T0329416

Translating Regenerative Medicine to the Clinic

Translating Regenerative Medicine to the Clinic

Edited by:

Jeffrey Laurence
Division of Hematology-Medical Oncology
Weill Cornell Medical College and New York Presbyterian Hospital
New York, NY, USA

Guest Editors:

Pedro Baptista
Aragon's Health Sciences Research Institute (IIS Aragon)
and CIBERehd, Zaragoza, Spain

Anthony Atala
Wake Forest Institute for Regenerative Medicine
Wake Forest University School of Medicine
Winston-Salem, NC, USA

Mary Van Beusekom
HealthPartners Institute for Education and Research
and Synapse Writing and Editing
Excelsior, MN, USA

AMSTERDAM • BOSTON • HEIDELBERG • LONDON • NEW YORK • OXFORD • PARIS
SAN DIEGO • SAN FRANCISCO • SINGAPORE • SYDNEY • TOKYO

Academic Press is an imprint of Elsevier

Academic Press is an imprint of Elsevier
125 London Wall, London EC2Y 5AS, UK
525 B Street, Suite 1800, San Diego, CA 92101-4495, USA
225 Wyman Street, Waltham, MA 02451, USA
The Boulevard, Langford Lane, Kidlington, Oxford OX5 1GB, UK

Notices

Knowledge and best practice in this field are constantly changing. As new research and experience broaden our understanding, changes in research methods, professional practices, or medical treatment may become necessary.

Practitioners and researchers may always rely on their own experience and knowledge in evaluating and using any information, methods, compounds, or experiments described herein. In using such information or methods they should be mindful of their own safety and the safety of others, including parties for whom they have a professional responsibility.

To the fullest extent of the law, neither the Publisher nor the authors, contributors, or editors, assume any liability for any injury and/or damage to persons or property as a matter of products liability, negligence or otherwise, or from any use or operation of any methods, products, instructions, or ideas contained in the material herein.

ISBN: 978-0-12-800548-4

British Library Cataloguing-in-Publication Data
A catalogue record for this book is available from the British Library

Library of Congress Cataloging-in-Publication Data
A catalog record for this book is available from the Library of Congress

For information on all Academic Press publications
visit our website at http://store.elsevier.com/

Working together
to grow libraries in
developing countries

www.elsevier.com • www.bookaid.org

Publisher: Mica Haley
Acquisition Editor: Mica Haley
Editorial Project Manager: Sam W. Young
Production Project Manager: Chris Wortley
Designer: Mark Rogers

Typeset by TNQ Books and Journals
www.tnq.co.in

Printed and bound in the United States of America

Contents

Contributors

Graça Almeida-Porada Wake Forest Institute for Regenerative Medicine, Winston-Salem, NC, USA

Greg Asatrian Department of Surgery, David Geffen School of Medicine, UCLA, Los Angeles, CA, USA; Department Orthopaedic Surgery, David Geffen School of Medicine, UCLA, Los Angeles, CA, USA

Anthony Atala Wake Forest Institute for Regenerative Medicine, Wake Forest University Health Sciences, Winston-Salem, NC, USA

Stephen F. Badylak McGowan Institute for Regenerative Medicine, University of Pittsburgh, Pittsburgh, PA, USA; Department of Surgery, School of Medicine, University of Pittsburgh, Pittsburgh, PA, USA

Pedro M. Baptista Aragon's Health Sciences Research Institute (IIS Aragon) and CIBERehd, Zaragoza, Spain

Cameron Best Tissue Engineering Program and Surgical Research, Nationwide Children's Hospital, Columbus, OH, USA

Khalil N. Bitar Wake Forest Institute for Regenerative Medicine, Wake Forest School of Medicine, Winston-Salem, NC, USA; Department of Molecular Medicine and Translational Sciences, Wake Forest School of Medicine, Winston Salem, NC, USA; Virginia Tech-Wake Forest School of Biomedical Engineering and Sciences, Winston-Salem, NC, USA

Christopher K. Breuer Tissue Engineering Program and Surgical Research, Nationwide Children's Hospital, Columbus, OH, USA

Bryan N. Brown McGowan Institute for Regenerative Medicine, University of Pittsburgh, Pittsburgh, PA, USA; Department of Bioengineering, Swanson School of Engineering, University of Pittsburgh, Pittsburgh, PA, USA

Raphaël F. Canadas 3B's Research Group – Biomaterials, Biodegradables and Biomimetics, University of Minho, Headquarters of the European Institute of Excellence on Tissue Engineering and Regenerative Medicine, Taipas, Guimarães, Portugal; ICVS/3B's – PT Government Associate Laboratory, Braga, Guimarães, Portugal

Arnold I. Caplan Skeletal Research Center, Department of Biology, Case Western University, Cleveland, OH, USA

Irene Cervelló Fundación IVI-Instituto Universitario IVI, Universidad de Valencia, INCLIVA, Valencia, Spain

William C.W. Chen Department of Bioengineering, University of Pittsburgh, PA, USA; Department of Orthopedic Surgery, University of Pittsburgh, PA, USA; Stem Cell Research Center, University of Pittsburgh, PA, USA; McGowan Institute for Regenerative Medicine, University of Pittsburgh, PA, USA

Dong F. Chen Schepens Eye Research Institute, Massachusetts Eye and Ear, Department of Ophthalmology, Harvard Medical School, Boston, MA, USA; VA Boston Healthcare System, Boston, MA, USA

Martin K. Childers Department of Rehabilitation Medicine, University of Washington, Seattle, WA, USA; Institute for Stem Cell and Regenerative Medicine, University of Washington, Seattle, WA, USA

Kin-Sang Cho Schepens Eye Research Institute, Massachusetts Eye and Ear, Department of Ophthalmology, Harvard Medical School, Boston, MA, USA

Claudio J. Conti Department of Bioengineering, Universidad Carlos III de Madrid (UC3M), Madrid, Spain; Instituto de Investigaciones Sanitarias Fundación Jiménez Díaz (IIS-FJD), Madrid, Spain

A. Crespo-Barreda Facultad de Ciencias Biosanitarias, Universidad Francisco de Vitoria, Madrid, Spain

Marcela Del Río Department of Bioengineering, Universidad Carlos III de Madrid (UC3M), Madrid, Spain; Regenerative Medicine Unit and Epithelial Biomedicine Division, CIEMAT, Madrid, Spain; Centre for Biomedical Research on Rare Diseases (CIBERER), Madrid, Spain; Instituto de Investigaciones Sanitarias Fundación Jiménez Díaz (IIS-FJD), Madrid, Spain

Giacomo Della Verde Department of Biomedicine, University Hospital Basel, University of Basel, Basel, Switzerland

Abritee Dhal Wake Forest Institute for Regenerative Medicine, Wake Forest University Health Sciences, Winston-Salem, NC, USA

Albert Donnenberg University of Pittsburgh Cancer Institute, Pittsburgh, PA, USA; Division of Hematology/Oncology, Department of Medicine, University of Pittsburgh, Pittsburgh, PA, USA

M.M. Encabo-Berzosa Gene and Cell Therapy Group, Instituto Aragonés de Ciencias de la Salud-IIS Aragón, Zaragoza, Aragón, Spain

Ignacio Giménez Aragon's Health Science Institutes, Zaragoza, Spain; Department of Pharmacology and Physiology, University of Zaragoza, Zaragoza, Spain

Melissa A. Goddard Department of Physiology and Pharmacology, School of Medicine, Wake Forest University Health Sciences, Winston-Salem, NC, USA; Department of Rehabilitation Medicine, University of Washington, Seattle, WA, USA; Institute for Stem Cell and Regenerative Medicine, University of Washington, Seattle, WA, USA

R. González-Pastor Gene and Cell Therapy Group, Instituto Aragonés de Ciencias de la Salud-IIS Aragón, Zaragoza, Aragón, Spain

Riccardo Gottardi Center for Cellular and Molecular Engineering, Department of Orthopaedic Surgery, University of Pittsburgh, Pittsburgh, PA, USA; RiMED Foundation, Palermo, Italy

Xuan Guan Department of Physiology and Pharmacology, School of Medicine, Wake Forest University Health Sciences, Winston-Salem, NC, USA; Department of Rehabilitation Medicine, University of Washington, Seattle, WA, USA; Institute for Stem Cell and Regenerative Medicine, University of Washington, Seattle, WA, USA

Sara Guerrero-Aspizua Department of Bioengineering, Universidad Carlos III de Madrid (UC3M), Madrid, Spain; Regenerative Medicine Unit and Epithelial Biomedicine Division, (CIEMAT), Madrid, Spain; Centre for Biomedical Research on Rare Diseases (CIBERER), Madrid, Spain; Instituto de Investigaciones Sanitarias Fundación Jiménez Díaz (IIS-FJD), Madrid, Spain

Chenying Guo Schepens Eye Research Institute, Massachusetts Eye and Ear, Department of Ophthalmology, Harvard Medical School, Boston, MA, USA

Winters Hardy Department of Surgery, David Geffen School of Medicine, UCLA, Los Angeles, CA, USA; Department Orthopaedic Surgery, David Geffen School of Medicine, UCLA, Los Angeles, CA, USA

Ira M. Herman Department of Developmental, Chemical and Molecular Biology, The Sackler School of Graduate Biomedical Sciences, The Center for Innovations in Wound Healing Research, Tufts University School of Medicine, Boston, MA, USA

Catalina K. Hwang Department of Developmental, Chemical and Molecular Biology, The Sackler School of Graduate Biomedical Sciences, The Center for Innovations in Wound Healing Research, Tufts University School of Medicine, Boston, MA, USA

M. Iglesias Facultad de Ciencias Biosanitarias, Universidad Francisco de Vitoria, Madrid, Spain

Glicerio Ignacio Wake Forest Institute for Regenerative Medicine, Winston-Salem, NC, USA

Aaron W. James Department of Surgery, David Geffen School of Medicine, UCLA, Los Angeles, CA, USA; Department Orthopaedic Surgery, David Geffen School of Medicine, UCLA, Los Angeles, CA, USA; Department of Pathology and Laboratory Medicine, UCLA, Los Angeles, CA, USA

Thi H. Khanh Vu Schepens Eye Research Institute, Massachusetts Eye and Ear, Department of Ophthalmology, Harvard Medical School, Boston, MA, USA; Departments of Ophthalmology, Leiden University Medical Center, Leiden, The Netherlands

Nobuaki Kikyo Stem Cell Institute, Department of Genetics, Cell Biology and Development, University of Minnesota, Minneapolis, MN, USA

Joanne Kurtzberg Department of Pediatrics, Duke University Medical Center, Durham, NC, USA

Gabriela M. Kuster Department of Biomedicine, University Hospital Basel, University of Basel, Basel, Switzerland; Division of Cardiology, University Hospital Basel, Basel, Switzerland

Angel Lanas University of Zaragoza, Zaragoza, Spain; IIS Aragón, CIBERehd, Zaragoza, Spain

Mark T. Langhans Center for Cellular and Molecular Engineering, Department of Orthopaedic Surgery, University of Pittsburgh, Pittsburgh, PA, USA

Fernando Larcher Department of Bioengineering, Universidad Carlos III de Madrid (UC3M), Madrid, Spain; Regenerative Medicine Unit and Epithelial Biomedicine Division, CIEMAT, Madrid, Spain; Centre for Biomedical Research on Rare Diseases (CIBERER), Madrid, Spain; Instituto de Investigaciones Sanitarias Fundación Jiménez Díaz (IIS-FJD), Madrid, Spain

Avione Y. Lee Tissue Engineering Program and Surgical Research, Nationwide Children's Hospital, Columbus, OH, USA

Yong-Ung Lee Tissue Engineering Program and Surgical Research, Nationwide Children's Hospital, Columbus, OH, USA

Ronglih Liao Cardiovascular Division, Department of Medicine, Brigham and Women's Hospital, Harvard Medical School, Boston, MA, USA

Jr-Jiun Liou Department of Bioengineering, Swanson School of Engineering, University of Pittsburgh, Pittsburgh, PA, USA

David L. Mack Department of Rehabilitation Medicine, University of Washington, Seattle, WA, USA; Institute for Stem Cell and Regenerative Medicine, University of Washington, Seattle, WA, USA

Nathan Mahler Tissue Engineering Program and Surgical Research, Nationwide Children's Hospital, Columbus, OH, USA

Alexandra P. Marques 3B's Research Group – Biomaterials, Biodegradables and Biomimetics, University of Minho, Headquarters of the European Institute of Excellence on Tissue Engineering and Regenerative Medicine, Taipas, Guimarães, Portugal; ICVS/3B's – PT Government Associate Laboratory, Braga, Guimarães, Portugal

Kacey G. Marra Department of Plastic Surgery, University of Pittsburgh, Pittsburgh, PA, USA; McGowan Institute of Regenerative Medicine, Pittsburgh, PA, USA

P. Martin-Duque Facultad de Ciencias Biosanitarias, Universidad Francisco de Vitoria, Madrid, Spain; Gene and Cell Therapy Group, Instituto Aragonés de Ciencias de la Salud-IIS Aragón, Zaragoza, Aragón, Spain; Fundación Araid, Zaragoza, Aragón, Spain

Patricia Meade Aragon's Health Science Institutes, Zaragoza, Spain; Department of Biochemistry and Cell Biology, University of Zaragoza, Zaragoza, Spain

Jose Vicente Medrano Fundación IVI-Instituto Universitario IVI, Universidad de Valencia, INCLIVA, Valencia, Spain; Fundación Instituto de Investigación Sanitaria La Fe, Valencia, Spain

Emma Moran Wake Forest Institute for Regenerative Medicine, Wake Forest University Health Sciences, Winston-Salem, NC, USA

Joaquim M. Oliveira 3B's Research Group – Biomaterials, Biodegradables and Biomimetics, University of Minho, Headquarters of the European Institute of Excellence on Tissue Engineering and Regenerative Medicine, Taipas, Guimarães, Portugal; ICVS/3B's – PT Government Associate Laboratory, Braga, Guimarães, Portugal

P. Ortíz-Teba Facultad de Ciencias Biosanitarias, Universidad Francisco de Vitoria, Madrid, Spain

Bruno Péault Department of Surgery, David Geffen School of Medicine, UCLA, Los Angeles, CA, USA; Department Orthopaedic Surgery, David Geffen School of Medicine, UCLA, Los Angeles, CA, USA; Center for Cardiovascular Science and MRC Center for Regenerative Medicine, University of Edinburgh, Edinburgh, UK

Otmar Pfister Department of Biomedicine, University Hospital Basel, University of Basel, Basel, Switzerland; Division of Cardiology, University Hospital Basel, Basel, Switzerland

Sandra Pina 3B's Research Group – Biomaterials, Biodegradables and Biomimetics, University of Minho, Headquarters of the European Institute of Excellence on Tissue Engineering and Regenerative Medicine, Taipas, Guimarães, Portugal; ICVS/3B's – PT Government Associate Laboratory, Braga, Guimarães, Portugal

Christopher D. Porada Wake Forest Institute for Regenerative Medicine, Winston-Salem, NC, USA

Rui L. Reis 3B's Research Group – Biomaterials, Biodegradables and Biomimetics, University of Minho, Headquarters of the European Institute of Excellence on Tissue Engineering and Regenerative Medicine, Taipas, Guimarães, Portugal; ICVS/3B's – PT Government Associate Laboratory, Braga, Guimarães, Portugal

J. Peter Rubin Department of Plastic Surgery, University of Pittsburgh, Pittsburgh, PA, USA; McGowan Institute of Regenerative Medicine, Pittsburgh, PA, USA

Keith Sabin Stem Cell Institute, Department of Genetics, Cell Biology and Development, University of Minnesota, Minneapolis, MN, USA

Natalia Sánchez-Romero Aragon's Health Science Institutes, Zaragoza, Spain

Alvaro Santamaria Department of Surgery and Orthopaedic Surgery, David Geffen School of Medicine, UCLA, Los Angeles, CA, USA

J.L. Serrano Institute of Nanosciences of Aragon Zaragoza, Aragón, Spain

Anthony R. Sheets Department of Developmental, Chemical and Molecular Biology, The Sackler School of Graduate Biomedical Sciences, The Center for Innovations in Wound Healing Research, Tufts University School of Medicine, Boston, MA, USA

Carlos Simón Fundación IVI-Instituto Universitario IVI, Universidad de Valencia, INCLIVA, Valencia, Spain; Reproductive Medicine Department, Instituto Valenciano Infertilidad (IVI) Valencia, Valencia, Spain; Department of Ob/Gyn, Stanford University School of Medicine, Stanford University, Stanford, CA, USA

Shay Soker Wake Forest Institute for Regenerative Medicine, Wake Forest University Health Sciences, Winston-Salem, NC, USA

Chia Soo Department of Surgery, David Geffen School of Medicine, UCLA, Los Angeles, CA, USA; Department Orthopaedic Surgery, David Geffen School of Medicine, UCLA, Los Angeles, CA, USA

Jessica M. Sun Department of Pediatrics, Duke University Medical Center, Durham, NC, USA

Mays Talib Schepens Eye Research Institute, Massachusetts Eye and Ear, Department of Ophthalmology, Harvard Medical School, Boston, MA, USA; Departments of Ophthalmology, Leiden University Medical Center, Leiden, The Netherlands

Shuhei Tara Tissue Engineering Program and Surgical Research, Nationwide Children's Hospital, Columbus, OH, USA

Kang Ting Dental and Craniofacial Research Institute and Section of Orthodontics, School of Dentistry, UCLA, Los Angeles, CA, USA

Rocky S. Tuan Department of Bioengineering, Swanson School of Engineering, University of Pittsburgh, Pittsburgh, PA, USA; Center for Cellular and Molecular Engineering, Department of Orthopaedic Surgery, University of Pittsburgh, Pittsburgh, PA, USA; McGowan Institute of Regenerative Medicine, School of Medicine, University of Pittsburgh, Pittsburgh, PA, USA

Jolene E. Valentin Department of Plastic Surgery, University of Pittsburgh, Pittsburgh, PA, USA

Dipen Vyas Wake Forest Institute for Regenerative Medicine, Wake Forest University Health Sciences, Winston-Salem, NC, USA

Bo Wang Comprehensive Transplant Center, Northwestern University Feinberg School of Medicine, Northwestern University, Chicago, IL, USA; Department of Surgery, Northwestern University Feinberg School of Medicine, Northwestern University, Chicago, IL, USA

Jason A. Wertheim Comprehensive Transplant Center, Northwestern University Feinberg School of Medicine,

Northwestern University, Chicago, IL, USA; Simpson Querrey Institute for BioNanotechnology, Northwestern University, Chicago, IL, USA; Chemistry of Life Processes Institute, Northwestern University, Evanston, IL, USA; Department of Biomedical Engineering, Northwestern University, Evanston, IL, USA; Department of Surgery, Jesse Brown VA Medical Center, Chicago, IL, USA; Department of Surgery, Northwestern University Feinberg School of Medicine, Northwestern University, Chicago, IL, USA

Honghua Yu Department of Ophthalmology, Liuhuaqiao Hospital, Guangzhou, PR China; Schepens Eye Research Institute, Massachusetts Eye and Ear, Department of Ophthalmology, Harvard Medical School, Boston, MA, USA

Elie Zakhem Wake Forest Institute for Regenerative Medicine, Wake Forest School of Medicine, Winston-Salem, NC, USA; Department of Molecular Medicine and Translational Sciences, Wake Forest School of Medicine, Winston Salem, NC, USA

Elisabeth Zapatero-Solana Regenerative Medicine Unit and Epithelial Biomedicine Division, (CIEMAT), Madrid, Spain; Centre for Biomedical Research on Rare Diseases (CIBERER), Madrid, Spain; Instituto de Investigaciones Sanitarias Fundación Jiménez Díaz (IIS-FJD), Madrid, Spain

Nan Zhang Wake Forest Institute for Regenerative Medicine, Wake Forest School of Medicine, Winston-Salem, NC, USA

Yuanyuan Zhang Wake Forest Institute for Regenerative Medicine, Wake Forest School of Medicine, Winston-Salem, NC, USA

Part I

Introduction

Chapter 1

Regenerative Medicine: The Hurdles and Hopes

Pedro M. Baptista, Anthony Atala

Key Concepts

1. Regenerative medicine is a subdivision of translational research in biomaterials science, tissue and organ engineering, and molecular and cell biology, which deals with the "process of replacing, engineering, or regenerating human cells, tissues, or organs to restore or reestablish normal function." This field holds the promise of engineering damaged tissues and organs via stimulating the body's own repair mechanisms to functionally heal previously irreparable tissues or organs.
2. Tissue and organ bioengineering is an emerging field driven by the shortage of donor tissue and organs for transplantation that relies on the conjugation of biomaterials with cells in order to produce functional tissue/organ constructs.
3. Cellular therapies are a form of therapeutic strategy in which cellular material is injected into a patient with the objective to provide a cellular replacement in damaged tissues or to release soluble factors such as cytokines, chemokines, and growth factors, which act in a paracrine or endocrine fashion.
4. Gene therapy relies on the use of nucleic acids as "drugs" to treat disease by therapeutically delivering them into patient's cells. Once inside the cells, they will either be translated as proteins, interfere with the expression of other proteins, or correct genetic mutations.
5. There are a few general limitations common to all these technologies, including cell sourcing and expansion, inconsistent therapeutic efficacy among different research groups, elevated price for most countries' healthcare systems, regulatory issues and the difficulty in scaling up some of these technologies.

1. INTRODUCTION

Regenerative medicine is characterized as the process of replenishing or restoring human cells, tissues, or organs to restore or reestablish normal function. This field holds the promise of transforming human medicine, by actually curing or treating diseases once poorly managed with conventional drugs and medical procedures.

It informally started almost 60 years ago with the first successful organ transplant performed in Boston by a team led by Dr Joseph Murray, John Merrill, and J. Hartwell Harrison.[1] This landmark accomplishment marked a new era in the emerging field of organ transplantation and allowed for the first time for the complete cure of a patient with end-stage organ disease.

The first effective cell therapies with bone marrow transplants followed in the late 1950s and 1960s. A team led by Dr Don Thomas was the first to treat leukemic patients with allogeneic marrow transplants in Seattle.[2,3] This was later followed by Dr Robert Good in 1968, where an immunodeficient patient was successfully treated with an allogeneic bone marrow transplant from his sibling at the University of Minnesota.[4]

Throughout these decades, many attempts on organ transplantation, cell therapies, and gene therapy ended in failure, but this vigorous scientific and clinical interest established the basis of the first wave of successes that regenerative medicine experienced and delivered to the clinic.[5–8]

With this paradigm change in medicine, came the first challenges of organ shortage and higher demand for matching bone marrow donors. Organ shortages established a driving force for novel advancements in molecular and cell biology that opened new avenues in several areas in regenerative medicine. The fields of cell transplantation and tissue engineering were proposed as alternatives to tissue and organ shortage by de novo reconstitution of functional tissues and organs in the laboratory for transplantation, and the use of cells for therapy.

Translating Regenerative Medicine to the Clinic. http://dx.doi.org/10.1016/B978-0-12-800548-4.00001-2

The present book you have just started to explore is an introduction for the translational and basic researcher as well as the clinician to the vast field of regenerative medicine technologies. It is the second book in a new series, Advances in Translational Medicine and presents 23 key chapters that describe in detail some of the contemporary regenerative medicine advances in different medical fields. These chapters review the state-of-the-art experimental data available from the bench, along with vital information provided by multiple clinical trials, giving a broad view of current and near future strategies to treat or cure human disease.

2. ON THE ORIGINS OF REGENERATIVE MEDICINE

It is hard to trace the origins of such a field of medicine that throughout the centuries has been part of legend and fact. From the ever regenerating liver of the titan Prometheus, the "first organ transplant" by Saints Cosmas and Damian after grafting a leg from a recently deceased Ethiopian to replace a patient's ulcerated or cancerous leg, to the first documented iron hand prosthesis in 1504, there are innumerous examples of body regeneration throughout history. Across the centuries, restoration of lost bodily functions has always been the focus of many shamans, apothecaries, and alchemist, who embraced the quest to save or improve human life. Some of them have been true pioneers that in parallel to the theological and mythological reports have progressively developed the fields of clinical medicine, surgery, anatomy, and biology, and with it, regenerative medicine.

Of particular interest is the adoption of the mechanical substitution of body parts by inanimate prosthesis (wooden legs, iron hands, metallic and ivory dentures, etc.), which can be seen as an early attempt to use the available materials of the time in reconstructive medicine.[9,10] Blood transfusions, repair of human skull with canine cranial bone, and skin transplants followed in the next centuries.[10]

The former, skin grafts, are actually considered a true landmark in the contemporary view of regenerative medicine. This is closely related to the work of the surgeon Johann Friedrich Dieffenbach who performed experimental and clinical work in skin transplantation described in "*Nonnulla de Regeneratione et Transplantatione.*"[11] He was also the pioneer that established the use of pedicled skin flaps (since most of the clinical skin transplants failed), one of the modern founding fathers of plastic and reconstructive surgery and an early specialist in transplantation medicine. Heinrich Christian Bünger followed with the first successful autologous skin transplantation in 1822.[12]

Followed by further experimental work and surgical advances on tissue transplants and reconstructive surgery during the nineteenth and beginning of twentieth century, the field moved on.

Clarification into the biological processes that determined the fate of transplants was first presented by the fundamental biological work of Rudolf Virchow that described in his "*Cellularpathologie*" that tissue regeneration is determined by cell proliferation.[13] His work led not only to the investigation of the cellular effects responsible for tissue healing, but also to the cultivation of cells outside the body. The groundbreaking achievement of in vitro cell cultivation was first reached by R.G. Harrison in 1910, demonstrating active growth of cells in culture.[14] Since then, cell biology and particularly in vitro cell culture developed the backbone of what can be termed classical tissue engineering.

These in vitro cell culture methods were followed by cell transplantation, contemporary tissue engineering, and regenerative medicine, all directly connected to microsurgery, which Alexis Carrel is considered the founding father. His development of microvascular surgery enabled organ transplantation and plastic surgery due to his work developing the methods of vascular anastomosis still used today.[15–18]

After this, organ transplantation, cell therapies, and many other procedures followed, shaping regenerative medicine, as we know it today.

3. FROM CELLS AND SCAFFOLDS TO TISSUES AND ORGANS

Tissue engineering as a field was casually "coined" in 1993 in a review authored by Dr Robert Langer and Joseph Vacanti that finally crystallized several concepts enunciated by many others in the previous years[19,20]: "Tissue engineering is an interdisciplinary field that applies the principles of engineering and life sciences toward the development of biological substitutes that restore, maintain, or improve tissue function." By then, both definitions (tissue engineering and regenerative medicine) were virtually identical, with a common expressed purpose to restore or reestablish normal bodily function. To the wide range of concurrent lines of research encompassed by tissue engineering, regenerative medicine eventually expands upon these by including additional methods and strategies from other areas of science not comprised by what was then defined as tissue engineering. It is this broadened view that more easily accommodates gene therapy, molecular medicine, and stem cell therapies into the same medical field, with the exact same goal in mind.

4. BIOMATERIALS, TISSUE AND ORGAN BIOENGINEERING

In this quest, biomaterials have played a critical role for surgical reconstructive purposes, and are also able to provide a physical carrier for cells—the scaffold.[21] Tissue engineered skin—one of the first bioengineered tissues—consisting of cultured epithelial sheets or fibroblast gels seeded onto polymer scaffolds, were first generated in the late 1970s and early 1980s with successful transplantation into burnt patients by Dr Howard Green in Boston in 1981.[22] Since then, combination of cells with a scaffold to generate a tissue prior to implantation became common. In the first years, synthetic polymeric materials were the most commonly used,[22,23] but experimentation with other types of biologically derived materials was also being sought.[24] In fact, naturally derived materials have been effectively used centuries before (canine cranial bones, teeth, etc.), but one of the first documented uses with this type of materials lies on the experimental work of Dr Guthrie that successfully used a fixed segment of *vena cava* to reconstitute the common carotid artery of a dog.[25] Tissue-derived extracellular matrices (ECM) were introduced in the 1960s, with one of the first reports of the use of intestinal submucosa for vascular grafts in Germany.[26]

Approximately at the same time, in 1967, occurred the very first successful heart transplantation by Christiaan Barnard in South Africa. This accomplishment in transplantation initiated the debate on the many ethical issues in transplantation medicine and later on in genetic engineering that followed in the next decades.[27] The field of transplantation and regenerative medicine evolved with the introduction of better immunosuppressants and novel therapeutic approaches in the following years. However, the shortage of hearts prompted for the creation of artificial hearts that could bridge patients to a heart transplant. This same approach prompted others in distinct fields for the creation of additional bioartificial organs.

In this book, several in-depth chapters cover the growing use and multiple applications of ECM-derived products in regenerative medicine. From the dissection of the dual role that ECM provides by serving as a mechanical framework for each tissue and organ and a substrate for cell signaling, to the available synthetic and naturally derived vascular, valvular, and heart tissue replacement strategies in cardiovascular disease and the development of whole organ decellularization, a wide range of subjects is thoroughly addressed. These advances in tissue and organ engineering are not isolated from the tremendous progress made in stem cell biology and enabling technologies in bioreactors and cell culture. Examples of these are now the deep understanding we have of lung, liver, and heart native mechanisms of development and regeneration that helped push our bioengineering boundaries forward.

5. GENE THERAPY

With the advances produced in molecular biology and DNA technology in the 1960s and 1970s, this same level of far-reaching innovation was observed with the creation of gene therapy as a novel research field. Theodore Friedmann and Richard Roblin first hypothesized it in 1972, in a coauthored paper titled "Gene therapy for human genetic disease?"[28] They cited Stanfield Rogers for proposing in 1970 that *exogenous good DNA* could be employed to replace the defective DNA in patients affected with genetic defects. The field gained traction with the development of new gene transfer vectors in the 1980s and the first approved gene therapy case in the United States occurred on September 14, 1990 at the National Institute of Health under the guidance of Professor William Anderson. It targeted severe combined immunodeficiency caused by adenosine deaminase deficiency on a 4-year-old girl. The treatment was successful, even if only transient.

In this book, a broad view of the current status of gene therapy vector types and clinical translation strategies is presented. Furthermore, in-depth analysis is provided of ongoing efforts of gene therapy correction of factor VIII deficiency and muscular dystrophies.

6. STEM CELL THERAPIES

Pericytes or mesenchymal stem cells derived from bone marrow, adipose, or other tissue sources, have been extensively used in tissue engineering and in the clinic due to their differentiation and immunomodulatory capabilities in many diseases. The molecular mechanisms by which pericyte and endothelial cells communicate are proving critical not only in cellular therapies but also in the bioengineering of vascularized tissues.

In some chapters of this book, a careful description of the phenotype, function, endothelial cell cross talk, molecular biology, and disease involvement of pericytes is provided. Furthermore, the biology of adipose-derived pericytes and their regenerative properties are presented with an extensive view of their clinical applications and future challenges for broad implementation.

Other cell populations are also in extensive experimentation to treat a myriad of other pathological conditions. From skeletal muscle disorders, to cardiovascular disease, to diabetes, or liver cirrhosis, cellular therapies are enabling the treatment of prior poorly manageable diseases. Hence, different types of stem cells are described here that have been tested in muscle wasting disorders. Additionally, a review of the advances within the field of β-cell regeneration and potential of establishing a future regenerative therapy for diabetes from adult tissues is presented here.

Even diseases previously incurable before, like retinal degenerative diseases, have now cell therapies leading the way in clinical trials, and the exciting findings in both human and animal models point to the potential of restoring vision through a cell replacement regenerative approach using endogenous- or differentiated-induced pluripotent stem cells. Hence, the emerging evidence of a subpopulation of stemlike cells resident in the mammalian retina that maintains the potential for retinal regeneration under certain conditions is fully described in this book.

7. FUTURE DIRECTIONS

It is hard to predict the future in a field so dynamic and broad as regenerative medicine is today. Novel advances are communicated almost on a daily basis and this makes it particularly hard to define where the best solution to a clinical challenge will rise. Nevertheless, the latest developments in molecular and cell biology point to a much profounder understanding of the molecular mechanisms of regeneration. Not only at the signaling level, but also from the cellular point of view, where microvesicles seem to play a center role.

In this book, the nature of microvesicles and their known functions and effects are carefully addressed. Moreover, data from animal models and in vitro studies are presented that suggest great applicability for microvesicle-based regenerative therapies, debating the current need for proof of efficacy and feasibility in clinical medicine.

The harnessing of these molecular mechanisms, as well as many of the described tissue/organ bioengineering and stem cell therapies, will have a pivotal role in what one can envision as the future of regenerative medicine. Presently, the fine details of what lies ahead might be hidden, but a future where chronic disability and disease seems to be alleviated by regenerative medicine seems to be finally taking shape.

REFERENCES

1. Guild WR, Harrison JH, Merrill JP, Murray J. Successful homotransplantation of the kidney in an identical twin. *Trans Am Clin Climatol Assoc* 1955;**67**:167–73.
2. Thomas ED, Lochte Jr HL, Lu WC, Ferrebee JW. Intravenous infusion of bone marrow in patients receiving radiation and chemotherapy. *N Engl J Med* 1957;**257**:491–6. http://dx.doi.org/10.1056/NEJM195709122571102.
3. Thomas ED, Lochte Jr HL, Cannon JH, Sahler OD, Ferrebee JW. Supralethal whole body irradiation and isologous marrow transplantation in man. *J Clin Invest* 1959;**38**:1709–16. http://dx.doi.org/10.1172/JCI103949.
4. Gatti RA, Meuwissen HJ, Allen HD, Hong R, Good RA. Immunological reconstitution of sex-linked lymphopenic immunological deficiency. *Lancet* 1968;**2**:1366–9.
5. Morris PJ. Transplantation—a medical miracle of the 20th century. *N Engl J Med* 2004;**351**:2678–80. http://dx.doi.org/10.1056/NEJMp048256.
6. Gage FH. Cell therapy. *Nature* 1998;**392**:18–24.
7. Platt JL. New directions for organ transplantation. *Nature* 1998;**392**:11–7. http://dx.doi.org/10.1038/32023.
8. Anderson WF. Human gene therapy. *Nature* 1998;**392**:25–30. http://dx.doi.org/10.1038/32058.
9. Ulrich Meyer T, Jorg Handshel M, Wiesmann HP. *Fundamentals of tissue engineering and regenerative medicine*. Berlin Heidelberg: Springer-Verlag; 2009. p. 5–11.
10. Historical highlights in bionics and related medicine. *Science* 2002;**295**:995.
11. Dieffenbach J. *Nonnulla de regeneratione et transplantatione*. Richter; 1822.
12. Bünger H. Gelungener Versuch einer Nasenbildung aus einem völligen getrennten Hautstück aus dem Beine. *J Chir Augenhkd* 1823;**4**:569.
13. Virchow R. *Die Cellularpathologie in ihrer Begründung auf physiologische und pathologische Gewebelehre*. Hirschwald; 1858.
14. Harrison RG. The outgrowth of the nerve fiber as a mode of protoplasmic extension. *J Exp Zool* 1910;**9**:787–846.
15. Witkowski JA. Alexis Carrel and the mysticism of tissue culture. *Med Hist* 1979;**23**:279–96.
16. Carrel A, Burrows MT. Cultivation in vitro of malignant tumors. *J Exp Med* 1911;**13**:571–5.
17. Carrel A, Burrows MT. Cultivation of tissues in vitro and its technique. *J Exp Med* 1911;**13**:387–96.
18. Carrel A. Landmark article, Nov 14, 1908: results of the transplantation of blood vessels, organs and limbs. By Alexis Carrel. *JAMA* 1983;**250**:944–53.
19. Heineken FG, Skalak R. Tissue engineering: a brief overview. *J Biomech Eng* 1991;**113**:111–2.
20. Langer R, Vacanti JP. Tissue engineering. *Science* 1993;**260**:920–6.
21. Fuchs JR, Nasseri BA, Vacanti JP. Tissue engineering: a 21st century solution to surgical reconstruction. *Ann Thorac Surg* 2001;**72**:577–91.
22. O'Connor NE, Mulliken JB, Banks-Schlegel S, Kehinde O, Green H. Grafting of burns with cultured epithelium prepared from autologous epidermal cells. *Lancet* 1981;**1**:75–8.
23. Vacanti JP, et al. Selective cell transplantation using bioabsorbable artificial polymers as matrices. *J Pediatr Surg* 1988;**23**:3–9.

24. Bell E, Ivarsson B, Merrill C. Production of a tissue-like structure by contraction of collagen lattices by human fibroblasts of different proliferative potential in vitro. *Proc Natl Acad Sci USA* 1979;**76**:1274–8.

25. Guthrie CC. End-results of arterial restitution with devitalized tissue. *J Am Med Assoc* 1919;**73**:186–7.

26. Rotthoff G, Haering R, Nasseri M, Kolb E. Artificial replacements in heart and blood vessel surgery by freely transplanted small intestines. *Langenbecks Arch Klin Chir Ver Dtsch Z Chir* 1964;**308**:816–20.

27. Barnard C, Pepper CB. *One life*. Timmins; 1969.

28. Friedmann T, Roblin R. Gene therapy for human genetic disease? *Science* 1972;**175**:949–55.

Part II

Biomaterials and Tissue/Organ Bioengineering

Chapter 2

Extracellular Matrix as an Inductive Scaffold for Functional Tissue Reconstruction

Bryan N. Brown, Stephen F. Badylak

Key Concepts

1. Mammalian extracellular matrix (ECM) represents the ideal biologic scaffold for cell, tissue, and organ development.
2. The ECM contains signaling molecules that support diverse physiologic functions including the recruitment of endogenous stem cells and modulation of the innate immune response.
3. Bioscaffolds composed of ECM can circumvent the default mammalian response to injury and promote constructive tissue remodeling.

1. INTRODUCTION

The extracellular matrix (ECM) is a composite of the secreted products of resident cells in every tissue and organ. The matrix molecules represent a diverse mixture of structural and functional proteins, glycoproteins, and glycosaminoglycans among other molecules that are arranged in an ultrastructure that is unique to each anatomic location. The ECM exists in a state of dynamic reciprocity with the resident cells. That is, the matrix composition and organization change as a function of the metabolic adaptations of the cells, which in turn respond to shifts in the mechanical properties, pH, oxygen concentration, and other variables in the microenvironment.[1] This constantly adapting structure–function relationship therefore represents the ideal scaffold for the resident cell population.

Although the ECM is a known repository for a variety of growth factors, it also represents a source of bioactive cryptic peptides.[2–4] Fragments of parent molecules such as collagen and fibronectin have been shown to exert biologic activities including angiogenesis,[5] antiangiogenesis,[6] antimicrobial effects,[7,8] and chemotactic effects,[9–12] among others. These growth factors and bioactive peptides play important roles in defining the microenvironmental niche within which cells function in both normal homeostasis and in response to injury. The matrix has also been shown to be important in fetal development[13] and plays a critical role in determination of stem/progenitor cell differentiation fate.[14,15]

The tremendous complexity of the composition and ultrastructure of the ECM is only partially understood. Therefore, it is hardly possible to design and engineer a mimic of this complex structure. However, the ECM can be harvested from parent tissues through decellularization. Attempts to harvest ECM for utilization as a tissue repair scaffold would ideally remove all potentially immunogenic cell products while minimizing damage to the remaining ECM. Many medical device products composed of allogeneic and xenogeneic ECM currently exist (Table 1), but their ultimate performance varies depending upon source of material, methods of preparation, and clinical application. These naturally occurring materials are generally considered as devices by most regulatory authorities. However, depending upon the formulation and indications for use, these materials may be regulated as a biologic in the future. Regardless of application or regulatory status, optimal clinical outcomes will be obtained if surgeons and other health-care providers understand the mechanisms of action and potential of ECM scaffolds to help define the microenvironment of an injury site.

The purpose of the present chapter is to briefly review the rationale for the selection of ECM as an "inductive" scaffold for regenerative medicine applications and the preparation of ECM scaffolds for such applications. Three recent translational applications of ECM in regenerative medicine are presented, and the known mechanisms by which ECM scaffolds promote "constructive remodeling" outcomes are discussed.

Translating Regenerative Medicine to the Clinic. http://dx.doi.org/10.1016/B978-0-12-800548-4.00002-4

TABLE 1 Partial List of Commercially Available Scaffold Materials Composed of Extracellular Matrix

Product	Company	Material	Form		Use
AlloDerm	LifeCell	Human skin	Cross-linked	Dry sheet	Abdominal wall, breast, ENT/head and neck reconstruction, grafting
AlloPatch®	Musculoskeletal Transplant Foundation	Human fascia lata	Cross-linked	Dry sheet	Orthopedic applications
Axis™ dermis	Mentor	Human dermis	Natural	Dry sheet	Pelvic organ prolapse
CollaMend®	C. R. Bard	Porcine dermis	Cross-linked	Dry sheet	Soft tissue repair
CuffPatch™	Arthrotek	Porcine SIS	Cross-linked	Hydrated sheet	Reinforcement of soft tissues
DurADAPT™	Pegasus Biologicals	Horse pericardium	Cross-linked		Repair dura matter after craniotomy
Dura-Guard®	Synovis Surgical	Bovine pericardium		Hydrated sheet	Spinal and cranial repair
Durasis®	Cook SIS	Porcine SIS	Natural	Dry sheet	Repair dura matter
Durepair®	TEI Biosciences	Fetal bovine skin	Natural	Dry sheet	Repair of cranial or spinal dura
FasLata®	C. R. Bard	Cadaveric fascia lata	Natural	Dry sheet	Soft tissue repair
Graft Jacket®	Wright Medical Tech	Human skin	Cross-linked	Dry sheet	Foot ulcers
MatriStem®	ACell, Inc.	Porcine urinary bladder	Natural	Dry sheet, powder	Soft tissue repair and reinforcement, burns, gynecologic
Oasis®	Healthpoint	Porcine SIS	Natural	Dry sheet	Partial and full thickness wounds; superficial and second degree burns
OrthADAPT™	Pegasus Biologicals	Horse pericardium	Cross-linked	Dry Sheet	Reinforcement, repair and reconstruction of soft tissue in orthopedics
Pelvicol®	C. R. Bard	Porcine dermis	Cross-linked	Hydrated sheet	Soft tissue repair
Peri-Guard®	Synovis Surgical	Bovine pericardium			Pericardial and soft tissue repair
Permacol™	Tissue Science Laboratories	Porcine skin	Cross-linked	Hydrated sheet	Soft connective tissue repair
PriMatrix™	TEI Biosciences	Fetal bovine skin	Natural	Dry sheet	Wound management
Restore™	DePuy	Porcine SIS	Natural	Sheet	Reinforcement of soft tissues
Stratasis®	Cook SIS	Porcine SIS	Natural	Dry sheet	Treatment of urinary incontinence
SurgiMend™	TEI Biosciences	Fetal bovine skin	Natural	Dry sheet	Surgical repair of damaged or ruptured soft tissue membranes
Surgisis®	Cook SIS	Porcine SIS	Natural	Dry sheet	Soft tissue repair and reinforcement
Suspend™	Mentor	Human fascia lata	Natural	Dry sheet	Urethral sling

TABLE 1 Partial List of Commercially Available Scaffold Materials Composed of Extracellular Matrix—cont'd

Product	Company	Material		Form	Use
TissueMend®	TEI Biosciences	Fetal bovine skin	Natural	Dry sheet	Surgical repair and reinforcement of soft tissue in rotator cuff
Vascu-Guard®	Synovis Surgical	Bovine pericardium			Reconstruction of blood vessels in neck, legs, and arms
Veritas®	Synovis Surgical	Bovine pericardium		Hydrated sheet	Soft tissue repair
Xelma™	Molnlycke	ECM protein, PGA, water		Gel	Venous leg ulcers
Xenform™	TEI Biosciences	Fetal bovine skin	Natural	Dry sheet	Repair of colon, rectal, urethral, and vaginal prolapse, pelvic reconstruction, urethral sling
Zimmer collagen Patch®	Tissue Science Laboratories	Procine dermis	Cross-linked	Dry sheet	Orthopedic applications

ECM, extracellular matrix; SIS, small intestinal submucosa; PGA, polyglycolic acid.

2. ECM AS A SCAFFOLD FOR REGENERATIVE MEDICINE

ECM-based substrates consisting of individual ECM components or of whole decellularized tissues have been used in a wide range of applications in both preclinical and clinical settings.[16–19] These materials, in their many forms, have been used in applications as basic as coatings for tissue culture plastic and as complex as inductive templates for tissue and organ reconstruction in regenerative medicine, a number of which are discussed in detail below. In more complex applications, ECM-based scaffold materials can promote a process termed "constructive remodeling"—the de novo in vivo formation of site-appropriate, functional tissue.[19] However, as will be discussed in more detail below, the ability to promote constructive remodeling is critically dependent upon the methods used to prepare the scaffold material. Regardless of the application or the outcome, the overall rationale for the use of ECM is similar. Simply stated, the ECM provides a naturally occurring and highly conserved substrate for cell viability and growth. As applications of ECM scaffolds in tissue engineering and regenerative medicine move toward the reconstruction of increasingly complex tissue structures and whole organs, it is important to understand the mechanisms by which ECM scaffolds promote constructive remodeling. While the full profile of mechanisms responsible for such outcomes is not known, these mechanisms extend beyond the role of the ECM as a mechanical substrate and include a number of processes which occur in development and tissue homeostasis.

2.1 The ECM as a Mechanical Substrate

The ECM provides a three-dimensional structural support occupying the space between cells, is a substrate for cell migration, and is a transmitter of biomechanical forces. The physical properties of the ECM, such as rigidity, porosity, insolubility, and topography that derive from composition of the matrix largely determine the mechanical behavior of each individual tissue as well as the behavior of the cells which reside within.[20,21] For example, the basement membrane is a dense ECM structure which serves as a selective barrier to migrating cells.[22–24] Migration of cells through this structure requires focal remodeling of the matrix. The compact ultrastructure of the basement membrane is in contrast to the more open and porous structure of the underlying connective tissues which allows greater cellular mobility. Numerous studies have demonstrated the potential effects of ECM ultrastructure and mechanics upon cell behavior, migration, and differentiation.[25–27]

ECM components also provide separation between distinct structures within a single tissue. For example, the basement membrane separates the mucosal epithelium of the intestine from the lamina propria. Each tissue compartment also serves a particular function within the organ as a whole. In addition to its role in separating the mucosal and submucosal compartments, the basement membrane provides a substrate for growth and maintenance of the intestinal mucosa and acts as a molecular sieve while the subjacent connective tissues provide mechanical support and serve as a conduit for vascular and lymphatic perfusion for the organ. The basement membrane is merely one example of a specialized form of the ECM which demarcates the boundary between mesenchymal and epithelial tissues. There are numerous other examples of boundaries within tissues, and in each example the transition from one tissue type to another is accompanied by a shift in the ECM composition and structure.

It is easy to appreciate the potential role of the ECM as a mechanical substrate for tissue engineering and regenerative medicine applications. ECM biomimetic approaches include attempts to recreate these structures by synthetic methods, electrospinning representing the most notable example.[28,29] Such approaches are capable of producing interconnected networks of randomly distributed fibers on the approximate scale of the fibrillar components of the ECM. However, no approach can account for the varied distribution of fiber diameters and orientation of these fibers, nor can they substitute for the three-dimensional distribution of biologically active molecular components within the ECM. While these studies have clearly demonstrated the potential role of topography, structure, and mechanics of the ECM in modulating cellular phenotype and migration, each tissue and organ contains a unique ECM composition which includes (at least) hundreds of component molecules—a target which is, practically speaking, beyond the capability of any existing engineering techniques.

2.2 ECM Composition

As stated above, the ECM is a combination of both structural and functional components arranged in a three-dimensional, tissue-specific architecture. These components of ECM include collagens, glycoproteins, proteoglycans, mucins, elastic fibers, and growth factors,[30] many of which are highly conserved across species.[31–35] As additional signaling pathways and mechanisms for ECM–cell interactions are identified, it is increasingly difficult to separate the mechanical and functional aspects of these components. This molecular multifunctionality is increasingly evident, as will be discussed in detail, when one considers the bioactivity of ECM degradation products during tissue remodeling.[36,37] As one would expect based upon varying tissue functions, the composition of the ECM varies greatly from tissue to tissue, and in some cases within the microstructure of a given tissue, based upon mechanical and metabolic requirements. For example, articular cartilage contains large amounts of collagen II and glycosaminoglycans which are specifically tailored to accommodate high water content and allow resistance to and recovery from compressive deformation. In contrast, tissues such as tendon contain much higher amounts of collagen I and an organization designed to withstand tensile loading. These tissues, by comparison, are quite dissimilar from organs such as the liver and kidney which serve few mechanical functions and are almost entirely physiologic in nature. Therefore, the ECM composition in these organs is quite dissimilar.

Again, the rationale for the use of a decellularized tissue (i.e., ECM-based) scaffold is clear. Removal of the cellular components will leave an intact meshwork of ECM components which are both highly conserved across mammalian species, arranged in a tissue-specific architecture, and with a composition that is functionally relevant to the native tissue.

2.3 Dynamic Reciprocity

In addition to its structural role, the pleiotropic effects of ECM upon tissue resident cells are known to include cell adhesion, migration, proliferation, differentiation, and death.[21,38] The mechanisms by which the ECM contributes to these processes are diverse. The ECM can transmit mechanical cues, and can provide signaling cues via direct cellular binding to ECM components, and through the sequestration of soluble growth factors and cytokines and regulation of access to these molecules.[20,38] Thus, to repeat, the ECM can be considered a highly specialized substrate for both spatial patterning and structural support as well as a functional substrate for cell growth and signaling. The ECM, even in fully developed tissues in adult mammals, is by no means static. Rather, the ECM is constantly subject to turnover through a process aptly termed "dynamic reciprocity."[1,39,40] That is, the ECM exerts effects upon cellular behavior and phenotype and the resident cells, in turn, produce, degrade, and remodel the ECM. This dynamic and reciprocal process is important to homeostasis of all tissue and organ form and function. The ability to rapidly and dynamically remodel the ECM is also an essential component of the wound healing process, allowing the host to effectively repair tissue damage and protect itself from further insult.

The ability of the ECM to modulate cellular activity while simultaneously being remodeled is particularly evident during tissue development and morphogenesis.[21,23] This process is highly regulated and cell signaling and patterning processes must be deployed promptly, transiently, and in a defined temporospatial sequence. The role of ECM remodeling in multiple developmental processes including epithelial branch morphogenesis and skeletal development and remodeling have been investigated in-depth.[21,23,41,42] In branch morphogenesis, both the basement membrane and other ECM components are in a constant state of dynamic remodeling leading to primary bud formation, branch formation, and branch reiteration. Cells participate through the degradation and remodeling the matrix in an exquisitely regulated process which relies heavily upon expression of matrix metalloproteinases (MMPs) and tissue inhibitor of metalloproteinases (TIMPs) concurrently with the production of fibronectin, collagen, and laminin. It is important to note that the role of the ECM in this process goes beyond spatial patterning and provision of a physical substrate. The ECM participates in the transmission of mechanical forces, regulation of cell migration, growth factor release and signaling, and tissue polarization.[21,23,38] The mechanisms which underlie each of these processes have been studied in-depth,[20,21,23,38] but are beyond the scope of the present chapter.

It is clear that ECM is intricately involved in the process of fetal development. The dynamic nature of the ECM is unique among the various biomaterials used in tissue engineering and regenerative medicine. Although synthetic materials can be finely tuned to degrade under specific conditions and at specific rates, the degradation process is not accompanied by the release of a variety of bioactive peptides, as will be discussed below, or by concurrent signaling to cells in the process of tissue remodeling. Disruption of the ability to degrade or blocking of cell–ECM interactions through chemical cross-linking can limit the ability to elicit a constructive remodeling response.[43,44]

2.4 Bioactive Degradation Products

All of the components of the ECM are degradable and subject to modification. The mechanisms by which ECM is degraded and remodeled have been reviewed in-depth elsewhere.[21] Briefly, the major families of proteinases which are responsible for degradation of the ECM are the MMP and metalloproteinase with thrombospondin motif families (ADAMTS).[45] There are at least 23 identified MMP family members[46] and 19 ADAMTS family members.[47] These proteinases target a wide variety of ECM components and are indispensable for maintenance, remodeling, and developmental processes.

Recent evidence shows that degradation or modification of the ECM by proteinase degradation can result in the exposure of new recognition sites with potent bioactivity. ECM degradation products include cryptic sites, termed matricryptins or matrikines, which have been shown to influence cell behavior through a number of mechanisms including integrin, toll-like receptor, and scavenger receptor signaling.[36,37,48] These cellular interactions result in a diverse array of bioprocesses including angiogenesis, antiangiogenesis, chemotaxis, adhesion, and antimicrobial effects, among others.[5,6,8–11,36,37,48] Exposure of matricryptic sites can play a role in ECM assembly and modification by influencing ECM multimerization and assembly of ECM–growth factor complexes.[37] Fibronectin, for example, has many functions including self-assembly, multimerization, and interactions with other ECM components and growth factors including vascular endothelial growth factor (VEGF). These processes have been shown to be controlled or affected by the exposure of matricryptic sites within fibronectin.[49–55] The degradation of fibronectin leads to the formation of peptides which can affect cellular behavior. There are now an increasingly large number of ECM fragments with recognized bioactivity (Table 2).

One of the best known examples of a matricryptic peptide in the tissue engineering and regenerative medicine field is the Arg-Gly-Asp (RGD) peptide present primarily not only within fibronectin, but also within collagen, vitronectin and

TABLE 2 Selected Examples of "Cryptic" Peptides within the Extracellular Matrix

Fragment	Parent Molecule	Activity
Endostatin	Collagen XVIII	Inhibits angiogenesis
Angiostatin	Plasminogen	Inhibits angiogenesis
Anastellin fragment III1c	Fibronectin	Inhibits angiogenesis
Canstatin	Collagen IV	Apoptosis, inhibits chemotaxis, and proliferation
Restin	Collagen XV	Inhibits migration
Tumstatin	Collagen IV	Inhibits angiogenesis, promotes apoptosis, antitumor activity
ABT-510	Thrombospondin-1	Inhibits angiogenesis
RGD	Fibronectin	Promotes adhesion
Hyaluronic acid fragments	Hyaluronic acid	Promotes angiogenesis, increased MMP production
VAVPG sites	Elastin	Promotes chemotaxis, increased MMP expression
C-terminal telopeptide of collagen III	Collagen III	Promotes chemotaxis, osteogenesis

MMP, matrix metalloproteinase.

osteopontin.[37,56–60] The RGD peptide has been used to promote cell adhesion to synthetic substrates.[61–65] Thus, an additional advantage of the use of an ECM-based biomaterial is that it acts not only as a reservoir of structural and functional proteins, but also as a degradable substrate with an additional reserve of "hidden" bioactive peptides released during context-dependent degradation processes.

2.5 ECM as an Instructive Niche for Stem Cells

Another important role of the ECM in tissue development and homeostasis is its ability to act as a niche for stem cell differentiation. The niche represents a specialized local microenvironment which contributes to the establishment and maintenance of stem cell phenotype and stem cell differentiation. Recent studies provide strong evidence that the niche is composed of both soluble factors and ECM macromolecules which direct cell fate.[27,66–68] The ECM composition and the biomechanical properties of the ECM within the niche have shown to play a role in cell fate.

It is now widely accepted that stem cells are present within all tissues of adult mammals and that such cells are associated with a unique niche. However, the anatomic niche has only been defined in a few select tissues. For example, neural stem cells are known to localize along blood vessels of the subventricular zone of the brain.[69] Within this environment the cells adhere to laminin on the vascular basement membrane, which has been suggested to be essential for the maintenance of stem cell properties within this niche. In another example, hematopoietic stem cells are found within the endosteum and their niche is rich in osteopontin secreted by osteoblasts.[70,71] Osteopontin has been shown to regulate both adherence and quiescence of the cells within the niche. The degradation and remodeling of the ECM within the stem cell niche is thought to mediate the activation and release of cells from the niche. A number of ECM properties, including composition, topography, and biomechanics, regulate the subsequent migration and differentiation of these stem cells in tissues following their release.[26,27]

The ability to deliver or recruit and then differentiate stem cell populations is considered a key aspect of regenerative medicine applications. In some applications, cells are delivered within a scaffold-based material into an injury site; in other applications stem cells are recruited from endogenous sources through a number of mechanisms including the activity of growth factors.

3. DECELLULARIZATION AND FABRICATION METHODS

As described above, the ECM can be thought of as a highly complex, tissue-specific reservoir of structural and functional proteins capable both of being remodeled by resident cells and of directing cellular behavior, phenotype, and survival. ECM is known to play a role in the development and maintenance of stem cell phenotype, and these processes are exquisitely regulated, potentially involving multiple ECM components. Therefore, it is desirable to maintain these components and their three-dimensional organization to the highest degree possible. The methods by which ECM is isolated from source tissue and is processed as a scaffold for tissue reconstruction applications should be specifically tailored to the tissue of interest.[72,73] However, the desire to maintain the ultrastructure and ligand landscape of the ECM must be balanced against the need to remove as much of the cellular content as possible to avoid a potentially adverse immune response.[74,75]

Generally, decellularization of source tissue involves a combination of physical, ionic, chemical, and enzymatic methods.[72,73,76] Each of these decellularization agents and the manner in which they are applied should be tailored to the characteristics of the source tissue of interest. These characteristics may include thickness, density, and intended clinical application of the matrix material. A full review of the methods commonly employed to achieve tissue decellularization is beyond the scope of this chapter, but has been reviewed in-depth elsewhere.[72,73,76] The effects of inefficient decellularization and/or the use of overly harsh methods to achieve decellularization have also been described.[75,77–79] Briefly, excessive remnant cellular constituents within an ECM scaffold or significant disruption of the native architecture and growth factor content by harsh processing methods have been shown to promote a proinflammatory process which adversely affects tissue remodeling upon implantation. Similarly, the chemical cross-linking used to mask cellular epitope and/or to increase mechanical properties in many commercially available products significantly disrupts the ligand landscape of the material and prevents the release of cryptic peptides from the matrix material. Such cross-linking limits constructive remodeling and promotes a foreign body and encapsulation type response following implantation.

The physical configuration of the ECM scaffold following decellularization is often dependent upon the three-dimensional shape of the source tissue and the mechanical processing methods used to remove excess or irrelevant tissue prior to decellularization. The majority of the clinically available ECM products are single or multilaminate sheets (Table 1). However, ECM materials can be processed into powders and three-dimensional constructs depending on the particular application of interest.[80,81] ECM hydrogels have also been described.[81–83] These materials have been harvested from a wide variety of tissues and organs and provide a tissue-specific, injectable, thermally sensitive, hydrogel which can be used in minimally invasive or space-filling applications. Such hydrogels have also been used to coat synthetic implants, which in turn modulate the host response following placement and leads to improved incorporation.[84,85] Additionally, though not a focus of the present chapter, the decellularization of many whole, intact organs has been performed largely via perfusion of the tissues with decellularization agents.[72,86,87] These decellularized organs maintain much of the native matrix including the vascular and lymphatic networks which are key for subsequent repopulation of the decellularized organ with host cells.

4. TRANSLATIONAL APPLICATIONS OF ECM IN REGENERATIVE MEDICINE

There are more than 30 commercially available ECM-based scaffold materials on the market as of the publication of this chapter. These materials vary in their source tissue and species, method of decellularization and sterilization, and three-dimensional form. These materials also vary in their indications for clinical use. Most commonly, ECM-based materials are regulated as surgical mesh materials (i.e., devices) and allowed for the reinforcement of soft tissues where weakness exists. However, it should be noted that as more complex forms and applications of ECM technology are developed, some materials may be regulated as biologics. We herein review three emerging applications of ECM-based scaffold materials in challenging anatomic sites where few clinically effective therapeutic solutions exist. The mechanisms by which ECM scaffolds may be capable of promoting the constructive remodeling outcomes observed in these applications are then reviewed.

4.1 Esophageal Disease

Barrett's esophagus and esophageal adenocarcinoma represent the sixth leading cause of cancer death worldwide and rates of esophageal cancer are increasing yearly.[88,89] Treatment of high-grade dysplasia (HGD) and cancer of the esophageal mucosa present significant challenges due to the high propensity of stricture of the esophagus. Stated differently, the default response of the esophagus to injury is fibrotic scar tissue with associated clinical stricture. Therefore esophagectomy with associated morbidity remains the standard of care for patients with HGD or early stage neoplasia.[90,91] Alternative effective methods for treatment of HGD and early stage cancer in the esophagus without the need for esophagectomy are desirable. Endoscopic resection has emerged as a promising treatment for HGD and early adenocarcinoma.[92–95] However, these methods are limited by disease recurrence, the frequent need for concurrent radiofrequency exposure, and ablation of the involved tissue. Endoscopic resection is also limited by the size of the nodule that can be removed, often requiring piecemeal resection and limiting the accuracy and utility of subsequent histologic evaluation. A recent study demonstrated that en bloc resection of a large, full circumference section of the mucosa was possible.[96] However, this method requires the prevention of stricture subsequent to large-scale disruption of the mucosal surface.

Preclinical canine studies utilizing ECM scaffold materials have shown that ECM is an effective material for reconstruction of the esophagus (Figure 1).[97–100] An approach to reconstruction of the esophagus included the placement of xenogeneic ECM derived from porcine urinary bladder and showed that full thickness defects that included approximately 40–50% of the circumference and 5cm of length could facilitate a constructive, nonstenotic healing response with formation of all layers of the esophageal wall in a preclinical dog model.[100] This neotissue was both functional and innervated.[101] Although similar remodeling was not observed when a full circumference, full thickness resection was performed, restoration of a functional mucosa was observed when the subjacent muscularis externa was left intact.[99]

Based upon these preclinical findings, five patients with early stage esophageal cancer were treated using long segment, circumferential resection of the mucosa and submucosa with subsequent placement of a tube-shaped ECM scaffold material derived from porcine small intestine over the resected surface.[102] The ECM material was held in place by an expandable stent which was removed between 9 and 14days posttreatment. Results showed that the ECM scaffold material remodeled rapidly resulting in the formation of a new epithelium and submucosal tissue layer. All patients required transient dilation for soft stricture, but did not experience recurrence of the disease or long-term stricture. A recent report in a small cohort of patients with localized esophageal defects showed that esophagoplasty with ECM augmented repair restored structure and function.[103]

4.2 Volumetric Muscle Loss

The incidence of volumetric muscle loss is increasing due to increased survival following en bloc resection of extremity tumors and increased incidence of battlefield injuries. While skeletal muscle has a limited capacity for regeneration following injury, it is generally accepted that a 20% or greater loss of muscle volume will result in the deposition of scar tissue, chronic weakness, and loss of function rather than regeneration.[104–113] There are no reproducible clinically effective options for reconstruction following large volumetric muscle loss. Current techniques may include autologous tissue transfer of vascularized or free muscle flaps. While these methods may provide potential cosmetic improvement, there is little restoration of function. Further, these methods are often not amenable to the reconstruction of large volumetric defects.

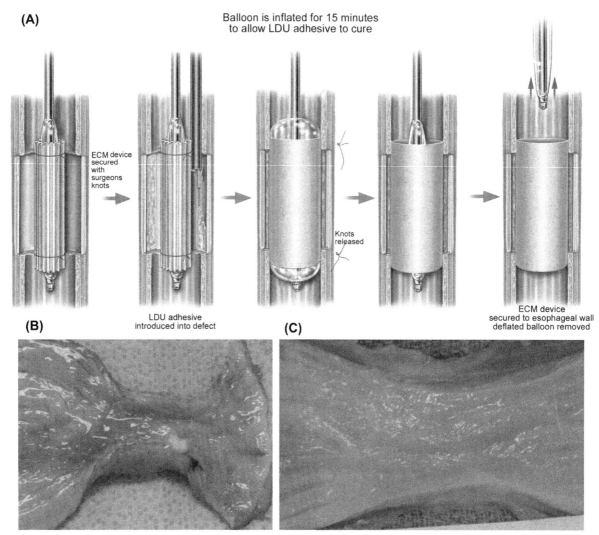

FIGURE 1 Use of extracellular matrix (ECM) scaffold material for reconstruction of postendomucosal resection (EMR) in a canine model. Schematic of the surgical deployment of ECM device with an achalasia balloon and delivery of the surgical adhesive (A). EMR is performed and tubular ECM scaffold is deployed using achalasia balloon. A lysine-derived urethane (LDU) surgical adhesive (TissuGlu, Cohera Medical) was used to secure ECM scaffold in place. Balloon was inflated and maintained for 15 min to allow adherence of the ECM scaffold to the esophagus prior to removal. Gross view of the remodeling EMR areas at 2 months after surgery (B, C). The control (B) shows pronounced stricture with reductions in the circumference and the length of the injury site. In contrast, the EMR site treated with ECM shows a smooth mucosal surface and limited circumferential and longitudinal reduction (C). *Reproduced from Ref. 97 with permission from Elsevier.*

ECM scaffold-based approaches to reconstruction of skeletal muscle have been shown to promote the formation of functional, innervated, and contractile skeletal muscle in preclinical models of volumetric muscle loss.[114–116] In one such study, a scaffold material composed of porcine small intestinal submucosa was fabricated into a three-dimensional shape and implanted into a canine model of musculotendinous junction repair.[116] In this model, the distal third of the gastrocnemius and musculotendinous junction were completely removed and replaced with the ECM implant. The results of this study showed that the implant promoted the formation of vascularized, functionally innervated skeletal muscle that was nearly indistinguishable from native muscle by 6 months postimplantation. These findings have now been translated into a treatment for human patients who have suffered volumetric muscle loss.[117] For example, placement of an ECM scaffold in a large quadriceps muscle injury in a 22-year-old male for whom all previous treatments had proven unsuccessful resulted in the restoration of new functional skeletal muscle and a significant increase in isokinetic performance and quality of life.

4.3 Temporomandibular Joint Meniscectomy

Temporomandibular joint disorders (TMJD) encompass a wide spectrum of clinical conditions involving the components of the temporomandibular joint (TMJ).[118–121] The etiology of TMJD is largely unknown due to the variety of suspected causes

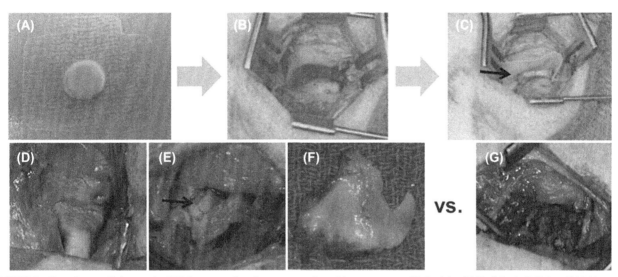

FIGURE 2 Use of extracellular matrix (ECM) as a template for reconstruction of the temporomandibular joint (TMJ) disk. An ECM scaffold (A) composed of porcine urinary bladder matrix was placed into the TMJ space (C, arrow indicates implant) following removal of the TMJ disk (B). Results demonstrated that the material was rapidly remodeled, acting as an interpositional material between the condyle and fossa (D). At explant the remodeled disk (E, arrow indicates explanted material) highly resembled native tissue (F). This tissue was also histologically and biomechanically similar to native tissues (not shown) and was shown to include integration with the lateral muscular and ligamentous attachments. This was in direct contrast to the contralateral side (G) which was left empty and resulted in the deposition of a small amount of granulation tissue and significant degenerative changes to the joint.

and their multifactorial nature. Symptoms of TMJD range from mild pain and clicking of the joint to chronic, intractable pain and limited jaw motion. It is estimated that TMJD affects 10–36 million Americans, 90% of which are women between the ages of 18 and 40. For a percentage of these patients, the only treatment which will relieve pain and restore motion to the jaw is to remove the fibrocartilaginous TMJ disk.[122–129]

Currently, no alloplastic alternatives exist to safely and effectively replace a degenerative, nonrepairable TMJ disk. Previous attempts to use alloplastic materials have resulted in unsatisfactory outcomes, including increased joint pathology, among other complications.[130–133] Several autogenous tissues, such as temporalis muscle, auricular cartilage, dermis, and abdominal adipose tissue, have been used as replacement materials, but only short-term success has been reported.[134–140] In addition, the use of these tissues has been associated with donor site morbidity, the eventual formation of scar tissue, decreased range of motion of the mandible, and additional joint pathology. Studies have documented a reduction of joint pain after discectomy without a replacement procedure; however, these patients inevitably experience varying degrees of subsequent degenerative changes.[123,124,141] Thus, the identification of a suitable off-the-shelf disk replacement material would obviate the associated donor site morbidity and avoid downstream degenerative changes to the condyle. Ideally, such a material would also act as a template for cellular ingrowth, integrate with the surrounding host tissues, and eventually restore the native morphology and function of the TMJ disk.

A number of tissue engineering and regenerative medicine approaches to the replacement of the TMJ disk have been suggested.[142–153] However, to date, these studies have focused primarily on the selection of ideal cell sources, growth factors, and scaffold materials for the engineering of tissues that recapitulate the TMJ disk in vitro. Many of these approaches involve long culture times and would be considered difficult to implement both from practical standpoint and from a regulatory standpoint. A recent study demonstrated that a device composed of decellularized porcine urinary bladder matrix alone was capable of providing an effective interpositional material while serving as an inductive template for reconstruction of the TMJ disk in vivo (Figure 2).[154] In that study, a device consisting of a powdered urinary bladder matrix (UBM) "pillow" encapsulated within sheets of the same material was placed as an interpositional graft after discectomy in a canine model. The implanted material was observed to progressively remodel from 3 weeks to 6 months after implantation, and the newly formed host tissues resembled the native fibrocartilage of the TMJ disk in both gross and histologic morphology. A follow-up study of 10 dogs demonstrated that the composition and mechanical properties of the remodeled tissue were also similar to that of the native disk.[155] Of note, the placement of the UBM device resulted in formation of fibrocartilage not only within the bulk of the implant, but also within muscular and ligamentous attachments resembling those found at the periphery of native menisci.

5. MECHANISMS OF CONSTRUCTIVE REMODELING

Each of the examples provided above demonstrates the phenomenon of ECM scaffold-mediated constructive remodeling. In each case, an acellular scaffold is populated by host cells, and gradually degrades and remodels, resulting in replacement

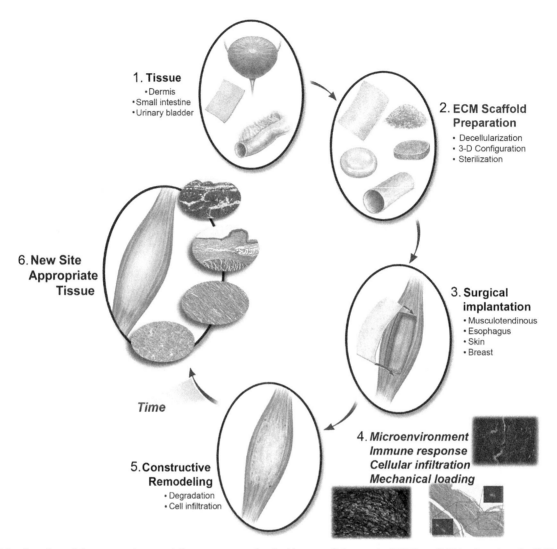

FIGURE 3 Overview of the constructive remodeling process associated with extracellular matrix (ECM) scaffold implantation. Scaffold materials obtained from tissue decellularization are processed into application-specific formulations. Upon implantation, the material provides a microenvironment for ingrowth of cells and mechanotransduction. The material is degraded rapidly resulting in modulation of the innate and adaptive immune response and recruitment of progenitor cells. Over time, these processes result in constructive remodeling—the formation of new, site-appropriate, functional host tissues. *Reproduced from Ref. 19 with permission from Elsevier.*

of the material with functional, site-appropriate host tissue. These outcomes are distinctly different from the default mammalian response to tissue injury which typically induces scar tissue formation and loss of function. In each example, the scaffold was derived from xenogeneic (porcine) small intestine or urinary bladder, despite the application in heterologous (i.e., nonintestine or bladder applications). The mechanisms by which such scaffold materials are capable of promoting constructive remodeling in diverse applications and heterologous anatomic locations is not yet fully understood, but have been shown to include exposure to mechanical forces, modulation of the host immune response, and degradation of the scaffold material with recruitment of endogenous stem cells. In the absence of any of these key factors, constructive remodeling does not occur and outcomes are undesirable. We review each of these physiologic variables below, with a particular focus upon the role of ECM degradation in the constructive remodeling process. An overview of the constructive remodeling process associated with ECM implantation is shown in Figure 3.

5.1 Mechanical Forces

Several studies have shown that early site-appropriate mechanical loading facilitates the remodeling of ECM scaffold materials into anatomically site-specific tissue.[156,157] In a canine preclinical study, partial cystectomies repaired with an

ECM scaffold material were exposed to long-term catheterization and prevention of bladder filling, with an associated lack of cyclic distention and decrease in maximal bladder distention, and results were compared to bladders that experienced an early return to normal micturition following ECM scaffold implantation.[156] The presence of physiologic mechanical loading in the early postsurgical period promoted remodeling of the ECM scaffold material into tissue with a highly differentiated transitional urothelium, vasculature, innervation, and islands of smooth muscle cells. Delayed return of normal mechanical loading was insufficient to overcome the lack of early mechanical signals and resulted in degradation of the ECM scaffold material, dense scar tissue deposition, and absence of constructive remodeling. Similar results have also been observed in a study of Achilles tendon repair in rabbits with and without postsurgical immobilization.[157]

The mechanisms by which ECM scaffolds promote site-specific remodeling in the presence of site-appropriate mechanical loading have only been partially elucidated. Static and cyclic stretching of cells seeded on an ECM scaffold in vitro modulated the collagen expression by the cells, enhanced the cell and collagen alignment, and improved the mechanical behavior of the scaffold.[158–162] During in vivo remodeling, it is thought that the progenitor cells recruited to the site of ECM remodeling will differentiate into site-appropriate cells in the presence of local mechanical cues.[12,163–167] Several in vitro studies have shown that mechanical loading can induce progenitor cells to differentiate into fibroblasts, smooth muscle cells, and osteoblasts.[168–170] Finally, and as described above, mechanical loading of the ECM may lead to the exposure of cryptic ligand motifs and thereby initiate ECM multimerization and cell responses which lead to constructive remodeling.

5.2 Modulation of the Host Response

Materials derived from mammalian tissue sources elicit a distinctly different host response than those composed of synthetic materials due to their unique surface topologies and naturally derived ligand landscapes. Further, naturally derived materials likely experience adsorption of a different repertoire of molecules following in vivo implantation than do synthetic materials and possess inherent surface functionality. The constructive remodeling response elicited by ECM-based scaffold materials has been shown to be dependent upon the ability to both elicit and modulate the host response.[44,74,75,171–174] In general, the host response to a nonchemically cross-linked ECM scaffold material consists of neutrophils at early time points (<48 h) following implantation changing to a prominent macrophage response by 72 h postimplantation.[43] The macrophage response to the scaffold material persists through much of the early remodeling process and as long as several months postimplantation depending on the clinical application. This intense and long-term macrophage response to a biomaterial implant is conventionally associated with negative implications, including chronic inflammation and fibrous encapsulation or scarring.[175] However, the presence of these cells, and macrophages in particular, is essential for the promotion of a constructive remodeling response.[44,75,173,174] A recent study that investigated the response to ECM scaffolds in animals treated with clodronate to remove circulating mononuclear phagocytes showed that the ECM scaffold was not degraded or remodeled, suggesting a prominent, if not necessary, role for circulating mononuclear cells in the constructive remodeling process.[44]

Subsequent investigations have revealed that ECM scaffolds both promote the host cellular response and modulate the phenotype of the cells which participate.[74,75,171–174,176] Briefly, scaffold materials composed of ECM promote a switch from a predominantly M1 macrophage (proinflammatory, cytotoxic) population immediately following implantation to a population enriched in M2 macrophages (anti-inflammatory and prohealing) by 7–14 days postimplantation.[75,173,174] In addition to eliciting and modulating the host macrophage response, ECM scaffolds have been shown to consistently evoke a Th2-type immune response,[171,172] which is generally associated with transplant acceptance.[177–179] The mechanisms by which ECM-based scaffold materials promote the M1 to M2 and Th1 to Th2 transition remain unknown. However, the phenotypic profile of the immune cells which respond to these scaffold materials at early time points has now been shown to be a strong statistical predictor of the downstream outcome associated with their implantation.[174] For example, modification of such scaffold materials with chemical cross-linking agents which delay or prevent macrophage-mediated degradation inhibits the formation of the beneficial M2 response, promotes the M1 response, and results in downstream scar tissue formation.[173,174,180] These results suggest that interactions of host cells with intact ligands on the surface of the material, or their degradation products, may be responsible for the observed phenomena. However, a recent study showed that both the structural and soluble components of ECM scaffold degradation products elicited a shift in macrophage phenotype toward an M2 response, albeit through different mechanisms,[181] suggesting that both the structural and soluble fractions of ECM scaffold degradation products play an important and interrelated role.

5.3 ECM Scaffold Degradation

ECM scaffolds are rapidly degraded in vivo. One study showed that 10-layer ^{14}C-labeled ECM scaffolds were 60% degraded at 30 days postimplantation and 100% at 90 days postsurgery in a model of canine Achilles tendon repair.[182] During this period, the scaffold was populated and degraded by host cells and resulted in the formation of site-specific functional host tissue. The major route of excretion of the degraded scaffold was found to be via hematogenous circulation and elimination by the kidneys and exhaled CO_2. The mechanisms of in vivo degradation of ECM scaffolds are complex and include both cellular and enzymatic pathways. The process is mediated by inflammatory cells, such as macrophages, which produce reactive oxygen species and proteolytic enzymes that aid in the degradation of the matrix. Another study in a rat model utilizing ^{14}C-labeled ECM scaffolds showed that peripheral blood monocytes are required for the early and rapid degradation of both ECM scaffolds, and that cross-linked ECM scaffolds are resistant to macrophage-mediated degradation.[44]

ECM scaffolds have also been degraded in vitro by chemical and physical methods. Recent findings suggest that the degradation products of ECM scaffolds are bioactive.[8,9,11,12,82,163,164,183–185] Studies have shown antimicrobial activity associated with the degradation products of ECM scaffolds;[7,8] however, in the absence of degradation, antimicrobial activity was not seen, suggesting that some of the bioactive properties of the ECM are derived from its degradation products, rather than from whole molecules present in the ECM.[186] Degradation products of ECM scaffolds have also been shown to be chemoattractants for progenitor and nonprogenitor cell populations.[9–12,163,183] An ECM scaffold that cannot degrade (i.e., is chemically cross-linked) may not release bioactive degradation products, including those bioactive molecules that may be responsible for modulating the host response toward constructive remodeling. Furthermore, surface characterization studies have shown that chemical cross-linking alters the ligand landscape of the material potentially affecting ligand–receptor interactions important in determining cell–scaffold interactions.[77,78]

One of the biological effects of ECM scaffold degradation is the recruitment of host stem and progenitor cells to the site of degradation. A study of ECM scaffold remodeling in a mouse Achilles tendon injury model examined the ability of ECM scaffolds and autograft tissue to recruit bone marrow-derived cells.[166] Bone marrow-derived cells were observed in the sites of remodeling associated with both ECM scaffolds and autograft control tissue, among the predominantly mononuclear cell population at early time points (1 and 2 weeks) postsurgery. Both scaffold types remodeled into tissue resembling the native Achilles tendon; however, by 16 weeks, the presence of bone marrow-derived cells was observed only in the ECM-treated group. Another study, also utilizing a model of mouse Achilles tendon injury, examined the ability of ECM scaffold explants to cause the chemotaxis of progenitor cells after 3, 7, and 14 days of in vivo remodeling.[164] The results of the study showed greater migration of progenitor cells toward tendons repaired with ECM scaffolds, compared to tendons repaired with autologous tissue and uninjured normal contralateral tendon.

While these studies have shown the effects of ECM scaffold implantation and subsequent degradation in vivo, additional studies have investigated the delivery of ECM scaffold degradation products in place of whole ECM scaffolds.[163,183–185] The injection of peptides derived by pepsin degradation of a porcine small intestinal ECM scaffold resulted in the recruitment of multipotent progenitor cells in a model of mouse digit amputation.[163,183] These cells expressed a number of markers of multipotency including Sox2, Sca1, and Rex1. When isolated from the site of injury, these cells were shown to be able to differentiate along mesodermal and neuroectodermal lineages.[163,183] A proteomics approach to identifying the peptides responsible showed that a single α subunit of the collagen III molecule could promote the chemotaxis of multiple progenitor cells in vitro and was able to recruit the Sox2, Sca1 positive progenitor cells in vivo.[163] Additional studies showed that this peptide was able to promote the osteogenic differentiation of human perivascular cells in vitro.[184]

5.4 Undesirable Responses to ECM Scaffold Materials

While the primary focus of this chapter is a description of the rationale for the use of ECM as a bioscaffold and the known mechanisms by which constructive remodeling is facilitated, it must be noted that variations in outcomes have also been reported.[43,174,187,188] As discussed above, although the mechanisms of constructive remodeling are only partially understood, there are certain aspects of preparation and use of ECM bioscaffolds that may account for variable results.

First, the ability of an ECM scaffold material to act as a highly conserved inductive template for constructive tissue remodeling is dependent upon thorough decellularization.[73,76] As is logical, a scaffold which has not been fully decellularized will contain remnant DNA and epitopes such as the α-gal epitope as well as other cell debris. These molecules are well characterized for their ability to promote an inflammatory or rejection type response.[189–191] ECM scaffold materials containing large amounts of cellular content following decellularization have been shown to promote a more inflammatory, M1 type, response and result in scar tissue deposition rather than constructive remodeling.[75] While the degree of inflammatory response appears to be linked to the quantity of remnant cellular content, an exact threshold beyond which the constructive remodeling response is affected is unknown.[79]

Second, the use of chemical cross-linking reagents such as carbodiimide or glutaraldehyde has been shown to alter the ligand landscape of the material and prevent ECM scaffold degradation.[43,44,77] The conversion to a nondegradable material is clearly associated with a foreign body reaction and downstream encapsulation rather than constructive remodeling.[43,174] Figure 4 demonstrates the host remodeling response to three different ECM scaffold materials following implantation into

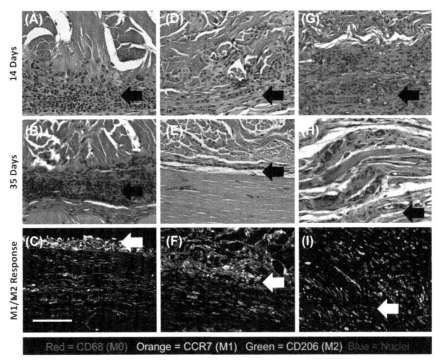

FIGURE 4 Outcomes following 14 and 35 day implantation of three different extracellular matrix-based biomaterials in a rodent abdominal wall defect model. Results demonstrated that those scaffold materials which were cross-linked (Collamend, BARD) were associated with a foreign body type response and an M1 type macrophage response (A–C). Scaffold materials which were not cross-linked, but degraded slowly (InteXen, American Medical Systems) were associated with a dense mononuclear cell response early, with reduction of the inflammatory response at later times (D–E). These materials were associated with a mixed M1/M2 macrophage phenotype (F). Scaffold materials which were noncross-linked and degraded rapidly (MatriStem, ACell) were associated with a more polarized M2 response and showed signs of early constructive remodeling (G–I). Hematoxylin and eosin (A–B, D–E, G–H) and immunofluorescent labeling of macrophages (14 days) are shown. Image magnification = 40×. Scale bar = 100 μm. CD68 (pan-macrophage) = red (light gray in print versions), CCR7 (M1) = orange (lighter gray in print versions), CD206 (M2) = green (gray in print versions), DRAQ5 (nuclei) = blue (dark gray in print versions). Arrow = interface between scaffold and underlying native tissue. Asterisk = new skeletal muscle bundle formation. *Reproduced from Ref. 174 with permission from Elsevier.*

a rodent partial thickness abdominal wall defect model.[174] Similarly, harsh processing methods which disrupt the ECM and remove selected growth factors and other components critical to a constructive remodeling process can account for poor clinical outcomes.[77] The use of strong detergents such as sodium dodecyl sulfate or deoxycholate can have disruptive effects upon basement membrane structures.[192]

Finally, it appears to be essential that materials are not only prepared in such a manner as to avoid an undesirable response, but also are used in an application appropriate manner. Each clinical application will pose specific challenges which must be considered in the use of an ECM scaffold material. Three such examples are provided above. Of particular importance is the exposure of the matrix material to the normal local tissue microenvironment as well as local mechanical forces. In the absence of the application of mechanical stimuli, remodeling of ECM scaffold materials has largely resulted in degradation of the scaffold material without remodeling.[156,157]

6. CONCLUSIONS

The ECM is a highly complex and highly dynamic structural and functional microenvironment which is both dependent upon and a critical determinant of cell phenotype and behavior. For these reasons, ECM represents an ideal biomaterial for tissue reconstruction. ECM is commonly used as a simple surgical mesh to bridge and reinforce tissues, and also has the potential to act as an inductive template for constructive remodeling. Successful use of ECM scaffolds as inductive templates in complex and challenging applications requires an understanding of the potential for ECM to define or modulate the injury microenvironment in an application-specific manner. Key factors leading to constructive remodeling outcomes include the use of appropriate processing in scaffold preparation, application of site-appropriate mechanical forces, modulation of the host immune response, and degradation of the scaffold with release of bioactive cryptic peptides. Disruption of any of these processes may negatively affect downstream outcomes. A better understanding of these factors will logically lead to improved clinical outcomes associated with the use of ECM-based scaffold materials.

ACKNOWLEDGMENTS

The authors do not have any financial or personal relationships which influenced the preparation of this chapter.

REFERENCES

1. Bissell MJ, Hall HG, Parry G. How does the extracellular matrix direct gene expression? *J Theor Biol* 1982;**99**:31–68.

2. Houghton AM, Grisolano JL, Baumann ML, Kobayashi DK, Hautamaki RD, Nehring LC, et al. Macrophage elastase (matrix metalloproteinase-12) suppresses growth of lung metastases. *Cancer Res* 2006;**66**:6149–55.

3. O'Reilly MS, Boehm T, Shing Y, Fukai N, Vasios G, Lane WS, et al. Endostatin: an endogenous inhibitor of angiogenesis and tumor growth. *Cell* 1997;**88**:277–85.

4. Colorado PC, Torre A, Kamphaus G, Maeshima Y, Hopfer H, Takahashi K, et al. Anti-angiogenic cues from vascular basement membrane collagen. *Cancer Res* 2000;**60**:2520–6.

5. Vlodavsky I, Goldshmidt O, Zcharia E, Atzmon R, Rangini-Guatta Z, Elkin M, et al. Mammalian heparanase: involvement in cancer metastasis, angiogenesis and normal development. *Semin Cancer Biol* 2002;**12**:121–9.

6. Ramchandran R, Dhanabal M, Volk R, Waterman MJ, Segal M, Lu H, et al. Antiangiogenic activity of restin, NC10 domain of human collagen XV: comparison to endostatin. *Biochem Biophys Res Commun* 1999;**255**:735–9.

7. Sarikaya A, Record R, Wu CC, Tullius B, Badylak S, Ladisch M. Antimicrobial activity associated with extracellular matrices. *Tissue Eng* 2002;**8**:63–71.

8. Brennan EP, Reing J, Chew D, Myers-Irvin JM, Young EJ, Badylak SF. Antibacterial activity within degradation products of biological scaffolds composed of extracellular matrix. *Tissue Eng* 2006;**12**:2949–55.

9. Brennan EP, Tang XH, Stewart-Akers AM, Gudas LJ, Badylak SF. Chemoattractant activity of degradation products of fetal and adult skin extracellular matrix for keratinocyte progenitor cells. *J Tissue Eng Regen Med* 2008;**2**:491–8.

10. Haviv F, Bradley MF, Kalvin DM, Schneider AJ, Davidson DJ, Majest SM, et al. Thrombospondin-1 mimetic peptide inhibitors of angiogenesis and tumor growth: design, synthesis, and optimization of pharmacokinetics and biological activities. *J Med Chem* 2005;**48**:2838–46.

11. Li F, Li W, Johnson S, Ingram D, Yoder M, Badylak S. Low-molecular-weight peptides derived from extracellular matrix as chemoattractants for primary endothelial cells. *Endothelium* 2004;**11**:199–206.

12. Reing JE, Zhang L, Myers-Irvin J, Cordero KE, Freytes DO, Heber-Katz E, et al. Degradation products of extracellular matrix affect cell migration and proliferation. *Tissue Eng Part A* 2009;**15**:605–14.

13. Calve S, Odelberg SJ, Simon HG. A transitional extracellular matrix instructs cell behavior during muscle regeneration. *Dev Biol* 2010;**344**:259–71.

14. Murry CE, Keller G. Differentiation of embryonic stem cells to clinically relevant populations: lessons from embryonic development. *Cell* 2008;**132**:661–80.

15. Cortiella J, Niles J, Cantu A, Brettler A, Pham A, Vargas G, et al. Influence of acellular natural lung matrix on murine embryonic stem cell differentiation and tissue formation. *Tissue Eng Part A* 2010;**16**:2565–80.

16. Badylak SF. The extracellular matrix as a scaffold for tissue reconstruction. *Semin Cell Dev Biol* 2002;**13**:377–83.

17. Badylak SF. Xenogeneic extracellular matrix as a scaffold for tissue reconstruction. *Transpl Immunol* 2004;**12**:367–77.

18. Badylak SF. The extracellular matrix as a biologic scaffold material. *Biomaterials* 2007;**28**:3587–93.

19. Badylak SF, Brown BN, Gilbert TW, Daly KA, Huber A, Turner NJ. Biologic scaffolds for constructive tissue remodeling. *Biomaterials* 2011;**32**:316–9.

20. Daley WP, Peters SB, Larsen M. Extracellular matrix dynamics in development and regenerative medicine. *J Cell Sci* 2008;**121**:255–64.

21. Lu P, Takai K, Weaver VM, Werb Z. Extracellular matrix degradation and remodeling in development and disease. *Cold Spring Harb Perspect Biol* 2011;**3**.

22. Egeblad M, Rasch MG, Weaver VM. Dynamic interplay between the collagen scaffold and tumor evolution. *Curr Opin Cell Biol* 2010;**22**:697–706.

23. Rozario T, DeSimone DW. The extracellular matrix in development and morphogenesis: a dynamic view. *Dev Biol* 2010;**341**:126–40.

24. Brown B, Lindberg K, Reing J, Stolz DB, Badylak SF. The basement membrane component of biologic scaffolds derived from extracellular matrix. *Tissue Eng* 2006;**12**:519–26.

25. Engler AJ, Humbert PO, Wehrle-Haller B, Weaver VM. Multiscale modeling of form and function. *Science* 2009;**324**:208–12.

26. Engler AJ, Sen S, Sweeney HL, Discher DE. Matrix elasticity directs stem cell lineage specification. *Cell* 2006;**126**:677–89.

27. Reilly GC, Engler AJ. Intrinsic extracellular matrix properties regulate stem cell differentiation. *J Biomech* 2010;**43**:55–62.

28. Kumbar SG, James R, Nukavarapu SP, Laurencin CT. Electrospun nanofiber scaffolds: engineering soft tissues. *Biomed Mater* 2008;**3**:034002.

29. Barnes CP, Sell SA, Boland ED, Simpson DG, Bowlin GL. Nanofiber technology: designing the next generation of tissue engineering scaffolds. *Adv Drug Deliv Rev* 2007;**59**:1413–33.

30. Hynes RO, Naba A. Overview of the matrisome—an inventory of extracellular matrix constituents and functions. *Cold Spring Harb Perspect Biol* 2012;**4**:a004903.

31. Bernard MP, Chu ML, Myers JC, Ramirez F, Eikenberry EF, Prockop DJ. Nucleotide sequences of complementary deoxyribonucleic acids for the pro alpha 1 chain of human type I procollagen. Statistical evaluation of structures that are conserved during evolution. *Biochemistry* 1983;**22**:5213–23.

32. Bernard MP, Myers JC, Chu ML, Ramirez F, Eikenberry EF, Prockop DJ. Structure of a cDNA for the pro alpha 2 chain of human type I procollagen. Comparison with chick cDNA for pro alpha 2(I) identifies structurally conserved features of the protein and the gene. *Biochemistry* 1983;**22**:1139–45.

33. Constantinou CD, Jimenez SA. Structure of cDNAs encoding the triple-helical domain of murine alpha 2 (VI) collagen chain and comparison to human and chick homologues. Use of polymerase chain reaction and partially degenerate oligonucleotide for generation of novel cDNA clones. *Matrix* 1991;**11**:1–9.

34. Exposito JY, D'Alessio M, Solursh M, Ramirez F. Sea urchin collagen evolutionarily homologous to vertebrate pro-α2(I) collagen. *J Biol Chem* 1992;**267**:15559–62.

35. Brown NH. Extracellular matrix in development: insights from mechanisms conserved between invertebrates and vertebrates. *Cold Spring Harb Perspect Biol* 2011;**3**.

36. Davis GE. Matricryptic sites control tissue injury responses in the cardiovascular system: relationships to pattern recognition receptor regulated events. *J Mol Cell Cardiol* 2010;**48**:454–60.

37. Davis GE, Bayless KJ, Davis MJ, Meininger GA. Regulation of tissue injury responses by the exposure of matricryptic sites within extracellular matrix molecules. *Am J Pathol* 2000;**156**:1489–98.

38. Hynes RO. The extracellular matrix: not just pretty fibrils. *Science* 2009;**326**:1216–9.

39. Boudreau N, Myers C, Bissell MJ. From laminin to lamin: regulation of tissue-specific gene expression by the ECM. *Trends Cell Biol* 1995;**5**:1–4.

40. Ingber D. Extracellular matrix and cell shape: potential control points for inhibition of angiogenesis. *J Cell Biochem* 1991;**47**:236–41.

41. Fata JE, Werb Z, Bissell MJ. Regulation of mammary gland branching morphogenesis by the extracellular matrix and its remodeling enzymes. *Breast Cancer Res* 2004;**6**:1–11.

42. Sternlicht MD, Kouros-Mehr H, Lu P, Werb Z. Hormonal and local control of mammary branching morphogenesis. *Differentiation* 2006;**74**:365–81.

43. Valentin JE, Badylak JS, McCabe GP, Badylak SF. Extracellular matrix bioscaffolds for orthopaedic applications. A comparative histologic study. *J Bone Joint Surg Am* 2006;**88**:2673–86.

44. Valentin JE, Stewart-Akers AM, Gilbert TW, Badylak SF. Macrophage participation in the degradation and remodeling of extracellular matrix scaffolds. *Tissue Eng Part A* 2009;**15**:1687–94.

45. Cawston TE, Young DA. Proteinases involved in matrix turnover during cartilage and bone breakdown. *Cell Tissue Res* 2010;**339**:221–35.

46. Page-McCaw A, Ewald AJ, Werb Z. Matrix metalloproteinases and the regulation of tissue remodelling. *Nat Rev Mol Cell Biol* 2007;**8**:221–33.

47. Apte SS. A disintegrin-like and metalloprotease (reprolysin-type) with thrombospondin type 1 motif (ADAMTS) superfamily: functions and mechanisms. *J Biol Chem* 2009;**284**:31493–7.

48. Maquart FX, Bellon G, Pasco S, Monboisse JC. Matrikines in the regulation of extracellular matrix degradation. *Biochimie* 2005;**87**:353–60.

49. Hocking DC, Kowalski K. A cryptic fragment from fibronectin's III1 module localizes to lipid rafts and stimulates cell growth and contractility. *J Cell Biol* 2002;**158**:175–84.

50. Hocking DC, Titus PA, Sumagin R, Sarelius IH. Extracellular matrix fibronectin mechanically couples skeletal muscle contraction with local vasodilation. *Circ Res* 2008;**102**:372–9.

51. Smith ML, Gourdon D, Little WC, Kubow KE, Eguiluz RA, Luna-Morris S, et al. Force-induced unfolding of fibronectin in the extracellular matrix of living cells. *PLoS Biol* 2007;**5**:e268.

52. Vakonakis I, Staunton D, Rooney LM, Campbell ID. Interdomain association in fibronectin: insight into cryptic sites and fibrillogenesis. *EMBO J* 2007;**26**:2575–83.

53. Vogel V. Mechanotransduction involving multimodular proteins: converting force into biochemical signals. *Annu Rev Biophys Biomol Struct* 2006;**35**:459–88.

54. Gao M, Craig D, Lequin O, Campbell ID, Vogel V, Schulten K. Structure and functional significance of mechanically unfolded fibronectin type III1 intermediates. *Proc Natl Acad Sci USA* 2003;**100**:14784–9.

55. Mitsi M, Forsten-Williams K, Gopalakrishnan M, Nugent MA. A catalytic role of heparin within the extracellular matrix. *J Biol Chem* 2008;**283**:34796–807.

56. Ugarova TP, Zamarron C, Veklich Y, Bowditch RD, Ginsberg MH, Weisel JW, et al. Conformational transitions in the cell binding domain of fibronectin. *Biochemistry* 1995;**34**:4457–66.

57. Krammer A, Lu H, Isralewitz B, Schulten K, Vogel V. Forced unfolding of the fibronectin type III module reveals a tensile molecular recognition switch. *Proc Natl Acad Sci USA* 1999;**96**:1351–6.

58. Seiffert D, Smith JW. The cell adhesion domain in plasma vitronectin is cryptic. *J Biol Chem* 1997;**272**:13705–10.

59. Davis GE. Affinity of integrins for damaged extracellular matrix: $\alpha_v\beta_3$ binds to denatured collagen type I through RGD sites. *Biochem Biophys Res Commun* 1992;**182**:1025–31.

60. Smith LL, Cheung HK, Ling LE, Chen J, Sheppard D, Pytela R, et al. Osteopontin N-terminal domain contains a cryptic adhesive sequence recognized by $\alpha_9\beta_1$ integrin. *J Biol Chem* 1996;**271**:28485–91.

61. Hirano Y, Okuno M, Hayashi T, Goto K, Nakajima A. Cell-attachment activities of surface immobilized oligopeptides RGD, RGDS, RGDV, RGDT, and YIGSR toward five cell lines. *J Biomater Sci Polym Ed* 1993;**4**:235–43.

62. Hern DL, Hubbell JA. Incorporation of adhesion peptides into nonadhesive hydrogels useful for tissue resurfacing. *J Biomed Mater Res* 1998;**39**:266–76.

63. Hsiong SX, Huebsch N, Fischbach C, Kong HJ, Mooney DJ. Integrin-adhesion ligand bond formation of preosteoblasts and stem cells in three-dimensional RGD presenting matrices. *Biomacromolecules* 2008;**9**:1843–51.

64. LeBaron RG, Athanasiou KA. Extracellular matrix cell adhesion peptides: functional applications in orthopedic materials. *Tissue Eng* 2000;**6**:85–103.

65. Vidal G, Blanchi T, Mieszawska AJ, Calabrese R, Rossi C, Vigneron P, et al. Enhanced cellular adhesion on titanium by silk functionalized with titanium binding and RGD peptides. *Acta Biomater* 2013;**9**:4935–43.

66. Brizzi MF, Tarone G, Defilippi P. Extracellular matrix, integrins, and growth factors as tailors of the stem cell niche. *Curr Opin Cell Biol* 2012;**24**:645–51.

67. Kazanis I, ffrench-Constant C. Extracellular matrix and the neural stem cell niche. *Dev Neurobiol* 2011;**71**:1006–17.

68. Votteler M, Kluger PJ, Walles H, Schenke-Layland K. Stem cell microenvironments – unveiling the secret of how stem cell fate is defined. *Macromol Biosci* 2010;**10**:1302–15.

69. Shen Q, Wang Y, Kokovay E, Lin G, Chuang SM, Goderie SK, et al. Adult SVZ stem cells lie in a vascular niche: a quantitative analysis of niche cell–cell interactions. *Cell Stem Cell* 2008;**3**:289–300.

70. Kollet O, Dar A, Shivtiel S, Kalinkovich A, Lapid K, Sztainberg Y, et al. Osteoclasts degrade endosteal components and promote mobilization of hematopoietic progenitor cells. *Nat Med* 2006;**12**:657–64.

71. Lymperi S, Ferraro F, Scadden DT. The HSC niche concept has turned 31. Has our knowledge matured? *Ann NY Acad Sci* 2010;**1192**:12–8.

72. Crapo PM, Gilbert TW, Badylak SF. An overview of tissue and whole organ decellularization processes. *Biomaterials* 2011;**32**:3233–43.

73. Gilbert TW, Sellaro TL, Badylak SF. Decellularization of tissues and organs. *Biomaterials* 2006;**27**:3675–83.

74. Badylak SF, Gilbert TW. Immune response to biologic scaffold materials. *Semin Immunol* 2008;**20**:109–16.

75. Brown BN, Valentin JE, Stewart-Akers AM, McCabe GP, Badylak SF. Macrophage phenotype and remodeling outcomes in response to biologic scaffolds with and without a cellular component. *Biomaterials* 2009;**30**:1482–91.

76. Gilbert TW. Strategies for tissue and organ decellularization. *J Cell Biochem* 2012;**113**:2217–22.

77. Reing JE, Brown BN, Daly KA, Freund JM, Gilbert TW, Hsiong SX, et al. The effects of processing methods upon mechanical and biologic properties of porcine dermal extracellular matrix scaffolds. *Biomaterials* 2010;**31**:8626–33.

78. Barnes CA, Brison J, Michel R, Brown BN, Castner DG, Badylak SF, et al. The surface molecular functionality of decellularized extracellular matrices. *Biomaterials* 2011;**32**:137–43.

79. Keane TJ, Londono R, Turner NJ, Badylak SF. Consequences of ineffective decellularization of biologic scaffolds on the host response. *Biomaterials* 2012;**33**:1771–81.

80. Gilbert TW, Stolz DB, Biancaniello F, Simmons-Byrd A, Badylak SF. Production and characterization of ECM powder: implications for tissue engineering applications. *Biomaterials* 2005;**26**:1431–5.

81. Freytes DO, Martin J, Velankar SS, Lee AS, Badylak SF. Preparation and rheological characterization of a gel form of the porcine urinary bladder matrix. *Biomaterials* 2008;**29**:1630–7.

82. Medberry CJ, Crapo PM, Siu BF, Carruthers CA, Wolf MT, Nagarkar SP, et al. Hydrogels derived from central nervous system extracellular matrix. *Biomaterials* 2013;**34**:1033–40.

83. Zhang L, Zhang F, Weng Z, Brown BN, Yan H, Ma XM, et al. Effect of an inductive hydrogel composed of urinary bladder matrix upon functional recovery following traumatic brain injury. *Tissue Eng Part A* 2013;**19**:1909–18.

84. Faulk DM, Londono R, Wolf MT, Ranallo CA, Carruthers CA, Wildemann JD, et al. ECM hydrogel coating mitigates the chronic inflammatory response to polypropylene mesh. *Biomaterials* 2014;**35**:8585–95.

85. Wolf MT, Daly KA, Brennan-Pierce EP, Johnson SA, Carruthers CA, D'Amore A, et al. A hydrogel derived from decellularized dermal extracellular matrix. *Biomaterials* 2012;**33**:7028–38.

86. Badylak SF, Taylor D, Uygun K. Whole-organ tissue engineering: decellularization and recellularization of three-dimensional matrix scaffolds. *Annu Rev Biomed Eng* 2011;**13**:27–53.

87. Baptista PM, Orlando G, Mirmalek-Sani SH, Siddiqui M, Atala A, Soker S. Whole organ decellularization – a tool for bioscaffold fabrication and organ bioengineering. *Conf Proc IEEE Eng Med Biol Soc* 2009;**2009**:6526–9.

88. Enzinger PC, Mayer RJ. Esophageal cancer. *N Engl J Med* 2003;**349**:2241–52.

89. Ries LA, Wingo PA, Miller DS, Howe HL, Weir HK, Rosenberg HM, et al. The annual report to the nation on the status of cancer, 1973–1997, with a special section on colorectal cancer. *Cancer* 2000;**88**:2398–424.

90. Orringer MB, Marshall B, Iannettoni MD. Transhiatal esophagectomy for treatment of benign and malignant esophageal disease. *World J Surg* 2001;**25**:196–203.

91. Gawad KA, Hosch SB, Bumann D, Lubeck M, Moneke LC, Bloechle C, et al. How important is the route of reconstruction after esophagectomy: a prospective randomized study. *Am J Gastroenterol* 1999;**94**:1490–6.

92. Ell C, May A, Pech O, Gossner L, Guenter E, Behrens A, et al. Curative endoscopic resection of early esophageal adenocarcinomas (Barrett's cancer). *Gastrointest Endosc* 2007;**65**:3–10.

93. Namasivayam V, Wang KK, Prasad GA. Endoscopic mucosal resection in the management of esophageal neoplasia: current status and future directions. *Clin Gastroenterol Hepatol* 2010;**8**:743–54. quiz e96.

94. Chennat J, Konda VJ, Ross AS, de Tejada AH, Noffsinger A, Hart J, et al. Complete Barrett's eradication endoscopic mucosal resection: an effective treatment modality for high-grade dysplasia and intramucosal carcinoma – an American single-center experience. *Am J Gastroenterol* 2009;**104**:2684–92.

95. Pouw RE, Seewald S, Gondrie JJ, Deprez PH, Piessevaux H, Pohl H, et al. Stepwise radical endoscopic resection for eradication of Barrett's oesophagus with early neoplasia in a cohort of 169 patients. *Gut* 2010;**59**:1169–77.

96. Witteman BP, Foxwell TJ, Monsheimer S, Gelrud A, Eid GM, Nieponice A, et al. Transoral endoscopic inner layer esophagectomy: management of high-grade dysplasia and superficial cancer with organ preservation. *J Gastrointest Surg* 2009;**13**:2104–12.

97. Nieponice A, McGrath K, Qureshi I, Beckman EJ, Luketich JD, Gilbert TW, et al. An extracellular matrix scaffold for esophageal stricture prevention after circumferential EMR. *Gastrointest Endosc* 2009;**69**:289–96.

98. Nieponice A, Gilbert TW, Badylak SF. Reinforcement of esophageal anastomoses with an extracellular matrix scaffold in a canine model. *Ann Thorac Surg* 2006;**82**:2050–8.

99. Badylak SF, Vorp DA, Spievack AR, Simmons-Byrd A, Hanke J, Freytes DO, et al. Esophageal reconstruction with ECM and muscle tissue in a dog model. *J Surg Res* 2005;**128**:87–97.

100. Badylak S, Meurling S, Chen M, Spievack A, Simmons-Byrd A. Resorbable bioscaffold for esophageal repair in a dog model. *J Pediatr Surg* 2000;**35**:1097–103.

101. Agrawal V, Brown BN, Beattie AJ, Gilbert TW, Badylak SF. Evidence of innervation following extracellular matrix scaffold-mediated remodelling of muscular tissues. *J Tissue Eng Regen Med* 2009;**3**:590–600.

102. Badylak SF, Hoppo T, Nieponice A, Gilbert TW, Davison JM, Jobe BA. Esophageal preservation in five male patients after endoscopic inner-layer circumferential resection in the setting of superficial cancer: a regenerative medicine approach with a biologic scaffold. *Tissue Eng Part A* 2011;**17**:1643–50.

103. Nieponice A, Ciotola FF, Nachman F, Jobe BA, Hoppo T, Londono R, et al. Patch esophagoplasty: esophageal reconstruction using biologic scaffolds. *Ann Thorac Surg* 2014;**97**:283–8.

104. Beiner JM, Jokl P. Muscle contusion injuries: current treatment options. *J Am Acad Orthop Surg* 2001;**9**:227–37.

105. Garrett Jr WE. Muscle strain injuries. *Am J Sports Med* 1996;**24**:S2–8.

106. Lehto MU, Jarvinen MJ. Muscle injuries, their healing process and treatment. *Ann Chir Gynaecol* 1991;**80**:102–8.

107. Jarvinen TA, Kaariainen M, Jarvinen M, Kalimo H. Muscle strain injuries. *Curr Opin Rheumatol* 2000;**12**:155–61.

108. Jarvinen TA, Jarvinen TL, Kaariainen M, Kalimo H, Jarvinen M. Muscle injuries: biology and treatment. *Am J Sports Med* 2005;**33**:745–64.

109. Aarimaa V, Kaariainen M, Vaittinen S, Tanner J, Jarvinen T, Best T, et al. Restoration of myofiber continuity after transection injury in the rat soleus. *Neuromuscul Disord* 2004;**14**:421–8.

110. Crow BD, Haltom JD, Carson WL, Greene WB, Cook JL. Evaluation of a novel biomaterial for intrasubstance muscle laceration repair. *J Orthop Res* 2007;**25**:396–403.

111. Garrett Jr WE, Seaber AV, Boswick J, Urbaniak JR, Goldner JL. Recovery of skeletal muscle after laceration and repair. *J Hand Surg* 1984;**9**:683–92.

112. Menetrey J, Kasemkijwattana C, Fu FH, Moreland MS, Huard J. Suturing versus immobilization of a muscle laceration. A morphological and functional study in a mouse model. *Am J Sports Med* 1999;**27**:222–9.

113. Terada N, Takayama S, Yamada H, Seki T. Muscle repair after a transsection injury with development of a gap: an experimental study in rats. *Scand J Plast Reconstr Surg Hand Surg* 2001;**35**:233–8.

114. Sicari BM, Agrawal V, Siu BF, Medberry CJ, Dearth CL, Turner NJ, et al. A murine model of volumetric muscle loss and a regenerative medicine approach for tissue replacement. *Tissue Eng Part A* 2012;**18**:1941–8.

115. Turner NJ, Badylak JS, Weber DJ, Badylak SF. Biologic scaffold remodeling in a dog model of complex musculoskeletal injury. *J Surg Res* 2012;**176**:490–502.

116. Valentin JE, Turner NJ, Gilbert TW, Badylak SF. Functional skeletal muscle formation with a biologic scaffold. *Biomaterials* 2010;**31**:7475–84.

117. Mase Jr VJ, Hsu JR, Wolf SE, Wenke JC, Baer DG, Owens J, et al. Clinical application of an acellular biologic scaffold for surgical repair of a large, traumatic quadriceps femoris muscle defect. *Orthopedics* 2010;**33**:511.

118. Rollman GB, Gillespie JM. The role of psychosocial factors in temporomandibular disorders. *Curr Rev Pain* 2000;**4**:71–81.

119. Oakley M, Vieira AR. The many faces of the genetics contribution to temporomandibular joint disorder. *Orthod Craniofac Res* 2008;**11**:125–35.

120. Tanaka E, Detamore MS, Mercuri LG. Degenerative disorders of the temporomandibular joint: etiology, diagnosis, and treatment. *J Dent Res* 2008;**87**:296–307.

121. Farrar WB, McCarty Jr WL. The TMJ dilemma. *J Ala Dent Assoc* 1979;**63**:19–26.

122. Hall HD, Indresano AT, Kirk WS, Dietrich MS. Prospective multicenter comparison of 4 temporomandibular joint operations. *J Oral Maxillofac Surg* 2005;**63**:1174–9.

123. Nyberg J, Adell R, Svensson B. Temporomandibular joint discectomy for treatment of unilateral internal derangements – a 5 year follow-up evaluation. *Int J Oral Maxillofac Surg* 2004;**33**:8–12.

124. Krug J, Jirousek Z, Suchmova H, Cermakova E. Influence of discoplasty and discectomy of the temporomandibular joint on elimination of pain and restricted mouth opening. *Acta Med (Hradec Kralove)* 2004;**47**:47–53.

125. Vazquez-Delgado E, Valmaseda-Castellon E, Vazquez-Rodriguez E, Gay-Escoda C. Long-term results of functional open surgery for the treatment of internal derangement of the temporomandibular joint. *Br J Oral Maxillofac Surg* 2004;**42**:142–8.

126. McCain JP, Sanders B, Koslin MG, Quinn JH, Peters PB, Indresano AT. Temporomandibular joint arthroscopy: a 6-year multicenter retrospective study of 4,831 joints. *J Oral Maxillofac Surg* 1992;**50**:926–30.

127. Kaneyama K, Segami N, Sato J, Murakami K, Iizuka T. Outcomes of 152 temporomandibular joints following arthroscopic anterolateral capsular release by holmium: YAG laser or electrocautery. *Oral Surg Oral Med Oral Pathol Oral Radiol Endod* 2004;**97**:546–51. Discussion 52.

128. Nitzan DW, Samson B, Better H. Long-term outcome of arthrocentesis for sudden-onset, persistent, severe closed lock of the temporomandibular joint. *J Oral Maxillofac Surg* 1997;**55**:151–7. Discussion 7–8.

129. Reston JT, Turkelson CM. Meta-analysis of surgical treatments for temporomandibular articular disorders. *J Oral Maxillofac Surg* 2003;**61**:3–10. Discussion 10–2.

130. Alonso A, Kaimal S, Look J, Swift J, Fricton J, Myers S, et al. A quantitative evaluation of inflammatory cells in human temporomandibular joint tissues from patients with and without implants. *J Oral Maxillofac Surg* 2009;**67**:788–96.

131. Ferreira JN, Ko CC, Myers S, Swift J, Fricton JR. Evaluation of surgically retrieved temporomandibular joint alloplastic implants: pilot study. *J Oral Maxillofac Surg* 2008;**66**:1112–24.

132. Fricton JR, Look JO, Schiffman E, Swift J. Long-term study of temporomandibular joint surgery with alloplastic implants compared with nonimplant surgery and nonsurgical rehabilitation for painful temporomandibular joint disc displacement. *J Oral Maxillofac Surg* 2002;**60**:1400–11. Discussion 11-2.

133. Dolwick MF, Aufdemorte TB. Silicone-induced foreign body reaction and lymphadenopathy after temporomandibular joint arthroplasty. *Oral Surg Oral Med Oral Pathol* 1985;**59**:449–52.

134. Dimitroulis G. A critical review of interpositional grafts following temporomandibular joint discectomy with an overview of the dermis-fat graft. *Int J Oral Maxillofac Surg* 2011;**40**:561–8.

135. Dimitroulis G. The use of dermis grafts after discectomy for internal derangement of the temporomandibular joint. *J Oral Maxillofac Surg* 2005;**63**:173–8.

136. Matukas VJ, Lachner J. The use of autologous auricular cartilage for temporomandibular joint disc replacement: a preliminary report. *J Oral Maxillofac Surg* 1990;**48**:348–53.

137. Meyer RA. The autogenous dermal graft in temporomandibular joint disc surgery. *J Oral Maxillofac Surg* 1988;**46**:948–54.

138. Pogrel MA, Kaban LB. The role of a temporalis fascia and muscle flap in temporomandibular joint surgery. *J Oral Maxillofac Surg* 1990;**48**:14–9.

139. Thyne GM, Yoon JH, Luyk NH, McMillan MD. Temporalis muscle as a disc replacement in the temporomandibular joint of sheep. *J Oral Maxillofac Surg* 1992;**50**:979–87. Discussion 87–8.

140. Dimitroulis G. Macroscopic and histologic analysis of abdominal dermis-fat grafts retrieved from human temporomandibular joints. *J Oral Maxillofac Surg* 2011;**69**(9):2329–33.

141. McKenna SJ. Discectomy for the treatment of internal derangements of the temporomandibular joint. *J Oral Maxillofac Surg* 2001;**59**:1051–6.

142. Almarza AJ, Athanasiou KA. Seeding techniques and scaffolding choice for tissue engineering of the temporomandibular joint disk. *Tissue Eng* 2004;**10**:1787–95.

143. Bean AC, Almarza AJ, Athanasiou KA. Effects of ascorbic acid concentration on the tissue engineering of the temporomandibular joint disc. *Proc Inst Mech Eng H* 2006;**220**:439–47.

144. Detamore MS, Athanasiou KA. Structure and function of the temporomandibular joint disc: implications for tissue engineering. *J Oral Maxillofac Surg* 2003;**61**:494–506.

145. Detamore MS, Athanasiou KA. Evaluation of three growth factors for TMJ disc tissue engineering. *Ann Biomed Eng* 2005;**33**:383–90.

146. Elder BD, Athanasiou KA. Synergistic and additive effects of hydrostatic pressure and growth factors on tissue formation. *PLoS One* 2008;**3**:e2341.

147. Elder BD, Eleswarapu SV, Athanasiou KA. Extraction techniques for the decellularization of tissue engineered articular cartilage constructs. *Biomaterials* 2009;**30**:3749–56.

148. Grayson WL, Frohlich M, Yeager K, Bhumiratana S, Chan ME, Cannizzaro C, et al. Regenerative Medicine Special Feature: engineering anatomically shaped human bone grafts. *Proc Natl Acad Sci USA* 2009;**107**:3299–304.

149. Johns DE, Athanasiou KA. Improving culture conditions for temporomandibular joint disc tissue engineering. *Cells Tissues Organs* 2007;**185**:246–57.

150. Johns DE, Wong ME, Athanasiou KA. Clinically relevant cell sources for TMJ disc engineering. *J Dent Res* 2008;**87**:548–52.

151. Lumpkins SB, McFetridge PS. Regional variations in the viscoelastic compressive properties of the temporomandibular joint disc and implications toward tissue engineering. *J Biomed Mater Res A* 2009;**90**:784–91.

152. Lumpkins SB, Pierre N, McFetridge PS. A mechanical evaluation of three decellularization methods in the design of a xenogeneic scaffold for tissue engineering the temporomandibular joint disc. *Acta Biomater* 2008;**4**:808–16.

153. Wang L, Lazebnik M, Detamore MS. Hyaline cartilage cells outperform mandibular condylar cartilage cells in a TMJ fibrocartilage tissue engineering application. *Osteoarthritis Cartilage* 2009;**17**:346–53.

154. Brown BN, Chung WL, Pavlick M, Reppas S, Ochs MW, Russell AJ, et al. Extracellular matrix as an inductive template for temporomandibular joint meniscus reconstruction: a pilot study. *J Oral Maxillofac Surg* 2011;**69**:e488–505.

155. Brown BN, Chung WL, Almarza AJ, Pavlick MD, Reppas SN, Ochs MW, et al. Inductive, scaffold-based, regenerative medicine approach to reconstruction of the temporomandibular joint disk. *J Oral Maxillofac Surg* 2012;**70**:2656–68.

156. Boruch AV, Nieponice A, Qureshi IR, Gilbert TW, Badylak SF. Constructive remodeling of biologic scaffolds is dependent on early exposure to physiologic bladder filling in a canine partial cystectomy model. *J Surg Res* 2010;**161**:217–25.

157. Hodde JP, Badylak SF, Shelbourne KD. The effect of range of motion on remodeling of small intestinal submucosa (SIS) when used as an Achilles tendon repair material in the rabbit. *Tissue Eng* 1997;**3**:27–37.

158. Almarza AJ, Yang G, Woo SL, Nguyen T, Abramowitch SD. Positive changes in bone marrow-derived cells in response to culture on an aligned bioscaffold. *Tissue Eng Part A* 2008;**14**:1489–95.

159. Androjna C, Spragg RK, Derwin KA. Mechanical conditioning of cell-seeded small intestine submucosa: a potential tissue-engineering strategy for tendon repair. *Tissue Eng* 2007;**13**:233–43.

160. Nguyen TD, Liang R, Woo SL, Burton SD, Wu C, Almarza A, et al. Effects of cell seeding and cyclic stretch on the fiber remodeling in an extracellular matrix-derived bioscaffold. *Tissue Eng Part A* 2009;**15**:957–63.

161. Wallis MC, Yeger H, Cartwright L, Shou Z, Radisic M, Haig J, et al. Feasibility study of a novel urinary bladder bioreactor. *Tissue Eng Part A* 2008;**14**:339–48.

162. Gilbert TW, Stewart-Akers AM, Sydeski J, Nguyen TD, Badylak SF, Woo SL. Gene expression by fibroblasts seeded on small intestinal submucosa and subjected to cyclic stretching. *Tissue Eng* 2007;**13**:1313–23.

163. Agrawal V, Tottey S, Johnson SA, Freund JM, Siu BF, Badylak SF. Recruitment of progenitor cells by an extracellular matrix cryptic peptide in a mouse model of digit amputation. *Tissue Eng Part A* 2011;**17**:2435–43.

164. Beattie AJ, Gilbert TW, Guyot JP, Yates AJ, Badylak SF. Chemoattraction of progenitor cells by remodeling extracellular matrix scaffolds. *Tissue Eng Part A* 2009;**15**:1119–25.

165. Crisan M, Yap S, Casteilla L, Chen CW, Corselli M, Park TS, et al. A perivascular origin for mesenchymal stem cells in multiple human organs. *Cell Stem Cell* 2008;**3**:301–13.

166. Zantop T, Gilbert TW, Yoder MC, Badylak SF. Extracellular matrix scaffolds are repopulated by bone marrow-derived cells in a mouse model of Achilles tendon reconstruction. *J Orthop Res* 2006;**24**:1299–309.

167. Badylak SF, Park K, Peppas N, McCabe G, Yoder M. Marrow-derived cells populate scaffolds composed of xenogeneic extracellular matrix. *Exp Hematol* 2001;**29**:1310–8.

168. Altman GH, Horan RL, Martin I, Farhadi J, Stark PR, Volloch V, et al. Cell differentiation by mechanical stress. *FASEB J* 2002;**16**:270–2.

169. Matziolis G, Tuischer J, Kasper G, Thompson M, Bartmeyer B, Krocker D, et al. Simulation of cell differentiation in fracture healing: mechanically loaded composite scaffolds in a novel bioreactor system. *Tissue Eng* 2006;**12**:201–8.

170. Nieponice A, Maul TM, Cumer JM, Soletti L, Vorp DA. Mechanical stimulation induces morphological and phenotypic changes in bone marrow-derived progenitor cells within a three-dimensional fibrin matrix. *J Biomed Mater Res A* 2007;**81**:523–30.

171. Allman AJ, McPherson TB, Badylak SF, Merrill LC, Kallakury B, Sheehan C, et al. Xenogeneic extracellular matrix grafts elicit a TH2-restricted immune response. *Transplantation* 2001;**71**:1631–40.

172. Allman AJ, McPherson TB, Merrill LC, Badylak SF, Metzger DW. The Th2-restricted immune response to xenogeneic small intestinal submucosa does not influence systemic protective immunity to viral and bacterial pathogens. *Tissue Eng* 2002;**8**:53–62.

173. Badylak SF, Valentin JE, Ravindra AK, McCabe GP, Stewart-Akers AM. Macrophage phenotype as a determinant of biologic scaffold remodeling. *Tissue Eng Part A* 2008;**14**:1835–42.

174. Brown BN, Londono R, Tottey S, Zhang L, Kukla KA, Wolf MT, et al. Macrophage phenotype as a predictor of constructive remodeling following the implantation of biologically derived surgical mesh materials. *Acta Biomater* 2012;**8**:978–87.

175. Anderson JM. Inflammatory response to implants. *ASAIO Trans* 1988;**34**:101–7.

176. Palmer EM, Beilfuss BA, Nagai T, Semnani RT, Badylak SF, van Seventer GA. Human helper T cell activation and differentiation is suppressed by porcine small intestinal submucosa. *Tissue Eng* 2002;**8**:893–900.

177. Bach FH, Ferran C, Hechenleitner P, Mark W, Koyamada N, Miyatake T, et al. Accommodation of vascularized xenografts: expression of "protective genes" by donor endothelial cells in a host Th2 cytokine environment. *Nat Med* 1997;**3**:196–204.

178. Chen N, Field EH. Enhanced type 2 and diminished type 1 cytokines in neonatal tolerance. *Transplantation* 1995;**59**:933–41.

179. Piccotti JR, Chan SY, VanBuskirk AM, Eichwald EJ, Bishop DK. Are Th2 helper T lymphocytes beneficial, deleterious, or irrelevant in promoting allograft survival? *Transplantation* 1997;**63**:619–24.

180. Turner NJ, Pezzone MA, Brown BN, Badylak SF. Quantitative multispectral imaging of Herovici's polychrome for the assessment of collagen content and tissue remodelling. *J Tissue Eng Regen Med* 2013;**7**:139–48.

181. Slivka PF, Dearth CL, Keane TJ, Meng FW, Medberry CJ, Riggio RT, et al. Fractionation of an ECM hydrogel into structural and soluble components reveals distinctive roles in regulating macrophage behavior. *Biomater Sci* 2014;**2**:1521–34.

182. Gilbert TW, Stewart-Akers AM, Badylak SF. A quantitative method for evaluating the degradation of biologic scaffold materials. *Biomaterials* 2007;**28**:147–50.

183. Agrawal V, Johnson SA, Reing J, Zhang L, Tottey S, Wang G, et al. Epimorphic regeneration approach to tissue replacement in adult mammals. *Proc Natl Acad Sci USA* 2010;**107**:3351–5.

184. Agrawal V, Kelly J, Tottey S, Daly KA, Johnson SA, Siu BF, et al. An isolated cryptic peptide influences osteogenesis and bone remodeling in an adult mammalian model of digit amputation. *Tissue Eng Part A* 2011;**17**:3033–44.

185. Agrawal V, Siu BF, Chao H, Hirschi KK, Raborn E, Johnson SA, et al. Partial characterization of the Sox2+ cell population in an adult murine model of digit amputation. *Tissue Eng Part A* 2012;**18**:1454–63.

186. Holtom PD, Shinar Z, Benna J, Patzakis MJ. Porcine small intestine submucosa does not show antimicrobial properties. *Clin Orthop Relat Res* 2004:18–21.

187. de Castro Bras LE, Shurey S, Sibbons PD. Evaluation of crosslinked and non-crosslinked biologic prostheses for abdominal hernia repair. *Hernia* 2012;**16**:77–89.

188. Shah BC, Tiwari MM, Goede MR, Eichler MJ, Hollins RR, McBride CL, et al. Not all biologics are equal!. *Hernia* 2011;**15**:165–71.

189. Collins BH, Chari RS, Magee JC, Harland RC, Lindman BJ, Logan JS, et al. Mechanisms of injury in porcine livers perfused with blood of patients with fulminant hepatic failure. *Transplantation* 1994;**58**:1162–71.

190. Cooper DK, Good AH, Koren E, Oriol R, Malcolm AJ, Ippolito RM, et al. Identification of alpha-galactosyl and other carbohydrate epitopes that are bound by human anti-pig antibodies: relevance to discordant xenografting in man. *Transpl Immunol* 1993;**1**:198–205.

191. Galili U, Macher BA, Buehler J, Shohet SB. Human natural anti-alpha-galactosyl IgG. II. The specific recognition of alpha (1–3)-linked galactose residues. *J Exp Med* 1985;**162**:573–82.

192. Faulk DM, Carruthers CA, Warner HJ, Kramer CR, Reing JE, Zhang L, et al. The effect of detergents on the basement membrane complex of a biologic scaffold material. *Acta Biomater* 2014;**10**:183–93.

Chapter 3

Whole-Organ Bioengineering—Current Tales of Modern Alchemy

Emma Moran*, Abritee Dhal*, Dipen Vyas*, Angel Lanas, Shay Soker, Pedro M. Baptista

Key Concepts

1. Organ bioengineering is an emerging field driven by the shortage of donor organs for transplantation.
2. Whole-organ bioengineering uses a technique called decellularization to remove all of the cells of an organ, leaving behind an intact extracellular matrix and vascular network.
3. The acellular scaffolds are then seeded with the organ-specific cell type in a bioreactor to create the bioengineered tissue.
4. Whole-organ bioengineered has been studied for liver, intestine, kidney, heart, pancreas, and lung tissue, among others.
5. Although human transplantation is not yet realized, these bioengineered organs have numerous applications such as research in drug metabolism, tissue physiology, developmental biology, and stem cell differentiation studies.
6. Current limitations of this technology include cell sourcing and expansion, assembling a functional vascular network for transplantation, bioreactor technologies to grow the tissue, regulatory issues, and scaling up the technology.

1. INTRODUCTION

Organ transplantation has come a long way since 1954 that celebrated the first successful solid organ transplant, performed in Boston by a team led by Joseph Murray, John Merrill, and J. Hartwell Harrison. This living donor kidney transplant between homozygotic twins, rendering immunosuppression unnecessary, showed that renal failure could be reversed completely. This allowed the first successful solid organ transplant procedure, reshaping the field of transplantation medicine from an experimental science into a clinical discipline.[1]

Nonetheless, it was only with the advent of a new generation of immunosuppressant drugs in the 1960s and 1970s that this area of medicine really started to become more prominent and more effective.[2] Azathioprine in combination with low-dose steroids carried the whole field of renal transplantation to 90% patient survival, with 60% graft survival per year, by the end of 1970s. However, it was with the introduction of cyclosporine, a calcineurin inhibitor, in the beginning of 1980s that renal transplantation changed dramatically, and with it, the whole field of transplantation medicine.[3–5] By 1984, and in only a few years, more than two-thirds of all cadaveric heart, kidney, and liver transplant patients retained their graft for more than 1 year.[6,7] Gradually, with the development of better combinatorial immunosuppressive regimens and with the introduction of new immunosuppressant agents (tacrolimus, mycophenolic acid, etc.) organ transplants became commonplace, limited only by the number of available donors. As the rising success rate of transplants and modern immunosuppression made transplants more common, the need for more organs became critical.

The increasing shortage of available organs in the late 1980s and 1990s urged for the development of alternative approaches that could quickly provide a complementary source for patients in the waiting list. Transplants from living donors, especially relatives, became increasingly common, as well as significant research into xenotransplantation, or transgenic organs.[7,8] The creation of specialized organ procurement and allocation networks (United Network for Organ Sharing—UNOS, Eurotransplant, etc.) and the introduction of more efficient prioritizing allocation methods (Model for End-Stage Liver Disease (MELD) score for the liver, Calculated Panel Reactive Antibody (CPRA) for the kidney, Low Attenuation Density (LAD) for lungs, etc.) has additionally attenuated or stabilized the wide gap between donors and recipients.[9,10] Nevertheless, and despite the massive advancement and innovation experienced by the organ transplantation field in the past decades, the goal of reaching all patients in a meaningful time is still elusive, at best.[10]

---

* These authors contributed equally to this work.

Translating Regenerative Medicine to the Clinic. http://dx.doi.org/10.1016/B978-0-12-800548-4.00003-6

2. CURRENT STATUS OF ORGAN TRANSPLANTATION

When looking at organ transplants internationally, transplant totals and rates differ significantly between countries around the world. Multiple factors account for this: (1) disparities in the incidence of end-stage organ disease, (2) different economical capabilities of the national health-care systems to provide transplants, (3) cultural acceptance or rejection of organ donation and transplant, and (4) reporting of transplants.[11]

Hence, by the end of 2012, there were nearly 120,000 people in the waiting list for organ transplantation in the USA. From these, 28,053 received an organ transplant (total transplant rate ~23%).[9] In Central Europe the scenario is moderately better, with 15,027 reported candidates on the waiting list in January 1, 2013 and 7357 patients transplanted by the end of 2012 (transplant rate ~49%), displaying a higher total transplant rate.[12] Nevertheless, a more accurate reading of transplant rates for different organs (transplants per million population) show that for many countries with advanced economies and modern health-care systems the rates have been reasonably high (e.g., >40 transplants per million—kidney) and stable for several years. However, for Third World countries or emerging economies with far less proficient procurement agencies and trained transplant teams, the total transplant rates are much lower (<10 transplants per million—kidney).[11] Regardless of these differences, what seems to be equally true for all countries is that any mechanism that could provide more organs, and sooner, to all patients that at some point make it to the waiting list for an organ transplant, would give a critical contribution to tackle the shortage of available organs. In this particular point, the emerging field of regenerative medicine holds the promise to help solve this problem. Regrowing organs in the lab, using patient's own cells (stem cells or healthy cells harvested from the donor site), or by using allogeneic human tissue, embryonic or human induced pluripotent stem (iPS) cells, is now a possibility. Especially, since the development of whole-heart decellularization by Ott et al., the field of organ bioengineering has gained incredible pace.[13] The generation of truly vascularized organ scaffolds from native solid organs has been the main catalyst for this, and it is this particularly critical development, now applied to all major solid organs (Figure 1), that we are going to describe in detail.

3. CURRENT STATUS ON ORGAN BIOENGINEERING

3.1 Liver

Liver organ bioengineering has become possible due to recent advances in tissue decellularization techniques.[13–17] Decellularized liver scaffolds with intact vasculature and hepatic microarchitecture can provide specific hepatic cues for cell survival, proliferation, differentiation, and function. Several groups have reported proof-of-principle studies showing effective recellularization of decellularized liver scaffolds with fetal human liver cells and rat hepatocytes.[15,16,18] These bioengineered livers also performed several important metabolic functions such as albumin synthesis, urea secretion, and drug metabolism. Although these reports have shown promising results, major challenges still need to be overcome. These challenges include identifying appropriate species for providing decellularized livers, selecting ideal cell sources, achieving robust vascularization, optimizing bioreactor perfusion technology along with scalability, and preventing graft rejection.[17,19–22]

3.2 Intestine

Bioengineering of small intestine has been of particular interest for treatment of short bowel syndrome (SBS), a condition resulting from 70% to 80% resection of small bowel. This is a severe condition associated with high mortality rates, and

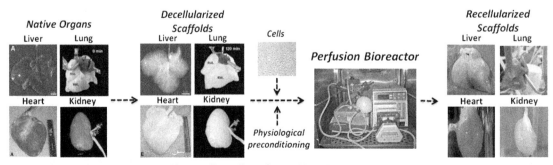

FIGURE 1   Organ Bioengineering Paradigm. The overall paradigm of organ bioengineering is based on the generation of whole-organ scaffolds by perfusion decellularization. These organ scaffolds can then be recellularized with multiple cell types, and under appropriate physiological conditions in a perfusion bioreactor, generate bioengineered organs with numerous applications. *Liver: Baptista et al.*[69]*; Lung: Ott et al.*[51]*; Heart: Wainwright et al.*[42]*; Kidney: Sullivan et al.*[36]

the current therapeutic options, including total parenteral nutrition (TPN) or surgical reconstruction, have proven to be suboptimal.[23,24] Grikscheit et al. were the first to report development of tissue-engineered small intestine (TESI), by seeding intestinal organoid units on polymer scaffolds which where then transplanted into rat omentum.[25] In recent years, with the emergence of decellularized tissue/organ scaffolds, various groups have reported decellularization approaches to create acellular intestinal matrix to be used as scaffold for small intestine tissue engineering.[26–29] These studies showed preserved intestinal microarchitecture with functional units, and crypts with a villus, using currently available decellularization techniques. Patil et al. have recently demonstrated development of TESI using decellularized human small intestine reseeded with cells (epithelial, endothelial, and smooth muscle) derived from allogeneic bone marrow.[28] These TESI had intact regenerated mucosa, intact villi, and crypts lined with mucin-positive goblet cells and repopulated blood vessels with endothelial cells (ECs). Although major strides have been made in development of TESI recently, a significant effort is required in evaluation of decellularized intestinal scaffolds, identification of ideal cell source, and efficient strategy for reseeding the scaffolds and more importantly, functional assessment of TESI.

3.3 Kidney

The need for an alternative treatment for patients with end-stage renal disease is highlighted by the fact that 100,000 Americans currently wait a donor kidney while only 18,000 transplants are performed annually.[30] Regenerative medicine approaches may alleviate some of this burden through the delivery of stem and progenitor cells to the site of renal failure.[31] However, evidence suggests that cell therapy alone may not be successful in chronic renal failure and that a whole-organ approach may be necessary. Successful decellularization of rat,[32–34] porcine,[35,36] rhesus monkey,[37,38] and human[39] kidney scaffolds have been reported by various groups. Ross et al. successfully recellularized rat scaffolds with murine embryonic stem cells (mESCs) through the renal artery and ureter. The mESCs successfully proliferated within the glomerular, vascular, and tubular structures and expressed markers indicating differentiation.[33] Further work by this group demonstrated differentiation of mESCs toward mature ECs, by expression of BsLB4 and VEGFR2 in the arterial vessels and glomeruli, an important milestone for transplantation.[34] Using rhesus monkey decellularized scaffolds, Nakayama et al. demonstrated successful differentiation and migration of fetal liver cells[38] and more recently, the expression of cytokeratin-positive epithelial tubule phenotypes when scaffolds were seeded with human ESCs.[37] The most promising progress in terms of successful transplantation and obtaining renal function was achieved in recent work by Song et al.[32] Rat scaffolds were recellularized with ECs and rat neonatal kidney cells (NKCs) and maintained in a perfusion bioreactor for 3–5 days for endothelialization and 12 days for mature NKC function. These scaffolds were orthotopically transplanted in rats and provided urine production and clearance of metabolites, despite functional immaturity. While the exhibition of urine production is promising, demonstration of more advanced functions including blood filtration and blood pressure maintenance is needed before these efforts can translate clinically.

3.4 Heart

Causing 600,000 deaths per year, heart disease is the leading cause of death for both men and women in the United States.[40] While tissue engineering of myocardial muscle has successfully been used to improve function in infarcted rat hearts,[41] human tissue requires much thicker tissue which can pose a problem due to the high metabolic demands of cardiomyocytes. Additionally, this "cardiac patch" approach is only feasible if a small region of the heart is damaged. In 2008, Ott et al. developed a whole-organ decellularization technique to create a bioengineered heart that contained intact vascular tree, heart valves, and atrial and ventricular geometry.[13] Rat hearts were decellularized via perfusion techniques and reseeded with the injection of rat neonatal cardiac cells and perfusion of ECs. After 8 days of culture in a bioreactor that mimicked physiological flow properties, the recellularized scaffold demonstrated contractile function and electric responses to stimulation. Following this work, others have shown decellularization of whole porcine heart[42] and sections of human myocardial tissue.[43] Decellularized rat scaffolds supported the differentiation of human embryonic stem cells (hESCs) and human mesendodermal cells derived from hESCs, an important milestone in cardiac bioengineering, however, the cells lacked beating functionality.[44] While research to date has been successful in decellularization and reseeding of scaffolds with various cell types, much more needs to be achieved in whole heart bioengineering, including physiological mechanical function of seeded cells and transplantation.

3.5 Pancreas

According to the World Health Organization, at least 285 million people worldwide suffer from diabetes.[45] The treatment for patients suffering from diabetes continues to be quite unsatisfactory. Even though exogenous insulin therapy is effective at preventing acute decompensation in type I diabetes, less than 40% of these patients reach and maintain therapeutic targets. In other words, the treatment does not provide a cure and also has the potential to lead to long-term complications.[46,47] The only available treatments capable of establishing long-term, stable euglycemia in type I diabetes patients is β-cell replacement, which can be done through pancreas or islet cell transplantation.[46] It has also been found that pancreatic islet transplantation has limited engraftment potential and short-term therapeutic effect.[48] Hence, researchers have been seeking novel approaches to tackle this shortcome. Microencapsulation technology has been one of the platforms in development for islet cell transplantation. Encapsulation involves the packaging of islets within a semipermeable and bioinert membrane, which selectively allows the passage of oxygen, glucose, nutrients, waste products, and insulin and prevents the penetration of immune cells. In other terms, encapsulation isolates foreign antigens from allogeneic or xenogeneic islets from the host immune system.[49] However, issues like poor biocompatibility of capsule materials, immune isolation due to penetration of small immune mediators such as chemokines, cytokines, and nitric oxide, or hypoxia secondary to failed revascularization remain major challenges.[48]

The use of organ scaffolds is emerging as cutting edge new technology for whole-organ bioengineering. The development of a bioengineered pancreas by appropriate combination of cells, decellularized pancreas scaffold, and biologically active molecules has great potential to provide an alternative avenue for diabetes therapy.[47] Goh et al. demonstrated successful cellular engraftment within decellularized pancreas. The resulting graft gave rise to strong upregulation of insulin gene expression. Their findings support the use of whole-pancreas scaffolds as a biomaterial for the support and enhancement of pancreatic cell functionality, particularly with enhanced insulin function upon seeding of β-cells within the pancreas scaffold. Mirmalek-Sani et al. also showed the utility of decellularized pancreas scaffolds,[46] in this particular case, generated from a porcine pancreas. The scaffold was seeded with human stem cells and porcine pancreatic islets, and like in the rodent model, the acellular porcine pancreas also supported cell adhesion and maintenance. The islets seeded on the scaffolds had statistically significant increased insulin secretion when compared to control islets plated in petri dishes. This study shows the potential of using acellular porcine pancreas extracellular matrix (ECM) as a platform for bioengineered pancreas for transplantation into humans, but like the previous study, lack true scaffold revascularization and transplantation.

3.6 Lungs

Lung disease continues to remain a significant cause of morbidity and mortality worldwide. To date, there is no long-term cure for patients suffering from chronic lung disease except lung transplantation. However, like most organs for transplantation there is a limited availability of donor lungs, graft rejections are frequent, and the long-term need for immune suppression limits the wide use and effectiveness of lung transplantation.[50] Therefore, lung bioengineering is an exciting emerging solution for patients suffering from end-stage lung disease. Ott et al., as well as Petersen et al., generated decellularized rat lungs as a platform to bioengineer lungs[51,52] and seeded these with pneumocytes and ECs. As a means to establish function of the lungs, the bioengineered lungs were perfused and ventilated in a bioreactor simulating the physiologic environment of developing lung. After bioreactor preconditioning the constructs were perfused with blood and ventilated using physiologic pressures, generating gas exchange comparable to that of isolated native lungs. When orthotopically transplanted into rats, these constructs were perfused by the recipient's circulation and ventilated by means of the recipient's airway and respiratory muscles, providing short-term gas exchange in vivo, for up to 6 h. In 2011, Song et al. were able to further enhance the ability of the bioengineered lung to function in their rat model in vivo for up to 2 weeks.[53] Ott et al. and Song et al., along with Petersen et al., obtained very encouraging results; however, scale-up is necessary for translation into the clinic.

O'Neil et al. were able to successfully decellularize pig and human lungs and found that the decellularized pig lung scaffolds had similar properties as those of human origin. Thus concluded the use of porcine decellularized lung scaffolds as a platform for eventually bioengineered human lungs.[54] However, there has not been any report of a bioengineered lung with the size of a pig or human lung so far.

4. CURRENT APPLICATIONS FOR BIOENGINEERED ORGANS

Presently, with multiple solid organs already successfully bioengineered and in further development by several groups around the world, there is a huge promise and potential on these technologies. However, due to the current intrinsic limitations (described ahead), these bioengineered organs are still not ready for prime time in transplantation medicine. Nevertheless, there are already multiple applications for the current generation of bioengineered organs and their acellular scaffolds that go much further than just organ transplantation (Figure 2).

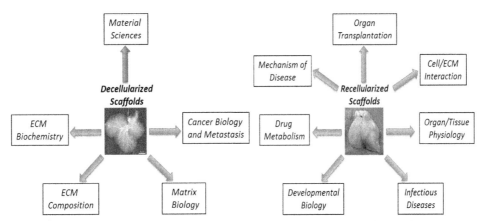

FIGURE 2   **Current applications of organ scaffolds and bioengineered organs.** There are multiple applications for the current status of organ decellularization and recellularization technology that go much further than organ transplantation. These bioengineered organs/tissues have the potential to be "customized" with particular features (with a subtracted cell type present in the native tissue; added cell type not present in the native tissue; all cells from a particular genetic background or gender; etc.).

Most of these applications, like drug metabolism,[16] organ/tissue physiology,[13,16,32,52] matrix biology,[18] developmental biology,[18,55] stem cell differentiation studies,[26] just to name a few, are gaining relevance as new platforms for these areas of research. Besides the applications described in Figure 2, these bioengineered organs/tissues have also the potential to be truly "customizable" with particular features in order to answer or investigate a specific problem (by subtracting a cell type present in native tissue; by adding a particular cell type not present in the native tissue; all cells from a specific genetic background susceptible to a certain disease; etc.). This will certainly enhance researchers' arsenal and allow them to answer challenging questions in novel ways previously impossible.

5. CURRENT LIMITATIONS

5.1 Cell Sources and Expansion

The source and type of cells used for repopulating decellularized scaffolds is critical for developing optimally functional and eventually clinically successful bioengineered tissues/organs. As a general idea, a perfect cell type for tissue engineering should have several important hallmark characteristics such as the ability to self-renew/proliferate and differentiate into multiple cell types found within the tissue/organ of interest. They should also be readily available, well characterized, and able to expand to large number of cells to repopulate a whole-organ scaffold. Stem or progenitor cells like hESCs, adult or fetal progenitors, and iPS cells have become an attractive choice as they have the aforementioned characteristics. The ethical debate and teratogenic properties have slowed down the use of hESCs in tissue engineering. iPS cells on the other hand, present epigenetic and safety issues and are also capable of forming teratomas in vivo. Adult stem/progenitor cells derived from bone marrow and several organs or tissues have been utilized quite extensively for cellular therapy and could be an ideal autologous cell source for tissue engineering.[56] Fetal stem cells have shown extensive expansion ability along with ease of differentiation to mature cell types and no teratoma formation in vivo.[57,58] Nonetheless, numerous hurdles remain to be cleared for identifying the best cell source tailored for a specific tissue or organ. The expansion of billions of cells, along with maintaining a stable genotype and the required phenotype, so as to render them to be feasible for organ engineering, remains the biggest hurdles of all.

5.2 Assembly of Patent Vascular Networks

Achieving full vascularization of bioengineered solid organs of human size continues to be one of the greatest challenges in organ engineering and regenerative medicine. It is a widely known fact that oxygen is a limiting factor in cell survival in most grafts and that oxygen transport in tissues is limited to a diffusion of 150–200 μm from the source.[59] When cells are not in proximity to their oxygen source, cells become metabolically inactive and the tissue becomes necrotic. Consequently, and without a doubt, there is a well-defined urgency and need for full vascularization of thicker bioengineered organs such as kidney, liver, heart, etc.

Many strategies for enhanced vascularization are currently under investigation. Some include scaffold design, the inclusion of angiogenic factors, along with both in vivo and in vitro prevascularization.[60] The architecture and design of a scaffold have a great effect on the rate of vascularization after implantation. The use of decellularized whole-organs is especially useful not only because of the ECM proteins needed for cell adhesion and proliferation within a graft upon scaffold reseeding, but also because thick organ scaffolds upon decellularization contain an intact vascular network which can be coated with vascular cells upon reseeding.[16,61] Angiogenic factor delivery, such as vascular endothelial growth factor (VEGF), relies on the ingrowth of host vessels into the entire implanted construct. The use of decellularized scaffolds along with angiogenic factor delivery has the ability to increase the rate of vascularization within a tissue. However, it would still take several days to weeks for the center of the implant to become perfused.[60] In vivo prevascularization could result in the instantaneous perfusion of a construct after implantation at the final site since the construct is connected to the host vasculature. However, a preimplantation period is needed, during which time the implant has to rely on spontaneous angiogenesis from the surrounding vessels into the construct. Therefore, nutrient limitations are likely to occur during this stage leading to tissue degeneration. In vitro prevascularization does not result in the instantaneous perfusion of a construct since vessels have to grow from the host into the construct until they reach the vascular network formed in vitro. Compared to scaffold design and angiogenic factor delivery, this method can dramatically decrease the time that is needed to vascularize the implant because host vessels do not have to grow into the entire construct but only into its outer regions, that is, until the ingrowing vessels meet the preformed vascular network.[60]

The generation of mature and stable capillaries and larger vascular structures depends on the rapid proliferation of ECs. Therefore, one of the first steps to assemble a patent vascular network within a decellularized scaffold is to repopulate a large portion of its vascular tree with ECs, but not exclusively with EC, but also with proper mural cells such as pericytes and mature vascular smooth muscle cells. The heavy recruitment of these mural cells is a hallmark of revascularization processes in native tissue, but also of stable vascular networks within bioengineered organs.[62]

In our view, the combinatorial use of a 3D organ scaffold, preconditioned with angiogenic growth factors, and subsequently reseeded with ECs and pericytes could all but help with the assembly of a robust and patent vascular network within a bioengineered organ.

5.3 Enabling Bioreactor Technologies

In order to create complex tissues, cells must be cultured in ways that provide the physical and mechanical conditions that cells experience in their physiological environment. This process preconditions and matures the tissue for eventual transplantation. Perfusion bioreactors are most commonly used in whole-organ bioengineering to fill this role and take advantage of the intact vascular networks remaining after the decellularization procedure.[63] To date, researchers have used mainly two methods to seed organ scaffolds with cells: injection seeding of cells directly into the parenchyma of the tissues[13,17,33] or delivering cells via the inherent vascular network of the decellularized organ.[15–17,32,51] This latter approach has been shown to provide a better seeding efficiency than direct injection in liver bioreactors[17] and has the added advantage of seeding cells in the same bioreactor setup in which the organ will be maintained for tissue growth. Generally, near physiological flow types (pulsatile vs steady), flow rates, and pressures are successful at long-term culturing of the organ in perfusion bioreactors.[13,15,16,51,64] Nonetheless, the system must be adapted to the most favorable conditions for the particular cell type that is being used. Certain organs require additional components to the perfusion bioreactor that prime the cells for their in vivo function. For example, heart bioreactors are designed to deliver physiological preload, afterload, and intraventricular pressure in addition to electrical stimulation to prime the cells for contractile functions.[13] Lung bioreactors may contain both media perfusion and ventilation components[52] in addition to breathing-induced stretch[64] to prime different cell types for their physiological function.

Bioreactor design will also need to include precise control over environmental and operating conditions, including sensors to monitor pH, temperature, pressure, waste, oxygen content, and nutrient supply.[65] These controls are necessary to achieve reproducible biological results and will allow large-scale translation of this research.

5.4 Regulatory Challenges, Scaling Up, and Other Issues

One of the current issues for the successful clinical translation of whole-organ bioengineering is scaffold sourcing. It is widely known that ECM molecules are highly conserved among species.[55] Therefore, in the field of organ bioengineering scientists have been developing novel ways to generate scaffolds from animals such as pig or sheep, since these animals have organs with a similar size of human organs. Thus, porcine organs and tissues are highly sought, not only because of their comparable size to human organs, but also because of their similarities in specific details, such as their native microarchitecture. Use of porcine-derived ECM scaffolds, analogous to aortic valve repair with decellularized pig valves or the use of decellularized small intestine submucosa in patients with chronic ulcers or hernias, with proper regulation could significantly increase the number of available organs for decellularization, and consequently for transplantation, when reseeded with appropriate human cells.[66] Human organs that are not suitable for transplantation are also widely available for decellularization.[57] However, since these organs are rejected for transplantation, there is a credible possibility that they may be damaged, raising additional technical issues. Both animal and human decellularized organs will have to be subject to standardized protocols and sterilization for clinical translation, a topic that still requires much research. The addition of cells to the scaffolds (combinatorial product) will also provide additional layers of technical and safety validation with the regulatory agencies, something that will certainly slow clinical translation.[67]

Clinical translation of whole-organ bioengineering will require scaled-up, high-throughput, and standardized methodologies for both decellularization and organ culture. Researchers have begun addressing such realities, including work on the development of a rapid and high-throughput kidney decellularization protocol that may be adapted for other organs[36] and the evaluation of culturing cells on decellularized organs following long-term storage.[68]

6. CONCLUSIONS

Generally, there are still many outstanding questions in the clinical translation of whole-organ bioengineering. The search for a robust cell source that can be expanded to staggering numbers, along with the generation of a complex and patent vascular network within the bioengineered organs, are just but some of the most prominent challenges that need to be resolved in the near future. However, it is important to keep in mind that research in organ bioengineering and regeneration using decellularized tissues is providing novel tools to study organ development and disease. This area of research might not be ready for prime time in transplantation medicine, but is certainly starting to contribute with novel tissue and organ models to a research community starving for more meaningful research platforms and tools, while more ambitious goals of clinical translation are slowly, but progressively being pushed forward.

REFERENCES

1. Guild WR, Harrison JH, Merrill JP, Murray J. Successful homotransplantation of the kidney in an identical twin. *Trans Am Clin Climatol Assoc* 1955;**67**:167–73.
2. Morris PJ. Transplantation–a medical miracle of the 20th century. *N Engl J Med* 2004;**351**:2678–80.
3. Calne RY, et al. Cyclosporin A in patients receiving renal allografts from cadaver donors. *Lancet* 1978;**2**:1323–7.
4. Cosimi AB, et al. Prolongation of allograft survival by cyclosporin A. *Surg Forum* 1979;**30**:287–9.
5. White DJ, Calne RY. The use of cyclosporin A immunosuppression in organ grafting. *Immunol Rev* 1982;**65**:115–31.
6. Cohen DJ, et al. Cyclosporine: a new immunosuppressive agent for organ transplantation. *Ann Intern Med* 1984;**101**:667–82.
7. Linden PK. History of solid organ transplantation and organ donation. *Crit care Clin* 2009;**25**:165–84. ix.
8. Hammer C, Linke R, Wagner F, Diefenbeck M. Organs from animals for man. *Int Arch Allergy Immunol* 1998;**116**:5–21.
9. Sharing, U.N.F.O., vol. 2013. http://optn.transplant.hrsa.gov; 2013.
10. Israni AK, Zaun DA, Rosendale JD, Snyder JJ, Kasiske BL. OPTN/SRTR 2011 Annual Data Report: deceased organ donation. *Am J Transplant* 2013;**13**(Suppl. 1):179–98.
11. Kasiske BL, et al. OPTN/SRTR 2011 Annual Data Report: international data. *Am J Transplant* 2013;**13**(Suppl. 1):199–225.
12. Eurotransplant, vol. 2013. http://www.eurotransplant.org/cms/; 2013.
13. Ott HC, et al. Perfusion-decellularized matrix: using nature's platform to engineer a bioartificial heart. *Nat Med* 2008;**14**:213–21.
14. Baptista PM, et al. Whole-organ decellularization – a tool for bioscaffold fabrication and organ bioengineering. *Conf Proc IEEE Eng Med Biol Soc* 2009;**2009**:6526–9.
15. Uygun BE, et al. Organ reengineering through development of a transplantable recellularized liver graft using decellularized liver matrix. *Nat Med* 2010;**16**:814–20.
16. Baptista PM, et al. The use of whole organ decellularization for the generation of a vascularized liver organoid. *Hepatology* 2011;**53**:604–17.
17. Soto-Gutierrez A, et al. A whole-organ regenerative medicine approach for liver replacement. *Tissue Eng Part C Methods* 2011;**17**:677–86.
18. Wang Y, et al. Lineage restriction of human hepatic stem cells to mature fates is made efficient by tissue-specific biomatrix scaffolds. *Hepatology* 2010;**53**:293–305.
19. Duncan AW, Soto-Gutierrez A. Liver repopulation and regeneration: new approaches to old questions. *Curr Opin Organ Transplant* 2013;**18**:197–202.
20. Fukumitsu K, Yagi H, Soto-Gutierrez A. Bioengineering in organ transplantation: targeting the liver. *Transplant Proc* 2011;**43**:2137–8.
21. Uygun BE, Yarmush ML, Uygun K. Application of whole-organ tissue engineering in hepatology. *Nat Rev Gastroenterol Hepatol* 2012;**9**:738–44.
22. Wertheim JA, Baptista PM, Soto-Gutierrez A. Cellular therapy and bioartificial approaches to liver replacement. *Curr Opin Organ Transplant* 2012;**17**:235–40.
23. Orlando G, et al. Cell and organ bioengineering technology as applied to gastrointestinal diseases. *Gut* 2013;**62**:774–86.
24. Levin DE, Grikscheit TC. Tissue-engineering of the gastrointestinal tract. *Curr Opin Pediatr* 2012;**24**:365–70.
25. Grikscheit TC, et al. Tissue-engineered small intestine improves recovery after massive small bowel resection. *Ann Surg* 2004;**240**:748–54.
26. Nowocin AK, Southgate A, Gabe SM, Ansari T. Biocompatibility and potential of decellularized porcine small intestine to support cellular attachment and growth. *J Tissue Eng Regen Med* 2013.

27. Park KM, Woo HM. Systemic decellularization for multi-organ scaffolds in rats. *Transplant Proc* 2012;**44**:1151–4.

28. Patil PB, et al. Recellularization of acellular human small intestine using bone marrow stem cells. *Stem Cells Transl Med* 2013;**2**:307–15.

29. Totonelli G, et al. A rat decellularized small bowel scaffold that preserves villus-crypt architecture for intestinal regeneration. *Biomaterials* 2012;**33**:3401–10.

30. OPTN/SRTR 2011 Annual Report: kidney. Department of Health and Human Services Administration, Healthcare Systems Bureau. *Division of Transplantation* 2012:11–47.

31. Morigi M, et al. Life-sparing effect of human cord blood-mesenchymal stem cells in experimental acute kidney injury. *Stem Cells* 2010;**28**:513–22.

32. Song JJ, et al. Regeneration and experimental orthotopic transplantation of a bioengineered kidney. *Nat Med* 2013;**19**:646–51.

33. Ross EA, et al. Embryonic stem cells proliferate and differentiate when seeded into kidney scaffolds. *J Am Soc Nephrol* 2009;**20**:2338–47.

34. Ross EA, et al. Mouse stem cells seeded into decellularized rat kidney scaffolds endothelialize and remodel basement membranes. *Organogenesis* 2012;**8**:49–55.

35. Orlando G, et al. Production and implantation of renal extracellular matrix scaffolds from porcine kidneys as a platform for renal bioengineering investigations. *Ann Surg* 2012;**256**.

36. Sullivan DC, et al. Decellularization methods of porcine kidneys for whole organ engineering using a high-throughput system. *Biomaterials* 2012;**33**:7756–64.

37. Nakayama KH, Lee CCI, Batchelder CA, Tarantal AF. Tissue specificity of decellularized rhesus monkey kidney and lung scaffolds. *PLoS One* 2013;**8**:10.

38. Nakayama KH, Batchelder CA, Lee CI, Tarantal AF. Decellularized rhesus monkey kidney as a three-dimensional scaffold for renal tissue engineering. *Tissue Eng Part A* 2010;**16**:2207–16.

39. Orlando G, et al. Discarded human kidneys as a source of ECM scaffold for kidney regeneration technologies. *Biomaterials* 2013;**34**:5915–25.

40. Kochanek KD, Xu J, Murphy SL, Minino AM, Kung H-C. Deaths: Final Data for 2009. Wyattsville, MD:Center for Health Statistics; 2012.

41. Zimmermann WH, et al. Engineered heart tissue grafts improve systolic and diastolic function in infarcted rat hearts. *Nat Med* 2006;**12**:452–8.

42. Wainwright JM, et al. Preparation of cardiac extracellular matrix from an intact porcine heart. *Tissue Eng Part C Methods* 2010;**16**:525–32.

43. Oberwallner B, et al. Preparation of cardiac extracellular matrix scaffolds by decellularization of human myocardium. *J Biomed Mater Res A* 2014;**102**(9):3263–72.

44. Ng SLJ, Narayanan K, Gao S, Wan ACA. Lineage restricted progenitors for the repopulation of decellularized heart. *Biomaterials* 2011;**32**:7571–80.

45. Shaw JE, Sicree RA, Zimmet PZ. Global estimates of the prevalence of diabetes for 2010 and 2030. *Diabetes Res Clin Pract* 2010;**87**:4–14.

46. Mirmalek-Sani S-H, et al. Porcine pancreas extracellular matrix as a platform for endocrine pancreas bioengineering. *Biomaterials* 2013;**34**:5488–95.

47. Goh S-K, et al. Perfusion-decellularized pancreas as a natural 3D scaffold for pancreatic tissue and whole organ engineering. *Biomaterials* 2013;**34**:6760–72.

48. Vaithilingam V, Tuch BE. Islet transplantation and encapsulation: an update on recent developments. *Rev Diabet Stud* 2011;**8**:51–67.

49. Opara EC, Mirmalek-Sani S-H, Khanna O, Moya ML, Brey EM. Design of a bioartificial pancreas. *J Investig Med* 2010;**58**:831–7.

50. Prakash YS, Stenmark KR. Bioengineering the lung: molecules, materials, matrix, morphology, and mechanics. *Am J Physiol Lung Cell Mol Physiol* 2012;**302**:L361–2.

51. Ott HC, et al. Regeneration and orthotopic transplantation of a bioartificial lung. *Nat Med* 2010;**16**:927–33.

52. Petersen TH, et al. Tissue-engineered lungs for in vivo implantation. *Science* 2010;**329**:538–41.

53. Song JJ, et al. Enhanced in vivo function of bioartificial lungs in rats. *Ann Thorac Surg* 2011;**92**:998–1006.

54. O'Neill JD, et al. Decellularization of human and porcine lung tissues for pulmonary tissue engineering. *Ann Thorac Surg* 2013;**96**:1046–56.

55. Badylak SF. Regenerative medicine and developmental biology: the role of the extracellular matrix. *Anat Rec Part B New Anat* 2005;**287**:36–41.

56. Koh CJ, Atala A. Tissue engineering, stem cells, and cloning: opportunities for regenerative medicine. *J Am Soc Nephrol* 2004;**15**:1113–25.

57. Badylak SF, Taylor D, Uygun K. Whole-organ tissue engineering: decellularization and recellularization of three-dimensional matrix scaffolds. *Annu Rev Biomed Eng* 2011;**13**:27–53.

58. Guillot PV, Cui W, Fisk NM, Polak DJ. Stem cell differentiation and expansion for clinical applications of tissue engineering. *J Cell Mol Med* 2007;**11**:935–44.

59. Soker S, Machado M, Atala A. Systems for therapeutic angiogenesis in tissue engineering. *World J Urol* 2000;**18**:10–8.

60. Rouwkema J, Rivron NC, van Blitterswijk CA. Vascularization in tissue engineering. *Trends Biotechnol* 2008;**26**:434–41.

61. Crapo PM, Gilbert TW, Badylak SF. An overview of tissue and whole organ decellularization processes. *Biomaterials* 2011;**32**:3233–43.

62. Auger FA, Gibot L, Lacroix D. The pivotal role of vascularization in tissue engineering. *Annu Rev Biomed Eng* 2013;**15**(15):177–200.

63. Bijonowski BM, Miller WM, Wertheim JA. Bioreactor design for perfusion-based, highly-vascularized organ regeneration. *Curr Opin Chem Eng* 2013;**2**:32–40.

64. Price AP, England KA, Matson AM, Blazar BR, Panoskaltsis-Mortari A. Development of a decellularized lung bioreactor system for bioengineering the lung: the matrix reloaded. *Tissue Eng Part A* 2010;**16**:2581–91.

65. Martin I, Wendt D, Heberer M. The role of bioreactors in tissue engineering. *Trends Biotechnol* 2004:80–6.

66. Cooper DKC, Gollackner B, Sachs DH. Will the pig solve the transplantation backlog? *Annu Rev Med* 2002;**53**:133–47.

67. Lee MH, et al. Considerations for tissue-engineered and regenerative medicine product development prior to clinical trials in the United States. *Tissue Eng Part B Rev* 2010;**16**:41–54.

68. Crawford B, et al. Cardiac decellularisation with long-term storage and repopulation with canine peripheral blood progenitor cells. *Can J Chem Eng* 2012;**90**:1457–64.

69. Baptista PM, Siddiqui MM, Lozier G, Rodriguez SR, Atala A, Soker S. The use of whole organ decellularization for the generation of a vascularized liver organoid. *Hepatology* 2011;**53**(2):604–17

Chapter 4

Regenerative Implants for Cardiovascular Tissue Engineering

Avione Y. Lee, Yong-Ung Lee, Nathan Mahler, Cameron Best, Shuhei Tara, Christopher K. Breuer

Key Concepts

In this chapter, we summarize the current state of regenerative implants in cardiovascular tissue engineering by covering the following aspects: (1) brief historical background, (2) types of regenerative implants, (3) the function of implants, and (4) clinical applications in cardiovascular repair.

1. A fundamental problem that affects the field of cardiovascular surgery is the paucity of autologous tissue available for surgical reconstructive procedures. The field of tissue engineering represents a promising new option for replacement surgical procedures. Throughout the years, intensive interdisciplinary, translational research into cardiovascular regenerative implants has been undertaken in an effort to improve surgical outcome and better quality of life for patients with cardiovascular defects.
2. Biocompatibility, biodegradability, mechanical profile, and manufacturability are several variables to be considered in the development of tissue engineering scaffolds. To meet those guidelines, three broad classes of materials have been pursued: (1) biodegradable synthetic polymers, (2) biologic polymers, (3) hybrid polymer blends.
3. Implants used in regenerative medicine serve one task, restoring healthy function of the injured tissue or organ. It can be accomplished in a variety of ways. Implants can be used as an in situ structure or scaffold for cells to grow into for tissue formation, as an in vitro cell delivery vehicle, and as a way to manipulate the tissues environment through drugs and genes.
4. The first reported synthetic vascular patches and grafts date back to the 1950s. Since then, tremendous advancements have been achieved in cardiovascular repair. This section focuses on the most promising and widely used clinical applications to date.

List of Abbreviations

B-FGF Basic Fibroblast Growth Factor
BiVAD Biventricular assist devices
BM-MNC Bone marrow mononuclear cell
BM-MSC Bone marrow mesenchymal cell
DDS Drug delivery systems
EC Endothelial cell
ECM Extracellular matrix
EPC Endothelial progenitor cell
hMSC Human mesenchymal stem cell
iPS Inducible pluripotent stem
LVAD Left ventricular assist devices
MSC Mesenchymal stem cell
P(CL/LA) Polycaprolactone polylactic acid
PCL Poly(ε-caprolactone)
PDGF Platelet Derived Growth Factor
PET Polyethylene terephthalate
PGA Polyglycolic acid
PGS Polyglycerol sebacate
PHB Polyhydroxyalkanoate
PLA Polylactic acid
PLCL Poly(lactide-co-ε-caprolactone)

Translating Regenerative Medicine to the Clinic. http://dx.doi.org/10.1016/B978-0-12-800548-4.00004-8

PLGA Polylactic-co-glycolic acid
PLLA Poly-L-lactide acid
PTFE Polytetrafluoroethylene
RVAD Right ventricular assist devices
SBTE Sheet-based tissue engineering
SIS Small intestinal submucosa
SIS-ECM Small intestinal submucosa—extracellular matrix
SMC Smooth muscle cell
TAH Total artificial heart
TEHV Tissue-engineered heart valve
TESA Self-assembled tissue-engineered vascular grafts
TEVG Tissue-engineered vascular graft
VAD Ventricular assist devices
VEGF Vascular Endothelial Growth Factor

1. INTRODUCTION

1.1 The Path to a Regenerative Approach

A fundamental problem that affects all fields of surgery is the paucity of autologous tissue available for surgical reconstructive procedures.[1] When a surgeon removes a tissue that is diseased or damaged, or when a surgeon replaces a tissue that is congenitally absent, the best results are obtained when an individual's own tissues are used for the surgical repair. When this is not possible the surgeon is forced to use alternative biomaterials and typically select from either prosthetic, artificial synthetic materials or from biological materials typically derived from allografts or xenografts (Figure 1). Prosthetic materials have the advantage of ready "off-the-shelf" availability, but frequently suffer from problems related to biocompatibility and the fact that they never become completely integrated into the host. Biological materials are typically more biocompatible than synthetic prosthetic materials, however, they are still a source of rejection and require treatment with immunosuppressive agents, as in the case of organ transplantation, or they are treated in order to reduce their immunogenicity and allow implantation without immunosuppressive agents. This is typically accomplished using either cryopreservation techniques or tissue fixation methods. Alternatively, decellularization methods can also be used. Such treatments either remove the cellular component of the tissue or render the cells nonviable which worsens the durability of these products. The use of autologous tissue outperforms currently available prosthetic or biological materials designed for use in surgery and is always preferable when autologous tissue is available in adequate supply.[1]

Tissue engineering is a multidisciplinary science that attempts to create living biomaterials from a patient's own cells. One method of tissue engineering uses a three-dimensional (3D) scaffold that serves as a site for cell attachment and provides space for neotissue formation.[2] The scaffold can serve as a template for neotissue development. Tissue engineering attempts to exploit the cells reproductive potential and harness the body's innate capacity for healing and regeneration (Figure 2). The goal of tissue engineering is to create living, autologous neotissues that can be used to repair or replace tissues that are diseased, damaged, or congenitally absent.[2] The central hypothesis of tissue engineering is that the tissue-engineered construct will perform more like an individual's own tissue and less like a prosthetic or biological material.

Departing from more traditional approaches to treat lost organ or tissue function, tissue engineering seeks to replace or restore function to diseased or damaged tissues and organs through implantable devices. From its onset, tissue engineering has been a multidisciplinary field that combines efforts of basic scientists, engineers, and clinicians.[2] Through trial and error, these researchers established three major approaches toward tissue engineering implants for regeneration: implanting neotissues derived from cells, implanting engineered matrices, or implanting cells combined within matrices.[3] The first tissue-engineered cells were implanted in the belly of a pig in 1933.[3] Tissue-engineered skin matrices, consisting of cultured epithelial sheets or fibroblast gels seeded onto polymer scaffolds, were created in the late 1970s and in the early 1980s.[4–6] In the late 1980s, the first tissue-engineered implantation studies were conducted consisting of the seeding of pancreatic islet cells onto a synthetic polymer and subsequently implanted into animals.[7] By the 1990s, tissue engineering established itself as a recognized field.[2]

Tissue engineering techniques have been used to create a host of tissue types with varying degrees of success. From a translational perspective, dermatological applications in the form of tissue-engineered skin substitutes are the furthest along and a variety of commercially available products are commonly used in the clinic.[8] Other tissues such as nervous tissues have proven significantly more challenging to create using tissue engineering methodology and are therefore not yet available for clinical use. For the purpose of this review, we will focus our discussions on tissue-engineered cardiovascular

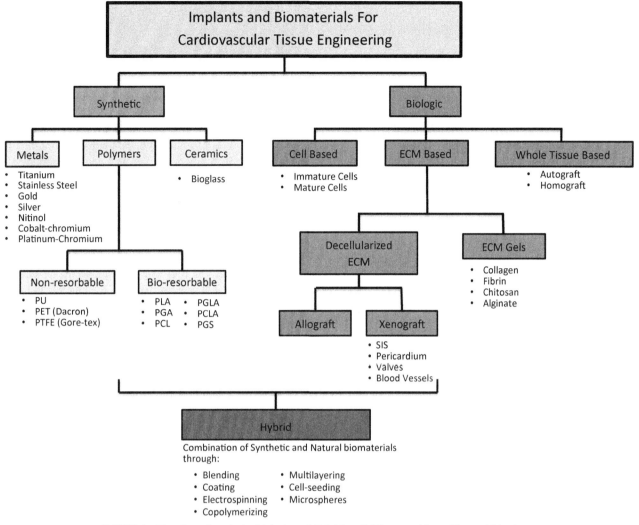

FIGURE 1 Flowchart of synthetic, Biologic, and Hybrid scaffolding materials used in arterial implants.

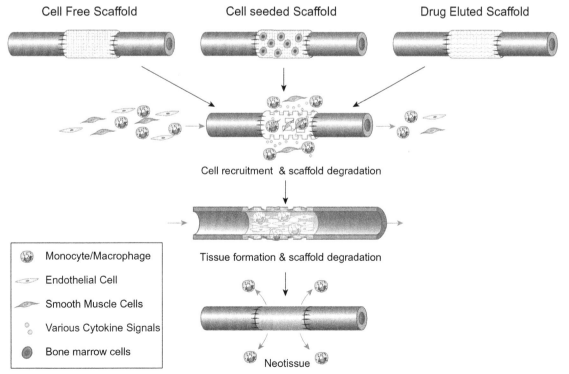

FIGURE 2 Proposed mechanism of neovessel formation after implantation of a 1) cell-free scaffold, 2) cell-seeded scaffold, or 3) a drug eluted scaffold. After implantation, there is early monocyte recruitment to the scaffold due to either 1) the scaffold inherent biocompatibility, 2) monocyte chemoattractant protein-1 (MCP-1) and related cytokines from seeded bone marrow-derived mononuclear cells (BM-MNCs) or 3) drugs created to mimic the cytokines from cells. The monocytes in turn recruit smooth muscle cells and endothelial cells. The scaffold simultaneously degrades as the cells develop into neotissue. The neotissue continues its growth and remodeling as the patient ages.

implants, which provide examples of a variety of tissue engineering applications along the translational spectrum. In this way, we can provide an accurate snapshot of the current state-of-the-art technologies and in the process provide an overview of the use of implants in the field of tissue engineering.

The field of tissue engineering is an applied science that approaches complex problems by deconstructing them into multiple simpler components. The development of most tissue engineering applications utilizes this paradigm beginning by creating small pieces of tissue, then developing more complex functional tissue components, before finally attempting to create entire bioartificial organs. This "pieces, to parts, to whole organ" experimental design motif is a common thread that will run throughout most tissue engineering projects discussed in this review. It provides a rational framework for designing any tissue engineering application.

Tissue engineering of cardiovascular structures from blood vessels to heart valves, and even whole heart has undergone great strides in the last decade. Historically, the foundation for regenerative cardiovascular implants can be traced back to CC Guthrie who, in 1919, stated that for repairing a blood vessel "an implanted segment need only temporarily restore mechanical continuity and serve as a scaffolding or bridge for the laying down of an ingrowth of tissue derived from the host."[9] The defining characteristics of regenerative cardiovascular implants have not strayed from that statement, but have taken it one step further. Not only does the tissue-engineered implant restore continuity and serve as a scaffolding device for tissue in growth, it also has the ability to degrade, leaving only healthy neotissue behind; thus enabling full integration into the host. In this review, we seek to provide a brief historical perspective on regenerative cardiovascular devices, insight on types and functions of regenerative cardiovascular implants, and the most recent clinical breakthroughs in cardiovascular regenerative medicine.

For the purpose of this review, we will define an implant as any tissue-engineered construct that is placed in the body and used to repair or replace tissue that is either diseased, damaged, or absent.

1.2 Brief Historical Overview of Vascular, Heart Valve, and Heart Implants

1.2.1 Vascular Implants

Atherosclerotic cardiovascular disease is the major cause of morbidity and mortality in the western world.[10] Vascular bypass with an autologous conduit remains the gold standard and treatment of choice for end-stage cardiovascular disease. Autologous vein grafting was first pioneered by Nobel laureate Alex Carrel, who is considered the father of modern cardiovascular surgery.[11,12] The technical aspects of vascular bypass have been optimized and are performed routinely throughout the United States. Autologous vascular conduits in the form of a saphenous vein or mammary artery are the conduits of choice for coronary bypass grafting procedures due to their superior clinical performance.[13] Unfortunately, atherosclerosis is a progressive systemic disease, so the use of these autologous grafts is frequently not possible when their supply has been exhausted. When autologous vascular grafts are not available the surgeon is forced to use either prosthetic or biologic vascular grafts.

The first clinical study using an artificial vascular graft was reported in 1954 by Blakemore et al.[14] This report described a small clinical series in which Vinyon-N was used to make conduits which were used in the surgical repair of patient's with symptomatic abdominal aortic aneurysms, a disease with 100% mortality at the time of the study. Subsequent materials used to fabricate synthetic vascular grafts included grafts made of polyethylene terephthalate (PET) (Dacron®) which was subsequently replaced by polytetrafluoroethylene (Gore-Tex®). These artificial, prosthetic grafts have many advantages, the greatest of which is their ready availability. Prosthetic vascular graft function is inversely related to graft diameter with an increase in thrombotic occlusion as graft size decreases.[15] When used for bypass procedures in the arterial circulation, prosthetic vascular graft function is adequate for use in large-diameter grafts (>8 mm, such as for use in aortic surgery), intermediate (between 6 and 8 mm, such as femoral artery bypass surgery), and poor when used in procedures requiring the use of small-diameter vascular grafts (<6 mm, such as coronary artery bypass surgery or peripheral arterial bypass surgery).[15,16] Synthetic vascular grafts have several disadvantages related to biocompatibility issues. Most notably they are prone to thromboembolic complications, and have poor long-term durability related to the development of neointimal hyperplasia.[16] Biological vascular grafts are more biocompatible than synthetic vascular grafts, however, in order to reduce their immunogenicity, most biological grafts are processed using either cryopreservation or various forms of tissue fixation. This processing renders the cellular component of the vessel not viable, which in turn significantly reduces the durability of the graft. Biological grafts are also a potential source of donor-derived infections. Compared to prosthetic grafts, biological grafts tend to be more expensive due to the complex processing required to make and store these grafts, thus their use tends to be quite limited compared to synthetic vascular grafts.

One strategy that has been explored for improving the function of small-diameter vascular grafts is seeding the luminal surface of the conduit with autologous endothelial cells (ECs), which would create a thromboresistant surface.[17] Spontaneous reendothelialization of prosthetic grafts occurs in some species; however, it does not occur to any great extent in humans.[18–20] Several groups developed methods to harvest, expand, and seed autologous ECs onto the luminal surface of a prosthetic vascular graft in an attempt to improve graft function.[21–24] Results of preclinical studies were promising and ultimately led to a prospective clinical trial.[25] Despite the fact that the clinical trial demonstrated improved efficacy, it has still not been widely adopted presumably due to the added time, expense, and complexity associated with using this technology.

In 1986, a collagen gel was used by Weinberg and Bell to create the first tissue-engineered vascular graft (TEVG),[26] however, the resulting neovessel did not have sufficient burst strength to withstand the hemodynamic forces of the cardiovascular circulation. Over the ensuing years, several investigators attempted to create biocompatible TEVGs, by combining biologic materials with nondegradable materials such as PET in order to create a hybrid tissue-engineered prosthetic vascular graft.[27–29] In 1998, Shinoka et al. first reported the use of a biodegradable synthetic scaffold and ovine cells with promising long-term result in low-pressure pulmonary artery system.[30] Similarly, in 1998, L'Heureux reported the first "scaffold-free" self-assembled tissue-engineered vascular graft (TESA).[31] This graft was fabricated using a method called sheet-based tissue engineering (SBTE).[31] This technique utilizes the monolayer of tissue that forms when cells are grown on a cell culture dish in vitro. The thin sheets of tissue are rolled creating a multilayered laminated tube. In 1999, Niklason et al. reported the use of TEVGs grown using a bioreactor.[32] In this methodology, smooth muscle cells (SMCs) are seeded onto a biodegradable tubular scaffold and are cultured in a device that exposes the cells to gradually increasing hemodynamic forces which induce the cells to produce a more robust extracellular matrix (ECM) compared to statically grown neovessels. The luminal surface of the scaffold was seeded with ECs, thus creating the tissue-engineered conduit.[32] These experiments demonstrated the feasibility of creating a TEVG from an individual's own cells that could be implanted and function as a vascular interposition graft (Table 3).

1.2.2 Heart Valve

Valvular heart disease is a significant source of morbidity and mortality in the United States.[33] The treatment of choice for end-stage valvular heart disease is heart valve replacement.[34] The first human valve implantation was performed by Hufnagel in 1952 using a caged ball valve implanted into the descending aorta of six patients.[35] The first orthotopic valve transplant was reported in 1960 by Harken et al.[36] Mechanical valves are advantageous because they are readily available, offer an unlimited supply, and are the most durable replacement heart valve. However, despite various advances in design, they are associated with significant problems related to biocompatibility. Thromboembolic complications are the most common serious valve-related event. The risk of thromboembolic complications is so high as to justify the use of anticoagulation therapy despite its 1% per year cumulative mortality. Mechanical heart valves also cause hemolysis and are susceptible to infectious complications.[37] Bioprosthetic replacement heart valves are more biocompatible than mechanical heart valves, less thrombogenic, and typically do not require anticoagulation therapy. The first bioprosthetic replacement heart valve used in humans was an allogeneic cadaveric valve placed in the descending aorta by Murray et al. (1956).[37] The clinical use of a glutaraldehyde-fixed xenogeneic replacement heart valve was first reported by Carpentier et al.[38] They used glutaraldehyde fixation to overcome the immunogenicity of the xenograft. Cryopreservation is also applied to bioprosthetic valves for the same reason. Though both of these models provide excellent short-term treatment, they are prone to ectopic calcification and have poor durability resulting in the need for replacement.[37] Additionally, bioprosthetic valves do not possess growth capacity which results in somatic overgrowth and the need for replacement when used in the pediatric population.[39]

The tissue-engineered heart valve was developed as an alternative to prosthetic and bioprosthetic replacement heart valves. Shinoka et al. reported the development of the first tissue-engineered heart valve leaflet which was created by seeding autologous fibroblasts and ECs onto a biodegradable polyglycolic acid (PGA) fiber-based matrix.[40,41] Hoerstrup and colleagues began making whole heart valves using this methodology and ultimately replaced the stiff PGA matrix with a biodegradable elastomer, polyhydroxyalkanoate.[42–44] This resulted in giving the tissue-engineered valve construct a more physiologic biomechanical profile.[42] Several advances have been made in biodegradable valve material, seeding, and insertion techniques, all designed to improve the clinical utility of this technology.[45–50] However, no biodegradable valve has been placed in humans to date. Extensive work has also been done with xenogeneic and allogeneic decellularized valves, beginning with Brendel and Duhamel's decellularized xenograft in 1984.[51–56] These valves are created by decellularizing allogeneic or xenogeneic valves using either detergent-based methods and/or enzymatic methods. Removal of the cellular component of the heart valve tissue should render it inert enabling implantation in humans with less risk of rejection. It also preserves the unique ECM of the heart valve, which enables it to completely open and rapidly close under minimal pressure gradients, thus preventing stenosis and regurgitation.[57] The decellularized heart valve does not recellularize spontaneously when implanted in vivo. Due to the dense, nonporous nature of the heart valve ECM, it is difficult to recellularize these constructs which compromises their long-term durability.[51]

1.2.3 Whole Heart

Heart failure is a significant cause of death in the United States. The incidence of heart failure is increasing and medical management does little to slow the progression of disease.[58,59] Currently, the treatment of choice for end-stage heart failure is allogeneic transplantation, with a 10-year survival rate of nearly 50%.[60] In 1967, Barnard and colleagues reported the first known human heart transplant in which a cadaveric heart was successfully transplanted into a 54-year-old male.[61] Although a life-saving intervention to refractory heart failure, since its inception heart transplantation has been plagued by donor organ scarcity and transplant rejection.[60,62,63] As a response to the paucity of donor organs, cardiac prostheses and total artificial hearts (TAHs) were developed to sustain circulation while a suitable heart for transplant was obtained.[64] Cooley et al. reported the first successful application of dual ventricle prosthesis in 1969 (the Liotta artificial heart).[64,65] In spite of the recent refinement and improvement of TAH technology, these devices, although mechanically durable, are prone to infection and thromboembolism, which necessitate the use of anticoagulation, and current lack of clinical reliability limits their widespread use as an alternative to allogeneic heart transplantation.[65,66] Recently, extracorporeal left, right, and biventricular assist devices (LVAD, RVAD, and BiVAD, respectively) have been used as a "bridge-to-transplantation" and in some cases as destination therapy with relative clinical success.[60,62] However, ventricular assist devices (VAD) share the disadvantages of TAHs along with the need for additional surgery once a donor heart is available for transplant.[60,62,65,67]

Tissue engineering presents a potential future alternative to transplantation or the use of TAH and VAD technology in the treatment of heart failure because it offers the possibility of a nonimmunogenic and fully functional bioartificial heart. A recent advancement reported by Ott and colleagues suggests that advances are being made toward that end.[63] In that study, cadaveric rat hearts were decellularized by coronary diffusion and subsequently reseeded with neonatal cardiac cells and aortic ECs to culture contractile and pharmacologically responsive bioartificial hearts after 8 days in vitro.[63] Current research is aimed at upscaling and refining the decellularization, seeding, and in vitro culture processes reported by Ott et al. to larger animal models with the ultimate goal being a tissue-engineered human heart as a clinical solution to end-stage heart failure.[63]

2. TYPES OF REGENERATIVE IMPLANTS—THE CONTINUITY BRIDGE

Successful tissue-engineered implants must be rationally designed as biologically compatible, nonimmunogenic tissue replacements that mimic the morphology and function of the native tissue.[68] An initial consideration in the development of a tissue-engineered implant is the ECM.[69,70] Like the tissue engineering scaffold, the native ECM is a complex structure that provides a platform for cell adhesion, mechanoreception, and paracrine signaling, large contribution to the biomechanical properties of the tissue, and acts as a reservoir for growth factors and biomolecules essential to cell function, proliferation, and differentiation.[69] Thus, in order to support neotissue formation, a tissue-engineered scaffold needs to provide or induce an extracellular environment that closely resembles native ECM in both structure and function.[71]

Other variables to be considered in the development of tissue engineering scaffolds are biodegradability, biocompatibility, mechanical profile, and manufacturability. To meet those guidelines, three broad classes of materials have been pursued: (1) degradable synthetic polymers, (2) biologic polymers, and (3) hybrid polymer blends (Figure 1). In the following discussion, current advances and the relative utility of each are reviewed.

2.1 Synthetic

2.1.1 Degradable Synthetic Polymers

Degradable scaffold materials are advantageous in tissue engineering applications since they provide initial biomechanical integrity as cells adhere, mature to the desired phenotype, and secrete natural ECM, and most importantly, they are designed to be replaced by neotissues upon biodegradation and bioresorption.[69,72,73] Polymers that have been extensively used in tissue engineering applications are PGA, polylactic acid (PLA), and poly(ε-caprolactone) (PCL) (Figure 1).[73–76] These polymers are widely used due to their biocompatible degradation properties. Hydrolysis of ester bonds in situ is thought to release nontoxic monomers that are excreted or absorbed by surrounding tissue.[69,73,75] However, some groups report that the release of acidic degradation products can be a source of toxicity causing adverse local and systemic reactions that could result in implant rejection.[77] For example, PGA has a fast degradation rate and is reported to lose all mechanical strength within 4 weeks of in vivo implantation and is completely degraded after 6 months, whereas PCL can require up to 2–3 years for in vivo degradation to occur depending on how it is attached to the graft.[77] For instance, a PCL coating degrades faster than if it is electrospun onto the graft (Figure 3). In fact, it is important to note that the degradation timeline is vast for all materials and depends on their fabrication and use (i.e., coating, 3D printing, electrospinning, etc.). Use of these polymers provides a versatile platform for making scaffolds for tissue engineering applications. Altering polymer concentrations and using polymer blends and copolymers can yield a scaffold with a degradation rate specifically tailored to mirror the rate of neotissue

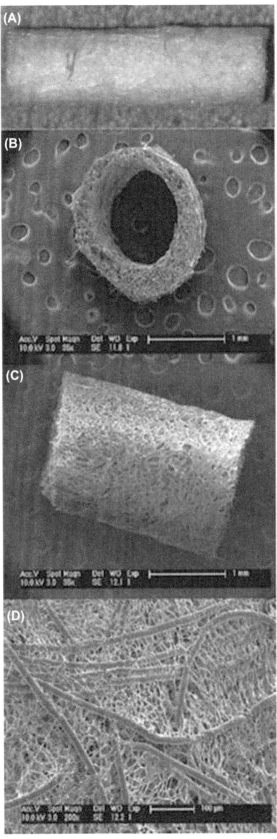

FIGURE 3 Gross and scanning electron microscopy (SEM) images of biodegradable polyglycolic acid (PGA) synthetic scaffold coated with polycap-rolactonepolylactic acid for mouse model of tissue engineered vascular graft. (A) Gross axial image of scaffold. (B) SEM of scaffold showing thickness across wall. The average luminal diameter is 0.9 mm and wall thickness is 0.15 mm. (C) Axial SEM view of scaffold. (D) SEM of PGA fibers and PCL coating in scaffold. The average pore size is 45.4 ± 17.6 mm and porocity is 78.5%.

formation. Examples of copolymers under current investigation are polylactic-co-glycolic acid (PLGA), polycaprolactone polylactic acid (P(CL/LA)), or poly(lactide-co-ε-caprolactone) (PLCL).[69,73,78,79] The popularity of PGA, PLA, PCL and their copolymers in tissue engineering scaffold research is due in part to their approval for use in humans by the US Food and Drug Administration; however, individual approval of each new tissue engineering applications is ultimately required before clinical implementation.[68,69,77,78] Recently, an innovative, rationally designed polymer, polyglycerol sebacate (PGS), has been reported by Wang and colleagues that shows promise in cardiovascular tissue engineering applications.[80,81]

While most of the commonly used biodegradable polymers such as PGA, PLA, and PCA were originally designed for use outside of biomedical applications, PGS, or biorubber, was specifically designed as a scaffold for making tissue-engineered cardiovascular structures.[81] PGS is a tough, compliant, 3D, inexpensive, biocompatible, and biodegradable elastomer designed to provide an optimal environment for cellular attachment, proliferation, and the deposition of ECM that mimics native morphology. In addition, mechanical analysis of PGS scaffolds illustrates Young's moduli that approach native artery, vein, and myocardium.[81–83] As opposed to other polymers mentioned herein that were originally synthesized for uses outside of tissue engineering, PGS was developed with a "bottom-up" approach. It was rationally designed to be an ideal tissue engineering scaffold material. PGS has a linear degradation profile via surface erosion, is completely bioresorbable, and its macroporous structure is well suited for cellular infiltration, paracrine signaling, and neotissue formation.[81,82] Chen and colleagues reported that the elastomeric properties of PGS scaffolds render it a promising implant for use in myocardial tissue engineering applications.[83] Similarly, exciting findings by Wu et al. have demonstrated that heparin-coated PGS scaffolds reinforced with an electrospun PCL sheath support host remodeling and yield small-diameter tissue-engineered arterial grafts that approach the biomechanical characteristics of native vessels after 3 months in a rat model.[84]

2.1.2 Fabrication Methods

In developing viable scaffold constructs for regenerative cardiovascular applications, researchers have employed a host of innovative fabrication technologies in order to design clinically relevant secondary characteristics such as gross size and shape, surface morphology, porosity, mechanical profile, and degradation rate. Three general approaches to prefabricated scaffold (scaffolds created before implantation) manufacture are: (1) physical methods, (2) rapid prototyping, and (3) textile methods.[69,85] Physical methods most commonly employed include phase separation, gas foaming, freeze-drying, solvent casting, particulate leaching, thermal processing, and molding.[69,85] Rapid prototyping methods comprise technologies such as 3D printing, stereolithography, selective laser sintering, 3D shape deposition manufacturing, fused deposition modeling, and electrospinning.[69,85] Recently, electrospinning has become widely used and is a promising low-cost rapid-prototyping method due to the wide range of polymers that can be electrospun; the ability to fine-tune degradation characteristics and mechanical properties; and the ability to control fiber size, orientation, and alignment.[86] Finally, prefabricated scaffolds are often created by traditional textile methods, including wet, dry, or melt spinning, nonwoven bonding, weaving, knitting, or braiding.[85] An alternative to prefabricated scaffolds is in situ peptide-based hydrogel constructs, also referred to as self-assembling scaffolds, whose curing occurs in situ to produce biomimetic scaffold constructs.[77,85] Self-assembled constructs are produced by genetic and chemical engineering to yield biomimetic and functional solutions by methods such as molecular entanglement via cross-linking and polymer–polymer interactions or by use of specifically designed hydroregulators or oligomeric precursors (e.g., synthetic polymers, polysaccharides, proteins, and polyamino acids) that direct molecular self-assembly.[87]

2.2 Biologic

Biologic materials used in regenerative medicine comprise naturally occurring polymers typically derived from the ECM. The ECM is an ideal scaffold for tissue engineering because it is inherently biocompatible and bioresorbable, nonimmunogenic, and provides a natural home for cellular adhesion, proliferation, and differentiation.[88,89] The ECM components are most widely used in two ways: the individual proteins are isolated and used in gel form or the whole ECM is decellularized and used as a decellularized matrix.[88]

2.2.1 ECM Gels

Commonly used ECM proteins are collagen, fibrin, and elastin because they are the major components of the blood vessel wall and myocardium.[90,91] Collagen is one of the largest proteins in the body and helps mediate cell adhesion, migration, proliferation, and differentiation. It has a wide range of applications and has been used in the heart and blood vessels. Collagen gel was used by Weinberg and Bell to create the first TEVG in 1986[26] and since the early 1990s, it has been used as a matrix to study both myocardial contraction and electrophysiology.[92] Fibrin is a protein produced from fibrinogen and has been investigated as a substrate for cell adhesion, migration, and proliferation. It is an autocrine regulator of SMCs and has a controlled degradation, which makes it a natural drug delivery substrate. ECM gene expression correlated with in vitro tissue growth in fibrin gel[93] and elastogenesis could be achieved better on fibrin gels than collagen gels.[94] A fibrin gel patch has been used to deliver mesenchymal stem cells (MSCs) to an infarct region in swine[91] and fibrin has also been used to produce small-diameter elastic blood vessels.[95,96] Elastin is an autocrine SMC regulator, has a controlled degradation, and is critical in conferring elasticity to cardiovascular structures, especially the vascular wall. Unfortunately, gels often do not have the mechanical strength to support the physical load needed in the hemodynamic environment of the cardiovascular system. However, in recent years several researchers have created grafts made of fibrin gels that approach physiological mechanical strength and compliance and are sufficient for implantation in animal models.[97–99]

2.2.2 Decellularized Scaffolds

A response to the poor mechanical properties of protein gels for use in various tissue engineering applications is to use decellularized ECM scaffolds. The method of decellularization has been used since the mid-1980s.[100] The most commonly employed methods for decellularizing tissues use either detergent-based methods and/or enzymatic methods; in addition, cryopreservation techniques also have utility. The decellularizing process produces a weaker ECM than native vessels, but cross-linking of the ECM proteins can increase the mechanical properties. The cells are removed to reduce the immunogenicity of the cellular component of the allogeneic or xenogeneic tissues and once removed, only the protein matrix remains with the natural cross-linking and mechanical properties inherent in it. It can then be reseeded with autologous cells and implanted when needed. Rejection still occurs, however, mainly due to remnants of the α Gal antigen, which is an antigen responsible for rejection of xenotransplants,[101,102] due to a natural antibody response in humans to pig antigens.[103] Cryopreservation can assist in improving the durability of xenografts due to immunomodulation of the xenografic tissue, which could prevent rejection.[104]

Nevertheless, the advantage of using decellularized scaffolds is the potential use of animal tissue, which is in greater supply than human. For vascular applications, porcine and canine arteries and veins are the most widely used because of their similar geometry to humans.[105] Decellularized porcine small intestinal submucosa (SIS) has been applied to a variety of cardiovascular applications for more than two decades.[106] Intriguing independent research has come out of the Niklason and L'Heureux groups where allogeneic vessels are grown in vitro, decellularized/devitalized for storage, then reseeded with allogeneic cells prior to implantation.[107,108]

2.2.3 Hybrid

Hybrid scaffolds are the combination of synthetic and biologic materials to form a scaffold that uses the advantages of both. These scaffolds can be made in a variety of ways such as coating biologic material onto a synthetic one, electrospinning a biologic and synthetic material together, or using biologic microspheres with a synthetic patch. Iwai et al. showed good in situ cellularization, no thrombus formation, and 100% patency 2 months after implanting a cell-free collagen micro sponge with a PGA/PLA knitted patch onto the porcine descending aorta, pulmonary artery trunk, and canine right ventricle.[109] Similarly, Takahashi et al. showed in situ cell repopulation after implanting a cell-free hybrid PGA, PLLA, and collagen microsphere patches onto canine pulmonary artery trunks.[110] For high-pressure applications, Koch et al. developed a small caliber vascular composite graft that combined fibrin gel with a synthetic mesh that demonstrated good mid-term patency, and no thrombus, aneurysm, or calcification formation after 6 months in vivo in the sheep carotid artery.[111]

3. FUNCTION OF IMPLANTS

Implants used in regenerative medicine serve one task, restoring healthy function of the injured tissue or organ. It can be accomplished in a variety of ways. Implants can be used as an in situ structure or scaffold for cells to grow into for tissue formation, as an in vitro cell delivery vehicle, and as a way to manipulate the tissues environment through drugs and genes.

3.1 Implants as a Cell Delivery Vehicle Grown In Vitro

A commonly used method in tissue engineering is cell-seeded grafts cultured in vitro. Several commonly used cell types are shown in Figure 4. Cells are vitally important in regeneration and cell therapy has remarkable promise in regenerative medicine, especially as it applies to the cardiovascular field. Cell seeding of regenerative implants is important because seeding can promote endothelialization and neotissue formation, which helps in long-term patency of the graft. For that reason, cells have undergone an immense amount of study to determine which cells are the best to use in the cardiovascular system. The earliest reports of cell-seeded TEVGs used mature cells commonly seen in the vasculature such as ECs, fibroblasts, and SMCs.[26,112,113] Promising results were seen, but the process of harvesting the cells required invasive procedures, long-term

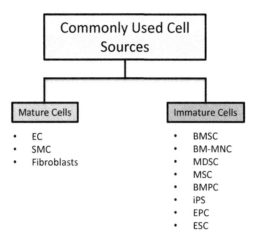

FIGURE 4 Flowchart of various common mature and immature cell sources for tissue-engineered grafts.

incubation, and an increased risk of contamination.[114] In addition, increased donor age directly correlated to decreased cell proliferation, ECM synthesis, and mechanical properties.[115] Given those limitations, it was necessary to seek alternative autologous cell sources that could be better candidates for regeneration. Autologous cells are the preferred cell type due to their lack of rejection, and therefore, are widely used in clinical TEVG applications.[116,117] However, allogeneic cell sourcing could provide off-the-shelf availability if the biocompatibility issue is solved. Recent research by the Niklason group has shown great strides in growing and decellularizing allogeneic tissues in vitro as a push toward off-the-shelf availability, which is critical in commercialization of TEVGs.[107,118]

3.1.1 Endothelial Progenitor Cells and MSCs

Endothelial progenitor cells (EPCs) are an attractive alternative to mature ECs because they can easily be isolated from the peripheral blood, umbilical cord blood, or amniotic fluid, share similarities to mature ECs, and promote angiogenesis.[115,119] In a study by Kaushal et al., EPC-seeded vascular grafts stayed patent for 130 days in an animal model.[120] The delivery of EPC-seeded scaffolds to infarcted hearts improved cardiac function in animal models.[121] MSCs are attractive because they can expand into multiple tissue lineages, including ECs, and can be readily isolated from a variety of sources such as bone marrow (BM), adipose tissue, amniotic fluid, and umbilical cord blood. Recent studies from Hoerstrup and colleagues showed that freshly isolated, as well as cryopreserved MSCs isolated from amniotic fluid, can be used to fabricate endothelialized tissue-engineered heart valves and vascular grafts.[122–124] Limitations of these cells are that they have low numbers in vivo (1 out of 10,000 cells for EPCs) and have cell line contamination problems when grown in vitro.

3.1.2 Embryonic and Inducible Stem Cells

Continued advances in our understanding of stem cell biology have dramatically altered the field of tissue engineering. Embryonic stem cells and inducible pluripotent stem (iPS) cells provide an unparalleled cell source due to their tremendous capacity to be expanded and their ability to differentiate down alternative cell lineages, thus providing an abundant supply of cells for any tissue engineering application. The use of iPS cells provides an unlimited source of autologous cells for these applications and is a potentially useful tool for cardiovascular regenerative medicine. Hibino et al. created iPS cell sheets for construction of TEVGs and successfully implanted them into mice.[125] These grafts did not produce calcification, thrombosis, or aneurysm formation and exerted a paracrine effect that induced neotissue formation. These results are promising; however, they recommended additional study on iPS purification procedures and prevention of apoptosis. Additionally, the development of the iPS cell provides the potential to create an abundant source of autologous cardiomyocytes in situ that could enable their clinical use in treating severe heart failure.[126] However, many hurdles still remain including the development of improved techniques to mitigate the risk of teratoma formation and to stably and efficiently redifferentiate the iPS cells to the various lineages necessary to make tissue-engineered heart tissue.[125,126]

3.1.3 BM Cells

Recently, bone marrow-derived mononuclear cells (BM-MNCs) and mesenchymal stem cells (BM-MSCs) are utilized in cardiovascular regenerative medicine to regenerate or replace heart valves, myocardium, and blood vessels in vivo.[127]

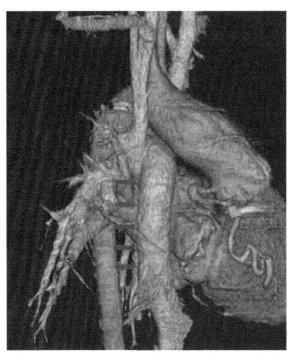

FIGURE 5 Computed tomography image of 1-year follow-up of tissue-engineered vascular graft (TEVG). The TEVG is patent with no aneurysmal dilatation. Arrows denotes the grafts. *Adopted with permission from Hibino et al.*[117]

There are several advantages to using BM-derived cells: first is the abundance of the cells which greatly contribute to shortening the preparation time because there is minimal need for cell expansion, second, BM-MSCs are shown to modulate immunological response, which contributes to endothelialization and SMC and myofibroblast migration from adjacent blood vessel, and lastly BM cells can differentiate into various lineages including smooth muscle and ECs.[74,76,127,128,129] Ongoing research using BM-derived cells can be divided along two lines: cells differentiated in vitro then implanted versus directly implanting undifferentiated cells in vivo.[127]

In vitro, BM cells are typically differentiated into SMCs and ECs with the purpose of either creating a confluent EC layer on the scaffold or a fully formed TEVG before implantation. Zhao et al. was able to differentiate BM-MSC into SMCs and ECs in vitro, seed them onto a decellularized graft, then implant the graft using an ovine model.[105,130] These grafts remained stable for 5 months, whereas the unseeded controls occluded in just 2 weeks. Gong and Niklason also created a TEVG in vitro by differentiating autologous human BM-MSCs into SMCs and ECs.[131] These TEVGs showed similar morphology, histology, and protein synthesis to native vessels, but were lacking in mechanical strength. Cho et al. differentiated SMCs and ECs from autologous ovine BM-MNC, seeded them onto decellularized scaffolds, and implanted them into the abdominal aorta of the donor pigs. These TEVGs were patent after 18 weeks with no signs of thrombus or aneurysm formation.[132] As enticing as the in vitro controlled cell differentiation model is, there are still drawbacks such as the time to culture and differentiate the cells and the possibility of contamination. These issues can be lessened by implanting BM-MNC- or BM-MSC-seeded grafts directly into the body, thus using the body's ability to differentiate the cells into the appropriate cell type.

There are two main approaches using undifferentiated BM cells, one is culturing them into a cell sheet, and the other is seeding the cells directly into a scaffold without culturing. Zhao et al. created TEVGs through culturing undifferentiated autologous BM-MSCs into cell sheets and rolling them around a mandrel. They implanted the TEVG into a rabbit common carotid artery where it integrated with the native vessel and remained patent after 4 weeks with no aneurysm or thrombus formation.[133] Piao et al. cultured BM-MNC into a monolayered cell sheet on a PGCL scaffold for the treatment of myocardial infarction in a rat model.[134] This model attenuated left ventricle dysfunction by inducing neovascularization and differentiation of BM-MNCs into cardiomyocytes. Injectable BM cells have also been used in the treatment of myocardial infarction.[135] Instead of culturing BM cells into sheets or injectables for TE applications, Hibino et al. seeded BM-MNC into a biodegradable scaffold and implanted it directly into patients undergoing Fontan operations. This approach has been used with great clinical success because it limits contamination and waiting time of culturing cells and is most advantageous for surgical procedures (Figure 5).[116,117]

TABLE 1 Advantages and Disadvantages of Cell Seeding Techniques

Seeding Techniques	Passive	Dynamic	Tissue Seeding	Other
Types	Static	Rotational Centrifugal Vacuum	Sheet seeding Bioreactor seeding	Magnetic Electrostatic
Efficiency	10–25%	38–90%	75–94%	90–99%
Advantages	Simple, quick, established	High seeding efficiency, cell penetrates into scaffold	Good seeding efficiency, morphological maturation	High seeding efficiency, fast
Disadvantages	Low seeding efficiency, operator dependent	Complex, adverse effects on cell morphology, culture time	Very complex, long culture time (weeks/months), expensive, requires very sterile environment, contamination hazard, must be decellularized or dehydrolyzed for storage	Only for specific cell types, long culture time

3.1.4 Cell Seeding Techniques

There are a wide variety of methods available for seeding cells onto scaffolds (Table 1). However, there are several questions to address when approaching the proper seeding technique. What types of cells are the best for the particular application? What is the most optimal sell seeding efficiency? Is the method safe? Is it cost-effective? How long are the culture times? The seeding technique that works for each lab and applications are different. For instance, in the case of TEVGs, our team wants to address several issues: what is the most optimal cell source for the best clinical outcome? what is the optimal sell seeding numbers? which technique maximizes cell attachment and cell seeding efficiency? does it minimize laboratory processing? is it safe, cost-effective, and reliable? and most importantly, does it promote long-term patency? In order to understand which technique works for your lab, it is important to have an understanding of the available seeding techniques, quantification techniques, and assessment of cell viability. In the following paragraphs, we will outline the most common cell seeding, quantification, and viability methods used clinically today.

Passive seeding: Passive seeding is the simplest cell seeding method, which is often called static seeding. This method entails pipetting the cell solution directly onto the scaffold or onto biological glue that is coated onto the scaffold (Figure 6); the scaffold is then incubated for several hours to several days to maximize cell attachment. Though the easiest method, this method has several shortcomings compared to other methods, including smallest cell attachment efficiency (10–25%), an increased incidence of contamination, and a potential for thrombogenicity. Though the most inferior method of all seeding types, it remains the simplest which makes it the most widely used.

Dynamic seeding: Dynamic seeding increases cell seeding efficiency, uniformity, and penetration into the scaffold through the use of hydraulic forces (rotational seeding) or pressure differentials (vacuum seeding) (Figure 6). Rotational seeding seeds cells while the graft is rotated or spun in a cell-medium suspension (Figure 6). Rotational conditions can range from 0.2 to 2500 rpm with culture periods from 12 to 72 h and has a cell seeding efficiency of 38–90% depending on the rotational speed. Increased rotational speeds allow better cell infiltration and shorter incubation times, however, the increased speeds can also change the cells morphology. Vacuum seeding relies on either internal or external pressure forces to force a cell through the pores of a scaffold. This method is quick and efficient with 60–90% seeding efficiencies. In addition, contamination is reduced in these systems due to the simplicity of some vacuum apparatus that allows for its use as a disposable seeding device. This method is advantageous because it has a shorter incubation time, 1–2 h, which can then be used in same day operations on patients. In addition, this system has been used clinically in pediatric patients with good short- and long-term results.

Tissue-based seeding and others: Sheet-based cell seeding and bioreactor cell seeding involve growing a living tissue in a laboratory environment (Figure 6). With sheet-based cell seeding, the cells are grown in sheets for 1–2 months; these sheets are then separated from their culture plate and wrapped around a cylindrical mandrel to form a confluent concentric cylindrical shape such as with an artery. Bioreactor seeding is similar except that in this case, the cells are grown on a biodegradable scaffold in a bioreactor, which simulates physiologic pressure and flow conditions. The scaffold degrades over the period of 1–2 months leaving a vessel behind. Both methods are interesting in that they both produce a vessel from the initial cell seeding. However, these techniques are costly and time-consuming. There are other types of seeding techniques in addition to the ones mentioned above, such as electrostatic cell seeding, magnetic cell seeding, polymerized hydrogel

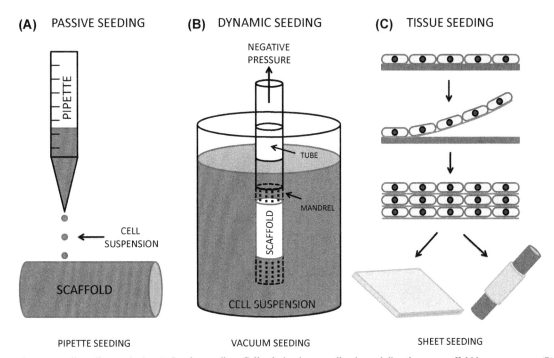

FIGURE 6 Common cell seeding methods. (A) Passive seeding. Cell solution is manually pipetted directly onto scaffold by an operator. (B) Dynamic seeding. Vacuum seeding is shown where the scaffold is on a perforated mandrel and submersed in a cell suspension. The mandrel is connected to a tube where a negative pressure can be applied to it to draw the cells into the scaffold. (C) Tissue seeding. The method shown is sheet seeding. In this method, one layer of cells are grown into a sheet. The sheets can be layered on top of each other and further cultured. They can then be used as a sheet of vascular tissue for ischemic myocardium treatment or rolled on to a spindle to further form into a blood vessel.

cell seeding, and hybrid methods that combine two or more of the aforementioned types. Though these systems improve cell efficiency, they are more complicated and costly and they are not yet used clinically.

Quantification and cell viability: Once a scaffold is seeded by any of the aforementioned methods, seeding efficiency can be quantified by determining the number of cells that are attached to the scaffold. This can be accomplished by two methods: (1) direct cell count and (2) assays. Through direct cell count, the cells can be manually counted, which provides evidence of superficial cell attachment. However, it does not address cells that have penetrated the scaffold and cannot be seen through observation alone. Assays, on the other hand, account for the entire cell population (both superficial and penetrated cells), are generally faster than manual counting, and may account for living or dead cells. Several quantification methods are shown in Table 2. For cell viability, a (3-(4,5-dimethylthiazol-2-yl)-2,5-diphenyl tetrasodium bromide) or live-dead assay can be run to ensure that the cells attached to the scaffold can contribute to long-term growth and remodeling of the graft.

3.2 Implants as an In Situ "Cell-Free" Tissue Forming Device

Many researchers choose to utilize the body's ability to regenerate and culture their implants in situ, suggesting that the host is a more efficient bioreactor than any in vitro strategy. In this approach, a cell-free degradable implant is surgically placed in the impacted area. This method trusts that neighboring cells will migrate into the implant area and proliferate to form neotissue. In an ideal case, the implant will degrade at the same rate as the formation of neotissue, leaving a new vessel, valve, or myocardium behind providing perfect integration with the host tissue.[84] In situ tissue engineering provides multiple advantages to in vitro culturing: it eliminates the need for pretreatments involving complicated or invasive procedures, there is no need for cell culture which provides off-the-shelf capabilities, and the graft can be stored much easier and longer than living tissue.[136]

This method seems like a promising regenerative strategy, but cell-free approaches are still scarce in tissue engineering. This may be due to the unpredictability of the hosts' response to the implants degradation timeline. If the implant degrades too slowly then it could limit the cells penetration and cause an immune response due to prolonged presence of foreign materials. If the implant degrades too fast it may structurally fail before any new tissue is formed. There is also the issue of myocardium which lacks that ability to regenerate, which makes this method better suited for blood vessels or heart valves.

TABLE 2 Advantages and Disadvantages of Quantification Methods Used in cell Seeding

Methods	Hemocytometer	Histology	SEM	Picogreen (DNA)	MTS and MTT
Time	<1 h	1 day	1 day	2 days	2 days
Technique	Counts cells before and after seeding and back calculates possible number of cells.	Hematoxylin and Eosin or similar staining. Operator manually counts cells per high magnification histological image.	Counts cells on magnified scaffold surface.	Measures DNA content on scaffold sections by generating a standard curve.	Determines enzymatic reduction in the mitochondria by measuring cellular metabolic activity.
Advantages	Simple, fast	Accurate	Good surface analysis	Accurate, reliable	Quantitative and qualitative
Disadvantages	Inaccurate	Time consuming, operator dependent	Does not count penetrated cells	Counts alive and dead cells	Cell line dependent

SEM, scanning electron microscopy; MTS (3-(4,5-dimethylthiazol-2-yl)-5-(3-carboxymethoxyphenyl)-2-(4-sulfophenyl)-2H-tetrazolium); MTT, (3-(4,5-dimethylthiazol-2-yl)-2,5-diphenyl tetrasodium bromide).

Despite those shortcomings, there have been recent advances in the field of in situ "cell-free" tissue engineering. Two studies conducted in Japan showed complete endothelialization in an inferior vena cava within a month postimplantation and in a pulmonary artery 6 month after implantation in a canine model with no stenosis or calcification.[137,138] The Wang group has seen promising results with a rat model that showed almost complete host remodeling within 3 months, good integration, and no stenosis.[84] Their philosophy departs from mainstream thought in that they feel a rapidly degrading implant is the most critical factor in regeneration, because it quickly creates room to promote cell migration, proliferation, and matrix production. It also reduces the exposure to degradation materials, thus limiting the inflammatory response.

There are also several strategies to promote in situ regeneration that take a hybrid approach to in vitro and in situ tissue engineering. One method is to seed cells directly onto the implant and implant it in the body without a prolonged culture period. This method can be time-consuming to harvest the cells and seed them, but depending on the cell type, it can be done in one surgical procedure and has shown great clinical promise.[117]

3.3 Implants as a Drug Delivery Vehicle

Most of the biodegradable polymers used to make tissue engineering scaffolds have also been used as drug delivery systems (DDS) for various facilitators of tissue growth factors and/or drugs.[139] Thus the use of the scaffold as a drug delivery platform that can be used to deliver growth factors designed to accelerate or direct neotissue formation or the delivery of drugs selected to prevent the development of pathological change in the neotissue is a logical and natural extension of this methodology. Although moderate success has been made with complete in situ repopulation of naked scaffolds, the majority of cardiovascular tissue engineering uses cell seeding, a DDS, or a combination thereof.[139] Various growth factors including VEGF, PDGF, and b-FGF are used for their varying beneficial effects on grafts.[140–142] Heparin is primarily used to prevent fibrin deposition, and has also been found to block intimal hyperplasia.[107,143] Rapamycin inhibits proliferation of vascular SMCs.[144] Alternatively, cells seeded onto the scaffold can release substances that facilitate regeneration, thus serving as DDS themselves. For example, Roh et al. demonstrated that cytokines released by seeded cells have a paracrine effect on surrounding native cells.[145] Human mesenchymal stem cells (hMSCs) seeded onto a PGA-P(CL/LA) scaffold in a mouse model secreted high levels of MCP-1, a cytokine that induces angiogenesis. After 24 weeks the grafts were completely repopulated with mouse cells, showing that the hMSCs only participated in a paracrine fashion to aid the inflammation and regeneration process.[145]

4. CLINICAL APPLICATIONS IN CARDIOVASCULAR REPAIR

The first reported synthetic vascular patches and grafts date back to the 1950s by Blakemore et al. and Voorhees et al.[14] Since then, tremendous advancements have been achieved in cardiovascular repair. Here, we discuss the most promising and widely used clinical applications to date.

TABLE 3 Notable Tissue Engineering Applications in the Cardiovascular Field

	Approach	Scaffold Material	Cell Type	Application	Notes	References
Heart valve	Biodegradable scaffold	PGA	Endothelial cells, fibroblasts	Ovine, PV	First in vivo placement of a heart valve	40
		PGA/P4HB	Vascular cells	Ovine, PV	Material more closely resembles native valve properties than PGA	46
		Hyaluronan	Cardiac valvular interstitial cells	In vitro culture	Simplifies seeding process and complex shape formation	47
		Fibrin gel	Human venous myofibroblasts	In vitro culture	Gel could be derived autologous from patient	171
		PGA/PLLA	BM-MSC	Ovine, PV	Noted regurgitation due to decreased valve cusp length	45
	Decellularized tissue	Porcine	Unseeded	Ovine, human	Difficulties in making valves antigen free has led to undesirable complications	55,172
		Intestinal submucosa	Unseeded	Porcine, PV	Intestinal ECM is reabsorbed and replaced by fibrous CT	54
		Porcine + PEGylation	Unseeded	In vitro culture	Optimal for effective drug delivery	173
	Decellularization/ recellularization	PGA/P4HB	Fibroblasts, MSC	In vitro culture	Reduces risk of regurgitation due to length decrease	174
		Fibrin gel	MSC	In vitro culture	Decellularized fibrin scaffold favorably facilitates recellularization	175
Myocardium	Biodegradable scaffold	Collagen matrix	Embryonic chick cardiomyocytes	In vitro culture	First in vitro culture of contractile heart tissue	176
		Collagen matrix	Neonatal rat cardiomyocytes	In vitro culture	First culture of mammalian heart tissue	177
		Gelatin mesh (Gelfoam)	Cardiomyocytes	In vitro culture, in vivo application, rat myocardium	Attenuated LV dilation, no effect on LV contractility	178
		Collagen, Matrigel matrix	Neonatal rat cardiomyocytes	In vitro culture		179
		Collagen matrix	Neonatal rat cardiomyocytes	In vitro culture, in vivo application as rat myocardium	Injured heart model, graft improved LV function	180
		Collagen matrix	BMMNC	In vivo, human myocardium		167
		Collagen matrix	Human embryonic stem cells, human iPS cells	In vitro culture, in vivo application, rat myocardium		181

Continued

TABLE 3 Notable Tissue Engineering Applications in the Cardiovascular Field—cont'd

Approach	Scaffold Material	Cell Type	Application	Notes	References
Scaffold free	Polystyrene tissue culture plates and microcarrier beads	Neonatal rat cardiomyocytes	In vitro, rotational bioreactor	One of the first 3D cardiac tissues engineered by cell self-assembly	182
	PIPAAm culture surface	Neonatal rat cardiomyocytes	In vitro, cell sheet method	Novel thermosensitive culture surface	183
	PIPAAm culture surface	Human iPS cells	In vitro, cell sheet method		184
Cell therapy	N/A	CD34+ stem cells	In vivo, human myocardium	MAGNUM clinical trial	185
Vascular	Biodegradable scaffold	Collagen, Dacron (PET) mesh	Bovine endothelial cells, smooth muscle cells, fibroblasts	In vitro	First report of TEVG	26
	PG/PGA	Ovine endothelial, smooth muscle cells, fibroblasts	In vivo, ovine pulmonary artery	First report of completely biodegradable scaffold	30
	PGA	Bovine aortic smooth muscle and endothelial cells	In vitro culture, in vivo application as saphenous artery in miniature swine	Introduction of novel pulsatile bioreactor	32
	PGA/P(CL/LA)	Autologous human cells isolated from peripheral vein	In vitro culture/seeding, in vivo application as human pulmonary artery	First report of clinical TEVG implant	151
	PGA/P(CL/LA)	Autologous BM-MNCs	In vivo, human extracardiac cavopulmonary conduit	Long-term follow-up of TEVG clinical trial	117
Cell-sheet method	N/A	Human umbilical cord, endothelial, and smooth muscle cells, human dermal fibroblasts	In vivo, femoral IVC in canine model	First report of novel cell-sheet method based on vessel self-assembly	31
	N/A	Autologous human fibroblasts and endothelial cells	In vitro culture, in vivo application as arteriovenous hemodialys is shunt	Clinical use of cell-sheet method in 10 patients	155
Decellularization/recellularization	PGA	Human allogeneic smooth muscle cells	In vitro culture, in vivo application as baboon arteriovenous shunt		156

PV, pulmonary valve; PEG, polyethylene glycol; BM-MSC, bone marrow mesynchymal stem cells; P4HB, poly-4-hydroxybutyrate; PIPAAm, poly (N-isopropylacrylamide); CT, connective tissue; CD34, cluster of differentiation molecule #34; TEVG, tissue-engineered vascular graft; ECM, extracellular matrix; LV, left ventricle; PGA, polyglycolic acid; PLLA, poly-L-lactide acid; MSC, mesenchymal stem cells; BM-MNC, bone marrow mononuclear cells; iPS, inducible pluripotent stem; PET, polyethylene terephthalate; P(CL/LA), polycaprolactone polylactic acid; IVC, inferior vena cava; PG, polyglactine; MAGNUM, Myocardial Assistance by Grafting a New bioartificial Upgraded Myocardium.

4.1 Vascular

4.1.1 Vascular Patch

CorMatrix® Cardiovascular, Inc. (CorMatrix Alpharetta, GA) is a commercially available vascular patch based on the work of Badylack on SIS-ECM research.[146–148] CorMatrix® has several FDA-approved products for vascular and cardiac applications, including the CorMatrix ECM® for Carotid Repair, the CorMatrix ECM for Pericardial Closure, and the CorMatrix ECM for Cardiac Tissue Repair and Suture Line Buttressing. The three products are derived from the same SIS-ECM material, but are of different sizes and have different indications for use. CorMatrix ECM helps to remodel, regrow, and restore damaged tissue through signaling the body to start producing healthy tissue, providing resistance to infection, providing permanent repair to the site, and supporting the mechanical environment through providing long-term strength. The benefits of these CorMatrix products include its off-the-shelf availability, being able to remodel into native tissue, and leaving no foreign material to induce an immune response. Potential drawbacks of this patch are the significant variability of the material properties between batches and the lack of a long-term follow-up in clinical cases.[149] Quarti et al. described their preliminary experience using the CorMatrix patch with 26 patients in both cardiac tissue repair and pericardial closure. Their results demonstrated encouraging results using this patch with no patch-related complications and, no cases of patch failure or calcification after 13.2 months.[150] However, they recommended a further follow-up period to fully understand the potential of the CorMatrix patch. Multiple other off-label uses are being investigated with no approval and reporting is haphazard.

4.1.2 Vascular Grafts

In 1998, Shinoka et al. first reported the use of a cell-seeded biodegradable synthetic scaffold in a canine model, with promising long-term result in low-pressure pulmonary artery system.[30] These preclinical results included continued growth capacity with a reasonable safety profile. This was a breakthrough research, especially in the pediatric cardiovascular field where implant failure correlated with patients' continued growth. In 2001, Shin'oka et al., reported the first clinical use of a tissue-engineered vascular construct (Figure 7).[151] In that case, the application of a biodegradable PCL-PLA scaffold

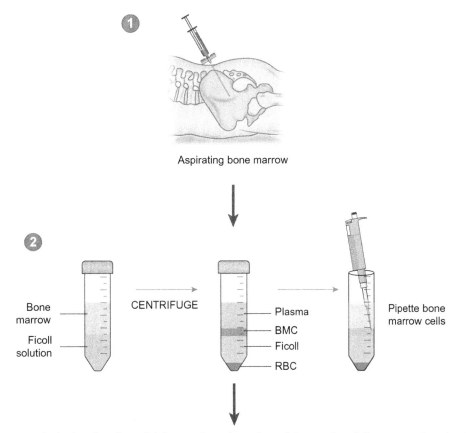

FIGURE 7   Current strategy for implantation of a seeded tissue engineered vascular graft into a patient. 1) Bone marrow is aspirated from pelvis of the patient; 2) The bone marrow derived mononuclear cells are isolated through centrifugation with Ficoll solution; 3) Bone marrow derived mononuclear cells are seeded onto the scaffold via a closed loop seeding method; 4) The tissue engineered vascular graft is implanted into the patient during the surgical Operation.

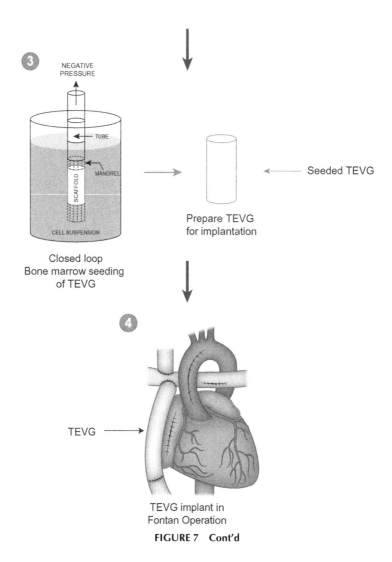

FIGURE 7 Cont'd

seeded with the patient's own cells was approved for use in a 4-year-old girl who had total occlusion of the right intermediate pulmonary artery. Seven months later, the TEVG showed no evidence of occlusion. That led to a pilot clinical trial in Japan where an evolving method of TEVGs composed of woven PGA and ε-caprolactone or L-lactide were seeded with BM-MNC in 25 patients from 1 to 24 years old. The same group released a follow-up report on the success of the TEVG over a range of 4.3–7.3 years. Of the 25 patients, six demonstrated graft narrowing and four of the six patients, with severe stenosis, underwent successful balloon angioplasty. One patient demonstrated thrombosis formation. Four patients died of causes that were unrelated to TEVG placement.[117] Even with the great successes of this treatment, two areas of improvement were needed: first in the duration of cell culturing, 8 weeks, which increased exposure to contamination and second, in decreasing the occurrence of stenosis. Recently, the culture time was tremendously shortened to a same day surgical procedure that only required 2 h incubation of the BM cells. In August 2011, the FDA approved a human clinical trial in the USA investigating the safety and growth capacity of this synthetic biodegradable TEVG use in pediatric patients undergoing a modified Fontan procedure for treatment of single ventricle cardiac anomaly.[152]

Currently there are two methods for making scaffold-free TEVGs, one is the sheet seeding method and the other is the decellularizing/recellularizing method.[31,32] In 1998, the L'Heureux lab created the first sheet-based "scaffold-free" TESAs using SBTE.[31] These implants were grown as cell sheets and displayed three distinct biological layers: an internal membrane, a media of SMCs, and an adventitia of living human skin fibroblast cells. These sheets were then rolled around a cylindrical support and cultured in vitro until they formed a confluent vessel.[31,108,153] In 2000, Cytograft Tissue Engineering, Inc. was formed from this research with the purpose of driving the technology to the clinic and encouraging clinical results were reported in 2007 with expanded findings in 2009.[154,155] These clinical results were in using the Lifeline graft as clinical shunts for dialysis treatment for 10 patients with end-stage renal disease (ESRD). Of these patients, one had immediate aneurysmal failure of the graft and two other grafts failed in the first 90 days to thrombus and aneurysm. In total, the Lifeline graft has had 13 implantations and saw 4 failures in the first 90 days after implantation.[108,155] The remaining eight grafts needed three interventions to maintain patency. Despite those complications, these grafts were used for a clinically challenging application and hold promise for use in dialysis treatment.

In 1999, concurrent research by the Niklason group reported the first use of in vitro culture, of TEVGs using biodegradable scaffolds seeded with porcine or ovine vascular cells.[32] They are also the first group to decellularize and then recellularize an in vitro grown scaffold for clinical use. Their method created a scaffold-free vessel by culturing allogeneic cells onto a highly degradable scaffold in vitro.[156] The scaffold degrades leaving a robust tissue behind. This vessel can then be decellularized and refrigerated for future use for "off-the-shelf" availability.[118] Once needed, the vessel can either be directly implanted (i.e., for use as a large caliber vascular graft) or reseeded with the patient's ECs and implanted (i.e., for use as a small caliber vascular graft), thus having, ideally, no immune response. To prove this method, nine allogeneic TEVG, grown in vitro, were implanted into baboons as arteriovenous conduits.[118] Of these grafts, only one showed thrombosis and no calcification or aneurysm was observed in any graft, but they did exhibit neointimal hyperplasia. They also evaluated this method for use in small-diameter vessels in a canine model where five TEVGs were implanted as carotid artery bypass grafts and two as coronary artery bypass grafts. In the carotid study, four grafts remained patent and one occluded after 1 week. In the coronary study, all remained patent with no aneurysm dilation. However, there was wide variation in endothelial cell coverage (0–60%). With the success of the late term preclinical results, Humacyte, Inc. was founded to bring this method toward clinical application. Initial European trials are underway in Poland, with the first human TEVGs implanted in 2012 and used for dialysis in February 2013. Use in US clinical trials has been approved for a pilot study in vascular access for dialysis in patients with ESRD, to evaluate the safety and efficacy of the vessels after 6 months. Clinical reports are pending.

4.2 Heart Valve

Clinical implementation of tissue-engineered heart valves has shown promising success in recent years, though early trials resulted in unnecessary patient mortality. Early applications of the SynerGraft valve illustrate the importance of assessing quality control and assurance before introducing products to the clinical setting. The SynerGraft decellularized porcine valve experienced initial short-term success in an adult human trial.[157] Soon thereafter, valves were implanted into the right ventricular outflow tract of four pediatric patients for evaluation of long-term function.[158] However, no long-term animal studies were conducted prior to the commencement of this second human trial.[51] Not only did the grafts fail to repopulate in situ as planned, but they triggered an immune response leading to fibrosis and death in three of the patients. Upon further observation, the method employed to remove porcine antigens and cells from valves was not sufficient, and an adverse immune reaction ensued when the valves were implanted in vivo.[51] Future models of the SynerGraft design were approached with more caution.[159] This unfortunate episode has highlighted the need for significant quality control and quality assurance measures to be in place before moving products into the clinical realm.

Multiple clinical studies have since been conducted using various allograft and xenograft decellularized valves placed in right ventricular outflow tracts[160,161] and some work has been done placing decellularized human allografts into the aortic position, however, placement of decellularized xenografts remains challenging.[162,163] Dohmen et al. showed the first successful placement of a xenograft valve into systemic circulation in an animal model.[162] Seven sheeps received decellularized stentless porcine xenografts into the aortic position and then explanted at 4 months. The grafts showed minimal calcification and repopulation of ECs on the surface and fibroblasts underneath. These promising results implicate the possibility of moving toward a clinical trial using xenografts in systemic circulation.[162]

Despite improvements made in the field of TEHVs, their future off-the-shelf clinical use is impeded by technological difficulties of production and approval by medical regulations because of "ex vivo cell processing procedures," or the complications that come with preparing a seeded scaffold using in vitro methods while maintaining adequate sterility for clinical use.[164] Recent advances are in overcoming these obstacles and making them more feasible for future clinical application. Weber et al. demonstrated that TEHVs can be decellularized and used as "off-the-shelf" constructs for valve replacement.[164] A PGA/P4HB-based scaffold was seeded with human fibroblasts for ECM fabrication, following subsequent placement in a bioreactor for 4 weeks. The valve was then decellularized and implanted into orthotopic pulmonary valve position. Although mild-to-moderate regurgitation was observed due to leaflet valve shortening, the valves demonstrated optimal recellularization capacity in comparison to decellularized human native heart valves. This model removes complications that come with ex vivo seeding while still providing homologous ECM optimal for in situ host repopulation. These valves could be mass produced and ready for engraftment without extensive presurgical preparation.[164] TEHVs have also shown compatibility with the most current methodology for heart valve replacement surgery. During a one-step surgical approach, autologous marrow stromal cells from sheep were seeded onto a biodegradable TEHV and placed into orthotopic aortic valve position using a transcatheter, transapical approach. The valve placements were successful and show that TEHVs are compatible with current surgical techniques.[165]

4.3 Heart

The ultimate goal in cardiovascular tissue engineering is the creation of an entire bioartificial heart construct for the treatment of end-stage heart failure as an alternative to allogeneic transplantation, and although promising studies in a rat model have recently been completed, the technology is far from clinical application.[63] On the other hand, both scaffold-based and scaffold-free myocardial tissue engineering has immediate clinical potential as a regenerative solution to the repair of dilated cardiomyopathy, impaired contractibility, and/or infarcted myocardium after an ischemic event.[166–168]

Cell therapy (cellular cardiomyoplasty) is a promising approach, but is plagued by unreliable cell delivery, inconsistent cell viability, and poor cell retention.[167] Ten clinical trials are currently underway to explore the administration of autologous hMSCs, BM mononuclear stem cells, BM-derived mesenchymal cardiopoietic cells, CD34+ stems cells, and allogeneic cardiosphere-derived stem cells.[169] Of note are recently published results of a 5-year follow-up in which 55 patients with nonischemic dilated cardiomyopathy received intracoronary administration of CD34+ stem cells and showed marked improvement in ventricular function, exercise tolerance, and long-term survival.[170] Similarly, the scaffold-based approach has demonstrated clinical promise. Chachques et al. implanted a collagen matrix seeded with BM-MNCs in 10 patients with left ventricular postischemic myocardial scarring and report improvement in left ventricular ejection fraction and a reduction in adverse ventricular remodeling.[167] Although sustained cell regeneration of the heart remains the ultimate goal, obtaining that vision is difficult. However, ongoing research utilizing tissue-engineered patches in combination with cell therapy approaches may alleviate adverse remodeling and improve the mechanical properties of the infarct.

5. CONCLUSION

Over the years, intensive interdisciplinary, translational research into cardiovascular regenerative implants has been undertaken in an effort to improve surgical outcome and better quality of life for patients with cardiovascular defects. These collaborative efforts have resulted in a variety of innovative approaches for vessel, heart valves, and myocardium regenerative medicine, eventually, moving diligently toward whole heart replacement. These current models and promising clinical trials are great achievements in regenerative medicine that are progressing toward more functional tissue-engineered cardiovascular implants with off-the-shelf capability. However, with all the great achievement this field had in the last 10 years, there are still areas that need to be addressed, namely, minimizing immunological response and having true off-the-shelf capability for patients who need immediate treatment. With the growth and accomplishments this field has had, meeting those goals in the future should be attainable.

ACKNOWLEDGMENT

All authors have read the journal's policy on disclosure of potential conflicts of interest. There are no conflicts of interest to disclose.

REFERENCES

1. Vacanti JP. Beyond transplantation. Third annual Samuel Jason Mixter lecture. *Arch Surg (Chicago, Ill: 1960)* 1988;**123**(5):545–9. Epub 1988/05/01.
2. Langer R, Vacanti JP. Tissue engineering. *Science* 1993;**260**(5110):920–6. Epub 1993/05/14.
3. Fuchs JR, Nasseri BA, Vacanti JP. Tissue engineering: a 21st century solution to surgical reconstruction. *Ann Thorac Surg* 2001;**72**(2):577–91. Epub 2001/08/23.
4. Bell E, Ivarsson B, Merrill C. Production of a tissue-like structure by contraction of collagen lattices by human fibroblasts of different proliferative potential in vitro. *Proc Natl Acad Sci USA* 1979;**76**(3):1274–8. Epub 1979/03/01.
5. Oconnor NE, Mulliken JB, Banksschlegel S, Kehinde O, Green H. Grafting of burns with cultured epithelium prepared from autologous epidermal-cells. *Lancet* 1981;**1**(8211):75–8.
6. Burke JF, Yannas IV, Quinby WC, Bondoc CC, Jung WK. Successful use of a physiologically acceptable artificial skin in the treatment of extensive burn injury. *Ann Surg* 1981;**194**(4):413–28.
7. Vacanti JP, Morse MA, Saltzman WM, Domb AJ, Perez-Atayde A, Langer R. Selective cell transplantation using bioabsorbable artificial polymers as matrices. *J Pediatr Surg* 1988;**23**(1 Pt 2):3–9. Epub 1988/01/01.
8. Priya SG, Jungvid H, Kumar A. Skin tissue engineering for tissue repair and regeneration. *Tissue Eng Part B, Rev* 2008;**14**(1):105–18. Epub 2008/05/06.

9. Guthrie CC. End-results of arterial restitution with devitalized tissue. *J Am Med Assoc* 1919;**73**:186–7.

10. Kavey RE, Daniels SR, Lauer RM, et al. American Heart Association guidelines for primary prevention of atherosclerotic cardiovascular disease beginning in childhood. *J Pediatr* 2003;**142**(4):368–72. Epub 2003/04/25.

11. Orlando G, Soker S, Stratta RJ, Atala A. Will regenerative medicine replace transplantation? *CSH Perspect Med* 2013;**3**(8).

12. Benveniste GL. Alexis Carrel: the good, the bad, the ugly. *ANZ J Surg* 2013;**83**(9):609–11. Epub 2013/04/27.

13. Bello SO, Peng EW, Sarkar PK. Conduits for coronary artery bypass surgery: the quest for second best. *J Cardiovasc Med* 2011;**12**(6):411–21. Epub 2011/03/25.

14. Blakemore AH, Voorhees Jr AB. The use of tubes constructed from Vinyon N cloth in bridging arterial defects; experimental and clinical. *Ann Surg* 1954;**140**(3):324–34. Epub 1954/09/01.

15. Hoerstrup SP, Zund G, Sodian R, Schnell AM, Grunenfelder J, Turina MI. Tissue engineering of small caliber vascular grafts. *Eur J Cardiothorac Surg Off J Eur Assoc Cardiothorac Surg* 2001;**20**(1):164–9. Epub 2001/06/26.

16. Cittadella G, de Mel A, Dee R, De Coppi P, Seifalian AM. Arterial tissue regeneration for pediatric applications: inspiration from up-to-date tissue-engineered vascular bypass grafts. *Artif Organs* 2013;**37**(5):423–34. Epub 2013/04/05.

17. Herring MB, Dilley R, Jersild Jr RA, Boxer L, Gardner A, Glover J. Seeding arterial prostheses with vascular endothelium. The nature of the lining. *Ann Surg* 1979;**190**(1):84–90. Epub 1979/07/01.

18. Tassiopoulos AK, Greisler HP. Angiogenic mechanisms of endothelialization of cardiovascular implants: a review of recent investigative strategies. *J Biomater Sci Polym Ed* 2000;**11**(11):1275–84. Epub 2001/03/27.

19. Dixit P, Hern-Anderson D, Ranieri J, Schmidt CE. Vascular graft endothelialization: comparative analysis of canine and human endothelial cell migration on natural biomaterials. *J Biomed Mater Res* 2001;**56**(4):545–55. Epub 2001/06/16.

20. Rupnick MA, Hubbard FA, Pratt K, Jarrell BE, Williams SK. Endothelialization of vascular prosthetic surfaces after seeding or sodding with human microvascular endothelial cells. *J Vasc Surg* 1989;**9**(6):788–95. Epub 1989/06/01.

21. Graham LM, Burkel WE, Ford JW, Vinter DW, Kahn RH, Stanley JC. Immediate seeding of enzymatically derived endothelium in Dacron vascular grafts. Early experimental studies with autologous canine cells. *Arch Surg (Chicago, Ill: 1960)* 1980;**115**(11):1289–94. Epub 1980/11/01.

22. Zilla P, Fasol R, Deutsch M, et al. Endothelial-cell seeding of polytetrafluoroethylene vascular grafts in humans - a preliminary-report. *J Vasc Surg* 1987;**6**(6):535–41.

23. Zilla P, Fasol R, Preiss P, et al. Use of fibrin glue as a substrate for in vitro endothelialization of PTFE vascular grafts. *Surgery* 1989;**105**(4):515–22. Epub 1989/04/01.

24. Graham LM, Vinter DW, Ford JW, Kahn RH, Burkel WE, Stanley JC. Cultured autogenous endothelial cell seeding of prosthetic vascular grafts. *Surg Forum* 1979;**30**:204–6. Epub 1979/01/01.

25. Magometschnigg H, Kadletz M, Vodrazka M, et al. Prospective clinical study with in vitro endothelial cell lining of expanded polytetrafluoroethylene grafts in crural repeat reconstruction. *J Vasc Surg* 1992;**15**(3):527–35. Epub 1992/03/01.

26. Weinberg CB, Bell E. A blood vessel model constructed from collagen and cultured vascular cells. *Science* 1986;**231**(4736):397–400. Epub 1986/01/24.

27. Drury JK, Ashton TR, Cunningham JD, Maini R, Pollock JG. Experimental and clinical experience with a gelatin impregnated Dacron prosthesis. *Ann Vasc Surg* 1987;**1**(5):542–7. Epub 1987/12/01.

28. Freischlag JA, Moore WS. Clinical experience with a collagen-impregnated knitted Dacron vascular graft. *Ann Vasc Surg* 1990;**4**(5):449–54. Epub 1990/09/01.

29. Massia SP, Hubbell JA. Tissue engineering in the vascular graft. *Cytotechnology* 1992;**10**(3):189–204. Epub 1992/01/01.

30. Shinoka T, Shum-Tim D, Ma PX, et al. Creation of viable pulmonary artery autografts through tissue engineering. *J Thorac Cardiovasc Surg* 1998;**115**(3):536–45. Discussion 45-6. Epub 1998/04/16.

31. L'Heureux N, Paquet S, Labbe R, Germain L, Auger FA. A completely biological tissue-engineered human blood vessel. *FASEB J Off Publ Fed Am Soc Exp Biol* 1998;**12**(1):47–56. Epub 1998/01/23.

32. Niklason LE, Gao J, Abbott WM, et al. Functional arteries grown in vitro. *Science* 1999;**284**(5413):489–93.

33. Rosamond W, Flegal K, Friday G, et al. Heart disease and stroke statistics–2007 update: a report from the American Heart Association Statistics Committee and Stroke Statistics Subcommittee. *Circulation* 2007;**115**(5):e69–171. Epub 2006/12/30.

34. Hammermeister K, Sethi GK, Henderson WG, Grover FL, Oprian C, Rahimtoola SH. Outcomes 15 years after valve replacement with a mechanical versus a bioprosthetic valve: final report of the Veterans Affairs randomized trial. *J Am Coll Cardiol* 2000;**36**(4):1152–8. Epub 2000/10/12.

35. Cooley DA, Frazier OH. The past 50 years of cardiovascular surgery. *Circulation* 2000;**102**(Suppl. 4):IV-87–93.

36. Harken DE, Soroff HS, Taylor WJ, Lefemine AA, Gupta SK, Lunzer S. Partial and complete prostheses in aortic insufficiency. *J Thorac Cardiovasc Surg* 1960;**40**:744–62. Epub 1960/12/01.

37. Rahimtoola SH. Choice of prosthetic heart valve in adults—an update. *J Am Coll Cardiol* 2010;**55**(22):2413–26. Epub 2010/06/01.

38. Carpentier A, Lemaigre G, Robert L, Carpentier S, Dubost C. Biological factors affecting long-term results of valvular heterografts. *J Thorac Cardiovasc Surg* 1969;**58**(4):467–83. Epub 1969/10/01.

39. Kanter KR, Budde JM, Parks WJ, et al. One hundred pulmonary valve replacements in children after relief of right ventricular outflow tract obstruction. *Ann Thorac Surg* 2002;**73**(6):1801–6. discussion 6-7. Epub 2002/06/25.

40. Shinoka T, Breuer CK, Tanel RE, et al. Tissue engineering heart valves: valve leaflet replacement study in a lamb model. *Ann Thorac Surg* 1995;**60**(6 Suppl.):S513–6. Epub 1995/12/01.

41. Shinoka T, Ma PX, Shum-Tim D, et al. Tissue-engineered heart valves. Autologous valve leaflet replacement study in a lamb model. *Circulation* 1996;**94**(9 Suppl.):II164–8. Epub 1996/11/01.

42. Sodian R, Hoerstrup SP, Sperling JS, et al. Evaluation of biodegradable, three-dimensional matrices for tissue engineering of heart valves. *ASAIO J Am Soc Artif Intern Organs 1992* 2000;**46**(1):107–10. Epub 2000/02/10.

43. Sodian R, Hoerstrup SP, Sperling JS, et al. Tissue engineering of heart valves: in vitro experiences. *Ann Thorac Surg* 2000;**70**(1):140–4. Epub 2000/08/02.

44. Sodian R, Hoerstrup SP, Sperling JS, et al. Early in vivo experience with tissue-engineered trileaflet heart valves. *Circulation* 2000;**102**(19 Suppl. 3): III22–9. Epub 2000/11/18.

45. Gottlieb D, Kunal T, Emani S, et al. In vivo monitoring of function of autologous engineered pulmonary valve. *J Thorac Cardiovasc Surg* 2010;**139**(3):723–31. Epub 2010/02/24.

46. Hoerstrup SP, Sodian R, Sperling JS, Vacanti JP, Mayer Jr JE. New pulsatile bioreactor for in vitro formation of tissue engineered heart valves. *Tissue Eng* 2000;**6**(1):75–9. Epub 2000/08/15.

47. Masters KS, Shah DN, Leinwand LA, Anseth KS. Crosslinked hyaluronan scaffolds as a biologically active carrier for valvular interstitial cells. *Biomaterials* 2005;**26**(15):2517–25. Epub 2004/12/09.

48. Schmidt D, Dijkman PE, Driessen-Mol A, et al. Minimally-invasive implantation of living tissue engineered heart valves: a comprehensive approach from autologous vascular cells to stem cells. *J Am Coll Cardiol* 2010;**56**(6):510–20. Epub 2010/07/31.

49. Sutherland FW, Perry TE, Yu Y, et al. From stem cells to viable autologous semilunar heart valve. *Circulation* 2005;**111**(21):2783–91. Epub 2005/06/02.

50. Weber B, Scherman J, Emmert MY, et al. Injectable living marrow stromal cell-based autologous tissue engineered heart valves: first experiences with a one-step intervention in primates. *Eur Heart J* 2011;**32**(22):2830–40. Epub 2011/03/19.

51. Vesely I. Heart valve tissue engineering. *Circulation Res* 2005;**97**(8):743–55. Epub 2005/10/15.

52. Cheung DT, Weber PA, Grobe AC, et al. A new method for the preservation of aortic valve homografts. *J Heart Valve Dis* 2001;**10**(6):728–34. Discussion 34-5. Epub 2002/01/05.

53. Dohmen PM, Ozaki S, Verbeken E, Yperman J, Flameng W, Konertz WF. Tissue engineering of an auto-xenograft pulmonary heart valve. *Asian Cardiovasc Thorac Ann* 2002;**10**(1):25–30. Epub 2002/06/25.

54. Matheny RG, Hutchison ML, Dryden PE, Hiles MD, Shaar CJ. Porcine small intestine submucosa as a pulmonary valve leaflet substitute. *J Heart Valve Dis* 2000;**9**(6):769–74. discussion 74-5. Epub 2000/12/29.

55. O'Brien MF, Goldstein S, Walsh S, Black KS, Elkins R, Clarke D. The SynerGraft valve: a new acellular (nonglutaraldehyde-fixed) tissue heart valve for autologous recellularization first experimental studies before clinical implantation. *Semin Thorac Cardiovasc Surg* 1999;**11**(4 Suppl. 1): 194–200. Epub 2000/02/05.

56. Steinhoff G, Stock U, Karim N, et al. Tissue engineering of pulmonary heart valves on allogenic acellular matrix conduits: in vivo restoration of valve tissue. *Circulation* 2000;**102**(19 Suppl. 3):III50–5. Epub 2000/11/18.

57. Breuer CK, Mettler BA, Anthony T, Sales VL, Schoen FJ, Mayer JE. Application of tissue-engineering principles toward the development of a semilunar heart valve substitute. *Tissue Eng* 2004;**10**(11–12):1725–36. Epub 2005/02/03.

58. Vincent J. Heart failure: a disease complex with challenging therapeutics. *Clin Pharmacol Ther* 2013;**94**(4):415–21. Epub 2013/09/21.

59. Heidenreich PA, Albert NM, Allen LA, et al. Forecasting the impact of heart failure in the United States: a policy statement from the American Heart Association. *Circ Heart Fail* 2013;**6**(3):606–19. Epub 2013/04/26.

60. Westaby S. Cardiac transplantation and ventricular assist devices. *Medicine* 2010;**38**(9):490–5.

61. Barnard CN. The operation. A human cardiac transplant: an interim report of a successful operation performed at Groote Schuur Hospital, Cape Town. *S Afr Med J (Suid-Afrikaanse tydskrif vir geneeskunde)* 1967;**41**(48):1271–4. Epub 1967/12/30.

62. La Franca E, Iacona R, Ajello L, Sansone A, Caruso M, Assennato P. Heart failure and mechanical circulatory assist devices. *Glob J Health Sci* 2013;**5**(5):11–9. Epub 2013/08/30.

63. Ott HC, Matthiesen TS, Goh SK, et al. Perfusion-decellularized matrix: using nature's platform to engineer a bioartificial heart. *Nat Med* 2008;**14**(2):213–21. Epub 2008/01/15.

64. Cooley DA, Liotta D, Hallman GL, Bloodwell RD, Leachman RD, Milam JD. Orthotopic cardiac prosthesis for two-staged cardiac replacement. *Am J Cardiol* 1969;**24**(5):723–30. Epub 1969/11/01.

65. Meyer A, Slaughter M. The total artificial heart. *Panminerva Medica* 2011;**53**(3):141–54. Epub 2011/07/22.

66. Ward RA, Wellhausen SR, Dobbins JJ, Johnson GS, DeVries WC. Thromboembolic and infectious complications of total artificial heart implantation. *Ann N Y Acad Sci* 1987;**516**:638–50. Epub 1987/01/01.

67. Parameshwar J, Wallwork J. Left ventricular assist devices: current status and future applications. *Int J Cardiol* 1997;**62**(Suppl. 1):S23–7. Epub 1998/02/17.

68. Zimmermann WH, Cesnjevar R. Cardiac tissue engineering: implications for pediatric heart surgery. *Pediatr Cardiol* 2009;**30**(5):716–23. Epub 2009/03/26.

69. Chung S, King MW. Design concepts and strategies for tissue engineering scaffolds. *Biotechnol Appl Biochem* 2011;**58**(6):423–38. Epub 2011/12/17.

70. Naderi H, Matin MM, Bahrami AR. Review paper: critical issues in tissue engineering: biomaterials, cell sources, angiogenesis, and drug delivery systems. *J Biomater Appl* 2011;**26**(4):383–417. Epub 2011/09/20.

71. Naito Y, Williams-Fritze M, Duncan DR, et al. Characterization of the natural history of extracellular matrix production in tissue-engineered vascular grafts during neovessel formation. *Cells Tissues Organs* 2012;**195**(1–2):60–72. Epub 2011/10/15.

72. Chen G, Ushida T, Tateishi T. Scaffold design for tissue engineering. *Macromol Biosci* 2002;**2**(2):67–77.

73. Cleary MA, Geiger E, Grady C, Best C, Naito Y, Breuer C. Vascular tissue engineering: the next generation. *Trends Mol Med* 2012;**18**(7):394–404. Epub 2012/06/15.

74. Patterson JT, Gilliland T, Maxfield MW, et al. Tissue-engineered vascular grafts for use in the treatment of congenital heart disease: from the bench to the clinic and back again. *Regen Med* 2012;**7**(3):409–19. Epub 2012/05/19.

75. Ravi S, Chaikof EL. Biomaterials for vascular tissue engineering. *Regen Med* 2010;**5**(1):107–20. Epub 2009/12/19.

76. Hibino N, Yi T, Duncan DR, et al. A critical role for macrophages in neovessel formation and the development of stenosis in tissue-engineered vascular grafts. *FASEB J Off Publ Fed Am Soc Exp Biol* 2011;**25**(12):4253–63. Epub 2011/08/26.

77. Gunatillake PA, Adhikari R. Biodegradable synthetic polymers for tissue engineering. *Eur Cells Mater* 2003;**5**:1–16. discussion Epub 2003/10/17.

78. Dean EW, Udelsman B, Breuer CK. Current advances in the translation of vascular tissue engineering to the treatment of pediatric congenital heart disease. *Yale J Biol Med* 2012;**85**(2):229–38. Epub 2012/06/28.

79. Naito Y, Lee YU, Yi T, et al. Beyond burst pressure: Initial evaluation of the natural history of the biaxial mechanical properties of tissue engineered vascular grafts in the venous circulation using a murine model. *Tissue Eng. Part A* 2014;**20**(1-2):346–55. Epub 2013/08/21.

80. Wang Y, Kim YM, Langer R. In vivo degradation characteristics of poly(glycerol sebacate). *J Biomed Mater Res Part A* 2003;**66**(1):192–7. Epub 2003/07/02.

81. Wang Y, Ameer GA, Sheppard BJ, Langer R. A tough biodegradable elastomer. *Nat Biotechnol* 2002;**20**(6):602–6. Epub 2002/06/04.

82. Gao J, Crapo PM, Wang Y. Macroporous elastomeric scaffolds with extensive micropores for soft tissue engineering. *Tissue Eng* 2006;**12**(4):917–25. Epub 2006/05/06.

83. Chen QZ, Bismarck A, Hansen U, et al. Characterisation of a soft elastomer poly(glycerol sebacate) designed to match the mechanical properties of myocardial tissue. *Biomaterials* 2008;**29**(1):47–57. Epub 2007/10/05.

84. Wu W, Allen RA, Wang Y. Fast-degrading elastomer enables rapid remodeling of a cell-free synthetic graft into a neoartery. *Nat Med* 2012;**18**(7):1148–53. Epub 2012/06/26.

85. Kim TG, Shin H, Lim DW. Biomimetic scaffolds for tissue engineering. *Adv Funct Mater* 2012;**22**(12):2446–68.

86. Ingavle G, Leach JK. Advancements in electrospinning of Polymeric nanofibrous scaffolds for tissue engineering. *Tissue Eng Part B Rev* 2014;**20**(4):277–93. Epub 2013/09/06.

87. Fichman G, Gazit E. Self-assembly of short peptides to form hydrogels: design of building blocks, physical properties and technological applications. *Acta Biomater* 2014;**10**(4):1671–82. Epub 2013/08/21.

88. Badylak SF, Freytes DO, Gilbert TW. Extracellular matrix as a biological scaffold material: structure and function. *Acta Biomater* 2009;**5**(1):1–13. Epub 2008/10/22.

89. Sreejit P, Verma RS. Natural ECM as biomaterial for scaffold based cardiac regeneration using adult bone marrow derived stem cells. *Stem Cell Rev* 2013;**9**(2):158–71. Epub 2013/01/16.

90. Zhang WJ, Liu W, Cui L, Cao YL. Tissue engineering of blood vessel. *J Cell Mol Med* 2007;**11**(5):945–57.

91. Ye L, Zimmermann WH, Garry DJ, Zhang J. Patching the heart: cardiac repair from within and outside. *Circ Res* 2013;**113**(7):922–32. Epub 2013/09/14.

92. Souren JE, Schneijdenberg C, Verkleij AJ, Van Wijk R. Factors controlling the rhythmic contraction of collagen gels by neonatal heart cells. *In Vitro Cell Dev Biol J Tissue Cult Assoc* 1992;**28A**(3 Pt 1):199–204. Epub 1992/03/01.

93. Ross JJ, Tranquillo RT. ECM gene expression correlates with in vitro tissue growth and development in fibrin gel remodeled by neonatal smooth muscle cells. *Matrix Biol J Int Soc Matrix Biol* 2003;**22**(6):477–90. Epub 2003/12/12.

94. Long JL, Tranquillo RT. Elastic fiber production in cardiovascular tissue-equivalents. *Matrix Biol J Int Soc Matrix Biol* 2003;**22**(4):339–50. Epub 2003/08/26.

95. Yao L, Swartz DD, Gugino SF, Russell JA, Andreadis ST. Fibrin-based tissue-engineered blood vessels: differential effects of biomaterial and culture parameters on mechanical strength and vascular reactivity. *Tissue Eng* 2005;**11**(7–8):991–1003. Epub 2005/09/08.

96. Swartz DD, Russell JA, Andreadis ST. Engineering of fibrin-based functional and implantable small-diameter blood vessels. *Am J Physiol Heart Circ Physiol* 2005;**288**(3):H1451–60. Epub 2004/10/16.

97. Syedain ZH, Meier LA, Bjork JW, Lee A, Tranquillo RT. Implantable arterial grafts from human fibroblasts and fibrin using a multi-graft pulsed flow-stretch bioreactor with noninvasive strength monitoring. *Biomaterials* 2011;**32**(3):714–22. Epub 2010/10/12.

98. Yao L, Liu J, Andreadis ST. Composite fibrin scaffolds increase mechanical strength and preserve contractility of tissue engineered blood vessels. *Pharm Res* 2008;**25**(5):1212–21. Epub 2007/12/20.

99. Weber M, Heta E, Moreira R, et al. Tissue-engineered fibrin-based heart valve with a tubular leaflet design. *Tissue Eng Part C Methods* 2014;**20**(4):265–75. Epub 2013/07/09.

100. Malone JM, Brendel K, Duhamel RC, Reinert RL. Detergent-extracted small-diameter vascular prostheses. *J Vasc Surg* 1984;**1**(1):181–91. Epub 1984/01/01.

101. Wang B, Tedder ME, Perez CE, et al. Structural and biomechanical characterizations of porcine myocardial extracellular matrix. *J Mater Sci Mater Med* 2012;**23**(8):1835–47. Epub 2012/05/16.

102. Kasimir MT, Rieder E, Seebacher G, Wolner E, Weigel G, Simon P. Presence and elimination of the xenoantigen gal (alpha1, 3) gal in tissue-engineered heart valves. *Tissue Eng* 2005;**11**(7–8):1274–80. Epub 2005/09/08.

103. Sandrin MS, McKenzie IF. Gal alpha (1,3)Gal, the major xenoantigen(s) recognised in pigs by human natural antibodies. *Immunol Rev* 1994;**141**:169–90. Epub 1994/10/01.

104. Nagasaka S, Taniguchi S, Nakayama Y, et al. In vivo study of the effects of cryopreservation on heart valve xenotransplantation. *Cardiovasc Pathol Off J Soc Cardiovasc Pathol* 2005;**14**(2):70–9. Epub 2005/03/23.

105. Swartz DD, Andreadis ST. Animal models for vascular tissue-engineering. *Curr Opin Biotechnol* 2013;**24**(5):916–25. Epub 2013/06/19.

106. Badylak S, Gilbert T, Myers-Irvin J. The extracellular matrix as a biologic scaffold for tissue engineering. *Tissue Eng* 2008:121–43.

107. Quint C, Kondo Y, Manson RJ, Lawson JH, Dardik A, Niklason LE. Decellularized tissue-engineered blood vessel as an arterial conduit. *Proc Natl Acad Sci USA* 2011;**108**(22):9214–9. Epub 2011/05/17.

108. Peck M, Gebhart D, Dusserre N, McAllister TN, L'Heureux N. The evolution of vascular tissue engineering and current state of the art. *Cells Tissues Organs* 2012;**195**(1–2):144–58. Epub 2011/10/15.

109. Iwai S, Sawa Y, Ichikawa H, et al. Biodegradable polymer with collagen microsponge serves as a new bioengineered cardiovascular prosthesis. *J Thorac Cardiovasc Surg* 2004;**128**(3):472–9.

110. Takahashi H, Yokota T, Uchimura E, et al. Newly developed tissue-engineered material for reconstruction of vascular wall without cell seeding. *Ann Thorac Surg* 2009;**88**(4):1269–76.

111. Koch S, Flanagan TC, Sachweh JS, et al. Fibrin-polylactide-based tissue-engineered vascular graft in the arterial circulation. *Biomaterials* 2010;**31**(17):4731–9. Epub 2010/03/23.

112. L'Heureux N, Germain L, Labbe R, Auger FA. In vitro construction of a human blood vessel from cultured vascular cells: a morphologic study. *J Vasc Surg* 1993;**17**(3):499–509. Epub 1993/03/01.

113. Hirai J, Kanda K, Oka T, Matsuda T. Highly oriented, tubular hybrid vascular tissue for a low pressure circulatory system. *ASAIO J Am Soc Artif Intern Organs 1992* 1994;**40**(3):M383–8. Epub 1994/07/01.

114. Kurobe H, Maxfield MW, Breuer CK, Shinoka T. Concise review: tissue-engineered vascular grafts for cardiac surgery: past, present, and future. *Stem Cells Transl Med* 2012;**1**(7):566–71. Epub 2012/12/01.

115. Bajpai VK, Andreadis ST. Stem cell sources for vascular tissue engineering and regeneration. *Tissue Eng Part B Rev* 2012;**18**(5):405–25. Epub 2012/05/11.

116. Hibino N, Nalbandian A, Devine L, et al. Comparison of human bone marrow mononuclear cell isolation methods for creating tissue-engineered vascular grafts: novel filter system versus traditional density centrifugation method. *Tissue Eng Part C Methods* 2011;**17**(10):993–8. Epub 2011/05/26.

117. Hibino N, McGillicuddy E, Matsumura G, et al. Late-term results of tissue-engineered vascular grafts in humans. *J Thorac Cardiovasc Surg* 2010;**139**(2):431–6. 6 e1-2. Epub 2010/01/29.

118. Dahl SLM, Blum JL, Niklason LE. Bioengineered vascular grafts: can we make them off-the-shelf? *Trends Cardiovasc Med* 2011;**21**(3):83–9.

119. Weber B, Emmert MY, Hoerstrup SP. Stem cells for heart valve regeneration. *Swiss Med Wkly* 2012;**142**:w13622. Epub 2012/07/18.

120. Kaushal S, Amiel GE, Guleserian KJ, et al. Functional small-diameter neovessels created using endothelial progenitor cells expanded ex vivo. *Nat Med* 2001;**7**(9):1035–40. Epub 2001/09/05.

121. Hiesinger W, Frederick JR, Atluri P, et al. Spliced stromal cell-derived factor-1alpha analog stimulates endothelial progenitor cell migration and improves cardiac function in a dose-dependent manner after myocardial infarction. *J Thorac Cardiovasc Surg* 2010;**140**(5):1174–80. Epub 2010/10/19.

122. Weber B, Kehl D, Bleul U, et al. In vitro fabrication of autologous living tissue-engineered vascular grafts based on prenatally harvested ovine amniotic fluid-derived stem cells. *J Tissue Eng Regen Med* 2013. Epub 2013/07/25.

123. Schmidt D, Achermann J, Odermatt B, et al. Prenatally fabricated autologous human living heart valves based on amniotic fluid derived progenitor cells as single cell source. *Circulation* 2007;**116**(11 Suppl.):I64–70. Epub 2007/09/14.

124. Schmidt D, Achermann J, Odermatt B, Genoni M, Zund G, Hoerstrup SP. Cryopreserved amniotic fluid-derived cells: a lifelong autologous fetal stem cell source for heart valve tissue engineering. *J Heart Valve Dis* 2008;**17**(4):446–55. Discussion 55. Epub 2008/08/30.

125. Hibino N, Duncan DR, Nalbandian A, et al. Evaluation of the use of an induced puripotent stem cell sheet for the construction of tissue-engineered vascular grafts. *J Thorac Cardiovasc Surg* 2012;**143**(3):696–703. Epub 2012/01/17.

126. Kawamura M, Miyagawa S, Fukushima S, et al. Enhanced survival of transplanted human induced pluripotent stem cell-derived cardiomyocytes by the combination of cell sheets with the pedicled omental flap technique in a porcine heart. *Circulation* 2013;**128**(26 Suppl. 1):S87–94. Epub 2013/09/18.

127. Krawiec JT, Vorp DA. Adult stem cell-based tissue engineered blood vessels: a review. *Biomaterials* 2012;**33**(12):3388–400.

128. Park JS, Huang NF, Kurpinski KT, Patel S, Hsu S, Li S. Mechanobiology of mesenchymal stem cells and their use in cardiovascular repair. *Front Biosci A J Virtual Libr* 2007;**12**:5098–116. Epub 2007/06/16.

129. Hashi CK, Zhu Y, Yang GY, et al. Antithrombogenic property of bone marrow mesenchymal stem cells in nanofibrous vascular grafts. *Proc Natl Acad Sci USA* 2007;**104**(29):11915–20. Epub 2007/07/07.

130. Zhao YL, Zhang S, Zhou JY, et al. The development of a tissue-engineered artery using decellularized scaffold and autologous ovine mesenchymal stem cells. *Biomaterials* 2010;**31**(2):296–307.

131. Gong ZD, Niklason LE. Small-diameter human vessel wall engineered from bone marrow-derived mesenchymal stem cells (hMSCs). *FASEB J* 2008;**22**(6):1635–48.

132. Cho SW, Kim IK, Kang JM, et al. Evidence for in vivo growth potential and vascular remodeling of tissue-engineered artery. *Tissue Eng Part A* 2009;**15**(4):901–12. Epub 2008/09/12.

133. Zhao J, Liu L, Wei J, et al. A novel strategy to engineer small-diameter vascular grafts from marrow-derived mesenchymal stem cells. *Artif Organs* 2012;**36**(1):93–101. Epub 2011/07/28.

134. Piao H, Kwon JS, Piao S, et al. Effects of cardiac patches engineered with bone marrow-derived mononuclear cells and PGCL scaffolds in a rat myocardial infarction model. *Biomaterials* 2007;**28**(4):641–9. Epub 2006/10/19.

135. Ravichandran R, Venugopal JR, Sundarrajan S, Mukherjee S, Ramakrishna S. Minimally invasive cell-seeded biomaterial systems for injectable/epicardial implantation in ischemic heart disease. *Int J Nanomedicine* 2012;**7**:5969–94. Epub 2012/12/29.

136. Torikai K, Ichikawa H, Hirakawa K, et al. A self-renewing, tissue-engineered vascular graft for arterial reconstruction. *J Thorac Cardiovasc Surg* 2008;**136**(1):37–45. e1. Epub 2008/07/08.

137. Matsumura G, Nitta N, Matsuda S, et al. Long-term results of cell-free biodegradable scaffolds for in situ tissue-engineering vasculature: in a canine inferior vena cava model. *PLoS One* 2012;**7**(4):e35760. Epub 2012/04/26.

138. Matsumura G, Isayama N, Matsuda S, et al. Long-term results of cell-free biodegradable scaffolds for in situ tissue engineering of pulmonary artery in a canine model. *Biomaterials* 2013;**34**(27):6422–8. Epub 2013/06/12.

139. Spadaccio C, Chello M, Trombetta M, Rainer A, Toyoda Y, Genovese JA. Drug releasing systems in cardiovascular tissue engineering. *J Cell Mol Med* 2009;**13**(3):422–39. Epub 2009/04/22.

140. Conklin BS, Richter ER, Kreutziger KL, Zhong DS, Chen C. Development and evaluation of a novel decellularized vascular xenograft. *Med Eng Phys* 2002;**24**(3):173–83. Epub 2002/06/14.

141. Peters MC, Isenberg BC, Rowley JA, Mooney DJ. Release from alginate enhances the biological activity of vascular endothelial growth factor. *J Biomater Sci Polym Ed* 1998;**9**(12):1267–78. Epub 1998/12/22.

142. Wei G, Jin Q, Giannobile WV, Ma PX. Nano-fibrous scaffold for controlled delivery of recombinant human PDGF-BB. *J Control Release Off J Control Release Soc* 2006;**112**(1):103–10. Epub 2006/03/07.

143. Rosenberg RD. Biochemistry of heparin antithrombin interactions, and the physiologic role of this natural anticoagulant mechanism. *Am J Med* 1989;**87**(3B):2S–9S. Epub 1989/09/11.

144. Edelman ER, Nathan A, Katada M, Gates J, Karnovsky MJ. Perivascular graft heparin delivery using biodegradable polymer wraps. *Biomaterials* 2000;**21**(22):2279–86. Epub 2000/10/12.

145. Roh JD, Sawh-Martinez R, Brennan MP, et al. Tissue-engineered vascular grafts transform into mature blood vessels via an inflammation-mediated process of vascular remodeling. *Proc Natl Acad Sci USA* 2010;**107**(10):4669–74. Epub 2010/03/09.

146. Lantz GC, Badylak SF, Hiles MC, et al. Small intestinal submucosa as a vascular graft: a review. *J Invest Surg Off J Acad Surg Res* 1993;**6**(3):297–310. Epub 1993/05/01.

147. Crapo PM, Gilbert TW, Badylak SF. An overview of tissue and whole organ decellularization processes. *Biomaterials* 2011;**32**(12):3233–43. Epub 2011/02/08.

148. Badylak SF. Xenogeneic extracellular matrix as a scaffold for tissue reconstruction. *Transpl Immunol* 2004;**12**(3–4):367–77. Epub 2004/05/26.

149. Kalfa D, Bacha E. New technologies for surgery of the congenital cardiac defect. *Rambam Maimonides Med J* 2013;**4**(3):e0019. Epub 2013/08/03.

150. Quarti A, Nardone S, Colaneri M, Santoro G, Pozzi M. Preliminary experience in the use of an extracellular matrix to repair congenital heart diseases. *Interact Cardiovasc Thorac Surg* 2011;**13**(6):569–72. Epub 2011/10/08.

151. Shin'oka T, Imai Y, Ikada Y. Transplantation of a tissue-engineered pulmonary artery. *N Engl J Med* 2001;**344**(7):532–3.

152. Udelsman BV, Maxfield MW, Breuer CK. Tissue engineering of blood vessels in cardiovascular disease: moving towards clinical translation. *Heart (Br Cardiac Soc)* 2013;**99**(7):454–60. Epub 2013/02/01.

153. L'Heureux N, Stoclet JC, Auger FA, Lagaud GJ, Germain L, Andriantsitohaina R. A human tissue-engineered vascular media: a new model for pharmacological studies of contractile responses. *FASEB J Off Publ Fed Am Soc Exp Biol* 2001;**15**(2):515–24. Epub 2001/02/07.

154. L'Heureux N, Dusserre N, Marini A, Garrido S, de la Fuente L, McAllister T. Technology insight: the evolution of tissue-engineered vascular grafts—from research to clinical practice. *Nat Clin Pract Cardiovasc Med* 2007;**4**(7):389–95. Epub 2007/06/26.

155. McAllister TN, Maruszewski M, Garrido SA, et al. Effectiveness of haemodialysis access with an autologous tissue-engineered vascular graft: a multicentre cohort study. *Lancet* 2009;**373**(9673):1440–6. Epub 2009/04/28.

156. Dahl SL, Kypson AP, Lawson JH, et al. Readily available tissue-engineered vascular grafts. *Sci Transl Med* 2011;**3**(68):68ra9. Epub 2011/02/04.

157. Elkins RC, Dawson PE, Goldstein S, Walsh SP, Black KS. Decellularized human valve allografts. *Ann Thorac Surg* 2001;**71**(5 Suppl. 1):S428–32.

158. Simon P, Kasimir MT, Seebacher G, et al. Early failure of the tissue engineered porcine heart valve SYNERGRAFT in pediatric patients. *Eur J Cardiothorac Surg Off J Eur Assoc Cardiothorac Surg* 2003;**23**(6):1002–6. discussion 6. Epub 2003/06/28.

159. Sharp MA, Phillips D, Roberts I, Hands L. A cautionary case: the SynerGraft vascular prosthesis. *Eur J Vasc Endovasc Surg Off J Eur Soc Vasc Surg* 2004;**27**(1):42–4. Epub 2003/12/04.

160. Dohmen PM. Clinical results of implanted tissue engineered heart valves. *HSR Proc Intensive Care Cardiovasc Anesth* 2012;**4**(4):225–31. Epub 2013/02/27.

161. Dohmen PM, Lembcke A, Holinski S, Pruss A, Konertz W. Ten years of clinical results with a tissue-engineered pulmonary valve. *Ann Thorac Surg* 2011;**92**(4):1308–14. Epub 2011/10/01.

162. da Costa FD, Costa AC, Prestes R, et al. The early and midterm function of decellularized aortic valve allografts. *Ann Thorac Surg* 2010;**90**(6):1854–60. Epub 2010/11/26.

163. Zehr KJ, Yagubyan M, Connolly HM, Nelson SM, Schaff HV. Aortic root replacement with a novel decellularized cryopreserved aortic homograft: postoperative immunoreactivity and early results. *J Thorac Cardiovasc Surg* 2005;**130**(4):1010–5. Epub 2005/10/11.

164. Weber B, Dijkman PE, Scherman J, et al. Off-the-shelf human decellularized tissue-engineered heart valves in a non-human primate model. *Biomaterials* 2013;**34**(30):7269–80. Epub 2013/07/03.

165. Emmert MY, Weber B, Behr L, et al. Transcatheter aortic valve implantation using anatomically oriented, marrow stromal cell-based, stented, tissue-engineered heart valves: technical considerations and implications for translational cell-based heart valve concepts. *Eur J Cardiothorac Surg Off J Eur Assoc Cardiothorac Surg* 2014;**45**:61–68. Epub 2013/05/10.

166. Assmus B, Zeiher AM. Early cardiac retention of administered stem cells determines clinical efficacy of cell therapy in patients with dilated cardiomyopathy. *Circulation Res* 2013;**112**(1):6–8. Epub 2013/01/05.

167. Chachques JC, Trainini JC, Lago N, Cortes-Morichetti M, Schussler O, Carpentier A. Myocardial assistance by grafting a new bioartificial upgraded myocardium (MAGNUM trial): clinical feasibility study. *Ann Thorac Surg* 2008;**85**(3):901–8. Epub 2008/02/23.

168. Jawad H, Lyon AR, Harding SE, Ali NN, Boccaccini AR. Myocardial tissue engineering. *Br Med Bull* 2008;**87**:31–47. Epub 2008/09/16.

169. Sanganalmath SK, Bolli R. Cell therapy for heart failure: a comprehensive overview of experimental and clinical studies, current challenges, and future directions. *Circulation Res* 2013;**113**(6):810–34. Epub 2013/08/31.

170. Vrtovec B, Poglajen G, Haddad F. Stem cell therapy in patients with heart failure. *Methodist DeBakey Cardiovasc J* 2013;**9**(1):6–10. Epub 2013/03/23.

171. Mol A, van Lieshout MI, Dam-de Veen CG, et al. Fibrin as a cell carrier in cardiovascular tissue engineering applications. *Biomaterials* 2005;**26**(16):3113–21. Epub 2004/12/18.

172. Dohmen PM, Costa F, Lopes SV, et al. Results of a decellularized porcine heart valve implanted into the juvenile sheep model. *Heart Surg Forum* 2005;**8**(2):E100–4. discussion E4. Epub 2005/03/17.

173. Zhou J, Hu S, Ding J, Xu J, Shi J, Dong N. Tissue engineering of heart valves: PEGylation of decellularized porcine aortic valve as a scaffold for in vitro recellularization. *Biomed Eng Online* 2013;**12**(1):87. Epub 2013/09/07.

174. Dijkman PE, Driessen-Mol A, Frese L, Hoerstrup SP, Baaijens FP. Decellularized homologous tissue-engineered heart valves as off-the-shelf alternatives to xeno- and homografts. *Biomaterials* 2012;**33**(18):4545–54. Epub 2012/04/03.

175. Syedain ZH, Bradee AR, Kren S, Taylor DA, Tranquillo RT. Decellularized tissue-engineered heart valve leaflets with recellularization potential. *Tissue Eng Part A* 2013;**19**(5–6):759–69. Epub 2012/10/24.

176. Eschenhagen T, Fink C, Remmers U, et al. Three-dimensional reconstitution of embryonic cardiomyocytes in a collagen matrix: a new heart muscle model system. *FASEB J Off Publ Fed Am Soc Exp Biol* 1997;**11**(8):683–94. Epub 1997/07/01.

177. Zimmermann WH, Fink C, Kralisch D, Remmers U, Weil J, Eschenhagen T. Three-dimensional engineered heart tissue from neonatal rat cardiac myocytes. *Biotechnol Bioeng* 2000;**68**(1):106–14. Epub 2000/03/04.

178. Li RK, Jia ZQ, Weisel RD, Mickle DA, Choi A, Yau TM. Survival and function of bioengineered cardiac grafts. *Circulation* 1999;**100**(19 Suppl.):II63–9. Epub 1999/11/24.

179. Zimmermann WH, Didie M, Wasmeier GH, et al. Cardiac grafting of engineered heart tissue in syngenic rats. *Circulation* 2002;**106**(12 Suppl. 1):I151–7. Epub 2002/10/02.

180. Zimmermann WH, Melnychenko I, Wasmeier G, et al. Engineered heart tissue grafts improve systolic and diastolic function in infarcted rat hearts. *Nat Med* 2006;**12**(4):452–8. Epub 2006/04/04.

181. Tulloch NL, Muskheli V, Razumova MV, et al. Growth of engineered human myocardium with mechanical loading and vascular coculture. *Circulation Res* 2011;**109**(1):47–59. Epub 2011/05/21.

182. Akins RE, Boyce RA, Madonna ML, et al. Cardiac organogenesis in vitro: reestablishment of three-dimensional tissue architecture by dissociated neonatal rat ventricular cells. *Tissue Eng* 1999;**5**(2):103–18. Epub 1999/06/08.

183. Shimizu T, Yamato M, Isoi Y, et al. Fabrication of pulsatile cardiac tissue grafts using a novel 3-dimensional cell sheet manipulation technique and temperature-responsive cell culture surfaces. *Circulation Res* 2002;**90**(3):e40. Epub 2002/02/28.

184. Matsuura K, Wada M, Shimizu T, et al. Creation of human cardiac cell sheets using pluripotent stem cells. *Biochem Biophysic Res Commun* 2012;**425**(2):321–7. Epub 2012/07/31.

185. Vrtovec B, Poglajen G, Lezaic L, et al. Effects of intracoronary CD34+ stem cell transplantation in nonischemic dilated cardiomyopathy patients: 5-year follow-up. *Circulation Res* 2013;**112**(1):165–73. Epub 2012/10/16.

Chapter 5

Tissue Engineering and Regenerative Medicine: Gastrointestinal Application

Elie Zakhem, Khalil N. Bitar

Key Concepts

This chapter focuses on the different cell sources used in tissue engineering and the challenges in using them. It also covers the different biomaterials available to support gastrointestinal (GI) regeneration along with their characteristics. The last part of this chapter covers the recent advances in the regeneration of different parts of the GI tract in terms of cells and biomaterials. This includes the esophagus, stomach, small intestine, and large intestine.

1. THE GASTROINTESTINAL TRACT: OVERVIEW

The gastrointestinal tract (GI) is a continuous tubular system that extends from the mouth to the anus. The GI tract exhibits diverse motility patterns that aid in performing a variety of functions, including ingestion, digestion, absorption of nutritive elements, and excretion of waste. The organs that make up the GI tract are not limited to, but include mouth, esophagus, stomach, small intestine, and large intestine. The food enters through the mouth where it gets ingested, chemically and mechanically. The food bolus then passes into the esophagus, where alternating contractions and relaxations, a process defined as peristalsis, initiate to help propelling the food down to the stomach. Additional peristaltic pattern in the stomach coordinates digestion and propulsion of the food into the small intestine. The food is then transferred into the large intestine and any waste products are excreted from the body. Digested materials are absorbed into the blood through the epithelium. Waste substances get excreted from the body through the anus. Discontinuity exists along the tract between the different organs. The discontinuities serve as checkpoints to ensure unidirectional flow of the food down the tract.

Coordinated chemical and electrical interactions exist between smooth muscle, intramural innervation, interstitial cells, and mucosal epithelial layers. Those interactions are essential for proper GI motility. The inability of the GI tract to perform its function due to malformations or defects along the GI tract may lead to surgical resection which could cause loss of functionality. Surgical intervention serves as a short-term cure to many of the diseases. Major drawbacks of surgical resections are classified into psychological, economic, and social distress.[1–6] Currently, tissue engineering provides a promising field that can offer treatment for most parts of the GI tract. The goal is to obtain a biopsy from the patient for cell isolation and expansion in vitro. The cells are then seeded onto biodegradable, bioactive polymer scaffolds. The result is an engineered autologous tissue that can be implanted back into the same patient to regenerate a defective segment of the GI tract. The GI is composed of multiple cell layers including smooth muscle, enteric neurons, interstitial cells of Cajal (ICCs), epithelial cells, and skeletal muscle which populate one-third of the esophagus. This innate anatomical and physiological complexity dictates the requirement for a multidisciplinary approach to regeneration of functional tissue replacements. The main challenge lies in the ability to recapitulate the architecture and function of different components of the GI tract.

Engineered scaffolds are designed to recreate the anatomic complexity of the microenvironments associated with the native tissue. Design strategies for scaffolds include physical as well as chemical modifications.[7] After the design and manufacturing process, scaffolds are characterized by their porosities and stiffnesses. Cell-seeded scaffolds help the process of remodeling postimplantation and result in a functional final bioengineered product. The primary goal of tissue engineering is to be able to develop physiologically functional replacement tissues that have similar architecture and composition as the native tissue. Preconditioning the engineered tissues using bioreactors or mechanical cycling is one way of inducing maturation to the tissue prior to implantation. The goal is to obtain a functional engineered replacement while maintaining its integrity. Although the GI tract is a hollow tubular organ, it requires sophisticated design processes in order to generate functional replacements.

Translating Regenerative Medicine to the Clinic. http://dx.doi.org/10.1016/B978-0-12-800548-4.00005-X

This chapter will focus on the main steps in engineering the neuromusculature of the GI tract. These steps are considered as the basic requirements for successful outcomes. An overview on the regeneration of the esophagus, stomach, small intestine, and colon including the sphincters will be discussed in this chapter. In the last couple of decades, tissue engineering evolved quickly with the aim of reconstructing functional tissues and organs. Regenerative medicine approaches seek the reconstruction of tissues using cells and scaffolds.

2. NEURODEGENERATIVE DISEASES OF THE GI TRACT

The integrity of the neuronal circuitry in the GI tract is essential for coordination of cell function. Loss or defects of neuronal integrity leads to diseases related to motility and secretion. The smooth muscle is the basic musculature unit in the gastrointestinal tract that is responsible for contraction and relaxation. The smooth muscle receives regulatory signals from different levels including the enteric nervous system (ENS) and ICCs.[8] The ENS is arranged into two ganglionated plexi. The myenteric plexus exists between the two muscle layers and contains excitatory and inhibitory motor neurons. Neurons secrete neurotransmitters responsible for neural transmission to the smooth muscle. The submucosal plexus is located between the circular muscle layer and the submucosa. Neuro-neuronal transmission is mediated by ascending and descending interneurons. Neural dysfunction affects different parts of the GI tract and result in diseases that alter motility.

The inability of the sphincters to relax due to loss of inhibitory motor neurons, more specifically the nitrergic neurons, is a condition known as achalasia. The lower esophageal sphincter (LES) mediates the transport of food from the esophagus into the stomach. Esophageal achalasia is a neurodegenerative condition where the LES fails to relax. This condition is characterized by the absence of nitrergic neurons, however the excitatory cholinergic neurons are intact.[9,10] Absence of nitrergic neurons and presence of cholinergic neurons result in tonic contraction of the sphincter. Similarly, achalasia of the internal anal sphincter (IAS) is another form of neuropathies of the GI tract. In normal conditions, the distention of the rectum by luminal content induces rectoanal inhibitory reflexes (RAIR) that allow the relaxation of the IAS. This is followed by excretion of rectal luminal content. In achalasia, RAIR is lost due to the loss of nitrergic neurons. Achalasia is characterized by the absence of nitrergic neurons.[11] Both forms of achalasia can be diagnosed by manometric pressure measurement. Loss of intrinsic inhibitory innervation to the pyloric sphincter leads to pyloric stenosis, which is associated with gastric obstruction and altered gastric emptying. Current treatments for sphincteric achalasia include injection of botulinum toxin in order to reduce the tonic contraction, and thus facilitating the relaxation of the sphincter.[12,13]

Intestinal pseudoobstruction is another form of neurodegenerative diseases resulting from loss of intrinsic innervation. This disease is characterized by altered motility in the GI tract.[14] Pharmacological treatments are considered as the most effective way of relieving the symptoms and restoring intestinal motility.[15] Surgical correction does not provide long-term satisfactory outcomes.[16]

Hirschsprung's disease is another well characterized enteric neuropathy. It is characterized by either partial or total loss of enteric neurons in the gut. Areas of aganglionosis lack intrinsic inhibitory motor neurons and are dominated by the excitatory motor neurons. Aganglionated lengths are tonically contracted which leads to obstruction of the gut.[17] The most common and effective surgery involves a pull through. Although the surgery is helpful, it often causes disruption of the anal sphincter and loss of innervation of the rectum. Patients with pull-through surgeries suffer from fecal incontinence.[18]

Given the limited success of surgical intervention, there is an urgent need for functional replacements of the GI tract. Tissue engineering and regenerative medicine provide a therapeutic option for diseases that affect the GI tract. The concept of tissue engineering is to recapitulate the architecture and function of the tissue using the patient's own cells in combination with a scaffolding biomaterial. In the following sections of this chapter, we will discuss the different sources of cells and materials used in tissue engineering. The last section will also discuss the recent advances in tissue engineering of different regions of the GI tract.

3. CELL SOURCE IN REGENERATING THE NEUROMUSCULATURE OF THE GI TRACT

Cell source is an essential component in the design of bioengineering tissues/organs. The ability to find an appropriate cell source using minimally invasive surgeries has been recently the focus of research. This becomes more challenging when the organ is complex in its cellular composition, such as the GI tract. Ideally, biopsies obtained from patients are used to isolate the type of cells that is needed. It is critical to be able to culture the cells, expand them in vitro to obtain adequate number for reseeding. Depending on the tissue source and the type of cell needed, expansion might be problematic. Another challenge resides in the fact that some tissues, in their diseased state, do not provide a good source of cells. Finding other sources for biopsies might become an alternative approach. In the case of regenerating the GI tract, finding an efficient source of highly proliferative cells is a major limitation and is the focus of ongoing research. Stem cells derived from different parts of the body are being investigated to overcome the concern of cell source. Mesenchymal stem cells (from different sources), induced pluripotent stem cells (iPSCs), and organoid units are all examples of potential cell sources in tissue engineering.[19–23]

Tissue-specific cells obtained from biopsies are usually considered as the ideal source, given the fact that they have a determined physiological function. Tissue-specific smooth muscle cells have the contractile properties when cultured and used in the regeneration process. Smooth muscle cells can be isolated from gut segments using well-established enzymatic digestion protocols. The cells have the potential to be expanded and passaged in vitro. A major concern with tissue-specific cells is their ability to maintain their phenotype after isolation and expansion, and following their implantation.[24,25] Muscle-derived stem cells have the advantage of self-renewal and multilineage differentiation where they have been shown to differentiate into myotubes as well as smooth muscle.[26–28]

The GI tract is known to have its own nervous system which is referred to as the ENS. The ENS is considered as the intrinsic innervation of the GI tract and it is a major player in the regulatory apparatus that controls the smooth muscle function. Neural stem cells derived from both the central nervous system and the ENS are sources for neurons.[29] Well-established protocols have been developed to isolate enteric neural stem cells from embryonic, fetal, postnatal, and adult rodent as well as human gut tissue.[30–36] Enteric neural stem cells reinstated the innervation of denervated colon.[37] Enteric neural stem cells have the potential to migrate, proliferate, and differentiate into different subtypes of motor neurons.[38–40] Additionally, neural stem cells isolated from the central nervous system have the potential to differentiate into enteric-like neurons and treat gastrointestinal disorders.[35,41–44]

4. SCAFFOLDS AS SUPPORT FOR NEUROMUSCULATURE REGENERATION

The GI tract is a hollow tubular organ that extends from the esophagus down to the rectum. Natural and/or synthetic bio-materials can be used to develop tubular scaffolds to mimic the GI tract. Biomaterials are tested for their ability to support cell attachment, survival, proliferation, and differentiation. While designing scaffolds for tissue engineering applications, it is essential to characterize the scaffolds in terms of biocompatibility, porosity and pore sizes, mechanical properties, and biodegradability.[7,45] Determining all these characteristics of scaffolds will help in modulating the interaction of the cells with the scaffold, including cell attachment, alignment, and maintenance of phenotype. Taken all together, these elements are key for successful outcomes of the implant. A major hurdle in GI tissue engineering is the failure of the scaffold to support cell alignment.[46] A wide spectrum of natural and synthetic biomaterials is under investigation for the support of GI tissue engineering.

4.1 Natural Materials

The most common natural material is collagen. It is the most abundant and major component of the extracellular matrix. Collagen scaffolds were developed as support in GI tissue engineering. Collagen was tested using smooth muscle cells.[47–50] In addition to collagen scaffolds, collagen was used as a coating material to favor the regeneration process.[51] Studies have already demonstrated the excellent biocompatibility of collagen by supporting cell attachment and differentiation. Some of the weaknesses of using collagen as a scaffold are the low mechanical properties and its inability to support the native architecture of the tissue. Another common natural material is chitosan.[52] Chitosan is a widely used natural polymer in tissue engineering and it has gained special attention due to its biocompatibility. Chitosan has been tested in several applications including skin and bone tissue engineering. Our group demonstrated the biocompatibility of chitosan in GI applications.[53] Smooth muscle cells derived from the gut demonstrated attachment, proliferation, and maintenance of phenotype in vitro. Fibrin is another natural biomaterial used in GI tissue engineering. Circular smooth muscle tissue constructs were engineered using fibrin gel.[54] The smooth muscle cells demonstrated concentric alignment similar to the orientation of circular smooth muscle cells seen in native gut. Our group developed hollow tubular chitosan scaffolds for intestinal tissue engineering applications. Characterization of the scaffolds demonstrated high porosity. Engineered smooth muscle tissues maintained functionality around the tubular chitosan scaffolds. In order to obtain a scaffold that closely mimics the native tissue, physical and chemical modifications have been employed. Unidirectional porous structures were formed on OptiMaix collagen scaffolds.[55] The scaffolds were designed to allow unidirectional smooth muscle growth. Longitudinal smooth muscle tissues were engineered using substrate microtopography.[56] The tissues contracted and relaxed in a similar manner to native longitudinal smooth muscle layer. On the other hand, decellularization protocols have been well documented to remove cellular component of the tissue while preserving the native architecture.[57] The decellularized tissue is then reseeded with the appropriate cell types. The maintenance of architecture in the decellularized tissue guides the cells to orient themselves along the grooves.

4.2 Synthetic Materials

Apart from natural materials, synthetic materials are potential candidates for GI tissue engineering applications. Synthetic materials display strong biocompatibility and biodegradability properties. Synthetic materials lack binding domains, which makes them suitable for modifications.[58] Electrospun poly(L-lactide-co-caprolactone) (PLLC) scaffolds were immobilized with fibronectin as an attempt to enhance cell attachment.[59] Polyglycolic acid (PGA) scaffolds are also synthetic scaffolds that were coated with collagen to enhance the cell seeding process.[51,60]

5. TISSUE ENGINEERING OF DIFFERENT PARTS OF THE GI TRACT: CURRENT CONCEPTS

The GI tract has multiple complex functions that include transporting and digesting food, absorbing nutrients, and excreting waste. These functions are the result of coordinated interactions by the different cell types that make up the GI tract. Different layers of the GI tract consist of different cell types, arranged in a specific alignment. Smooth muscle cells are considered as the main effectors of motility, performing contraction and relaxation to propel luminal content. The outer smooth muscle layer, or also known as the longitudinal smooth muscle, is responsible for shortening the length of the intestine when contracted. The contraction of the inner circular smooth muscle layer is responsible for narrowing the lumen of the intestine. A primary regulator of the function of smooth muscle is the ENS which contains neurons (sensory, motor, secretory, etc.) and glia.[61] Regulation of function is also provided by the ICCs, which have pacemaker activity within the gut.[62] Unidirectional flow of luminal content is ensured by sphincters along the GI tract. Intestinal epithelial cells mediate absorption and secretion within the gut.

Postnatal diseases, damages, injuries, surgical or obstetric trauma, and age are causes of motility disorders. Additionally, motility can be altered due to congenital defects such as Hirschsprung's disease, intestinal pseudoobstruction, and achalasia.[63] Surgical intervention does not provide a long-lasting solution. Tissue engineering provides an approach to regenerate different parts of the GI tract using a combination of cells and scaffolds.[64,65]

5.1 Esophagus

The esophagus is a muscular tubular organ that extends from the pharynx to the stomach. A series of peristaltic waves mediate the transport of food bolus into the stomach through the LES. Diseases affecting the esophagus such as esophageal atresia and tracheoesophageal fistula result in esophageal stricture and gastroesophageal reflux disease, which lead to impaired motility.[66] Esophageal cancer is one of the most leading causes of deaths. Patients with esophageal cancer have low quality of life. Any engineered esophagus must take into consideration, the gravitational characteristic and the coordinated rhythmic activity of the tissue itself.

The basic requirements for a successful engineered esophagus include the following properties. First, the engineered esophagus must be a tubular structure with luminal patency. Second, all types of cells that make up the esophagus must be taken into account. The cells include epithelial cells along the lumen, the muscle component (striated muscle and smooth muscle). Intrinsic and extrinsic neural pathways are responsible for swallowing and for peristalsis. Third, analysis of the basement membrane of the esophagus is a prerequisite for designing scaffolds in tissue engineering.[67] The basement membrane composition of the esophagus dictates cell growth and differentiation. Defects or loss of function of any of the mentioned cell types will impair the process of motility.

The regeneration of the esophageal tissue has been studied using different natural-based scaffolds. OptiMaix collagen scaffolds with unidirectional porous structures were seeded with smooth muscle. This resulted in successful alignment of regenerated smooth muscle.[55] Removal of the cells from specific tissues using different detergents results in a matrix that lacks cells but maintains the native architecture of the tissue. This process of decellularization allows the reseeding of the tissue with healthy functional cells. A wide range of organs were decellularized and tested for their potential to regenerate esophageal tissue. These organs include the esophagus, urinary bladder, stomach, and small intestinal submucosa. Studies have shown that decellularized ovine esophagi exhibited rough surfaces with preservation of the extracellular matrix component. Reseeding these matrices with specific cells can be one approach to regenerate the esophagus.[68]

Porcine urinary bladders were also decellularized and resulted in matrices suitable for implantation in dog models. Results demonstrated the potential of using these tissues to reconstruct the esophagus. The mechanical properties of the regenerated tissue were similar to native esophagus. Additionally, histological analysis demonstrated the presence of organized layers that constitute the esophagus.[69] Small intestine submucosa (SIS) was seeded with bone marrow-derived mesenchymal stem cells. The seeded SIS was used as autologous replacement in dogs. Results have shown reepithelialization, revascularization, and muscle regeneration.[70]

First attempts to reconstruct the esophagus using synthetic materials have tested nitinol and silicon rubber. These approaches were associated with complications such as anastomosis and shedding problems. Further modifications of the model by adding polyester connecting rings reduced the risk of anastomotic leakage and improved the shedding time.[71] Another synthetic material polycaprolactone (PCL) was used as a matrix to support esophageal epithelial cells. PCL was coupled to type IV collagen in order to accelerate the regeneration process, cell spreading, and cell–cell interaction. The epithelial phenotype of the cells was confirmed via histological analysis.[58]

Synthetic materials often require chemical modifications in order to enhance the performance of the cells. Epithelial cells adhesion and growth on electrospun PLLC scaffolds was improved by fibronectin immobilization. Cells demonstrated morphological characteristics of epithelium with maintenance of function.[59] Adipose-derived smooth muscle cells seeded onto PGA/PLGA scaffolds were implanted in rat models. Results have shown reepithelialization and muscularis regeneration.[72]

5.2 Stomach

The stomach is the site of food mixing through cycles of coordinated rhythmic contractile activities. The luminal content then leaves the stomach and goes into the small intestine through the pylorus sphincter. Several diseases alter gastric motility. Gastroparesis is characterized by delayed gastric emptying due to autonomic neuropathy.[73,74] Current solutions include gastric electrical stimulation.[75,76] Gastric electrical stimulation requires surgical intervention which is associated with infection, pain, and device relocation.

Gastric cancer is another cause of dysmotility. The main solution is surgical removal of part of the stomach. This is also associated with high rates of morbidities, including malnutrition, anemia, and weight loss. Tissue engineering remains a promising field to regenerate the stomach.

Organoid units seeded onto different scaffolds have been the most common approach for regenerating the stomach. Composite scaffolds composed of PGA, poly-L-lactic acid, and collagen have been tested using organoid units. Following their implantation, the organoid units resulted into differentiated epithelium as well as muscularis layer.[51] One of the drawbacks was the lack of muscularis architecture, which is critical for proper function. Gastric patch was developed by seeding organoid units onto PGA scaffolds. A defect in the stomach was created and then replaced with engineered patch. Results have shown integration of the patch with the host tissue. Regeneration of the epithelium as well as the muscularis layer was also demonstrated by immunostaining.[77] Although the organoid units have resulted in successful regeneration of the epithelium, the architecture and function of the regenerated muscle remains a challenge.[78]

5.3 Small Intestine

As the food gets processed in the stomach, it gets emptied into the small intestine. Nutrient absorption mainly takes place in the small intestine. The microvilli structures in the epithelium enhance nutrient absorption. The smooth muscle performs peristalsis to propel food through the intestine. Inflammation or cancer of the intestine requires massive truncation of regions of the intestine which results in short bowel syndrome. Patients often suffer from malnutrition, malabsorption, and motility dysfunction. Patients experience weight loss, vitamin deficiency, and potential infections.[2] The most common therapies include chronic parenteral nutrition and intestinal transplantation. High costs of transplantation, availability of donors, and lifetime administration of immunosuppressive drugs are some of the limitations.[1–3]

A main requirement for successful regeneration of the muscularis layer of the intestine is the exact recapitulation of the alignment of the native smooth muscle. Additionally, providing the innervation using neuronal cells is essential for smooth muscle function. Pacemaker cells, ICCs, also provide additional regulation for motility. A tissue-engineered intestine must take into account the complexity of the native tissue.

Regeneration of the muscularis layer is a challenge in tissue engineering. This may be due to the fact that smooth muscle has a dynamic phenotype, switching between synthetic and contractile phenotype. Mesenchymal stem cells failed to regenerate the smooth muscle components.[46] Recent studies reported successful differentiation of iPSCs into smooth muscle sheets with peristaltic features.[79] This characteristic is promising to tissue engineer physiologically functional intestine.

Organoid units have also been tested in regenerating the small intestine. Following implantation of scaffolds seeded with organoid units, evidence of muscle and neuronal regeneration was observed.[80,81] The architecture of the muscularis layer was also regenerated. PGA scaffolds coated with poly-L-lactide acid and type I collagen were developed. Organoid units were seeded onto the scaffolds and were implanted into the omentum of mice. Organoid units generated the epithelium with villi structures. Regeneration of the smooth muscle layer was confirmed by positive stains for smooth muscle markers.[82] The regenerated muscle lacked the native orientation.

Attempts to reconstruct the small intestine employed decellularized tissues. In one study, smooth muscle cells were derived from the intestine and seeded onto SIS sheets. Following implantation, the implant became vascularized. Partial epithelialization was documented. There was a variation in the expression of smooth muscle markers.[83] Recently, a new decellularization protocol was developed using rat small bowel. The study had optimized the protocol in terms of number of cycles of detergents needed to completely remove all cellular components. The advantage of using decellularized matrices is the preservation of both the native architecture and the mechanical properties of the tissue.[57]

5.4 Large Intestine

The last part of the GI tract is the large intestine, where water absorption and waste excretion take place. In terms of cellular composition, the large intestine is considered as a continuation of the neuromusculature layer of the small intestine. Several diseases can affect the large intestine and therefore cause dysmotility in the large intestine. Diseases include inflammation (Crohn's disease) and loss/lack of innervation (Hirschsprung's disease). This is associated with either constipation or diarrhea. Engineered functional replacements must be available in order to restore motility.

Well-established protocols have successfully isolated organoid units directly from the large intestine. Organoid units were seeded onto polymer scaffolds. Results have shown regeneration of the musculature with a similar architecture to the native large intestine.[81] Those studies have not evaluated peristalsis in the engineered large intestine. Another approach to regenerate the innervation of the large intestine involved the use of neural crest progenitor cells. The large intestines of rats were denervated using benzalkonium chloride.[37] Neural progenitor cells were delivered to the site of denervation and differentiated into neurons and glia. Results have shown restoration of motility. Our group has successfully regenerated the neuromusculature of the GI tract using a combination of cells and biomaterials. We developed techniques to engineer the circular and longitudinal smooth muscle layers with the appropriate alignment. We were also successful in innervating the engineered smooth muscle using primary isolated neural progenitor cells derived from the gut. The engineered circular smooth muscle constructs maintained smooth muscle phenotype and contractile function around tubular chitosan scaffolds. We developed a novel approach to neoinnervate bioengineered smooth muscle constructs (Figure 1). Our approach involved the use of scaffold as support for the constructs. Noninnervated smooth muscle constructs and innervated smooth muscle constructs were bioengineered. The constructs were placed next to each other around tubular chitosan scaffolds as shown in the figure. The initially innervated constructs maintained neural differentiation and function. Neoinnervation of the abutting constructs around the scaffolds was demonstrated by immunohistochemistry and physiological functionality assays.[53] Longitudinal smooth muscle layer was engineered using molds with longitudinal grooves.[56,84] The influence of the extracellular matrix composition on the differentiation of the neural progenitor cells in the engineered longitudinal smooth muscle tissue was studied. Changing the composition of the extracellular matrix modulated the extent of differentiation of neural progenitor cells into different neuronal subtypes.

Fecal incontinence is caused by the degeneration or loss of function of the IAS.[85] This could be due to injury or aging. The sphincteric nature of the IAS smooth muscle is responsible for keeping the muscle in a constant contracture state until

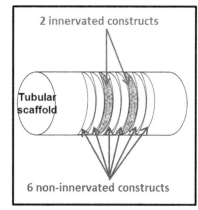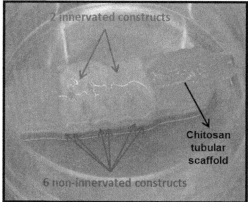

FIGURE 1 Schematic for neo-innervation concept in tissue engineering: Intrinsically innervated, concentrically aligned, smooth muscle constructs were bioengineered using smooth muscle cells and enteric neural progenitor cells (innervated constructs). Three-dimensional smooth muscle constructs were bioengineered using smooth muscle cells only (noninnervated constructs). The bioengineered constructs were placed next to each other around tubular chitosan scaffolds as shown in the figure. After 14 days in culture in vitro, the noninnervated constructs became neo-innervated as demonstrated by immunohistochemistry and physiological functionality assays. The intrinsically innervated construct maintained neuronal differentiation.

stimulated to relax. Current approaches involve the injection of silicone material or the use of mechanical devices.[86,87] However, these approaches fail to fully restore continence. We were successful in developing IAS constructs using primary isolated sphincteric smooth muscle and neural progenitor cells.[54,88] The engineered IAS constructs generated a spontaneous tone, which is a characteristic of native IAS. Following their implantation in situ in rodents, the engineered IAS constructs became neovascularized and maintained their myogenic and neurogenic phenotype and function.[89,90]

6. CONCLUSION

Tissue engineering and regenerative medicine provide an alternative approach as potential therapy for disorders affecting the GI tract. Current treatments for dysmotility in the GI tract involve surgical intervention associated with high rates of complications and limited success. Ongoing research has demonstrated promising results in generating three-dimensional tissues using autologous cells. The ultimate goal of tissue engineering and regenerative medicine is to recapitulate the architecture and function of the native tissue. The GI tract is highly regulated. In a system as complex as the GI tract, it is important to keep in mind all the different cell types and the specific cell alignment in each layer. Maintenance of cell phenotype is also critical for proper function. These characteristics are essential for peristalsis.

ACKNOWLEDGMENT

This work was supported by NIH/NIDDK R01DK071614.

REFERENCES

1. Levitt MA, Dickie B, Pena A. Evaluation and treatment of the patient with Hirschsprung disease who is not doing well after a pull-through procedure. *Semin Pediatr Surg* May 2010;**19**(2):146–53. PubMed PMID: 20307851.
2. Beyer-Berjot L, Joly F, Maggiori L, Corcos O, Bouhnik Y, Bretagnol F, et al. Segmental reversal of the small bowel can end permanent parenteral nutrition dependency: an experience of 38 adults with short bowel syndrome. *Ann Surg* November 2012;**256**(5):739–45. PubMed PMID: 23095617.
3. Sampietro GM, Corsi F, Maconi G, Ardizzone S, Frontali A, Corona A, et al. Prospective study of long-term results and prognostic factors after conservative surgery for small bowel Crohn's disease. *Clin Gastroenterol Hepatol* February 2009;**7**(2):183–91. quiz 25. PubMed PMID: 19118641.
4. Braghetto I, Korn O, Valladares H, Debandi A, Diaz JC, Brunet L. Laparoscopic surgical treatment for patients with short- and long-segment Barrett's esophagus: which technique in which patient?. *Int Surg* April–June 2011;**96**(2):95–103. PubMed PMID: 22026298.
5. Songun I, Putter H, Kranenbarg EM, Sasako M, van de Velde CJ. Surgical treatment of gastric cancer: 15-year follow-up results of the randomised nationwide Dutch D1D2 trial. *Lancet Oncol* May 2010;**11**(5):439–49. PubMed PMID: 20409751.
6. Prasad GA, Wu TT, Wigle DA, Buttar NS, Wongkeesong LM, Dunagan KT, et al. Endoscopic and surgical treatment of mucosal (T1a) esophageal adenocarcinoma in Barrett's esophagus. *Gastroenterology* September 2009;**137**(3):815–23. PubMed PMID: 19524578.
7. Bitar KN, Zakhem E. Design strategies of biodegradable scaffolds for tissue regeneration. *Biomed Eng Comput Biol* 2014;**6**:13–20. PubMed PMID: 25288907. Pubmed Central PMCID: 4147780.
8. Hansen MB. The enteric nervous system I: organisation and classification. *Pharmacol Toxicol* March 2003;**92**(3):105–13. PubMed PMID: 12753424. Epub 2003/05/20. eng.
9. Holloway RH, Dodds WJ, Helm JF, Hogan WJ, Dent J, Arndorfer RC. Integrity of cholinergic innervation to the lower esophageal sphincter in achalasia. *Gastroenterology* April 1986;**90**(4):924–9. PubMed PMID: 3949120. Epub 1986/04/01. eng.
10. De Giorgio R, Di Simone MP, Stanghellini V, Barbara G, Tonini M, Salvioli B, et al. Esophageal and gastric nitric oxide synthesizing innervation in primary achalasia. *Am J Gastroenterol* September 1999;**94**(9):2357–62. PubMed PMID: 10483991. Epub 1999/09/14. eng.
11. Hirakawa H, Kobayashi H, O'Briain DS, Puri P. Absence of NADPH-diaphorase activity in internal anal sphincter (IAS) achalasia. *J Pediatr Gastroenterol Nutr* January 1995;**20**(1):54–8. PubMed PMID: 7533833. Epub 1995/01/01. eng.
12. Messineo A, Codrich D, Monai M, Martellossi S, Ventura A. The treatment of internal anal sphincter achalasia with botulinum toxin. *Pediatr Surg Int* September 2001;**17**(7):521–3. PubMed PMID: 11666049. Epub 2001/10/23. eng.
13. Annese V, Bassotti G, Coccia G, D'Onofrio V, Gatto G, Repici A, et al. Comparison of two different formulations of botulinum toxin A for the treatment of oesophageal achalasia. The Gismad Achalasia Study Group. *Aliment Pharmacol Ther* October 1999;**13**(10):1347–50. PubMed PMID: 10540051. Epub 1999/10/30. eng.
14. Antonucci A, Fronzoni L, Cogliandro L, Cogliandro RF, Caputo C, De Giorgio R, et al. Chronic intestinal pseudo-obstruction. *World J Gastroenterol* May 21, 2008;**14**(19):2953–61. PubMed PMID: 18494042. Pubmed Central PMCID: 2712158. Epub 2008/05/22. eng.
15. Rudolph CD, Hyman PE, Altschuler SM, Christensen J, Colletti RB, Cucchiara S, et al. Diagnosis and treatment of chronic intestinal pseudo-obstruction in children: report of consensus workshop. *J Pediatr Gastroenterol Nutr* January 1997;**24**(1):102–12. PubMed PMID: 9093995. Epub 1997/01/01. eng.
16. Bitar KN, Raghavan S. Intestinal tissue engineering: current concepts and future vision of regenerative medicine in the gut. *Neurogastroenterol Motil* January 2012;**24**(1):7–19. PubMed PMID: 22188325. Pubmed Central PMCID: 3248673. Epub 2011/12/23. eng.
17. Arshad A, Powell C, Tighe MP. Hirschsprung's disease. *BMJ* 2012;**345**:e5521. PubMed PMID: 23028095. Epub 2012/10/03. eng.

18. Catto-Smith AG, Coffey CM, Nolan TM, Hutson JM. Fecal incontinence after the surgical treatment of Hirschsprung disease. *J Pediatr* December 1995;**127**(6):954–7. PubMed PMID: 8523196. Epub 1995/12/01. eng.

19. Wang A, Tang Z, Park IH, Zhu Y, Patel S, Daley GQ, et al. Induced pluripotent stem cells for neural tissue engineering. *Biomaterials* August 2011;**32**(22):5023–32. PubMed PMID: 21514663. Pubmed Central PMCID: 3100451.

20. Rodrigues MT, Lee SJ, Gomes ME, Reis RL, Atala A, Yoo JJ. Amniotic fluid-derived stem cells as a cell source for bone tissue engineering. *Tissue Eng Part A* August 14, 2012;**18**(23–24):2518–27. PubMed PMID: 22891759.

21. Williams C, Xie AW, Emani S, Yamato M, Okano T, Emani SM, et al. A comparison of human smooth muscle and mesenchymal stem cells as potential cell sources for tissue-engineered vascular patches. *Tissue Eng Part A* May 2012;**18**(9–10):986–98. PubMed PMID: 22145703.

22. Fuller MK, Faulk DM, Sundaram N, Shroyer NF, Henning SJ, Helmrath MA. Intestinal crypts reproducibly expand in culture. *J Surg Res* November 2012;**178**(1):48–54. PubMed PMID: 22564827.

23. Wagner W, Wein F, Seckinger A, Frankhauser M, Wirkner U, Krause U, et al. Comparative characteristics of mesenchymal stem cells from human bone marrow, adipose tissue, and umbilical cord blood. *Exp Hematol* November 2005;**33**(11):1402–16. PubMed PMID: 16263424.

24. Halayko AJ, Camoretti-Mercado B, Forsythe SM, Vieira JE, Mitchell RW, Wylam ME, et al. Divergent differentiation paths in airway smooth muscle culture: induction of functionally contractile myocytes. *Am J Physiol* January 1999;**276**(1 Pt 1):L197–206. PubMed PMID: 9887072.

25. Brittingham J, Phiel C, Trzyna WC, Gabbeta V, McHugh KM. Identification of distinct molecular phenotypes in cultured gastrointestinal smooth muscle cells. *Gastroenterology* September 1998;**115**(3):605–17. PubMed PMID: 9721158.

26. Deasy BM, Jankowski RJ, Huard J. Muscle-derived stem cells: characterization and potential for cell-mediated therapy. *Blood Cells Mol Dis* September–October 2001;**27**(5):924–33. PubMed PMID: 11783957. Epub 2002/01/11. eng.

27. Cao B, Huard J. Muscle-derived stem cells. *Cell Cycle* February 2004;**3**(2):104–7. PubMed PMID: 14712064. Epub 2004/01/09. eng.

28. Hwang JH, Yuk SH, Lee JH, Lyoo WS, Ghil SH, Lee SS, et al. Isolation of muscle derived stem cells from rat and its smooth muscle differentiation [corrected]. *Mol Cells* February 29, 2004;**17**(1):57–61. PubMed PMID: 15055528. Epub 2004/04/02. eng.

29. Kulkarni S, Becker L, Pasricha PJ. Stem cell transplantation in neurodegenerative disorders of the gastrointestinal tract: future or fiction?. *Gut* August 4, 2011;**61**(4):613–21. PubMed PMID: 21816959. Epub 2011/08/06. eng.

30. Suárez–Rodríguez R, Belkind–Gerson J. Cultured nestin–positive cells from postnatal mouse small bowel differentiate ex vivo into neurons, glia, and smooth muscle. *Stem Cells* 2004;**22**(7):1373–85.

31. Schafer KH, Hagl CI, Rauch U. Differentiation of neurospheres from the enteric nervous system. *Pediatr Surg Int* July 2003;**19**(5):340–4. PubMed PMID: 12845455.

32. Belkind-Gerson J, Carreon-Rodriguez A, Benedict LA, Steiger C, Pieretti A, Nagy N, et al. Nestin-expressing cells in the gut give rise to enteric neurons and glial cells. *Neurogastroenterol Motil* January 2013;**25**(1):61–9.e7. PubMed PMID: 22998406. Pubmed Central PMCID: 3531577.

33. Silva AT, Wardhaugh T, Dolatshad NF, Jones S, Saffrey MJ. Neural progenitors from isolated postnatal rat myenteric ganglia: expansion as neurospheres and differentiation *in vitro*. *Brain Res* 2008;**1218**:47–53.

34. Lindley RM, Hawcutt DB, Connell MG, Almond SL, Vannucchi MG, Faussone-Pellegrini MS, et al. Human and mouse enteric nervous system neurosphere transplants regulate the function of aganglionic embryonic distal colon. *Gastroenterology* July 2008;**135**(1):205–16.e6. PubMed PMID: 18515088.

35. Metzger M, Caldwell C, Barlow AJ, Burns AJ, Thapar N. Enteric nervous system stem cells derived from human gut mucosa for the treatment of aganglionic gut disorders. *Gastroenterology* June 2009;**136**(7):2214–25.e1–3. PubMed PMID: 19505425.

36. Rauch U, Hänsgen A, Hagl C, Holland-Cunz S, Schäfer K-H. Isolation and cultivation of neuronal precursor cells from the developing human enteric nervous system as a tool for cell therapy in dysganglionosis. *Int J Colorectal Dis* 2006;**21**(6):554–9.

37. Pan WK, Zheng BJ, Gao Y, Qin H, Liu Y. Transplantation of neonatal gut neural crest progenitors reconstructs ganglionic function in benzalkonium chloride-treated homogenic rat colon. *J Surg Res* May 15, 2011;**167**(2):e221–30. PubMed PMID: 21392806.

38. Hotta R, Stamp LA, Foong JP, McConnell SN, Bergner AJ, Anderson RB, et al. Transplanted progenitors generate functional enteric neurons in the postnatal colon. *J Clin Invest* March 1, 2013;**123**(3):1182–91. PubMed PMID: 23454768. Pubmed Central PMCID: 3582137. Epub 2013/03/05. eng.

39. Anitha M, Joseph I, Ding X, Torre ER, Sawchuk MA, Mwangi S, et al. Characterization of fetal and postnatal enteric neuronal cell lines with improvement in intestinal neural function. *Gastroenterology* May 2008;**134**(5):1424–35. PubMed PMID: 18471518. Pubmed Central PMCID: 2612783. Epub 2008/05/13. eng.

40. Tsai YH, Murakami N, Gariepy CE. Postnatal intestinal engraftment of prospectively selected enteric neural crest stem cells in a rat model of Hirschsprung disease. *Neurogastroenterol Motil* April 2011;**23**(4):362–9. PubMed PMID: 21199176. Pubmed Central PMCID: 3105196.

41. Kulkarni S, Zou B, Hanson J, Micci M-A, Tiwari G, Becker L, et al. Gut-derived factors promote neurogenesis of CNS-neural stem cells and nudge their differentiation to an enteric-like neuronal phenotype. *Am J Physiol Gastrointest Liver Physiol* October 2011;**301**(4):G644–55. PubMed PMID: 21817062. Pubmed Central PMCID: 3191554.

42. Micci MA, Kahrig KM, Simmons RS, Sarna SK, Espejo-Navarro MR, Pasricha PJ. Neural stem cell transplantation in the stomach rescues gastric function in neuronal nitric oxide synthase-deficient mice. *Gastroenterology* December 2005;**129**(6):1817–24. PubMed PMID: 16344050.

43. Dong YL, Liu W, Gao YM, Wu RD, Zhang YH, Wang HF, et al. Neural stem cell transplantation rescues rectum function in the aganglionic rat. *Transplant Proc* December 2008;**40**(10):3646–52. PubMed PMID: 19100458.

44. Metzger M, Bareiss PM, Danker T, Wagner S, Hennenlotter J, Guenther E, et al. Expansion and differentiation of neural progenitors derived from the human adult enteric nervous system. *Gastroenterology* December 2009;**137**(6):2063–73.e4. PubMed PMID: 19549531. Epub 2009/06/25. eng.

45. Yang S, Leong K-F, Du Z, Chua C-K. The design of scaffolds for use in tissue engineering. Part I. Traditional factors. *Tissue Eng* 2001;**7**(6):679–89.

46. Hori Y, Nakamura T, Kimura D, Kaino K, Kurokawa Y, Satomi S, et al. Experimental study on tissue engineering of the small intestine by mesenchymal stem cell seeding. *J Surg Res* February 2002;**102**(2):156–60. PubMed PMID: 11796013.

47. Lee M, Wu BM, Stelzner M, Reichardt HM, Dunn JC. Intestinal smooth muscle cell maintenance by basic fibroblast growth factor. *Tissue Eng Part A* August 2008;**14**(8):1395–402. PubMed PMID: 18680389.

48. Hori Y, Nakamura T, Kimura D, Kaino K, Kurokawa Y, Satomi S, et al. Functional analysis of the tissue-engineered stomach wall. *Artif Organs* October 2002;**26**(10):868–72. PubMed PMID: 12296927.

49. Araki M, Tao H, Sato T, Nakajima N, Hyon SH, Nagayasu T, et al. Development of a new tissue-engineered sheet for reconstruction of the stomach. *Artif Organs* October 2009;**33**(10):818–26. PubMed PMID: 19839991.

50. Nakase Y, Hagiwara A, Nakamura T, Kin S, Nakashima S, Yoshikawa T, et al. Tissue engineering of small intestinal tissue using collagen sponge scaffolds seeded with smooth muscle cells. *Tissue Eng* 2006;**12**(2):403–12.

51. Speer AL, Sala FG, Matthews JA, Grikscheit TC. Murine tissue-engineered stomach demonstrates epithelial differentiation. *J Surg Res* November 2011;**171**(1):6–14. PubMed PMID: 21571313.

52. Madihally SV, Matthew HW. Porous chitosan scaffolds for tissue engineering. *Biomaterials* June 1999;**20**(12):1133–42. PubMed PMID: 10382829.

53. Zakhem E, Raghavan S, Gilmont RR, Bitar KN. Chitosan-based scaffolds for the support of smooth muscle constructs in intestinal tissue engineering. *Biomaterials* June 2012;**33**(19):4810–7. PubMed PMID: 22483012. Pubmed Central PMCID: 3334429.

54. Somara S, Gilmont RR, Dennis RG, Bitar KN. Bioengineered internal anal sphincter derived from isolated human internal anal sphincter smooth muscle cells. *Gastroenterology* 2009;**137**(1):53–61.

55. Saxena AK, Kofler K, Ainodhofer H, Hollwarth ME. Esophagus tissue engineering: hybrid approach with esophageal epithelium and unidirectional smooth muscle tissue component generation in vitro. *J Gastrointest Surg* June 2009;**13**(6):1037–43. PubMed PMID: 19277795.

56. Raghavan S, Lam MT, Foster LL, Gilmont RR, Somara S, Takayama S, et al. Bioengineered three-dimensional physiological model of colonic longitudinal smooth muscle in vitro. *Tissue Eng Part C Methods* October 2010;**16**(5):999–1009. PubMed PMID: 20001822. Pubmed Central PMCID: 2943406.

57. Totonelli G, Maghsoudlou P, Garriboli M, Riegler J, Orlando G, Burns AJ, et al. A rat decellularized small bowel scaffold that preserves villus-crypt architecture for intestinal regeneration. *Biomaterials* April 2012;**33**(12):3401–10. PubMed PMID: 22305104.

58. Zhu Y, Ong WF. Epithelium regeneration on collagen (IV) grafted polycaprolactone for esophageal tissue engineering. *Mater Sci Eng C* 2009;**29**(3):1046–50.

59. Zhu Y, Leong MF, Ong WF, Chan-Park MB, Chian KS. Esophageal epithelium regeneration on fibronectin grafted poly(L-lactide-co-caprolactone) (PLLC) nanofiber scaffold. *Biomaterials* February 2007;**28**(5):861–8. PubMed PMID: 17081604.

60. Levin DE, Barthel ER, Speer AL, Sala FG, Hou X, Torashima Y, et al. Human tissue-engineered small intestine forms from postnatal progenitor cells. *J Pediatr Surg* 2013;**48**(1):129–37.

61. Furness JB. The enteric nervous system and neurogastroenterology. *Nat Rev Gastroenterol Hepatol* 2012;**9**(5):286–94.

62. Sanders KM, Koh SD, Ro S, Ward SM. Regulation of gastrointestinal motility—insights from smooth muscle biology. *Nat Rev Gastroenterol Hepatol* 2012;**9**(11):633–45.

63. Chumpitazi B, Nurko S. Pediatric gastrointestinal motility disorders: challenges and a clinical update. *Gastroenterol Hepatol* 2008;**4**(2):140.

64. Bitar KN, Raghavan S, Zakhem E. Tissue engineering in the gut: developments in neuromusculature. *Gastroenterology* June 2014;**146**(7):1614–24. PubMed PMID: 24681129. Pubmed Central PMCID: 4035447.

65. Bitar KN, Zakhem E. Tissue engineering and regenerative medicine as applied to the gastrointestinal tract. *Curr Opin Biotechnol* October 2013;**24**(5):909–15. PubMed PMID: 23583170. Pubmed Central PMCID: 3723710.

66. Kovesi T, Rubin S. Long-term complications of congenital esophageal atresia and/or tracheoesophageal fistula. *Chest* September 2004;**126**(3):915–25. PubMed PMID: 15364774.

67. Li Y, Zhu Y, Yu H, Chen L, Liu Y. Topographic characterization and protein quantification of esophageal basement membrane for scaffold design reference in tissue engineering. *J Biomed Mater Res B Appl Biomater* January 2012;**100**(1):265–73. PubMed PMID: 22102566.

68. Ackbar R, Ainoedhofer H, Gugatschka M, Saxena AK. Decellularized ovine esophageal mucosa for esophageal tissue engineering. *Technol Health Care* 2012;**20**(3):215–23. PubMed PMID: 22735736.

69. Badylak SF, Vorp DA, Spievack AR, Simmons-Byrd A, Hanke J, Freytes DO, et al. Esophageal reconstruction with ECM and muscle tissue in a dog model. *J Surg Res* September 2005;**128**(1):87–97. PubMed PMID: 15922361.

70. Tan B, Wei RQ, Tan MY, Luo JC, Deng L, Chen XH, et al. Tissue engineered esophagus by mesenchymal stem cell seeding for esophageal repair in a canine model. *J Surg Res* August 17, 2012;**182**(1):40–8. PubMed PMID: 22925499.

71. Liang JH, Zhou X, Zheng ZB, Liang XL. Long-term form and function of neoesophagus after experimental replacement of thoracic esophagus with nitinol composite artificial esophagus. *ASAIO J* May–June 2010;**56**(3):232–4. PubMed PMID: 20449897.

72. Basu J, Mihalko KL, Payne R, Rivera E, Knight T, Genheimer CW, et al. Extension of bladder-based organ regeneration platform for tissue engineering of esophagus. *Med Hypotheses* February 2012;**78**(2):231–4. PubMed PMID: 22100629.

73. Camilleri M, Vazquez-Roque M. Gastric dysmotility at the organ level in gastroparesis. In: Parkman HP, McCallum RW, editors. *Gastroparesis. Clinical gastroenterology*. Humana Press; 2012. p. 37–46.

74. Sachdeva P, Malhotra N, Pathikonda M, Khayyam U, Fisher RS, Maurer AH, et al. Gastric emptying of solids and liquids for evaluation for gastroparesis. *Dig Dis Sci* April 2011;**56**(4):1138–46. PubMed PMID: 21365240.

75. Guerci B, Bourgeois C, Bresler L, Scherrer ML, Böhme P. Gastric electrical stimulation for the treatment of diabetic gastroparesis. *Diabetes Metab* November 2012;**38**(5):393–402.

76. Chu H, Lin Z, Zhong L, McCallum RW, Hou X. Treatment of high-frequency gastric electrical stimulation for gastroparesis. *J Gastroenterol Hepatol* June 2012;**27**(6):1017–26. PubMed PMID: 22128901.

77. Maemura T, Kinoshita M, Shin M, Miyazaki H, Tsujimoto H, Ono S, et al. Assessment of a tissue-engineered gastric wall patch in a rat model. *Artif Organs* April 2012;**36**(4):409–17. PubMed PMID: 22040317.

78. Sala FG, Kunisaki SM, Ochoa ER, Vacanti J, Grikscheit TC. Tissue-engineered small intestine and stomach form from autologous tissue in a preclinical large animal model. *J Surg Res* October 2009;**156**(2):205–12. PubMed PMID: 19665143.

79. Yoshida A, Chitcholtan K, Evans JJ, Nock V, Beasley SW. In vitro tissue engineering of smooth muscle sheets with peristalsis using a murine induced pluripotent stem cell line. *J Pediatr Surg* February 2012;**47**(2):329–35. PubMed PMID: 22325385.

80. Grikscheit TC, Siddique A, Ochoa ER, Srinivasan A, Alsberg E, Hodin RA, et al. Tissue-engineered small intestine improves recovery after massive small bowel resection. *Ann Surg* November 2004;**240**(5):748–54. PubMed PMID: 15492554. Pubmed Central PMCID: 1356478.

81. Grikscheit TC, Ochoa ER, Ramsanahie A, Alsberg E, Mooney D, Whang EE, et al. Tissue-engineered large intestine resembles native colon with appropriate in vitro physiology and architecture. *Ann Surg* July 2003;**238**(1):35–41. PubMed PMID: 12832963. Pubmed Central PMCID: 1422658.

82. Sala FG, Matthews JA, Speer AL, Torashima Y, Barthel ER, Grikscheit TC. A multicellular approach forms a significant amount of tissue-engineered small intestine in the mouse. *Tissue Eng Part A* July 2011;**17**(13–14):1841–50. PubMed PMID: 21395443. Pubmed Central PMCID: 3118603.

83. Qin HH, Dunn JC. Small intestinal submucosa seeded with intestinal smooth muscle cells in a rodent jejunal interposition model. *J Surg Res* November 2011;**171**(1):e21–6. PubMed PMID: 21937060. Pubmed Central PMCID: 3195903.

84. Raghavan S, Bitar KN. The influence of extracellular matrix composition on the differentiation of neuronal subtypes in tissue engineered innervated intestinal smooth muscle sheets. *Biomaterials* August 2014;**35**(26):7429–40. PubMed PMID: 24929617. Pubmed Central PMCID: 4086147.

85. Bharucha AE. Fecal incontinence. *Gastroenterology* May 2003;**124**(6):1672–85. PubMed PMID: 12761725.

86. Nelson R, Norton N, Cautley E, Furner S. Community-based prevalence of anal incontinence. *JAMA* August 16, 1995;**274**(7):559–61. PubMed PMID: 7629985.

87. Frenckner B, Ihre T. Influence of autonomic nerves on the internal and sphincter in man. *Gut* April 1976;**17**(4):306–12. PubMed PMID: 773793. Pubmed Central PMCID: 1411095.

88. Gilmont RR, Raghavan S, Somara S, Bitar KN. Bioengineering of physiologically functional intrinsically innervated human internal anal sphincter constructs. *Tissue Eng Part A* June 2014;**20**(11–12):1603–11. PubMed PMID: 24328537. Pubmed Central PMCID: 4029137.

89. Raghavan S, Gilmont RR, Miyasaka EA, Somara S, Srinivasan S, Teitelbaum DH, et al. Successful implantation of bioengineered, intrinsically innervated, human internal anal sphincter. *Gastroenterology* July 2011;**141**(1):310–9. PubMed PMID: 21463628. Pubmed Central PMCID: 3129458.

90. Raghavan S, Miyasaka EA, Gilmont RR, Somara S, Teitelbaum DH, Bitar KN. Perianal implantation of bioengineered human internal anal sphincter constructs intrinsically innervated with human neural progenitor cells. *Surgery* April 2014;**155**(4):668–74. PubMed PMID: 24582493. Pubmed Central PMCID: 4017655.

Chapter 6

Injury and Repair of Tendon, Ligament, and Meniscus

Jr-Jiun Liou, Mark T. Langhans, Riccardo Gottardi, Rocky S. Tuan

Key Concepts

1. Current clinical interventions for tendon, ligament, and meniscus repair only target short-term pain but cause additional injuries and unfavorable readmission rate.
2. Tissue engineering and regenerative approaches, which usually incorporate cells and scaffolds with bioactive factors, are developed to form functional neotissues.
3. A biocompatible scaffold should promote optimal cell differentiation and extracellular matrix production as well as support mechanical stress of the joint.
4. Several clinical trials are ongoing to investigate safety and feasibility of native or synthetic biomaterials in combination of adult stem cells and platelet-rich plasma.

List of Abbreviations

ACL Anterior cruciate ligament
ASC Adipose-derived stem cells
LCL Lateral collateral ligament
MCL Medial collateral ligament
MSC Mesenchymal stem cell
PCL Posterior cruciate ligament/poly-ε-caprolactone
PLA Polylactic acid
PRP Platelet-rich plasma
PLGA Polylactic glycolic acid

1. INTRODUCTION

Musculoskeletal injuries, including those of the soft tissues such as tendon, ligament, and meniscus, affect more than 10 million people in the United States every year.[1] These injuries remain a significant clinical problem not only due to the intense pain and potential for mobility loss, but also because they represent the primary reason for visits to the doctor's office and contribute to the majority of physical disabilities.[1] The burden of musculoskeletal diseases is well recognized, afflicting more than 20% of the population, higher than cancer and cardiovascular disease combined.[1] Understanding the nature and cause of musculoskeletal soft tissue injuries is thus highly relevant. This chapter reviews the nature and current treatments for such injuries and the knowledge gaps in effective therapies. The potential of tissue engineering and regenerative medicine approaches for such injuries is highlighted with examples featuring both promises and challenges for future investigations.

2. PREVALENT INJURIES OF TENDON, LIGAMENT, AND MENISCUS

2.1 Anatomy of the Knee

The knee represents a paradigmatic example to illustrate the structural and functional importance of skeletal soft tissues in an articular joint (Figure 1). The osseous portions of the knee joint are the femur, tibia, patella, and fibula. Tendons and ligaments

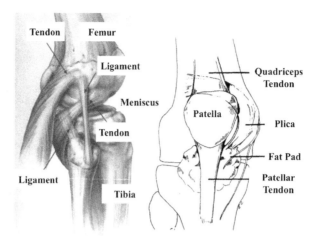

FIGURE 1 Anatomy of the knee. The interconnection between bones, articular cartilage, menisci, ligament, and tendons is illustrated. *Modified from LaPrade et al.*[60] *and Blackburn et al.*[61]

connect these bones to form a functional joint. Tendon connects muscle to the bone, such that upon muscular contraction, tendons pull on bone. Ligaments connect bone to bone, for example, the cruciate ligaments between femur and tibia. Between the bony structures, articular cartilage covers the surface end of the long bones and two horseshoe-shaped menisci serve as intermediate structures between femur and tibia. These components function together to provide a dynamically stable joint.

2.2 Tendon

Tendon is a viscoelastic connective tissue composed of an extracellular matrix of mainly collagen type I, and serves to connect muscle to bone at joints.[2] Common tendon injuries include those of the shoulder rotator cuff, hand flexor tendon, and Achilles tendon (Table 1).[3] The rotator cuff is a group of four muscles with tendons in the shoulder that insert onto the proximal humeral head; the flexor tendons are the extensions of the flexor digitorum superficialis and profundus; and the Achilles tendon is the extension of the gastrocnemius, plantaris, and soleus muscles that insert onto the calcaneus. The type and prevalence of tendon injuries vary depending on the gender, age, daily activity, and overall health of the individual,[2] and are often the result of an internal tensile overloading event or an underuse mechanism. Tendinopathy is characterized by a loss of normal tendon architecture, changes to the normal tenocyte morphology, cellular apoptosis, alterations in collagen fibril distribution profile, and neovascularization (Figure 2).[2] However, reconstructed tendons often fail to restore the native anatomy at the tendon–bone insertion site, resulting in tissue reinjury and continued disability.[4]

2.3 Development of Tendon–Bone Junction

Tendon–bone insertion site is the transition site between tendon and bone. During embryonic development, the formation of the bone and tendon attachment point, known as the enthesis, originates from a unique population of cells expressing the genes, sex-determining region Y box 9 (Sox9), and scleraxis (Scx).[4] The enthesis is structurally represented by four zones of tendon/ligament, nonmineralized fibrocartilage (NF), mineralized fibrocartilage (MF), and bone.[4] Within the NF region, chondrocytes are embedded in a collagen type II- and type III-rich extracellular matrix, while in the MF region, hypertrophic chondrocytes are associated with a collagen type X-rich matrix.[4] The balance between expression of Sox9 and Scx determines the fate of progenitor cells to different cell lineages.[4] If more Scx is expressed, the progenitor cells tend to develop tendon/ligament, as opposed to cartilage.[4] Signaling pathways involved in enthesis development include Indian hedgehog (Ihh) and parathyroid hormone-related peptide (PTHrP), acting in a feedback loop. Transforming growth factor-β (TGF-β) and bone morphogenetic protein also play important signaling roles in enthesis patterning.[4]

Unfortunately, these biological requirements of the tendon–bone insertion site are not satisfactorily addressed by most of the current surgical reconstruction strategies, thus contributing to the high failure rates of surgery.[4,5]

2.4 Ligament

Ligament is the connective tissue that connects bone to bone. Compared to tendon, the organization of collagen fibrils in ligament is more random, with a lower collagen but higher elastin and proteoglycan contents. Collagen type I, collagen type III, and elastin are the main extracellular matrix components of ligament.[6]

TABLE 1 Common Sites of Tendon, Ligament, and Meniscus Injuries

	Tissue Involved	Injuries	Incidence
Tendon	Achilles	Achilles tendinopathy, Achilles paratendinopathy, tendon rupture, calcaneal apophysitis (Sever's disease)	80%
	Patella	Patellar tendinopathy, patellar peritendinopathy, patellar apicitis (jumper's knee), Osgood–Schlatter lesion, Sinding-Larsen–Johansson lesion	
	Posterior tibial	Medial tibial syndrome	
	Iliotibial tract	Iliotibial tract syndrome	
	Biceps femoris, semitendinosus, semimembranosus	Hamstring syndrome	
	Supraspinatus	Supraspinatus syndrome (impingement syndrome, swimmer's shoulder)	
	Other rotator cuff tendons	Rotator cuff tendinopathy or tear	
	Wrist extensors	Lateral epicondylitis (tennis elbow)	20%
	Wrist flexors	Medial epicondylitis (thrower's elbow, golfer's elbow, little league elbow)	
Ligament	ACL	ACL tear	46%
	MCL	MCL tear	29%
	ACL+MCL complex, PCL, LCL	Ligament tear	25%
Meniscus	Medial meniscus	Medial meniscus tear	N/A
	Lateral meniscus	Lateral meniscus tear	N/A

ACL, anterior cruciate ligament; PCL, posterior cruciate ligament; LCL, lateral collateral ligament; MCL, medial collateral ligament.
Modified from Galloway et al.[2] and Maffulli et al.[65]

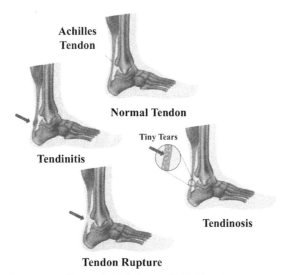

FIGURE 2 Common injuries of tendon. Common Achilles tendon injuries are divided into three groups: tendinitis, tendinosis, and rupture. *Modified from Sharma and Maffulli.*[62]

In the upper extremity, injuries to the acromioclavicular joint of the shoulder include ruptures of the coraco-, acromio-, and coracoacromial ligaments.[7] In the elbow, rupture of the ulnar collateral ligament is a common sports-related injury that requires reconstructive surgery.[8] Commonly injured wrist ligaments include the scapholunate and lunotriquetral ligaments.[9]

In the lower extremity, the lateral ankle ligaments, including the anterior talofibular ligament and the lateral collateral fibular ligament, are the most commonly injured ligaments that can be effectively managed with conservative treatment

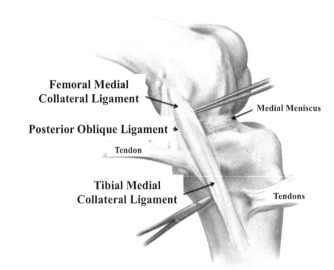

FIGURE 3 Common injuries of ligament. Ligament tears include mild ligament tear (first degree), moderate ligament tear (second degree), and severe/ruptured ligament tear (third degree) based on injury level. Common injury sites are medial collateral ligament (MCL), anterior cruciate ligament (ACL), and posterior cruciate ligament (PCL). *Modified from LaPrade et al.*[63]

in all but the most extreme cases (complete multiligament tears with gross instability).[10] Common injuries to ligaments of the knee include the anterior and posterior cruciate ligaments (ACL and PCL) and medial and lateral collateral ligaments (MCL and LCL) (Figure 3).

The most frequently repaired ligament is the ACL of the knee. In the United States, roughly 250,000 individuals suffer from ACL injuries annually, with more than 80% electing to undergo reconstruction for partial or complete tears. The costs associated with diagnostic magnetic resonance imaging, reconstructive surgery, and 6 months of rehabilitation for this injury are approximately $2 billion annually in the United States.[11] While return-to-activity rates are high, ACL reconstruction is ineffective in reducing the incidence of osteoarthritis (OA).[12]

While a number of studies have examined and compared the effects of surgical reconstruction techniques including anatomical ACL, double-bundle, and autograft or allograft options, recent efforts have also focused on developing regenerative medicine approaches to augment ligament healing of partially torn ligaments and to enhance performance and longevity of the reconstructed ligaments.

Together, tendon and ligament injuries represent 45% of the 32 million musculoskeletal injuries reported annually in the United States. In 2004, a study attempting to enumerate the intervention of these injuries reported 51,000 shoulder rotator cuff repairs, 44,000 Achilles tendon repairs, and 42,000 patellar tendon repairs being performed every year in the United States.[13]

2.5 Meniscus

Menisci, or meniscal cartilages, consist of two semilunar fibrocartilage discs. As the soft tissue of the superior tibia that absorbs shock at the knee joint, the meniscus is composed of approximately 30% of various collagen types (mainly collagen types I and II) and 70% of water. Common meniscal injuries include longitudinal, radial, and oblique tears (Figure 4). Longitudinal tear is a tear along inner layer of meniscus, whereas a radial tear is perpendicular to the inner or outer layer. Oblique tears, also known as flap tears, are defined by a small portion of meniscus separating from the meniscal tissue. Meniscal tears can be classified as acute or degenerative, the former resulting from excessive force applied to a normal knee and meniscus, while the latter results from repetitive normal forces acting upon a worn down meniscus. A prospective study in 2007 reported that the presence of Outerbridge II chondral degeneration was associated with degenerative meniscal tears 85% of the time, compared to 12% for flap tears and 0% for longitudinal tears.[14,15]

Table 1 tabulates the common injury sites of tendon, ligament, and meniscus, listed based on the clinical data as of 2003. As mentioned above, the type and prevalence of injuries are dependent on daily activity levels of the affected individuals.

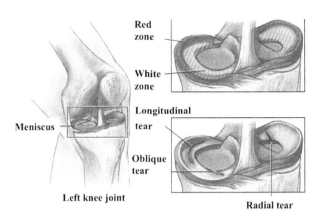

FIGURE 4   Common injuries of meniscus. The structure of the meniscus includes three zones: highly vascularized red zone (outer layer), red-to-white zone, and white zone (inner layer). Common meniscus injury sites are longitudinal, radial, and oblique tears. *Modified from Mesiha et al.*[14]

TABLE 2 Composition and Mechanical Characteristics of Tendon, Ligament, and Meniscus

		Tendon		Ligament		Meniscus	
Cells		Tenocytes (fibroblasts)		Ligamentocytes (fibroblasts)		Fibrochondrocytes	
Components	Water	60–80%		60–80%		72%	
	Collagen type I	20–40%	95–99%	20–40%	90%	22%	90%
	Collagen type III		1–5%		10%		10%
	Elastin		~0%		<1%		<1%
Young's modulus		1200–1800 MPa		100–300 MPa		0.2 MPa	

Modified from Fox et al.[16] and Hasan et al.[66]

3. TENDON, LIGAMENT, AND MENISCUS INJURIES AND JOINT FUNCTION

Musculoskeletal soft tissues usually contain large amounts of water for flexibility and elasticity. These tissues generally contain fibroblasts residing in highly collagenous matrices, with the fibroblasts in menisci referred to as fibrochondrocytes because of their chondrocyte-like characteristics.

Adequate and characteristic mechanical properties are critical for the function of these soft tissues since they need to withstand constant pressure and continuous, intense impacts on the joint. A typical tendon fails at 12–15% strain and 100–150 MPa stress. When the strain increases from 2% to 8%, the collagen fibers of tendon transit from straightening to rupture.

Viscoelasticity is a characteristic of a viscous and elastic material when undergoing deformation; viscoelastic responses to load depend on the magnitude and duration of the load and tissue composition. Typically, the higher the glycosaminoglycan or proteoglycan content of the tissue, the more stress is required for compression at the same strain. Compared to hyaline cartilage, the tensile modulus of knee meniscus is much higher due to its higher proteoglycan content. The tensile strength of ligament varies but is usually less than that of tendons. The elastic modulus of the ligament ranges from 100 to 300 MPa, whereas that of the tendon ranges from 1200 to 1800 MPa. Table 2 summarizes current information on the cell types, the extracellular matrix components, and some mechanical characteristics of tendons, ligaments, and menisci.[16]

Following a tendon injury, the natural healing process forms scar tissue via inflammation, matrix production, remodeling, and maturation. At first, fibrin starts to clot and immune cells, such as neutrophils and macrophages, migrate to the injury site in response of injury, termed as the inflammation stage.[2] The matrix production stage follows immediately via fibroblast localization, collagen, and extracellular matrix protein synthesis and continues for approximately 4 weeks.[2] Eventually, in the remodeling stage, the microenvironment change as the collagen fibrils rearrange, cell density and tissue vascularization decrease, and scar tissue forms.[2]

4. CURRENT CLINICAL INTERVENTIONS

Current intervention of soft tissue injuries can be nonsurgical or surgical. Nonsurgical techniques for tendon, ligament, and meniscus repair include RICE (rest, ice, compression, elevation),[17] physical therapy (strengthening exercise), the use of braces and crutches, and the administration of nonsteroidal anti-inflammatory drugs. Commonly used disease-modifying antirheumatic drugs (DMARDs) for degenerative joint diseases include methotrexate (Rheumatrex, Trexall),[18–20] hydroxychloroquine (Plaquenil),[20] and sulfasalazine (Azulfidine). However, side effects of these DMARDs, such as rash and nausea, are often reported.[21]

4.1 Tendon and Ligament

Surgical techniques for tendon or ligament repair include direct repair via sutured reattachment, augmentation (autografts, allografts, or xenografts), fixation (endobuttons), and biological or synthetic scaffolds embedded with cells. Autograft implantations are common but not preferred because of the propensity for additional tendon injuries as well as marked pain and neurologic damage.[22] Allografts, such as GraftJacket® (Wright Medical Technology) are derived from human cadaveric dermis. The US Food and Drug Administration (FDA)-approved Restore™ (DePuy Synthes)[23] and CuffPatch® (Arthrotek) are xenografts from porcine small intestine submucosa.[24] Fixation methods, similar to direct sutures, are used to stabilize the joints.[25] Opus® AutoCuff® device (ArthCare Sports Medicine) is an example of fixation methods for rotator cuff repair.[26]

   Last, natural and synthetic biomaterial scaffolds are also available for clinical applications. A large number of commercial products are available for ACL repair or reconstruction. Some examples include the following. Biological SeriScaffold™ (Serica Technologies) is a silk-based scaffold for ACL reconstruction.[27] BioCleanse® anterior and posterior tibialis tendons (RTI Surgical; NCT00975845, recruiting for Phase IV clinical trial) are tendon allografts from 18–65 years old donors. However, these products, including BioCleanse® and Allowash XG®(LifeNet Health), are produced using disinfection and sterilization methods, and may still have immunogenic concern after implantation. Smith and Nephew bioabsorbable ACL interference screws BIORCI® are made of polylactic acid (PLA) and range from 23 to 25 mm in length and are 7 mm wide. Bilok® screws (ArthroCare Sports Medicine) are made of PLA/β-tricalcium phosphate, and their use has been based on radiographic evaluation assessed 3 years after surgery.[28] Ligament Advanced Reinforcement System (LARS, Arc sur Tille), a polyethylene terephthalate graft, was popular in 1980s–1990s, but some synovitis and graft rupture cases were reported after surgery.[29]

   The limitations of the commercially available natural biomaterial scaffolds are low mechanical properties, nonspecific induction ability, and undefined degradation rate; on the other hand, those comprised of synthetic scaffolds have poor biocompatibility, poor cell infiltration, and risk of infection/immune response.[24]

4.2 Menisci

Surgical techniques for meniscus repair include menisectomy, sutures, or screws. Menisectomy involves the removal of all or part of an injured meniscus. During menisectomy, an arthroscope is inserted into the joint and damaged portions of the meniscus are ablated (Figure 5). However, high 5-year readmission rates remain a problem following menisectomy due to

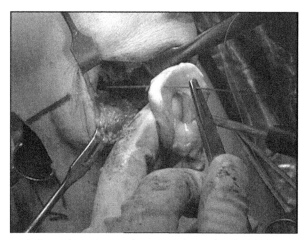

FIGURE 5   Interventions for meniscus repair. In menisectomy, an arthroscope is inserted to the joint to remove damaged meniscus after a small skin incision. If a meniscus transplant is available, the new meniscus will be sutured and anchored onto the knee bone. *Modified from Paxton et al.*[64]

joint space narrowing and reduction in meniscal thickness. Human meniscal allograft transplantations are also available. In these procedures, healthy cartilage or ligament are excised and sutured or screwed into replace diseased meniscus. This technique can reduce pain immediately, but the risk of meniscus tear after a few years remains high. Some collagen-based scaffolds are approved for meniscus repair by the FDA, but the natural biodegradability of collagen results in unfavorable rate of readmission. Some surgeons have been reported to decline to use collagen-based scaffolds for meniscal repair, particularly when involving total menisectomy, due to low level of clinical and research evidence.

5. TISSUE ENGINEERING AND REGENERATIVE THERAPEUTIC APPROACHES FOR INJURIES OF TENDON, LIGAMENT, AND MENISCUS

The field of tissue engineering was introduced more than two-and-a-half decades ago to describe the combination of biological principles and engineering technologies for the purpose of developing functional replacement of degenerated, injured, or diseased tissues and organs. Regenerative medicine refers to clinical applications of these therapeutic procedures. Successful tissue engineering and regenerative medicine approaches require cells, scaffolds or matrices, and bioactive factors, that when combined can facilitate the productive, conductive, and inductive activities, respectively, of tissue regeneration. Table 3 summarizes current clinical trials with cells, scaffolds, bioactive factors, and combination approaches for musculoskeletal tissue repair.

5.1 Cells

A combinatorial platform of cells, scaffolds, and biofactors has been conceptualized for tissue engineering and regenerative medicine in the past two decades and is described here. While endogenous cells of the target tissues are obvious candidate cells for regenerative and reparative applications, the intrinsic cytopaucity of soft skeletal tissues poses serious challenges to such approaches. Tendon cells or tenocytes are elongated fibroblasts. Cells in the ligament are also fibroblasts. Ligament stem cells from the periodontal ligament have been identified as having the capacity of self-renewal and differentiation.[30] Cells of the meniscus are fibrochondrocytes, exhibiting properties of both fibroblasts and chondrocytes.[31] However, cell densities in these skeletal connective tissues are intrinsically low, and it is thus difficult to harvest sufficient number of cells. Expansion of cell number by in vitro culture also has undesirable consequences, primarily because of the loss of cell phenotype as a function of culture passage.

Adult multipotent stem/progenitor cells, by virtue of their extensive proliferative activity and multilineage differentiation potential, are thus considered prime candidate cells for tissue engineering and regenerative medicine.[32] For musculoskeletal applications, adult stem cells, known as mesenchymal stem cells (MSCs) and most commonly derived from bone marrow and adipose tissue, are of particular interest due to the relative ease of their isolation and enrichment, safety, and freedom from ethical issues. MSCs cultured in the presence of specific growth factors and inductive supplements may be induced to differentiate into specific musculoskeletal tissue lineages, including chondrocytes and tenocytes. For the induction and enhancement of MSC chondrogenesis, a number of growth factors have been shown to be highly active, in particular, TGF-β. Of particular interest is the recent attention on platelet-enriched products, including plasma platelet-rich plasma (PRP) and platelet-rich fibrin (PRF) as bioactive additives to enhance the repair of cartilage, meniscus, ligament, and tendon (see below).

Adipose-derived stem cells (ASCs) have been widely studied because of the abundance of fat disposed after surgery. A Phase II clinical trial for tendon regeneration is recruiting to determine the effect of ASCs compared to placebo group and no treatment group (NCT02298023). Another Phase I/II clinical trial is focused on recovery after partial medial menisectomy with injection of different dosages of MSCs to intra-articular sites, with a placebo control group (NCT00702741).

MSCs may be derived from either autologous or allogeneic sources. Open surgery, ex vivo cell expansion, and iatrogenic damage caused by autologous stem cell harvesting are not preferable. The other disadvantage of using autologous stem cells, particularly from elderly individuals, relates to the age-associated reduction in proliferative and differentiation activities of MSCs. A number of clinical trials are currently testing the applicability of human allogeneic stem cells, some of which are listed in Table 4.

5.2 Bioactive Factors

Bioactive factors are critical in regulating the differentiation of stem cells to target cell lineages. For applications in skeletal soft tissues, the most common bioactive factors are PRP, PRF, and the orthokines. PRP, isolated from human whole blood as a cocktail of growth factors released from activated platelets, has been used in clinics since 1990s. Although the use of PRP treatment for tissue healing and regeneration has become clinically popular in the last decade, scientific

TABLE 3 Examples of Current Clinical Trials on Skeletal Soft Tissues (as of March, 2015) (www.clinicaltrials.gov)

	Condition	Intervention	Phase and Status	Comments	Identifier
Cells					
Meniscus	Recovery following partial medial menisectomy	Low dose MSC vs high dose MSC vs placebo	Phase 1/2, completed in 2011, no results posted	Higher dosage (150 million) of hMSCs could possibly provide a better wound healing than lower dosage or control.	NCT00702741
Ligament	ACL injury OA	MSC + HA vs HA	Phase 1/2, ongoing but not recruiting	This trial is to evaluate the safety and efficacy of MSC injection after ACL reconstruction.	NCT01088191
Tendon	Rotator cuff tear	ASC vs placebo vs no treatment	Phase 2, recruiting	This is the only trial that includes no treatment group for comparison	NCT02298023
Bioactive Factors					
Meniscus	Meniscal injury	PRF vs PRF + suture vs no PRF	Phase unknown, enrolling	PRF + suture group is expected to perform better than other groups.	NCT01211119
Meniscus	Cartilage disease	PRP vs no PRP	Phase unknown, withdrawn	The study has been withdrawn prior to enrollment in December 2014.	NCT00961597
Ligament	Postoperative pain	PRP vs placebo	Phase 4, recruiting	PRP may improve tendon healing after ACL reconstructive surgery using their own patellar tendons.	NCT01765712
Tendon	Knee stability	PRP vs control	Phase 4, completed in 2010, no results posted	27 subjects were evaluated for patellar tendon regeneration 6 months after surgery, but no results have been posted.	NCT01111747
Scaffolds					
Meniscus	Other tear of medial meniscus, current injury	Meniscus scaffold only	Phase unknown, not recruiting	FibroFix™ meniscus from Orthox Limited, UK was evaluated for its initial safety, and estimated enrollment is 10.	NCT02205645
Ligament	Complete tear, knee, ACL	Resorbable PLA implant only	Phase 2, ongoing but not recruiting	L-C Ligament® is a product of Soft Tissue Regeneration, Inc. There is another phase 2/3 trial of L-C Ligament versus hamstring autograft (NCT02183727).	NCT01634711
Tendon	Diabetes foot ulcer	Bovine collagen matrix on wound bed vs on wound bed and injected subcutaneously	Phase 4, terminated as the wound sizes decreased in 12 weeks	Integra LifeSciences Co.'s product INTEGRA™, made of cross-linked collagen and GAG, was tested in this trial. Matrix distributed locally and subcutaneously is expected to have better performance.	NCT01108263
Combination					
Tendon	Rotator cuff tear	MSCs + collagen membrane vs collagen membrane	Phase 2, status unknown	OrthADAPT® (collagen type I membrane) is expected to provide support for MSCs and improve rotator cuff tear restoration.	NCT01687777
All	Rheumatoid arthritis (RA)	MSCs + DMARD vs MSCs vs DMARD	Phase 1/2, status unknown	Inclusion of umbilical cord-derived MSCs is expected to have more anti-inflammatory effects and potentially alleviate the progression of RA, compared to disease-modifying antirheumatic drugs alone.	NCT01547091
All	OA, knee	MSC + PRP vs PRP	Phase 1/2, recruiting	This trial is to compare the efficacy of MSCs enhanced with PRP in early stages of knee OA. For more stem cell treatments for OA, see Table 4.	NCT01985633

MSC, mesenchymal stem cell; ACL, anterior cruciate ligament; MSC, mesenchymal stem cell; PRP, platelet-rich plasma; PRF, platelet-rich fibrin; PLA, polylactic acid; DMARD, disease-modifying antirheumatic drug; hMSCs, human mesenchymal stem cells; OA, osteoarthritis; HA, hyaluronic acid.

TABLE 4 Examples of Current Clinical Trials of MSCs for the Treatment of OA and Related Joint Defects (as of March, 2015) (www.clinicaltrials.gov)

Trial	Intervention	Phase; Current Stage	Comment	Identifier
Treatment of knee OA with allogeneic MSCs	Allogeneic MSC vs HA	Phase 1/2, completed	Results showed that MSC group had effective and durable pain relief and objective cartilage improvement, compared to HA group.	NCT01586312
Intra-articular autologous bone marrow MSC transplantation to treat mild to moderate OA	Autologous MSC + HA vs HA	Phase 2, unknown	Implantation of autologous bone marrow-derived MSCs is expected to be more effective than control group.	NCT01459640
Intra-articular injection of MSCs in knee joint OA	Autologous MSCs vs placebo	Phase 2, completed, no results posted	Intra-articular injection of MSCs is expected to have better improvement than placebo group six months after first injection.	NCT01504464
Allogeneic MSCs in OA	Allogeneic MSCs + Plasmalyte-A + HA vs Plasmalyte-A	Phase 2, unknown	Single intra-articular dose of Plasmalyte-A serves as placebo comparison in this trial, and allogeneic MSCs group is expected to improve OA two years after injection.	NCT01453738
Adult stem cell therapy for repairing articular cartilage in gonarthrosis	MSC group only	Phase 1/2, completed, no results posted	Bone marrow-derived MSCs	NCT01227694
Autologous adipose tissue-derived MSCs transplantation in patient with degenerative arthritis	Autologous MSC group only	Phase 1/2, completed, no results posted	Adipose tissue-derived MSCs	NCT01300598
Compare the efficacy and safety of Cartistem® and microfracture in patients with knee articular cartilage injury or defect	MSCs vs microfracture	Phase 3, completed, no results posted	Cartistem® is a commercial MSC product from Medipost Co Ltd. MSC group is expected to have better healing for articular cartilage defects.	NCT01041001
ADIPOA—clinical study	Adipose stem cells: 2 million vs 10 million vs 50 million cells/5 ml	Phase 1, completed, no results posted	Intra-articular injection of different adipose stem cell dosages to see the efficacy for OA therapy.	NCT01585857
Transplantation of bone marrow stem cells stimulated by protein scaffold to heal defects articular cartilage of the knee	MSC group only	Phase 0, unknown	Phase 0 is to evaluate pharmacodynamics and pharmacokinetics and half-life of the drug, i.e., the MSC group.	NCT01159899

MSC, mesenchymal stem cell; OA, osteoarthritis; HA, hyaluronic acid.

evidence remains insufficient with many conflicting results reported on its efficacy.[33] A number of reasons contribute to the observed variability, including variations in PRP composition, cellularity of the damaged or diseased target tissue, and the lack of controlled studies.[34] Several trials have reportedly focused on assessment of PRP effects compared to placebo controls. For example, a Phase IV clinical trial for patellar tendon regeneration in Brazil used PRP at the end of ACL reconstruction (NCT01111747). The trial was completed in 2010, but no results were posted. Another trial, currently recruiting, is attempting to determine if PRP improves postoperative pain after ACL surgery compared to a placebo group (NCT01765712).

PRF, also called fibrin clot, is used to improve tissue regeneration as well. The use of PRF is considered more desirable than the use of PRP since bovine thrombin and anticoagulants are not required. However, PRF applied to tendon–bone resulted in no significant tendon healing for rotator cuff repair.[35]

Recently, leukocytes were reported to be an inhibitory factor in PRP, acting to suppress chondrogenesis. A research study compared PRP with leukocyte-reduced PRP, and concluded that leukocytes in PRP downregulated the synthesis of collagen types I and III.[37] PRP also inhibited the mechanically induced catabolic and inflammatory responses in chondrocytes in vitro, with early addition of PRP showing the most significant benefits.[38]

5.3 Scaffolds

In addition to cells and bioactive factors, a biocompatible scaffold is beneficial in promoting cell growth and differentiation. Materials used as scaffolds include synthetic materials that degrade over time and native biomaterials derived from tissues. Poly(lactic-co-glycolic acid) (PLGA), poly(L-lactic acid) (PLLA), and poly(ε-caprolactone) (PCL) are the most common synthetic biodegradable polymers in use. Kuo et al. summarized current synthetic polymeric scaffolds and native biomaterials in tendon and ligament tissue engineering.[39] PLLA showed better performance compared to PLGA scaffold in terms of degradation profiles.[39] A recent study adopted a rational design of nanofiber scaffolds for tendon regeneration using PLGA and seeded with stem cells and when implanted into different animal models reported favorable outcomes; however, the prevention of adhesion after repair was not investigated.[5]

Native biomaterials derived from the extracellular matrix, such as collagen and fibrin, are also of common use. Silk fibroin, a native biomaterial, has favorable mechanical properties supporting cell and tissue ingrowth.[39] Decellularized tendon/ligament has also been studied to provide a more homogenous and porous scaffold.[40] A water-soluble extracellular matrix extract prepared from decellularized bovine tendons has been reported to promote adipose stem cell proliferation and tenogenic differentiation.[41]

Some products from soft tissue engineering companies, for example, FibroFix™ (Orthox; meniscus scaffold), were tested in a clinical trial to treat medial meniscus tears (NCT02205645). L-C ligament® from Soft Tissue Regeneration, Inc. used resorbable PLLA scaffold to treat ligament tears (NCT01634711). INTEGRA™ (Integra LifeSciences), made of cross-linked collagen and glycosaminoglycan, was tested in a Phase IV clinical trial; however, the trial was terminated because the wound size decreased after 12 weeks (Table 5).

5.4 Combination

For optimal tissue engineering, multipotent stem cells are exposed to appropriate bioactive factors and molecular environments to enhance optimal and mature matrix production. The most common approach is to seed stem cells into a biodegradable and biocompatible scaffold, with concomitant or subsequent treatment with growth factors, thereby promoting differentiation towards specific cell phenotypes and production of tissue-specific extracellular matrix that will compensate for the scaffold degradation, and ultimately resulting in a functional neotissue.

As a biomaterial, PRP may be considered both as a depot of bioactive factor as well as a biodegradable scaffold. In a recent study, PRP scaffolds embedded with bone marrow or adipose-derived MSCs were evaluated in a rabbit joint defect model. Twelve weeks after surgery, chondrogenic markers such as collagen type II and aggrecan strongly suggested PRP scaffold as a model capable of releasing endogenous growth factors as it degrades. However, the limitation includes its rapid degradation rate and low mechanical stiffness.[34]

Reliable means to assess the delivery of bioactive factors is critical. A number of studies have examined MSCs seeded in PLLA or polyethylene glycol scaffolds and cultured in conditioned medium for up to 4 weeks in terms of induction to undergo chondrogenic or osteogenic differentiation.[41–46] Saito et al. have reported the positive effects of injections of PRP in gelatin microspheres in ACL transection rabbit models.[47] Lin et al. have also developed a photocross-linkable and injectable gelatin hydrogel capable of encapsulating bone marrow-derived MSCs for cell-based repair of articular cartilage defects.[48] Such cartilage-based test platforms may be applied to tendon or ligament repair.

TABLE 5 Review of Current Common Biomaterials for Tendon, Ligament, and Meniscus Repair

	Scaffold	Experiment	Result and Comment	References
Synthetic	PLA	Autologous ASCs in PGA/PLA scaffolds in rabbit model for patellar tendon defect regeneration	ASCs-embedded matrix had tensile strength of 50 MPa and more matrix production. However, the scaffold degradation test, cell infiltration test, and inflammatory profile did not include.	67
	PLGA	MSCs in PEG hydrogel wrapped in meshes of PLGA and poly(ester-urethane urea) (PEUUR) in vitro for ligament regeneration	The scaffold improved the porosity and cellularization but the mechanical strength was weaker than previous reported scaffolds. PEUUR was not FDA approved yet.	68
	PCL	Fibrochondrocytes in PCL in vitro for meniscus regeneration	Fiber alignment and porosity was evaluated in a cellularized scaffold and it provided a potential of off-the-shelf tissue-engineered meniscus.	69
Native	Collagen	ACL fibroblasts in collagen scaffold in vitro for ACL regeneration	A thorough characterization of MMP1, MMP3, MMP13 synthesis was performed for future scaffold application.	70
	Collagen	PRP injections to collagen scaffold in minipig model for ACL reconstruction	The results showed that increased platelet concentration in collagen-based implants did not improve ACL regeneration after ACL transection.	71
Combination	PLA/Collagen	Tendon-derived stem cells (TSCs) in poly(lactide-co-caprolactone)/collagen (PLACL/Col) scaffolds in rabbit model for Achilles tendon repair	TSCs showed significant potential to develop tissue-engineered tendons. However, TSCs are not as practically as autologous ASCs or BM-MSCs.	72

PGA, polyglycolic acid; PLA, polylactic acid; MSCs, mesenchymal stem cells; PEG, polyethylene glycol; PLGA, polylactic glycolic acid; FDA, US Food and Drug Administration; PCL, poly-ε-caprolactone; ACL, anterior cruciate ligament; PRP, platelet-rich plasma; ASCs, adipose-derived stem cells; BM-MSCs, bone marrow-derived mesenchymal stem cells; MMP, matrix metalloproteinase.

Zhang et al. investigated the anti-inflammatory effects of PRP on tendon inflammation using both tendon stem/progenitor cell culture and a rat model. The results showed that with PRP treatment, tendon stem/progenitor cells differentiated into active tenocytes and nontenocytes, and did not undergo chondrogenesis, adipogenesis, or osteogenesis. In addition, PRP treatment resulted in reduced prostaglandin E2 reduction, which could be beneficial for tendon healing.[49]

Regarding scaffold design and fabrication, a widely used method is electrospinning for the production of fibrous scaffolds.[50–52] For bulk polymeric constructs, particulate leaching and phase separation are generally used to form scaffolds. Three-dimensional printing technologies have also recently gained significant attention.[53] Three-dimensional printing mimics the structure of defects or target organs. Lin et al. used projection stereolithographic methods to develop stem cell-based cartilaginous constructs[54] and to fabricate microphysiological osteochondral tissues.[55] Brown et al. reconstructed upper airway cartilage using computed tomography and cell-seeded construct fabrication, resulting in a structurally similar extracellular matrix composition compared to native tissue.[56]

6. CONCLUSIONS AND FUTURE DIRECTIONS

Current clinical treatments for tendon, ligament, and meniscus injuries mostly only target short-term pain and provide temporary restoration of joint function. Surgical procedures, in addition to being destructive in nature, are also associated with risks of infection, inflammation, postsurgery failure, and need for surgical revision. We have reviewed here, the principles and progress of current tissue engineering approaches that have the potential to repair or regenerate skeletal soft tissues, specifically tendon, ligament, and meniscus. Critical to the success of a regenerative medicine approach is the development of a biomaterial scaffold that, with the appropriate delivery of requisite growth and inductive factors, can provide cells with the optimal environment to grow and differentiate, and to regenerate a neotissue that is able to mature and remodel in response to the physiology of the native tissue.

A number of native and synthetic scaffolds are under investigation for tissue regenerative applications. Native biopolymeric scaffolds have natural extracellular matrix microstructure, which promotes cell adhesion, migration, infiltration, proliferation, and differentiation. However, poor mechanical properties and nonspecific and unregulated inductive activities are the limitations of current biological scaffolds. Synthetic, or polymeric, scaffolds have defined mechanical properties with quality consistency, but are not as biocompatible as native matrix-derived scaffolds. Future development in scaffold technology thus needs to be grounded in the natural history of soft tissue development and biology.[3,39,57]

In terms of bioactive factors, the use of PRP remains controversial. Although PRP is a natural product, obtained autologously and rapidly implemented, and requires no regulatory restriction, more investigations, are clearly needed to yield evidence that indicates its effectiveness and defines the most appropriate procedure for tissue in tendon, ligament and meniscus regeneration. In comparison, the bioactive factor, TGF-β, while having shown efficacy in improving tissue differentiation and regeneration in cell culture and animal models, does not have FDA approval for clinical use.

The most promising cell type for skeletal soft tissue engineering and regeneration is the MSC, derived from bone marrow or other tissues, such as adipose. MSCs that have undergone minimal or no culture expansion are preferred as they possess more robust proliferative and differentiation potential. MSCs are optimally delivered in 3D scaffolds that are formed in a manner that allows cell encapsulation, and recent developments in 3D bioprinting present interesting options in scaffold design and fabrication. The potential of using allogeneic MSCs should also be explored, particularly given the extensive literature supporting the low, intrinsic immunogenicity of MSCs.[58,59]

In conclusion, contemporary advances in stem cell biology, developmental biology, and scaffold technologies have provided exciting possibilities in developing regenerative therapeutic approaches for the repair of injuries to tendon, ligament, and meniscus. The potential of functional restoration will significantly alleviate the disease burden of these debilitating injuries and improve the quality of life of a significant segment of society. Sustaining the breakthroughs through continuous convergence of life science and engineering is critical to reach these goals.

ACKNOWLEDGMENTS

Supported in part by National Institutes of Health (5U18 TR000532, 5R01AR062947, 5T32HL076124), the Commonwealth of Pennsylvania Department of Health (SAP4100050913), the US Department of Defense (W81XWH-10-1-0850, W81XWH-13-2-0030, W81XWH-14-2-0003), and the Ri.MED Foundation (Italy).

REFERENCES

1. National Research Council and the Institute of Medicine. *Musculoskeletal disorders and the workplace: low back and upper extremities. Panel on musculoskeletal disorders and the workplace. commission on behavioral and social sciences and education.* Washington, DC: National Academy Press; 2001.
2. Galloway MT, Lalley AL, Shearn JT. The role of mechanical loading in tendon development, maintenance, injury, and repair. *J Bone Joint Surg Am* 2013;**95**(17):1620–8.
3. Yang G, Rothrauff BB, Tuan RS. Tendon and ligament regeneration and repair: clinical relevance and developmental paradigm. *Birth Defects Res C Embryo Today Rev* 2013;**99**(3):203–22.
4. Rothrauff BB, Tuan RS. Cellular therapy in bone-tendon interface regeneration. *Organogenesis* 2014;**10**(1):13–28.
5. Ma B, Xie J, Jiang J, Shuler FD, Bartlett DE. Rational design of nanofiber scaffolds for orthopedic tissue repair and regeneration. *Nanomedicine (London)* 2013;**8**(9):1459–81.
6. Freedman BR, Gordon JA, Soslowsky LJ. The Achilles tendon: fundamental properties and mechanisms governing healing. *Muscles Ligaments Tendons J* April 2014;**4**(2):245–55.
7. Stucken C, Cohen SB. Management of acromioclavicular joint injuries. *Orthop Clin North Am* January 2015;**46**(1):57–66.
8. Bruce JR, Andrews JR. Ulnar collateral ligament injuries in the throwing athlete. *J Am Acad Orthop Surg* May 2014;**22**(5):315–25.
9. Tanaka T, Ogino S, Yoshioka H. Ligamentous injuries of the wrist. *Semin Musculoskelet Radiol* December 2008;**12**(4):359–77.
10. Freymann U, Petersen W, Kaps C. Cartilage regeneration revisited: entering of new one-step procedures for chondral cartilage repair. *OA Orthop* 2013;**1**(1):1–6.
11. Silvers HJ, Mandelbaum BR. Prevention of anterior cruciate ligament injury in the female athlete. *Br J Sports Med* 2007;**41**(Suppl. 1):i52–9.
12. Lohmander LS, Englund PM, Dahl LL, Roos EM. The long-term consequence of anterior cruciate ligament and meniscus injuries: osteoarthritis. *Am J Sports Med* 2007;**35**(10):1756–69.
13. Butler DL, Juncosa N, Dressler MR. Functional efficacy of tendon repair processes. *Annu Rev Biomed Eng* 2004;**6**:303–29.

14. Mesiha M, Zurakowski D, Soriano J, Nielson JH, Zarins B, Murray MM. Pathologic characteristics of the torn human meniscus. *Am J Sports Med* January 2007;**35**(1):103–12.

15. Howell R, Kumar NS, Patel N, Tom J. Degenerative meniscus: pathogenesis, diagnosis, and treatment options. *World J Orthop* November 18, 2014;**5**(5):597–602.

16. Fox AJS, Bedi A, Rodeo SA. The basic science of human knee menisci: structure, composition, and function. *Sport Heal* 2012;**4**(4):340–51.

17. Bass E. Tendinopathy: why the difference between tendinitis and tendinosis matters. *Int J Ther Massage Bodyw* January 2012;**5**(1):14–7.

18. Atzeni F, Benucci M, Sallì S, Bongiovanni S, Boccassini L, Sarzi-Puttini P. Different effects of biological drugs in rheumatoid arthritis. *Autoimmun Rev* March 2013;**12**(5):575–9.

19. Roubille C, Richer V, Starnino T, McCourt C, McFarlane A, Fleming P, et al. The effects of tumour necrosis factor inhibitors, methotrexate, nonsteroidal anti-inflammatory drugs and corticosteroids on cardiovascular events in rheumatoid arthritis, psoriasis and psoriatic arthritis: a systematic review and meta-analysis. *Ann Rheum Dis* March 2015;**74**(3):480–9.

20. Goodman SM. Rheumatoid arthritis: perioperative management of biologics and DMARDs. *Semin Arthritis Rheum* January 30, 2015;**44**(6).

21. García-Hernández MH, González-Amaro R, Portales-Pérez DP. Specific therapy to regulate inflammation in rheumatoid arthritis: molecular aspects. *Immunotherapy* January 2014;**6**(5):623–36.

22. Makhni EC, Steinhaus ME, Mehran N, Schulz BS, Ahmad CS. Functional outcome and graft retention in patients with septic arthritis after anterior cruciate ligament reconstruction: a systematic review. *Arthroscopy* February 26, 2015. http://dx.doi.org/10.1016/j.arthro.2014.12.026.

23. Namdari S, Melnic C, Huffman GR. Foreign body reaction to acellular dermal matrix allograft in biologic glenoid resurfacing. *Clin Orthop Relat Res* August 2013;**471**(8):2455–8.

24. Longo UG, Lamberti A, Maffulli N, Denaro V. Tendon augmentation grafts: a systematic review. *Br Med Bull* 2010;**94**(1):165–88.

25. Li Q, Hsueh P, Chen Y. Coracoclavicular ligament reconstruction: a systematic review and a biomechanical study of a triple endobutton technique. *Medicine (Baltimore)* December 2014;**93**(28):e193.

26. Redziniak DE, Hart J, Turman K, Treme G, Hart J, Lunardini D, et al. Arthroscopic rotator cuff repair using the Opus knotless suture anchor fixation system. *Am J Sports Med* June 2009;**37**(6):1106–10.

27. Horan RL, Toponarski I, Boepple HE, Weitzel PP, Richmond JC, Altman GH. Design and characterization of a scaffold for anterior cruciate ligament engineering. *J Knee Surg* 2009;**22**(1):82–92.

28. Barber FA, Boothby MH. Bilok interference screws for anterior cruciate ligament reconstruction: clinical and radiographic outcomes. *Arthrosc* 2007;**23**(5):476–81.

29. Newman SDS, Atkinson HDE, Willis-Owen CA. Anterior cruciate ligament reconstruction with the ligament augmentation and reconstruction system: a systematic review. *Int Orthop* 2013:321–6.

30. Nagatomo K, Komaki M, Sekiya I, Sakaguchi Y, Noguchi K, Oda S, et al. Stem cell properties of human periodontal ligament cells. *J Periodontal Res* 2006;**41**(4):303–10.

31. Mauck RL, Martinez-Diaz GJ, Yuan X, Tuan RS. Regional multilineage differentiation potential of meniscal fibrochondrocytes: implications for meniscus repair. *Anat Rec* 2007;**290**(1):48–58.

32. Xie X, Zhang C, Tuan RS. Biology of platelet-rich plasma and its clinical application in cartilage repair. *Arthritis Res Ther* 2014;**16**(1):204.

33. Chen FH, Tuan RS. Adult stem cells for cartilage tissue engineering and regeneration. *Curr Rheumatol Rev* 2008;**4**(3):161–70.

34. Xie X, Wang Y, Zhao C, Guo S, Liu S, Jia W, et al. Comparative evaluation of MSCs from bone marrow and adipose tissue seeded in PRP-derived scaffold for cartilage regeneration. *Biomaterials* 2012;**33**(29):7008–18.

35. Rodeo SA, Delos D, Williams RJ, Adler RS, Pearle A, Warren RF. The effect of platelet-rich fibrin matrix on rotator cuff tendon healing: a prospective, randomized clinical study. *Am J Sports Med* 2012:1234–41.

36. Fox BA, Stephens MM. Treatment of knee osteoarthritis with Orthokine-derived autologous conditioned serum. *Expert Rev Clin Immunol* May 2010;**6**(3):335–45.

37. Boswell SG, Schnabel LV, Mohammed HO, Sundman EA, Minas T, Fortier LA. Increasing platelet concentrations in leukocyte-reduced platelet-rich plasma decrease collagen gene synthesis in tendons. *Am J Sports Med* 2013;**42**(1):42–9.

38. Xie X, Ulici V, Alexander PG, Jiang Y, Zhang C, Tuan RS. Platelet-rich plasma inhibits mechanically induced. *Arthrosc* 2015:1–9.

39. Kuo CK, Marturano JE, Tuan RS. Novel strategies in tendon and ligament tissue engineering: advanced biomaterials and regeneration motifs. *Sports Med Arthrosc Rehabil Ther Technol* January 2010;**2**(1):20.

40. Cheng CW, Solorio LD, Alsberg E. Decellularized tissue and cell-derived extracellular matrices as scaffolds for orthopaedic tissue engineering. *Biotechnol Adv* March–April 2014;**32**(2):462–84.

41. Yang G, Rothrauff BB, Lin H, Gottardi R, Alexander PG, Tuan RS. Enhancement of tenogenic differentiation of human adipose stem cells by tendon-derived extracellular matrix. *Biomaterials* 2013;**34**(37):9295–306.

42. Hwang NS, Varghese S, Li H, Elisseeff J. Regulation of osteogenic and chondrogenic differentiation of mesenchymal stem cells in PEG-ECM hydrogels. *Cell Tissue Res* 2011;**344**(3):499–509.

43. Nuttelman CR, Tripodi MC, Anseth KS. In vitro osteogenic differentiation of human mesenchymal stem cells photoencapsulated in PEG hydrogels. *J Biomed Mater Res A* 2004;**68**(4):773–82.

44. Briggs T, Treiser MD, Holmes PF, Kohn J, Moghe PV, Arinzeh TL. Osteogenic differentiation of human mesenchymal stem cells on poly(ethylene glycol)-variant biomaterials. *J Biomed Mater Res A* 2009;**91**(4):975–84.

45. Mattii L, Battolla B, D'Alessandro D, Trombi L, Pacini S, Cascone MG, et al. Gelatin/PLLA sponge-like scaffolds allow proliferation and osteogenic differentiation of human mesenchymal stromal cells. *Macromol Biosci* 2008;**8**(9):819–26.

46. Alvarez-Barreto JF, Landy B, Vangordon S, Place L, Deangelis PL, Sikavitsas VI. Enhanced osteoblastic differentiation of mesenchymal stem cells seeded in RGD-functionalized PLLA scaffolds and cultured in a flow perfusion bioreactor. *J Tissue Eng Regen Med* 2011;**5**(6):464–75.

47. Saito M, Takahashi KA, Arail Y, Inoue A, Sakao K, Tonomura H, et al. Intraarticular administration of platelet-rich plasma with biodegradable gelatin hydrogel microspheres prevents osteoarthritis progression in the rabbit knee. *Clin Exp Rheumatol* 2009:201–7.

48. Lin H, Lozito TP, Alexander PG, Gottardi R, Tuan RS. Stem cell-based microphysiological osteochondral system to model tissue response to interleukin-1B. *Mol Pharm* 2014;**11**(7):2203–12.

49. Zhang J, Wang JH-C. PRP treatment effects on degenerative tendinopathy – an in vitro model study. *Muscles Ligaments Tendons J* January 2014;**4**(1):10–7.

50. Li W-J, Cooper JA, Mauck RL, Tuan RS. Fabrication and characterization of six electrospun poly(alpha-hydroxy ester)-based fibrous scaffolds for tissue engineering applications. *Acta Biomater* July 2006;**2**(4):377–85.

51. Li W-J, Tuan RS. Fabrication and application of nanofibrous scaffolds in tissue engineering. *Curr Protoc Cell Biol* March 2009. http://dx.doi.org/10.1002/0471143030.cb2502s42. Chapter 25: Unit 25.2.

52. Li W-J, Jiang YJ, Tuan RS. Cell-nanofiber-based cartilage tissue engineering using improved cell seeding, growth factor, and bioreactor technologies. *Tissue Eng Part A* May 2008;**14**(5):639–48.

53. Lin H, Zhang D, Alexander PG, Yang G, Tan J, Cheng AWM, et al. Application of visible light-based projection stereolithography for live cell-scaffold fabrication with designed architecture. *Biomaterials* 2013;**34**(2):331–9.

54. Lin H, Cheng AW, Alexander PG, Beck AM, Tuan RS. Cartilage tissue engineering application of injectable gelatin hydrogel with in situ visible-light-activated gelation capability in both air and aqueous solution. *Tissue Eng Part A* September 2014;**20**(17–18):2402–11.

55. Alexander PG, Gottardi R, Lin H, Lozito TP, Tuan RS. Three-dimensional osteogenic and chondrogenic systems to model osteochondral physiology and degenerative joint diseases. *Exp Biol Med (Maywood)* 2014:1080–95.

56. Brown BN, Siebenlist NJ, Cheetham J, Ducharme NG, Rawlinson JJ, Bonassar LJ. Computed tomography-guided tissue engineering of upper airway cartilage. *Tissue Eng Part C* June 2014;**20**(6):506–13.

57. Cleary MA, van Osch GJVM, Brama PA, Hellingman CA, Narcisi RFGF. TGFβ and Wnt crosstalk: embryonic to in vitro cartilage development from mesenchymal stem cells. *J Tissue Eng Regen Med* April 2015;**9**(4):332–42.

58. Petrie Aronin CE, Tuan RS. Therapeutic potential of the immunomodulatory activities of adult mesenchymal stem cells. *Birth Defects Res C Embryo Today* March 2010;**90**(1):67–74.

59. Knaän-Shanzer S. Concise review: the immune status of mesenchymal stem cells and its relevance for therapeutic application. *Stem Cells* March 2014;**32**(3):603–8.

60. LaPrade RF, Ly TV, Wentorf FA, Engebretsen L. The posterolateral attachments of the knee: a qualitative and quantitative morphologic analysis of the fibular collateral ligament, popliteus tendon, popliteofibular ligament, and lateral gastrocnemius tendon. *Am J Sports Med* January 2003;**31**(6):854–60.

61. Blackburn TA, Craig E. Knee anatomy: a brief review. *Phys Ther* December 1980;**60**(12):1556–60.

62. Sharma P, Maffulli N. Biology of tendon injury: healing, modeling and remodeling. *J Musculoskelet Neuronal Interact* January 2006;**6**(2):181–90.

63. LaPrade RF, Engebretsen AH, Ly TV, Johansen S, Wentorf FA, Engebretsen L. The anatomy of the medial part of the knee. *J Bone Joint Surg Am* September 2007;**89**(9):2000–10.

64. Paxton ES, Stock MV, Brophy RH. Meniscal repair versus partial meniscectomy: a systematic review comparing reoperation rates and clinical outcomes. *Arthroscopy* September 2011;**27**(9):1275–88.

65. Maffulli N, Wong J, Almekinders LC. Types and epidemiology of tendinopathy. *Clin Sports Med* 2003:675–92.

66. Hasan MDA, Ragaert K, Swieszkowski W, Selimović Š, Paul A, Camci-Unal G, et al. Biomechanical properties of native and tissue engineered heart valve constructs. *J Biomech* 2013;**47**:1949–63.

67. Deng D, Wang W, Wang B, Zhang P, Zhou G, Zhang WJ, et al. Repair of Achilles tendon defect with autologous ASCs engineered tendon in a rabbit model. *Biomaterials* 2014;**35**(31):8801–9.

68. Thayer PS, Dimling AF, Plessl DS, Hahn MR, Guelcher SA, Dahlgren LA, et al. Cellularized cylindrical fiber/hydrogel composites for ligament tissue engineering. *Biomacromolecules* 2014;**15**(1):75–83.

69. Ionescu LC, Mauck RL. Porosity and cell preseeding influence electrospun scaffold maturation and meniscus integration in vitro. *Tissue Eng Part A* 2012:538–47.

70. Attia E, Bohnert K, Brown H, Bhargava M, Hannafin JA. Characterization of total and active matrix metalloproteinases-1, -3, and -13 synthesized and secreted by anterior cruciate ligament fibroblasts in three-dimensional collagen gels. *Tissue Eng Part A* 2014;**20**(1–2):171–7.

71. Fleming BC, Proffen BL, Vavken P, Shalvoy MR, Machan JT, Murray MM. Increased platelet concentration does not improve functional graft healing in bio-enhanced ACL reconstruction. *Knee Surg Sports Traumatol Arthrosc* 2014.

72. Xu Y, Dong S, Zhou Q, Mo X, Song L, Hou T, et al. The effect of mechanical stimulation on the maturation of TDSCs-poly(L-lactide-co-ε-caprolactone)/collagen scaffold constructs for tendon tissue engineering. *Biomaterials* 2014;**35**(9):2760–72.

Chapter 7

Cartilage and Bone Regeneration— How Close Are We to Bedside?

Raphaël F. Canadas*, Sandra Pina*, Alexandra P. Marques, Joaquim M. Oliveira, Rui L. Reis

Key Concepts

This chapter begins by presenting bone and cartilage physiology and disorders concepts and currently applied treatments, including their limitations and the potential solutions proposed within tissue engineering and regenerative medicine. Next, biomimetic strategies for bone and cartilage tissue engineering, namely scaffolds, hydrogels, and fibers, are presented, as well as natural and synthetic biomaterials, such as polymers, bioactive inorganic materials and composites, employed for such processing methodologies. Lastly, a thorough review of ongoing clinical trials and commercial products that are already in the market for osteochondral tissue regeneration are also presented.

List of Abbreviations

ACI Autologous chondrocyte implantation
ACs Articular chondrocytes
AMIC Autologous matrix-induced chondrogenesis
ASCs Adipose-derived stem cells
β-TCP β-Tricalcium phosphate
BM Bone marrow
BMPs Bone morphogenetic proteins
CaP Calcium phosphates
CRD Cartilage repair device
ECM Extracellular matrix
FDA Food and Drug Administration
HAp Hydroxyapatite
hBMSCs Human bone marrow stromal cells
MACI Matrix-induced autologous chondrocyte implantation
MSC Mesenchymal stem cell
NCs Neuroectoderm-derived nasal chondrocytes
OA Osteoarthritis
OC Osteochondral
OCD Osteochondral defect
OP Osteoporosis
PCL Poly(ε-caprolactone)
PEG Polyethylene glycol
PGA Polyglycolic acid (or polyglycolide)
PLA Polylactic acid (or polylactide)
PLGA Polylactide-co-glycolide
RFE Radio frequency energy
SF Silk fibroin
TGF-β Transforming growth factor-β

---

*The authors contributed equally to this work.

Translating Regenerative Medicine to the Clinic. http://dx.doi.org/10.1016/B978-0-12-800548-4.00007-3

1. INTRODUCTION

Osteoarthritis (OA) and osteoporosis (OP) are among the most disabling degenerative diseases that may lead to severe complications affecting the neuromuscular system, thus significantly impairing patients' quality of life.[1,2] OA is the highest ranking disease among the musculoskeletal diseases and contributes to approximately 50% of the disease burden in this group.[3] Current clinical treatments for OA and OP involve nonsteroidal anti-inflammatory drug administration and surgery such as osteotomy, abrasion arthroplasty, microfracture, and autologous and allogeneic cartilage tissue grafts, and autologous chondrocytes.[4] These treatments are well established and effective for reducing the patients' pain, but are not able to completely restore the patient's mobility. Therefore, the demand for new therapeutic options for complete healing of bone, cartilage, and osteochondral defects (OCDs) is significant. Bone and cartilage diseases or defects are directly related with joint degeneration. Such disorders can be caused by rheumatism, joint dysplasia, and/or trauma and are particularly prevalent in countries with high life expectation. Articular cartilage damage can arise as a consequence of both acute and repetitive trauma resulting in pain, effusion, and/or mechanical symptoms, affecting directly the individual's lifestyle, as work, hobbies, and daily tasks.[5] Cartilage lesions in joints can have different degrees; superficial lesions, as fissures or cracks are classified as grade 1. A grade 2 abnormality is defined when cartilage is affected up to 50% of its thickness, while grade 3 lesions are characterized by defects in which more than 50% of the cartilage thickness, down to the subchondral bone but without bone penetration, is affected. The final grade is the commonly termed OCD (grade 4) that results from the cartilage damage with penetration into the subchondral bone.

Although articular cartilage comprises just one type of cells, chondrocytes become less active with age and injury. Furthermore, the avascular nature of cartilage together with the declining function of chondrocytes leads to the inability of full-thickness defects to heal spontaneously. If untreated, these lesions can progress to more serious degenerative joint conditions. Tissue engineering is a multidisciplinary field of research that employs principles of chemistry, biology, and engineering sciences toward growth, development, and regeneration of damaged tissues or organs.[6] It can involve the use of scaffolds combined with cells and suitable biochemical signals to design and create off-the-shelf organs and tissues substitutes. Despite the promise of tissue engineering, a better understanding of the composition, structure, and properties of bone and cartilage can guide scientists to achieve the adequate tissue-engineered grafts to ideally repair and regenerate bone and cartilage tissues. Bone and cartilage have a three-dimensional architecture with several levels of organization comprising micro- and nanostructures (Figure 1).[7] Cancellous bone is a porous structure, whereas cortical bone is composed of osteons

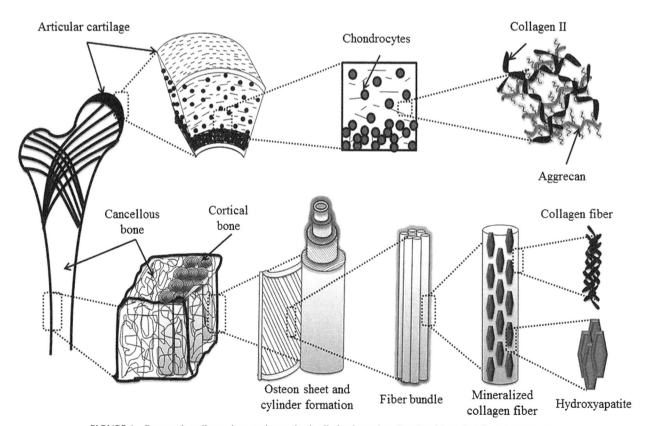

FIGURE 1 Bone and cartilage micro- and nanosized cylinder formation. *Reprinted from Ref. 7, with permission.*

that consist of a concentric series of layers (lamellae) of mineralized collagen Type I matrix. Articular cartilage presents a stratified architecture, which consists of three zones where in the superficial zone collagen Type II fibers are oriented tangentially to the articular surface, in the transitional zone have no predominant orientation and become aligned perpendicularly to the calcified cartilage, and anchored in subchondral bone in the deep zone. Articular cartilage has poor intrinsic ability for healing due to its isolation from vessels and nerve supply. On the other hand, bone is a vascular and innervated tissue. While bone is mineralized, normal cartilage tissue is not mineralized and is a highly hydrated tissue.

Herein, cartilage and bone physiology and disorders are discussed in terms of currently applied treatments. Their limitations and the potential solutions proposed within tissue engineering and regenerative medicine field are presented. Applied or tested materials, cells, bioactive molecules, and tissue-engineered techniques are briefly described. To address the question "How close are we to bedside?," the ongoing clinical trials and the related products that have already reach the market for osteochondral (OC) tissue regeneration are also reviewed.

2. CONCEPTS AND TREATMENT STRATEGIES

In order to understand what sort of bone and cartilage tissue engineering strategies could be best for repair/reconstruct defects, it is important first to recognize the structure, concepts, and current therapeutic approaches targeting the different bone and cartilage lesions.

2.1 Bone

Bone is a complex, highly organized, and specialized connective tissue with many functions. All bones have a mechanical function providing attachment to various muscle groups. In addition, in some parts of the body, bones provide a protective function to vital structures—skull (brain), ribs (lungs, heart), and pelvis (bladder, pelvic viscera). Some bones retain their hematopoietic function in adults—vertebrae, iliac crests, proximal parts of femur and humerus.[3,4] All bones serve as a reservoir of calcium and actively participate in the calcium homeostasis in the body.

Bone is composed of cortical (compact) (80%) and trabecular (cancellous or spongy) (20%) tissues. Cortical bone tissue forms the outer shell, or cortex, of the bone and has a dense structure with a porosity of about 5–10%.[3–5] It is the primary component of the long bones of the arm and leg and other bones, where its greater strength and rigidity are needed. Trabecular bone tissue typically occupies the interior region of bones and is composed of thin plates, or trabeculae, in a loose mesh structure with porosity of 50–90%. It is highly vascularized and frequently encloses the bone marrow (BM) with high proportions of mesenchymal and hematopoietic stem cells. Trabecular bone tissue has a higher surface area but is less dense and stiff, and weaker than cortical bone.

Bone has the ability to remodel, by altering its size, shape, and structure to meet the mechanical demands placed on it. Bone remodeling is a dynamic, lifelong process in which resorption is followed by formation, respectively, involving the activity of osteoclasts and osteoblasts.[6] During bone formation, also called osteogenesis, preosteogenic cells are stimulated to migrate through a provisional matrix, which in a therapeutic/regenerative approach could be represented by bone-graft substitutes or a blood clot.[7] The migrating cells then start a differentiation process that results in the secretion of the new bone matrix.

Bone defects are often associated to a disease state (e.g., OA, OP, osteomyelitis, and osteogenesis imperfect) and trauma-related injuries resulting from primary tumor resection and orthopedic surgeries (e.g., total joint arthroplasty and implant fixation). In addition, spinal fractures, called vertebral compression fractures, are the most common fracture in patients with OP, affecting nearly 700,000 people each year, typically postmenopausal women. However, others fractures like fractures of the hip, wrist, and proximal humerus are commonly observed in patients with OP.[8]

Treatments used by orthopedic surgeons for the reconstruction and repair of bone defects and fractures are mainly internal fixation and bone allografts and autografts.[9] Allografting involves the transplant of tissue from one individual to another with a different genotype but of the same species, carrying the risk of immune-mediated rejection or transmission of infectious diseases. By its turn, autograft is a piece of tissue that is transplanted from one part of the body to another in the same individual. This procedure ensures the long-term survival of graft and subsequent successful reconstruction due to intrinsic features such as osteoconductivity and histocompatibility. Autografting is thus considered the clinical gold standard method.[8,9] However, a number of complications including infection, vascular injuries, chronic donor site pain, and morbidity have been reported with the use of autografts.[10]

Other approaches, such as vascularized fibula autograft, Ilizarov bone transfer technique, and Masquelet technique, where autologous bone grafting alone is not recommended due to the risk of resorption, have been used particularly for long bone defects reconstruction.[11–13] Nevertheless, these techniques are mainly limited to cancer patients who have OP and suffer from impaired wound healing.

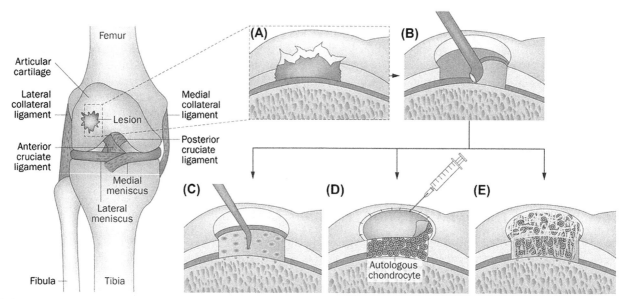

FIGURE 2 Representative images as summary of cartilage regeneration techniques: (A) full-thickness focal chondral lesion; (B) debridement; (C) microfracture; (D) autologous chondrocyte implantation; and (E) matrix-induced autologous chondrocyte implantation. *Reprinted from Ref. 17, with permission.*

2.2 Cartilage

Articular (hyaline) cartilage regeneration is a priority of orthopedic care because the clinical need is expanding with the aging population (mainly in developed countries). Articular cartilage enables the joints to tolerate shearing forces and absorb shock and loads up to 20 times the body's weight. As health care is evolving, people live longer and population ages. Moreover, societies are increasingly more dynamic, competitive, and physically more demanding. With time, articular cartilage increasingly bears more prolonged and cumulative skeletal stresses and shearing forces, increasing the potential to the development of degenerative diseases of cartilage as OA. Worldwide estimates indicate that 9.6% of men and 18% of women ≥60 years have symptomatic OA last decade.[2] Joint surface defects are ubiquitous, with reported prevalence of arthritis of about 31% in knees, 17% in hip, and 7% in hands.[10] In respect to prevalence of joint pain, 38% is incident in knees, 18% in the shoulders, 14% in hands and hip, and 16% in lower back.

While long-term research goals for cartilage regeneration focus on harnessing stem cell therapies alone or in combination with biodegradable materials, in the near term orthopedists choose from multiple treatment strategies to manage cartilage injuries. When injuries occur fracturing or damaging the tibia or knee, and the patient wants to continue to practice their normal lifestyle (or his physical activity in the case of sportsmen, for example) is usually indicated surgical intervention.[11] That type of surgery can be performed by arthroscopy, and at least from the standpoint of controlling pain, arthroscopic surgery has a main advantage of significant reduction of postoperative pain.[12] Literature describes that the minor the peripheral tissue damage, the lower the nociceptive stimulus at the surgery site, which will be crucial for the patient to have less pain after surgery.[13] Procedures such as the suprapatellar approach for nail insertion are seen as options to avoid late postoperative knee pain.[14] This minimally invasive approach uses an easy entry point, promoting lesions only in the Hoffa's body, which is usually removed during arthroscopy because it can be inflamed or damaged and to better visualize the knee also.[15]

Currently the methods for the treatment of cartilage defects (e.g., OA) include the insertion/transplantation of OC tissue, cells, scaffolds, or growth factors (GFs), alone or in combination, or even the use of radio frequency energy (RFE) methods.[16] Procedures that are normally used for treatment of small articular cartilage defects include also RFE, chondrocyte implantation, and BM stimulation techniques such as drilling, debridement, or microfracture. In the case of large defects, scaffolds and mosaicplasty are commonly used in the treatment (Figure 2).[17] Although current methods for articular cartilage defects are promising, no treatment has resulted in complete regeneration of the hyaline cartilage and the subchondral bone.

Drilling procedure aims to pierce the underlying subchondral bone, thereby inducing bleeding at the defect site and allowing the formation of a blood clot which contains BM mesenchymal stem cells (MSCs) that will differentiate helping in cartilage regeneration.[18] Drilling was described by Pridie and Gordon[19] and it is known to cause thermal necrosis of the subchondral bone, as well as result in an uneven repair surface and for these reasons it is not a favored method of treatment.[18,20] Debridement is the simple excision of the damaged cartilage and has been shown to improve symptoms for 5 years or more.[21] Opinion is divided as to whether arthroscopic debridement has any place in the treatment of established degenerative disorders, as OA, although this debate does not apply to the treatment of localized symptomatic chondral defects.

TABLE 1 Types of Collagen and Associated Genes and Cartilages Phenotypes Outcomes

Type	Outcome	Gene(s)	Disorders
I	Fibrocartilage	COL1A1, COL1A2	Osteogenesis imperfecta
II	Hyaline cartilage, makes up 50% of all cartilage protein	COL2A1	Collagenopathy, Types II and XI
X	Hypertrophic and mineralizing cartilage	COL10A1	Schmid metaphyseal dysplasia
XI	Cartilage	COL11A1, COL11A2	Collagenopathy, Types II and XI

Microfracture, which is based on marrow stimulation that creates fibrocartilage (as drilling and debridement) at the site of the procedure, with varied amounts of collagen Types I, II, and III,[20] was introduced by Steadman et al.[22] 20 years ago, and is one of the most used methods for cartilage repair. This type of cartilage is less durable, less resilient, and less able to withstand shearing forces than native articular cartilage, composed mainly by collagen Type II (Table 1). While this approach can have good results in smaller lesions, clinical studies reflect the lack of durability over a long-term follow-up. By its turn, treatment of OCD in the foot has also been considered a great challenge due to the different biomechanical features as interestingly reported elsewhere.[23]

Mosaicplasty, or OC cylinder transplantation, was first described in 1993.[24] In this procedure, OC plugs are taken with a cylindrical cutting device and used to fill an articular cartilage defect. Advantages of this technique are on one hand the immediate filling of the defects with mature, hyaline articular cartilage and on the other the simultaneous treatment of both chondral and OCDs. However, donor site morbidity is a concern and Hangody and Fules[25] recommend the limiting of the area to be treated to 1–4 cm². There are also technical difficulties in restoring the surfaces of both cartilage and bone to produce a smooth, convex joint surface. The thickness of the donor cartilage may differ from that of the area to be treated and the reconstitution of the important subchondral layer may not occur. In addition, lateral integration rarely occurs[26] raising the concern that synovial fluid may penetrate through the subchondral layer possibly causing cyst formation.

Autologous chondrocyte implantation (ACI) can result in more hyaline-like cartilage within the treated defect. The technique of ACI was first performed by Peterson et al.[27] in Gothenburg in 1987 and it was the first application of cell engineering in orthopedic surgery. In ACI, healthy cartilage cells are harvested, cultured, and then reimplanted into the defect under a patch, in a second-stage surgery. Brittberg et al.[28] presented the results of 23 patients with a mean follow-up of 39 months. Good or excellent clinical results were reported in 70% of cases (88% of femoral condylar defects). Of the biopsies from treated femoral condylar lesions, 11 of 15 had a hyaline-like appearance. A more recent publication from the same group showed durable results up to 11 years following the treatment of OC lesions.[29]

Current drawbacks to this procedure are hypertrophy of the patch that can lead to further surgery and unreliable biological potential of the reimplanted cartilage cells. Furthermore, histological analysis apparently shows that ACI is capable of producing hyaline-like tissue in some specimens, however, the best repaired tissue is not morphologically or histochemically identical to normal hyaline cartilage, and fibrocartilage can be found frequently.

A variation of the ACI technique using culture-expanded BM stromal cells (BMSCs) has the advantage of not requiring an additional arthroscopic procedure in order to harvest articular cartilage.[30]

All of these techniques encounter limited success due to issues which include fibrocartilage formation, chondrocyte dedifferentiation, and lack of tissue integration and mechanical support.

3. BIOMATERIALS FOR BONE AND CARTILAGE REGENERATION

The rationale of using biomaterials as scaffolds in tissue regeneration is to obtain a temporary three-dimensional structure for the in vitro growth of living cells and its subsequent implantation into the lesion area, followed by its biodegradation as new tissue is being formed. Several natural and synthetic polymers, bioactive inorganic materials, and their combinations have been employed for bone and cartilage tissue engineering and regeneration. Polymers have great stiffness and mechanical strength, and natural polymers add advantages such as their resemblance with the extracellular matrix (ECM), specific degradation rates due to the susceptibility to the action of enzymes, and improved recognition by the living body. Bioactive inorganic materials, such as calcium phosphates (CaP) and bioactive glasses, have good biocompatibility, osteoconductivity, and bioresorbability. Despite, they present poor mechanical properties that hinder its use in load-bearing applications. The combination of these different types of materials result in composite structures with significantly enhanced mechanical and biological properties for bone tissue engineering. In opposition, cartilage tissue is engineered using natural/synthetic polymers.

The most promising polymers, inorganic materials and composites, and their properties are briefly described as follows.

3.1 Polymers

Natural polymers, also known as biopolymers, have been extensively used owing to their ability to interact with cells and to be susceptible to enzymatic degradation providing space for tissue ingrowth.[31] Naturally occurring polymers most widely explored for bone and cartilage repair/regeneration are (1) proteins (e.g., silk fibroin (SF), collagen, and gelatin); (2) polysaccharides (e.g., chitosan, alginate, gellan gum, and derivatives); and (3) glycosaminoglycans (e.g., hyaluronic acid).[31]

In comparison to biopolymers, synthetic polymers have several advantages: excellent processing characteristics, excellent mechanical and physical properties (e.g., elastic modulus, strength, and degradation rates), and bioresorbability.[32] However, many of these polymers present several disadvantages, such as the possibility of causing persistent inflammatory reactions and not being capable to integrate with host tissues.[33] The most widely used polymers are polyglycolic acid (or polyglycolide (PGA)), polylactic acid (or polylactide (PLA)), polylactide-co-glycolide (PLGA), poly (D,L-lactic acid), polyethylene glycol (PEG), and poly(ε-caprolactone) (PCL). These polymers have received special attention since they can be self-reinforced to gain better strength properties.[34]

3.2 Bioactive Inorganic Materials

Inorganic materials often used for bone repair and regeneration are CaP, namely hydroxyapatite ($Ca_5(PO_4)_3OH$, HAp) and β-tricalcium phosphate ($Ca_3(PO_4)_2$, β-TCP), bioactive glasses, and glass-ceramics owing to their bioactivity, biocompatibility, and osteoconductivity.[35,36]

HAp is crystalline and is the most stable and least soluble CaP in an aqueous solution down to a pH of 4.2.[36] Aqueous precipitation;[37,38] hydrothermal synthesis;[39] solid-state reaction using calcium oxide, calcium hydroxide, or calcium carbonate; and hydrolysis of other CaP have been methods used to prepare HAp. A detailed information on HAp synthesis and preparation is well established.[40] β-TCP is a high temperature phase of CaP, which can only be obtained by its thermal decomposition at temperatures above 800 °C. β-TCP is biodegradable and has been extensively used as bone substitute, either as granules or blocks, or even in CaP bone cements.[41] The resorption capability of HAp and β-TCP is different though there is similarity in terms of chemical composition. It is believed that HAp has a slow resorption rate (1–2% per year) and may be integrated into the regenerated bone tissue, while β-TCP is completely reabsorbed.[42,43] Therefore, clinical applications have been performed by combining HAp with β-TCP, which forms the biphasic CaP, improving the bioresorbability and strength of the bone substitutes.[35,40,44] Nevertheless, these materials are limited to nonload-bearing applications due to their poor mechanical properties.

Bioactive glasses and glass-ceramics have been used in bone regeneration due to their capability to react with physiological fluids, thus bonding with bone through the formation of HAp layers at the implant interface and thereby stimulating bone growth.[45,46] These type of materials are osteogenetic and osteoconductive, while CaP exhibit only osteoconductive properties.[41] It has also been found that reactions on bioactive glass surfaces release concentrations of Si, Ca, P, and Na ions, thus inducing intracellular and extracellular responses.[47] They are also able to improve osteoblast adhesion, vascularization in vivo, enzymatic activity, and differentiation of MSCs.[48] In addition, glasses have shown great potential as reinforcing materials since they fully degrade in aqueous media.[49,50] Bioactive glasses are brittle materials; this limitation can be solved by the development of glass-ceramics or by the combination with an additional phase as a polymer, forming a composite.[51,52] There are different compositions of bioactive glasses, based on silicate, phosphate, and borate, which can be obtained through melt-quenching[53] and sol–gel process.[54] The most widely investigated bioactive glass for biomedical applications is the silicate-based glass designated 45S5, also known by its commercial name Bioglass®.[55] These types of glasses have higher chemical durability and durability limits as compared to other bioactive glasses.[56] By its turn, phosphate-based glasses have unique dissolution properties in aqueous fluids, while borate-based bioactive glasses have faster degradation rates and are able to completely convert into apatite.[57,58]

Bioactive inorganic materials can be doped with trace elements (e.g., strontium, zinc, magnesium, manganese, silicon), which can influence bone health and enhance biocompatibility, while strengthening the mechanical properties of the implants.[59] Besides, minerals and traces of metal elements may provide physicochemical modifications in the produced materials, which can accelerate bone formation and resorption in vivo.[60,61]

3.3 Composites

Composite materials embracing a natural/synthetic polymeric matrix and bioactive inorganic materials, as fillers, appeared as a strategy to mimic the human bone, which is a three-dimensional composite composed of organic, inorganic, and cellular phases, strictly assembled to form the natural bone tissue. Composites are the combination of two or more materials, with different compositions and properties, resulting in a single structure with significantly improved mechanical and biological properties. Special interest has been attributed to nanocomposites for bone tissue engineering and regeneration due to the nanosized features of the fillers which can intensely improve the tissue bonding capacity of the polymeric matrices, that the individual materials cannot attain thus allowing the production of better biomaterials.[7] The nanoparticles have large surface area when compared to the conventional microsized fillers, thus offering improved mechanical properties, while maintaining the osteoconductivity and biocompatibility of the fillers, as well as cell adhesion and differentiation.[62]

Many combinations of polymers and inorganic materials have been proposed for the production of nanocomposites which final properties will be dependent. As aforementioned, the most common polymers used are of natural origin (collagen, gelatin, silk, chitosan, alginate, hyaluronic acid, and gellan gum).[63–69] By its turn, some synthetic polymers (e.g., PEG, PLA, PGA, PLGA, and PCL) have been also used and applied in the clinics. On the other side, nanosized fillers include nanoparticles of CaP and bioactive glasses, carbon nanotubes, nanofibers, and nanoplatelets. These nanoparticles have been prepared through different processes, namely wet chemical precipitation,[38] sol–gel synthesis,[70] hydrothermal synthesis,[71] mechanochemical synthesis,[72] microwave processing,[73] spray drying methods,[74] and electrospinning,[75] while nanocomposites have been prepared by simple mechanical mixing or coprecipitation.

Further details on nanomaterials processing techniques and applications in bone tissue regeneration can be found elsewhere.[76]

4. BONE AND CARTILAGE TISSUE ENGINEERING

Biomimetic strategies to develop new bone and cartilage tissue-engineered constructs rely on bioactive structures able to mimic the natural tissue ECM in order to promote cell adhesion, migration, growth, and matrix deposition forming a tissuelike substitute. These structures embrace three-dimensional porous and fibrous scaffolds, and hydrogels, with specific design, controlled degradation rate, mechanical properties, and porosity for efficient gases, nutrients, and regulatory factors transport.

4.1 Bone Tissue Engineering

Bone tissue engineering focuses on alternative treatment strategies to reduce the shortcomings of the current clinical treatments (i.e., infection, vascular injuries, immune rejection, chronic donor-site pain, and morbidity) by using the combination of materials science, engineering principles, and cell biology. Hence, the fabrication of composite constructs hierarchically structured from nano- to macrosize ranges inspired by the nature of bone, has been followed.[77]

Conventional technologies, such as foam replica method,[78] solvent casting and particulate leaching,[79] freeze-drying,[80] phase separation,[81] and gas foaming,[82] often inexpensive, simple, and flexible to optimize physicochemical properties, have been used to fabricate scaffolds. Rapid prototyping[83] and electrospinning[84] are sophisticated techniques for the production of, respectively, 3D structures and fibers that allow the possibility of incorporating pharmaceutical agents. Molecular self-assembly is another strategy available for the production of nanofibers by creating supramolecular architectures.[85]

Several studies have reported the development of porous structures for bone tissue engineering using diverse materials and techniques.[65,78,86–88] For example, Oliveira et al.[78] developed macroporous HAp scaffolds with controlled morphology using the sponge replica method. The structures showed a porosity of ~70%, and highly interconnected macropores with a diameter in the range of 50–600 μm (Figure 3). Later, Barbani et al.[86] produced a gelatin/HAp nanocomposite scaffold with elastic modulus similar to natural bone, using freeze-drying technique. It was shown that HAp scaffolds supported the adhesion and proliferation of human MSCs onto the scaffolds.

Yan et al.[87] prepared a composite scaffold of SF and nanosized CaP combining solvent casting and freeze-drying methods that allowed the formation of a homogeneous macroporosity and porosity distribution (Figure 4). The scaffolds are also characterized by the good mechanical properties and stability, and self-mineralization capability which represents a major feature for bone tissue engineering.

Eftekhari et al.[89] developed a novel porous scaffold composed of cotton-sourced cellulose microcrystals, HAp nanoparticles, and PLLA with enhanced mechanical strength for bone tissue regeneration. A different preparation method, by applying cryogelation method as an alternative to freeze-drying, was used to prepare collagen/nano-HAp scaffolds for bone regeneration.[90] The scaffolds showed improved mechanical properties and allowed high cells proliferation.

Hydrogels have also been explored in the context of bone tissue engineering due to their structural and compositional similarities with the ECM that allow efficient mass transfer as well as the encapsulation of cells and biomolecules.[91] These structures comprise a hydrophilic porous network that can be controlled by solvent casting and particulate leaching, phase separation, gas foaming, solvent evaporation, freeze-drying, and blending with noncross-linkable linear polymers.[92]

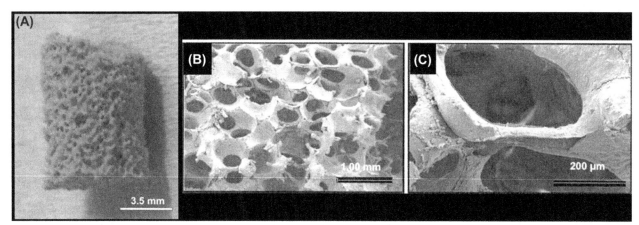

FIGURE 3   Hydroxyapatite scaffolds: (A) macroscopic image; (B and C) microstructure. *Reprinted from Ref. 78, with permission.*

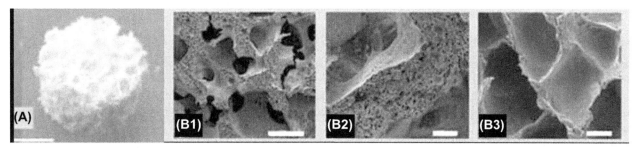

FIGURE 4   Silk fibroin/nanohydroxyapatite scaffolds: (A) macroscopic image (scale bar: 3 mm); (B1–B3) microstructure (scale bar: 500 nm). *Reprinted from Ref. 87, with permission.*

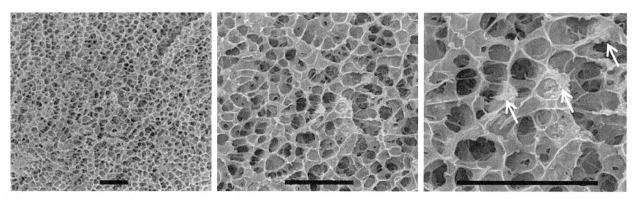

FIGURE 5   Microstructure of the hydrogel with 15% of nanosized hydroxyapatite concentration. Arrow indicates the polymer–nanoparticle aggregates (scale bar 2 μm). *Adapted from Ref. 95, with permission.*

Hydrogel networks can be engineered, into different sizes and shapes, as thin films, sheets, spheres, rods, hollow tubes, and bellows, due to their unique physical properties.[93] Excellent reviews regarding a deep description of hydrogels properties were recently reported.[94]

Hydrogels have been produced combining synthetic or/and biopolymers and inorganic biomaterials, with desired physical properties, reproducibility, and biological activity for use in bone tissue engineering. For example, Gaharwar et al.[95] developed hydrogels incorporating PEG and HAp presenting highly porous structures and interconnected porous structure with pore sizes of 100–300 nm (Figure 5). The results also showed osteoblast cell adhesion and bioactive attachment sites for the osteoblastic cells. An injectable and thermosensitive PEG-PCL-PEG copolymer, collagen, and nanosized HAp hydrogel for guided bone regeneration was developed by Fu et al.[96] The results revealed good biocompatibility, biodegradability, and new bone tissue formation after implanting the structures in rats. Gantar et al.[97] developed bioactive glass-reinforced gellan-gum hydrogels with an open and well-interconnected porosity of about 80% and a pore size of ~100–200 μm.

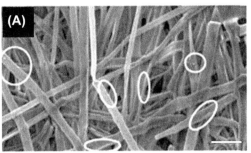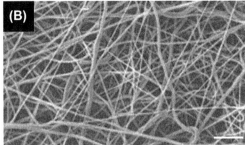

FIGURE 6   Microstructures of the composite fibrous scaffold gel/CaP/PCL (gel/calcium phosphates/poly(ε-caprolactone)): (A) gelatin side and (B) PCL side. Scale bar 10 μm. *Adapted from Ref. 99, with permission.*

Fiber-based scaffolds for bone tissue engineering are another good option to mimic the fibrous structure of ECM. Likewise, nanoscale fibrous scaffolds with similarities in the network of collagen fibrils of native ECM have received particular interest to enhance cell adhesion, proliferation, and differentiation.[98] Electrospun fibers have been explored as scaffolds similar to natural ECM to engineer and repair the bone tissue. Rajzer et al.[99] prepared composite fibrous scaffolds with electrospun PCL and gelatin/CaP fibers with diameter in the range of 2–6.5 μm and porosities of 74.3 ± 7.0% and 86.7 ± 2.3%, respectively, for the gelatin side and the PCL side of composite scaffold (Figure 6). In vitro tests proved the bioactivity of the scaffolds by the higher activity of alkaline phosphatase (ALP). Chae et al.[84] fabricated alginate/HAp fibrous scaffolds via electrospinning, composed of random nanofibers holding homogeneously distributed HAp nanocrystals.

4.2 Cartilage Tissue Engineering

The limited ability of articular cartilage to regenerate has prompted the development of cell-based tissue engineering techniques, such as ACI. However, the complexity of ACI and contraindications in wider clinical applications has driven the development of matrix-assisted chondrocyte implantation (MACI), which uses scaffolds to provide mechanical stability and to support chondrogenesis. Laboratory and clinical studies have examined the management of larger lesions using tissue-engineered cartilage.[100] To improve neotissue formation, cells can be cultured in vitro in 3D matrices with exogenous stimuli, such as GFs, to promote graft maturation and biomechanical integrity.

In order to outperform the currently used methods for treatment of grade 4 defects (penetration of the subchondral bone), novel tissue engineering approaches propose addressing OC regeneration by means of using bilayered scaffolds in combination with stem cells. That approach has in account the use of structures with two layers with different physical properties, usually a bioactive layer with a ceramic phase for the bony part, and a nonbioactive layer, composed by polysaccharides or proteins, or even biodegradable synthetic polymers for the cartilage-like layer.[33,101]

OC tissue is mainly composed by osteoblasts and chondrocytes. The two neighboring, but different cell lines have extremely different in vivo physiologic conditions, which have to be understood and replicated in vitro to obtain improvements in OC tissue regeneration.

Osteoblasts and chondrocytes are mononucleated cells that are derived from MSCs by GFs, via different signaling transcription pathways. Those factors can exhibit different and often opposite effects in the modulation of cells metabolism, depending on their maturation stage and phenotype. Many adult tissues contain cell niches that in response to injury, for example, provide stem cells that are able to differentiate into multiple cell linages, including chondrocytes. The adult stem cells have gained significant attention over the past decade and became frontline management for cartilage defects in the very recent past. BM is one of the main cell niches used to this end, presenting good potential and results.[102–104] However, there are more MSCs niches with potential to be used for cartilage and bone differentiation,[105] as umbilical cord- and adipose-derived stem cells (ASCs). In the case of ASCs, Hoffa's body is recently being explored as an interesting autologous source of cells to regenerate cartilage in knee-associated disorders.[106,107] Hoffa's body, which is a fat pad of ASCs, has to be removed during an arthroscopy to facilitate the visualization of the knee and surgery handling, and also to avoid tissue inflammation as was explained before. This way, this tissue can be considered too as a promising source of ASCs with great potential to differentiate into chondrocytes and osteoblasts.[108] Recently, also strategies using endothelial cells in coculture with osteoblasts or stem cells are being used to promote vascularization in bony part.[109,110]

To promote stem cell differentiation, GFs can be introduced in scaffolds in order to induce a faster host tissue response to the implanted matrix of material. In vitro osteogenic differentiation of ASCs could be induced by dexamethasone, L-ascorbic acid-2-phosphate, β-glycerophosphate, bone morphogenic proteins (BMPs), fibroblast growth factor, platelet-derived GF, and transforming growth factor-β (TGF-β).[111,112] Chondrogenic differentiation of ASCs requires GFs such as members of the TGF-β and BMPs families.

Steinwachs et al.[113] reported the technique of autologous matrix-induced chondrogenesis (AMIC). AMIC involves the joint use of the Chondro-Gide (Geistlich Biomaterials, Wallhausen, Germany) collagen Type I/III membrane as a scaffold over a defect treated by microfracture. Short-term results were encouraging, however, long-term follow-up data are needed to substantiate preliminary findings.[114] A novel approach to enhance cartilage repair with AMIC is to deliver GFs that selectively recruit and stimulate MSCs from the subchondral BM to invade cell-free scaffolds. Such GFs can be tailored also to activate chondrocytes in the surrounding healthy tissue to help filling the cartilage defect remodeling the tissue.

MACI (Genzyme, Oxford, United Kingdom) is a later surgical technique of cartilage repair and is a third generation variant of conventional ACI. Instead of injecting cultured chondrocytes underneath a periosteal or collagen Type I/III cover, the cells are preloaded onto a commercially produced porcine collagen patch. At the second stage of the operation, the patch is manually cut to cover the dimensions of the cartilage defect and held in place with tissue glue and, where necessary, sutures. Currently there is a limited data on the mid- to long-term follow-up success of such a technique.

Recently, a new promising cell niche for articular cartilage regeneration was investigated by Pelttari et al.[115] The authors showed that adult human neuroectoderm-derived nasal chondrocytes (NCs) can be constitutively distinguished from mesoderm-derived articular chondrocytes (ACs) by lack of expression of specific HOX genes, including HOXC4 and HOXD8. In contrast to ACs, serially cloned NCs could be continuously reverted from differentiated to dedifferentiated states, conserving the ability to form cartilage tissue in vitro and in vivo. NCs could also be reprogrammed to stably express HOX genes, typical of ACs, upon implantation into goat articular cartilage defects, directly contributing to cartilage regeneration.

The effect of the scaffolds pore size was studied by Zhang et al.[116] using porcine Type I collagen scaffolds applied for cartilage regeneration. The results obtained could help to establish the ideal conditions for future strategies for one of the most promising targets of regenerative medicine, the OC regeneration. The collagen porous scaffolds were prepared by means of using preprepared ice particulates that had diameters of 150–250, 250–355, 355–425, and 425–500 μm. The collagen porous scaffolds prepared with ice particulates 150–250 μm in size best promoted the expression and production of Type II collagen and aggrecan.[116]

Adachi et al.[117] evaluated the implantation of tissue-engineered cartilage-like tissue composed by autologous chondrocytes cultured in atelocollagen gel for the treatment for full-thickness cartilage defects of the knee. Arthroscopic analysis was performed 2 years after implantation. According to the International Cartilage Repair Society scale, in 64 of 73 knees (87.7%) the implanted constructs were graded normal or nearly normal. The authors concluded that the procedure can be suggested for repairing full-thickness cartilage defect of the knee.[117]

Lu et al.[118] used a baculovirus system that exploited FLPo/Frt-mediated transgene recombination and episomal minicircle formation to genetically engineer rabbit ASCs (rASCs). The baculovirus system conferred prolonged and robust TGF-β3/BMP-6 expression in rASCs cultured in porous scaffolds, which critically augmented rASCs chondrogenesis and suppressed osteogenesis/hypertrophy. Twelve weeks after implantation into full-thickness articular cartilage defects in rabbits, these engineered constructs displayed cartilage-specific zonal structures without signs of hypertrophy and degeneration, and eventually integrated with host cartilage.[118] Wang et al.[119] investigated the repair of articular cartilage defects with tissue-engineered cartilage constructed by acellular cartilage matrices from the rabbit ear and seeded with ASCs. After in vitro chondrogenic differentiation for 2 weeks, the constructs were implanted in 4 mm cartilage defects in rabbits. Articular cartilage defects of the rabbits implanted with tissue-engineered constructs were filled with chondrocyte-like tissue with smooth surface, while in the group implanted with acellular scaffolds, the defect was filled with fibrous tissue.[119] Forming a stable interface between the subchondral bone and tissue-engineered cartilage components remains a major challenge. Dua et al.[120] investigated the utility of HAp nanoparticles to promote controlled bone growth across the bone–cartilage interface in an in vitro engineered tissue model system using BM-derived stromal cells. Samples incorporated with HAp demonstrated significantly higher interfacial shear strength (at the junction between engineered cartilage and engineered bone) compared with the constructs without HAp, after 28 days of culture.

5. CLINICAL TRIALS

Significant strategies for regenerative medicine appeal to the use of cell therapy or tissue engineering. The first strategy applies the effect of cell signaling for clinical application, and the last approach combines scaffolds with bioactive signaling molecules and cells for tissue repair and regeneration. Among bioactive molecules, recombinant bone morphogenetic proteins (BMPs) GFs are most commonly used for bone growth and healing due to their osteoinduction ability. Nevertheless, some complications relating to the off-label use of BMP-2 in spinal and trauma surgery have been reported to result in the formation of ectopic epidural bone associated with severe neurological impairment in anterior interbody fusion surgery.[121,122]

Different stem cell sources such as human MSCs, human BMSCs, and human endometrial stem cells have been proposed for bone regeneration. It was reported in a clinical trial in which autologous BMSCs were seeded in macroporous HAp scaffolds showing a promising outcome of functional bone recovery, with good implant integration and host bone formation during 6–7 years postsurgery.[123] Other ongoing and complete clinical trials using several tissue engineering strategies for bone and cartilage repair/regeneration are well reported.[124–126] There is a variety of bone scaffolding products currently available in the market for clinical utility. On the contrary, few scaffolds are being commercialized for OC regeneration. This fact reflects the great challenge when addressing the simultaneous regeneration of two distinct tissues such as bone and cartilage. Considering all bone and cartilage tissue engineering and regenerative medicine strategies for clinical application, ongoing and completed clinical trials (with no reported results yet) for OC repair/regeneration using scaffolds or cell therapies, or even scaffolds combined with cells precultures in vitro are summarized in Table 2.

6. COMMERCIAL PRODUCTS

The process of commercialization of the scaffolds for implantation involves multiple stages of R&D replications before reaching the final approval from the governing bodies. R&D stages ensure safety and efficacy of the implants, which involve the production of medical grade scaffolds followed by animal testing under regulatory approved conditions. Scaffolds for bone tissue engineering are classified as biomedical devices under Class II Medium Risk.[127] For bone regeneration there are no tissue-engineered approaches fully approved for clinical application. Instead, just engineered materials/scaffolds already regulatory approved are arriving at the clinic as bone grafts (without the combination of cells), such as Infuse® Bone Graft (Medtronic Sofamor Danek) used for fusion of spinal cage, Osigraft (Stryker Biotech) for long bone nonunions applications, and Grafton® Orthoblend (OsteoTech) as a bone void filler for small and large defects, which have been successfully reported. Despite their efficacy in bone regeneration, clinical translation of scaffold-based bone therapies is limited to small defects due to insufficient mechanical integrity.

Food and Drug Administration (FDA)-approved scaffolds for craniofacial applications, such as Osteoplug and Osteomesh, are so far produced by Osteopore™. These scaffolds are composed of filaments of three-dimensional interwoven PCL polymer. They have higher mechanical strength and their architecture minimizes potential injuries to the exposed brain through the burr holes. Long-term clinical trials revealed significant bone regeneration with adequate resorption rate and no adverse reactions.[128]

Concerning cartilage tissue engineering, MACI is the better established technique for cartilage repair and the only FDA approved cell-based regenerative approach. Named Carticel, uses patient's own cartilage cells (chondrocytes) to treat and repair the articular cartilage damage in the knee.[28] However, nowadays there are some acellular approaches entering into the market. In fact, some of the products present in Table 2 are already being commercialized.

Bilayered or multilayered scaffolds, consisting of bone- and cartilage-like layers, seem to be the most promising strategy to achieve OC lesion regeneration.[129] Some studies have revealed promising results conjugating multilayer structures for OC regeneration and some acellular products are already being commercialized.[101,130] Among these few structures already in clinic, only three were reported in the literature (Figure 7). One is a bilayer PLGA-calcium-sulfate copolymer porous structure (Figure 7(A)).[131] The second OC scaffold is a nanostructured biomimetic HA-collagen scaffold with a porous 2D trilayer composite structure, mimicking the whole OC anatomy (Figure 7(B)).[130] The Kensey Nash cartilage repair device (CRD) is a biphasic (Figure 7(C)), bioresorbable scaffold intended to be implanted at the site of a focal articular cartilage lesion or OCD in the knee.[132] This CRD technology utilizes a biphasic design that contains two discrete layers. The chondral phase consists of a unique bovine collagen Type I matrix. The subchondral phase consists of β-TCP mineral suspended within a porous bioresorbable synthetic polymer scaffold.[132] Other OC scaffolds are still under preclinical investigation.[130]

7. CONCLUSIONS AND FUTURE DIRECTIONS

It is known that tissue-engineered technologies can take up to 20 years for reaching the market, and despite progress in many fields, this time frame has yet to be shorten. Accordingly, tissue engineering, which has officially given its first steps during the late 1980s, has not supplied many products to the bedside. Cell therapy strategies, as well as its first allogeneic stem cell therapy products, have been successfully applied for only few applications. Therefore, it is clear that, in spite of recent advances, tissue engineering has much to deliver in respect to combined products comprising biomaterials, GFs/bioactive molecules, and cells (differentiated or undifferentiated). Innovative strategies, such as the ones aforementioned, present out of the box solutions for some of the present challenges in the field and may constitute major breakthroughs in future in order to finally catalyze the translation of tissue-engineered products from bench to bedside.

TABLE 2 Overview of Ongoing and Complete Clinical Trials Using Strategies for OC Regeneration

NCT Number	Date and Phase	Name of the Clinical Trial	Patients Age	Follow-up	Procedure
NCT00891501	2006–2014 Phase 2 and 3	The Use of Autologous BM MSCs in the Treatment of Articular Cartilage Defects	15–55 years	n.d.	BM MSCs aspiration and implantation
NCT00560664	2007–2013 Phase 3	Comparison of ACI Versus Mosaicplasty	18–50 years	24 months	Autologous chondro-cytes transplantation and mosaicplasty
NCT00945399	2008–2011 Phase 3	Comparison of Micro-fracture Treatment and Cartipatch® Chondrocyte Graft Treatment in Femoral Condyle Lesions	18–45 years	18 months	ACI and microfracture
NCT00793104	2008–2012 Phase 3	Evaluation of the CR Plug (Allograft) for the Treatment of a Cartilage Injury in the Knee	≥18 years	24 months	Placement of allograft CR plug in primary injury site
NCT00821873	2008–2012 Phase 3	Evaluation of the CR Plug for Repair of Defects Created at the Harvest Site From an Autograft in the Knee	18–55 years	24 months	CR plug implantation in the harvest site
NCT01409447	2009–2011	Repair of Articular OCD	18–60 years	12 months	Biphasic OC composite implantation
NCT00984594	2009–2012 Phase 3	Evaluation of a Composite Cancellous and Demineral-ized Bone Plug (CR-Plug) for Repair of Knee OCDs	18–55 years	24 months	Autograft implantation in the primary defect site; CR plug implantation in the harvest site
NCT01183637	2010–2014 Phase 2	Evaluation of an Acellular OC Graft for Cartilage Lesions	≥21 years	24 months	Microfacture
NCT01159899	2010–2014 Phase 0	Transplantation of BM Stem Cells Stimulated by Proteins Scaffold to Heal Defects Articular Cartilage of the Knee	30–75 years	12 months	Transplantation of BM stem cells activated in knee arthrosis
NCT01209390	2010–2016	A Prospective, Post-marketing Registry on the Use of ChondroMimetic for the Repair of OCDs	18–65 years	36 months	Chondromimetic
NCT01473199	2011	BioPoly RS Knee Registry Study for Cartilage Defect Replacement	≥21 years	5 years	BioPoly RS partial resur-facing knee implantation
NCT01290991	2011–2012	A Study to Evaluate the Safety of Augment™ Bone Graft	18–40 years	12 months	Augment bone graft
NCT01410136	2011–2014	Chondrofix OC Allograft Prospective Study	18–70 years	24 months	Allogeneic OC grafting
NCT01477008	2011–2014 Phase 3	BiPhasic Cartilage Repair Implant	Up to 54 years	12 months	Marrow stimulation
NCT01282034	2011–2015 Phase 4	Study for the Treatment of Knee Chondral and OC Lesions	18–60 years	24 months	Marrow stimulation—drilling or microfractures

TABLE 2 Overview of Ongoing and Complete Clinical Trials Using Strategies for OC Regeneration—cont'd

NCT Number	Date and Phase	Name of the Clinical Trial	Patients Age	Follow-up	Procedure
NCT01471236	2011–2017 Phase 4	Evaluation of the Agili-C Biphasic Implant in the Knee Joint	18–55 years	24 months	Agili-C biphasic implantation and miniarthrotomy or arthroscopy
NCT01347892	2011–2019	DeNovo NT Ankle LDC Study	≥18 years	5 years	DeNovo NT natural tissue grafting
NCT01747681	2012–2013	Results at 10 to 14 years after Microfracture in the Knee	18–80 years	10 years	Microfracture
NCT01554878	2012–2014	Observational Study on the Treatment of Knee OC Lesions of Grade III–IV	30–60 years	12 months	Knee surgery
NCT01920373 (canceled)	2013 Phase 1	Platelet-Rich Plasma vs Corticosteroid Injection as Treatment for Degenerative Pathology of the Temporo-mandibular Joint	n.d.	6 months	Corticosteroid and platelet-rich plasma injection
NCT01799876	2013–2015	Use of Cell Therapy to Enhance Arthroscopic Knee Cartilage Surgery	18–68 years	12 months	Autologous cell and standard microfracture arthroscopic surgery
NCT02005861	2013–2016 Phase 3	"One-step" BM Mononu-clear Cell Transplantation in Talar OC Lesions	15–50 years	24 months	BM-derived cells trans-plantation on collagen scaffold
NCT02011295	2013–2017 Phase 4	BM Aspirate Concentrate Supplementation for OC Lesions	18–95 years	24 months	Ankle arthroscopy with debridement and micro-fracture

n.d., not defined; BM, bone marrow; MSCs, bone marrow stromal cells; ACI, autologous chondrocyte implantation; OCD, osteochondral defect; OC, osteo-chondral.
Information obtained from https://clinicaltrials.gov/

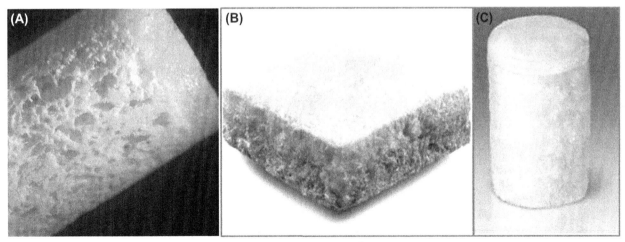

FIGURE 7 Scaffolds commercialized for osteochondral regeneration: (A) TruFit CB® implants in increasing diameters (7, 9, and 11 mm) *(Reprinted from Ref. 131 with permission.)*; (B) Maioregen® scaffold presenting three different gradient layers *(Reprinted from Ref. 130 with permission.)*; and (C) Cartilage repair device technology utilizes a biphasic design that contains two discrete layers. *(Reprinted from Ref. 132 with permission.)*

Using a product comprising a biomaterial seeded with cells from the own patient is appealing. Despite this, the culture conditions, mechanical stimuli, GFs cocktails, and environment conditions (as oxygen tension and tonicity) have to be better understood, correlated, and optimized for the ideal regenerative product to be obtained.

In a static cell culture system the environment is constantly changing due to accumulation of metabolites, nutrient consumption in the culture medium, and the consequent change in pH. This situation is clearly not representative of the in vivo state in which cells are maintained in equilibrium between the constant supply of nutrients and removal of products through the cellular secretory mechanism. Moreover, the metabolic requirements of a complex 3D environment are substantially larger than the necessary to maintain cell monolayers. To overcome these difficulties, several different bioreactor systems have been developed. A bioreactor can be described as a device or dynamic system for culturing cells or tissues under controlled conditions either biochemically or mechanically. Different bioreactors have been reported in the literature, including mixed flasks, rotating vessels, perfused cartridges, and bioreactors with different mechanical stimulation. For cartilage or bone tissue engineering separately there are some more options to improve the tissue maturation, as bioreactors with pneumatic compression, hypoxia chambers, and flow perfusion bioreactors. The main limitation of the current bioreactors for OC tissue engineering is that the newly formed tissue(s) is not homogeneously distributed within the constructs. Furthermore, there are just one bioreactor adapted for bilayered scaffolds (OC-related applications) that support different culture medium for each layer of the bilayered constructs, allowing inducing rotatory stimulus, compression, and vertical movement to avoid cell sedimentation and undesired tissue malformation, at the same time. Therefore, the developments in OC tissue regeneration and OC products in particular will be greatly dictated by the future developments in bioreactors and dynamic culture techniques/systems.

ACKNOWLEDGMENTS

The research leading to this work has received funding from the European Union's Seventh Framework Programme (FP7/2007–2013) under grant agreement n° REGPOT-CT2012-316331-POLARIS, and from QREN (ON.2—NORTE-01-0124-FEDER-000016) cofinanced by North Portugal Regional Operational Program (ON.2—O Novo Norte), under the National Strategic Reference Framework (NSRF), through the European Regional Development Fund (ERDF).

Thanks are due to the Portuguese Foundation for Science and Technology and POPH/FSE program for the fellowship grant of Raphaël Canadas (SFRH/BD/92565/2013). The FCT distinction attributed to J.M. Oliveira under the Investigator FCT program (IF/00423/2012) is also greatly acknowledged.

REFERENCES

1. Dequeker J, Aerssens J, Luyten F. Osteoarthritis and osteoporosis: clinical and research evidence of inverse relationship. *Aging Clin Exp Res* 2003;**15**:426–39.
2. Woolf AD, Pfleger B. Burden of major musculoskeletal conditions. *Bull World Health Organ* 2003;**81**:646–56.
3. Tanna S. *Osteoarthritis – opportunities to address pharmaceutical gaps. Priority medicines for Europe and the world. A public health approach to innovation.* 2004. P.6.121, 25.
4. Schindler O. Current concepts of articular cartilage repair. *Acta Orthop Belg* 2011;**77**:709–26.
5. Krych AJ, Stuart MJ. *Advances in articular cartilage defect management. ICRS Focus Meeting the Knee.* July 03–04, 2014. FIFA Auditorium Sonnenberg, Zurich, Switzerland.
6. Zaffagnini S, Giordano G, Vascellari A, Bruni D, Neri MP, Iacono F, et al. Arthroscopic collagen meniscus implant results at 6 to 8 years follow up. *Knee Surg Sports Traumatol Arthrosc* 2007;**15**(2):175–83.
7. Gaharwar AK, Schexnailder PJ, Schmidt G. Nanocomposite polymer biomaterials for tissue repair of bone and cartilage: a material science perspective. *Nanomaterials handbook*, vol. 24. Taylor and Francis Group, LLC; 2011.
8. Eastell R, Reid DM, Compston J, Cooper C, Fogelman I, Francis RM, et al. Secondary prevention of osteoporosis: when should a non-vertebral fracture be a trigger for action? *QJM* 2001;**94**(11):575–97.
9. Chapman MW, Rodrigo JJ. Bone grafting, bone grafts substitutes, and growth factors. In: *Chapman's orthopaedic surgery.* 3rd ed. Lippincott Williams & Wilkins; 2001.
10. Duncan R, Francis RM, Collerton J, Davies K, Jagger C, Kingston A, et al. Prevalence of arthritis and joint pain in the oldest old: findings from the Newcastle 85+ Study. *Age Ageing* 2011;**40**(6):752–5.
11. Wang S-Q, Gao Y-S, Wang J-Q, Zhang C-Q, Mei J, Rao Z-T. Surgical approach for high-energy posterior tibial plateau fractures. *Indian J Orthop* 2011;**45**(2):125–31.
12. Treuting R. Minimally invasive orthopedic surgery: arthroscopy. *Ochsner J* 2000;**2**(3):158–63.
13. Voscopoulos C, Lema M. When does acute pain become chronic? *Br J Anaesth* 2010;**105**(Suppl. 1):i69–85.
14. Cerqueira IS, Petersen PA, Mattar Júnior R, Silva JdS, Reis P, Gaiarsa GP, et al. Estudo anatômico da via de acesso suprapatelar lateral para a haste intramedular bloqueada na fratura da tíbia. *Rev Bras Ortop* 2012;**47**:169–72.
15. Maculé F, Sastre S, Lasurt S, Sala P, Segur JM, Mallofré C. Hoffa's fat pad resection in total knee arthroplasty. *Acta Orthop Belg* 2005;**71**:714–7.

16. Lubowitz JH. Partial-thickness articular cartilage defects: evaluation and treatment. *Oper Tech Orthop* 2006;**16**(4):227–31.

17. Makris EA, Gomoll AH, Malizos KN, Hu JC, Athanasiou KA. Repair and tissue engineering techniques for articular cartilage. *Nat Rev Rheumatol* 2014;**11**(1):21–34. Online publication.

18. Smith GD, Knutsen G, Richardson JB. A clinical review of cartilage repair techniques. *J Bone Joint Surg Br* 2005;**87-B**(4):445–9.

19. Pridie KH, Gordon G. A method of resurfacing osteoarthritic knee joints. *J Bone Joint Surg* 1959;**41**:618–9.

20. Johnson LL. Arthroscopic abrasion arthroplasty: a review. *Clin Orthop Relat Res* 2001;**391**:S306–17.

21. Hubbard MJS. Articular debridement versus washout for degeneration of the medial femoral condyle: a five-year study. *J Bone Joint Surg Br* 1996;**78-B**(2):217–9.

22. Steadman JR, Rodkey WG, Singleton SB, Briggs KK. Microfracture technique for full-thickness chondral defects: technique and clinical results. *Oper Tech Orthop* 1997;**7**(4):300–4.

23. Correia S, Pereira H, Silva-Correia J, Van Dijk C, Espregueira-Mendes J, Oliveira JM, et al. Current concepts: tissue engineering and regenerative medicine applications in the ankle joint. *J R Soc Interface* 2014;**11**(92).

24. Matsusue Y, Yamamuro T, Hama H. Arthroscopic multiple osteochondral transplantation to the chondral defect in the knee associated with anterior cruciate ligament disruption. *Arthroscopy* 1993;**9**(3):318–21.

25. Hangody L, Füles P. Autologous osteochondral mosaicplasty for the treatment of full-thickness defects of weight-bearing joints. *J Bone Joint Surg Am* 2003;**85-A**(Suppl. 2):25–32.

26. Horas U, Pelinkovic D, Herr G, Aigner T, Schnettler R. Autologous chondrocyte implantation and osteochondral cylinder transplantation in cartilage repair of the knee joint. *J Bone Joint Surg Am* 2003;**85-A**(2):185–92.

27. Peterson L, Minas T, Brittberg M, Lindahl A. Treatment of osteochondritis dissecans of the knee with autologous chondrocyte transplantation. *J Bone Joint Surg Am* 2003;**85-A**(Suppl. 2):17–24.

28. Brittberg M, Lindahl A, Nilsson A, Ohlsson C, Isaksson O, Peterson L. Treatment of deep cartilage defects in the knee with autologous chondrocyte transplantation. *N Engl J Med* 1994;**331**(14):889–95.

29. Peterson L, Brittberg M, Kiviranta I, Åkerlund EL, Lindahl A. Autologous chondrocyte transplantation: biomechanics and long-term durability. *Am J Sports Med* 2002;**30**(1):2–12.

30. Wakitani S, Imoto K, Yamamoto T, Saito M, Murata N, Yoneda M. Human autologous culture expanded bone marrow mesenchymal cell transplantation for repair of cartilage defects in osteoarthritic knees. *Osteoarthritis Cartilage* 2002;**10**(3):199–206.

31. Mano J, Silva G, Azevedo H, Malafaya P, Sousa R, Silva S, et al. Natural origin biodegradable systems in tissue engineering and regenerative medicine: present status and some moving trends. *J R Soc Interface* 2007;**4**:999–1030.

32. Pina S, Ferreira J. Bioresorbable plates and screws for clinical applications: a review. *J Healthc Eng* 2012;**3**(2):243–60.

33. Pereira D, Canadas R, Silva-Correia J, Marques A, Reis R, Oliveira J. Gellan gum-based hydrogel bilayered scaffolds for osteochondral tissue engineering. *Key Eng Mater* 2014;**587**:255–60.

34. Gunja NJ, Athanasiou KA. Biodegradable materials in arthroscopy. *Sports Med Arthrosc* 2006;**14**:112–9.

35. Daculsi G, Laboux O, Malard O, Weiss P. Current state of the art of biphasic calcium phosphate bioceramics. *J Mater Sci Mater Med* 2003;**14**(3):195–200.

36. Bohner M. Calcium orthophosphates in medicine: from ceramics to calcium phosphate cements. *Inj Int J Care Inj* 2000;**31**:37–47.

37. Kannan S, Goetz-Neunhoeffer F, Neubauer J, Ferreira JMF. Ionic substitutions in biphasic hydroxyapatite and beta-tricalcium phosphate mixtures: structural analysis by rietveld refinement. *J Am Ceram Soc* 2008;**91**(1):1–12.

38. Kannan S, Lemos AF, Ferreira JMF. Synthesis and mechanical performance of biological-like hydroxyapatites. *Chem Mater* 2006;**18**(8):2181–6.

39. Elliott JC. *Structure and chemistry of the apatites and other calcium orthophosphates*. London: Elsevier; 1994.

40. LeGeros RZ, LeGeros JP, Daculsi G, Kijkowska R. *Encyclopedia handbook of biomaterials and bioengineering*. New York: Marcel Dekker; 1995.

41. Dorozhkin S. Calcium orthophosphates in nature, biology and medicine. *Materials* 2009;**2**:399–498.

42. Ginebra MP, Traykova T, Planell JA. Calcium phosphate cements as bone drug delivery systems: a review. *J Control Release* 2006;**113**(2):102–10.

43. Takahashi Y, Yamamoto M, Tabata Y. Osteogenic differentiation of mesenchymal stem cells in biodegradable sponges composed of gelatin and beta-tricalcium phosphate. *Biomaterials* 2005;**26**:3587–96.

44. Metzger DS, Driskell TD, Paulsrud JR. Tricalcium phosphate ceramic: a resorbable bone implant: review and current status. *J Am Dent Assoc* 1982;**105**:1035–48.

45. Jones JR. Review of bioactive glass: from Hench to hybrids. *Acta Biomater* 2013;**9**(1):4457–86.

46. Kokubo T, Takadama H. How useful is SBF in predicting in vivo bone bioactivity? *Biomaterials* 2006;**27**(15):2907–15.

47. Xynos I, Edgar A, Buttery L, Hench L, Polak M. Gene expression profiling of human osteoblasts following treatment with the ionic products of Bioglass® 45S5 dissolution. *J Biomed Mater Res* 2001;**55**:151–7.

48. Lobel KD, Hench LL. In-vitro protein interactions with a bioactive gel-glass. *J Sol–Gel Sci Technol* 1996;**7**(1–2):69–76.

49. Wang G, Yang H, Li M, Lu S, Chen X, Cai X. The use of silk fibroin/hydroxyapatite composite co-cultured with rabbit bone-marrow stromal cells in the healing of a segmental bone defect. *J Bone Joint Surg Br* 2010;**92**:320–5.

50. Wang J, Zhou W, Hu W, Zhou L, Wang S, Zhang S. Collagen/silk fibroin bi-template induced biomimetic bone-like substitutes. *J Biomed Mater Res A* 2011;**99**:327–34.

51. Chen QZ, Thompson ID, Boccaccini AR. 45S5 Bioglass-derived glass ceramic scaffolds for bone tissue engineering. *Biomaterials* 2006;**27**:2414–25.

52. Yunos DM, Bretcanu O, Boccaccini AR. Polymer-bioceramic composites for tissue engineering scaffolds. *J Mater Sci Mater Med* 2008;**43**:4433–42.

53. Kashif I, Soliman AA, Sakr EM, Ratep A. Effect of different conventional melt quenching technique on purity of lithium niobate (LiNbO₃) nano crystal phase formed in lithium borate glass. *Results Phys* 2012;**2**:207–11.

54. Balamurugan A, Rebelo A, Kannan S, Ferreira JMF, Michel J, Balossier G, et al. Characterization and in vivo evaluation of sol–gel derived hydroxy-apatite coatings on Ti6Al4V substrates. *J Biomed Mater Res B Appl Biomater* 2007;**81B**(2):441–7.

55. Hench LL. Bioceramics. *J Am Ceram Soc* 1998;**81**:1705–28.

56. Huang W, Day D, Kittiratanapiboon K, Rahaman M. Kinetics and mechanisms of the conversion of silicate (45S5), borate, and borosilicate glasses to hydroxyapatite in dilute phosphate solutions. *J Mater Sci Mater Med* 2006;**17**(7):583–96.

57. Fu Q, Rahaman MN, Fu H, Liu X. Silicate, borosilicate, and borate bioactive glass scaffolds with controllable degradation rate for bone tissue engineering applications. I. Preparation and in vitro degradation. *J Biomed Mater Res A* 2010;**95**(1):164–71.

58. Knowles JC. Phosphate based glasses for biomedical applications. *J Mater Chem* 2003;**13**(10):2395–401.

59. Pina S, Ferreira J. Brushite-forming Mg-, Zn- and Sr-substituted bone cements for clinical applications. *Materials* 2010;**3**:519–35.

60. Mestres G, Le Van C, Ginebra M-P. Silicon-stabilized α-tricalcium phosphate and its use in a calcium phosphate cement: characterization and cell response. *Acta Biomater* 2012;**8**(3):1169–79.

61. Pina S, Vieira SI, Rego P, Torres PMC, Goetz-Neunhoeffer F, Neubauer J, et al. Biological responses of brushite-forming Zn- and ZnSr-substituted b-TCP bone cements. *Eur Cells Mater* 2010;**20**:162–77.

62. Bonfield W, Grynpas M, Tully A, Bowman J, Abram J. Hydroxyapatite reinforced polyethylene – a mechanically compatible implant material for bone replacement. *Biomaterials* 1981;**2**:185–6.

63. Yoshida T, Kikuchi M, Koyama Y, Takakuda K. Osteogenic activity of MG63 cells on bone-like hydroxyapatite/collagen nanocomposite sponges. *J Mater Sci Mater Med* 2010;**21**:1263–72.

64. Azami M, Samadikuchaksaraei A, Poursamar S. Synthesis and characterization of a laminated hydroxyapatite/gelatin nanocomposite scaffold with controlled pore structure for bone tissue engineering. *Int J Art Organs* 2010;**33**:86–95.

65. Yan L, Salgado A, Oliveira J, Oliveira A, Reis R. De novo bone formation on macro/microporous silk and silk/nano-sized calcium phosphate scaffolds. *J Bioact Comp Polym* 2013;**28**:439–52.

66. Tanase C, Sartoris A, Popa M, Verestiuc L, Unger R, Kirkpatrick C. In vitro evaluation of biomimetic chitosan-calcium phosphate scaffolds with potential application in bone tissue engineering. *Biomed Mater* 2013;**8**:025002.

67. Lee G, Park J, Shin U, Kim H. Direct deposited porous scaffolds of calcium phosphate cement with alginate for drug delivery and bone tissue engineering. *Acta Biomater* 2011;**7**:3178–86.

68. Heris H, Rahmat M, Mongeau L. Characterization of a hierarchical network of hyaluronic acid/gelatin composite for use as a smart injectable biomaterial. *Macromol Biosci* 2012;**12**:202–10.

69. Manda-Guiba G, Oliveira M, Mano J, Marques A, Oliveira J, Correlo V, et al. Gellan gum – hydroxyapatite composite hydrogels for bone tissue engineering. *J Tissue Eng Reg Med* 2012;**6**:15.

70. Costa DO, Dixon SJ, Rizkalla AS. One- and three-dimensional growth of hydroxyapatite nanowires during sol–gel–hydrothermal synthesis. *ACS Appl Mater Interfaces* 2012;**4**(3):1490–9.

71. Qi C, Zhu YJ, Zhao XY, Lu BQ, Tang QL, Zhao J, et al. Highly stable amorphous calcium phosphate porous nanospheres: microwave-assisted rapid synthesis using ATP as phosphorus source and stabilizer, and their application in anticancer drug delivery. *Chemistry* 2013;**19**(3):981–7.

72. Iwasaki T, Nakatsuka R, Murase K, Takata H, Nakamura H, Watano S. Simple and rapid synthesis of magnetite/hydroxyapatite composites for hyperthermia treatments via a mechanochemical route. *Int J Mol Sci* 2013;**14**(5):9365–78.

73. Zhou H, Bhaduri S. Novel microwave synthesis of amorphous calcium phosphate nanospheres. *J Biomed Mater Res B Appl Biomater* 2012;**100**(4):1142–50.

74. Sun L, Chow L, Frukhtbeyn S, Bonevich J. Preparation and properties of nanoparticles of calcium phosphates with various Ca/P ratios. *J Res Natl Inst Stand Technol* 2010;**115**:243–55.

75. Martins A, Reis RL, Neves NM. Electrospinning: processing technique for tissue engineering scaffolding. *Int Mater Rev* 2008;**53**.

76. Pina S, Oliveira JM, Reis RL. Natural polymer/calcium phosphates nanocomposites for bone tissue engineering and regenerative medicine: a review. *Adv Mater* 2015;**27**:1143–69.

77. Aizenberg J, Weaver J, Thanawala M, Sundar V, Morse D, Fratzl P. Skeleton of *Euplectella* sp.: structural hierarchy from the nanoscale to the macroscale. *Science* 2005;**309**:275–8.

78. Oliveira J, Silva S, Malafaya P, Rodrigues M, Kotobuki N, Hirose M, et al. Macroporous hydroxyapatite scaffolds for bone tissue engineering applications: physicochemical characterization and assessment of rat bone marrow stromal cell viability. *Inc J Biomed Mater Res A* 2009;**91**:175–86.

79. Hou Q, Grijpma D, Feijen J. Porous polymeric structures for tissue engineering prepared by a coagulation, compression moulding and salt leaching technique. *Biomaterials* 2003;**24**:1937–47.

80. Liapis A, Pikal M, Bruttini R. Research and development needs and opportunities in freeze drying. *Dry Technol* 1996;**14**:1265–300.

81. van de Witte P, Dijkstra P, van den Berg J, Feijen J. Phase separation processes in polymer solutions in relation to membrane formation. *J Memb Sci* 1996;**117**:1–31.

82. Dehghani F, Annabi N. Engineering porous scaffolds using gas-based techniques. *Curr Opin Biotechnol* 2011;**22**:661–6.

83. Abdelaal O, Darwish S. Fabrication of tissue engineering scaffolds using rapid prototyping techniques. In: *World academy of science, engineering and technology*, vol. 59. 2011. p. 577–85.

84. Chae T, Yang H, Leung V, Ko F, Troczynski T. Novel biomimetic hydroxyapatite/alginate nanocomposite fibrous scaffolds for bone tissue regeneration. *J Mater Sci Mater Med* 2013;**24**:1885–94.

85. Wang X, Ding B, Li B. Biomimetic electrospun nanofibrous structures for tissue engineering. *Mater Today* 2013;**16**(6):229–41.

86. Barbani N, Guerra G, Cristallini C, Urciuoli P, Avvisati R, Sala A. Hydroxyapatite/gelatin/gellan sponges as nanocomposite scaffolds for bone reconstruction. *J Mater Sci Mater Med* 2012;**23**:51–61.

87. Yan LP, Silva-Correia J, Correia C, Caridade SG, Fernandes EM, Sousa RA, et al. Bioactive macro/micro porous silk fibroin/nano-sized calcium phosphate scaffolds with potential for bone-tissue-engineering applications. *Nanomedicine (Lond)* 2013;**8**(3):359–78.

88. Yan LP, Oliveira JM, Oliveira AL, Caridade SG, Mano JF, Reis RL. Macro/microporous silk fibroin scaffolds with potential for articular cartilage and meniscus tissue engineering applications. *Acta Biomater* 2012;**8**(1):289–301.

89. Eftekhari S, El Sawi I, Bagheri ZS, Turcotte G, Bougherara H. Fabrication and characterization of novel biomimetic PLLA/cellulose/hydroxyapatite nanocomposite for bone repair applications. *Mater Sci Eng C* 2014;**39**(1):120–5.

90. Rodrigues SC, Salgado CL, Sahu A, Garcia MP, Fernandes MH, Monteiro FJ. Preparation and characterization of collagen-nanohydroxyapatite biocomposite scaffolds by cryogelation method for bone tissue engineering applications. *J Biomed Mater Res A* 2013;**101**(4):1080–94.

91. Slaughter BV, Khurshid SS, Fisher OZ, Khademhosseini A, Peppas NA. Hydrogels in regenerative medicine. *Adv Mater* 2009;**21**(32–33):3307–29.

92. Annabi N, Nichol J, Zhong X, Ji C, Koshy S, Khademhosseini A, et al. Controlling the porosity and microarchitecture of hydrogels for tissue engineering. *Tissue Eng Part B Rev* 2010;**16**:371–83.

93. Haraguchi K. Nanocomposite gels: new advanced functional soft materials. *Macromol Symp* 2007;**256**:120–30.

94. Zhu J, Marchant R. Design properties of hydrogel tissue-engineering scaffolds. *Expert Rev Med Devices* 2011;**8**:607–26.

95. Gaharwar AK, Dammu SA, Canter JM, Wu CJ, Schmidt G. Highly extensible, tough, and elastomeric nanocomposite hydrogels from poly(ethylene glycol) and hydroxyapatite nanoparticles. *Biomacromolecules* 2011;**12**(5):1641–50.

96. Fu S, Ni P, Wang B, Chu B, Luo F, Luo J, et al. Injectable and thermo-sensitive PEG-PCL-PEG copolymer/collagen/n-HA hydrogel composite for guided bone regeneration. *Biomaterials* 2012;**33**:4801–9.

97. Gantar A, da Silva LP, Oliveira JM, Marques AP, Correlo VM, Novak S, et al. Nanoparticulate bioactive-glass-reinforced gellan-gum hydrogels for bone-tissue engineering. *Mater Sci Eng C* October 1, 2014;**43**:27–36.

98. Nair LS, Bhattacharyya S, Laurencin CT. Development of novel tissue engineering scaffolds via electrospinning *Expert Opin Biol Ther* 2004;**4**: 659–68.

99. Rajzer I, Menaszek E, Kwiatkowski R, Planell JA, Castano O. Electrospun gelatin/poly(ε-caprolactone) fibrous scaffold modified with calcium phosphate for bone tissue engineering. *Mater Sci Eng C* 2014;**44**:183–90.

100. Marcacci M, Kon E, Zaffagnini S, Iacono F, Filardo G, Delcogliano M. Autologous chondrocytes in a hyaluronic acid scaffold. *Oper Tech Orthop* 2006;**16**(4):266–70.

101. Oliveira JM, Rodrigues MT, Silva SS, Malafaya PB, Gomes ME, Viegas CA, et al. Novel hydroxyapatite/chitosan bilayered scaffold for osteochondral tissue-engineering applications: scaffold design and its performance when seeded with goat bone marrow stromal cells. *Biomaterials* 2006;**27**:6123–37.

102. Gupta P, Das A, Chullikana A, Majumdar A. Mesenchymal stem cells for cartilage repair in osteoarthritis. *Stem Cell Res Ther* 2012;**3**(4):25.

103. Richter W. Mesenchymal stem cells and cartilage in situ regeneration. *J Int Med* 2009;**266**(4):390–405.

104. Wakitani S, Nawata M, Tensho K, Okabe T, Machida H, Ohgushi H. Repair of articular cartilage defects in the patello-femoral joint with autologous bone marrow mesenchymal cell transplantation: three case reports involving nine defects in five knees. *J Tissue Eng Reg Med* 2007;**1**(1):74–9.

105. Barry F, Murphy M. Mesenchymal stem cells in joint disease and repair. *Nat Rev Rheumatol* 2013;**9**(10):584–94.

106. English A, Jones EA, Corscadden D, Henshaw K, Chapman T, Emery P, et al. A comparative assessment of cartilage and joint fat pad as a potential source of cells for autologous therapy development in knee osteoarthritis. *Rheumatology* 2007;**46**(11):1676–83.

107. Marsano A, Millward-Sadler SJ, Salter DM, Adesida A, Hardingham T, Tognana E, et al. Differential cartilaginous tissue formation by human synovial membrane, fat pad, meniscus cells and articular chondrocytes. *Osteoarthr Cartil* 2007;**15**(1):48–58.

108. Khan WS, Adesida AB, Tew SR, Longo UG, Hardingham TE. Fat pad-derived mesenchymal stem cells as a potential source for cell-based adipose tissue repair strategies. *Cell Prolif* 2012;**45**(2):111–20.

109. Pirraco RP, Iwata T, Yoshida T, Marques AP, Yamato M, Reis RL, et al. Endothelial cells enhance the in vivo bone-forming ability of osteogenic cell sheets. *Lab Invest* 2014;**94**(6):663–73.

110. Fu WL, Xiang Z, Huang FG, Gu ZP, Yu XX, Cen SQ, et al. Coculture of peripheral blood-derived mesenchymal stem cells and endothelial progenitor cells on strontium-doped calcium polyphosphate scaffolds to generate vascularized engineered bone. *Tissue Eng A* 2014;**21**(5–6):948–59.

111. Zuk P, Zhu M, Mizuno H, Huang J, Futrell J, Katz A, et al. Multilineage cells from human adipose tissue: implications for cell-based therapies. *Tissue Eng* 2001;**7**(2):211–28.

112. Kyllonen L, Haimi S, Mannerstrom B, Huhtala H, Rajala K, Skottman H, et al. Effects of different serum conditions on osteogenic differentiation of human adipose stem cells in vitro. *Stem Cell Res Ther* 2013;**4**(1):17.

113. Steinwachs MR, Guggi T, Kreuz PC. Marrow stimulation techniques. *Injury* 2008;**39**(Suppl. 1):26–31.

114. Steinwachs MR, Kreuz PC, Guhlke-Steinwachs U, Niemeyer P. *Aktuelle Behandlung des Knorpelschadens im Patellofemoralgelenk*. Springer-Verlag; 2008;841–7.

115. Pelttari K, Pippenger B, Mumme M, Feliciano S, Scotti C, Mainil-Varlet P, et al. Adult human neural crest–derived cells for articular cartilage repair. *Sci Transl Med* 2014;**6**(251):25.

116. Zhang Q, Lu H, Kawazoe N, Chen G. Pore size effect of collagen scaffolds on cartilage regeneration. *Acta Biomater* 2014;**10**(5):2005–13.

117. Adachi N, Ochi M, Deie M, Nakamae A, Kamei G, Uchio Y, et al. Implantation of tissue-engineered cartilage-like tissue for the treatment for full-thickness cartilage defects of the knee. *Knee Surg Sports Traumatol Arthrosc* 2014;**22**(6):1241–8.

118. Lu CH, Yeh TS, Yeh CL, Fang YHD, Sung LY, Lin SY, et al. Regenerating cartilages by engineered ASCs: prolonged TGF-[beta]3/BMP-6 expression improved articular cartilage formation and restored zonal structure. *Mol Ther* 2014;**22**(1):186–95.

119. Wang ZJ, An RZ, Zhao JY, Zhang Q, Yang J, Wang JB, et al. Repair of articular cartilage defects by tissue-engineered cartilage constructed with adipose-derived stem cells and acellular cartilaginous matrix in rabbits. *Genet Mol Res* 2014;**13**(2):4599–606.

120. Dua R, Centeno J, Ramaswamy S. Augmentation of engineered cartilage to bone integration using hydroxyapatite. *J Biomed Mater Res B Appl Biomater* 2014;**102**(5):922–32.

121. Boraiah S, Paul O, Hawkes D, Wickham M, Lorich D. Complications of recombinant human BMP-2 for treating complex tibial plateau fractures: a preliminary report. *Clin Orthop Relat Res* 2009;**467**(12):3257–62.

122. Wong D, Kumar A, Jatana S, Ghiselli G, Wong K. Neurologic impairment from ectopic bone in the lumbar canal: a potential complication of off-label PLIF/TLIF use of bone morphogenetic protein-2 (BMP-2). *Spine J* 2008;**8**:1011–8.

123. Marcacci M, Kon E, Moukhachev V, Lavroukov A, Kutepov S, Quarto R, et al. Stem cells associated with macroporous bioceramics for long bone repair: 6- to 7-year outcome of a pilot clinical study. *Tissue Eng* 2007;**13**(5):947–55.

124. Liu Y, Lim J, Teoh S-H. Review: development of clinically relevant scaffolds for vascularised bone tissue engineering. *Biotechnol Adv* 2013;**31**(5):688–705.

125. Orth P, Rey-Rico A, Venkatesan JK, Madry H, Cucchiarini M. Current perspectives in stem cell research for knee cartilage repair. *Stem Cells Cloning* 2014;**7**:1–17.

126. Chen FH, Rousche KT, Tuan RS. Technology insight: adult stem cells in cartilage regeneration and tissue engineering. *Nat Clin Pract Rheum* 2006;**2**(7):373–82.

127. Sullivan F. *Advances in tissue engineering*. 2011.

128. Int O. *PCL Scaffolds for orbital reconstruction*. 2002.

129. Yan LP, Silva-Correia J, Oliveira M, Vilela C, Pereira H, Sousa RA, et al. Bilayered silk/silk-nanoCaP scaffolds for osteochondral tissue engineering: in vitro and in vivo assessment of biological performance. *Acta Biomater* 2014;**12**(2015):227–41.

130. Kon E, Delcogliano M, Filardo G, Pressato D, Busacca M, Grigolo B, et al. A novel nano-composite multi-layered biomaterial for treatment of osteochondral lesions: technique note and an early stability pilot clinical trial. *Injury* 2010;**41**:693–701.

131. Melton JTK, Wilson AJ, Chapman-Sheath P, Cossey AJ. TruFit CB® bone plug: chondral repair, scaffold design, surgical technique and early experiences. *Expert Rev Med Dev* 2010;**7**(3):333–41.

132. Ostrovsky G. *Bioresorbable, acellular, biphasic scaffold gets EU approval for knee cartilage repair: medGadget*. [updated February 17, 2010; cited November 25, 2014]. 2010.

Chapter 8

Current Applications for Bioengineered Skin

Sara Guerrero-Aspizua, Claudio J. Conti, Elisabeth Zapatero-Solana, Fernando Larcher, Marcela Del Río

Key Concepts

Bioengineered skin, stem cells, regenerative medicine.

1. INTRODUCTION

The formidable protective barrier function of the skin can be subverted by traumatic loss or disease. Self-repair of small and moderate cutaneous damage is an efficient process which, without any underlying disease, seldom requires external aid. However, extensive skin loss (e.g., severe burns) demands clinical intervention to achieve proper tissue restoration. Chirurgical techniques involving autologous transplantation of healthy skin to damaged areas (not considered in this review) remain the preferred approach to cover affected areas. Unfortunately, donor-site skin is often insufficient or its supply may inflict further suffering to the patient. Skin bioengineering provides an alternative source for repairing tissue. The skin has been one of the first organs in which several approaches of regenerative medicine have been successfully achieved. Easy access, as well as optimized methods to culture keratinocytes and fibroblasts, has played a pivotal role for this early success. Expanded epidermal and mesenchymal cells from 2D culture can be reassembled in three-dimensional (3D) skinlike structures. The optimization of these scaffolds using natural or synthetic materials has led to the development of graftable bioengineered 3D skin equivalents that fairly reproduce the morphology and physiology of skin.

Our laboratory has developed a whole autologous bioengineered skin, which is based on the use of fibroblasts embedded into a fibrin dermal scaffold that supports the growth and differentiation of keratinocytes. This autologous bioengineered skin has been widely used in our laboratory to develop preclinical models and more recently have also been used in the clinical scenario. This technology is particularly suitable to treat patients with severe skin loss that requires large-scale production of bioengineered skin.

The objective of this chapter is to present a review of prominent results achieved with bioengineered human skin in the clinic and for experimental purposes with the aim of permanent skin regeneration. We also touch on yet-to-come developments that can be envisioned from the use of accessory somatic cell types and those derived from pluripotent stem cells and new technological fabrication processes. Although different tissue engineering strategies will be considered, a thorough discussion of commercially available systems is beyond the scope of this review, and comprehensive discussions of the different products currently marketed can be found elsewhere.[1–4]

1.1 The Skin

The use of tissue engineering for regenerative purposes requires a deep understanding of the structure and functional characteristics of the native tissue and its development.

The skin is a complex organ that covers the surface of the body. It is the largest organ in terms of both weight and surface area and its main function is to act as a barrier to the surrounding environment preventing chemicals, bacteria, viruses, and ultraviolet light from entering the body. The skin also prevents water loss and regulates body temperature by blood flow and evaporation of sweat.[5]

The skin is formed by two anatomically, functionally, and developmentally distinct tissues: the epidermis and the dermis. In addition to these two distinct layers, the skin contains complex structures such as hair follicles, sebaceous

Translating Regenerative Medicine to the Clinic. http://dx.doi.org/10.1016/B978-0-12-800548-4.00008-5

glands, and sweat glands. These structures, known as skin appendages, are derived embryologically from the epidermis and represent a key component for skin functionality.[6]

1.1.1 Epidermis

The epidermis is the outermost component of the skin formed mostly by a particular kind of epithelial cells known as keratinocytes. Other epidermis resident cells also include melanocytes, Merkel cells, and Langerhans cells, responsible for pigmentation, sensory, and immunological functions, respectively.

The epidermis is morphologically divided into different layers or strata. From the bottom (innermost), these layers are stratum basale (basal cell layer), stratum spinosum (prickle cell layer), stratum granulosum (granular cell layer), stratum lucidum (clear layer), and stratum corneum (horny cell layer).[7] Keratinocytes produced in the basal layer, where cell proliferation is confined, move upward to the outer surface through a process of differentiation in which keratinocytes change their structures and physiological functions.

The differentiation process involves morphological and biochemical changes with temporal and spatial changes in gene expression. Specific proteins are characteristic of the cells at different layers of the skin. Thus, the proliferating basal keratinocytes, which express keratin 5 (K5) and keratin 14 (K14), adhere to the basal membrane (BM). Basal keratinocytes mature into suprabasal keratinocytes. This transition is characterized by loss of contact with the BM, proliferation arrest, and downregulation of K5 and K14 accompanied by upregulation of K1 and K10.[8] Finally, suprabasal keratinocytes undergo an apoptosis-related process called terminal differentiation which results in the formation of a layer of dead cornified cells, the stratum corneum. This layer is a main component of the protective skin barrier and its deficient formation is a cause enough to the development of barrier defective pathologies.[9] The terminal differentiation process is accompanied by the expression of marker proteins such as involucrin, fillagrin, and loricrin among many others.[10] It is important to notice that the epidermis is an avascular tissue and depends for its nourishing and oxygenation on the dermis.

1.1.2 Dermis

The dermis is a connective tissue layer that acts as a substrate and a support network for the epidermis. As every connective tissue, its embryological origin is the mesenchyme and its main cell type is the fibroblast which is responsible for the production and maintenance of the extracellular matrix (ECM). The ECM is formed by collagen and elastin, embedded in a nonfibrous ground substance composed of proteoglycans and glycosaminoglycans (GAGs).[11] Epidermal adnexa such as hair follicles and eccrine glands reside in the dermis. Cutaneous nervous and vascular tissues are also located in the dermis.

The epidermis and the dermis are separated by a complex BM composed by specialized proteins that serves as an epidermal and dermal anchoring structure but also establishes a boundary between epithelial and mesenchymal territories.[12] Mutations in the genes coding for the BM proteins (e.g., Type VII collagen or laminin 332) are responsible for rare inherited cutaneous diseases.

2. SKIN REGENERATIVE MEDICINE

A modern approach to tissue engineering for the purpose of regenerative medicine has to take into account three key components: cells, ECM, and growth factors (GFs) as well as the cross talk between them (Figure 1).[13] During the last years, significant efforts have been made in the development of engineered skin substitutes that try to mimic human skin. The basic premise of skin bioengineering to generate products aimed at permanent skin regeneration is to combine appropriate epidermal and mesenchymal cells with adequate biomaterial to produce a skin equivalent that is both functional and durable. This involves the integration and manipulation of the cell biology as well as the multitude of signals that control cell behavior,[14] including the maintenance of the stem cell population in a healthy, regenerative manner. The focus on developing precise functional objectives for the design and fabrication of skin bioengineering has led to significant advances in specific areas. These include, but are not limited to, cell sources, material design, and the creation of biomimic environments.[15] However, the knowledge of the components of bioengineered skin and their interplay is still incomplete to achieve a full organotypic skin, capable of restoring not only the barrier function but the whole variety of cutaneous appendages.

2.1 Cell Source

When developing a bioengineering skin substitute, it is essential to identify suitable sources of cells and to understand the mechanisms that allow them to properly maintain, proliferate, differentiate, and interact with the other components of the tissue.

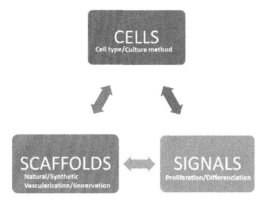

FIGURE 1 The three key components of skin tissue engineering and their interactions.

2.1.1 Epidermal Stem Cells

These cells reside in the skin and they are in charge of the maintenance of adult skin homeostasis and hair regeneration as well as participation in the repair and regeneration of injured skin.[15] The regeneration of the epidermis throughout life, in physiological conditions is achieved by specialized populations of basal keratinocytes known as epidermal stem cells (ESCs). Recent evidence suggests that multiple and dynamic stem cell populations exist.[16] Knowledge of the basic biology of the ESCs and the mechanisms that govern their capacity to differentiate into epidermis as well as skin appendages (multipotentiality) underlie the basis of cutaneous cell and genetic therapeutic strategies. In contrast with hematopoietic stem cells, the identification of bona fide functional markers for human ESCs is still controversial. Markers including, α6bri/CD71dim, Lrig1+, CD34, and others were suggested to be useful to enrich population of keratinocytes to obtain highly clonogenic subpopulations.[17,18] However, a criterion established more than 25 years ago, based on the in vitro proliferative capacity of keratinocytes, remains the most reliable way to identify the putative stem cells of human interfollicular epidermis.[19] Thus, clonogenic assays appear to be the best predictors of "stemness," at least in terms of the extensive proliferative capacity of the putative interfollicular ESCs.[19,20] These in vitro studies have demonstrated that cultured human epidermal cell clones are endowed with an extraordinary replicative potential. In vivo experiments from our group have made possible to evaluate the putative stem cell behavior of single genetically modified human clones (holoclones) through the use of bioengineering and improved surgical techniques.[21] Although the presence of functional stem cells is generally assumed in the keratinocyte batches (isolated from skin biopsies) used to produce bioengineered skin, failure of engraftment of these products may well be due to the lack of or highly reduced numbers of true stem cells, as has been shown in some genodermatoses.[22] Reliable and straightforward methods to identify these cells before seeding them on skin equivalents are still required.

2.1.2 Human Embryonic Stem Cells and Induced Pluripotent Stem Cell Derivatives

Much of the current enthusiasm for the study of human embryonic stem cells (hESCs) comes from the possible therapeutic use of somatic cells derived from them. Since hESCs are totipotent, they are able to differentiate into many different cell types. Many controversial ethical and technical problems need to be overcome before the full potential of hESC can be a reality. Several groups have shown the possibility of keratinocyte derivation from hESC. However, thus far only one group (which includes researchers of our team) succeeded in obtaining a homogenous population of keratinocytes derived from hESC with stem cell properties. Following assembly in a proper scaffold and grafting to immunodeficient mice, these keratinocytes retained their ability to regenerate a fully differentiated self-renewing epidermis.[23]

In the last few years, dedifferentiation of adult cells into a pluripotent embryonic stage has been achieved, and its interest for regenerative medicine has grown exponentially.[24–27] These cells are known as induced pluripotent stem cells (iPSCs).[28] While iPSCs have been generated from somatic cells, through genetic manipulation, optimization of the process is still underway. The obtaining of differentiated cells from iPSC or hESC is still a major challenge, but the potential use of patient-specific autologous cells appears as a very exciting strategy which eliminates the possibility of immunological rejection and avoids ethical dilemmas surrounding hESC.[29] Employing similar protocols than those used for hESC, several laboratories have attained differentiation of iPSC to fully functional keratinocytes.[25] Moreover, generation of 3D skin equivalents that includes iPSC-derived fibroblasts has also been reported.[30,31] iPSCs and keratinocyte derivatives from rare skin diseases such as dystrophic epidermolysis bullosa (EB) have also been produced.[24,32,33] Autologous iPSCs have the potential to provide an almost unlimited source of cells for skin regenerative therapies. However, pure iPSC-derived human skin keratinocytes capable of long-term skin regeneration have not been reported yet. Recently, derivation of melanocytes from iPSC has also been reported.[31,34]

2.1.3 Mesenchymal Stem Cells

Stem cells with the capacity to regenerate mesodermal tissues (connective tissue, bone, cartilage) can be easily isolated from blood, bone marrow, and fat.[35] These cells known as mesenchymal stem cells (MSCs) are a very attractive alternative, especially for the regeneration of the dermal compartment of the skin.[36] The interest of using MSCs in the regeneration of dermis is based on the immunomodulatory and antiapoptotic properties of MSCs. Different studies have demonstrated that MSCs can improve wound closure by controlling the inflammatory response, promoting the development of a well-vascularized granulation matrix, inducing the migration of keratinocytes, and inhibiting apoptosis during wound healing.[37–41] While the above-mentioned studies with MSC have been performed with locally applied cells, the challenge remains either to preserve their "stemness" or to achieve their differentiation in a controlled, efficient, and reproducible manner within graftable skin equivalents. Although some work demonstrate a good performance of skin equivalents containing MSC,[42] further studies are needed to understand the way to control the stem cell environment both chemically and physically in three dimensions. This new knowledge may lead to the development of a new generation of skin substitutes and replacements.

2.1.4 Miscellaneous Cells

The dermis is the place of residence of vascular and neural structures which perform cutaneous nourishing and sensory functions, respectively. Thus, it makes sense to include these cell types in 3D skin equivalents. Angiogenesis is a major event during skin wound healing. To mimic that physiological response, skin equivalents were manufactured containing, in the dermal compartment, endothelial cells capable of generating capillary structures competent for vasculature anastomosis with the recipient tissue after grafting.[43,44] This approach, which remains in an experimental stage, can be very important since the lack of vascularization of the tissue under regeneration post grafting is a major cause of cultured skin graft failure. Faulty cutaneous nerve regeneration leading to loss of pain, temperature, and touch perceptions is currently also a problem after grafting of skin substitutes which only restore barrier functions.[45] Thus, restoration of nerve fiber regeneration/migration is a major challenge in skin regenerative medicine. In this regard, Canadian researchers showed that the addition of Schwann cells to tissue-engineered skin not only enhanced nerve migration but also promoted myelin sheath formation in vitro and nerve function recovery in vivo in experimental models.[46]

An additional cell type that could eventually be included into the dermal equivalent is differentiated adipocytes. These cells present in the hypodermis functionally contribute with some of the mechanical and thermoregulatory properties of normal skin and could be obtained from MSC. In fact, a trilayer skin equivalent containing these cells has been described.[47]

2.2 Extracellular Matrix: Scaffolds

Originally, scaffolds were conceived to serve just as a basic support for cultured cells, but since then, several improvements have been achieved that provide a better physical and chemical properties for skin tissue engineering. The importance of 3D systems for cellular functions, such as migration and differentiation, has been widely discussed.[48]

Materials for skin bioengineering to date can be classified into two big groups, those derived from naturally occurring materials and those manufactured synthetically. In both groups, a large variety of materials with different porosities, permeabilities, and mechanical characteristics to address the diverse needs in skin regenerative medicine are available.[49]

Recent advances in the engineering of materials, such as microfabrication and nanotechnology, are particularly relevant for mimicking the natural ECM (biomimetic materials). These allow recreating the principles that govern whether cells grow, migrate, die, or differentiate. Similarly, findings in the interactions between cell surface receptors and ECM ligands will continue to provide inspiration for biomimetic surface modifications of scaffolds.[50]

One of the scaffolds used more in skin bioengineering is hydrogels. In general, gels have a soft tissuelike stiffness and aim to mimic the ECM. Gels made from ECM mixtures of natural origin, such as collagen, and alginate have been used for decades as substrates for 3D cell culture, especially for skin recreation. Others such as those prepared with GAGs are also popular.[51] The bioengineered skin equivalent used in our group is based on a natural fibrin hydrogel matrix and has been proved to be a perfect scaffold for skin bioengineering not only in preclinical models, but also in its clinical approach for burn patients and wound healing.[52,53] Its application is detailed below.

Given the possibility of producing robust artificial dermal scaffolds, novel tissue engineering techniques such as organ decellularization[54,55] are not practical in skin bioengineering aimed at producing large surfaces of skin equivalents for clinical purposes. However, a commercial acellular cadaveric skin matrix product (Alloderm®/Strattice®) exists and is suitable of seeding autologous keratinocytes to produce a full skin equivalent.[4]

The scaffold research represents an advance not only because it shows the profound influence of ECMs on stem cell differentiation but also because it ignites a pursuit toward identifying other properties of the different materials that can potentially control cell physiology.

2.3 Growth Factors and Their Delivery Systems

In addition to cells and scaffolds, molecular and biological signals such as cytokines or GFs are a key component for proper tissue regeneration/wound healing, a process intimately linked to the tissue embryonic development.[56] Thus, understanding the messages that trigger cellular proliferation, differentiation, and death to rationally incorporate these metabolic cues to the engineered tissue will provide a boost to man-driven skin regeneration.[57] More specifically, the delivery of biological molecules in a controlled way within the scaffold in the context of skin bioengineering appears highly attractive. The development of new technologies such as microspheres and nanospheres allows the delivery of different molecules inside the scaffolds in a spatially and temporally controlled way.[58] The choice of using a given combination from the numerous possible depends on what is the regenerative objective and is particularly important if nonnatural-sourced scaffolds are used, in order to mimic more deeply the conditions that are intrinsically present in the natural-sourced matrices. For instance, nanospheres or nanofibrous scaffolds that allow creating in vitro specific patterns are used for the study of cell migration and proliferation. On the contrary, if the goal is forming skinlike tissue within scaffolds in vivo, the choice will probably be hydrogels or microspheres.[59]

3. BIOENGINEERED SKIN SYSTEMS

Four decades ago, the methods for serial culture and large expansions of human epidermal keratinocytes were established.[60] Subsequently, methods to detach whole sheets of cultured epithelia were developed.[61] Although grafting of these pure epithelial sheets has been useful to reduce mortality of seriously burned patients all over the world,[62–64] this strategy showed important limitations including product fragility, limited engraftment efficacy, and abnormal ultrastructure of the dermoepidermal junction leading to ineffective clinical outcome.[65,66] It thus became evident that much more robust skin replacement systems containing dermal structures also were desirable leading to the advent of modern skin bioengineering.

When speaking about skin bioengineering, it should be important to differentiate between the generality of tissue-engineered skin substitutes, which encompass a great variety of cellular and acellular products, many of them in the market,[4] from the specificity of cultured (3D) skin equivalents or bioengineered skin that contain cellular epidermal and dermal structures. Thus, the most current commercial tissue-engineered products consist of sheets of a scaffold containing allogeneic cells (keratinocytes, fibroblasts, or both), which are typically derived from neonatal foreskin, a convenient tissue source with the added advantages of having a higher content of putative ESCs, robust cell growth, and metabolic activity. Remarkably, however, the allogeneic nature of these products makes them only suitable to promote endogenous wound healing as they cannot engraft. Therefore they are mainly used as transient healing aids for the treatment of chronic wounds. Many recent reviews have summarized the history and current status of matrices and skin substitutes.[1,2,67–71]

An ideal bioengineered skin should fulfill the following characteristics: it should mimic the structure and physiology of normal skin, being highly effective in achieving tissue regeneration and wound repair during the whole life of the individual; it should be autologous or not subject to immune rejection; and ready to be produced in a short period of time and in an inexpensive way.

The steps in creating and combining the components of bioengineered skin have been comprehensively discussed elsewhere.[15,67] While skin equivalents have been prepared in the laboratory by expert trained personnel, there is current interest in the automation of the process using 3D bioprinting techniques.[72]

Bioengineered skin equivalents have been growing in complexity, introducing different cell types, soluble mediators, scaffolds, and layers trying to mimic more completely native skin. Nowadays, only one autologous bioengineered skin product that uses collagen as a scaffolding material, PermaDerm® (Regenicin, Inc.), is in a premarket approval phase.

Although an ideal skin equivalent aiming at permanent skin regeneration is still a matter of study,[73] we have developed a fibroblast-containing fibrin scaffold-based bioengineered skin product (patent WO/2002/072800) that does fulfill many of the clinical requirements. We chose fibrin because in the wound healing process, fibrin is the primary and temporary wound healing matrix formed during blood clotting. Fibrin represents an excellent substrate for migration of both, epithelial and mesenchymal cellular elements that will repair the damaged tissue. As a consequence of keratinocyte–fibroblast–fibrin interactions, the right molecular signals and GFs for proliferation, survival, and homeostatic skin development are provided. In addition to fibrin, our matrix also contains other factors such as fibronectin or thrombospondin that may contribute to keratinocyte adherence and survival.

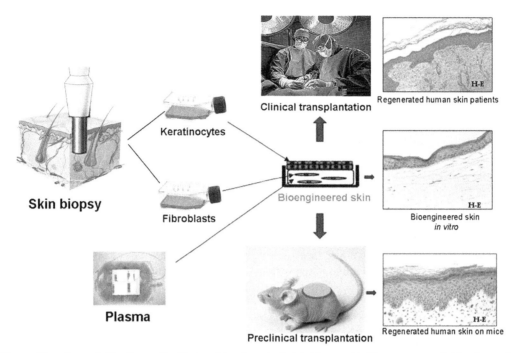

FIGURE 2 Human bioengineered skin. Human fibroblasts and keratinocytes isolated from a skin biopsy are expanded in vitro, and assembled in a fibrin-based matrix that acts as a dermal component. This bioengineered human skin has been successfully transplanted to patients and also represents a useful preclinical platform to model physiopathological process and to test innovative therapeutics. H–E, hematoxylin–eosin staining.

Our group has been working on a fully autologous skin equivalent with the development of a plasma-based matrix, since keratinocytes, fibroblasts, and fibrin may come from the same individual (Figure 2).[74] In this system, patient fibroblasts are embedded in human blood plasma that is clotted by the addition of calcium ions. The resulting fibrin hydrogel containing live fibroblasts is seeded with human keratinocytes that have been previously grown using the Rheinwald and Green culture method. When keratinocytes on top of the plasma dermal equivalent reach confluence, the bioengineered skin is manually detached from the culture flask or plate and is ready for grafting.

3.1 Clinical Applications of Bioengineered Skin

Here, we focus on the results obtained by our group using the above-mentioned fibrin-based bioengineered skin (Figure 3). Further information on skin tissue engineering applications can be found in several recent review articles.[70,75]

3.1.1 Permanent Skin Replacement

The plasma scaffold-based autologous system has been used successfully, for permanent skin regeneration in different situations such as extensive burns, necrotizing fascitis, removal of giant nevi and graft-versus-host disease.[74,76] In the case of burns a recent article compiled the satisfactory results achieved after grafting of the plasma-based autologous bioengineered skin in five Spanish burn units from 1999 to 2007 for 25 burned patients with an average total body surface area burned of 74%. The average graft take was 49%. In terms of patient satisfaction, cosmetic/functional outcomes (general appearance, texture, flexibility, sensitivity, and color) of the bioengineered skin behaved similarly to the split-thickness autografts.[77]

3.1.2 Transient Skin Replacement

Chronic wounds of various etiologies represent a chief medical challenge. The increasing aging population, prone to develop such wounds, will certainly require better products to cope with this problem. Allogeneic bioengineered skin substituted transplantation is widely used for both acute wounds and chronic ulcers.[76–79] The treatment of chronic wounds with allogeneic bioengineered skin does not produce acute rejection of the epidermis, but instead, the recipient's own cells gradually replace the keratinocytes of the transplant (silent rejection). These allogeneic skin substitutes have been useful because they not only provide temporary biologic dressing but also stimulate skin repair by recipient's cells. We have used our plasma-based bioengineered skin, in more than 100 patients. We observed a healing rate of 80% when fresh allogeneic bioengineered skin is applied weekly.[80]

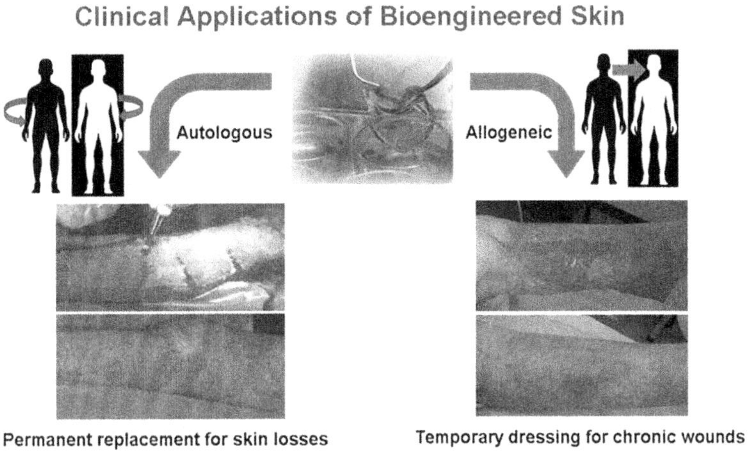

Clinical Applications of Bioengineered Skin

Permanent replacement for skin losses   Temporary dressing for chronic wounds

FIGURE 3 Clinical applications of fibrin-based bioengineered skin. Autologous: permanent skin regeneration on burn patients after autologous transplantation. Allogeneic: skin regeneration on chronic ulcers by allogeneic temporary coverage.

Two major causes are distinguished as responsible for healing failure: (1) poor vascularized wound beds and (2) infections of the graft-recipient bed. In this regard, recent studies from our and other laboratories showed that cutaneous tissue engineering in combination with gene therapies may provide a strategy to promote angiogenesis.[81,82] Antimicrobial peptides, such as the cathelicidin LL-37, have shown to be effective against a broad spectrum of pathogens, have low rates of bacterial resistance, and improve the repair process in part due to its additional angiogenic effect.[83,84] Skin equivalent systems able to deliver LL-37 or equivalent molecules emerge as powerful alternatives to conventional tissue-engineered skin products for recalcitrant ulcers.

3.1.3 Skin Bioengineering for EB

Within the large group of rare inherited skin diseases or genodermatosis, EB comprises an important family of bullous disorders characterized by skin fragility due to deficient epidermal–dermal adhesion (Table 1). The different forms of EB are due to mutation in the genes encoding various proteins with adhesive properties. Different grafting strategies have been used in EB although they do not treat the primary defect.[85–87] Cutaneous gene therapy appears as one of the therapeutic modalities explored to treat EB.[88] In fact, one patient with junctional epidermolysis bullosa (JEB), a form of EB caused by blistering at the level of BM has been successfully treated by ex vivo gene therapy using epithelial sheets of genetically corrected keratinocytes.[22] Similarly, the long-term correction of recessive dystrophic epidermolysis bullosa (RDEB), another form of EB, using a bioengineered skin equivalent containing gene-corrected keratinocytes in preclinical assays has also been achieved.[89,90] These approaches have concluded with the proposal of a European clinical trial (www.genegraft.eu) using our bioengineered skin system with genetically corrected RDEB keratinocytes and fibroblasts that obtained the orphan drug designation for the medicinal product: "Skin equivalent graft genetically corrected with a COL7A1-encoding SIN retroviral vector" (EU/3/09/630).

Other current tissue engineering approaches considered for EB can exploit revertant mosaicism. "Somatic revertant mosaicism" is a natural phenomenon involving spontaneous genetic correction of a pathogenic mutation in a somatic cell.[91] The incidence of this phenomenon, also known as natural gene therapy, appears to be more common than expected.[92,93] In the case of EB, the reversion has been found in several JEB patients[94] and also in RDEB patients[95,96] who display patches of unaffected skin, in which the presence of the corresponding adhesion protein (e.g., Type VII collagen), previously absent, has been reexpressed.

Transplantation of autologous "revertant" bioengineered skin may be a therapeutic option for patients with somatic mosaicism. This strategy is currently explored in our laboratory in collaboration with Marcel Jonkman's group in the Netherlands. In fact, the feasibility of the approach has been recently proven using revertant cells from a JEB patient (expressing Type XVII collagen) forming part of bioengineered skin transplanted to immunodeficient mice.[97] The ultimate goal of this strategy is the production of sufficient adhesion proteins from grafted revertant ESCs to ensure adequate and long-term phenotype recovery.

3.2 Preclinical Studies: The Skin-Humanized Mouse Model

In vivo studies in the skin in human beings are obviously limited by ethical and practical constrains. Pigs and murine models have been widely used to understand skin physiology and pathology, but its use is limited by elevated costs and their

TABLE 1 Review of Different Skin-Humanized Mouse Models

Pathway/Disease		Characteristics	Applications	References
Normal human skin		Skin-humanized mouse model that regenerates bioengineered normal human skin in an in vivo approach.	• To conduct studies in normal human skin to gain further insight into physiological processes. • To perform regenerative medicine, gene therapy, genomics, and pathology studies in a human context on homogeneous samples.	53,74,88
UV light response **Physiological and pathological**		Recapitulates the human phenotype in a physiological context: sunburn cell formation and p53 induction, two well-described surrogate markers of UV action.	• To measure protective effect of skin pigmentation. • To perform mutagenesis and carcinogenesis studies. • To test topic photoprotective agents.	53,108
Wound healing	**Physiological**	Recapitulates native skin wound-healing features: tissue architecture, cell proliferation, epidermal differentiation, dermal remodeling, and basement membrane regeneration.	• To evaluate pharmacological, cell and gene therapy strategies for impaired wound healing.	103
		Promoting wound healing through KGF gene therapy	• To compare the efficacy of different in vivo gene transfer strategies.	104
	Pathological	Promoting wound healing through LL-37 gene therapy in control and Ob/ob mice.	• To support the potential therapeutic applications of LL-37 in skin regeneration.	84
		Recapitulates delayed wound healing in a diabetic context and the improvement using dermal scaffolds containing fibroblasts.	• To shed light on the biological processes responsible for the wound healing and the improvement achieved. • To design new therapeutic approaches with clinical relevance.	105

Skin diseases			Recapitulated features	Objectives	Ref.
Simple monogenic	Ampullous disorders	EBs	Recapitulate general blistering caused by COL7A1 deficiency.	• To perform gene therapy studies in deficient COL7A1 keratinocytes and establish translational research toward individualized therapies in long-term treatments of EBs.	53,90,106
	Differentiation disorders	PC	Two different models with PC patients: 1. Unaffected skin biopsies: PC features after an acute hyperproliferative stimulus. 2. Affected skin biopsies: full recapitulation of the phenotype (marked acanthosis and epidermal blistering after minor trauma).	• To test novel pharmacological or gene-based therapies in order to establish translational research toward individualized therapies in long-term treatments of PC.	110
		NS	Recapitulates NS features and its correction by a lentiviral vector (SPINK5 gene).	• To study paracrine effects of gene therapy (benefits of LEKTI as a secreted protein) and establish translational research toward individualized therapies in long-term treatments of NS.	109
		LI	Recapitulates LI phenotype. Correction of the ichtyosis phenotype by enzyme replacement.	• To understand the pathology at the molecular level and to evaluate novel therapeutic approaches.	111,114
	Cancer-prone disorders	XPC	Recapitulates XPC features: severe deficiency in DNA repair mechanisms.	• To assess UVB-mediated DNA repair responses. • To provide a strong platform to test novel and individualized therapeutic strategies.	108
		GS	Recapitulates GS features: typical epithelial budding, a phenotypic change observed in premalignant human lesions.	• To establish translational research toward individualized therapies in long-term treatments of GS.	53
Complex inflammatory **Psoriasis**			Recapitulates psoriatic phenotype: hyperproliferation, alterations in the differentiation process and characteristic immune infiltration.	• To assess the contribution of potential susceptibility factors to the pathogenesis of psoriasis. • To provide a platform for developing new therapies.	112
Nonepidermal pathologies **Leptin deficiency**			Leptin replacement through genetically engineered human keratinocytes grafts induced body weight reduction.	• To establish a valuable therapeutic alternative for permanent treatment of human leptin deficiency conditions.	115

EB, epidermolysis bullosa; PC, pachyonychia congenita; NS, Netherton syndrome; LI, lamellar ichtyosis; XPC, xeroderma pigmentosum; GS, Gorlin syndrome.

structural differences, respectively.[98,99] It is necessary to search models that are closer to the human context and in a scale-up approach. Xenogeneic skin transplantation to immunocompromised mice is a valid alternative that also presents some drawbacks such as limited donor tissue and heterogeneity between samples.[100] A possibility to overcome all these problems involves the stable regeneration of normal or diseased human skin in appropriate hosts (e.g., immunodeficient mice) after bioengineered tissue engraftment.[101]

The preclinical model approach developed by our team (skin-humanized mouse model) allows the efficient regeneration of human skin in vivo and has been widely used in order to recapitulate the characteristics of native skin even with genetically modified cells[53,74,102] as well as to study physiological processes of the skin such as wound repair.[103–105] The approach has also been used to obtain experimental humanized models for a large number of pathologies such as different rare human monogenic skin diseases, including fragility disorders such as RDEB,[89,106] Kindler syndrome[107] (and unpublished results); the cancer-prone diseases xeroderma pigmentosum[108] and Gorlin syndrome[53]; and epidermal differentiation disorders such as Netherton syndrome,[109] pachyonichia congenita,[110] and lamellar ichtyosis.[111] In addition to genodermatosis, our team was able to generate a skin-humanized mouse model of a common chronic inflammatory disease such as psoriasis, where the immune component plays a pivotal role.[112]

Many of these humanized mouse models have been a unique platform to evaluate innovative therapeutic strategies for diseases that have no current treatment alternative, such as gene therapy,[81,102–104,113] enzyme replacement,[114] and even correcting a noncutaneous disease such as leptin defficiency.[115]

In Table 1, we present a list of the preclinical model of different diseases generated in our laboratory.

4. CHALLENGES AND FUTURE DIRECTIONS

The production of a full-thickness tissue-engineered skin, featuring both the epidermis and the dermis, is a major accomplished goal helping to improve healing quality and cosmetic outcome after grafting. The prospect is that new skin substitutes should be further engineered to offer the complete regeneration of functional skin compartments and the establishment of a functional vascular and nerve network with the surrounding host tissue. Recent and ongoing studies are addressing in part these challenging issues. Controlled incorporation of melanocytes, within the epidermal compartment, able to correct absent or irregular pigmentation is feasible.[116] Also, incorporation in the dermal equivalent of endothelial and Schwann cells has been attempted to initiate angiogenesis and neural regeneration in bioengineered skin grafts.[43,117,118] A bigger challenge remains in the regeneration of skin appendages (hair follicles, eccrine glands, and sensory organs). In contrast to that achieved with neonatal murine skin cells, the development of human bioengineered skin containing pilosebaceous units is still incomplete.[119] Further understanding of embryonic development mechanisms may be the key to tackle this unsolved problem. Concerning the autologous nature of all the cellular components, the hopes are put on the controlled derivation of every building block from iPSC.

A comprehensive understanding of bioengineered skin substitutes could lead to smart combinations with genetic modification of the cells to express various GFs and cytokines. As many other young therapeutic modalities, skin bioengineering and cutaneous cell therapy will certainly meet further improvements as basic and applied science continue to nourish the field.

The ultimate challenge of building a completely functional skin seems ostensibly insurmountable. Nevertheless, rapid progress in tissue engineering to design a skin substitute, including the use of stem cells, biomimic materials, and the right combination of GFs, may give us hope that such a product will be developed in the near future (Figure 4).

5. CONCLUSIONS

Even though skin tissue engineering is still a relatively young discipline, it has no choice but to mature in order to be translated into clinical applications. Systems based on bioengineered skin like those described in this chapter offer several advantages. These comprise the ability to grow cells that retain their functionality, including stemness, needed for long-term tissue regeneration. This bioengineered skin not only provides a clinically relevant alternative in the field of skin regeneration but has also led to the development of a robust preclinical platform for human pathologies in which innovative strategies such as gene and cell therapies can be evaluated.

Despite the fact that currently available bioengineered skin substitutes display a range of drawbacks discussed above, they are still more than a valuable tool for their clinical applications and potential development in the biomedical field. Although the challenge of building a completely functional skin is not yet a reality, multidisciplinary researchers, including biologists, clinicians, and engineers continue to work together to meet this challenge.

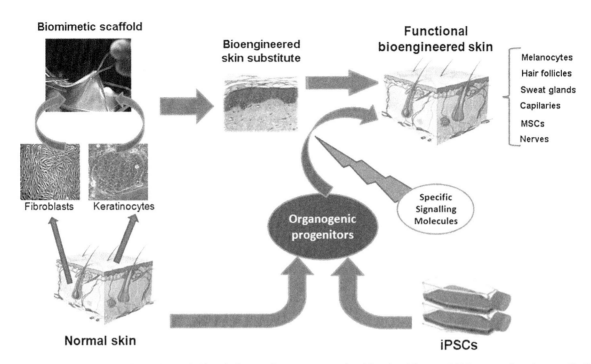

FIGURE 4 Future development of bioengineered skin substitutes to improve structural and functional features. MSCs, mesenchymal stem cells; iPSCs, induced pluripotent stem cells.

ACKNOWLEDGMENTS

FL was supported by grants from Instituto de Salud Carlos III (PI11/01225) and Comunidad de Madrid (S2010/BMD- 2359; SKINMODEL). MDR was supported by grants from the Science and Innovation Ministry of Spain (SAF2010-16976), Comunidad de Madrid (S2010/BMD-2420; CELLCAM), GENEGRAFT—contract N° HEALTH-F2-2011-261392 and CIBERER ACCI 13-714/172.04.

REFERENCES

1. Ehrenreich M, Ruszczak Z. Update on tissue-engineered biological dressings. *Tissue Eng* 2006;**12**(9):2407–24.
2. Horch RE, et al. Tissue engineering of cultured skin substitutes. *J Cell Mol Med* 2005;**9**(3):592–608.
3. Huang S, Fu X. Tissue-engineered skin: bottleneck or breakthrough. *Int J Burns Trauma* 2011;**1**(1):1–10.
4. Zhang Z, Michniak-Kohn BB. Tissue engineered human skin equivalents. *Pharmaceutics* 2012;**4**(1):26–41.
5. Ross MH, Pawlina W. *Histology: a text and atlas with correlated cell and molecular biology.* 6th ed. Lippincott Williams and Wilkins; 2010.
6. Mescher AL. *Junqueira's basic histology.* 13th ed. China: Lange: McGraw-Hill; 2013.
7. Segre JA. Epidermal barrier formation and recovery in skin disorders. *J Clin Invest* 2006;**116**(5):1150–8.
8. Fuchs E, Green H. Changes in keratin gene expression during terminal differentiation of the keratinocyte. *Cell* 1980;**19**(4):1033–42.
9. Kalinin AE, Kajava AV, Steinert PM. Epithelial barrier function: assembly and structural features of the cornified cell envelope. *Bioessays* 2002;**24**(9):789–800.
10. Mehrel T, et al. Identification of a major keratinocyte cell envelope protein, loricrin. *Cell* 1990;**61**(6):1103–12.
11. Fawcett D, Jensh R. *Bloom and Fawcett: concise histology.* 2nd ed. McGraw-Hill Interamericana; 2002.
12. Gawkrodger DJ. *Dermatology: an illustrated colour text.* Edinburgh: Churchill Livingstone; 2002.
13. MacNeil S. Progress and opportunities for tissue-engineered skin. *Nature* 2007;**445**(7130):874–80.
14. Boyce ST. Design principles for composition and performance of cultured skin substitutes. *Burns* 2001;**27**(5):523–33.
15. Lu G, Huang S. Bioengineered skin substitutes: key elements and novel design for biomedical applications. *Int Wound J* 2012;**10**(4):365–71.
16. Kretzschmar K, Watt FM. Markers of epidermal stem cell subpopulations in adult mammalian skin. *Cold Spring Harb Perspect Med* 2014;**4**(10).
17. Li A, et al. Extensive tissue-regenerative capacity of neonatal human keratinocyte stem cells and their progeny. *J Clin Invest* 2004;**113**(3):390–400.
18. Jensen KB, Watt FM. Single-cell expression profiling of human epidermal stem and transit-amplifying cells: Lrig1 is a regulator of stem cell quiescence. *Proc Natl Acad Sci USA* 2006;**103**(32):11958–63.
19. Barrandon Y, Green H. Three clonal types of keratinocyte with different capacities for multiplication. *Proc Natl Acad Sci USA* 1987;**84**(8):2302–6.
20. Mathor MB, et al. Clonal analysis of stably transduced human epidermal stem cells in culture. *Proc Natl Acad Sci USA* 1996;**93**(19):10371–6.

21. Larcher F, et al. Long-term engraftment of single genetically modified human epidermal holoclones enables safety pre-assessment of cutaneous gene therapy. *Mol Ther* 2007;**15**(9):1670–6.

22. Mavilio F, et al. Correction of junctional epidermolysis bullosa by transplantation of genetically modified epidermal stem cells. *Nat Med* 2006;**12**(12):1397–402.

23. Guenou H, et al. Human embryonic stem-cell derivatives for full reconstruction of the pluristratified epidermis: a preclinical study. *Lancet* 2009;**374**(9703):1745–53.

24. Wenzel D, et al. Genetically corrected iPSCs as cell therapy for recessive dystrophic epidermolysis bullosa. *Sci Transl Med* 2014;**6**(264):264ra165.

25. Bilousova G, Roop DR. Induced pluripotent stem cells in dermatology: potentials, advances, and limitations. *Cold Spring Harb Perspect Med* 2014;**4**(11).

26. Savla JJ, et al. Induced pluripotent stem cells for the study of cardiovascular disease. *J Am Coll Cardiol* 2014;**64**(5):512–9.

27. Fujita J, Fukuda K. Future prospects for regenerated heart using induced pluripotent stem cells. *J Pharmacol Sci* 2014;**125**(1):1–5.

28. Takahashi K, et al. Induction of pluripotent stem cells from adult human fibroblasts by defined factors. *Cell* 2007;**131**(5):861–72.

29. Nelson TJ, et al. Induced pluripotent stem cells: advances to applications. *Stem Cells Cloning* 2010;**3**:29–37.

30. Kogut I, Roop DR, Bilousova G. Differentiation of human induced pluripotent stem cells into a keratinocyte lineage. *Methods Mol Biol* 2014;**1195**: 1–12.

31. Guo Z, et al. Building a microphysiological skin model from induced pluripotent stem cells. *Stem Cell Res Ther* 2013;**4**(Suppl. 1):S2.

32. Itoh M, et al. Generation of keratinocytes from normal and recessive dystrophic epidermolysis bullosa-induced pluripotent stem cells. *Proc Natl Acad Sci USA* 2011;**108**(21):8797–802.

33. Sebastiano V, et al. Human COL7A1-corrected induced pluripotent stem cells for the treatment of recessive dystrophic epidermolysis bullosa. *Sci Transl Med* 2014;**6**(264):264ra163.

34. Nissan X, et al. Functional melanocytes derived from human pluripotent stem cells engraft into pluristratified epidermis. *Proc Natl Acad Sci USA* 2011;**108**(36):14861–6.

35. Hocking AM, Gibran NS. Mesenchymal stem cells: paracrine signaling and differentiation during cutaneous wound repair. *Exp Cell Res* 2010;**316**(14):2213–9.

36. Jackson WM, Nesti LJ, Tuan RS. Concise review: clinical translation of wound healing therapies based on mesenchymal stem cells. *Stem Cells Transl Med* 2012;**1**(1):44–50.

37. Sasaki M, et al. Mesenchymal stem cells are recruited into wounded skin and contribute to wound repair by transdifferentiation into multiple skin cell type. *J Immunol* 2008;**180**(4):2581–7.

38. Chen L, et al. Analysis of allogenicity of mesenchymal stem cells in engraftment and wound healing in mice. *PLoS One* 2009;**4**(9):e7119.

39. Smith AN, et al. Mesenchymal stem cells induce dermal fibroblast responses to injury. *Exp Cell Res* 2010;**316**(1):48–54.

40. Fathke C, et al. Contribution of bone marrow-derived cells to skin: collagen deposition and wound repair. *Stem Cells* 2004;**22**(5):812–22.

41. Badiavas EV, et al. Participation of bone marrow derived cells in cutaneous wound healing. *J Cell Physiol* 2003;**196**(2):245–50.

42. Formigli L, et al. Dermal matrix scaffold engineered with adult mesenchymal stem cells and platelet-rich plasma as a potential tool for tissue repair and regeneration. *J Tissue Eng Regen Med* 2012;**6**(2):125–34.

43. Supp DM, Wilson-Landy K, Boyce ST. Human dermal microvascular endothelial cells form vascular analogs in cultured skin substitutes after grafting to athymic mice. *FASEB J* 2002;**16**(8):797–804.

44. Sanchez-Munoz I, et al. The use of adipose mesenchymal stem cells and human umbilical vascular endothelial cells on a fibrin matrix for endothelialized skin substitute. *Tissue Eng Part A* 2014;**21**(1–2):214–23.

45. Blais M, et al. Concise review: tissue-engineered skin and nerve regeneration in burn treatment. *Stem Cells Transl Med* 2013;**2**(7):545–51.

46. Blais M, Grenier M, Berthod F. Improvement of nerve regeneration in tissue-engineered skin enriched with Schwann cells. *J Invest Dermatol* 2009;**129**(12):2895–900.

47. Monfort A, et al. Production of human tissue-engineered skin trilayer on a plasma-based hypodermis. *J Tissue Eng Regen Med* 2013;**7**(6):479–90.

48. Baker BM, Chen CS. Deconstructing the third dimension: how 3D culture microenvironments alter cellular cues. *J Cell Sci* 2012;**125**(Pt 13): 3015–24.

49. Lee J, Cuddihy MJ, Kotov NA. Three-dimensional cell culture matrices: state of the art. *Tissue Eng Part B Rev* 2008;**14**(1):61–86.

50. Place ES, Evans ND, Stevens MM. Complexity in biomaterials for tissue engineering. *Nat Mater* 2009;**8**(6):457–70.

51. Zhang L, et al. The modification of scaffold material in building artificial dermis. *Artif Cells Blood Substit Immobil Biotechnol* 2002;**30**(4):319–32.

52. Guerrero-Aspizua S, Carretero M, del Rio M. Hot topics in cell biology. In: Becerra J, Santos-Ruiz L, editors. *Stem cells in skin tisular engineering*. Spanish Society for Cell Biology; 2012. Congress (2011: Malaga, Spain). Chartridge. Books Oxford. xiii, 254 p.

53. Garcia M, et al. Modeling normal and pathological processes through skin tissue engineering. *Mol Carcinog* 2007;**46**(8):741–5.

54. Baptista PM, et al. Human liver bioengineering using a whole liver decellularized bioscaffold. *Methods Mol Biol* 2013;**1001**:289–98.

55. Murphy SV, Atala A. Organ engineering–combining stem cells, biomaterials, and bioreactors to produce bioengineered organs for transplantation. *Bioessays* 2013;**35**(3):163–72.

56. Barrientos S, et al. Growth factors and cytokines in wound healing. *Wound Repair Regen* 2008;**16**(5):585–601.

57. Saltzman WM, Olbricht WL. Building drug delivery into tissue engineering. *Nat Rev Drug Discov* 2002;**1**(3):177–86.

58. Martin Y, et al. Microcarriers and their potential in tissue regeneration. *Tissue Eng Part B Rev* 2011;**17**(1):71–80.

59. Garg T, Goyal AK. Biomaterial-based scaffolds – current status and future directions. *Expert Opin Drug Deliv* 2014;**11**(5):767–89.

60. Rheinwald JG, Green H. Serial cultivation of strains of human epidermal keratinocytes: the formation of keratinizing colonies from single cells. *Cell* 1975;**6**(3):331–43.

61. Green H, Kehinde O, Thomas J. Growth of cultured human epidermal cells into multiple epithelia suitable for grafting. *Proc Natl Acad Sci USA* 1979;**76**(11):5665–8.

62. Carsin H, et al. Cultured epithelial autografts in extensive burn coverage of severely traumatized patients: a five year single-center experience with 30 patients. *Burns* 2000;**26**(4):379–87.

63. Compton CC. Current concepts in pediatric burn care: the biology of cultured epithelial autografts: an eight-year study in pediatric burn patients. *Eur J Pediatr Surg* 1992;**2**(4):216–22.

64. O'Connor N, Mulliken J, Banks-Schlegel S, Kehinde O, Green H. Grafting of burns with cultured epithelium prepared from autologous epidermal cells. *Lancet* 1981;**1**(8211):75–8.

65. Mommaas AM, et al. Ontogenesis of the basement membrane zone after grafting cultured human epithelium: a morphologic and immunoelectron microscopic study. *J Invest Dermatol* 1992;**99**(1):71–7.

66. Woodley DT, et al. Burn wounds resurfaced by cultured epidermal autografts show abnormal reconstitution of anchoring fibrils. *JAMA* 1988;**259**(17):2566–71.

67. Boyce ST, Warden GD. Principles and practices for treatment of cutaneous wounds with cultured skin substitutes. *Am J Surg* 2002;**183**(4):445–56.

68. Hansen SL, et al. Using skin replacement products to treat burns and wounds. *Adv Skin Wound Care* 2001;**14**(1):37–44. quiz 45–6.

69. Beele H. Artificial skin: past, present and future. *Int J Artif Organs* 2002;**25**(3):163–73.

70. Catalano E, et al. Tissue-engineered skin substitutes: an overview. *J Artif Organs* 2013;**16**(4):397–403.

71. Cook EA, et al. Bioengineered alternative tissues. *Clin Pediatr Med Surg* 2014;**31**(1):89–101.

72. Murphy SV, Atala A. 3D bioprinting of tissues and organs. *Nat Biotechnol* 2014;**32**(8):773–85.

73. Metcalfe AD, Ferguson MW. Tissue engineering of replacement skin: the crossroads of biomaterials, wound healing, embryonic development, stem cells and regeneration. *J R Soc Interface* 2007;**4**(14):413–37.

74. Llames SG, et al. Human plasma as a dermal scaffold for the generation of a completely autologous bioengineered skin. *Transplantation* 2004;**77**(3):350–5.

75. Sun BK, Siprashvili Z, Khavari PA. Advances in skin grafting and treatment of cutaneous wounds. *Science* 2014;**346**(6212):941–5.

76. Llames S, et al. Clinical results of an autologous engineered skin. *Cell Tissue Bank* 2006;**7**(1):47–53.

77. Gomez C, et al. Use of an autologous bioengineered composite skin in extensive burns: clinical and functional outcomes. A multicentric study. *Burns* 2011;**37**(4):580–9.

78. Coto-Segura P, et al. Letter: efficacy of a self-made artificial skin in the treatment of chronic ulcers. *Dermatol Surg* 2007;**33**(3):392–4.

79. Coto-Segura P, et al. Potent analgesic effect of tissue-engineered skin in a terminal patient with severe leg ulcer pain. *Dermatol Surg* 2008;**34**(10): 1414–6. discussion 1416.

80. Llames S, Garcia-Perez E, Garcia V, Pevida M, Camblor L, Llaneza J, et al. Treatment of diabetic foot trophic lesions using a tissue-engineered dermal graft. *Tissue Eng Part A* 2008;**14**.

81. Lasso JM, et al. Improving flap survival by transplantation of a VEGF-secreting endothelised scaffold during distal pedicle flap creation. *J Plast Reconstr Aesthet Surg* 2007;**60**(3):279–86.

82. Lugo LM, Lei P, Andreadis ST. Vascularization of the dermal support enhances wound re-epithelialization by in situ delivery of epidermal keratinocytes. *Tissue Eng Part A* 2011;**17**(5–6):665–75.

83. Carretero M, et al. A cutaneous gene therapy approach to treat infection through keratinocyte-targeted overexpression of antimicrobial peptides. *FASEB J* 2004;**18**(15):1931–3.

84. Carretero M, et al. In vitro and in vivo wound healing-promoting activities of human cathelicidin LL-37. *J Invest Dermatol* 2008;**128**(1):223–36.

85. Eisenberg M, Llewelyn D. Surgical management of hands in children with recessive dystrophic epidermolysis bullosa: use of allogeneic composite cultured skin grafts. *Br J Plast Surg* 1998;**51**(8):608–13.

86. Falabella AF, et al. Tissue-engineered skin (Apligraf) in the healing of patients with epidermolysis bullosa wounds. *Arch Dermatol* 2000;**136**(10): 1225–30.

87. Fivenson DP, et al. Graftskin therapy in epidermolysis bullosa. *J Am Acad Dermatol* 2003;**48**(6):886–92.

88. Del Rio M, Larcher F, Jorcano J. Recent advances in gene therapy with skin cells. *Eur Rev* 2002;**10**:369–88.

89. Spirito F, et al. Sustained phenotypic reversion of junctional epidermolysis bullosa dog keratinocytes: establishment of an immunocompetent animal model for cutaneous gene therapy. *Biochem Biophys Res Commun* 2006;**339**(3):769–78.

90. Del Rio M, et al. Current approaches and perspectives in human keratinocyte-based gene therapies. *Gene Ther* 2004;**11**(Suppl. 1):S57–63.

91. Davis BR, Candotti F. Genetics. Mosaicism—switch or spectrum? *Science* 2010;**330**(6000):46–7.

92. May M. Mutations to the rescue. *Nat Med* 2011;**17**(4):405–7.

93. Lai-Cheong JE, McGrath JA, Uitto J. Revertant mosaicism in skin: natural gene therapy. *Trends Mol Med* 2011;**17**(3):140–8.

94. Jonkman MF, Pasmooij AM. Revertant mosaicism—patchwork in the skin. *N Engl J Med* 2009;**360**(16):1680–2.

95. Pasmooij AM, et al. Revertant mosaicism due to a second-site mutation in COL7A1 in a patient with recessive dystrophic epidermolysis bullosa. *J Invest Dermatol* 2010;**130**(10):2407–11.

96. Kiritsi D, et al. Mechanisms of natural gene therapy in dystrophic epidermolysis bullosa. *J Invest Dermatol* 2014;**134**(8):2097–104.

97. Gostynski A, Pasmooij AM, Jonkman MF. Successful therapeutic transplantation of revertant skin in epidermolysis bullosa. *J Am Acad Dermatol* 2014;**70**(1):98–101.

98. Garlick JA. Engineering skin to study human disease – tissue models for cancer biology and wound repair. *Adv Biochem Eng Biotechnol* 2007;**103**:207–39.

99. Sullivan TP, et al. The pig as a model for human wound healing. *Wound Repair Regen* 2001;**9**(2):66–76.

100. Wrone-Smith T, Nickoloff BJ. Dermal injection of immunocytes induces psoriasis. *J Clin Invest* 1996;**98**(8):1878–87.

101. Khavari PA. Modelling cancer in human skin tissue. *Nat Rev Cancer* 2006;**6**(4):270–80.

102. Del Rio M, et al. A preclinical model for the analysis of genetically modified human skin in vivo. *Hum Gene Ther* 2002;**13**(8):959–68.

103. Escamez MJ, et al. An in vivo model of wound healing in genetically modified skin-humanized mice. *J Invest Dermatol* 2004;**123**(6):1182–91.

104. Escamez MJ, et al. Assessment of optimal virus-mediated growth factor gene delivery for human cutaneous wound healing enhancement. *J Invest Dermatol* 2008;**128**(6):1565–75.

105. Martinez-Santamaria L, et al. The regenerative potential of fibroblasts in a new diabetes-induced delayed humanised wound healing model. *Exp Dermatol* 2013;**22**(3):195–201.

106. Gache Y, et al. Construction of skin equivalents for gene therapy of recessive dystrophic epidermolysis bullosa. *Hum Gene Ther* 2004;**15**(10):921–33.

107. Zapatero-Solana E, et al. Oxidative stress and mitochondrial dysfunction in Kindler syndrome. *Orphanet J Rare Dis* 2014;**9**:211.

108. Garcia M, et al. In vivo assessment of acute UVB responses in normal and Xeroderma Pigmentosum (XP-C) skin-humanized mouse models. *Am J Pathol* 2010;**177**(2):865–72.

109. Di WL, et al. Ex-vivo gene therapy restores LEKTI activity and corrects the architecture of Netherton syndrome-derived skin grafts. *Mol Ther* 2011;**19**(2):408–16.

110. Garcia M, et al. Development of skin-humanized mouse models of pachyonychia congenita. *J Invest Dermatol* 2011;**131**(5):1053–60.

111. Aufenvenne K, et al. Long-term faithful recapitulation of transglutaminase 1-deficient lamellar ichthyosis in a skin-humanized mouse model, and insights from proteomic studies. *J Invest Dermatol* 2012;**132**(7):1918–21.

112. Guerrero-Aspizua S, et al. Development of a bioengineered skin-humanized mouse model for psoriasis: dissecting epidermal-lymphocyte interacting pathways. *Am J Pathol* 2010;**177**(6):3112–24.

113. Bergoglio V, et al. Safe selection of genetically manipulated human primary keratinocytes with very high growth potential using CD24. *Mol Ther* 2007;**15**(12):2186–93.

114. Aufenvenne K, et al. Topical enzyme-replacement therapy restores transglutaminase 1 activity and corrects architecture of transglutaminase-1-deficient skin grafts. *Am J Hum Genet* 2013;**93**(4):620–30.

115. Larcher F, et al. A cutaneous gene therapy approach to human leptin deficiencies: correction of the murine ob/ob phenotype using leptin-targeted keratinocyte grafts. *FASEB J* 2001;**15**(9):1529–38.

116. Swope VB, Supp AP, Boyce ST. Regulation of cutaneous pigmentation by titration of human melanocytes in cultured skin substitutes grafted to athymic mice. *Wound Repair Regen* 2002;**10**(6):378–86.

117. Huang S, et al. In vitro constitution and in vivo implantation of engineered skin constructs with sweat glands. *Biomaterials* 2010;**31**(21):5520–5.

118. Huang S, et al. Mesenchymal stem cells delivered in a microsphere-based engineered skin contribute to cutaneous wound healing and sweat gland repair. *J Dermatol Sci* 2012;**66**(1):29–36.

119. Sriwiriyanont P, et al. Characterization of hair follicle development in engineered skin substitutes. *PLoS One* 2013;**8**(6):e65664.

Chapter 9

Urologic Tissue Engineering and Regeneration

Nan Zhang, Yuanyuan Zhang, Anthony Atala

Key Concepts

Cell-based therapy offers promising treatments for urological diseases such as tissue regeneration for bladder and urethral reconstruction, urinary incontinence, erectile dysfunction and renal insufficiency. This chapter describes the current status of tissue engineering and regeneration technology for each disease, and future directions to improve areas such as vascularization and innervation.

1. INTRODUCTION

Significant progress has been made in urological tissue engineering and cell therapy over the past two decades. Urological tissue regeneration strategies include several approaches: cell-based, biomaterials-based, and combined strategies (tissue engineering).

The ideal cell source for urological tissue engineering should be nonimmunogenic, easy to harvest, allow rapid large-scale expansion, and be highly functional. Autologous cells, in general, are preferable because they will not cause immune rejection and are less likely to transmit disease.[1] Two major cell types are critical for urological tissue regeneration—urothelial cells (UCs) and smooth muscle cells (SMCs). However, it is difficult to culture UCs and SMCs when the bladder biopsy is taken from the patients with urological diseases such as inflammation, stone or foreign body. For urologic regeneration and cell therapy, more appealing types of cells include pluripotent stem cells, neonatal stem cells, and mesenchymal stem cells (MSCs). In addition to their relative ease of use, stem cells have multipotent differentiation capacity, paracrine effect, and immunomodulatory properties.

The ideal scaffold for urological tissue regeneration is comparable to the bladder in terms of porosity, biocompatibility, and mechanical and physical properties. Although scaffolds are considered essential for supporting the transition from two- to three-dimensional (3D) cell growth, we do not know enough about the specific requirements of a given biological system to allow a rational approach to advanced development and design of biomaterials and scaffolds.[2] Meanwhile, current options include synthetic biomaterials, collagen matrix, nanosphere beads, and hydrogels.

Tissue engineering combines cell and scaffold technologies, and could confer major benefit over existing therapies. It capitalizes on the complexity of cell–cell and cell–matrix interactions. Three-dimension bioreactor provides a dynamic culture environment to improve the metabolic and nutritional environment, and promote tissue maturation. In particular, this 3D dynamic culture system significantly enhances cell adhesion, proliferation, infiltration, differentiation, and homologous distribution, enhanced nutrition, gas exchange, and waste removal, increased extracellular matrix (ECM) deposition, and amplified mechanical properties compared to traditional static-culture condition.[3–5] Whether cells are needed in combination with scaffolds depends on the clinical situation, for example, the size of injured tissue.

Tissue regeneration offers promising treatments for urological diseases such as renal insufficiency, urinary incontinence, erectile dysfunction (ED), and bladder and urethral tissue reconstruction. This chapter describes the current status of tissue regeneration technology for each disease and future directions to improve areas such as vascularization and innervation.

2. CELLS FOR IMPLANTATION

There are two primary cell sources currently in use for urological tissue engineering, UCs and SMCs. Human UCs have been successfully isolated from bladder urothelium, and expanded as monolayers in serum-free conditions. However, cells may be damaged by electrocautery during biopsy, or the presence of catheters or preexisting stones. UCs are usually capable of 4–10 in vitro passages. To provide functional cells for bladder or urethral reconstruction and remodeling, cultured

Translating Regenerative Medicine to the Clinic. http://dx.doi.org/10.1016/B978-0-12-800548-4.00009-7

cells at an early passage (<P5) provide optimal results. As with UCs, SMCs can only be expanded for a limited number (<P5) of generations in vitro. As a result of normal cell aging in tissue culture, the cultured cells eventually lose their phenotype and/or function (called dedifferentiation), or become senescent in early passages.

Stem cells are an appealing cell source for regenerative medicine and tissue engineering applications due, in part, to their self-renewal, long-term ability to expand in vitro, differentiation potential, immune characteristics, and paracrine effects. Various types of stem cells have been used to treat urological diseases, including pluripotent embryonic and induced pluripotent stem cells (iPSCs), neonatal stem cells from amniotic fluid and placenta, and MSCs isolated from bone marrow, adipose tissue, skeletal muscle and peripheral blood, and urine, see Table 1.

2.1 Cell Expansion

Embryonic stem cells (ESCs) and iPSCs have remarkable self-renewal and diverse differentiation capacity. These cells can be maintained for over 80 passages and 200 population doublings (PDs) in vitro.[9,10] However, their application has been hindered by ethical issues and in vivo tumorigenicity. Fetal stem cells, such as amniotic fluid and placenta-derived stem cells, show promise for future clinical applications.[11,12] These cells can give rise to cells from the endoderm, mesoderm, and ectoderm, and can be maintained for over 250 passages. Long telomeres are retained and a normal karyotype without tumorigenicity is observed in vivo.[13] MSCs have been successfully isolated from various types of tissues. These cells usually reach a PD rate of 20–40 in 10 passages.[14] However, their differentiation capacities are usually limited to mesodermal cell lineages.

TABLE 1 Cells in Urological Tissue Engineering

Type of Cells	Type of Cells	Cell Renewal (Population Doubling/Passage)	Differentiation Capacity	Paracrine Effect	Immunomodulatory Properties
Pluripotent stem cell	ESC	200 PD/80P	Pluripotent with teratoma formation	Unknown	Unknown
	iPSC	120 PD/40P	Pluripotent with teratoma formation	Unknown	Unknown
Neonatal stem cells	PLSC	30PD/20P	Pluripotent without teratoma formation	IL-6, VEGF, HGF, FGF, TGF-β, IGF	IL-8, IL-10, and HGF
	AFSC		Pluripotent without teratoma formation	IL-8, IL-6, TGF-β, TNFRI, VEGF, and EGF[6]	IL-10, TGF-β, IDO, PEG-2[7]
Mesenchymal stem cell	USC	60–70 PD/20P	Multipotent differentiation potential; give rise to three dermal cell lineages	VEGF, IGF-1, FGF-1, PDGF, HGF, NGF and MMP9	IL-6 and IL-8
	BMSC	25–30 PD/10P	Multipotent, but mainly limited within mesodermal cell lineages	VEGF, HGF, FGF, MCP-1, G-CSF, M-CSF, GM-CSF, IL-7, PGF, IL-1 and IL-6, SDF-1, and MMP-9[8]	IFN-γ, TNF-α, and IL-1α or -1β[8]
	ASC	25–30 PD/10P	Multipotent, but mainly limited within mesodermal cell lineages	VEGF, HGF, G-CSF, M-CSF, GM-CSF, IL-7[8]	IFN-γ, TNF-α, and IL-1α or -1β[8]
	skMPC	70PD			
	HFSC				
Mature function cell	UC	4–10P	No	No	No
	SMC	10P	No	No	No

ADSC=Adipose-derived stem cells; AFSC=amniotic fluid stem cells; PLSC=placental stem cells; BMSC=bone marrow stem cells; ESC=embryonic stem cells; iPSC=induced pluripotent stem cells; SkMPC=skeletal muscle progenitor cells; USC=urinary stem cells; HFSC=hair follicle stem cells; UC=urothelial cells; SMC=smooth muscle cells.

Urine-derived stem cells (USCs) are reported to have a PD rate of 60–70 for up to 20 passages, whereas other USCs without telomerase activity can be maintained for 8–10 passages with 34 PDs.[15,16] Based on this ratio of cells and urine volume, two urine samples containing 20–30 USC clones could potentially yield at least 1.5×10^9 USCs at the end of passage four within 4–5 weeks.[1,17] Isolation of USCs is a separation- and digestion-free procedure. Urine samples are simply centrifuged, and cells are seeded in a mixed media composed of keratinocyte serum-free medium and embryonic fibroblast medium in a 1:1 ratio.[18] Expanded USCs are a relatively homogeneous population, and only require 2–5% serum to be maintained in vitro; in contrast, most MSCs require 10–20% serum.[15] When cells collected from voided urine are cultured in USC culture media, only USCs tend to attach to the culture container and continuously expand in culture.[15] This quick, easy, and economical process may also facilitate their large-scale expansion for potential clinical trials (Table 1).

To refine cell-based therapies to treat kidney, ureter, bladder, urethra, or penile diseases, the appropriate animal models are essential. For example, because canine models of renal disease share similarities to human beings, translational studies aimed at improving clinical outcomes have important implications for similar therapies in patients. Table 2 summarizes the currently available animal models used for studying specific urological diseases. Use of the most appropriate animal model will expedite the understanding of disease mechanisms, onset, and progression, and ultimately the discovery of treatments.

2.2 Multipotentiality of MSCs

During tissue repair, stem cells work in multiple ways to accelerate tissue repair and regeneration (Table 3). Although ESCs and iPSCs are pluripotent and capable of differentiating into multiple specialized cell types,[21] their ability to differentiate into urologic-specific cell types is relatively low.[14] Under appropriate conditions, bone marrow stromal cells (BMSCs) can be successfully induced into cells with bladder SMC characteristics, both in vitro and in multiple animal models.[22–24] Induced BMSCs proliferate at similar rate as bladder SMCs, and possess a similar histological appearance and contractile phenotype. However, only 5–10% of cells in BMSCs can be efficiently induced into either urethral endothelial or SMCs

TABLE 2 Preferred Types of Stem Cells and Animal Models for Studies for Different Urinary Tract System Sites and Disorders

	Kidney	Ureter	Bladder	Urethra	Urethra Sphincter for SUI	Penile Tissue for ED
Stem cell types	iPSC, ESC, BMSC, ADSC, AFSC, USC	ADSC,[19] BMSC,[20] USC	BMSC, ADSC, AFSC, USC	BMSC, ADSC, USC	SkMPC, BMSC, ADSC, AFSC, USC	EC, BMSC, ADSC, AFSC, USC
Experimental animals preferred	Mouse, rat, rabbit, canine, porcine	Canine, porcine	Canine, porcine	Rabbit, canine	Rat, canine, porcine	Rat

ADSC=Adipose-derived stem cells; AFSC=amniotic fluid stem cells; BMSC=bone marrow stem cells; ESC=embryonic stem cells; ED=erectile dysfunction; iPSC=induced pluripotent stem cells; SkMPC=skeletal muscle progenitor cells; SUI=stress urinary incontinence; USC=urinary stem cells.

TABLE 3 Multiple Modes of Action Attributed to Mesenchymal Stem Cells

Multiple Functions of Stem Cells	Outcomes and Potential Applications	References
Multipotent differentiation	Osteocytes, chondrocytes, adipocytes, myocytes, epithelial and endothelial cells for cell replacement	Bharadwaj et al.[15,17]
Secretion of trophic factors	Recruitment resident cells to tissue repair	Liu et al.[32]
Secretion of extracellular matrix	Prompt cell proliferation, rejuvenation, and differentiation of bone marrow-derived stem cells	Pei et al., 2014[33]
Immunomodulatory and anti-inflammatory	Inhibited T and B cell proliferation to decrease fibrosis	Wu et al.[34]
Gene delivery via angiogenic growth factor gene transfection	Soluble factor gene therapy to accelerate tissue regeneration at the site of chronic injury with extensive scarring	Liu et al.[36,49] and Ouyan et al.[35]

with specific lineage marker expression.[25–30] One reason for this could be their relative scarcity; only an estimated 1 in 10,000 or 15,000 cells in bone marrow are BMSCs.[31]

Although the effect of aging on MSCs remains controversial, some studies show an age-related decline in stromal vascular fraction number, proliferation rate, longevity, differentiation potential, and MSC immunophenotype expression in BMSCs from older versus younger donors.[37–41] This is an important question given that the aging population is increasing, as is the prevalence of urologic diseases among aging patients. Considering the large quantity of MSCs (up to 10^9) required for clinical applications, especially for bladder reconstruction, it is critical to find a stable cell source with potent MSCs.[1]

USCs can differentiate into a number of lineages in vitro. After exposure to certain conditions, USCs from a single clone can give rise to adipogenic, osteogenic, chondrogenic, and myogenic lineages, confirmed at the gene, protein, and cellular levels.[15] They also can transdifferentiate into SMCs (endoderm) and neurogenic cells (ectoderm). When differentiated USCs are seeded onto a porcine small intestinal submucosal scaffold, subcutaneously implanted constructs successfully generate functional adipose tissue, bone, cartilage, endothelium, and urothelium in nude mice.[15] Further, no teratomas were found in vivo, indicating that USCs are a promising source for stem cell-based urological treatment.[15]

More importantly, USCs can differentiate into urological tissue-specific cell lineages, including UCs, SMCs, and endothelial cells (ECs). They have higher differentiation frequency into UCs than BMSCs. Under the same induction conditions for 7 days, up to 60–70% USCs are induced into a uroepithelial lineage, compared to only 5% for BMSCs.[15,23] If induction extended to 14 days, up to 90% USCs are differentiated. Cells develop a cobblestone-like morphology, and express both urothelial-specific cell markers (uroplakin-III, uroplakin-Ia), and generic epithelial cell markers (CK7 and AE1/AE3). Differentiated USCs also show tight junction markers (ZO-1, E-cadherin, cingulin) in a dose-dependent and time-dependent manner.[15] Those cells display enhanced barrier function, with at least a 60% decrease in leakage compared to noninduced cells.[15]

In one study, urothelially differentiated USCs were seeded on a porcine small intestinal submucosal scaffold and cultured in vitro for 14 days before implantation into nude mice. Constructs generated stratified layers in vivo, and neotissue expressed urothelial-specific cell markers (uroplakin-III, uroplakin-Ia). USCs and BMSCs have comparable differentiation potency into SMCs.[25] Up to 80% of induced USCs express early differentiation markers (α-smooth muscle actin and calponin), contractile SMC markers (desmin, myosin), and smooth muscle-specific marker (smoothelin) after 14 days induction with smooth muscle differentiation media in vitro.[15] When implanted in vivo on a porcine small intestinal submucosal scaffold, those cells formed multiple layers of SMCs beneath UC layers, and expressed SMC markers (desmin, myosin, α-smooth muscle actin). USCs are also capable of endothelial differentiation with barrier function. EC-induced USCs display vessel-like structures on a solidified Matrigel surface after in vitro induction. Those cells express EC-specific genes (vWF, CD31), proteins (CD31, vWF, KDR, FLT1, eNOS), and a marker of tight endothelial junctions (VE-cadherin). When implanted subcutaneously into athymic mice, EC-induced USCs effectively form neovessel structures.[42]

2.3 Paracrine Effects

Tissue MSCs are thought to replace cells lost to age, disease, or trauma. They have a paracrine effect, producing local bioactive factors. The MSC production of cytokines and growth factors is reported to be relatively constant among different donors and not influenced by age or health status.[43] However, bioactive factors produced by MSCs are affected by differentiation and the local environment.[43] Effects of these factors can be direct (i.e., inducing intracellular signaling), indirect (i.e., stimulating contiguous cells to secrete other bioactive factors), or both. Indirect activity is referred as trophic, since the MSCs do not differentiate themselves, but their secretion of bioactive factors mediates cell behavior. Some trophic effects include local immune suppression, enhancing angiogenesis, inhibiting apoptosis and fibrosis, and inducing differentiation and mitosis of tissue-specific and tissue-intrinsic progenitors.[43]

Exogenous MSCs contribute to tissue repair by homing to damaged tissue, where they secrete paracrine and autocrine signals to influence local cellular dynamics. This effect has been reported in spinal cord injury; myocardial infarction; meniscus, cartilage, and bone repair; and Crohn's disease.[44–46] Chopp et al. showed that rat marrow MSCs promote intrinsic neural progenitor cells to regenerate functional neurological pathways in damaged brain tissue when applied directly or systemically. Notably, the exogenous cells do not differentiate into neurons or neuronal support cells to mediate these effects.[44–47]

When implanted BMSCs are placed on an appropriate biodegradable scaffold, they can act as antifibrotic, angiogenic, antiapoptotic, and mitotic agents.[48] When exposed in a favorable microenvironment, USCs can secrete angiogenic growth factors and cytokines.[35,49] When genetically modified via vascular endothelial growth factor (VEGF), USCs proficiently differentiated into myogenic-specific cell types, and further promoted angiogenesis and innervation.[50] However, the viral transfection required causes hyperemia, hemorrhage, and even death. An alternative approach is highly desirable to induce efficient gene transfection without severe side effects. An exogenous hydrogel, alginate, successfully supported tissue regeneration and healing.[51] Alginate microbeads stably released active FGF-1 for at least 3 weeks in vitro, and stimulated neovascularization in vivo without side effects.[52–54] Additionally, USCs locally released a combination of growth factors, including VEGF, IGF-1, FGF-1, PDGF, HGF, and NGF, which induced USCs myogenic differentiation, revascularization and innervation, and stimulated resident cell growth in vivo.[32]

2.4 Immunoregulatory Properties

MSCs play multiple roles in immunomodulation. The immunosuppressive effects of MSCs allow them to be used for allogeneic transplantation.[55] By definition, major histocompatibility complex (MHC) class I proteins are expressed by MSCs while MHC class II proteins are not.[56,57] However, this characteristic is influenced by the microenvironment. The MHC II expression can be either upregulated or downregulated by interferon-γ.[58–61] Therefore, the immunosuppressive effect of MSCs is independent of MHC I and II expressions. The MSCs modulate the functions of the major immune cells including T, B cells, natural killer (NK) cells, monocytes, and dendritic cells (DCs). They suppress T cell proliferation, cytotoxic T lymphocyte formation, and interferon-γ production, thus inducing the expansion of T_{reg} cells.[62–65] MSCs also inhibit B cell proliferation, differentiation, chemotactic functions, and IgG secretion.[66–68] Additionally, they inhibit NK cell proliferation, NK cell-mediated cytolysis, and NK cell production of interferon-γ.[69] They also inhibit both DC maturation and activation, and differentiation of monocytes to DCs, thereby increasing T cell anergy.[70] Thus, MSCs are currently being used to reduce immunological rejection.[55]

Immunosuppressive function in MSCs is mediated by soluble factors, since cellular communication is not contact dependent.[55,71] Several factors have been found in coculture systems of immune cells and MSCs (human and murine BMSCs, placenta and dental pulp MSCs). These include stem cell factor, multiple forms of interleukin (IL-1β, IL-6, IL-8, IL-10, IL-12), interferon-γ, transforming growth factor-β1, VEGF, prostaglandin E2, macrophage colony-stimulating factor, hepatocyte growth factor, and indoleamine 2,3-dioxygenase.[69,56,72–74] However, how MSCs interact with immune cells is still unclear.

USCs can impart profound immunomodulatory effects by inhibiting proliferation of peripheral blood mononuclear cells (PBMNCs), T cells and B cells, and by secreting IL-6 and IL-8. They are involved in induction of peripheral tolerance, inhibition of proinflammatory immune responses, and decreased immune reactions.[34] PBMNCs proliferate when mixed with other cells due to immune stimulation.[75] However, PBMNC concentrations in USC wells were much lower than in BMSC culture wells.[75] CD80 and CD86 expressed on the surface of antigen-presenting cells interact with cytotoxic T lymphocyte antigen-4 expressed on activated T cells and mediate critical T cell inhibitory signals. On flow cytometry, 3.35% of the BMSCs were positive for CD80 (vs 1.05% of USCs), and 1.3% of the BMSCs were positive for CD86 (vs 0.55% of USCs). Human cytokine released arrays showed that IL-6 and IL-8 concentrations were elevated after stimulation by PBMNCs in the USC supernatant compared to the BMSC supernatant. Thus, IL-6 and IL-8 might be the main immunomodulatory cytokines that will be target in future studies, aimed at preventing and treating diabetic bladder tissue lesions, other immune system disorders, or rejection of transplanted organs.

Due to the advantages of unlimited supply; simple, fast, and economical isolation; superior in vitro expansion and longevity; and high differentiation potential to urological cell types, USCs may represent one of the most promising cell sources for cell therapies in treatment of various urological diseases, such as renal insufficiency, urinary incontinence, and ED, and bladder and urethral tissue regeneration.

3. BIODEGRADABLE BIOMATERIALS

As a required component in tissue engineering, a biological scaffold that is tailored to stem cells is central to their capability to mimic the function of the ECM. The ECM provides structural support and a physical environment so that cells can attach, proliferate, migrate, differentiate, and function.[76] It confers mechanical properties to tissues and delivers bioactive cues for regulating activities of residing cells, and provides a dynamic environment for vascularization and new tissue formation. Scaffolds can be designed to stimulate and direct tissue formation to replace portions of tissues or whole tissue structures. The material would possess appropriate porosity and microporosity (interconnectivity between pores) to expedite cell attachment, migration, penetration, differentiation, tissue growth, and integration. The ideal replacement should be composed of materials with similar physical and mechanical properties as the native tissue, and should degrade at the same rate as the new tissue is generated. Porosity should be sufficient for nutrient transfer and cell adhesion without compromising mechanical strength. Four categories of scaffolds designed to carry cells include synthetic scaffolds, collagen matrix, nanosphere beads with slow-releasing growth factors, and hydrogel with controllable cytokine release.

3.1 Synthetic Scaffolds

Porous structures composed of natural or synthetic biodegradable and biocompatible materials are one of the most popular approaches to scaffold carriers. Certain biomaterials have been approved by the US Food and Drug Administration (FDA) for human use as medical devices, such as polyglycolic acid (PGA), polylactic acid (PLA), poly(lactic-co-glycolic acid)

(PLGA), and polycaprolactone. The degradation metabolites of these materials are confirmed to be nontoxic, and eventually eliminated from the body as carbon dioxide and water.[77] In addition, synthetic polymers can be manufactured on a large scale using various techniques, including electrospinning, phase separation, gas foaming, particulate leaching, inkjet printing, and chemical cross-linking. Strength, degradation rate, and microstructure may be adjusted during manufacturing. Scaffolds can be made in different shapes and porosity to facilitate cell engraftment, or further modified by incorporation, surface adsorption, or chemical attachment of bioactive factors. However, it is critical that synthetic scaffolds are biocompatible according to the US FDA regulations, meaning that they have minimum cytotoxicity, sensitization, hemocompatibility, pyrogenicity, genotoxicity, carcinogenicity, and reproductive and developmental toxicity.

3.2 Collagen Matrix

Naturally occurring matrix materials may also function as 3D scaffolds in tissue engineering and regenerative medicine, including both decellularized natural matrix and matrix produced from natural extracted polymers. Natural materials are widely used in tissue regeneration. For example, collagen—the most abundant structural protein in the body—is an FDA-approved biocompatible material for various types of medical uses. Natural ECM that is decellularized retains tissue-specific architecture, and provides a wide range of biological and physical material properties specified by the nature of the original tissue. Additionally, those matrix materials share highly conserved matrix proteins among species, such as collagen, laminin, and fibronectins, indicating that these matrix materials could be nonimmunogenic and attractive for use in recellularization and tissue integration. Under certain circumstances, UCs and SMCs differentiated from USCs can form multiple uniform layers on a porcine decellularized small intestine submucosa scaffold in vitro and in vivo, showing that this 3D cell matrix can develop into a multilayer mucosal structure similar to native urinary tract tissue.[78] In addition, ECM secreted by USCs greatly improves the chondrogenic capacity of BMSCs at later passages, possibly due to the trophic factors released by USCs.[33] When seeded with USCs, bacterial cellulose scaffolds represent a promising material for urinary conduits, with a multilayered urothelium and cell/matrix infiltration in vitro and in vivo.[16] When considering clinical applications, decellularized matrix could be contaminated by xenogeneic factors. There is also a risk of incomplete decellularization and residual cell bodies, as well as altered tissue properties, due to complete decellularization and deproteinization.[79,80]

3.3 Hydrogels and Nanosphere Beads

Hydrogel formulations can be made with both natural and synthetic materials by physical and chemical cross-linking.[81,82] Scaffolds are customized to the intended in vivo application. Tissue-specific properties and the amount of tissue to be regenerated are considerations when identifying scaffold carriers. Alginate, a polysaccharide isolated from seaweed, has been widely used as a natural hydrogel and cell immobilization matrix, due to its mild gelling conditions and tunable microbead characteristics. This polymer delivers molecules in a controlled fashion to promote regeneration and tissue healing.[51] Because of their negatively charged nature, alginate microspheres repel protein adsorption, stably release active FGF-1, and promote neovascularization without side effects in vivo.[52–54] Alginate microbeads are also reported to control local levels of myogenic, angiogenic, and neurogenic growth factors in vitro, and to induce USCs to differentiate into a myogenic lineage, with enhanced revascularization and innervation, and stimulated resident cell growth in vivo.[32] Given these promising results, this approach could potentially be used for cell therapy in the treatment of stress urinary incontinence (SUI). A family of peptide amphiphile molecules can apply self-assembling coatings for fiber-bonded PGA scaffolds, and facilitate SMC adherence to human bladder.[83] These nanostructured biomaterials have great potential to improve cell alignment and tissue formation.[84]

4. APPLICATIONS IN URINARY TRACT SYSTEM

4.1 Kidney Regeneration

4.1.1 Incidence

It is estimated that 20 million people in the US—over 10% of all adults—may have chronic kidney disease of varying degrees. It is associated with all cause and cardiovascular mortality, and may progress to end-stage renal disease (ESRD).[85] Other conditions such as glomerular diseases, malnutrition, infectious diseases, and acute kidney injury also may progress to ESRD. ESRD accounts for 6.3% of Medicare expenses in the US, and is predicted to increase to nearly 13% by 2015. Therefore, it is imperative to develop therapeutic treatments to prevent, alleviate, or decelerate the progression of renal failure. According to Centers for Disease Control and Prevention, in 2011 alone, over 113,000 patients in the US started treatment for ESRD. Diabetes or hypertension is considered the primary cause for 70% of new cases of ESRD.

4.1.2 Current Treatments

Chronic glomerular and tubulointerstitial fibrosis is a common pathway to ESRD, and often associated with apoptosis, oxidative damage, and microvascular rarefaction. ESRD is usually treated with dialysis or transplantation. Dialysis treatment can only partially replace kidney function. It also requires intermittent connection with an artificial means of renal replacement. Organ transplantation is highly limited to a large patient population by the shortage of donor organs. Based on reports from National Kidney Foundation, there are currently over 123,000 people waiting for a kidney transplant; of those, over 100,000 patients are waiting for the right kidney, and the waiting list increases by 3000 patients per month. In contrast, fewer than 17,000 kidney transplants took place in 2013. In 2013, around 4500 patients died while awaiting a suitable match for a kidney transplant. As the number of ESRD patients continues to rise disproportionate to the number of available donor kidneys, it is urgent to identify a new source of organs.

4.1.3 Cell Therapy

The kidney is a very complex organ; it has cortex, medulla, and papilla regions, with as many as 25 different types of cells. Recent advancements in cell therapy introduce the possibility of a regenerative approach to treat renal disease. Initially, stem cells induce regeneration of damaged kidney tissue. After tubular necrosis, BMSCs were recruited to the site of injury in mice, indicating a possible contribution to kidney regeneration.[86] Those injected BMSCs are also visible in mice with acute kidney injury but not normal animals.[87,88] BMSCs are associated with enhanced tubular proliferation and differentiation; protection from peritubular capillary changes (e.g., EC abnormalities, leukocyte infiltration, and low EC and lumen volume density); and prolonged survival of animals. Cord blood MSCs also contribute to similar improvement of renal function.[89] Even though these MSCs did not differentiate into kidney-specific cell types, they function as trophic factors that promote kidney tissue regeneration and repair. When seeded on a scaffold, murine ESCs accumulate within the glomerular, vascular, and tubular structures of the decellularized ECM.[90] These cells are capable of differentiating into renal cells, losing their embryonic markers, and expressing renal markers.

In swine models, intrarenally delivered allogeneic adipose-derived stem cells (ASCs) protected the stenotic kidney despite sustained hypertension with attenuated renal inflammation, endoplasmic reticulum stress, and apoptosis.[91] Further studies showed that those MSCs improved renal function and structure after renal revascularization and reduced inflammation, oxidative stress, apoptosis, microvascular remodeling, and fibrosis in the stenotic kidney.[92] Similar treatment also restored oxygen-dependent tubular function in the stenotic kidney medulla, which retained medullary structure and function under chronic ischemic conditions.[93] Further, a decellularization protocol for kidneys from rhesus monkeys has been established for future application of renal tissue engineering and repair. The decellularized scaffolds retain critical structure and functional properties.[94] After recellularization with age-matched kidney explants, the scaffold shows renal cell attachment and migration. This method could be applied as a xenogenic scaffold into patients.

Many clinical trials with stem cell therapies are currently underway, both with MSCs and other cell types.[95] The autologous transplant study shows some encouraging safety outcomes, which provides preliminary reassurance that the efficacy of administering mobilized MSCs (or other cells) to ameliorate human kidney disease may be testable in the near future.

4.2 Ureter

4.2.1 Incidence

Ureteral damage is mostly caused by injury or disorders that may lead to stricture formation, such as urolithiasis, chronic inflammation, and muscle-invasive ureteral cancer.[96,97] Most injuries are iatrogenic, due to surgical procedures or radiation therapy. Up to 73% of ureteral injuries occur during gynecological surgery, most often hysterectomy, but 33–87.5% of cases remain unrecognized at the time of surgery. If not repaired properly, ureteral damage can result in sepsis or loss of renal function.

4.2.2 Current Treatment

Ureteral defects are usually treated with endourological procedures and ureteroscopy. However, this is not always appropriate for long ureteral defects (e.g., Boari flap, psoas hitch, transureteroureterostomy, reimplantation, Blandy cystoplasty, or ileal interposition). In addition, surgical complications may occur, such as recurrent strictures, urinary leakage, metabolic side effects, and problems with harvesting donor tissue.

4.2.3 Cell-Seeded Tissue Engineering

Tissue engineering techniques can eliminate most of the complications related to currently used procedures. However, compared to other urological diseases, tissue engineering studies on urinary conduit construction have been sporadic.[98–100] In initial studies using a small intestinal submucosal scaffold seeded with the 3T3 fibroblast cell line, severe inflammatory reactions occurred.[98] In a porcine model, a scaffold built with type I collagen and Vyproll synthetic mesh was sufficient to enhance fibroblast deposition and tissue contraction.[101] When using a bladder acellular matrix (BAM) seeded with bladder epithelial cells in rabbits, the constructs successfully generated mature and functional epithelial cell layers on the lumen without obstruction. In comparison, in cell-free scaffolds, lack of epithelial layer regeneration, scar formation, atresia, severe hydronephrosis, and death occurred.[102,103] For small repairs, cell-seeded scaffolds may not be necessary, since urothelium can easily self-regenerate from surrounding tissue.[99] However, when large pieces of ureter tissue are reconstructed, UCs may be included to promote tissue regeneration.

4.3 Bladder

4.3.1 Incidence

Stem cell-based therapy for bladder repair is most relevant to congenital bladder anomalies (bladder exstrophy, myelomeningocele, posterior urethral valves) or conditions such as radiation damage, infection, interstitial cystitis, neuropathic small bladder disease, and bladder cancer.[1] Currently, bladder cancer is the second most common urologic malignancy and the sixth most common cancer in the United States. According to National Cancer Institute, in 2014, an estimated 74,690 people had bladder cancer in the US; estimated deaths from bladder cancer are about 15,580 annually. The 5-year survival rate for bladder cancer is around 77%. Based on 2009–2011 data, approximately 2.4% of the population will be diagnosed with bladder cancer at some point during their lifetime. According to the American Urological Association, bladder exstrophy occurs in approximately 2.07 per 100,000 live births. Neurogenic bladder occurs in 40–90% of patients in the US with multiple sclerosis, 37–72% of patients with Parkinsonism, and 15% of patients with stroke.[104] It is estimated that 70–84% of patients with spinal cord injuries have at least some degree of bladder dysfunction.[104,105] Chronic bladder diseases cause reduced contractility and compliance, form heavy scar tissue, and significantly reduce bladder volume (end-stage bladder disease).

4.3.2 Current Treatment

Current treatment for invasive malignancies or end-stage bladder diseases frequently requires cystectomy. Gastrointestinal segments or gastric flaps are used as donor tissues to create a neobladder or a continent urinary reservoir to restore bladder function and increase volume.[106] However, severe complications may occur, such as metabolic disturbances, urolithiasis, excess mucus secretion, and urinary tract infection.[107,108] Additionally, there are increased risks of malignancy, particularly adenocarcinoma, due to histological changes in the intestinal mucosa after long-term exposure to urine. Children with neurogenic bladder disease have a higher risk of bladder cancer regardless of exposure to intestinal mucosa.[109]

Many other materials and tissues have been investigated to search for an alternative to bowel tissue. Anastomoses between two sets of urological tissue are limited by the amount of autologous urological tissues. Tissues from other sources include fascia, skin, bladder submucosa, omentum, dura, peritoneum, placenta, seromuscular grafts, and small intestinal submucosa.[110–115] Synthetic materials include polyvinyl sponge, tetrafluoroethylene, gelation sponge, collagen matrices, Vicryl® matrices, resin-sprayed paper, and silicone.[116–118] However, most attempts to test these materials have resulted in mechanical, structural, functional, or biocompatibility failures. Therefore, bowel tissue remains the most frequent application for cystectomy.

4.3.3 Cell-Seeded Tissue Engineering

The ideal tissue-engineered bladder would reconstruct the substitute as native bladder ECM, and restore bladder function. The native bladder consists of epithelium on the lumen surrounded by a collagen-rich connective tissue (submucosa), and then a muscle layer that contains mainly SMCs. The submucosa mainly constitutes collagen types I and III fibers, elastic fibers, and unmyelinated nerve endings, joining the muscle and epithelial layers.[119] Both submucosal and muscle layers maintain the structural integrity of the organ and contract to transport or expel the urine.[120] The epithelial layer is transitional, and all UCs are attached on the basal lamina. The urothelium consists of fully differentiated cells, predifferentiated cells, and progenitor cells, from the outside to the inside of the bladder architecture. The fully differentiated UCs express uroplakin, a urothelium-specific marker. Uroplakins and tight junctions between cells assure the impermeability of the bladder.

Presently, both cell-free and cell-based scaffold systems are used for bladder tissue engineering.[121] Cell-seeded scaffolds generally are preferred due to their superior biocompatibility and adequate regenerative capacity. Cell-free scaffold implantation has failed to show full regeneration of the bladder wall, with issues such as scarring, reduced reservoir, incompetence of graft contraction, and cycling.[122,123] Only approximately 30% of the smooth muscle layer regenerated.[124] Cell-based technology is mainly focused on reconstruction of a 3D structure of the urinary tissue in vitro by seeding UCs and SMCs on the scaffold. The tissue-engineered product is expected to become more fully differentiated and phenotypically mature at implantation. Furthermore, this approach would accelerate neotissue formation and tissue integration, reduce host inflammatory or immune response, as well as prevent graft contracture and shrinkage.

Using this approach, engineered bladder tissue has developed anatomically and functionally neobladder tissue in several animal models.[125–129] In a canine model, up to 95% of the original precystectomy volume was achieved in tissue-engineered bladder replacement, with normal cellular organization in all three cell layers.[130]

Engineered bladder constructs were successfully implanted into seven patients with neurogenic bladder due to myelomeningocele.[1] Constructs containing collagen–PGA scaffolds with autologous UCs and SMCs implanted with omentum wrap displayed decreased bladder leak point pressure, increased volume and compliance, and prompt recovery of bowel function with normal mucus production and renal function, and without metabolic consequences or urinary calculi formation up to 4 years. Further biopsies showed an adequate structural architecture and phenotype. After implantation, 6 of the 10 pediatric patients and 4 of 6 adult patients showed clinical improvement based on urodynamic studies, radiography, and voiding diary results.[131]

The selection of scaffold is another challenge for bladder engineering. So far, most commonly used scaffolds are natural collagen materials and synthetic polymers. Different types of scaffolds have been applied to bladder regeneration, including submucosa, small intestinal submucosa, type I collagen matrix, PGA, PLGA, poly-l-lactide acid (PLLA), and bacterial cellulose polymer.[1,78,98,130,132–134] For instance, PLLA seems to be a good candidate for bladder reconstruction. It possesses a three-dimensional porous structure (50–200 μm) with adequate mechanical strength and biodegradability. It also promotes angiogenesis. It has been used in a number of tissue engineering studies, such as bone, cartilage, blood vessel, and wound healing.[135–138] Combined with USCs, it may generate promising neobladder tissue.

According to phase II clinical studies, patients who did not respond well to this therapy (without normal bladder cycles) had open bladder necks or other physiological issues.[131] This supports the idea that conditioning engineered bladder tissue within a specially designed bioreactor before implantation may improve clinical results.

4.4 Urethra

4.4.1 Incidence

Various urethral conditions, such as inflammatory and posttraumatic strictures, congenital defects, and malignancy, often require extensive urethral reconstruction. In the US, urethral strictures occur in up to 0.6% of susceptible men and result in over 5000 inpatient visits per year.[139] The total cost of treatment was almost $200 million, not including medication costs. Patients with urethral stricture disease appeared to have a high rate of urinary tract infection (41%) and incontinence (11%). Most commonly, strictures are caused by scarring process involving the vascular tissue of the corpus spongiosum, leading to ischemic spongiofibrosis in the urethra.[140] Occasionally, a specific cause such as lichen sclerosis may be identified.

4.4.2 Current Treatments

Depending on the situation, different surgical repair techniques are performed for urethral strictures, such as urethrotomy or anastomotic and substitution urethroplasty.[141] Urethrotomy and dilatation are standard procedures and minimally invasive, but more applicable in patients with relatively short strictures (<2 cm).[142] Recurrence rates of over 90% are reported within 7 months;[143] in up to 60% of patients, the failure rate can be 100% by 48 months.[144]

Anastomotic urethroplasty involves removal of stricture and simple anastomosis of the ends. It is also limited to short strictures, up to 6 cm. It also results in fibrosis and chronic inflammation, and these conditions can lead to stricture recurrence. Longer strictures still require autologous grafts or flaps from genital skin or buccal mucosa.[145] Usually, grafts are preferred to flaps, which may lead to greater morbidity.[146] Oral mucosa has been the graft of choice due to its ease of access, low donor-site morbidity, and immune status.[147] Combinations of buccal (cheek), lingual (tongue), and labial (lip) mucosa also have been used, but these grafts are restricted by availability. The risk of morbidity increases as the length of grafts increases. In addition, they are not appropriate in patients who are smokers or tobacco chewers, or those with poor oral hygiene.[148] Liver tissue grafts have also been tried, but complications such as bleeding and infection occurred. Based on these results, the tissue-engineered urethra would be a promising alternative method for urethral reconstruction.

4.4.3 Cell-Seeded Tissue Engineering for Urethral Reconstruction

The urethra is a tubular structure roughly similar to bladder, consisting of smooth muscles with connective tissue, submucosa with collagen fibers and microvascularization, and the urethral epithelium. Like bladder, tissue-engineered urethral tissue also has used both cell-free and cell-seeded constructs. Acellular grafts can be used only when a healthy part of urethral wall exists. Cadaveric bladder submucosa graft implantation has been successful in patients with urethral stricture and hypospadias.[149–151] Small intestinal submucosa is an established acellular matrix with long-term safety and efficacy. However, cell-free scaffolds can be used only as an alternative option in patients with short-to-medium urethral defects who have a healthy urethral bed and no or minimal spongiofibrosis. With cell-seeded constructs, three human studies have been reported, and only one showed successful results up to 6 years.[152–154] With UCs and SMCs seeded on PGA and PLGA scaffolds, constructs remained patent in all five patients with posterior defects.[153]

Urethral tissue engineering incorporates two major cell types, UCs and SMCs. However, tissue biopsy procedures required by cell isolation may increase hospital stays and lead to donor-site morbidity. Cell quality may be affected by taking cells from a less optimal microenvironment. The use of UCs isolated from urine or bladder washes may also be associated with a low success rate of cultures (55%) and limited expansion capability in vitro.[155,156] Immortalized UC cell lines have the advantage of large expansion quantities, but their clinical applications are extremely limited because of tumorigenicity. The cell seeding density for urethral tissue engineering is approximately 5×10^7 cells/cm^3.[1] Similar to bladder regeneration, USCs isolated from 200 mL of voided urine can create sufficient cells for a 6.5–10 cm^3 urethral construct.

Biomaterials such as PLA, polyethylene glycol (PEG), and type I collagen have been tested in several studies. For both small intestinal submucosa and BAM, the permeability is only direction dependent, which may cause urine leakage from the lumen into surrounding tissue. This side effect is less important in cell-based tissue engineering, where cells can fill in the gaps or pores within the matrix, making it suitable for tissue-engineered urethral products. Parallel cell seeding and bioreactor designs could be applied for seeding and culturing urethral tissue-engineered products such as bladder. In short, despite considerable progress in tissue engineering of bladders and urethras, tissues remain in creating fully optimized grafts, including stimulating the implanted muscles to contract, vascularity, and nerve supply.

4.5 Urethral Sphincter

4.5.1 Incidence of Stress Urinary Incontinence

Around 25 million adult Americans experience transient or chronic urinary incontinence.[157] It is estimated that 75–80% of patients are women, 9–13 million of whom have bothersome to severe symptoms. Incontinence frequently occurs from middle age onward and is associated with a reduced quality of life. SUI affects an estimated 15 million adult women in the US.[158,159] Almost 30% of individuals aged 60–70, experience leakage when coughing, sneezing, or laughing, compared to 17% of men and women ages 30–39. Up to half of men report leakage due to SUI in the first few weeks after prostate surgery; in about 20% of them, some degree of SUI will continue to be a significant problem one year postsurgery.

4.5.2 Current Treatments

SUI arises when bladder pressure exceeds the urethral pressure in a sudden increase in intra-abdominal forces. This may be caused by the anatomical change or intrinsic sphincter deficiency. Current treatments for SUI include nonsurgical and surgical options. Nonsurgical treatment, usually pelvic floor muscle training, results in only limited improvement.[160] Surgical interventions show more promising results.[161,162] However, in the case of more invasive sling procedures, the use of bulking agents sometimes lead to complications such as chronic inflammatory reactions, a foreign-body giant-cell response, particle migration, periurethral abscess, erosion of the bladder or urethra, or obstruction with urinary retention.[163–165] Cell-based therapy could be a potential option for treating SUI.

4.5.3 Cell Therapy

Different types of cells have been tested in clinical studies of SUI. One study involved 42 women and 21 men undergoing autologous myoblast and fibroblast injection, and 28 patients with collagen injection.[166] Skeletal muscle biopsies were taken from the left arm of the patients. The myoblasts were directly injected into the omega-shaped rhabdosphincter and the fibroblast/collagen into the submucosa. Only 10% of patients with collagen treatment showed improvement in SUI, while 85% of patients treated with myoblasts and fibroblasts were considered cured at 12-month follow-up. Other studies also show a 65–90% success rate after a year of follow-up.[167–169]

MSCs also demonstrate potency for treating SUI. Five of eight patients who received muscle-derived stem cell injection show improvement and cure at a median of 10 months with no severe adverse effects.[170] Previously, we reported that USCs expressed SMC markers when cultured in media with serum.[18] VEGF expression by USCs, along with concurrent endothelial cell implantation, promoted angiogenesis, significantly improved in vivo cell survival and myogenic differentiation of USCs, and enhanced nerve regeneration within the graft, which maintained its size.[42] Further studies demonstrated that the angiogenic potential of USCs in collagen-I hydrogel appreciably increased cell viability, resident cell recruitment, myogenic regeneration, and innervation via angiogenic effects. Thus, this treatment might have potential in treatment of patients with moderate-to-severe SUI.[49] Considering that 5×10^7 skeletal muscle-derived stem cells would be needed to treat SUI, one urine sample would be sufficient for treatment.[171]

4.6 Erectile Dysfunction

4.6.1 Incidence

Although not life-threatening by itself, ED is a strong predictor of coronary artery disease and cardiovascular disease.[172–174] ED can be a complication of other diseases, such as diabetes mellitus or Peyronie's disease.[175,176] It can occur after prostatectomy or radiotherapy for prostate cancer. Half to three-quarters of men with diabetes mellitus have ED, regardless of age.[177] Diabetic men also tend to have ED 10–15 years earlier and are three times more likely to have ED than nondiabetic men.[178,179] Prostate cancer is the most common malignancy in men, with estimated numbers of new cases and related deaths being 217,730 and 32,050, respectively, in the USA in 2010.[180] In up to 80% of cases, prostatectomy and subsequent radiotherapy are the typical treatments.[181] Incidence of ED ranges from 60.8% to 93.9% up to 15 years after treatment.[182]

4.6.2 Current Treatment

ED can be classified as psychogenic, organic, or mixed psychogenic/organic. Most organic types of ED can be treated with intracavernous injection of erectogenic agents, a transurethral prostaglandin suppository, a vacuum device, and/or PDE5 inhibitors.[183] However, up to 50% of patients discontinue or never take medication due to noneffectiveness, concerns about cardiovascular effects, cost, or inconvenience.[184] In addition, about 20% of current ED patients do not respond to current treatments.[185]

4.6.3 Cell Therapy

The corpora cavernosa are comprised of two parallel cylinders that contain interconnected sinusoids surrounded by SMCs. Each cylinder is encased in a fibrous sheath known as the tunica albuginea. This unique tissue architecture controls inflow of blood to the penis to maintain erections. During sexual stimulation, nitric oxide released from the nerve endings and endothelium causes smooth muscle relaxation, which permits an influx of blood into the cavernosal sinusoids. The expanded sinusoidal walls press against one another within the tunica albuginea, resulting in an erection. Corporal tissue is highly specialized to perform this function, and engineered erectile tissue must mimic it to be effective.

A number of cell-based studies of potential ED treatments have been reported. Choices of cell types vary among ASC, BMSC, ESC, bone marrow mononuclear cells, skeletal MSC, endothelial progenitor cells, and others. Because cell-based therapy for ED is not yet approved for clinical trials in most countries, there is only one published clinical study.[186] All studies reported improved erectile function with subcutaneous transplantation in patients with ED or animal models. Intracavernous pressure was increased by 70% compared to non-ED controls. However, it is unclear whether this level of improvement is sufficient for ED patients to regain erectile function.

USCs or USCs genetically modified with FGF2 enhanced the expression of endothelial cell markers, smooth muscle contents, and improve neurogenic-mediated erectile responses in type 2 diabetic ED rats.[35] Improvements in diabetic ED in a rodent model after administration of USCs or USCs–FGF2 was similar to those in cell therapy with other types of MSCs. Paracrine action of USCs may play an important role in recruiting resident endothelial and SMCs to participate in tissue repair within the cavernous tissue. All these results indicate the potential of USCs for ED treatment. Further investigations are needed in other types of EDs and large animal models.

5. FUTURE DIRECTIONS

5.1 Efficient Revascularization

So far, the use of tissue-engineered products is hindered by issues related to limited nutrient and gas exchange. To achieve functional complex urological tissue in vivo, revascularization of regenerated products is essential. To overcome the current problems, different approaches have been used to enhance revascularization. Angiogenic factors, such as VEGF, have been included in implanted cells with enhanced grafted cell survival; they recruited resident cells and promoted myogenic phenotype differentiation of USCs and innervation.[49] Endothelial cells may be involved in the seeding process. Finally, prevascularization of matrix may be considered before cell seeding to increase cell viability and functionality.

5.2 Innervation

For use as an in vitro model for testing, engineered tissues need to mimic key morphological, physiological, and biochemical properties of natural tissue as closely as possible. Tissue-engineered products also require a sensory nerve supply to mechanistically mimic natural tissue.[187] To achieve recovery, engineered tissues require integration of a host nerve to facilitate coordinated function.[188] USCs expressing VEGF lead to enhanced nerve regeneration,[49] due to increased angiogenesis and thus increased nutrition in native nerve tissue. This nerve regeneration may be beneficial in restoring function of regenerated bladder and other urological tissues.

5.3 Antifibrotic Effects

Bladder fibrosis can adversely affect composition of the bladder wall by reducing the smooth muscle component, resulting in low bladder compliance.[189] The main mechanisms for improved bladder function may be prevention or resolution of fibrosis and restoration of bladder architecture. MSCs themselves cannot substitute for damaged tissue directly or totally, but they can secrete growth factors or cytokines that contribute to reducing fibrosis, such as hepatic growth factor, nerve growth factor, brain-derived growth factor, glial-derived growth factor, IGF, VEGF, and ciliary neurotrophic growth factor. Further studies are needed to determine how best to maintain or enhance the efficacy of MSCs as a therapeutic option for bladder fibrosis.[190]

5.4 Optimal Biomaterials

The ideal biomaterial scaffold is not necessarily a mimic of the natural tissue matrix, but must provide a suitable biophysical microenvironment to direct cell behavior toward functional tissue generation. In addition, appropriate biomaterials should be biodegradable and bioresorbable without inflammation. Each type of biomaterial should be selected according to the specific therapeutic application.

6. CONCLUSION

The field of urological tissue engineering and regenerative medicine is developing rapidly in (stem) cell therapy and biomaterial research. Multiple cell types offer the promise of clinical applications within the near future, and are being tested in both cell therapy and tissue engineering techniques. Among those, USCs exhibit extreme capabilities as a source in a series of urological diseases, with easy accessibility, great longevity, compelling multipotentiality (especially to urological tissue-specific cells), trophic effects, and favorable immunomodulatory characteristics. Appropriate biomaterial and bioreactor design is also essential in engineering genitourinary tissues. The translation of tissue engineering into urological practice still anticipates new developments and progress in biomaterial and stem cell research.

REFERENCES

1. Atala A, Bauer SB, Soker S, Yoo JJ, Retik AB. Tissue-engineered autologous bladders for patients needing cystoplasty. *Lancet* 2006;**367**(9518): 1241–6. Epub 2006/04/25.
2. Wezel F, Southgate J, Thomas DF. Regenerative medicine in urology. *BJU Int* 2011;**108**(7):1046–65.
3. Martin I, Wendt D, Heberer M. The role of bioreactors in tissue engineering. *Trends Biotechnol* 2004;**22**(2):80–6. Epub 2004/02/06.
4. Grayson WL, Frohlich M, Yeager K, Bhumiratana S, Chan ME, Cannizzaro C, et al. Engineering anatomically shaped human bone grafts. *Proc Natl Acad Sci USA* 2010;**107**(8):3299–304. Epub 2009/10/13.
5. Korossis S, Bolland F, Ingham E, Fisher J, Kearney J, Southgate J. Review: tissue engineering of the urinary bladder: considering structure-function relationships and the role of mechanotransduction. *Tissue Eng* 2006;**12**(4):635–44.
6. Yoon BS, Moon JH, Jun EK, Kim J, Maeng I, Kim JS, et al. Secretory profiles and wound healing effects of human amniotic fluid-derived mesenchymal stem cells. *Stem Cells Dev* 2010;**19**(6):887–902. Epub 2009/08/19.
7. Insausti CL, Blanquer M, Garcia-Hernandez AM, Castellanos G, Moraleda JM. Amniotic membrane-derived stem cells: immunomodulatory properties and potential clinical application. *Stem Cells Cloning Adv Appl* 2014;**7**:53–63. Epub 2014/04/20.
8. Baraniak PR, McDevitt TC. Stem cell paracrine actions and tissue regeneration. *Regen Med* 2010;**5**(1):121–43. Epub 2009/12/19.

9. Thomson JA, Itskovitz-Eldor J, Shapiro SS, Waknitz MA, Swiergiel JJ, Marshall VS, et al. Embryonic stem cell lines derived from human blastocysts. *Science* 1998;**282**(5391):1145–7.

10. Xue Y, Cai X, Wang L, Liao B, Zhang H, Shan Y, et al. Generating a non-integrating human induced pluripotent stem cell bank from urine-derived cells. *PLoS One* 2013;**8**(8):e70573. Epub 2013/08/14.

11. De Coppi P, Bartsch Jr G, Siddiqui MM, Xu T, Santos CC, Perin L, et al. Isolation of amniotic stem cell lines with potential for therapy. *Nat Biotechnol* 2007;**25**(1):100–6. Epub 2007/01/09.

12. De Coppi P, Callegari A, Chiavegato A, Gasparotto L, Piccoli M, Taiani J, et al. Amniotic fluid and bone marrow derived mesenchymal stem cells can be converted to smooth muscle cells in the cryo-injured rat bladder and prevent compensatory hypertrophy of surviving smooth muscle cells. *J Urol* 2007;**177**(1):369–76. Epub 2006/12/13.

13. Mosquera A, Fernandez JL, Campos A, Goyanes VJ, Ramiro-Diaz J, Gosalvez J. Simultaneous decrease of telomere length and telomerase activity with ageing of human amniotic fluid cells. *J Med Genet* 1999;**36**(6):494–6. Epub 2000/06/30.

14. Qin D, Long T, Deng J, Zhang Y. Urine-derived stem cells for potential use in bladder repair. *Stem Cell Res Ther* 2014;**5**(3):69. Epub 2014/08/27.

15. Bharadwaj S, Liu G, Shi Y, Wu R, Yang B, He T, et al. Multipotential differentiation of human urine-derived stem cells: potential for therapeutic applications in urology. *Stem Cells* 2013;**31**(9):1840–56. Epub 2013/05/15.

16. Bodin A, Bharadwaj S, Wu S, Gatenholm P, Atala A, Zhang Y. Tissue-engineered conduit using urine-derived stem cells seeded bacterial cellulose polymer in urinary reconstruction and diversion. *Biomaterials* 2010;**31**(34):8889–901. Epub 2010/08/31.

17. Bharadwaj S, Liu G, Shi Y, Markert C, Andersson KE, Atala A, et al. Characterization of urine-derived stem cells obtained from upper urinary tract for use in cell-based urological tissue engineering. *Tissue Eng Part A* 2011;**17**(15–16):2123–32. Epub 2011/04/26.

18. Zhang Y, McNeill E, Tian H, Soker S, Andersson KE, Yoo JJ, et al. Urine derived cells are a potential source for urological tissue reconstruction. *J Urol* 2008;**180**(5):2226–33. Epub 2008/09/23.

19. Zhao Z, Yu H, Xiao F, Wang X, Yang S, Li S. Differentiation of adipose-derived stem cells promotes regeneration of smooth muscle for ureteral tissue engineering. *J Surg Res* 2012;**178**(1):55–62. Epub 2012/04/10.

20. Shen J, Fu X, Ou L, Zhang M, Guan Y, Wang K, et al. Construction of ureteral grafts by seeding urothelial cells and bone marrow mesenchymal stem cells into polycaprolactone-lecithin electrospun fibers. *Int J Artif Organs* 2010;**33**(3):161–70. Epub 2010/04/13.

21. Brivanlou AH, Gage FH, Jaenisch R, Jessell T, Melton D, Rossant J. Stem cells. Setting standards for human embryonic stem cells. *Science* 2003;**300**(5621):913–6. Epub 2003/05/10.

22. Chung SY, Krivorov NP, Rausei V, Thomas L, Frantzen M, Landsittel D, et al. Bladder reconstitution with bone marrow derived stem cells seeded on small intestinal submucosa improves morphological and molecular composition. *J Urol* 2005;**174**(1):353–9. Epub 2005/06/11.

23. Zhang Y, Lin HK, Frimberger D, Epstein RB, Kropp BP. Growth of bone marrow stromal cells on small intestinal submucosa: an alternative cell source for tissue engineered bladder. *BJU Int* 2005;**96**(7):1120–5. Epub 2005/10/18.

24. Kanematsu A, Yamamoto S, Iwai-Kanai E, Kanatani I, Imamura M, Adam RM, et al. Induction of smooth muscle cell-like phenotype in marrow-derived cells among regenerating urinary bladder smooth muscle cells. *Am J Pathol* 2005;**166**(2):565–73. Epub 2005/02/01.

25. Tian H, Bharadwaj S, Liu Y, Ma PX, Atala A, Zhang Y. Differentiation of human bone marrow mesenchymal stem cells into bladder cells: potential for urological tissue engineering. *Tissue Eng Part A* 2010;**16**(5):1769–79. Epub 2009/12/22.

26. Tian H, Bharadwaj S, Liu Y, Ma H, Ma PX, Atala A, et al. Myogenic differentiation of human bone marrow mesenchymal stem cells on a 3D nano fibrous scaffold for bladder tissue engineering. *Biomaterials* 2010;**31**(5):870–7. Epub 2009/10/27.

27. Kovanecz I, Rivera S, Nolazco G, Vernet D, Segura D, Gharib S, et al. Separate or combined treatments with daily sildenafil, molsidomine, or muscle-derived stem cells prevent erectile dysfunction in a rat model of cavernosal nerve damage. *J Sex Med* 2012;**9**(11):2814–26.

28. Qiu X, Villalta J, Ferretti L, Fandel TM, Albersen M, Lin G, et al. Effects of intravenous injection of adipose-derived stem cells in a rat model of radiation therapy-induced erectile dysfunction. *J Sex Med* 2012;**9**(7):1834–41. Epub 2012/05/03.

29. Sun C, Lin H, Yu W, Li X, Chen Y, Qiu X, et al. Neurotrophic effect of bone marrow mesenchymal stem cells for erectile dysfunction in diabetic rats. *Int J Androl* 2012;**35**(4):601–7. Epub 2012/03/21.

30. Huang YC, Ning H, Shindel AW, Fandel TM, Lin G, Harraz AM, et al. The effect of intracavernous injection of adipose tissue-derived stem cells on hyperlipidemia-associated erectile dysfunction in a rat model. *J Sex Med* 2010;**7**(4 Pt 1):1391–400. Epub 2010/02/10.

31. Weissman IL. Stem cells: units of development, units of regeneration, and units in evolution. *Cell* 2000;**100**(1):157–68. Epub 2000/01/27.

32. Liu G, Pareta RA, Wu R, Shi Y, Zhou X, Liu H, et al. Skeletal myogenic differentiation of urine-derived stem cells and angiogenesis using microbeads loaded with growth factors. *Biomaterials* 2013;**34**(4):1311–26. Epub 2012/11/10.

33. Pei M, Li J, Zhang Y, Liu G, Wei L. Expansion on a matrix deposited by nonchondrogenic urine stem cells strengthens the chondrogenic capacity of repeated-passage bone marrow stromal cells. *Cell Tissue Res* 2014;**356**(2):391–403. Epub 2014/04/08.

34. Wu RP, Soland M, Liu G, Shi YA, Bharadwaj S, Atala A, et al. *Immunomodulatory properties of urine derived stem cells. The 3rd Annual Regenerative Medicine Foundation Conference 2012 Abstract Book Charlotte, NC, USA October 18–19*. 2012.

35. Ouyang B, Sun X, Han D, Chen S, Yao B, Gao Y, et al. Human urine-derived stem cells alone or genetically-modified with FGF2 improve type 2 diabetic erectile dysfunction in a rat model. *PLoS One* 2014;**9**(3):e92825. Epub 2014/03/26.

36. Liu G, Sun X, Bian J, Wu R, Guan X, Ouyang B, et al. Correction of diabetic erectile dysfunction with adipose derived stem cells modified with the vascular endothelial growth factor gene in a rodent diabetic model. *PLoS One* 2013;**8**(8):e72790. Epub 2013/09/12.

37. Baxter MA, Wynn RF, Jowitt SN, Wraith JE, Fairbairn LJ, Bellantuono I. Study of telomere length reveals rapid aging of human marrow stromal cells following in vitro expansion. *Stem Cells* 2004;**22**(5):675–82. Epub 2004/09/03.

38. Zhou S, Greenberger JS, Epperly MW, Goff JP, Adler C, Leboff MS, et al. Age-related intrinsic changes in human bone-marrow-derived mesenchymal stem cells and their differentiation to osteoblasts. *Aging Cell* 2008;**7**(3):335–43. Epub 2008/02/06.

39. Mareschi K, Ferrero I, Rustichelli D, Aschero S, Gammaitoni L, Aglietta M, et al. Expansion of mesenchymal stem cells isolated from pediatric and adult donor bone marrow. *J Cell Biochem* 2006;**97**(4):744–54. Epub 2005/10/18.

40. Stolzing A, Jones E, McGonagle D, Scutt A. Age-related changes in human bone marrow-derived mesenchymal stem cells: consequences for cell therapies. *Mech Ageing Dev* 2008;**129**(3):163–73. Epub 2008/02/05.

41. Dexheimer V, Mueller S, Braatz F, Richter W. Reduced reactivation from dormancy but maintained lineage choice of human mesenchymal stem cells with donor age. *PLoS One* 2011;**6**(8):e22980. Epub 2011/08/19.

42. Wu S, Wang Z, Bharadwaj S, Hodges SJ, Atala A, Zhang Y. Implantation of autologous urine derived stem cells expressing vascular endothelial growth factor for potential use in genitourinary reconstruction. *J Urol* 2011;**186**(2):640–7. Epub 2011/06/21.

43. Caplan AI, Dennis JE. Mesenchymal stem cells as trophic mediators. *J Cell Biochem* 2006;**98**(5):1076–84. Epub 2006/04/19.

44. Li Y, Chen J, Zhang CL, Wang L, Lu D, Katakowski M, et al. Gliosis and brain remodeling after treatment of stroke in rats with marrow stromal cells. *Glia* 2005;**49**(3):407–17.

45. Murphy JM, Fink DJ, Hunziker EB, Barry FP. Stem cell therapy in a caprine model of osteoarthritis. *Arthritis Rheumatism* 2003;**48**(12):3464–74.

46. Laflamme MA, Murry CE. Regenerating the heart. *Nat Biotechnol* 2005;**23**(7):845–56.

47. Chen J, Li Y, Katakowski M, Chen X, Wang L, Lu D, et al. Intravenous bone marrow stromal cell therapy reduces apoptosis and promotes endogenous cell proliferation after stroke in female rat. *J Neurosci Res* 2003;**73**(6):778–86.

48. Caplan AI. Adult mesenchymal stem cells for tissue engineering versus regenerative medicine. *J Cell Physiol* 2007;**213**(2):341–7. Epub 2007/07/11.

49. Liu G, Wang X, Sun X, Deng C, Atala A, Zhang Y. The effect of urine-derived stem cells expressing VEGF loaded in collagen hydrogels on myogenesis and innervation following after subcutaneous implantation in nude mice. *Biomaterials* 2013;**34**(34):8617–29. Epub 2013/08/13.

50. Albersen M, Fandel TM, Lin G, Wang G, Banie L, Lin CS, et al. Injections of adipose tissue-derived stem cells and stem cell lysate improve recovery of erectile function in a rat model of cavernous nerve injury. *J Sex Med* 2010;**7**(10):3331–40. Epub 2010/06/22.

51. Camarata PJ, Suryanarayanan R, Turner DA, Parker RG, Ebner TJ. Sustained release of nerve growth factor from biodegradable polymer microspheres. *Neurosurgery* 1992;**30**(3):313–9. Epub 1992/03/01.

52. Moya ML, Lucas S, Francis-Sedlak M, Liu X, Garfinkel MR, Huang JJ, et al. Sustained delivery of FGF-1 increases vascular density in comparison to bolus administration. *Microvasc Res* 2009;**78**(2):142–7. Epub 2009/06/27.

53. Moya ML, Garfinkel MR, Liu X, Lucas S, Opara EC, Greisler HP, et al. Fibroblast growth factor-1 (FGF-1) loaded microbeads enhance local capillary neovascularization. *J Surg Res* 2010;**160**(2):208–12. Epub 2009/12/05.

54. Moya ML, Cheng MH, Huang JJ, Francis-Sedlak ME, Kao SW, Opara EC, et al. The effect of FGF-1 loaded alginate microbeads on neovascularization and adipogenesis in a vascular pedicle model of adipose tissue engineering. *Biomaterials* 2010;**31**(10):2816–26. Epub 2010/01/19.

55. Djouad F, Plence P, Bony C, Tropel P, Apparailly F, Sany J, et al. Immunosuppressive effect of mesenchymal stem cells favors tumor growth in allogeneic animals. *Blood* 2003;**102**(10):3837–44. Epub 2003/07/26.

56. Brooke G, Tong H, Levesque JP, Atkinson K. Molecular trafficking mechanisms of multipotent mesenchymal stem cells derived from human bone marrow and placenta. *Stem Cells Dev* 2008;**17**(5):929–40. Epub 2008/06/20.

57. Abumaree M, Al Jumah M, Pace RA, Kalionis B. Immunosuppressive properties of mesenchymal stem cells. *Stem Cell Rev* 2012;**8**(2):375–92. Epub 2011/09/06.

58. Chan WK, Lau AS, Li JC, Law HK, Lau YL, Chan GC. MHC expression kinetics and immunogenicity of mesenchymal stromal cells after short-term IFN-gamma challenge. *Exp Hematol* 2008;**36**(11):1545–55. Epub 2008/08/22.

59. Deuse T, Stubbendorff M, Tang-Quan K, Phillips N, Kay MA, Eiermann T, et al. Immunogenicity and immunomodulatory properties of umbilical cord lining mesenchymal stem cells. *Cell Transplant* 2011;**20**(5):655–67. Epub 2010/11/09.

60. Dickhut A, Schwerdtfeger R, Kuklick L, Ritter M, Thiede C, Neubauer A, et al. Mesenchymal stem cells obtained after bone marrow transplantation or peripheral blood stem cell transplantation originate from host tissue. *Ann Hematol* 2005;**84**(11):722–7. Epub 2005/09/01.

61. Le Blanc K, Tammik C, Rosendahl K, Zetterberg E, Ringden O. HLA expression and immunologic properties of differentiated and undifferentiated mesenchymal stem cells. *Exp Hematol* 2003;**31**(10):890–6. Epub 2003/10/11.

62. Nauta AJ, Fibbe WE. Immunomodulatory properties of mesenchymal stromal cells. *Blood* 2007;**110**(10):3499–506. Epub 2007/08/01.

63. Tse WT, Pendleton JD, Beyer WM, Egalka MC, Guinan EC. Suppression of allogeneic T-cell proliferation by human marrow stromal cells: implications in transplantation. *Transplantation* 2003;**75**(3):389–97. Epub 2003/02/18.

64. Schurgers E, Kelchtermans H, Mitera T, Geboes L, Matthys P. Discrepancy between the in vitro and in vivo effects of murine mesenchymal stem cells on T-cell proliferation and collagen-induced arthritis. *Arthritis Res Ther* 2010;**12**(1):R31. Epub 2010/02/24.

65. Selmani Z, Naji A, Zidi I, Favier B, Gaiffe E, Obert L, et al. Human leukocyte antigen-G5 secretion by human mesenchymal stem cells is required to suppress T lymphocyte and natural killer function and to induce CD4+CD25highFOXP3+ regulatory T cells. *Stem Cells* 2008;**26**(1):212–22. Epub 2007/10/13.

66. Volarevic V, Al-Qahtani A, Arsenijevic N, Pajovic S, Lukic ML. Interleukin-1 receptor antagonist (IL-1Ra) and IL-1Ra producing mesenchymal stem cells as modulators of diabetogenesis. *Autoimmunity* 2010;**43**(4):255–63. Epub 2009/10/23.

67. Asari S, Itakura S, Ferreri K, Liu CP, Kuroda Y, Kandeel F, et al. Mesenchymal stem cells suppress B-cell terminal differentiation. *Exp Hematol* 2009;**37**(5):604–15. Epub 2009/04/21.

68. Corcione A, Benvenuto F, Ferretti E, Giunti D, Cappiello V, Cazzanti F, et al. Human mesenchymal stem cells modulate B-cell functions. *Blood* 2006;**107**(1):367–72. Epub 2005/09/06.

69. Aggarwal S, Pittenger MF. Human mesenchymal stem cells modulate allogeneic immune cell responses. *Blood* 2005;**105**(4):1815–22. Epub 2004/10/21.

70. Abdi R, Fiorina P, Adra CN, Atkinson M, Sayegh MH. Immunomodulation by mesenchymal stem cells: a potential therapeutic strategy for type 1 diabetes. *Diabetes* 2008;**57**(7):1759–67. Epub 2008/07/01.

71. Groh ME, Maitra B, Szekely E, Koc ON. Human mesenchymal stem cells require monocyte-mediated activation to suppress alloreactive T cells. *Exp Hematol* 2005;**33**(8):928–34. Epub 2005/07/26.

72. Jiang XX, Zhang Y, Liu B, Zhang SX, Wu Y, Yu XD, et al. Human mesenchymal stem cells inhibit differentiation and function of monocyte-derived dendritic cells. *Blood* 2005;**105**(10):4120–6. Epub 2005/02/05.

73. Tomic S, Djokic J, Vasilijic S, Vucevic D, Todorovic V, Supic G, et al. Immunomodulatory properties of mesenchymal stem cells derived from dental pulp and dental follicle are susceptible to activation by toll-like receptor agonists. *Stem Cells Dev* 2011;**20**(4):695–708. Epub 2010/08/25.

74. Munn DH, Zhou M, Attwood JT, Bondarev I, Conway SJ, Marshall B, et al. Prevention of allogeneic fetal rejection by tryptophan catabolism. *Science* 1998;**281**(5380):1191–3. Epub 1998/08/26.

75. Nawa Y, Teshima T, Sunami K, Hiramatsu Y, Yano T, Shinagawa K, et al. Responses of granulocyte colony-stimulating factor-mobilized peripheral blood mononuclear cells to alloantigen stimulation. *Blood* 1997;**90**(4):1716–8. Epub 1997/08/15.

76. Chan BP, Leong KW. Scaffolding in tissue engineering: general approaches and tissue-specific considerations. *Eur Spine J Off Publ Eur Spine Soc Eur Spinal Deformity Soc Eur Sect Cerv Spine Res Soc* 2008;**17**(Suppl. 4):467–79. Epub 2008/11/14.

77. Atala A. Tissue engineering of human bladder. *Br Med Bull* 2011;**97**:81–104.

78. Wu S, Liu Y, Bharadwaj S, Atala A, Zhang Y. Human urine-derived stem cells seeded in a modified 3D porous small intestinal submucosa scaffold for urethral tissue engineering. *Biomaterials* 2011;**32**(5):1317–26. Epub 2010/11/09.

79. Kuo YR, Kuo MH, Chou WC, Liu YT, Lutz BS, Jeng SF. One-stage reconstruction of soft tissue and Achilles tendon defects using a composite free anterolateral thigh flap with vascularized fascia lata: clinical experience and functional assessment. *Ann Plastic Surg* 2003;**50**(2):149–55. Epub 2003/02/05.

80. Crossett LS, Sinha RK, Sechriest VF, Rubash HE. Reconstruction of a ruptured patellar tendon with Achilles tendon allograft following total knee arthroplasty. *J Bone Joint Surg Am Vol* 2002;**84-A**(8):1354–61. Epub 2002/08/15.

81. Grohn P, Klock G, Zimmermann U. Collagen-coated Ba(2+)-alginate microcarriers for the culture of anchorage-dependent mammalian cells. *Biotechniques* 1997;**22**(5):970–5. Epub 1997/05/01.

82. Huaping Tan KGM. Injectable, biodegradable hydrogels for tissue engineering applications. *Materials* 2010;**3**(3):1746–67.

83. Harrington DA, Cheng EY, Guler MO, Lee LK, Donovan JL, Claussen RC, et al. Branched peptide-amphiphiles as self-assembling coatings for tissue engineering scaffolds. *J Biomed Mater Res Part A* 2006;**78**(1):157–67. Epub 2006/04/19.

84. Choi JS, Lee SJ, Christ GJ, Atala A, Yoo JJ. The influence of electrospun aligned poly(epsilon-caprolactone)/collagen nanofiber meshes on the formation of self-aligned skeletal muscle myotubes. *Biomaterials* 2008;**29**(19):2899–906. Epub 2008/04/11.

85. Jha V, Garcia-Garcia G, Iseki K, Li Z, Naicker S, Plattner B, et al. Chronic kidney disease: global dimension and perspectives. *Lancet* 2013;**382**(9888):260–72.

86. Yen TH, Alison MR, Cook HT, Jeffery R, Otto WR, Wright NA, et al. The cellular origin and proliferative status of regenerating renal parenchyma after mercuric chloride damage and erythropoietin treatment. *Cell Prolif* 2007;**40**(2):143–56. Epub 2007/05/03.

87. Morigi M, Introna M, Imberti B, Corna D, Abbate M, Rota C, et al. Human bone marrow mesenchymal stem cells accelerate recovery of acute renal injury and prolong survival in mice. *Stem Cells* 2008;**26**(8):2075–82. Epub 2008/05/24.

88. Herrera MB, Bussolati B, Bruno S, Fonsato V, Romanazzi GM, Camussi G. Mesenchymal stem cells contribute to the renal repair of acute tubular epithelial injury. *Int J Mol Med* 2004;**14**(6):1035–41. Epub 2004/11/18.

89. Morigi M, Rota C, Montemurro T, Montelatici E, Lo Cicero V, Imberti B, et al. Life-sparing effect of human cord blood-mesenchymal stem cells in experimental acute kidney injury. *Stem Cells* 2010;**28**(3):513–22. Epub 2010/01/06.

90. Ross EA, Williams MJ, Hamazaki T, Terada N, Clapp WL, Adin C, et al. Embryonic stem cells proliferate and differentiate when seeded into kidney scaffolds. *J Am Soc Nephrol* 2009;**20**(11):2338–47. Epub 2009/09/05.

91. Zhu XY, Urbieta-Caceres V, Krier JD, Textor SC, Lerman A, Lerman LO. Mesenchymal stem cells and endothelial progenitor cells decrease renal injury in experimental swine renal artery stenosis through different mechanisms. *Stem Cells* 2013;**31**(1):117–25.

92. Eirin A, Zhu XY, Krier JD, Tang H, Jordan KL, Grande JP, et al. Adipose tissue-derived mesenchymal stem cells improve revascularization outcomes to restore renal function in swine atherosclerotic renal artery stenosis. *Stem Cells* 2012;**30**(5):1030–41.

93. Ebrahimi B, Eirin A, Li Z, Zhu XY, Zhang X, Lerman A, et al. Mesenchymal stem cells improve medullary inflammation and fibrosis after revascularization of swine atherosclerotic renal artery stenosis. *PLoS One* 2013;**8**(7):e67474.

94. Nakayama KH, Batchelder CA, Lee CI, Tarantal AF. Decellularized rhesus monkey kidney as a three-dimensional scaffold for renal tissue engineering. *Tissue Eng Part A* 2010;**16**(7):2207–16. Epub 2010/02/17.

95. Trounson A, Thakar RG, Lomax G, Gibbons D. Clinical trials for stem cell therapies. *BMC Med* 2011;**9**:52.

96. Brandt AS, von Rundstedt FC, Lazica DA, Roth S. Ureteral reconstruction after ureterorenoscopic injuries. *Der Urol Ausg A* 2010;**49**(7):812–21. Epub 2010/06/19. Harnleiterrekonstruktion nach ureterorenoskopischen Verletzungen.

97. Roupret M, Babjuk M, Comperat E, Zigeuner R, Sylvester R, Burger M, et al. European guidelines on upper tract urothelial carcinomas: 2013 update. *Eur Urol* 2013;**63**(6):1059–71. Epub 2013/04/02.

98. Drewa T. The artificial conduit for urinary diversion in rats: a preliminary study. *Transplant Proc* 2007;**39**(5):1647–51. Epub 2007/06/21.

99. Kloskowski T, Jundzill A, Kowalczyk T, Nowacki M, Bodnar M, Marszalek A, et al. Ureter regeneration-the proper scaffold has to be defined. *PLoS One* 2014;**9**(8):e106023. Epub 2014/08/28.

100. Kloskowski T, Kowalczyk T, Nowacki M, Drewa T. Tissue engineering and ureter regeneration: Is it possible?. *Int J Artif Organs* 2013;**36**(6):392–405. Epub 2013/05/07.

101. Geutjes P, Roelofs L, Hoogenkamp H, Walraven M, Kortmann B, de Gier R, et al. Tissue engineered tubular construct for urinary diversion in a preclinical porcine model. *J Urol* 2012;**188**(2):653–60. Epub 2012/06/19.

102. Liao WB, Song C, Li YW, Yang SX, Meng LC, Li XH. Tissue-engineered conduit using bladder acellular matrix and bladder epithelial cells for urinary diversion in rabbits. *Chin Med J* 2013;**126**(2):335–9. Epub 2013/01/18.

103. Liao W, Yang S, Song C, Li Y, Meng L, Li X, et al. Tissue-engineered tubular graft for urinary diversion after radical cystectomy in rabbits. *J Surg Res* 2013;**182**(2):185–91. Epub 2012/11/13.

104. Dorsher PT, McIntosh PM. Neurogenic bladder. *Adv Urol* 2012;**2012**:816274.

105. Manack A, Motsko SP, Haag-Molkenteller C, Dmochowski RR, Goehring Jr EL, Nguyen-Khoa BA, et al. Epidemiology and healthcare utilization of neurogenic bladder patients in a US claims database. *Neurourol Urodyn* 2011;**30**(3):395–401.

106. Mitchell ME. Bladder augmentation in children: where have we been and where are we going? *BJU Int* 2003;**92**(Suppl. 1):29–34. Epub 2003/09/13.

107. McDougal WS. Metabolic complications of urinary intestinal diversion. *J Urol* 1992;**147**(5):1199–208.

108. Soergel TM, Cain MP, Misseri R, Gardner TA, Koch MO, Rink RC. Transitional cell carcinoma of the bladder following augmentation cystoplasty for the neuropathic bladder. *J Urol* 2004;**172**(4 Pt 2):1649–51. discussion 1651–2.

109. Higuchi TT, Granberg CF, Fox JA, Husmann DA. Augmentation cystoplasty and risk of neoplasia: fact, fiction and controversy. *J Urol* 2010;**184**(6):2492–6. Epub 2010/10/22.

110. Cheng E, Rento R, Grayhack JT, Oyasu R, McVary KT. Reversed seromuscular flaps in the urinary tract in dogs. *J Urol* 1994;**152**(6 Pt 2):2252–7.

111. Gleeson MJ, Griffith DP. The use of alloplastic biomaterials in bladder substitution. *J Urol* 1992;**148**(5):1377–82.

112. Kelami A, Ludtke-Handjery A, Korb G, Rolle J, Schnell J, Danigel KH. Alloplastic replacement of the urinary bladder wall with lyophilized human dura. *Eur Surg Res Eur Chir Forschung Recherches Chir Eur* 1970;**2**(3):195–202.

113. Probst M, Dahiya R, Carrier S, Tanagho EA. Reproduction of functional smooth muscle tissue and partial bladder replacement. *Br J Urol* 1997;**79**(4):505–15.

114. Tsuji I, Ishida H, Fujieda J. Experimental cystoplasty using preserved bladder graft. *J Urol* 1961;**85**:42–4.

115. Vaught JD, Kropp BP, Sawyer BD, Rippy MK, Badylak SF, Shannon HE, et al. Detrusor regeneration in the rat using porcine small intestinal sub-mucosal grafts: functional innervation and receptor expression. *J Urol* 1996;**155**(1):374–8.

116. Monsour MJ, Mohammed R, Gorham SD, French DA, Scott R. An assessment of a collagen/Vicryl composite membrane to repair defects of the urinary bladder in rabbits. *Urol Res* 1987;**15**(4):235–8.

117. Rohrmann D, Albrecht D, Hannappel J, Gerlach R, Schwarzkopp G, Lutzeyer W. Alloplastic replacement of the urinary bladder. *J Urol* 1996;**156**(6):2094–7.

118. Tsuji I, Kuroda K, Fujieda J, Shiraishi Y, Kunishima K. Clinical experiences of bladder reconstruction using preserved bladder and gelatin sponge bladder in the case of bladder cancer. *J Urol* 1967;**98**(1):91–2.

119. Wilson CB, Leopard J, Cheresh DA, Nakamura RM. Extracellular matrix and integrin composition of the normal bladder wall. *World J Urol* 1996;**14**(Suppl. 1):S30–7. Epub 1996/01/01.

120. Orabi H, Bouhout S, Morissette A, Rousseau A, Chabaud S, Bolduc S. Tissue engineering of urinary bladder and urethra: advances from bench to patients. *Sci World J* 2013;**2013**:154564. Epub 2014/01/24.

121. Orlando G, Wood KJ, De Coppi P, Baptista PM, Binder KW, Bitar KN, et al. Regenerative medicine as applied to general surgery. *Ann Surg* 2012;**255**(5):867–80. Epub 2012/02/15.

122. Atala A. Tissue engineering in urologic surgery. *Urologic Clin North Am* 1998;**25**(1):39–50. Epub 1998/04/08.

123. Atala A. This month in investigative urology: commentary on the replacement of urologic associated mucosa. *J Urol* 1996;**156**(2 Pt 1):338–9. Epub 1996/08/01.

124. Adamowicz J, Juszczak K, Bajek A, Tworkiewicz J, Nowacki M, Marszalek A, et al. Morphological and urodynamic evaluation of urinary bladder wall regeneration: muscles guarantee contraction but not proper function–a rat model research study. *Transplant Proc* 2012;**44**(5):1429–34. Epub 2012/06/06.

125. Fraser M, Thomas DF, Pitt E, Harnden P, Trejdosiewicz LK, Southgate J. A surgical model of composite cystoplasty with cultured urothelial cells: a controlled study of gross outcome and urothelial phenotype. *BJU Int* 2004;**93**(4):609–16.

126. Kanematsu A, Yamamoto S, Noguchi T, Ozeki M, Tabata Y, Ogawa O. Bladder regeneration by bladder acellular matrix combined with sustained release of exogenous growth factor. *J Urol* 2003;**170**(4 Pt 2):1633–8.

127. Nuininga JE, van Moerkerk H, Hanssen A, Hulsbergen CA, Oosterwijk-Wakka J, Oosterwijk E, et al. A rabbit model to tissue engineer the bladder. *Biomaterials* 2004;**25**(9):1657–61.

128. Sievert KD, Fandel T, Wefer J, Gleason CA, Nunes L, Dahiya R, et al. Collagen I: III ratio in canine heterologous bladder acellular matrix grafts. *World J Urol* 2006;**24**(1):101–9.

129. Lai JY, Yoon CY, Yoo JJ, Wulf T, Atala A. Phenotypic and functional characterization of in vivo tissue engineered smooth muscle from normal and pathological bladders. *J Urol* 2002;**168**(4 Pt 2):1853–7. discussion 8.

130. Oberpenning F, Meng J, Yoo JJ, Atala A. De novo reconstitution of a functional mammalian urinary bladder by tissue engineering. *Nat Biotechnol* 1999;**17**(2):149–55. Epub 1999/03/03.

131. Yoo JJ, Olson J, Atala A, Kim B. Regenerative medicine strategies for treating neurogenic bladder. *Int Neurourol J* 2011;**15**(3):109–19. Epub 2011/11/17.

132. Yoo JJ, Meng J, Oberpenning F, Atala A. Bladder augmentation using allogenic bladder submucosa seeded with cells. *Urology* 1998;**51**(2):221–5.

133. Becker C, Olde Damink L, Laeufer T, Brehmer B, Heschel I, Jakse G. 'UroMaix' scaffolds: novel collagen matrices for application in tissue engineering of the urinary tract. *Int J Artif Organs* 2006;**29**(8):764–71.

134. Kobashi LI, Raible DA. Biocarbon urinary conduit: laboratory experience and clinical applications. *Urology* 1980;**16**(1):27–32. Epub 1980/07/01.

135. Woo KM, Chen VJ, Jung HM, Kim TI, Shin HI, Baek JH, et al. Comparative evaluation of nanofibrous scaffolding for bone regeneration in critical-size calvarial defects. *Tissue Eng Part A* 2009;**15**(8):2155–62. Epub 2009/04/08.

136. Gong Y, He L, Li J, Zhou Q, Ma Z, Gao C, et al. Hydrogel-filled polylactide porous scaffolds for cartilage tissue engineering. *J Biomed Mater Res B Appl Biomater* 2007;**82**(1):192–204. Epub 2006/11/16.

137. Hu J, Sun X, Ma H, Xie C, Chen YE, Ma PX. Porous nanofibrous PLLA scaffolds for vascular tissue engineering. *Biomaterials* 2010;**31**(31):7971–7. Epub 2010/08/03.

138. Spadaccio C, Rainer A, De Porcellinis S, Centola M, De Marco F, Chello M, et al. A G-CSF functionalized PLLA scaffold for wound repair: an in vitro preliminary study. *Conf Proc IEEE Eng Med Biol Soc* 2010;**2010**:843–6.

139. Santucci RA, Joyce GF, Wise M. Male urethral stricture disease. *J Urol* 2007;**177**(5):1667–74. Epub 2007/04/18.

140. Mangera A, Chapple CR. Tissue engineering in urethral reconstruction–an update. *Asian J Androl* 2013;**15**(1):89–92. Epub 2012/10/09.

141. Mangera A, Chapple C. Management of anterior urethral stricture: an evidence-based approach. *Curr Opin Urol* 2010;**20**(6):453–8. Epub 2010/09/10.

142. Steenkamp JW, Heyns CF, de Kock ML. Internal urethrotomy versus dilation as treatment for male urethral strictures: a prospective, randomized comparison. *J Urol* 1997;**157**(1):98–101. Epub 1997/01/01.

143. Santucci R, Eisenberg L. Urethrotomy has a much lower success rate than previously reported. *J Urol* 2010;**183**(5):1859–62.

144. Heyns CF, Steenkamp JW, De Kock ML, Whitaker P. Treatment of male urethral strictures: is repeated dilation or internal urethrotomy useful? *J Urol* 1998;**160**(2):356–8. Epub 1998/07/29.

145. McAninch JW. Urethral reconstruction: a continuing challenge. *J Urol* 2005;**173**(1):7. Epub 2004/12/14.

146. Dubey D, Vijjan V, Kapoor R, Srivastava A, Mandhani A, Kumar A, et al. Dorsal onlay buccal mucosa versus penile skin flap urethroplasty for anterior urethral strictures: results from a randomized prospective trial. *J Urol* 2007;**178**(6):2466–9. Epub 2007/10/17

147. Barbagli G, Vallasciani S, Romano G, Fabbri F, Guazzoni G, Lazzeri M. Morbidity of oral mucosa graft harvesting from a single cheek. *Eur Urol* 2010;**58**(1):33–41. Epub 2010/01/29.

148. Sinha RJ, Singh V, Sankhwar SN, Dalela D. Donor site morbidity in oral mucosa graft urethroplasty: implications of tobacco consumption. *BMC Urol* 2009;**9**:15. Epub 2009/09/24.

149. el-Kassaby A, AbouShwareb T, Atala A. Randomized comparative study between buccal mucosal and acellular bladder matrix grafts in complex anterior urethral strictures. *J Urol* 2008;**179**(4):1432–6. Epub 2008/02/26.

150. El-Kassaby AW, Retik AB, Yoo JJ, Atala A. Urethral stricture repair with an off-the-shelf collagen matrix. *J Urol* 2003;**169**(1):170–3. discussion 3. Epub 2002/12/13.

151. Atala A, Guzman L, Retik AB. A novel inert collagen matrix for hypospadias repair. *J Urol* 1999;**162**(3 Pt 2):1148–51. Epub 1999/08/24.

152. Bhargava S, Patterson JM, Inman RD, MacNeil S, Chapple CR. Tissue-engineered buccal mucosa urethroplasty-clinical outcomes. *Eur Urol* 2008;**53**(6):1263–9. Epub 2008/02/12.

153. Raya-Rivera A, Esquiliano DR, Yoo JJ, Lopez-Bayghen E, Soker S, Atala A. Tissue-engineered autologous urethras for patients who need reconstruction: an observational study. *Lancet* 2011;**377**(9772):1175–82. Epub 2011/03/11.

154. Fossum M, Svensson J, Kratz G, Nordenskjold A. Autologous in vitro cultured urothelium in hypospadias repair. *J Pediatr Urol* 2007;**3**(1):10–8. Epub 2008/10/25.

155. Fossum M, Gustafson CJ, Nordenskjold A, Kratz G. Isolation and in vitro cultivation of human urothelial cells from bladder washings of adult patients and children. *Scand J Plast Reconstr Surg Hand Surg* 2003;**37**(1):41–5. Epub 2003/03/11.

156. Nagele U, Maurer S, Feil G, Bock C, Krug J, Sievert KD, et al. In vitro investigations of tissue-engineered multilayered urothelium established from bladder washings. *Eur Urol* 2008;**54**(6):1414–22. Epub 2008/02/19.

157. Resnick NM. Improving treatment of urinary incontinence. *JAMA* 1998;**280**(23):2034–5. Epub 1998/12/24.

158. Hampel C, Wienhold D, Benken N, Eggersmann C, Thuroff JW. Definition of overactive bladder and epidemiology of urinary incontinence. *Urology* 1997;**50**(Suppl. 6A):4–14. discussion 5–7. Epub 1998/01/14.

159. Neveus T, von Gontard A, Hoebeke P, Hjalmas K, Bauer S, Bower W, et al. The standardization of terminology of lower urinary tract function in children and adolescents: report from the Standardisation Committee of the International Children's Continence Society. *J Urol* 2006;**176**(1):314–24. Epub 2006/06/07.

160. Holroyd-Leduc JM, Straus SE. Management of urinary incontinence in women: scientific review. *JAMA* 2004;**291**(8):986–95. Epub 2004/02/26.

161. Ross J. Two techniques of laparoscopic Burch repair for stress incontinence: a prospective, randomized study. *J Am Assoc Gynecol Laparosc* 1996;**3**(3):351–7. Epub 1996/05/01.

162. Rezapour M, Ulmsten U. Tension-Free vaginal tape (TVT) in women with mixed urinary incontinence–a long-term follow-up. *Int Urogynec J Pelvic Floor Dysfunct* 2001;**12**(Suppl. 2):S15–8. Epub 2001/07/14.

163. Sweat SD, Lightner DJ. Complications of sterile abscess formation and pulmonary embolism following periurethral bulking agents. *J Urol* 1999;**161**(1):93–6. Epub 1999/02/26.

164. Papa Petros PE. Tissue reaction to implanted foreign materials for cure of stress incontinence. *Am J Obstet Gynecol* 1994;**171**(4):1159. Epub 1994/10/01.

165. Pannek J, Brands FH, Senge T. Particle migration after transurethral injection of carbon coated beads for stress urinary incontinence. *J Urol* 2001;**166**(4):1350–3. Epub 2001/09/08.

166. Strasser H, Marksteiner R, Margreiter E, Mitterberger M, Pinggera GM, Frauscher F, et al. Transurethral ultrasonography-guided injection of adult autologous stem cells versus transurethral endoscopic injection of collagen in treatment of urinary incontinence. *World J Urol* 2007;**25**(4):385–92. Epub 2007/08/19.

167. Mitterberger M, Marksteiner R, Margreiter E, Pinggera GM, Colleselli D, Frauscher F, et al. Autologous myoblasts and fibroblasts for female stress incontinence: a 1-year follow-up in 123 patients. *BJU Int* 2007;**100**(5):1081–5. Epub 2007/09/01.

168. Mitterberger M, Pinggera GM, Marksteiner R, Margreiter E, Fussenegger M, Frauscher F, et al. Adult stem cell therapy of female stress urinary incontinence. *Eur Urol* 2008;**53**(1):169–75. Epub 2007/08/09.

169. Mitterberger M, Marksteiner R, Margreiter E, Pinggera GM, Frauscher F, Ulmer H, et al. Myoblast and fibroblast therapy for post-prostatectomy urinary incontinence: 1-year followup of 63 patients. *J Urol* 2008;**179**(1):226–31. Epub 2007/11/16.

170. Carr LK, Steele D, Steele S, Wagner D, Pruchnic R, Jankowski R, et al. 1-year follow-up of autologous muscle-derived stem cell injection pilot study to treat urinary incontinence. *Int Urogynec J Pelvic Floor Dysfunct* 2008;**19**(6):881–3. Epub 2008/01/22.

171. Smaldone MC, Chen ML, Chancellor MB. Stem cell therapy for urethral sphincter regeneration. *Minerva urologica e nefrologica = Italian J Urol Nephrol* 2009;**61**(1):27–40. Epub 2008/11/13.

172. Inman BA, Sauver JL, Jacobson DJ, McGree ME, Nehra A, Lieber MM, et al. A population-based, longitudinal study of erectile dysfunction and future coronary artery disease. *Mayo Clin Proc* 2009;**84**(2):108–13. Epub 2009/02/03.

173. Dong JY, Zhang YH, Qin LQ. Erectile dysfunction and risk of cardiovascular disease: meta-analysis of prospective cohort studies. *J Am Coll Cardiol* 2011;**58**(13):1378–85. Epub 2011/09/17.

174. Jackson G, Boon N, Eardley I, Kirby M, Dean J, Hackett G, et al. Erectile dysfunction and coronary artery disease prediction: evidence-based guidance and consensus. *Int J Clin Pract* 2010;**64**(7):848–57. Epub 2010/06/30.

175. Garaffa G, Trost LW, Serefoglu EC, Ralph D, Hellstrom WJ. Understanding the course of Peyronie's disease. *Int J Clin Pract* 2013;**67**(8):781–8. Epub 2013/07/23.

176. Lopez JA, Jarow JP. Penile vascular evaluation of men with Peyronie's disease. *J Urol* 1993;**149**(1):53–5. Epub 1993/01/01.

177. Lewis RW, Fugl-Meyer KS, Corona G, Hayes RD, Laumann EO, Moreira Jr ED, et al. Definitions/epidemiology/risk factors for sexual dysfunction. *J Sex Med* 2010;**7**(4 Pt 2):1598–607. Epub 2010/04/15.

178. Ponholzer A, Temml C, Mock K, Marszalek M, Obermayr R, Madersbacher S. Prevalence and risk factors for erectile dysfunction in 2869 men using a validated questionnaire. *Eur Urol* 2005;**47**(1):80–5. discussion 5–6. Epub 2004/12/08.

179. Fonseca V, Seftel A, Denne J, Fredlund P. Impact of diabetes mellitus on the severity of erectile dysfunction and response to treatment: analysis of data from tadalafil clinical trials. *Diabetologia* 2004;**47**(11):1914–23. Epub 2004/12/16.

180. Jemal A, Siegel R, Xu J, Ward E. Cancer statistics, 2010. *CA Cancer J Clin* 2010;**60**(5):277–300. Epub 2010/07/09.

181. Brandeis J, Pashos CL, Henning JM, Litwin MS. A nationwide charge comparison of the principal treatments for early stage prostate carcinoma. *Cancer* 2000;**89**(8):1792–9. Epub 2000/10/24.

182. Resnick MJ, Koyama T, Fan KH, Albertsen PC, Goodman M, Hamilton AS, et al. Long-term functional outcomes after treatment for localized prostate cancer. *N Engl J Med* 2013;**368**(5):436–45. Epub 2013/02/01.

183. Shamloul R, Ghanem H. Erectile dysfunction. *Lancet* 2013;**381**(9861):153–65. Epub 2012/10/09.

184. Carvalheira AA, Pereira NM, Maroco J, Forjaz V. Dropout in the treatment of erectile dysfunction with PDE5: a study on predictors and a qualitative analysis of reasons for discontinuation. *J Sex Med* 2012;**9**(9):2361–9. Epub 2012/05/24.

185. Melman A, Davies K. Gene therapy for erectile dysfunction: what is the future?. *Curr Urol Rep* 2010;**11**(6):421–6. Epub 2010/09/15.

186. Bahk JY, Jung JH, Han H, Min SK, Lee YS. Treatment of diabetic impotence with umbilical cord blood stem cell intracavernosal transplant: preliminary report of 7 cases. *Exp Clin Transplant* 2010;**8**(2):150–60. Epub 2010/06/23.

187. Suuronen EJ, McLaughlin CR, Stys PK, Nakamura M, Munger R, Griffith M. Functional innervation in tissue engineered models for in vitro study and testing purposes. *Toxicol Sci Off J Soc Toxicol* 2004;**82**(2):525–33. Epub 2004/09/03.

188. Dhawan V, Lytle IF, Dow DE, Huang YC, Brown DL. Neurotization improves contractile forces of tissue-engineered skeletal muscle. *Tissue Eng* 2007;**13**(11):2813–21. Epub 2007/09/08.

189. Elbadawi A, Yalla SV, Resnick NM. Structural basis of geriatric voiding dysfunction. IV. Bladder outlet obstruction. *J Urol* 1993;**150**(5 Pt 2): 1681–95. Epub 1993/11/01.

190. Sakuma T, Matsumoto T, Kano K, Fukuda N, Obinata D, Yamaguchi K, et al. Mature, adipocyte derived, dedifferentiated fat cells can differentiate into smooth muscle-like cells and contribute to bladder tissue regeneration. *J Urol* 2009;**182**(1):355–65. Epub 2009/05/22.

Chapter 10

Regenerative Medicine and Tissue Engineering in Reproductive Medicine: Future Clinical Applications in Human Infertility

Irene Cervelló*, Jose Vicente Medrano*, Carlos Simón

Key Concepts

Endometrium, Gametes, Ovary, Uterus, Tissue Engineering, Stem Cells.

List of Abbreviations

ABCG2, Bcrp1 ATP-binding cassette subfamily G member 2
ASCs Adult stem cells
BMDCs Bone marrow-derived cells
BMT Bone marrow transplantation
CD9 Cluster differentiation 9
c-Kit, CD117, SCFR Mast/stem cell growth factor receptor
Dazl Deleted in azoospermia like
DNA Deoxyribonucleic acid
ECM Extracellular matrix
ESCs Embryonic stem cells
GDNF Glial-derived neural factor
GFP Green fluorescent protein
GFRA1 Gdnf receptor alpha 1
HLA Human leukocyte antigen
ICSI Intracytoplasmic sperm injection
iPSCs Induced pluripotent stem cells
Itga6, CD49f Integrin alpha-6
KLF4 Kruppel-like factor 4
Mdr Multi-drug resistance
MHC Major histocompatibility complex
MRKHS Mayer–Rokitansky–Küster–Hauser syndrome
MSCs Mesenchymal stem cells
Mvh Mouse vasa homologue
Neurog3 Neurogenin-3
Oct4, Pou5f1 Octamer-binding transcription factor 4
OLCs Oocyte-like cells
OSE Ovarian surface epithelium
PDGFR-β Platelet-derived growth factor receptor beta
PGA Polyglycolic acid

---

* These authors contributed equally.

Translating Regenerative Medicine to the Clinic. http://dx.doi.org/10.1016/B978-0-12-800548-4.00010-3

Plzf, ZBTB16 Zinc finger and BTB domain containing 16
RNA Ribonucleic acid
Sall4 Sal-like protein 4
SDS Sodium dodecyl sulfate
SOX2 SRY (sex-determining region Y)-box 2
SP Side population
SRY Sex-determining region Y
SSCs Spermatogonial stem cells
Ssea-1 Stage-specific embryonic antigen 1
Thy1, CD90 THYmocyte differentiation antigen 1
UTF1 Undifferentiated embryonic cell transcription factor 1
VSEL Very small embryonic-like
W5C5, SUSD2 Sushi domain containing 2
WHO World Health Organization

1. INTRODUCTION

"A brave new Medicine termed Regenerative Medicine" was presented by William Haseltine at a conference on Lake Como, Italy in 1999. W. Haseltine was a professor at Harvard Medical School and Harvard School of Public Health, where he did important work on HIV/AIDS and on the human genome. He is the founder of Human Genome Sciences Inc. and other biotechnology companies, and now his career is focused on philanthropy. He was listed by Time Magazine as one of the world's 25 most influential business people in 2001.

Regenerative medicine is an emerging multidisciplinary field that incorporates cell biology, medicine, and engineering, and which focuses on restoring tissues and/or organs. Based on novel therapeutic approaches using stem cell technology and tissue engineering, regenerative medicine may soon offer new hope to patients who suffer from diseases which still remain untreatable in the twenty-first century (Box 1).

Stem cells are an attractive cell source for use in combination with tissue engineering in regenerative therapeutics. Several different studies have described the use of this methodology over the last 10 years in different pathologies and organs such as macular degeneration,[1] heart diseases,[2] diabetes,[3] and muscular dystrophy.[4]

According to the World Health Organization (WHO), infertility affects up to 14% of couples in reproductive age. This percentage tends to be higher in eastern countries, mainly due to the higher percentage of toxic habits, uterine pathologies, and the delaying of motherhood in eastern countries. Based on the 2005 National Survey on Family Growth American report, approximately 12% of American couples experienced impaired fertility in 2002. This implies a 20% increase from 6.1 million couples who have reported fertility difficulties to have children problems in 1995. In this chapter, we review and explain the putative uses of regenerative medicine to enhance our knowledge of reproductive medicine.

2. CELL THERAPY APPROACHES/STEM CELL TECHNOLOGY IN REPRODUCTIVE MEDICINE

Cell therapy is a concept, which encompasses all the possible technical advances related to cell transplantation that may eventually repair a damaged tissue or organ. The transplanted cells may have various origins and can be used as replacement cells in damaged tissues (i.e., transplanted cardiomyocytes could be used to regenerate ischemic areas of heart after a myocardial infarction[5,6]), or may act indirectly to help resident cells to repair the injury by releasing soluble factors (i.e., transplanted mesenchymal cells can act as immune-regulators in injured tissues, promoting self-repair[7–10]). In the following sections we will focus on describing the main cell therapy strategies for untreatable conditions affecting human reproduction.

Box 1 Recent Classification of Regenerative Medicine Based on the Use of Scaffolds

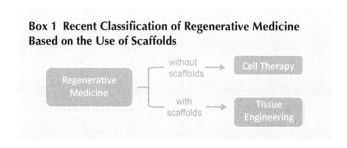

2.1 The Female Side: Endometrial and Ovarian Stem Cells

2.1.1 The Regenerative Capacity of Human Endometrium

The endometrium, the mucosal layer that lines the inside of the uterus, is a very interesting "organ," not only because of its pivotal role in the blastocyst implantation process and ensuing pregnancy, but also because of its exceptional remodeling and renewal ability. The endometrium follows a precisely programmed series of morphological and physiological changes associated with growth, differentiation, and in the absence of conception, degeneration, and regeneration. For this reason it has been described in the medical literature as one of the most regenerative tissues present in the human body.[11,12] Like any tissue that undergoes self-renewal and subsequent differentiation (e.g., the hematopoietic system, epidermis, intestines, or muscle), it must have a specific population of tissue-specific stem cells, adult stem cells (ASCs), or somatic stem cells.[13–16]

2.1.1.1 Autologous Sources of Endometrial ASCs

Despite early work in the 1970s suggesting the presence of endometrial stem cells, their existence was not demonstrated until the beginning of 2004.[17] Initial cell cloning studies showed human epithelial (0.22%) and stromal (1.25%) cell colony-forming activity in normal cycling and menopausal endometrium. These clonogenic cells demonstrated characteristics of ASCs: the ability to divide, proliferate, and self-renew while maintaining the potential of their originator cells, and thus supported the theory that the endometrium possesses an ASC subpopulation. Due to the absence of specific markers, the side population (SP) method, which is well described in other adult tissues, including bone marrow,[18] was also used to identify this subset of cells. This approach is based on enriching cells with specific cell membrane transporters such as ABCG2 and MDR which extrude the Hoechst vital dye and are implicated in protecting immature cells in vivo. Two independent groups[19–21] used flow cytometry to isolate the endometrial SP, and both demonstrated that these cells are capable of multipotential differentiation in vitro, giving rise to mesenchymal lineage cells including osteocytes, adipocytes, and chondrocytes.[22] The final proof-of-concept was corroborated by both groups by generating endometrial-like tissue in vivo in a murine model.[19–21]

Currently, there are no publications describing universal markers for human endometrial ASCs, although several candidate endometrial ASCs surface markers are being investigated by different groups, including CD146, CD140b/PDGFR-β, W5C5, Musashi-1, Lgr5, Ssea-1, and ABCG2.[23–28]

ASCs are the main cell type responsible for severe endometrial disorders, for example, endometriosis, endometrial cancer, and Asherman syndrome.[20] Notably, bone marrow progenitors (described in the following section) have already been used to regenerate human endometrium,[29,30] however, endometrial ASCs could be considered as potentially safer candidates for such therapeutic purposes.

2.1.1.2 Exogenous Sources of Endometrial ASCs

The role of bone marrow-derived cells (BMDCs) in the physiological reconstitution of endometrium after bone marrow transplantation (BMT) was demonstrated by Taylor et al. (2004).[31] They analyzed the origin of endometrial cells in four patients who underwent BMT by using the specific human leukocyte antigen of the donors as a marker and thus demonstrated that BMDCs can differentiate into the human endometrium. Our study in 2012 also showed the presence of XY donor-derived cells in the endometrium of five women after male-donor BMT, suggesting that these cells have the plasticity to renew endometrial tissue despite their differing exogenous source.[32]

At the clinical level several studies suggest the possible application of stem cell therapies for Asherman syndrome, which causes endometrial destruction. Nagori et al. published a case report describing a 33-year-old woman with Asherman syndrome who was treated with adult autologous bone marrow stem cells: CD9+, CD90+, and CD133+ candidate cells were isolated and placed into the endometrial cavity with a catheter. Four months later endometrial thickness, morphology, and vascularity improved and a pregnancy was successfully obtained after in vitro fertilization treatment.[29]

Indeed, very recent works (published in 2014) based on mouse models demonstrated the specific functional role of BMDCs in uterine repair, suggesting that they are essential for endometrial regeneration and that their transplant contributes to improving fertility by increasing endometrial thickness.[33–36] Our group is also performing similar work, which has demonstrated the beneficial effect of BMDCs and their ability to home and to colonize murine endometrium, where they engraft mainly around the vascular space (unpublished data). These cells, probably acting by paracrine effects, were able to provoke endometrial regeneration in an animal model of Asherman syndrome.

2.1.2 Stem Cells in the Ovaries of Female Mammals: Fallacy or Hidden Truth?

It is generally accepted that follicular regeneration occurs in most nonvertebrate animals (reviewed in Ref. 37) and in some vertebrates.[38] However, one of the central dogmas in reproductive medicine is that female mammals are born with a finite number of germ cells (primordial follicles). This follicular or ovarian reserve decreases over an individual's life span until is completely depleted at the start of menopause.[39] However, this view was questioned few years ago when John Tilly's group at Harvard University postulated the existence of ovarian stem cells in postnatal mammalian ovaries.[40] Tilly's group hypothesized that, based on the amount of atretic follicles they found in mouse ovaries, the ovarian reserve would become depleted at a much faster rate than what actually occurs. This implies that there had to be some kind of germinal stem cells within mammalian ovaries that replace atretic follicles by newly formed ones. To test their hypothesis, the authors transplanted ovarian fragments from wild-type mice into the ovarian bursal cavity of transgenic green fluorescent protein (GFP) mice and observed the appearance of newly formed chimeric follicles with GFP positive oocytes surrounded by non-GFP expressing granulose cells.[40] The issue became controversial, and other groups criticized the methods they used to calculate follicle atresia and the ovarian graft results.[41,42]

Only 1 year later, Tilly's group presented a new report suggesting the bone marrow as the putative source of germ stem cells for adult neofolliculogenesis in mammals.[43] The authors based their hypothesis on the observation that Ssea-1 (Stage-Specific Embryonic Antigen 1) positive cells isolated from either disassociated mouse ovaries or bone marrow express early germ line markers but not oocyte-specific markers. To test this hypothesis, both bone marrow and peripheral blood transplants from GFP transgenic mice into chemotherapy-ablated wild-type females were performed, resulting in the formation of new chimeric follicles with GFP positive oocytes. Once again, this report generated intense discussion and there were several criticisms.[44]

This led Eggan's group to conduct parabiosis experiments in which the circulatory systems of wild-type and GFP transgenic mice were joined for up to 6 months. These results demonstrated a high degree of circulating cell chimerism, and even some GFP positive granulosa cells associated with ovulated oocytes in wild-type mice. However, no evidence of GFP positive oocytes was observed in wild-type females. Finally, the authors transplanted bone marrow from a GFP donor into wild-type females treated with cyclophosphamide/busulfan and, once again, found no evidence of oocyte chimerism. In summary, these results indicated that neither peripheral blood nor bone marrow transplants seemed to provide germ stem cells that could help to regenerate the ovarian reserve. The authors suggested that the recovery of ovulation in chemically sterilized mice could be explained by incomplete depletion of endogenous follicles by the alquilant drugs and the presence of some GFP positive cells within the granulosa cell layer, or infiltration of white cells from the GFP donor.

These respective reports subsequently divided the scientific community, with publications supporting[45–49] or arguing against[50–52] the neofolliculogenesis hypothesis in mammalian ovaries. A new report in 2009[53] by Zou et al. established germ stem cell lines in vitro from mouse vasa homolog (Mvh) positive cells isolated from mouse ovaries. In agreement with Tilly's previous observations, Mvh lines expressed pluripotency-associated markers as well as early, but not late, germ cell markers. Once again, this work was received with skepticism because, to isolate ovarian stem cells, the authors used an antibody which binds to a surface region of the Mvh protein, however, this RNA helicase is classically localized in the cytoplasm/nucleus of germ cells.

Tilly's group published a new report in 2012 describing how Mvh/VASA-based isolation of germ cells from both mouse and human ovaries allows germ stem cells to be established in vitro.[54] The authors also reported the spontaneous detachment of some oocyte-like cells (OLCs) that expressed meiotic markers and were even able to complete meiosis. Finally, transplantation of putative GFP positive mouse germ stem cells into mouse ovaries resulted in the formation of GFP oocytes after 6 months and these were able to form healthy embryos. In a similar way, putative human germ stem cells were also marked with GFP and cocultured with disassociated adult human cortical tissues and transplanted into immunosuppressed mice, resulting in the formation of GFP positive oocytes.

In relation to this debate, new opinions on the possible origin of these putative germinal stem cells have arisen in recent years. One of the possible explanations for the existence of germinal stem cells in mammalian ovaries is an ovarian cell dedifferentiation process that, under specific in vitro conditions, produces cells which are able to develop into germ-like cells in the same way as previously described for other cell types.[55–58] Related to this hypothesis, some authors have even postulated a relationship between the existence of stem cells in the ovarian surface epithelium (OSE) and the incidence of ovarian cancer of epithelial cell origin, a theory that is supported by the deregulation of cell cycle and cell signaling genes which has previously been observed.[59] Alternatively, some authors also propose the presence of very small embryonic-like stem cells resident in the ovaries which have the potential to differentiate into both germ and somatic cells.[60] Taking all of this evidence together, it is clear that further research is needed to determine the true identity of these putative germinal stem cells in mammalian ovaries.

2.2 The Male Side: Spermatogonial Stem Cells

Fertility preservation in men undergoing alquilant chemotherapy and/or pelvic irradiation as part of a cancer treatment is necessary since they are in serious danger of becoming infertile due to the toxicity of such treatments for germ cells especially.[61] In the case of adult patients this issue is quite easily resolved by cryopreserving their sperm for future fertility treatments. However, for prepubertal patients unable to produce sperm this solution is not an option. From this point of view, the use of spermatogonial stem cells (SSCs) is presented as a possible solution.

SSCs are the origin of the process of spermatogenesis in the testes and are comprised by a subset of spermatogonia attached to the basal layer of the seminiferous.[62] SSCs were first described in 1994 by Brinster's lab using a murine germ cell transplant model, and were defined as "the spermatogonial subpopulation of spermatogenic stem cells with the ability to colonise the seminiferous epithelium of chemically sterilised mice and reconstitute their spermatogenesis."[63,64] Because prepubertal seminiferous tubules are mainly composed by Sertoli cells and quiescent SSCs,[62,65,66] one possible solution for fertility preservation in these patients could be the use of these SSCs. Indeed, some groups have started to cryopreserve testicular biopsies from cancer patients with a risk of becoming infertile, just before they begin their cancer treatments[67–71] (Figure 1). However, it is important that several issues be resolved before trying to restore fertility in humans using SSCs because they are still far from being a well-defined cell population. The lack of consensus about appropriate markers to isolate SSCs has hampered their study, leaving their ability to restore spermatogenesis following their transplant into mouse testes the best approach to identify them. By using this approach as a read out, several surface markers such as GFRA1, Thy1, Itga6, CD9, and Cdh1 and some intracellular markers such as Plzf, Sall4, UTF1, Neurog3, Nanos2, and Lin28 have been identified as putative SSCs markers in both mouse and human. However, all of these markers are also expressed in other undifferentiated spermatogonial cohorts that represent transient amplifying progenitors.[72]

Moreover, in vitro propagation of human SSCs before any transplantation back to the germ cell-depleted testes of patients is essential for two main reasons: the need to obtain a high number of cells, and because of in vitro propagation of SSC purification and effective elimination of malignant cells must be ensured, especially in patients with nonsolid tumors such as leukemia that may infiltrate the testis and cause a new cancer if transplanted back into the patient.[61] Studying the SSC niche has led to the identification of several factors that play important roles in SSC self-renewal. The SSC niche is located on the basal layer of seminiferous tubules and is tightly regulated by Sertoli cells, usually in areas close to the interstitial tissue. The testicular pool of SSCs (estimated at between 0.01% and 1% of the total number of cells in testis) is principally maintained by glial-derived neural factor secreted by Sertoli cells.[73,74] Although in vitro propagation of mouse SSCs is a well-established model used to study their biology, in vitro culture of human SSCs has proven to be more difficult and was not accomplished until 2008.[75] This study, and

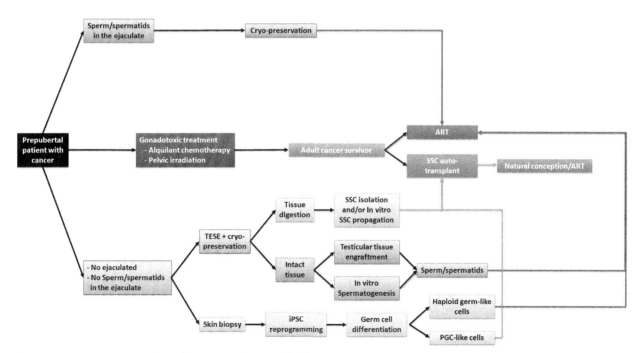

FIGURE 1 Schematic diagram of the different approaches described in this review to preserve human male fertility. ART: assisted reproduction technique; SSC: spermatogonial stem cell; TESE: testicular sperm extraction; iPSC: induced pluripotent stem cell; PGC: primordial germ cell.

subsequent reports on human SSC culture, has paved the way for the development of better technologies for in vitro propagation. However, these tools will still require further work before their effective use in regenerative therapies can be contemplated.[76–78]

Finally, one last issue that must be resolved before SSCs can be used in clinical applications is the need for appropriate models for spermatogenic restoration. Transplantation of human SSCs into germ cell-depleted mouse testes results in colonization of the seminiferous epithelium, but fails to restore spermatogenesis because of phylogenetic discrepancies such as differences in the biology of human/murine spermatogenesis, as previously described in this chapter. However, recent reports on fertility restoration in primates[79] and transplantation of cells obtained from cadaver donors into human testes[80] are beginning to shed some light on this issue.

2.3 In Vitro Derivation of Gametes for the Study of Human Germ Line Development

Donation of sperm and eggs is one solution when fertility problems are caused by bad quality of gametes, or their absence. However, gamete donation raises ethical, legal, and personal concerns, which has led to increasing scientific interest in the study of human germ line development. Our knowledge in this area is limited, mainly due to ethical and technical limitations, which complicate access to germ cells in early human developmental stages for molecular and genetic analysis. Therefore many groups are using pluripotent stem cells as an in vitro model to provide some insight into human germ cell development in vitro.[81]

Several groups have now reported evidence of the in vitro generation of germ cells, which express specific germ line-related markers, from pluripotent stem cells such as embryonic stem cells (ESCs) and induced pluripotent stem cells (iPSCs). However, in 2003, the scientific community was surprised by the first report providing evidence of germ line differentiation from mouse ESCs.[82] In this study, ESC lines, expressing GFP driven by a germ cell-specific distal Oct4 enhancer, were spontaneously differentiated and GFP positive cells were selected. Surprisingly, the authors observed that some follicle-like structures spontaneously started to detach from the monolayer and extrude OLCs; the group parthenogenetically activated these, and they subsequently formed pseudoblastocysts. This fact was a breakthrough in regenerative medicine because it opened up the possibility to obtaining germ cells in vitro in order to study their genetics and gain insight into their development. Since the publication of this work in 2003, several reports have used different strategies to obtain putative mouse germ cells in vitro with different success rates.[83–85] However, to date only one group has reported efficiently obtaining offspring using such in vitro-derived germ cells.[86–88]

Unfortunately, germ line differentiation seems to be different in humans compared to mice. The first evidence of obtaining such cells from human ESCs (hESCs) in vitro was reported in 2004 by spontaneous differentiation.[89] This study consisted of the exhaustive characterization of stage-specific expression of different germ cell markers in both undifferentiated and spontaneously differentiated ESC lines and established a reference model for in vitro germ line differentiation from ESCs. Among their findings, the authors showed that undifferentiated hESCs expressed some early germ-specific markers, but not late markers, and hypothesized that ESCs could actually be a heterogeneous pluripotent population in which some cells have a predisposition to a germ cell fate. Following this work, several groups have reported the in vitro formation of human germ cells using different techniques,[90,91] however, most of these reports showed problems in accomplishing the maturation of germ cells derived in vitro, although the initiation of meiosis was achieved in most cases.

It has been reported that ectopic expression of the DAZ gene family members DAZ2, DAZL, and BOULE in both hESC and human iPSC (hiPSC) lines subjected to spontaneous differentiation can drive complete meiotic progression of in vitro-derived germ cells.[92–94] The role of RNA-binding proteins has also been tested in vivo. In recent reports, the ability of iPSCs to colonize the lumen of the seminiferous tubules of sterilized immunodeficient mice improved when VASA was added to the cocktail of reprogramming factors (OCT4, SOX2, KLF4, and c-MYC). These results highlight the role VASA plays in making cells competent to form germ cells and in controlling the pluripotency state in a mixed in vitro/in vivo model.[95] This same model has been used to analyze the capability of hiPSCs derived from azoospermic men with different deletions in the Y chromosome, demonstrating how genetic background affects their capacity to differentiate into germ cells and colonize the seminiferous niche in vivo.[96] Therefore this line of research may pave the way for better understanding germ line development, and may even provide the tools to create "artificial gametes" which can eventually help infertile patients. However, although amazing progress has been made since the first report on germ cell derivation from mouse ESCs, as shown, there are still several questions to be answered before this possibility can become a reality.

3. TISSUE ENGINEERING IN REPRODUCTIVE MEDICINE

3.1 Regenerating the Uterus

Uterine reconstruction currently focuses mainly on cell transplantation assays (see Section 2.1.2), and very recently on tissue engineering (Figure 2)[97–101] (Table 1).

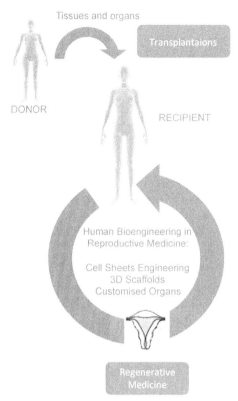

FIGURE 2   Diagram showing the advances described here and related with bioengineering in the human uterus.

TABLE 1 List of Recent Publications Related to Uterine Regenerative Medicine

Title of Publication	Model	Journal and Month/Year	Impact Factor	Characteristics
Reconstruction of functional endometrium-like tissue in vitro and in vivo using cell sheet engineering	Mice	*Biomedical and Biophysical Research Communications* March 2014	2.2	Reconstruction of mouse and rat functional endometrial tissues in vitro and in vivo using cell sheet engineering
Partial regeneration and reconstruction of the rat uterus through recellularization of a decellularized uterine matrix	Rat	*Biomaterials* July 2014	8.3	Maruyama shows the regeneration and reconstruction of uterine tissues permitting the achievement of pregnancy in this elegant study using rats
Tissue-engineered autologous vaginal organs in patients: a pilot cohort study	Human	*The Lancet* July 2014	39.3	First study describing the use of biodegradable scaffolds for the generation of human vaginal organs
Towards the development of a bioengineered uterus: Comparison of different protocols for rat uterus decellularization	Rat	*Acta Biomaterialia* August 2014	5.6	Brännström's group describe several decellularization protocols for rat uteruses, and makes suggestions for their future use in in vivo applications

3.1.1 Cell Sheet Engineering for Endometrial Reconstruction

In experimental assays with small animals (rats, rabbits, mice, etc.), injection of single cell suspensions can be sufficient, but this technique does not seem to be suitable for large-scale tissue reconstructions because only a low percentage of the injected cells are capable of correctly homing and engrafting into host tissues. To avoid this problem, single sheets of multiple layers of candidate cells with intact cell–cell junctions can be obtained by cell sheet engineering, as first proposed by Okano.[102] This technique is based on the properties of a temperature-responsive thermoresponsive polymer (which undergoes dramatic changes in physical properties with changes in temperature), which was used to precoat the surface of the tissue culture dishes. When the temperature drops below 32 °C the cells spontaneously detach, thus eliminating the requirement for enzymatic products and the consequent degradation of cell adhesion molecules, extracellular matrix (ECM), and protein interactions. The first clinical applications of these cell sheets were for ocular surface regeneration,[103–105] periodontal diseases, and as cardiac patches.[106,107] Furthermore, scientists also demonstrated that transplantation of these multilayered cell sheets into the endometrial compartment of immunocompromised mice generates endometrium-like tissues with a three-dimensional (3D) structure.[98] Thus this study was a proof-of-concept that endometrial tissue can be reconstructed in vitro and of ectopic regeneration in vivo in animal models using cell sheet engineering.

3.1.2 Vaginal Reconstruction from 3D Scaffolds

Creation of 3D scaffolds with autologous cells with the purpose of reconstructing organs such as the vagina has already been attempted. The main objective of using an artificial structure is to guide the proliferation and spread of seeded cells both in vitro and in vivo. Bioengineering techniques used to create a 3D scaffold were described by Dr Atala's group.[100,108,109] A biodegradable composite scaffold made of collagen and polyglycolic acid (PGA) was used to create a specific template for each organ and patient. PGA belongs to a group of biodegradable aliphatic polyesters which are currently exploited in a variety of medical applications and which have favorable properties such as biocompatibility, bioresorbability, and high tensile strength.

After successfully obtaining bladder and urethra tissues, vagina was created for patients with vaginal aplasia caused by Mayer–Rokitansky–Küster–Hauser syndrome (MRKHS) who participated in a pilot study coordinated by Dr Atala. Human vaginas were bioengineered according to the individual patients' morphological features using PGA scaffolds into which autologous muscle and vaginal epithelial cells were seeded; finally these matrices were implanted into four patients.[100] All the patients showed normal structural and functional characteristics in the reconstructed vaginas, and in all cases presented adequate vascularization and innervation. In conclusion, the use of engineered organs using autologous cells may be a viable option for vaginal reconstruction and for the repair of other genital and gynecological abnormalities.

3.1.3 Customized Uterus

Whole-organ decellularization with perfusion protocols using detergents provides some key advantages compared to the techniques previously described in this chapter (Sections 3.1.1 and 3.1.2): mainly the preservation of tissue-specific ECM and the maintenance of a 3D structure of the original organ. Successful decellularization has been reported in several organs, including bladder,[110] arteries,[111] esophagus,[112] trachea,[113] liver,[114] and lung,[115] as well as in rat uterus.[99,101] Importantly, both groups highlight the importance of appropriate surgical hysterectomy procedures to preserve the vasculature.

Seventy-two hours of SDS detergent treatment yielded a translucent acellular structure which maintained its uterine shape.[99] Seventy-nine million primary uterine cells and mesenchymal stem cells were introduced in vitro, in a perfusion system which gave rise to uterine-like tissue after 6 days. However, the reconstituted in vitro tissue gradually became atrophic around days 6–10. In contrast, rats transplanted with a piece of recellularized uterine horn, after excising the native horn and replacing it with a recellularized matrix, showed normal endometrial organization on day 90. Immunohistochemical analysis demonstrated positive expression for typical endometrial markers such as vimentin (stroma), cytokeratin (epithelium), α-smooth muscle actin (myometrium), and progesterone receptors. Moreover, six out of eight excision model animals were able to conceive. The authors demonstrated that the 3D architecture of the rat uterus could be maintained in vitro, and that the partial excised and replaced uterus in a rat model could give rise to a structural and functional uterus.

Exhaustive uterine decellularization in the rat model is based on three protocols which were published by Brännström's group. They demonstrated the elimination of immunoreactive major histocompatibility complex class I and class II elements in all three protocols, suggesting that the in vivo use of these decellularized uteruses may be possible. Also they showed that some intracellular protein or DNA remains in uterine matrix and may have an important role in autologous stem cell homing during the recellularization process.

Uterine transplantation involves complicated surgery and requires strong immunosuppression to prevent rejections. However, it is hoped that the tissue engineering applications described in this section may eventually replace the need for a live donor, and that they circumvent the need for immunosuppressive treatment.

3.2 In Vitro Spermatogenesis and Tissue Engraftment

Creating tissue engineering approximations of male spermatogenesis is complicated because of the complexity of spermatogenic tissue which comprises different, and very specialized, cell types. However, apart from the SSC transplant approach described above, other techniques such as the testicular engraftment technique reported by Schlatt's group in 2002[116] have also appeared. The authors describe the engraftment of small pieces of mouse testicular tissue under the skin of immunosuppressed mice that were previously castrated in order to prevent the feedback inhibition of gonadotropins by their own testes. After a few weeks, some of the engrafted tissues showed active spermatogenesis and sperm production. Researchers used the sperm retrieved to perform intracytoplasmic sperm injection (ICSI) which produced normal offspring, thus demonstrating the feasibility of using this technique to produce functional sperm.

Following on from this approach, several reports have documented the production of sperm from different species such as rat, rabbit, horse, donkey, and monkey in testicular grafts implanted under the skin of castrated immunodeficient mice.[117,118] Additionally, slight modifications to the technique, such as performing scrotal rather than subcutaneous engraftment, have further improved results.[67] However, as several reports have described, in the case of xenografts from testicular tissue from humans, and other species with higher than average daily sperm production such as pig and goat, the survival rates of the engrafted tissue are extremely poor, and show a high degree of fibrosis and degenerated spermatogenesis.[67,68,117,119,120] This is likely caused by fibrosis resulting from the ischemic period the engrafted tissue endures before it becomes vascularized. It is possible that using immature testicular tissue for grafts can help to solve this problem as evidence suggests that spermatogonia are quiescent in these tissues, making them more resistant to ischemia. However, so far, human grafts designed to preserve spermatogenesis have not been able to develop beyond the spermatocyte stage,[68] thus highlighting the need for further improvements to the technique before its eventual possible clinical application to produce functional human sperm in azoospermic patients (Figure 1).

Finally, a recent alternative tissue engineering possibility for producing sperm in vitro has emerged. In 2011, a Japanese group demonstrated, for the first time, that functional sperm can be produced in vitro from pieces of mouse testicular tissue.[121,122] In this new approach, small testicular pieces taken from new born mice, carrying GFP gene expression under the control of the postmeiotic markers Gsg2 (haspin) and acrosin, were placed in a liquid–gas interphase on agarose soaked in a medium containing knockout serum replacement and were placed at 34 °C to mimic in vivo testicular conditions. After around 3 weeks in these conditions, GFP spermatids and sperm started to appear within the testicular pieces. The sperm retrieved were used for ICSI and gave raise to healthy offspring. Furthermore, following this technique, and adding exogenous kit ligand to the culture medium, the same group has recently reported spermatogenic recovery in tissue fragments from sterile c-kit ligand mutant mice.[123] Even more interesting is that the efficiency of the technique using frozen/thawed tissue has also recently been demonstrated.[124] Taking these findings together, in vitro spermatogenesis using an organ culture approach represents an encouraging alternative solution to help solve human spermatogenic deficiency problems (Figure 1). However, the model from Ogawa's group has not yet been replicated by other groups, and no data regarding the feasibility of this technique with human tissue have so far been reported.

4. CONCLUSIONS AND FUTURE DIRECTIONS

These novel findings should help to facilitate the development of organ-specific tissue engineering and stem cell technologies, not only for reproductive organs but also for other organs with similar tissue structures. We believe that regenerative medicine is very promising and has the potential to become a fundamental discipline in the next generation of treatments for both the male and female reproductive systems. However, it is clear that many further studies will be required to determine how the novel techniques described here can be optimized for human use.

ACKNOWLEDGMENTS

This work was supported by a research project grant PI13/00546 from ISCIII, cofounded by the European Regional Development Fund "A way to make Europe" and a Sara Borrell grant given to JVM (CD12/00568) by the ISCIII. This work was also financed by the Prometeo II/2013/018 project (CS, IC), the *Generalitat Valencia* GV/2013/120 (IC), and SAF2012/31017 from MINECO (CS).

REFERENCES

1. Huang Y, Enzmann V, Ildstad ST. Stem cell-based therapeutic applications in retinal degenerative diseases. *Stem Cell Rev* June 2011;**7**(2):434–45.
2. Fox IJ, Daley GQ, Goldman SA, Huard J, Kamp TJ, Trucco M. Stem cell therapy. Use of differentiated pluripotent stem cells as replacement therapy for treating disease. *Science* August 22, 2014;**345**(6199):1247391.
3. Ryan EA, Paty BW, Senior PA, Bigam D, Alfadhli E, Kneteman NM, et al. Five-year follow-up after clinical islet transplantation. *Diabetes* July 2005;**54**(7):2060–9.
4. De Angelis L, Berghella L, Coletta M, Lattanzi L, Zanchi M, Cusella-De Angelis MG, et al. Skeletal myogenic progenitors originating from embryonic dorsal aorta coexpress endothelial and myogenic markers and contribute to postnatal muscle growth and regeneration. *J Cell Biol* November 15, 1999;**147**(4):869–78.
5. Jackson KA, Majka SM, Wang H, Pocius J, Hartley CJ, Majesky MW, et al. Regeneration of ischemic cardiac muscle and vascular endothelium by adult stem cells. *J Clin Invest* June 2001;**107**(11):1395–402.
6. Kawada H, Fujita J, Kinjo K, Matsuzaki Y, Tsuma M, Miyatake H, et al. Nonhematopoietic mesenchymal stem cells can be mobilized and differentiate into cardiomyocytes after myocardial infarction. *Blood* December 1, 2004;**104**(12):3581–7.
7. Yagi H, Soto-Gutierrez A, Parekkadan B, Kitagawa Y, Tompkins RG, Kobayashi N, et al. Mesenchymal stem cells: mechanisms of immunomodulation and homing. *Cell Transplant* 2010;**19**(6):667–79.
8. Nauta AJ, Fibbe WE. Immunomodulatory properties of mesenchymal stromal cells. *Blood* November 15, 2007;**110**(10):3499–506.
9. Salem HK, Thiemermann C. Mesenchymal stromal cells: current understanding and clinical status. *Stem Cells* March 31, 2010;**28**(3):585–96.
10. Devine SM, Cobbs C, Jennings M, Bartholomew A, Hoffman R. Mesenchymal stem cells distribute to a wide range of tissues following systemic infusion into nonhuman primates. *Blood* April 15, 2003;**101**(8):2999–3001.
11. Prianishnikov VA. On the concept of stem cell and a model of functional-morphological structure of the endometrium. *Contraception* September 1978;**18**(3):213–23.
12. Padykula HA, Coles LG, Okulicz WC, Rapaport SI, McCracken JA, King Jr NW, et al. The basalis of the primate endometrium: a bifunctional germinal compartment. *Biol Reprod* March 1989;**40**(3):681–90.
13. Bjerknes M, Cheng H. Clonal analysis of mouse intestinal epithelial progenitors. *Gastroenterology* January 1999;**116**(1):7–14.
14. Jankowski RJ, Deasy BM, Huard J. Muscle-derived stem cells. *Gene Ther* May 2002;**9**(10):642–7.
15. Alonso L, Fuchs E. Stem cells of the skin epithelium. *Proc Natl Acad Sci USA* September 30, 2003;**100**(Suppl. 1):11830–5.
16. Spangrude GJ, Smith L, Uchida N, Ikuta K, Heimfeld S, Friedman J, et al. Mouse hematopoietic stem cells. *Blood* September 15, 1991;**78**(6):1395–402.
17. Chan RW, Schwab KE, Gargett CE. Clonogenicity of human endometrial epithelial and stromal cells. *Biol Reprod* June 2004;**70**(6):1738–50.
18. Goodell MA, Brose K, Paradis G, Conner AS, Mulligan RC. Isolation and functional properties of murine hematopoietic stem cells that are replicating in vivo. *J Exp Med* April 1, 1996;**183**(4):1797–806.
19. Cervello I, Gil-Sanchis C, Mas A, Delgado-Rosas F, Martinez-Conejero JA, Galan A, et al. Human endometrial side population cells exhibit genotypic, phenotypic and functional features of somatic stem cells. *PLoS One* 2010;**5**(6):e10964.
20. Cervello I, Mas A, Gil-Sanchis C, Peris L, Faus A, Saunders PT, et al. Reconstruction of endometrium from human endometrial side population cell lines. *PLoS One* 2011;**6**(6):e21221.
21. Masuda H, Matsuzaki Y, Hiratsu E, Ono M, Nagashima T, Kajitani T, et al. Stem cell-like properties of the endometrial side population: implication in endometrial regeneration. *PLoS One* 2010;**5**(4):e10387.
22. Dominici M, Le Blanc K, Mueller I, Slaper-Cortenbach I, Marini F, Krause D, et al. Minimal criteria for defining multipotent mesenchymal stromal cells. The International Society for Cellular Therapy position statement. *Cytotherapy* 2006;**8**(4):315–7.
23. Gargett CE, Schwab KE, Zillwood RM, Nguyen HP, Wu D. Isolation and culture of epithelial progenitors and mesenchymal stem cells from human endometrium. *Biol Reprod* June 2009;**80**(6):1136–45.
24. Masuda H, Anwar SS, Buhring HJ, Rao JR, Gargett CE. A novel marker of human endometrial mesenchymal stem-like cells. *Cell Transplant* 2012;**21**(10):2201–14.
25. Gil-Sanchis C, Cervello I, Mas A, Faus A, Pellicer A, Simon C. Leucine-rich repeat-containing G-protein-coupled receptor 5 (Lgr5) as a putative human endometrial stem cell marker. *Mol Hum Reprod* July 2013;**19**(7):407–14.
26. Zhou S, Schuetz JD, Bunting KD, Colapietro AM, Sampath J, Morris JJ, et al. The ABC transporter Bcrp1/ABCG2 is expressed in a wide variety of stem cells and is a molecular determinant of the side-population phenotype. *Nat Med* September 2001;**7**(9):1028–34.
27. Gotte M, Wolf M, Staebler A, Buchweitz O, Kelsch R, Schuring AN, et al. Increased expression of the adult stem cell marker Musashi-1 in endometriosis and endometrial carcinoma. *J Pathol* July 2008;**215**(3):317–29.
28. Valentijn AJ, Palial K, Al-Lamee H, Tempest N, Drury J, Von Zglinicki T, et al. SSEA-1 isolates human endometrial basal glandular epithelial cells: phenotypic and functional characterization and implications in the pathogenesis of endometriosis. *Hum Reprod* October 2013;**28**(10):2695–708.
29. Gargett CE, Healy DL. Generating receptive endometrium in Asherman's syndrome. *J Hum Reprod Sci* January 2011;**4**(1):49–52.
30. Nagori CB, Panchal SY, Patel H. Endometrial regeneration using autologous adult stem cells followed by conception by in vitro fertilization in a patient of severe Asherman's syndrome. *J Hum Reprod Sci* January 2011;**4**(1):43–8.
31. Taylor HS. Endometrial cells derived from donor stem cells in bone marrow transplant recipients. *JAMA* July 7, 2004;**292**(1):81–5.
32. Cervello I, Gil-Sanchis C, Mas A, Faus A, Sanz J, Moscardo F, et al. Bone marrow-derived cells from male donors do not contribute to the endometrial side population of the recipient. *PLoS One* 2012;**7**(1):e30260.
33. Jing Z, Qiong Z, Yonggang W, Yanping L. Rat bone marrow mesenchymal stem cells improve regeneration of thin endometrium in rat. *Fertil Steril* February 2014;**101**(2):587–94.

34. Alawadhi F, Du H, Cakmak H, Taylor HS. Bone Marrow-Derived Stem Cell (BMDSC) transplantation improves fertility in a murine model of Asherman's syndrome. *PLoS One* 2014;**9**(5):e96662.

35. Zhao J, Zhang Q, Wang Y, Li Y. Uterine infusion with bone marrow mesenchymal stem cells improves endometrium thickness in a rat model of thin endometrium. *Reprod Sci* June 19, 2014;**22**(2).

36. Kilic S, Yuksel B, Pinarli F, Albayrak A, Boztok B, Delibasi T. Effect of stem cell application on Asherman syndrome, an experimental rat model. *J Assist Reprod Genet* August 2014;**31**(8):975–82.

37. Lin H. The stem-cell niche theory: lessons from flies. *Nat Rev Genet* December 2002;**3**(12):931–40.

38. Nakamura S, Kobayashi K, Nishimura T, Higashijima S, Tanaka M. Identification of germline stem cells in the ovary of the teleost medaka. *Science* June 18, 2010;**328**(5985):1561–3.

39. Green SH, Zuckerman S. The number of oocytes in the mature rhesus monkey (*Macaca mulatta*). *J Endocrinol* June 1951;**7**(2):194–202.

40. Johnson J, Canning J, Kaneko T, Pru JK, Tilly JL. Germline stem cells and follicular renewal in the postnatal mammalian ovary. *Nature* March 11, 2004;**428**(6979):145–50.

41. Gosden RG. Germline stem cells in the postnatal ovary: is the ovary more like a testis? *Hum Reprod Update* May–June 2014;**10**(3):193–5.

42. Bristol-Gould SK, Kreeger PK, Selkirk CG, Kilen SM, Mayo KE, Shea LD, et al. Fate of the initial follicle pool: empirical and mathematical evidence supporting its sufficiency for adult fertility. *Dev Biol* October 1, 2006;**298**(1):149–54.

43. Johnson J, Bagley J, Skaznik-Wikiel M, Lee HJ, Adams GB, Niikura Y, et al. Oocyte generation in adult mammalian ovaries by putative germ cells in bone marrow and peripheral blood. *Cell* July 29, 2005;**122**(2):303–15.

44. Eggan K, Jurga S, Gosden R, Min IM, Wagers AJ. Ovulated oocytes in adult mice derive from non-circulating germ cells. *Nature* June 29, 2006;**441**(7097):1109–14.

45. Lee HJ, Selesniemi K, Niikura Y, Niikura T, Klein R, Dombkowski DM, et al. Bone marrow transplantation generates immature oocytes and rescues long-term fertility in a preclinical mouse model of chemotherapy-induced premature ovarian failure. *J Clin Oncol* August 1, 2007;**25**(22): 3198–204.

46. Bukovsky A, Caudle MR, Svetlikova M, Upadhyaya NB. Origin of germ cells and formation of new primary follicles in adult human ovaries. *Reprod Biol Endocrinol* April 28, 2004;**2**:20.

47. Bukovsky A, Svetlikova M, Caudle MR. Oogenesis in cultures derived from adult human ovaries. *Reprod Biol Endocrinol* 2005;**3**:17.

48. Virant-Klun I, Zech N, Rozman P, Vogler A, Cvjeticanin B, Klemenc P, et al. Putative stem cells with an embryonic character isolated from the ovarian surface epithelium of women with no naturally present follicles and oocytes. *Differentiation* October 2008;**76**(8):843–56.

49. Szotek PP, Chang HL, Brennand K, Fujino A, Pieretti-Vanmarcke R, Lo Celso C, et al. Normal ovarian surface epithelial label-retaining cells exhibit stem/progenitor cell characteristics. *Proc Natl Acad Sci USA* August 26, 2008;**105**(34):12469–73.

50. Liu Y, Wu C, Lyu Q, Yang D, Albertini DF, Keefe DL, et al. Germline stem cells and neo-oogenesis in the adult human ovary. *Dev Biol* June 1, 2007;**306**(1):112–20.

51. Begum S, Papaioannou VE, Gosden RG. The oocyte population is not renewed in transplanted or irradiated adult ovaries. *Hum Reprod* October 2008;**23**(10):2326–30.

52. Veitia RA, Gluckman E, Fellous M, Soulier J. Recovery of female fertility after chemotherapy, irradiation, and bone marrow allograft: further evidence against massive oocyte regeneration by bone marrow-derived germline stem cells. *Stem Cells* May 2007;**25**(5):1334–5.

53. Zou K, Yuan Z, Yang Z, Luo H, Sun K, Zhou L, et al. Production of offspring from a germline stem cell line derived from neonatal ovaries. *Nat Cell Biol* May 2009;**11**(5):631–6.

54. White YA, Woods DC, Takai Y, Ishihara O, Seki H, Tilly JL. Oocyte formation by mitotically active germ cells purified from ovaries of reproductive-age women. *Nat Med* March 2012;**18**(3):413–21.

55. Lei L, Spradling AC. Female mice lack adult germ-line stem cells but sustain oogenesis using stable primordial follicles. *Proc Natl Acad Sci USA* May 21, 2013;**110**(21):8585–90.

56. Wei S, Zan L, Hausman GJ, Rasmussen TP, Bergen WG, Dodson MV. Dedifferentiated adipocyte-derived progeny cells (DFAT cells): potential stem cells of adipose tissue. *Adipocyte* July 1, 2013;**2**(3):122–7.

57. Herreros-Villanueva M, Zhang JS, Koenig A, Abel EV, Smyrk TC, Bamlet WR, et al. SOX2 promotes dedifferentiation and imparts stem cell-like features to pancreatic cancer cells. *Oncogenesis* 2013;**2**:e61.

58. Lee ST, Gong SP, Yum KE, Lee EJ, Lee CH, Choi JH, et al. Transformation of somatic cells into stem cell-like cells under a stromal niche. *FASEB J* July 2013;**27**(7):2644–56.

59. Bowen NJ, Walker LD, Matyunina LV, Logani S, Totten KA, Benigno BB, et al. Gene expression profiling supports the hypothesis that human ovarian surface epithelia are multipotent and capable of serving as ovarian cancer initiating cells. *BMC Med Genomics* 2009;**2**:71.

60. Bhartiya D, Unni S, Parte S, Anand S. Very small embryonic-like stem cells: implications in reproductive biology. *Biomed Res Int* 2013;**2013**:682326.

61. Geens M, Goossens E, De Block G, Ning L, Van Saen D, Tournaye H. Autologous spermatogonial stem cell transplantation in man: current obstacles for a future clinical application. *Hum Reprod Update* March–April 2008;**14**(2):121–30.

62. de Rooij DG. The spermatogonial stem cell niche. *Microsc Res Tech* August 2009;**72**(8):580–5.

63. Brinster RL, Avarbock MR. Germline transmission of donor haplotype following spermatogonial transplantation. *Proc Natl Acad Sci USA* November 22, 1994;**91**(24):11303–7.

64. Brinster RL, Zimmermann JW. Spermatogenesis following male germ-cell transplantation. *Proc Natl Acad Sci USA* November 22, 1994;**91**(24): 11298–302.

65. Hermann BP, Sukhwani M, Hansel MC, Orwig KE. Spermatogonial stem cells in higher primates: are there differences from those in rodents? *Reproduction* March 2010;**139**(3):479–93.

66. Phillips BT, Gassei K, Orwig KE. Spermatogonial stem cell regulation and spermatogenesis. *Philos Trans R Soc Lond B Biol Sci* May 27, 2010;**365**(1546):1663–78.

67. Wyns C, Curaba M, Martinez-Madrid B, Van Langendonckt A, Francois-Xavier W, Donnez J. Spermatogonial survival after cryopreservation and short-term orthotopic immature human cryptorchid testicular tissue grafting to immunodeficient mice. *Hum Reprod* June 2007;**22**(6):1603–11.

68. Wyns C, Van Langendonckt A, Wese FX, Donnez J, Curaba M. Long-term spermatogonial survival in cryopreserved and xenografted immature human testicular tissue. *Hum Reprod* November 2008;**23**(11):2402–14.

69. Baert Y, Van Saen D, Haentjens P, In't Veld P, Tournaye H, Goossens E. What is the best cryopreservation protocol for human testicular tissue banking? *Hum Reprod* July 2013;**28**(7):1816–26.

70. Goossens E, Van Saen D, Tournaye H. Spermatogonial stem cell preservation and transplantation: from research to clinic. *Hum Reprod* April 2013;**28**(4):897–907.

71. Poels J, Van Langendonckt A, Many MC, Wese FX, Wyns C. Vitrification preserves proliferation capacity in human spermatogonia. *Hum Reprod* March 2013;**28**(3):578–89.

72. de Rooij DG, Griswold MD. Questions about spermatogonia posed and answered since 2000. *J Androl* November–December 2012;**33**(6):1085–95.

73. Kubota H, Avarbock MR, Brinster RL. Growth factors essential for self-renewal and expansion of mouse spermatogonial stem cells. *Proc Natl Acad Sci USA* November 23, 2004;**101**(47):16489–94.

74. Oatley JM, Avarbock MR, Brinster RL. Glial cell line-derived neurotrophic factor regulation of genes essential for self-renewal of mouse spermatogonial stem cells is dependent on Src family kinase signaling. *J Biol Chem* August 31, 2007;**282**(35):25842–51.

75. Conrad S, Renninger M, Hennenlotter J, Wiesner T, Just L, Bonin M, et al. Generation of pluripotent stem cells from adult human testis. *Nature* November 20, 2008;**456**(7220):344–9.

76. Sadri-Ardekani H, Mizrak SC, van Daalen SK, Korver CM, Roepers-Gajadien HL, Koruji M, et al. Propagation of human spermatogonial stem cells in vitro. *JAMA* November 18, 2009;**302**(19):2127–34.

77. Sadri-Ardekani H, Akhondi MA, van der Veen F, Repping S, van Pelt AM. In vitro propagation of human prepubertal spermatogonial stem cells. *JAMA* June 15, 2011;**305**(23):2416–8.

78. Kossack N, Meneses J, Shefi S, Nguyen HN, Chavez S, Nicholas C, et al. Isolation and characterization of pluripotent human spermatogonial stem cell-derived cells. *Stem Cells* January 2009;**27**(1):138–49.

79. Hermann BP, Sukhwani M, Winkler F, Pascarella JN, Peters KA, Sheng Y, et al. Spermatogonial stem cell transplantation into rhesus testes regenerates spermatogenesis producing functional sperm. *Cell Stem Cell* November 2, 2012;**11**(5):715–26.

80. Faes K, Tournaye H, Goethals L, Lahoutte T, Hoorens A, Goossens E. Testicular cell transplantation into the human testes. *Fertil Steril* October 2013;**100**(4):981–8.

81. Marques-Mari AI, Lacham-Kaplan O, Medrano JV, Pellicer A, Simon C. Differentiation of germ cells and gametes from stem cells. *Hum Reprod Update* May–June 2009;**15**(3):379–90.

82. Hubner K, Fuhrmann G, Christenson LK, Kehler J, Reinbold R, De La Fuente R, et al. Derivation of oocytes from mouse embryonic stem cells. *Science* May 23, 2003;**300**(5623):1251–6.

83. Toyooka Y, Tsunekawa N, Akasu R, Noce T. Embryonic stem cells can form germ cells in vitro. *Proc Natl Acad Sci USA* September 30, 2003;**100**(20):11457–62.

84. Geijsen N, Horoschak M, Kim K, Gribnau J, Eggan K, Daley GQ. Derivation of embryonic germ cells and male gametes from embryonic stem cells. *Nature* January 8, 2004;**427**(6970):148–54.

85. Novak I, Lightfoot DA, Wang H, Eriksson A, Mahdy E, Hoog C. Mouse embryonic stem cells form follicle-like ovarian structures but do not progress through meiosis. *Stem Cells* August 2006;**24**(8):1931–6.

86. Hayashi K, Ohta H, Kurimoto K, Aramaki S, Saitou M. Reconstitution of the mouse germ cell specification pathway in culture by pluripotent stem cells. *Cell* August 19, 2011;**146**(4):519–32.

87. Hayashi K, Ogushi S, Kurimoto K, Shimamoto S, Ohta H, Saitou M. Offspring from oocytes derived from in vitro primordial germ cell-like cells in mice. *Science* November 16, 2012;**338**(6109):971–5.

88. Nakaki F, Hayashi K, Ohta H, Kurimoto K, Yabuta Y, Saitou M. Induction of mouse germ-cell fate by transcription factors in vitro. *Nature* August 4, 2013;**501**(7466).

89. Clark AT, Bodnar MS, Fox MS, Rodriquez RT, Abeyta MJ, Firpo MT, et al. Spontaneous differentiation of germ cells from human embryonic stem cells in vitro. *Hum Mol Genet* 2004;**13**:727–39.

90. Park TS, Galic Z, Conway AE, Lindgren A, van Handel BJ, Magnusson M, et al. Derivation of primordial germ cells from human embryonic and induced pluripotent stem cells is significantly improved by coculture with human fetal gonadal cells. *Stem Cells* April 2009;**27**(4):783–95.

91. Kee K, Gonsalves JM, Clark AT, Pera RA. Bone morphogenetic proteins induce germ cell differentiation from human embryonic stem cells. *Stem Cells Dev* December 2006;**15**(6):831–7.

92. Kee K, Angeles V, Flores M, Nguyen H, Pera RR. Human DAZL, DAZ and BOULE genes modulate primordial germ cell and haploid gamete formation. *Nature* 2009;**462**:222–5.

93. Panula S, Medrano JV, Kee K, Bergstrom R, Nguyen HN, Byers B, et al. Human germ cell differentiation from fetal- and adult-derived induced pluripotent stem cells. *Hum Mol Genet* February 15, 2011;**20**(4):752–62.

94. Medrano JV, Ramathal C, Nguyen HN, Simon C, Reijo-Pera RA. Divergent RNA-binding proteins, DAZL and vasa, induce meiotic progression in human germ cells derived in vitro. *Stem Cells* December 12, 2011;**30**(3).

95. Durruthy Durruthy J, Ramathal C, Sukhwani M, Fang F, Cui J, Orwig KE, et al. Fate of induced pluripotent stem cells following transplantation to murine seminiferous tubules. *Hum Mol Genet* January 24, 2014;**23**(12).

96. Ramathal C, Durruthy-Durruthy J, Sukhwani M, Arakaki JE, Turek PJ, Orwig KE, et al. Fate of iPSCs derived from azoospermic and Fertile men following xenotransplantation to murine seminiferous tubules. *Cell Rep* April 30, 2014;**7**(4).

97. Wang HB, Lu SH, Lin QX, Feng LX, Li DX, Duan CM, et al. Reconstruction of endometrium in vitro via rabbit uterine endometrial cells expanded by sex steroid. *Fertil Steril* May 1, 2010;**93**(7):2385–95.

98. Takagi S, Shimizu T, Kuramoto G, Ishitani K, Matsui H, Yamato M, et al. Reconstruction of functional endometrium-like tissue in vitro and in vivo using cell sheet engineering. *Biochem Biophys Res Commun* March 28, 2014;**446**(1):335–40.

99. Miyazaki K, Maruyama T. Partial regeneration and reconstruction of the rat uterus through recellularization of a decellularized uterine matrix. *Biomaterials* October 2014;**35**(31):8791–800.

100. Raya-Rivera AM, Esquiliano D, Fierro-Pastrana R, Lopez-Bayghen E, Valencia P, Ordorica-Flores R, et al. Tissue-engineered autologous vaginal organs in patients: a pilot cohort study. *Lancet* July 26, 2014;**384**(9940):329–36.

101. Hellstrom M, El-Akouri RR, Sihlbom C, Olsson BM, Lengqvist J, Backdahl H, et al. Towards the development of a bioengineered uterus: comparison of different protocols for rat uterus decellularization. *Acta Biomater* August 25, 2014;**10**(12).

102. Nishida K, Yamato M, Hayashida Y, Watanabe K, Maeda N, Watanabe H, et al. Functional bioengineered corneal epithelial sheet grafts from corneal stem cells expanded ex vivo on a temperature-responsive cell culture surface. *Transplantation* February 15, 2004;**77**(3):379–85.

103. Yamamoto Y, Ito A, Fujita H, Nagamori E, Kawabe Y, Kamihira M. Functional evaluation of artificial skeletal muscle tissue constructs fabricated by a magnetic force-based tissue engineering technique. *Tissue Eng Part A* January 2011;**17**(1–2):107–14.

104. Nishida K. Tissue engineering of the cornea. *Cornea* October 2003;**22**(7 Suppl.):S28–34.

105. Nishida K, Yamato M, Hayashida Y, Watanabe K, Yamamoto K, Adachi E, et al. Corneal reconstruction with tissue-engineered cell sheets composed of autologous oral mucosal epithelium. *N Engl J Med* September 16, 2004;**351**(12):1187–96.

106. Shimizu T, Yamato M, Akutsu T, Shibata T, Isoi Y, Kikuchi A, et al. Electrically communicating three-dimensional cardiac tissue mimic fabricated by layered cultured cardiomyocyte sheets. *J Biomed Mater Res* April 2002;**60**(1):110–7.

107. Shimizu T, Yamato M, Kikuchi A, Okano T. Cell sheet engineering for myocardial tissue reconstruction. *Biomaterials* June 2003;**24**(13):2309–16.

108. Atala A, Bauer SB, Soker S, Yoo JJ, Retik AB. Tissue-engineered autologous bladders for patients needing cystoplasty. *Lancet* April 15, 2006;**367**(9518):1241–6.

109. Raya-Rivera A, Esquiliano DR, Yoo JJ, Lopez-Bayghen E, Soker S, Atala A. Tissue-engineered autologous urethras for patients who need reconstruction: an observational study. *Lancet* April 2, 2011;**377**(9772):1175–82.

110. Yoo JJ, Meng J, Oberpenning F, Atala A. Bladder augmentation using allogenic bladder submucosa seeded with cells. *Urology* February 1998;**51**(2):221–5.

111. Dahl SL, Koh J, Prabhakar V, Niklason LE. Decellularized native and engineered arterial scaffolds for transplantation. *Cell Transplant* 2003;**12**(6):659–66.

112. Nieponice A, Gilbert TW, Badylak SF. Reinforcement of esophageal anastomoses with an extracellular matrix scaffold in a canine model. *Ann Thorac Surg* December 2006;**82**(6):2050–8.

113. Macchiarini P, Jungebluth P, Go T, Asnaghi MA, Rees LE, Cogan TA, et al. Clinical transplantation of a tissue-engineered airway. *Lancet* December 13, 2008;**372**(9655):2023–30.

114. Uygun BE, Soto-Gutierrez A, Yagi H, Izamis ML, Guzzardi MA, Shulman C, et al. Organ reengineering through development of a transplantable recellularized liver graft using decellularized liver matrix. *Nat Med* July 2010;**16**(7):814–20.

115. Ott HC, Clippinger B, Conrad C, Schuetz C, Pomerantseva I, Ikonomou L, et al. Regeneration and orthotopic transplantation of a bioartificial lung. *Nat Med* August 2010;**16**(8):927–33.

116. Honaramooz A, Snedaker A, Boiani M, Scholer H, Dobrinski I, Schlatt S. Sperm from neonatal mammalian testes grafted in mice. *Nature* August 15, 2002;**418**(6899):778–81.

117. Arregui L, Rathi R, Zeng W, Honaramooz A, Gomendio M, Roldan ER, et al. Xenografting of adult mammalian testis tissue. *Anim Reprod Sci* June 2008;**106**(1–2):65–76.

118. Shinohara T, Inoue K, Ogonuki N, Kanatsu-Shinohara M, Miki H, Nakata K, et al. Birth of offspring following transplantation of cryopreserved immature testicular pieces and in-vitro microinsemination. *Hum Reprod* December 2002;**17**(12):3039–45.

119. Geens M, De Block G, Goossens E, Frederickx V, Van Steirteghem A, Tournaye H. Spermatogonial survival after grafting human testicular tissue to immunodeficient mice. *Hum Reprod* February 2006;**21**(2):390–6.

120. Schlatt S, Honaramooz A, Ehmcke J, Goebell PJ, Rubben H, Dhir R, et al. Limited survival of adult human testicular tissue as ectopic xenograft. *Hum Reprod* February 2006;**21**(2):384–9.

121. Sato T, Katagiri K, Gohbara A, Inoue K, Ogonuki N, Ogura A, et al. In vitro production of functional sperm in cultured neonatal mouse testes. *Nature* March 24, 2011;**471**(7339):504–7.

122. Sato T, Katagiri K, Kubota Y, Ogawa T. In vitro sperm production from mouse spermatogonial stem cell lines using an organ culture method. *Nat Protoc* November 2013;**8**(11):2098–104.

123. Sato T, Yokonishi T, Komeya M, Katagiri K, Kubota Y, Matoba S, et al. Testis tissue explantation cures spermatogenic failure in c-Kit ligand mutant mice. *Proc Natl Acad Sci USA* October 16, 2012;**109**(42):16934–8.

124. Yokonishi T, Sato T, Komeya M, Katagiri K, Kubota Y, Nakabayashi K, et al. Offspring production with sperm grown in vitro from cryopreserved testis tissues. *Nat Commun* 2014;**5**:4320.

Part III

Gene Therapy and Molecular Medicine

Chapter 11

Viral and Nonviral Vectors for In Vivo and Ex Vivo Gene Therapies

A. Crespo-Barreda*, M.M. Encabo-Berzosa*, R. González-Pastor*, P. Ortíz-Teba, M. Iglesias, J.L. Serrano, P. Martin-Duque

Key Concepts

Nonintegrative vectors are vital for the short term gene expression. The lack of integration will avoid undesired side effects.
   On the contrary, integrative vectors are ideal when long-term expression is needed.
   Both kinds of expression could be achieved by using either viral or nonviral vectors (Figure 1).

1. VIRAL VECTORS

1.1 Nonintegrative Viral Vectors

Viral vectors are among the most common strategies to deliver genes for gene and cellular therapies and could be divided into integrative and nonintegrative vectors. Nonintegrative viral vectors have multiple advantages to be used as vehicles for gene therapy. Due to their ability to not integrate into the genome of the host cells, the risk of insertional mutagenesis is reduced and short-time expression of the therapeutic gene occurs. This is very useful in situations where short-term activity is required such as suicide therapy or to initiate cascade processes. However, it can be a disadvantage when long-term activity is needed, that is, for the treatment of chronic conditions as cystic fibrosis (CF). The most common nonintegrative viral vectors are adenovirus (Ad), herpes simplex virus (HSV), and vaccinia virus (VV). In this chapter, we will review both types of gene therapy vectors in detail.

1.1.1 Adenovirus

1.1.1.1 Generalities

There is high interest in the search for novel non integrative vectors,[1] however, Ads offer a great promise as gene vectors for diagnostic or therapeutic applications. Currently, adenoviral vectors are the most common vehicles being used for gene transfer in clinical trials for cancer, as they could be used just as simple vectors or as their oncolytic versions.

1.1.1.2 Ad as Vectors—Genetic Modifications

Adenoviral vectors are very attractive tools for gene therapy because of their ability to not integrate into the host genome and so, reducing the risk of insertional mutagenesis. They can be used to transduce genes to a wide range of tissues including dividing and nondividing cells.[2] Ad vectors are relatively easy to genetically engineer, accommodating large segments of DNA (until 7.5 kb), and consequently to modify the tissue tropism of the viral vector.

   In most of viruses, transcription of the virus genome is divided into three phases, early (E), intermediate (I), and late (L) according to the time at which the gene is expressed.[3] At the Ad genome, we can find two groups of genes; the *cis*-genes which must be carried by the virus for an efficient viral production, such as packing signals, and the *trans*-genes that can generally be replaced by exogenous DNA (being very useful when the virus should be used as a vector).

   E1a gene is the one which starts the viral replication[4] and so, if that gene is deleted, the virus is incapable to replicate. E1 deletion (and substitution by the therapeutic transgene) is the base for the *first generation* of Ad vectors. Other early genes are frequently deleted for the incorporation of larger transgenes, such as E3 or E4 regions.[5] Even with the E1 and E3 or E4

---

* Those authors have contributed equally.

Translating Regenerative Medicine to the Clinic. http://dx.doi.org/10.1016/B978-0-12-800548-4.00011-5

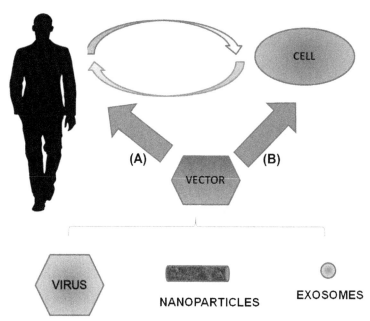

FIGURE 1 Gene therapy approaches. The vector used could be a virus or a nonviral vector (such as nanoparticles or exosomes). Moreover, gene therapy could be due to (A) the direct injection of the vector (in vivo gene therapy) or (B) the extraction of the patient's cells and external transduction by the vector before reinjection (ex vivo gene therapy or cell therapy).

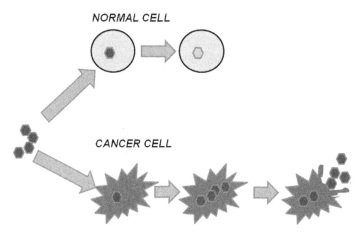

FIGURE 2 Principle of oncolytic vectors. When the virus infects a normal cell, it suffers an abortive infection and the virus is inactivated by the healthy cell. But when the virus infects a cancerous cell, the virus replicates and the cell lysates to release the virus.

removal, the infected cells still had low transcriptional levels of viral proteins, and a host immune response. To avoid those adverse effects, RACs or replication competent Ads (carrying the E1 genes) were developed, named as oncolytic vectors.[6]

Oncolytic Ads are genetically manipulated human Ads that acquired a replication phenotype in tumor cells, but showed a more restricted phenotype in normal cells (Figure 2). The most frequently tested modifications could be approached by three methods: (1) by the deletion of viral genes which interact with tumor suppressor genes, (2) by the modification of the tropism to infect specifically cancer cells, and (3) by the inclusion of transcriptional elements in the viral genome that are sensitive to transcription factors which are upregulated only in cancer cells.[7]

Second-generation vectors have further deletions in other early genes (as E2a, E2b, or E4) to decrease their immunogenicity and to increase the size of the exogenous transgene[8] but adenoviral producer cell lines are difficult to engineer and to grow leading to a poor viral titer.

Third-generation or gutless Ads have all viral genes replaced by therapeutic genes, except for the packing signals.[9] Therefore, they are nonimmunogenic and have long-term transgene expression, but they need a helper virus which carries all the genes needed for viral replication.

1.1.1.3 Strategies for the Targeting of Adenoviral Vectors

Adenoviral vectors have great advantages compared to other vehicles for gene transfer, but there are problems which limit their use. The broad tropism of Ad is also a limitation for therapeutic applications. Moreover, experiments done in animals have shown that Ads are capable to spread around the body causing toxic effects (mainly in the liver) and therefore systemic administration is not recommended. Moreover, usually the most interesting targets are refractory to adenoviral infections and higher doses are required to increase the efficiency of the infection leading to a stronger immune response and liver retention.[10] Therefore, to optimize the infection rate (by using transductional or/and transcriptional targeting modifications) is one important goal as it could increase the gene transfer to specific cells.

Transductional targeting might use different strategies: *Changing the structure of the knob fiber* on the adenoviral surface could be modified by using different strategies: (1) it can be modified by inserting peptides, such as the Arg-Gly-Asp (RGD) cell attachment site, that serve as receptor for integrins and that has improved the infection for esophagic cancer[11]; (2) the use of the knob fiber from other viral serotypes (with a diverse tissue tropism, that is, fiber from Ad 16 targets fibroblasts and chondrocytes but the fiber from Ad 50 targets myoblasts and hematopoietic stem cells)[12]; (3) the use of dispecific adapter molecules that bind to the knob fiber on one side (ablating the normal tropism) and they bind to the antigen of interest on the other side, and so, retargeting the specificity of Ad viral vectors (i.e., single-chain diabody scDb Mel was used to increase targeting against melanoma cells).[13] Other examples of adaptamers might be bispecific antibodies against the Ad knob fiber protein combined with epithelial cell adhesion molecules (anti-EpCAM/antiknob antibodies) that have been tested for gastric and esophageal cancer cell lines.[14]

Another way to target the viral expression to the desired tissue of interest might be triggered by transcriptional targeting by *using specific tissue promoters or even fusion promoters* with double targets to increase the level of specificity when commonly the second promoter is of general expression.[15]

Finally, another strategy to retarget Ad vectors is based on the *coating with nanoparticles*, such as a poly-[*N*-(2-hydroxypropyl)methacrylamide] (pHPMA). hHPMA is a hydrophilic and nonbiodegradable polymer that cooperatively binds to Ad being able to bind it by using reactive esters avoiding the host antibody neutralization and the recognition. New ligands could also be introduced chemically into the pHPMA-coated virus to be retargeted and to change the normal pathways of receptor binding.[16]

1.1.2 Herpes Simplex Virus

1.1.2.1 Generalities

HSV is a neurotropic virus which commonly infects humans. HSV has diverse characteristics that make it an important tool for gene and cellular therapy, such as their ability to carry large-sized inserts, long-term episomal expression in central nervous system (CNS) or a wide range of host cells and species. The most common herpes viruses used for gene therapy are serotypes 1 and 2 (HSV-1 and HSV-2) characterized by similar structures of their genomes but different prevalence.

1.1.2.2 HSV Vectors

HSV-1 are enveloped viruses of around 120–300 nm in diameter that have their genome in a double-stranded DNA (dsDNA) of 152 kb codifying for at least 34 viral proteins and mainly infect cells from the CNS and peripheral nervous system (PNS).

HSV-1 vectors have attracted the interest of the scientific community due to their ability to replace almost all of their genome by external DNA (inserting by homologous recombination a large cassette of 40–50 kb or multiple expression cassettes), their capacity to infect neurons, a high titter or long-term expression of the inserted therapeutic gene.[17]

Most of HSV-1 vectors used for gene therapy are designed to reactivate from their state of latency. Those manipulations made them lose the capacity to infect other hosts and they may need a complementary element (cell line, for example) to replicate. Removal of essential genes from the HSV-based vectors is important for genetic transference and the expression of the therapeutic gene. However, this removal might also reduce the cytotoxicity of the vectors and also affect the interaction between the vector and the host.[18] Essential ICP4 or ICP27 genes encode products needed for the transcription of early and late genes. For cancer therapy applications, retention of certain viral gene expression has several beneficial effects, and it might bring an increase in the levels of transgene expression, arrest the tumor cell division, and induce apoptosis in tumor cells. Similarly, the maintenance of other viral genes as the LAT promoter or LAP2 gene, to build HSV-based vectors, allows a longer expression of the infections of cells from CNS or PNS.[19]

Alternative vector systems might be the viral amplicons. Amplicons are plasmids with a bacterial origin for DNA replication in bacteria and HSV viral packing signals to propagate when cotransfected in cells with a defective HSV helper virus (in which packing signals have been removed). Amplicons are nontoxic "vectors" that transport very large transgenes and they can obtain a high titer.[20]

1.1.2.3 Modifications and Targeting

HSV-1 vectors have great advantages such as long-term expression of the therapeutic gene, high titer or the size of the transgene. Nevertheless, research on HSV-1-derived vectors is still ongoing to improve the vector distribution, to direct the targeting to improve the vector distribution, and to avoid the immune system.

Some of the modifications are resumed here: (1) *Viral distribution* has been improved by using simultaneously the viral vectors and MMP-9 (a matrix of metalloprotease with an essential role in the attachment of the vectors to the host cell surface), showing the spread of the viral particles to distant sites.[21] Another system used was the change in distribution by a small hydrostatic pressure differential system named convention-enhanced delivery and helped by the coadministration with heparin.[22]

(2) To engineer the HSV-1 vectors *to target specific surfaces*, the discovery of novel cell receptors has been of great use. The host cell surface contains heparin sulfate and other glycosaminoglycan molecules that are recognized by the external domains of the viral glycoproteins C and B. Modifications of these regions allow selective viral targeting as some examples showed: (1) Some authors have mutated the lysine-rich region of the viral gB sequence and the heparin sulfate binding domain and replacing them with the coding sequence for erythropoietin (EPO), and therefore EPO-expressing cell lines could be infected selectively; (2) another strategy could be followed by using adapter molecules (such as the epithelial growth factor receptor (EGFR)) that could interact simultaneously with the cell-specific receptor and with the virion gD and so, to allow a targeted HSV entry and they might favor the infection of EGFR-expressing tumor tissue in vivo.[23]

Other tools to engineer those viruses are the promoters. In general, HSV-1 promoters show a transient activity (as the intermediate-early HSV-1 promoter or IE-HSV-1 promoter) and they might be changed for other viral IE promoters, such as the IE human cytomegalovirus promoter, that has a vigorous expression (21 days after infection),[24] or by LAP1 and LAP2 promoters to reach long-term expression of the therapeutic transgenes.[25]

1.1.3 Vaccinia Virus

1.1.3.1 Generalities

VV is a poxvirus that was clinically used as vaccine for smallpox. The genome of VV is completely sequenced and facilitates the creation of recombinant viral vectors that could carry up to 25 kb of foreign DNA without the need for viral gene deletions.

1.1.3.2 Vaccinia Vectors and Genomic Modifications

VV is a very large, complex, and enveloped virus that contains a linear and dsDNA genome of about 180–190 kbp, with internal terminal repeats, and hairpin loops. The life cycle of the VV occurs in the cytoplasm and so it is a nonintegrative vector.[26]

An advantage of using vaccinia-derived vectors is that the vaccinia vectors may carry until 25 kb of foreign DNA without the need for viral deletions.[27] Vaccinia vectors present other advantages as a broad host range that permits the infections of primary cultures and many different cell lines, cytoplasmic replication, or the fact that the viral genome does not splice its primary transcripts.

To regulate the activity levels of the VV-derived vectors, diverse promoters are used, such as natural or synthetic promoters and also early or late expression promoters. Several promoters can be used and are relatively easy to make recombinant viruses, growth, and purify them[26] In order to improve the efficiency of the homologous recombination when insertion is desired, other strategies could be used such as the vaccinia–bacteria artificial chromosome technology.[28]

Nevertheless, there are problems associated with the host immune response. A vigorous immune response is desirable because it may enhance its potential as vaccine but it might lead to a premature clearance of the virus before adequate levels of replication have occurred. To avoid the early clearance of the virus, vaccinia has developed a wide range of strategies to evade the immune response and to survive in vivo.[29]

1.2 Integrative Viral Vectors

The integrative viral vectors have multiple advantages to be used as vehicles for gene therapy. Due to their ability to integrate into the genome of the host cells, long-term expression of the therapeutic gene is guaranteed. This is very useful in situations when the treatment of chronic conditions (as CF) is required. However, the use of integrative viral vectors has some risks associated as the insertional mutagenesis. The most common integrative viral vectors are retrovirus (RT), lentivirus (LV), and adeno-associated virus (AAV).

1.2.1 Retrovirus

1.2.1.1 Introduction

RTs have a great potential to be used as gene vectors for the treatment of chronic diseases due to their ability to integrate into the genome of the host. RTs infect a broad range of vertebrate host (and cells in all the division states) and are able to convert its RNA genome into DNA and integrate it into the host genome.

After the attachment and the entry of the RT to the host cell, the reverse transcriptase is activated, dsDNA synthesized and bound to the host genome through the action of the viral integrase and cellular enzymes. Viral DNA is transcribed by the host cell machinery synthesizing all the viral proteins required for the assembly of new viral particles that have to mature before realizing.[30]

The genome of the RT is divided into essential genes: *gag* that encodes internal proteins of the virus; *pro* that synthesizes a viral protease involved in the assembly of the viral particles; *pol* gene codifies the reverse transcriptase with DNA polymerase and RNase H activity; and *env* genes that encode transmembrane proteins and surface glycoproteins.[31]

1.2.1.2 Retroviral Vectors

RTs are enveloped viruses with a diameter of 80–130 nm that replicate through a DNA intermediate. The genome of RT is a single strain positive sense diploid RNA of 7–13 kb with long terminal repeats (LTRs) that flank coding regions and acts as regulatory domains.[32]

Retroviral vectors are commonly used for cancer gene therapy and mostly derived from the murine leukemia virus. They have several characteristics useful for gene therapy such as the fact that up to 8 kb of external genome can be inserted or their ability to integrate into the host genome, allowing long-term expression of the transgene and ensuring the maintenance of the genetic information in the self-renewing tissues or cells. The pathological genes are mostly removed and large-scale manufacturing of the therapeutic vector for clinical use is possible. However, there are some limitations such as the inability to infect nondividing cells, that can be also understand as an "inherent targeting" for diseases with a high rate of cell division (such as brain tumors) or the position-dependent transcription.

The RT vector has all the protein-coding sequences replaced by the therapeutic gene of interest and still carry the essential sequences as packing sequence, reverse transcriptase, and integration sequences. So they are replication deficient and the packing cell line must provide the viral proteins needed for infection. The self-inactivating (SIN) RT vectors have mutated the enhancer-promoter regions of the LTR and to initiate transcription they use an internal promoter.[33]

Nevertheless, there are some risks associated with the use of retroviral vectors as the insertional mutagenesis. In order to avoid it, retroviral vectors that act in a site-specific manner are being developed following different approaches.[34]

1.2.1.3 Modifications

Similar to other viral vectors, new targeting several items have been incorporated into the virion such as polypeptide ligands, small targeting motif, single-chain antibodies, or adapters.[35] Pseudotyping of RT-derived vectors are based on the transfer of viral proteins from other viruses which are used for the attachment to cells. Pseudotyping has been used to incorporate into the retroviral vectors glycoproteins from the vesicular stomatitis virus and so to confer a broad host range to the vector.[36]

Another common vector modification is the use of chemical agents to change its tropism coating the virus with polyethylene glycol (PEG), PEG-derived or avidin-biotin systems. Those compounds usually have linked specific antibodies for a therapeutic target binding in order to infect of a specific cell line (such as the streptavidin-bound antibodies used to achieve infection of human cells by a murine ecotropic RT).[37]

1.2.2 Lentivirus

1.2.2.1 Introduction

Lentiviruses (LV) are viruses of the RT family characterized by longer incubation periods. Lentiviral vectors are commonly used for basic biological research, functional genomics, and ex vivo and in vivo gene therapy due to their useful characteristics. LV vectors integrate stably into the host genome of dividing and nondividing cells, carrying large transgenes up to 10 kbp and making them suitable for many applications.[38]

They have an RNA genome that transcribes reversely into DNA and integrates into the host genome. Examples of lentiviruses are human immunodeficiency virus (HIV), simian immunodeficiency virus, feline immunodeficiency virus, and equine infection anemia virus.

1.2.2.2 Lentiviral Vectors

Lentiviral vectors (LVs) are enveloped particles of 80–100 nm in diameter and RNA positive sense genome of 9–10 kbp. After the attachment to specific surface cell receptors and the entry, the virus is partially uncoated and the RNA is copied by the viral RT into dsDNA that incorporates into the host genome mediated by viral integrase.

LVs have been modified in order to increase the safety, minimize the risk of unwanted viral transmission, optimize the infections, improve the expression of the therapeutic gene, and target viral infection. Three main modifications have been done: (1) insert a heterologous promoter, (2) remove the enhancer-promoter LTR sequences and addition of central polypurine tract, and (3) the central termination sequence that increases considerably transduction efficiency.[39]

1.2.2.3 Modifications and Targeting

Transcriptional targeting of lentivirus-derived vectors using heterologous promoters has several benefits. Using tissue-specific promoters, the efficiency and safety of the vectors increase allowing an improvement in the synthesis and the biological activity of the therapeutic transgene.[40] Moreover, the immune response against the infection is lower and the risk of the transgene silencing after insertion is reduced. For example, the use of an albumin gene promoter allows hepatocyte infection. The genetic incorporation of polypeptides ligands or single-chain antibodies (single-chain variable fragment—scFV) is also an efficient technique to retarget viral tropism as previously mentioned with other viruses.[41]

Besides, pseudotyping modification of the lentiviral vector is a powerful tool for targeting. As an example, pseudotyped LV vectors with enveloped glycoproteins that have previously coupled with a specific antibody from sindbis virus are successfully retargeted to metastatic melanoma cells in vivo.[42]

1.2.3 Adeno-Associated Viruses

1.2.3.1 Introduction

AAVs are nonpathogenic human parvoviruses that integrate into the host genome allowing long-term expression of the viral genes. AAVs are characterized by their replication incompetency, low immunogenicity, and a wide range of cell tropism. Their abilities have aroused a great interest toward the use of AAV-based recombinant vectors to gene therapy.

1.2.3.2 Adeno-Associated Viral Vectors and Genome Modifications

Adeno-associated viral vectors are promising tools for the transfer of therapeutic genes. They have major advantages compared to other viral vectors, such as the integration into the host genome, their replication incompetency, low immunogenicity, broad cell tropism, and long-term expression of the therapeutic transgene. Nevertheless, the use of AAV has some disadvantages such as the small size of its genome, which limits size of the foreign gene, and the requirement of helper functions for viral activation and lytic cycle begins.[43]

The improvement of the infection to targeted cells, optimization of the expression of the therapeutic gene, and the increase of the transgene size are the next important goals for AAVs.

Pseudotyping modifications of the AAV enable the use of the capsid from other isolated AAV strains without changing the rest of the viral genome. Genetic incorporation of small peptide ligands or single-chain antibodies (scFV) into the capsid surface also might be an efficient technique to redirect viral tropism.[44]

Furthermore, self-complementary vectors (scAAV) have been developed to avoid de novo DNA synthesis. scAAVs are based on a half-size DNA replicon in single-stranded dimeric genome with an inverted repeat configuration that allows the fold back into dsDNA and integration independently of DNA synthesis.[45]

2. NONVIRAL VECTORS FOR GENE THERAPY

Although viral vectors have better transfection efficacy, their preparation is complex and they can be highly immunogenic.[46] Nonviral vectors are synthesized in the laboratory from biocompatible precursors. In contrast to viral vectors, they show low cytotoxicity, low immunogenicity, and their chemical surface can be easily modified that could have advantages in terms of ease and large-scale production.[47]

Nanomaterials are structures with morphological properties smaller than 1 micron in at least one dimension, usually between 1 and 100 nm. Those used in gene therapy can be organic or inorganic and are designed to bind and protect DNA or RNA, simultaneously achieving high efficiency, prolonged gene expression, and low toxicity. Among organic vectors, we can find polymers, dendrimers, cationic liposomes, etc. Inorganic materials include silica, iron oxide, carbon nanotubes (CNTs), or gold nanoparticles (GNPs) among others.[48]

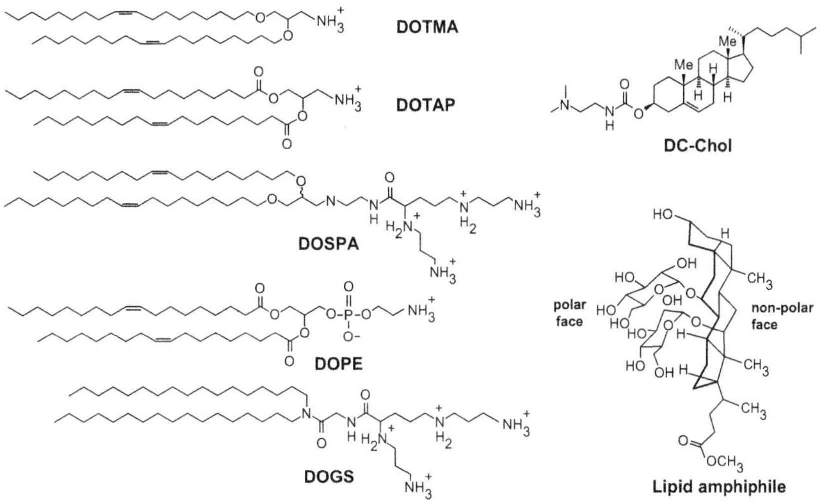

FIGURE 3 **Lipids chemical structure.** Some of the most common lipids used for gene delivery are shown in this figure.

Nanoparticles are good candidates for gene delivery; their surfaces are functionalized with cationic polymers to improve the nucleic acid union. They are very feasible alternatives against viral vectors but it is necessary to develop further in vivo studies before their clinical use.

2.1 Organic Vectors

Organic particles, natural or synthetic, may be composed of polymers, lipids, or peptides, and have the advantage of reduced toxicity, as their degradation leads to nontoxic products and avoidance of accumulation inside the cells.[49,50]

2.1.1 Lipid-Based Vectors

The first liposome-based gene delivery method was described by Felgner in 1987,[51,52] which contained the cationic lipid DOTMA (N-[1-(2,3-dioleyloxy)propyl]-N,N,N-trimethylammonium chloride). Cationic and neutral lipids are typically used for gene delivery because of the coulombic attraction between them and the polyanionic DNA/RNA (Figure 3). However, and mainly due to their toxicity, cationic lipids are more frequently formulated as *liposomes, lipid nanoemulsions, or solid lipid nanoparticles*.[53]

Liposomes typically contain a cationic lipid combined with a neutral or helper lipid forming a stable structure under physiological conditions. The composition of a dozen of commercial products is based on this combination, including *Lipofectin* (a mixture of DOTMA and DOPE (dioleoylphosphatidylethanolamine)) and *Lipofectamine* (a mixture of DOSPA (2,3-dioleyloxy-N-[2(sperminecarboxamido)ethyl]-N,N-dimethyl-l-propanaminium trifluoroacetate) and DOPE).[54]

Cationic lipids are amphiphilic molecules with three structural domains: a polar headgroup that commonly acquires a positive charge through one or more amines, a hydrophobic tail, and a linker between these two domains. Each lipid varies the structural aspects, such as the size of the headgroup and the length and cross section of the hydrocarbon tail. These characteristics confer distinct properties to the final lipid–DNA complex, which directly affects its association and uptake into the cell and the levels of transfection activity and cytotoxicity.[55]

When compared to other methods of gene transfer used in the late 1980s, such as the use of calcium phosphate,[56] DOTMA proved to facilitate up to 100-fold more efficient gene delivery. Once commercialized, many alterations in the main moieties of DOTMA have reflected the efforts to reduce toxicity and increase transfection efficiencies, producing first DOTAP,[57] then DOGS (dioctadecylamido-glycylspermine), followed by DC-Chol (3β-[N-(N′,N′-dimethylaminoethane) carbamoyl]-cholesterol hydrochloride)[54] (Figure 3).

2.1.1.1 Structure of Cationic Lipids

The *hydrophobic domain* represents the nonpolar part of the cationic lipid and the majority is made of aliphatic chains that commonly contain between 8 and 18 carbons. The length and degree of saturation of the lipid chain affect the phase transition temperature and the fluidity of the bilayer, which is fundamental in determining the stability and toxicity of the liposome.[58] They might assemble into different ordered microscopic structures (liposomes) with a posterior formation of macroscopically ordered phases (lipoplexes) (Figure 4).

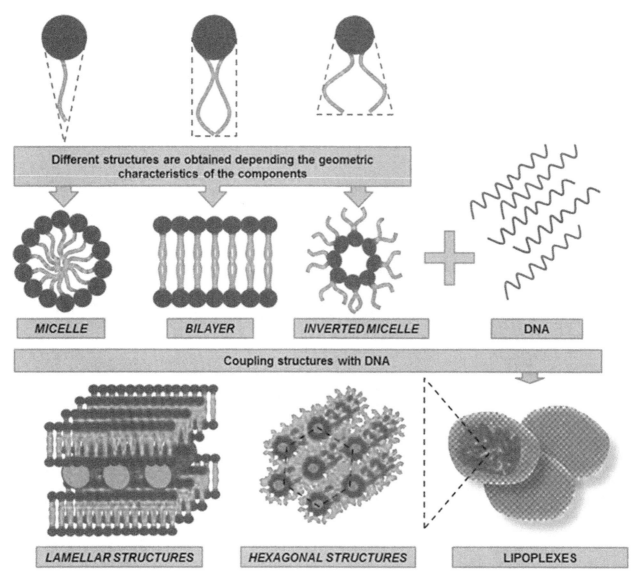

FIGURE 4 Self-assembly of lipids. Different ordered microscopic structures (liposomes) and posterior formation of macroscopically ordered phases (lipoplexes) are shown in this figure.

Several studies have also shown that incorporating aliphatic chains with a different number of C in the chains can improve transfection efficiency potentially by promoting endosomal escape.[59] As it occurs with tail length, several studies have shown that lipid mixtures containing both types of chains produce higher transfection efficiencies than with one type only.

In addition, various alternative hydrophobic chains have been used to promote gene transfer.[60] Cholesterol is most frequently used because of its rigidity, endogenous biodegradability, and fusion activity.[61] Other steroids used include vitamin D and bile acid derivates.[62]

The chemistry of the *linker* commonly includes amides, carbamates, esters, or ethers. Carbamates and amides are more biodegradable and so they allow for cleavage once inside the cell. A typical example of cationic lipid with carbamate linker is DC-Chol, which was the first lipid used in clinical trials because of its combined properties of transfection efficiency, stability, and low toxicity.[63]

The effect of transfection efficiency and cytotoxicity is mainly associated with the cationic nature of the vectors, which is fundamentally determined by the structure of its *hydrophilic headgroup*.[64] The positive charge generally comes from amine groups with different degrees of substitution (mostly tertiary amine or quaternary ammonium groups), but other groups like amino acids, guanidinium, or peptides have been reported.

The mechanism of entry into the cells is controversial. Early work suggested that the lipoplex-mediated delivery of genes occurs through the fusion of the lipids with the cell membrane and the direct release of DNA to the cytosol.[52] However, more recent reports clearly show a significant involvement of endocytosis in the uptake, suggesting that fusion with the cell membrane contributes minimally.[65] Physicochemical properties of lipoplexes such as particle size distribution, lipid composition, and charge ratio may also influence their uptake route.[66]A "flip-flop" mechanism was described[67] that starts inside the endosomes, where electrostatic interactions between the cationic lipoplexes and the negatively charged lipids of the endosomal membrane promote lateral diffusion of the anionic lipids into the lipoplexes and form charge-neutralized ion pairs. As a result, nucleic acids are displaced from the lipoplexes, allowing the release of the nucleic acids into the cytoplasm.[68]

Lipid-based vectors do not stimulate a cellular immune response, but may be recognized as foreign and initiate the production of cytokines such as tumor necrosis factor (TNF). Recent improvements in lipofection have facilitated protection from degradation in vivo, due to surface modifications with PEG. PEG presents many attractive qualities such as lack of toxicity and ready excretion by the kidneys.[68] Also it was proved that the PEGylated lipoplexes display improved stabilities and longer circulation times in the blood than the non-PEGylated version.

2.1.2 Polymer-Based Vectors

Wu and Wu introduced the use of cationic polymers for gene transfer in 1987, using a poly-L-lysine coupled to a galactose-terminal asialoglycoprotein to deliver a foreign gene to hepatocytes both in vitro and in vivo.[69] Cationic polymers include natural DNA-binding proteins, synthetic polypeptides, polyethylenimine (PEI), cationic dendrimers, polymethacrylate, and carbohydrate-based polymers (Figure 5). Natural biopolymers and proteins, such as histones or collagen, have recently attracted attention in gene delivery due to their unique properties, which are biodegradability, biocompatibility, and controlled release. However, most of the polymers used are synthetic compounds because it is relatively easy to tune their chemical and physical properties through polymer chemistry. In this case, it is possible to modify their molecular weight, structure or molecular composition by introducing different biofunctional groups. But their low gene transfection efficiency limits their clinical applications. Nowadays, commercially available polymers and polyamidoamine (PAMAM) dendrimers are typically used in gene-delivery studies, being poly(L-lysine) (PLL) and PEI among the most widely utilized.

The formulation of polyplexes plays an important role in both their transfection efficiency and stability. Polymers with linear, branched, or dendritic structures bind nucleic acids through the cationic amines (primary, secondary, tertiary, and quaternary). Nucleic acid can be entrapped into the polymeric matrix and also they can be adsorbed or conjugated on the polymeric surface. Moreover, the degradation of the polymer can be used as a tool to release the nucleic acid into the cytosol,[70] but then the molar ratio of nitrogens on the polymer/phosphates from the DNA (N:P) should be around 10. These results have been confirmed using different polycations and cell lines.

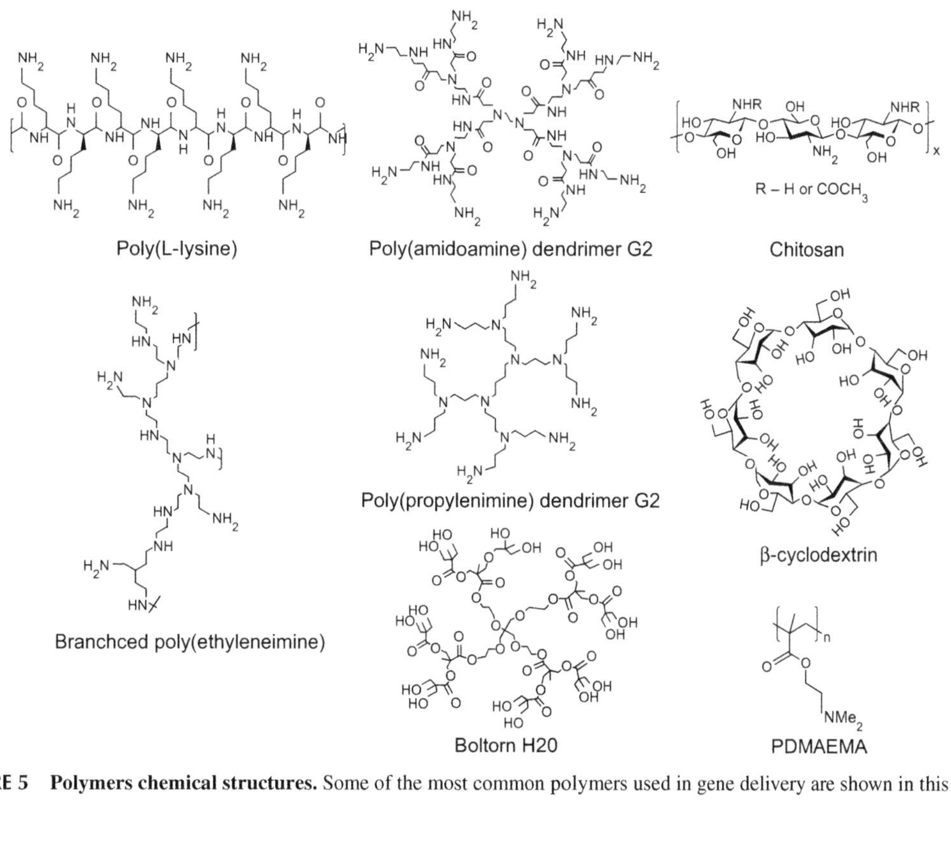

FIGURE 5   Polymers chemical structures. Some of the most common polymers used in gene delivery are shown in this figure.

The final size of the polyplex is controlled by the structure and complexation behavior of its components. Loosely condensed polymer–plasmid complexes have a larger size; in contrast, complete complexation results in condensation. In general, the use of a long polymer chain or a higher branched polymer promotes a better condensation of the nucleic acid, conferring protection against degradation in the extracellular environment and promoting cellular uptake.[71] Additionally, the molecular weight of the polymer also has a significant effect on cellular uptake, endosomal escape, DNA unpackaging, and nuclear internalization.[72]

In general, polyplexes have some advantages compared to lipoplexes, such as small size, narrow distribution, higher protection against enzymatic degradation, more stability, and easy control of the physical factors.[73]

2.1.2.1 Poly(L-lysine)

All primary amino groups of this linear polypeptide with L-lysine residues in repeat units[74] are protonated at physiological pH, forming a structure with no buffering capacity to aid in endosomal escape, which causes lower transfection efficiency than that obtained using PEI.[75] Various biodegradable polylysine conjugates have been synthesized in order to reduce the high cytotoxicity of high-molecular-weight PLL structures, for example, by incorporating terminal cysteine residues to form complexes with reducible disulfide linkages.[76] Recently, Kim's group developed a thermosensitive PEG–PLGA–PEG triblock copolymer, where DNA could be formulated in the polymer solution at room temperature, and in response to temperature changes slowly release DNA for prolonged transfection at the injection site.[77]

2.1.2.2 Polyethylenimine

PEI was the second polymeric transfection agent developed, which was synthesized by Boussif et al.[78] in 1995. It has been showed that the transfection ability of PEIs depends on their molecular weights, degree of branching, PEI nitrogen/nucleic acid phosphate charge ratios (N/P), and cell type. Specifically, PEI is the active species used in a number of commercial reagents, such as jet-PEI and TransIT-TKO.[79]

Sonawane et al.[80] proved that the transfection efficiency of PEI is closely related to its "proton sponge effect." In this case, the extensive buffering capacity of PEI promotes both the inhibition of lysosomal nuclease activity and a change in the osmolarity of acidic vesicles, which results in endosomal swelling and rupture. Describing the process, the presence of amines in PEI increases the endosomal proton concentration, which induces a massive influx of chloride ions triggering the entry of water molecules. The final effect is the osmotic swelling of the endosome and subsequent membrane disruption.

The most utilized modification to PEI structure to improve transfection efficiency is PEGylation. But PEI has also been modified by linking cholesterol to the amines[81] and by deacylation to boost delivery of DNA and small-interfering RNA (siRNA) by orders of magnitude in vitro and in vivo.[82] Synthesis of biodegradable PEI compounds has involved either the incorporation of reducible disulfide linkages, ester conjugation, hydrolyzable amide, or imine linkages.[83]

2.1.2.3 Carbohydrate-Based Polymers

Due to the hydrophilic nature of oligosaccharides and the fact that sugars are relatively well tolerated by the body, cationic polysaccharides have been explored for gene and nucleic acid delivery. MacLaughlin et al.[84] was the first to report chitosan as gene delivery vectors. The biodegradability, biocompatibility, cationic potential, mucoadhesive property, and the ability to open intercellular tight junctions of chitosan have helped it to become one of the most outstanding, naturally derived nonviral vectors for gene transfer. The commercial transfection reagent *Novafect* is based on chitosan.

Many studies have shown that pDNA-loaded chitosan nanocarriers are able to achieve high transfection levels in most cell lines, and the variation in gene transfer activity of different cell lines is attributed to differences in both cellular uptake due to the cell-specific plasma membrane composition and in chitosan-degrading enzymes present within the endosomal compartments of the cells.

2.1.2.4 Dendrimers

Dendrimers are three-dimensional polymers with spherical and highly branched structures. They are good candidates for an efficient gene transfection, as they have some advantages in comparison with their linear analogs: For example, they are monodisperse and have a well-defined structure. They usually present a low viscosity and good solubility. Depending on their generation (which represents the number of focal points when going from the core toward the dendrimer surface), dendrimers present a dense concentration of functional groups in their periphery; this characteristic allows the easy binding of selected molecules to create different nanostructures. Dendrimers consist of three distinct segments, which are the central core, the branching units, and the terminal functional groups. The core and the internal units determine the shape, the directionality, the multiplicity, the microenvironment of the nanocavities, and so their solubilization properties, whereas the number of the external groups and its nature characterize their chemical and biological behavior. The binding properties can be modulated by the modification of the surface using a large number of functional groups. Cationic functionalization, for example, introducing quaternary ammonium groups, promotes the formation of complexes between polymers and negatively charged nucleic acids. Additionally, targeting ligands such as guanidinium groups, folate, or galactose can be added onto the dendrimers.[85]

In general, increasing diameter, dendrimers enhance transfection efficiency. On the other hand, primary amine groups promote cellular uptake because of their participation in nucleic acid binding. Since pKa value is 3.9 for internal tertiary amines and 6.9 for terminal amines, the internal groups act as a proton-sponge in the endosomes and enhance the release of nucleic acid into the cytoplasm.

Polypropylenimine (PPI) dendrimers with diaminobutane and diaminoethane cores and PAMAM dendrimers with ammonium (NH_3) or ethylenediamine core have been widely used for their delivery potential. Although low-generation (G2, G3) complexes show great transfection efficiency, high generations are generally quite cytotoxic. Two kits commercially available based on intact (Polyfect) and fractured (Superfect) PAMAM have been successfully used with DNA and RNA.[86] Other frequently used dendrimers are based on phosphorous, carbosilane, poly-L-Lysine, and polyester.[87]

Compared to structurally well-defined dendrimers, hyperbranched polymers possess a lower degree of perfection, with a broad molecular-weight distribution and a randomly branched architecture. However, their synthesis is less tedious and less expensive and requires fewer steps of separation and purification. Hyperbranched polylysine and hyperbranched aliphatic polyesters based on poly-2,2-bis(methylol) propionic acid have transfection properties that are comparable to those of similar molecular weight of PEI, but leads to improved long-term cell viability during transfection, attributed to its partial biodegradability.[88]

2.1.2.5 Polymeric Hydrogels

Various groups have investigated cross-linking polycation chains using emulsification/solvent evaporation techniques to form a structure in which DNA is trapped. The use of these nanogels as delivery carriers is desirable, due to their degradable nature, facile encapsulation of materials in the core, release of biomolecules upon an external stimuli, and low toxicity. It has been reported that encapsulation of large DNA molecules in nanogels leads to significant aggregation, but siRNA loading is successfully achieved.[89] Nanogels were conjugated with antisense oligonucleotides, showing a higher cellular accumulation of the nanogel-antisense oligonucleotide complex and higher gene transfer activity.

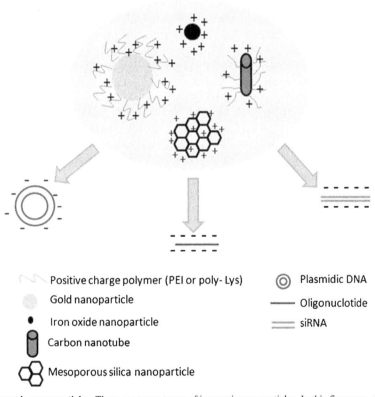

FIGURE 6 **Examples of inorganic nanoparticles.** There are many types of inorganic nanoparticles. In this figure we show several examples of them. In general, they act as nucleic acid vectors for in vitro or in vivo transfections.

2.2 Inorganic Vectors

Inorganic nanovectors have a simple, controllable, and scalable synthesis, good dispersion, large loading capacity, no immunogenicity, and low cytotoxicity. Both DNA and RNA can be bound to the inorganic nanovectors through electro or chemical interactions (Figure 6). These interactions permit to form a complex between the inorganic nanomaterial and the nucleotidic sequence. The complex protects the genetic material from enzymatic degradation and improves the transfection efficiency. Although endocytosis is the main entrance route of the nanovectors, specific molecules such as antibodies can be coupled on the nanomaterial surface, which enables the interaction with cell surface target molecules. Under this receptor mediation, gene transfection is more specific and effective.

In all cases, there are some variables that must be considered, such as the nanoparticles/nucleotidic material ratio, the surface charge coverage, or the hydrophobicity. The endosomal escape is the main critical step for the transfection efficacy. It is thought that the "proton sponge" has an important role in this process, as it was previously explained. For this process, it is important to functionalize the particle surface with cationic polymers as PEI or poly-L-Lysine (PLL).[90]

2.2.1 Magnetic Nanoparticles

One of the pioneers using magnetofection for in vitro applications was Lin et al.[91] There are various cationic magnetic nanoparticles types that have the capacity to bind nucleotidic material on their surface. With this method, the magnetic nanoparticles are concentrated in the target cells by the influence of an external magnetic field (EMF). Normally, the internalization is accomplished by endocytosis or pinocytosis, so the membrane architecture stays intact. This is an advantage over other physical transfection methods. Other advantages are the low vector dose needed to reach saturation yield and the short incubation time needed to achieve high transfection efficiency. Moreover, with the application of an EMF, cells transfected with magnetic nanoparticles can be used to target the region of interest in vivo.

2.2.1.1 Iron Oxide Nanoparticles

The magnetic nanoparticles most used in magnetofection include the iron oxide nanoparticles (IONPs). IONPs are biodegradable and not cytotoxic and can be easily functionalized with PEI, PEG, or PLL. Poly-L-lysine-modified iron oxide nanoparticles (IONP–PLL) are good candidates as DNA and microRNA (miRNA) vectors because they bind and protect nucleic acids and showed high transfection efficiency in vitro. In addition, they are highly biocompatible in vivo.

Chen et al.[92] used human vascular endothelial growth factor siRNA bound to superparamagnetic iron oxide nanoparticles (SPIONs) and it was capable of hepatocellular carcinoma growth inhibition in nude mice. Moreover, Li et al.[93] demonstrated that the intravenous injection of IONP–PLL carrying NM23-H1 (a tumor suppressor gene) plasmid DNA significantly extended the survival time of an experimental pulmonary metastasis mouse model.

Another advantage of this kind of nanoparticles is that they can be used as MRI agents. Chen et al.[94] bound siRNA to PEG-PEI SPIONs together to a gastric cancer-associated CD44v6 single-chain variable fragment. This bound permitted both cancer cell's transfection and their visualization by MRI.

But those complexes might be used for cell therapies as well. Schade et al.[95] used iron oxide magnetic nanoparticles (MNPs) to bind miRNA and transfect human mesenchymal stem cells. As the binding between the MNPs and PEI took place via biotin-streptavidin conjugation, these particles cannot pass the nuclear barrier, so they are good candidates to deliver miRNA, as it exerts its function in the cytosol. They functionalized the surface nanoparticles with PEI and were able to obtain a better transfection than PEI 72 h after transfection. Moreover, they demonstrated that magnetic polyplexes provided a better long-term effect, also when included inside of the stem cells.

2.2.2 Gold Nanoparticles

In general, gold nanoparticles (GNPs) are good candidates for biomedical applications. Essentially, they are practically inert, nontoxic, and biocompatible. In addition, their synthesis is simple and scalable, so their size is easily controllable, as well as their disparity. Other important aspect is their easy functionalization and modifications.

Moreover, GNPs can bind different types of nucleic acids and make different kinds of complexes. (1) Those nanoparticles could be employed *to bind therapeutic transgenes*. Tandon et al.[96] used PEI-GNPs to transfer BMP7 gene into corneal cells for the inhibition of the corneal fibrosis in vivo, delivering the transgene into cornea without causing cytotoxicity or inflammation (2) Also *gene silencing* could be achieved using GNPs. The use of siRNA for glioblastoma treatment was approached by Jensen et al.[97] as they targeted the oncoprotein BCl2L12, showing a reduction of Bcl2L12 mRNA and protein levels effectively and reduction of tumor progression in xenograft mice. Also, Lee Seung et al.[98] achieved ultralong in vitro and in vivo gene silencing by using GNPs which were synthesized sequentially by layers. When a PLL coating is used, the PLL cover is gradually degraded by proteases inside the cells and so, the siRNA is continuously released. After cell division, the nanoparticles are distributed to daughter cells and the siRNA continues to be liberated for a long time. (3) But also those nanocomplexes could be *formed by GNPs, DNA, and other elements* such as dexamethasone (Dexa).[99] Those studies showed that PEI-GNPs/pTRAIL/PEI-Dexa significantly inhibited tumor growth in vivo with minimal side effects.

Finally, other important characteristic of many nanoparticles, included GNPs, is their ability to reach tumors by the permeability and retention effect (EPR). The EPR is a unique phenomenon of solid tumors related to their anatomical and physiological differences from normal tissues. On tumoral tissues, angiogenesis leads to larger gaps between endothelial cells in tumor blood vessels. Therefore, certain sizes of molecules (liposomes, nanoparticles, and macromolecular drugs) tend to accumulate in tumor tissue more than in normal tissues. That accumulation is called the enhanced and permeability retention (EPR) effect and based on that, Kim et al.[100] developed an siRNA delivery vehicle formed by a PEI/PLL/siRNA complex and GNPs, that could reach the tumor, because of the size of the complex. They efficiently delivered siRNA into cultured cancer cells without apparent cytotoxicity, achieving a significant sequence-specific gene silencing. These results are important advances toward RNA-based cancer therapy.

2.2.3 Silica Nanoparticles

Silica nanoparticles are very promising tools in Nanomedicine. Mesoporous silica nanoparticles (MSNs) have mesopores between 2 and 50 nm pore size. They have hydrophilic surface that favors nucleic acid-protected circulation, easily modifiable and functionalizable silane chemistry surface, simple and scalable synthesis, and low cost of production.[101] Also, silica nanoparticles are more stable than organic materials like lipoplexes.

MSNs have been investigated for siRNA delivery. For example, Hom et al.[102] modified MSNs with PEI to increase the positive-charge of the nanoparticle surface. They showed that comparing the cytotoxicity of their PEI-MSNs with other transfection vehicles like Lipofectamine 2000 and PEI/siRNA complexes, their PEI-MSNs resulted less toxic but showing a high effect in silencing capability of EGFP, Akt, and K-*ras*. Similar results were obtained by Zhu et al.[103] who used PLL-NPSNs to bind and protect *c-myc* antisense oligonucleotides. They showed significant downregulation of c-myc mRNA levels in HNE1 and HeLa cell lines.

However, the combination of different types of nanoparticles could be very attractive. Hybrid NPs were developed by Hartono et al.[104] as they used MSNs loaded with iron oxide, and covalently modified with PEI for gene silencing into osteosarcoma cancer cells. These nanocomplexes improved the MSNs internalization due to the magnetic field. Nowadays, there are hybrid silica nanoparticles which can be used for diagnostic imaging and gene therapy. Silica nanoparticles doped with gadolinium oxide can deliver DNA and serve as contrast agents for magnetic resonance imaging.[105]

There are fewer studies in vivo than in vitro, and it is necessary to do more experiments in order to confirm the possible use of silica nanoparticles in the clinic, but at the moment there are good expectations. Preclinical studies by Bharali et al.[106] reported a high efficiency of transfection through stereotaxic injection of a DNA–silica NPs complex. Their results were equal or superior to that obtained with a viral vector.

2.2.4 Carbon Nanotubes

Carbon-based materials like graphene oxide or CNTs are interesting candidates for biomedical applications such as implant devices or scaffolds. CNTs are carbon atoms arranged in condensed atomic rings. They can be organized in one (single-walled carbon nanotubes, SWNTs) or more (multiwalled carbon nanotubes, MWNTs) concentric sheets rolled up into cylinders.[107] CNTs can be modified by functionalization to increase their solubility and to facilitate the nucleic acid binding (DNA, siRNA, etc.).[108]

Both types of CNTs have been used to bind nucleic acids, but the most notable success has been achieved with MWNTs where Siu et al.[109] were able to bind siRNA to SWNTs for treating melanoma, showing a reduction in the significant tumor progression reduction in a day 25 interval.

2.2.5 Quantum Dots

Quantum dots (QDs) are semiconductor nanoparticles which can be used as photoluminescent markers for bioimaging. They are crystalline particles that range from 2 to 10 nm. The electronic properties of these materials are intermediate between those of bulk semiconductors and of discrete molecules. The main problem in the use of QDs in biomedical applications is their toxicity, because of the cadmium release into the cells or the organs. But their surface functionalization may prevent this problem and in addition, permit the binding of different biomolecules as aptamers, peptides, or nucleic acids.

The matrix-degrading metalloproteinases (MMPs) are involved in the neuroinflammation and disruption of the blood–brain barrier (BBB) that occurs in many diseases. Bonoiu et al.[110] used QDs with MMP-9 siRNA for silencing the MMP-9

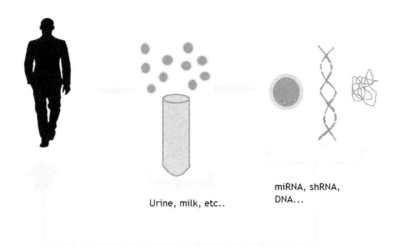

FIGURE 7 **Exosomes as gene vectors.** Exosomes could be extracted from most of the biological fluids and they could be used to transfer genetic material to the patients.

protein and maintain the BBB integrity. They achieved a QD-mediated gene silencing efficiency of almost 80% in brain microvascular endothelial cells (BMVECs). Nanoplexes provided a great advantage over other therapeutic options because of their ability to combine diagnosis and therapy.

3. EXOSOMES AS BIOLOGICAL VEHICLES

Exosomes are cell-derived vesicles that are believed to be present in all body fluids, including blood, urine, and cultured medium of cell cultures. They are small (30–150 nm) extracellular vesicles involved in cell signaling, although they were first discovered nearly 30 years ago, they were considered little more than garbage cans.

The evolution in the study of the exosomes in the last few years have allowed us to discover new opportunities in the transfer of genetic material used in gene therapy (Figure 7). The different attributes of exosomes, such as (1) low immunogenicity (compared to other plausible vectors as virus or nanoparticles), (2) It has been proved that exosomes can transfer siRNA made of the exosomes promising vectors for gene transfer, (3) Exosomes can go through the hematoencefalic membrane, but also they can target specific types of cells or regions, and (4) they can stay in action for a long period of time. All these attributes make of the exosomes a new and exciting way of gene delivering to different tissues.[111]

3.1 Vehicles for RNA Transfer

In their natural form, exosomes are loaded with different types of RNA molecules, and they can deliver it in a successful way to different types of cells. Exosome composition and function depend on the physiological or pathological situation of the cells that secrete them. That is the reason why exosomes could be useful in different treatments, especially by siRNA and miRNA transfer. In neurological pathologies such as Parkinson disease, it has been shown that exosomes can transport AlfaSyn siRNA in the midbrain, striatum and cortex in vivo, reducing the total AlfaSyn protein. That makes the administration of exosomes charged with this siRNA in vivo, a potential treatment of this degenerative disease among others.

Other uses of exosomes for Alzheimer disease are by the administration of exosomes containing the siRNA that silences the production of BACE1 (one of the most important proteins involved in this disease). It has shown a reduction of the 60%, and a reduction of the beta-amyloid (which is one of the markers of the pathological degeneration).[112]

Exosomes are also widely used for cardiovascular diseases, as they are involved in the progression of the pathology. In the last few years, the treatment of these diseases with miRNAs (such as miRNA-126) has been proved to be efficient, and it can promote vascular regeneration and regeneration of the endothelium in vivo.[113] Exosomes in vivo administration has shown an improvement in the endothelial function, by an increase of the NO liberation, which can be used in the treatment of ischemia and reperfusion.[113]

The use of exosomes in cancer treatment is spreading fast as they have been proved to migrate to the tumor microenvironment.[114] One of the many treatment possibilities is the administration of siRNAs against RAD51 loaded into exosomes, which could be useful because it is involved in the proliferation of aberrant cells. Studies showed that the injection of those exosomes was functional, and promoted cellular death in cancer cells, because of a decrease of the RAD51 recombinase.

3.2 As Vehicles for Protein Transfer

Classic vaccines have some disadvantages such as they are not efficient against intracellular antigens like virus or parasites. Tumor vaccines are being currently investigated, but the results in this field are not yet positive. Against a tumor, the immune system does not make a strong response (as if it was an exogenous antigen). There are several studies in which a tumoral antigen was expressed in exosomes, resulting in a stronger immune response against the tumor. It was proved by the slower growth of the tumors treated with ovalbumin bound to exosomes versus soluble albumin.

Acute respiratory syndrome and toxoplasmosis (caused by *Toxoplasma gondii*) are two diseases where exosomes might be used as potential vaccines.[115] The eosinophil level and the specific immune response producing IgE demonstrated the efficacy of the exosome treatment.

3.3 Vehicles for DNA Transfer

Very little is known about the DNA transfer through exosomes as the first evidence that tumor-derived exosomes could carry dsDNA is very recent.[116]

Recently, it was shown that exosomal DNA (exoDNA) contains the entire enome of the person from whom it was obtained, highlighting the translational value of exoDNA (1) as a potential biomarker and (2) for gene therapy.

4. CLINICAL APPLICATIONS

4.1 Clinical Trials Involving Viral Vectors

4.1.1 Adenovirus

Adenoviral vectors have been widely studied for many years. Based on their ability for short-term expression, adenovirus have been proved to be very useful for medical applications in which transitory expression of the therapeutic gene is required. 284 clinical trials using adenoviral vectors were performed in different areas such as oncology, genetic, or cardiovascular diseases.

4.1.1.1 Oncology

Adenoviral vectors can be powerful/useful tools to treat oncology-related diseases. There are four main antitumoral approaches: (1) tumor suppressor genes, (2) oncolytic therapy, (3) suicide gene therapy, and (4) immunotherapy.

1. The replenishment of the altered tumor suppressor genes was the first strategy that took place on clinical trials. P53 transfer was the first approach assayed, and it was indeed the first gene therapy product to be licensed and commercially available. It was licensed by the Chinese Food and Drug Agency in China in 2003 under the name Gendicine. However, a similar product was turned down by the US Food and Drug Administration (FDA) in 2008. It was due to the concerns about the safety of the adenoviral vectors after Jesse Gelsinger died in 1999 while participating in a clinical trial. Since then all the viruses assayed have safety mechanisms to avoid further problems.
2. Oncolytic viruses are viruses that infect selectively tumoral cells and kill them by lysis, releasing new infectious virus particles and so, destroying the remaining tumor. The first oncolytic virus approved by China's State Food and Drug Administration (SFDA) in 2005 was a genetically modified Ad named H101. It was very similar to the original Onyx-15 which was engineered to remove the interaction with the human gene *p53*, frequently altered in tumoral cells.[117] Short-term response rates are approximately doubled for H101 plus chemotherapy when compared to chemotherapy alone. The pharmaceutical market has commercialized it under the brand name Oncorine.

But oncolytic vectors could be also used as vaccines. As a result of this process, ColoAd1 (a novel chimeric hybrid of Ad serotypes Ad11p and Ad3) was generated, showing a higher potency and tumor selectivity than the control viruses. They could also be employed in conjunction with classical radio or chemotherapy. Onyx-015 underwent trials together with chemotherapy before it was abandoned in the early 2000s.[118]

In summer 2014, at least nine oncolytic viruses were in clinical trials and three are Ad based. (1) The previously mentioned Oncorine which is based on the Ad H101. (2) ColoAd1, developed by Psioxus Therapeutics Ltd which has successfully completed recruitment in a phase I clinical trial of ColoAd1. The second phase of the ColoAd1 study is planned to commence in 2014 by intravenous administration, and it will examine efficacy in patients with metastatic colorectal cancer. Unlike many other oncolytic viruses, ColoAd1 can be administered by intravenous injection rather than requiring intratumoral injection. (3) GL-ONC1, by Genelux, is used in phase I and they administer the virus intravenously.[119]

3. Suicide gene therapy consist in the incorporation of a foreign gene, which encodes for an enzyme, that metabolizes a non-toxic prodrug into a toxic metabolite capable to kill the tumoral cell. The phase I research studied a recombinant Ad that carries a suicide gene (TK, presumably) and a reporter gene and infects patients with ovarian and gynecological cancers. The simultaneous administration of this vector and ganciclovir induced the tumor death.[120]

4. Immunotherapy is based on the immunostimulation of genes that promote the natural immune surveillance and the elimination of tumors. Currently, there are 140 clinical trials based on immunotherapeutic approaches such as the ex vivo or in vivo transference of cytokine genes which enhance T-cell reactivity against tumors but only four of them are using Ad.

4.1.1.2 Genetic Diseases

Ongoing studies suggest that adenoviral vectors could be used to add a good copy of a defective gene in monogenic diseases, such as CF or ornithine transcarbamylase (OTC) deficiencies. OTC is an X-linked recessive metabolic disorder of nitrogen metabolism leading to ammonia accumulation in the blood. High ammonia concentrations might cause damage to the organism, leading to ataxia, lethargy, or even death. In adult patients with partial OTC deficiency, genetic function could be restored by the addition of an external copy of the OTC gene. A phase I clinical trial using a recombinant Ad containing the OTC gene is currently ongoing, with good perspectives.[121]

4.1.1.3 Cardiovascular

The idea of the use of adenoviral vectors to treat cardiovascular disorders is the secretion of growth factors that might induce a cascade of tissue remodeling factors that would not require continues expression of the therapeutic gene. A phase I/II clinical trial uses the vector AdVEGF-AII6A+ in the myocardium of individuals with diffuse CAD trying to induce the growth of collateral blood vessels, thereby improving the cardiac function.[122]

4.1.2 Herpes Simplex Virus

HSV-derived vectors have a great potential as long-term and/or short-term expression therapeutic gene vehicles. They have been used in several clinical trials for neuropathies, pain, and cancer.

4.1.2.1 Neuropathy and Pain

HSV-derived vectors have a great potential as tools for neuropathies and pain treatments as PNS is a natural target for HSV infections (specifically sensorial neurons with peripheral axonal terminal in the skin). Subcutaneous inoculation of replication-incompetent HSV vectors expressing the NT-3 gene was able to preserve the functionality of large myelinated fibers and prevents their degeneration (characteristics of neuropathy linked to diabetes or neuropathies related to cancer treatment with cisplatin).[123] Meanwhile, HSV vectors expressing nerve growth factor, vascular endothelial growth factor, or EPO are also used to prevent diabetic polyneuropathy.[124]

To treat pain, HSV vectors can express inhibitory neurotransmitters (anti-inflammatory peptides) that could modify the nociceptive neurotransmitter activation. A phase I clinical trial had studied the safety and the efficacy of NP2 in patients with intractable cancer pain. NP2 is a nonreplication HSV-based vector that expresses prepoenkephalin in order to interrupt pain signaling at the spinal level by blocking the opioid receptors. Treatment with NP2 was well tolerated and the dose-responsive analgesic effects suggest that NP2 may be effective to treat pain and requires further clinical investigation.[125]

4.1.2.2 Oncology

The vectors based on HSV follow the same strategies to treat cancer as the adenoviral vectors.

The immunotherapy is based on the vaccination with an HSV vector with a transgene that costimulates immunogenicity or provides tumor antigens to enhance the tumor suppressor. The therapeutic vector express cytokines on the tumor and that expression makes the neoplasic area to recruit natural killer cells to the area and start an immune response, decreasing the tumor size. A phase I clinical trial was focused on the study of recombinant hGM-CSF HSV also named OrienX010. OrienX010 is a recombinant virus that expresses a cytokine granulocyte-macrophage colony-stimulating factor (GM-CSF), and it selectively infects, replicates, and lyses tumor cells (as an oncolytic version). Moreover, GM-CSF attracts dendritic cells and may stimulate a cytotoxic T-cell response against the tumor cells.[126]

The suicide gene therapy also uses conditionally replicative viruses (or oncolytic viruses) to kill tumoral cells. The prodrug vectors express a transgene whose product induces cell death or increases the sensibility to chemo or radioactivity, such as the thymidine kinase gene that converts the ganciclovir into a toxic compound that causes cell death. An ongoing phase I/II clinical trial is studying the safety and effectiveness of a genetically engineered HSV-1, G207, in patients with recurrent brain cancer. G207 has been designed to harm and destroy tumor cells and do not affect normal brain cells.[127]

4.1.3 Vaccinia Virus

4.1.3.1 Oncology

Vaccinia-derived vector can be a powerful tool to treat oncologic diseases. The oncolytic therapy is based on the natural ability of vaccinia to destroy tissues. The double mutation of TK and vaccinia growth factor practically suppresses the viral replication in resting cells without modifying its ability to replicate in a tumoral environment and allows the tumor elimination.[128] Strategies such as the immunotherapy, uses vaccinia-derived vectors as a vaccine to stimulate the immune system response is another approach that could be combined. VV is engineered to carry antigens of tumors allowing an advantage of the stimulatory effects.[129]

4.1.4 Retrovirus

4.1.4.1 Oncology

Retroviral vectors are commonly used in cancer gene therapy although a major problem is that they only infect dividing cells (and not even tumoral cells are cycling in a coordinated manner). Approaches might be by carrying: (1) suicidal genes such as the TK of HSV or (2) tumor suppressor genes to restore the tumor suppressor function as p53 or (3) immune stimulatory genes that might activate the host immune response or bypass a defect of the immune system, such as interleukins, TNF, interferon-γ, or GM-CSF.[130]

4.1.4.2 Genetic Diseases

On a French clinical trial which took place in 1999 in Paris, involving almost a dozen very young patients with X-linked severe-combined immunodeficiency syndrome (X-SCID), the insertion of a RT-based vector was responsible for the development of leukemia in two children participating in the clinical trial.

Therefore, all the regulatory issues regarding RT and other integrative vectors were readjusted and so, retroviral vectors are mainly used to infect stem cells for ex vivo therapies. An ongoing phase I/II clinical trial for X-linked severe-combined immunodeficiency (SCID-X1) is focused on the study of an ex vivo gene therapy using autologous CD34$^+$ hematopoietic stem cells transduced with a therapeutic retroviral vector that are reinfused on the patients.[131] Another remarkable Phase I/II clinical trial is studying the safety and efficacy of autologous hematopoietic stem cells transduced with a retroviral vector MT-gp91 in patients with a defective chronic granuloma disease.[132]

4.1.5 Lentivirus

According to its versatility, lentiviral vectors are an interesting alternative for ex vivo and in vivo gene therapies, as they can infect dividing and nondividing cells. It leads to a wide range of applications for treatment of cancer, HIV, monogenic or neurodegenerative diseases.

4.1.5.1 Oncology

Several clinical trials using lentivirus have been approved. A phase 2 clinical trial for adults suffering of chemotherapy resistance in acute lymphoblastic leukemia is ongoing, and it is achieving a great success. The patient's autologous T cells were transduced with a lentiviral vector expressing anti-CD19 scFV TCR:41BB costimulatory domains.[133]

4.1.5.2 Neuropathies

Recent success in long-term correction of β-thalassemia and expression in transduced human hematopoietic stem cells has led to a phase I clinical trial with lentivirus. Lentiviral vectors encoding normal human β-globin gene are used to transduce aberrant CD34$^+$ hematopoietic progenitor cells from patients with β-thalassemia.[134]

Moreover, the outcomes previously obtained in correction of X-SCID with RT have resulted in a pilot feasibility study of gene transfer for X-SCID using SIN lentiviral vectors to transduce autologous CD34$^+$ hematopoietic stem cells by transferring a functional copy of the gene.[135]

4.1.6 Adeno-Associated Virus

The biological characteristics of AAV-derived vectors make them suitable for multiple applications where gene transfer would be required. Most of the AAV clinical applications are focused on chronic diseases that affect various organ systems or vaccination applications showing the great potential of AAV vectors.

4.1.6.1 Oncology

Recent results obtained using AAVs encourage further research focused on clinical trials. Of special interest is the study of the clinical safety and efficacy of antigen-specific cytotoxic T lymphocytes induced by dendritic cells infected by recombinant AAV with CEA gene in stage IV gastric cancer. The aim of this treatment is to enhance the immune response of the patient.[136]

4.1.6.2 Genetic Diseases

AAVs are also useful to treat genetic diseases like CF. Several clinical trials using AAV-CFTR by inhaled aerosol had been made.[137] AAV are also used to study an alternative therapy to treat hemophilia. Hemophilia is a disease caused by the absence or aberrant form of coagulating factor IX protein causing a noneffective clot of the blood and severe bleeding episodes. AAV expressing human clotting factor IX are used in vivo to correct the mutated gene in patients with severe hemophilia B.[138]

An AAV vector encoding for the enzyme that transforms l-dopa into dopamine (AAV-hAADC-2) has been used to transduce the enzyme on the damaged brain cells. Therefore, the damaged cells recovered their functionality.[139]

4.2 Clinical Trials Involving Nanoparticles

Nanoparticles are very popular for in vitro studies, but not many of them went into clinical trials. Lipofection is the second more popular nonviral system in clinical trials (after direct injection of naked DNA), used in 5.9% of all trials until 2012. A formulation composed of DOTAP:cholesterol liposome nanoparticles complexed with the plasmid C-VISA BiKDD (with potential antineoplastic activity) is used to find the highest tolerable dose of BikDD nanoparticles that can be given to patients with advanced cancer of the pancreas.[140]

siRNAs reached the clinic with successful results. TKM-080301 is a lipid nanoparticle formulation of an siRNA directed against PLK1, a serine/threonine kinase that regulates multiple critical aspects of cell cycle progression and mitosis. Antitumor activity, RNA interference, and pharmacodynamic effects of PLK1 inhibition have been conclusively demonstrated in preclinical models. Demonstration of pharmacodynamic effects of PLK1 inhibition in patients' biopsy is an exploratory objective of this phase 1 clinical trial.[141]

Other approaches delivering therapeutic genes have also been successful. DOTAP:Chol-fus1 is a drug that helps transfer a gene called fus1 into cancer cells. Researchers think that cells without this gene may be involved in the development of lung cancer tumors. The clinical phase would show if replacing the gene in these cells may keep the tissue from forming cancer cells.[142]

These systems have been proposed for intraperitoneal transfer of genes because of the toxicity they present when administered intravenously. Although Allovectin-7 (Vical, San Diego, CA, USA), which consists of a DNA plasmid bearing the genes for the allogenic MHC I protein, HLA-B7 and β2-microglobulin complexed with a DMRIE/DOPE cationic lipid mixture, failed to demonstrate and to improve a phase III clinical trial for treatment of advanced metastatic melanoma, other liposomal formulations are being developed clinically.

With regard to polymers, DermaVir consists of pDNA encoding 15 antigens of the HIV condensed with PEI mannose and dextrose. It induces long-lasting cytotoxic T cells capable of killing latently infected cells that remain in the reservoirs after highly active antiretroviral therapy, and it has now entered into phase III human trials.[143] Another recent example of a successful polymer is a transferrin-targeted, PEGylated CD-containing polymer for siRNA delivery for different types of cancers. It uses CDs for nucleic acid delivery, and it has finished a phase I clinical trial named as CALAA-01.

4.3 Clinical Trials Involving Exosomes

Despite the fact that research using exosomes is relatively novel, there are ongoing clinical trials using exosomes. Recently, it was shown that exosomal DNA (exoDNA) contains the entire genome of the person from whom it was obtained, highlighting the translational value of exoDNA (1) as a potential biomarker and (2) for gene therapy. One of the clinical trials (phase I) for stage III or IV melanoma patients, used exosomes expressing MAGE3 protein, showing an immunomodulatory effect with protein MAGE3. Patients showed an increase of the immune response against tumors, showing promising results.[144]

5. CONCLUSIONS AND FUTURE DIRECTIONS

Gene therapy is a field of research that has been under studies for decades with multiple clinical trials occurring worldwide (Figure 8). Although most of the vectors could be included on viral or nonviral vectors, the second group was not very successful for years. That is the reason why most of the clinical approaches were done by viral vectors. However, as nanotechnology is a field under expansion, many advances on the nonviral vectors have been made and progression on that area is quickly happening. Finally, the combination with patient's cells (specially stem cells) is a good tandem with an increasing number on clinical trials.

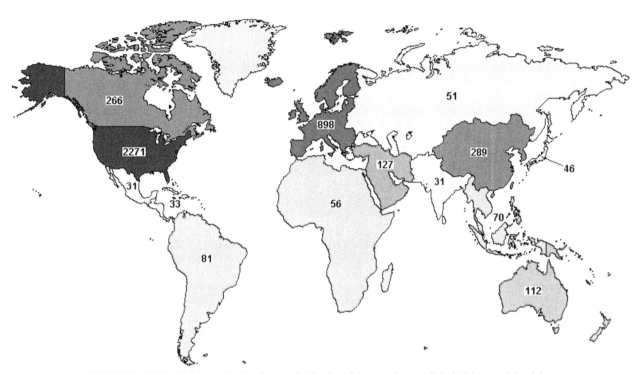

FIGURE 8 Clinical trials worldwide. Current distribution of the gene therapy clinical trials around the globe.

REFERENCES

1. Viecelli HM, Harbottle RP, Wong SP, Schlegel A, Chuah MK, VandenDriessche T, et al. Treatment of phenylketonuria using minicircle-based naked-DNA gene transfer to murine liver. *Hepatology* 2014;**60**(3):1035–43.
2. Reddy VS, Nemerow GR. Structures and organization of adenovirus cement proteins provide insights into the role of capsid maturation in virus entry and infection. *Proc Natl Acad Sci USA* 2014;**111**(32):11715–20.
3. San Martín C. Latest insights on adenovirus structure and assembly. *Viruses* May 2012;**4**(5):847–77. Erratum in Viruses 2012;**4**(12):3952.
4. Radko S, Koleva M, James KM, Jung R, Mymryk JS, Pelka P. Adenovirus E1A targets the DREF nuclear factor to regulate virus gene expression, DNA replication, and growth. *J Virol* 2014;**88**(22):13469–81.
5. Cheng CY, Gilson T, Wimmer P, Schreiner S, Ketner G, Dobner T, et al. Role of E1B55K in E4orf6/E1B55K E3 ligase complexes formed by different human adenovirus serotypes. *J Virol* 2013;**87**(11):6232–45.
6. Caravokyri C, Leppard KN. Human adenovirus type 5 variants with sequence alterations flanking the E2A gene: effects on E2 expression and DNA replication. *Virus Genes* 1996;**12**(1):65–75.
7. Nakashima H, Chiocca EA. Switching a replication-defective adenoviral vector into a replication-competent, oncolytic adenovirus. *J Virol* January 2014;**88**(1):345–53.
8. Amalfitano A, Hauser MA, Hu H, Serra D, Begy CR, Chamberlain JS. Production and characterization of improved adenovirus vectors with the e1, e2b, and e3 genes deleted. *J Virol* 1998;**72**(2):926–33.
9. Catalucci D, Sporeno E, Cirillo A, Ciliberto G, Nicosia A, Colloca S. An adenovirus type 5 (Ad5) amplicon-based packaging cell line for production of high-capacity helper-independent deltaE1-E2-E3-E4 Ad5 vectors. *J Virol* 2005;**79**(10):6400–9.
10. Brokhman I, Pomp O, Fishman L, Tennenbaum T, Amit M, Itzkovitz-Eldor J, et al. Genetic modification of human embryonic stem cells with adenoviral vectors: differences of infectability between lines and correlation of infectability with expression of the coxsackie and adenovirus receptor. *Stem Cells Dev* 2009;**18**(3):447–56.
11. Jiang H, Fueyo J. Eradication of brain tumor stem cells with an oncolytic adenovirus. *Discov Med* 2010;**10**(50):24–8.
12. Havenga MJ, Lemckert AA, Ophorst OJ, van Meijer M, Germeraad WT, Grimbergen J, et al. Exploiting the natural diversity in adenovirus tropism for therapy and prevention of disease. *J Virol* 2002;**76**(9):4612–20.
13. Nettelbeck DM, Rivera AA, Kupsch J, Dieckmann D, Douglas JT, Kontermann RE, et al. Retargeting of adenoviral infection to melanoma: combining genetic ablation of native tropism with a recombinant bispecific single-chain diabody (scDb) adapter that binds to fiber knob and HMWMAA. *Int J Cancer* 2004;**108**(1):136–45.
14. Heideman DA, Snijders PJ, Craanen ME, Bloemena E, Meijer CJ, Meuwissen SG, et al. Selective gene delivery toward gastric and esophageal adenocarcinoma cells via EpCAM-targeted adenoviral vectors. *Cancer Gene Ther* 2001;**8**(5):342–51.
15. Hogg R, Garcia JA, Gerard RD. Adenoviral targeting of gene expression to tumors. *Cancer Gene Ther* 2010;**17**(6):375–86.
16. Fisher K, Stallwood Y, Green NK, Ulbrich K, Mautner V, Seymour LW. Polymer-coated adenovirus permits efficient retargeting and evades neutralizing antibodies. *Gene Ther* March 2001;**8**(5):341–8.

17. Burton EA, Fink DJ, Glorioso JC. Gene delivery using herpes simplex virus vectors. *DNA Cell Biol* 2002;**21**(12):915–36.

18. Glorioso JC, Goins WF, Fink DJ, DeLuca NA. Herpes simplex virus vectors and gene transfer to brain. *Dev Biol Stand* 1994;**82**:79–87.

19. Wu N, Watkins SC, Schaffer PA, Deluca NA. Prolonged gene expression and cell survival after infection by a herpes simplex virus mutant defective in the immediate-early genes encoding icp4, icp27, and icp22. *J Virol* 1996;**70**(9):6358–69.

20. de Silva S, Mastrangelo MA, Lotta Jr LT, Burris CA, Federoff HJ, Bowers WJ. Extending the transposable payload limit of Sleeping Beauty (SB) using the herpes simplex virus (HSV)/SB amplicon-vector platform. *Gene Ther* 2010;**17**(3):424–31.

21. Hong CS, Fellows W, Niranjan A, Alber S, Watkins S, Cohen JB, et al. Ectopic matrix metalloproteinase 9 expression in human brain tumor cells enhances oncolytic HSV vector infection. *Gene Ther* 2010;**17**(10):1200–5.

22. Hadjipanayis CG, Fellows-Mayle W, DeLuca NA. Therapeutic efficacy of a herpes simplex virus with radiation or temozolomide for intracranial glioblastoma after convection-enhanced delivery. *Mol Ther* November 2008;**16**(11):1783–8.

23. Nakano K, Kobayashi M, Nakamura KI, Nakanishi T, Asano R, Kumagai I, et al. Mechanism of HSV infection through soluble adapter-mediated virus bridging to the EGF receptor. *Virology* April 2011;**413**(1):12–8.

24. Krisky DM, Wolfe D, Goins WF, Marconi PC, Ramakrishnan R, Mata M, et al. Deletion of multiple immediate-early genes from herpes simplex virus reduces cytotoxicity and permits long-term gene expression in neurons. *Gene Ther* December 1998;**5**(12):1593–603.

25. Goins WF, Lee KA, Cavalcoli JD, O'Malley ME, DeKosky ST, Fink DJ, et al. Herpes simplex virus type 1 vector-mediated expression of nerve growth factor protects dorsal root ganglion neurons from peroxide toxicity. *J Virol* January 1999;**73**(1):519–32.

26. Moss B. Poxviridae: the viruses and their replication. In: Knipe DM, Howley PM, editors. *Field's virology*. 5th ed. Philadelphia: Lippincott Williams and Wilkins; 2001. p. 2906–45.

27. Condit RC, Moussatche N, Traktman P. In a nutshell: structure and assembly of the vaccinia virion. *Adv Virus Res* 2006;**66**:31–124.

28. Domi A, Moss B. Engineering of a vaccinia virus bacterial artificial chromosome in *E. coli* by bacteriophage lambda-based recombination. *Nat Methods* 2005;**2**(2):95–7.

29. Smith GL, Benfield CT, Maluquer de Motes C, Mazzon M, Ember SW, Ferguson BJ, et al. Vaccinia virus immune evasion: mechanisms, virulence and immunogenicity. *J Gen Virol* 2013;**94**(Pt 11):2367–92.

30. Ramshaw IA, Ramsay AJ, Karupiah G, Rolph MS, Mahalingam S, Ruby JC. Cytokines and immunity to viral infections. *Immunol Rev* 1997;**159**:119–35.

31. Weiss RA. Retrovirus classification and cell interactions. *J Antimicrob Chemother* 1996;**37**(Suppl. B):1–11.

32. Nisole S, Saïb A. Early steps of retrovirus replicative cycle. *Retrovirology* 2004;**14**(1):9.

33. Yu SF, von Rüden T, Kantofft PW, Garbert C, Seiberg M, Rüthert L, et al. Self-inactivating retroviral vectors designed for transfer of whole genes into mammalian cells. *Proc Natl Acad Sci USA* 1986;**83**:3194–8.

34. Bushman FD. Targeting survival: integration site selection by retroviruses and LTR-retrotransposons. *Cell* 2003;**115**:135–8.

35. Waehler R, Russell SJ, Curiel DT. Engineering targeted viral vectors for gene therapy. *Nat Rev Genet* 2007;**8**(8):573–87.

36. Yang Y, Vanin EF, Whitt MA, Fornerod M, Zwart R, Schneiderman RD, et al. Inducible, high-level production of infectious murine leukemia retroviral vector particles pseudotyped with vesicular stomatitis virus G envelope protein. *Hum Gene Ther* 1995;**6**(9):1203–13.

37. Etienne-Julan M, Roux P, Carillo S, Jeanteur P, Piechaczyk M. The efficiency of cell targeting by recombinant retroviruses depends on the nature of the receptor and the composition of the artificial cell-virus linker. *J Gen Virol* 1992;**73**(Pt 12):3251–5.

38. Mátrai J, Chuah M, VandenDriessche T. Recent advances in lentiviral vector development and applications. *Mol Ther* 2010;**18**(3):477–90.

39. Durand S, Cimarelli A. The inside out of lentiviral vectors. *Viruses* 2011;**3**(2):132–59.

40. Sirven A, Pflumio F, Zennou V, Titeux M, Vainchenker W, Coulombel L, et al. The human immunodeficiency virus type-1 central DNA flap is a crucial determinant for lentiviral vector nuclear import and gene transduction of human hematopoietic stem cells. *Blood* December 15, 2000;**96**(13):4103–10.

41. Chang AH, Stephan MT, Sadelain M. Stem cell-derived erythroid cells mediate long-term systemic protein delivery systemic protein delivery. *Nat Biotechnol* 2006;**24**(8):1017–21.

42. Morizono K, Xie Y, Ringpis GE, Johnson M, Nassanian H, Lee B, et al. Lentiviral vector retargeting to P-glycoprotein on metastatic melanoma through intravenous injection. *Nat Med* 2005;**11**(3):346–52.

43. Kotterman MA, Schaffer DV. Engineering adeno-associated viruses for clinical gene therapy. *Nat Rev Genet* July 2014;**15**(7):445–51.

44. Gonçalves MA. Adeno-associated virus: from defective virus to effective vector. *Virol J* 2005;**6**(2):43.

45. McCarty DM, Monahan PE, Samulski RJ. Self-complementary recombinant adeno-associated virus (scAAV) vectors promote efficient transduction independently of DNA synthesis. *Gene Ther* 2001;**8**(16):1248–54.

46. Mukherjee S, Thrasher AJ. Gene therapy for PIDs: progress, pitfalls and prospects. *Gene* 2013;**525**(2):174–81.

47. Yin H, Kanasty RL, Eltoukhy AA, Vegas AJ, Dorkin JR, Anderson DG. Non-viral vectors for gene-based therapy. *Nat Rev Genet* 2014;**15**(8):541–55.

48. Sunshine JC, Corey J, Bishop JJG. Advances in polymeric and inorganic vectors for nonviral nucleic acid delivery. *Ther Deliv* 2014;**2**(4):493–521.

49. Nitta S, Numata K. Biopolymer-based nanoparticles for Drug/Gene delivery and tissue engineering. *Int J Mol Sci* 2013;**14**(1):1629–54.

50. Liu C, Zhang N. Nanoparticles in gene therapy: principles, prospects, and challenges. In: Antonio V, editor. *Progress in molecular biology and translational science*. 2011. p. 509–62.

51. Malone RW, Felgner PL, Verma IM. Cationic liposome-mediated RNA transfection. *Proc Natl Acad Sci USA* August 1, 1989;**86**(16):6077–81.

52. Felgner PL, Ringold GM. Cationic liposome-mediated transfection. *Nature* January 26, 1989;**337**(6205):387–8.

53. Gascón AR, Pozo-Rodríguez A, Solinís MÁ. Non-viral delivery systems in gene therapy. In: Martin DF, editor. *Gene therapy–tools and potential applications*. 2013.

54. Balazs DA, Godbey W. Liposomes for use in gene delivery. *J Drug Deliv* 2011;**2011**.

55. Niyomtham N, Apiratikul N, Chanchang K, Opanasopit P, Yingyongnarongkul BE. Synergistic effect of cationic lipids with different polarheads, central core structures and hydrophobic tails on gene transfection efficiency. *Biol Pharm Bull* 2014;**37**(9):1534–42.

56. Padeganeh A, Bakhshinejad B, Sadeghizadeh M, Khalaj-Kondori M. Non-viral vehicles: principles, applications, and challenges in gene delivery. In: Brown GG, editor. *Molecular cloning–selected applications in medicine and biology.* INTECH Open Access Publisher; 2011.

57. Behr JP, Demeneix B, Loeffler JP, Perez-Mutul J. Efficient gene transfer into mammalian primary endocrine cells with lipopolyamine-coated DNA. *Proc Natl Acad Sci* 1989;**86**(18):6982–6.

58. Zhao Y, Zhi D, Zhang S. Cationic liposomes in different structural levels for gene delivery. In: Yuan PX, editor. *Non-viral gene therapy.* InTech; 2011.

59. Zhi D, Zhang S, Wang B, Zhao Y, Yang B, Yu S. Transfection efficiency of cationic lipids with different hydrophobic domains in gene delivery. *Bioconjug Chem* 2010;**21**(4):563–77.

60. Misra SK, Biswas J, Kondaiah P, Bhattacharya S. Gene transfection in high serum levels: case studies with new cholesterol based cationic Gemini lipids. *PLoS ONE* 2013;**8**(7):e68305.

61. Zhang S, Xu Y, Wang B, Qiao W, Liu D, Li Z. Cationic compounds used in lipoplexes and polyplexes for gene delivery. *J Control Release* 2004;**100**(2):165–80.

62. Walker S, Sofia MJ, Kakarla R, Kogan NA, Wierichs L, Longley CB, et al. Cationic facial amphiphiles: a promising class of transfection agents. *Proc Natl Acad Sci USA* 1996;**93**(4):1585–90.

63. Zhi D, Zhang S, Cui S, Zhao Y, Wang Y, Zhao D. The headgroup evolution of cationic lipids for gene delivery. *Bioconjugate Chem* 2013;**24**(4):487–519.

64. Oh N, Park J-H. Endocytosis and exocytosis of nanoparticles in mammalian cells. *Int J Nanomedicine* 2014;**9**(Suppl. 1):51–63.

65. Zuhorn IS, Engberts JB, Hoekstra D. Gene delivery by cationic lipid vectors: overcoming cellular barriers. *Eur Biophys J* 2007;**36**(4–5):349–62.

66. Zelphati O, Szoka FC. Mechanism of oligonucleotide release from cationic liposomes. *Proc Natl Acad Sci USA* 1996;**93**(21):11493–8.

67. Liang W, Lam JKW. *Endosomal escape pathways for non-viral nucleic acid delivery systems.* 2012.

68. Gjetting T, Arildsen NS, Laulund C, et al. In vitro and in vivo effects of polyethylene glycol (PEG)-modified lipid in DOTAP/cholesterol-mediated gene transfection. *Int J Nanomedicine* 2010;**5**:371–83.

69. Wu GY, Wu CH. Receptor-mediated in vitro gene transformation by a soluble DNA carrier system. *J Biol Chem* 1987;**262**(10):4429–32.

70. Yue Y, Wu C. Progress and perspectives in developing polymeric vectors for in vitro gene delivery. *Biomaterials Sci* 2013;**1**(2):152–70.

71. Lin C, Lou B. Bioreducible cationic polymers for gene transfection. In: Lin C, editor. *Biomedicine.* InTech; 2012. p. 85–104.

72. Tros de Ilarduya C, Sun Y, Duzgunes N. Gene delivery by lipoplexes and polyplexes. *Eur J Pharm Sci* 2010;**40**(3):159–70.

73. Tschiche A, Staedtler AM, Malhotra S, Bauer H, Bottcher C, Sharbati S, et al. Polyglycerol-based amphiphilic dendrons as potential siRNA carriers for in vivo applications. *J Mater Chem B* 2014;**2**(15):2153–67.

74. Dizaj S, Jafari S, Khosroushahi A. A sight on the current nanoparticle-based gene delivery vectors. *Nanoscale Res Lett* 2014;**9**(1):252.

75. Laemmli UK. Characterization of DNA condensates induced by poly(ethylene oxide) and polylysine. *Proc Natl Acad Sci USA* 1975;**72**(11):4288–92.

76. Mintzer MA, Simanek EE. Nonviral vectors for gene delivery. *Chem Rev* 2009;**109**(2):259–302.

77. Kim T-i, Kim SW. Bioreducible polymers for gene delivery. *React Funct Polym* March 2011;**71**(3):344–9.

78. Boussif O, Lezoualc'h F, Zanta MA, Mergny MD, Scherman D, Demeneix B, et al. A versatile vector for gene and oligonucleotide transfer into cells in culture and in vivo: polyethylenimine. *Proc Natl Acad Sci USA* 1995;**92**(16):7297–301.

79. Demeneix B, Behr JP. Polyethylenimine (PEI). In: Leaf Huang M-CH, Ernst W, editors. *Advances in genetics.* Academic Press; 2005. p. 215–30.

80. Sonawane ND, Francis CS, Verkman AS. Chloride accumulation and swelling in endosomes enhances DNA transfer by polyamine-DNA polyplexes. *J Biol Chem* 2003;**278**(45):44826–31.

81. Navarro G, Essex S, Sawant RR, Biswas S, Nagesha D, Sridhar S, et al. Phospholipid-modified polyethylenimine-based nanopreparations for siRNA–mediated gene silencing: implications for transfection and the role of lipid components. *Nanomedicine Nanotechnol Biol Med* 2014;**10**(2):411–9.

82. Arima H. Dendrimers as DNA carriers. In: Taira K, Kataoka K, Niidome T, editors. *Non-viral gene therapy.* Tokyo: Springer; 2005. p. 75–86.

83. Thomas M, Ge Q, Lu JJ, Chen J, Klibanov A. Cross-linked small polyethylenimines: while still nontoxic, deliver DNA efficiently to mammalian cells in vitro and in vivo. *Pharm Res* 2005;**22**(3):373–80.

84. MacLaughlin FC, Mumper RJ, Wang J, Tagliaferri JM, Gill I, Hinchcliffe M, Rolland AP. Chitosan and depolymerized chitosan oligomers as condensing carriers for in vivo plasmid delivery. *J Control Release* 1998;**56**(1–3):259–72.

85. Dufès C, Uchegbu IF, Schätzlein AG. Dendrimers in gene delivery. *Adv Drug Deliv Rev* 2005;**57**(15):2177–202.

86. Fuentes JLJ, Ortega P, Ferrando-Martínez S, Gómez R, Leal M, Mata J, et al. *Dendrimers in RNAi delivery. Advanced delivery and therapeutic applications of RNAi.* John Wiley and Sons, Ltd; 2013. p. 163–85.

87. Wu J, Huang W, He Z. Dendrimers as carriers for siRNA delivery and gene silencing: a review. *Sci World J* 2013;**2013**:16.

88. Kadlecova Z, Rajendra Y, Matasci M, Baldi L, Hacker DL, Wurm FM, et al. DNA delivery with hyperbranched polylysine: a comparative study with linear and dendritic polylysine. *J Control Release* 2013;**169**(3):276–88.

89. Sunasee R, Wattanaarsakit P, Ahmed M, Lollmahomed FB, Narain R. Biodegradable and nontoxic nanogels as nonviral gene delivery systems. *Bioconjugate Chem* 2012;**23**(9):1925–33.

90. Benjaminsen RV, Mattebjerg Ma, Henriksen JR, Moghimi SM, Andresen TL. The possible "proton sponge" effect of polyethylenimine (PEI) does not include change in lysosomal pH. *Mol Ther* 2013;**21**(1):149–57.

91. Lin MM, Kim DK, El Haj AJ, Dobson J. Development of superparamagnetic iron oxide nanoparticles (SPIONS) for translation to clinical applications. *IEEE Trans Nanobioscience* 2008;**7**(4):298–305.

92. Chen J, Zhu S, Tong L, Li J, Chen F, Han Y, et al. Superparamagnetic iron oxide nanoparticles mediated (131)I-hVEGF siRNA inhibits hepatocellular carcinoma tumor growth in nude mice. *BMC Cancer* 2014.

93. Li Z, Xiang J, Zhang W, Fan S, Wu M, Li X, et al. Nanoparticle delivery of anti-metastatic NM23-H1 gene improves chemotherapy in a mouse tumor model. *Cancer Gene Ther* 2009;**16**(5):423–9.

94. Chen Y, Wang W, Lian G, Qian C, Wang L, Zeng L, et al. Development of an MRI-visible nonviral vector for siRNA delivery targeting gastric cancer. *Int J Nanomedicine* 2012;**7**:359–68.

95. Schade A, Delyagina E, Scharfenberg D, Skorska A, Lux C, David R, et al. Innovative strategy for MicroRNA delivery in human mesenchymal stem cells via magnetic nanoparticles. *Int J Mol Sci* 2013;**14**(6):10710–26.

96. Tandon A, Sharma A, Rodier JT, Klibanov AM, Rieger FG, Mohan RR. BMP7 gene transfer via gold nanoparticles into stroma inhibits corneal fibrosis in vivo. *PLoS One* 2013;**8**(6):e66434.

97. Jensen SA, Day ES, Ko CH, Hurley LA, Janina P, Kouri FM, et al. Spherical nucleic acid nanoparticle conjugates as an RNAi- based therapy for glioblastoma. *Sci Transl Med* 2013;**5**(209).

98. Lee Seung K, Tung C-H. A fabricated siRNA nanoparticle for ultra-long gene silencing in vivo. *Adv Funct Mater* 2014;**23**(28):3488–93.

99. Papasani MR, Wang G, Hill Ra. Gold nanoparticles: the importance of physiological principles to devise strategies for targeted drug delivery. *Nanomedicine* 2012;**8**(6):804–14. Elsevier Inc.

100. Kim HJ, Takemoto H, Yi Y, Zheng M, Maeda Y, Chaya H, et al. Precise engineering of siRNA delivery vehicles to tumors using polyion. 2014;23(8(9)):8979–8991.

101. Cheng J, Tang L. Nonporous silica nanoparticles for nanomedicine application. *Nano Today* 2013;**8**(3):290–312.

102. Hom C, Lu J, Liong M, Luo H, Zongxi L, Jeffrey I, et al. Mesoporous silica nanoparticles facilitate delivery of siRNA to shutdown signaling pathways in mammalian cells. *Small* 2011;**6**(11):1185–90.

103. Zhu SG, Xiang JJ, Li XL, Shen SR, Lu HB, Zhou J, et al. Poly(L-lysine)-modified silica nanoparticles for the delivery of antisense oligonucleotides. *Biotechnol Appl Biochem* 2004;**39**(2):179–87.

104. Hartono SB, Yu M, Gu W, Yang J, Strounina E, Wang X, et al. Synthesis of multi-functional large pore mesoporous silica nanoparticles as gene carriers. *Nanotechnology* 2014;**25**(5):055701.

105. Gupta N, Shrivastava A, Sharma RK. Silica nanoparticles coencapsulating gadolinium oxide and horseradish peroxidase for imaging and therapeutic applications. *Int J Nanomedicine* 2012;**7**:5491–500.

106. Bharali DJ, Klejbor I, Stachowiak EK, Dutta P, Roy I, Kaur N, et al. Organically modified silica nanoparticles: a nonviral vector for in vivo gene delivery and expression in the brain. *Proc Natl Acad Sci USA* 2005;**102**(32):11539–44.

107. Bates K, Kostarelos K. Carbon nanotubes as vectors for gene therapy: past achievements, present challenges and future goals. *Adv Drug Deliv Rev* 2013;**65**(15):2023–33. Elsevier B.V.

108. Hollanda LM, Lobo AO, Lancellotti M, Berni E, Corat EJ, Zanin H. Graphene and carbon nanotube nanocomposite for gene transfection. *Mater Sci Eng C Mater Biol Appl* 2014;**39**(1):288–98. Elsevier B.V.

109. Siu KS, Chen D, Zheng X, Zhang X, Johnston N, Liu Y, et al. Non-covalently functionalized single-walled carbon nanotube for topical siRNA delivery into melanoma. *Biomaterials* 2014;**35**(10):3435–42.

110. Bonoiu A, Mahajan SD, Ye L, Kumar R, Ding H, Roy I, et al. MMP-9 gene silencing by a quantum dot-siRNA nanoplex delivery to maintain the integrity of the blood–brain barrier. *Brain Res* 2010:142–55.

111. Lee Y, Andaloussi S, Wood M. Exosomes and microvesicles: extracellular vesicles for genetic information transfer and gene therapy. *Hum Mol Genet* 2012;**21**(1):R125–34.

112. Shtam T, Kovalev R, Varfolomeeva E, Makarov E, Kil Y, Filatov M. Exosomes are natural carriers of exogenous siRNA to human cells in vitro. *Cell Commun Signal* 2013;**11**:88.

113. Fleury A, Martinez M, Le Lay S. Extracellular vesicles as therapeutic tools in cardiovascular diseases. *Front Immunol* 2014;**5**:370.

114. Peinado H, Alečković M, Lavotshkin S, Matei I, Costa-Silva B, Moreno-Bueno G, et al. Melanoma exosomes educate bone marrow progenitor cells toward a pro-metastatic phenotype through MET. *Nat Med* 2012;**18**(6):883–91.

115. Hartman Z, Wei J, Glass O. Increasing vaccine potency through exosome antigen targeting. *Vaccine* 2011;**29**(50):9361–7.

116. Thakur BK, Zhang H, Becker A, Matei I, Huang Y, Costa-Silva B, et al. Double-stranded DNA in exosomes: a novel biomarker in cancer detection. *Cell Res* 2014;**24**:766–9.

117. Huang X, Jia R, Zhao X, Liu B, Wang H, Wang J, et al. Recombinant oncolytic adenovirus H101 combined with siBCL2: cytotoxic effect on uveal melanoma cell lines. *Br J Ophthalmol* 2012;**96**(10):1331–8.

118. Kuhn I, Harden P, Bauzon M, Chartier C, Nye J, Thorne S, et al. Directed evolution generates a novel oncolytic virus for the treatment of colon cancer. *PLoS One* June 18, 2008;**3**(6):e2409.

119. Kyula JN, Khan AA, Mansfield D, Karapanagiotou EM, McLaughlin M, Roulstone V, et al. Synergistic cytotoxicity of radiation and oncolytic Lister strain vaccinia in (V600D/E)BRAF mutant melanoma depends on JNK and TNF-α signaling. *Oncogene* March 27, 2014;**33**(13):1700–12.

120. A study of an infectivity enhanced suicide gene expressing adenovirus for ovarian cancer in patients with recurrent ovarian and other selected gynecologic cancer. ClinicalTrials.gov Identifier NCT 00964756. http://clinicaltrials.gov/ct2/show/record/NCT00964756?term=%22adenovirus%22+AND+%22cancer%22&rank=3.

121. A study of MM-121 combination therapy in patients with advanced non-small cell lung. ClinicalTrials.gov Identifier NCT01757223. http://clinicaltrials.gov/ct2/show/NCT01757223?term="cardiovascular"+AND+"adenovirus"&rank=1.

122. Wu CC, D'Argenio D, Asgharzadeh S, Triche T. TARGETgene: a tool for identification of potential therapeutic targets in cancer. *PLoS One* 2012;**7**(8):e43305.

123. Wolfe D, Mata M, Fink DJ. Targeted drug delivery to the peripheral nervous system using gene therapy. *Neurosci Lett* October 11, 2012;**527**(2):85–9.

124. Fink DJ, Wechuck J, Mata M, Glorioso JC, Goss J, Krisky D, et al. Gene therapy for pain: results of a phase I clinical trial. *Ann Neurol* 2011;**70**(2):207–12.

125. A phase I study of recombinant hGM-CSF herpes simplex virus to treat cancer. Clinical Trials.gov Identifier NCT01935453. http://clinicaltrials.gov/ct2/show/study/NCT0193545).

126. Safety and effectiveness study of G207, a tumor-killing virus, in patients with recurrent brain cancer. Clinical Trials.gov Identifier NCT00028158. http://clinicaltrials.gov/ct2/show/NCT00028158.

127. McCart GM, Kayser SR. Therapeutic equivalency of low-molecular-weight heparins. *Ann Pharmacother* 2002;**36**(6):1042–57.

128. Safety Study of Recombinant Vaccinia Virus to Treat Refractory Solid Tumors. ClinicalTrials.gov Identifier NCT00625456. http://www.clinicaltrials.gov/show/NCT00625456.

129. Barzon L, Pacenti M, Franchin E, Colombo F, Palù G. HSV-TK/IL-2 gene therapy for glioblastoma multiforme. *Methods Mol Biol* 2009;**542**:529–49.

130. Gene therapy for X-linked severe combined immunodeficiency (SCID2). ClinicalTrials.gov Identifier NCT01410019. http://clinicaltrials.gov/ct2/show/NCT01410019.

131. Gene Therapy for Chronic Granulomatous Disease in Korea. ClinicalTrials.gov Identifier NCT00778882. http://clinicaltrials.gov/ct2/show/NCT00778882.

132. Study of redirected autologous T cells engineered to contain anti-CD19 attached to TCR and 4-1BB signaling domains in patients with chemotherapy resistant or refractory acute lymphoblastic leukemia. ClinicalTrials.gov Identifier NCT02030847. http://clinicaltrials.gov/ct2/show/NCT02030847.

133. SS-Thalassemia major with autologous CD34$^+$ hematopoietic progenitor cells transduced with TNS9.3.55 a lentiviral vector encoding the normal human β-globin gene. ClinicalTrials.gov Identifier NCT01639690. http://clinicaltrials.gov/ct2/show/NCT01639690.

134. Gene transfer for X-linked severe combined immunodeficiency in newly diagnosed infants (LVXSCID-ND). ClinicalTrials.gov Identifier NCT01512888. http://clinicaltrials.gov/ct2/show/NCT01512888.

135. Clinical safety and Preliminary efficacy of AAV-DC-CTL treatment in stage IV gastric cancer. NCT01637805. http://clinicaltrials.gov/ct2/show/study/NCT01637805.

136. Moss AC, Farrell RJ. Adding fuel to the fire: GM-CSF for active Crohn's disease. *Gastroenterology* 2005;**129**(6):2115–7.

137. Safety of a new type of treatment called gene transfer for the treatment of severe Hemophilia B. ClinicalTrials.gov Identifier NCT00076557. http://clinicaltrials.gov/ct2/show/record/NCT00076557.

138. Dose-escalation study of a self complementary adeno-associated viral vector for gene transfer in hemophilia B. ClinicalTrials.gov Identifier NCT00979238. http://clinicaltrials.gov/ct2/show/NCT00979238.

139. A study of AAV-hAADC-2 in subjects with Parkinson's disease. ClinicalTrials.gov Identifier NCT00229736. http://clinicaltrials.gov/ct2/show/NCT00229736.

140. C-V.I.S.A. BikDD: liposome in advanced pancreatic cancer. ClinicalTrials.gov Identifier NCT00968604. https://clinicaltrials.gov/ct2/show/NCT00968604.

141. TKM 080301 for primary or secondary liver cancer. ClinicalTrials.gov Identifier NCT01437007. https://clinicaltrials.gov/ct2/show/NCT01437007.

142. FUS1-nanoparticles and Erlotinib in stage IV lung cancer. ClinicalTrials.gov Identifier NCT01455389. https://clinicaltrials.gov/ct2/show/NCT01455389.

143. Lori F, Calarota SA, Lisziewicz J. Nanochemistry-based immunotherapy for HIV-1. *Curr Med Chem* 2007;**14**(18):1911–9.

144. Pitt JM, Charrier M, Viaud S, André F, Besse B, Chaput N, et al. Dendritic cell-derived exosomes as immunotherapies in the fight against cancer. *J Immunol* August 1, 2014;**193**(3):1006–11.

Chapter 12

Treating Hemophilia by Gene Therapy

Christopher D. Porada, Glicerio Ignacio, Anthony Atala, Graça Almeida-Porada

Key Concepts

1. Gene therapy could promise a permanent cure for hemophilia A and B.
2. Current treatments are expensive and do not cure disease.
3. Many characteristics of the hemophilias make them ideal diseases for treating by gene therapy.
4. Very good preclinical animal models of hemophilia A and B exist for testing novel treatments.
5. Multiple gene therapy vectors are well-suited for delivering FVIII or FIX.
6. In addition to ongoing direct vector injection trials, several cell types would make very good cellular vehicles for delivering a FVIII or FIX transgene.
7. Performing gene therapy for the hemophilias prior to birth (in utero) would circumvent many of the issues that have thus far hindered clinical application of gene therapy to the hemophilias and many other genetic diseases.

1. RATIONALE FOR GENE THERAPY

Gene therapy is the transfer of a normal functional copy of a gene into the appropriate cells of an individual, with the intent to correct a disease caused by a defect within the individual's own copy of the gene in question.[1–3] This relatively new (<25 years) treatment promises to make it possible to permanently cure one day any of the over 4000 diseases that are caused by an error in only a single gene (monogenic). Some of the best known examples among these include hemophilia, hemoglobin disorders such as sickle cell and the thalassemias; lysosomal storage diseases like Gaucher's, Tay–Sachs, and the mucopolysaccharidoses (e.g., Hurler's); diseases of immune function such as adenosine deaminase deficiency or deficiencies in the shared γc receptor subunit; and cystic fibrosis. While one might imagine that many of these genetic diseases could be corrected by simply providing an exogenous source of the missing/defective protein, this is not always the case, since therapy of this type is only available for some of these disorders. Even when the required protein can be purified or produced in recombinant form in sufficient quantities to be therapeutically useful, one is still faced with the challenge of providing the missing/defective protein in a therapeutic fashion, which may require the delivery of the complex, and often fragile protein to the precise subcellular location in which it is normally expressed in order to achieve benefit. Even if these delivery hurdles can be overcome, protein-based treatments can never cure the underlying disease. Rather, they require a lifetime of regularly spaced infusions to keep the disease process at bay. Even after years of treatment, if the patient misses a single dose of replacement protein, the symptoms return with potentially life-threatening consequences.

In contrast to existing protein-based treatment approaches, performing gene therapy, by providing a normal copy of the defective gene to the affected tissues within the patient, would eliminate the problem of having to deliver the protein to the proper subcellular locale, since the protein would be synthesized within the cell, utilizing the cell's own translational and posttranslational modification machinery. This would ensure that the protein arrives at the appropriate target site. In addition, although the gene defect is present within every cell of the affected individual, in most cases, only selected cells within a limited number of organs in the body transcribe each given gene and synthesize the resultant protein. Therefore, only cells that express the product of the gene in question would be affected by the genetic abnormality. This greatly simplifies the task of delivering the defective gene to the patient and achieving therapeutic benefit, since the gene need to be delivered only to a fairly limited number of sites within the body. Furthermore, if the gene could somehow be specifically targeted to the organs which are most affected by the disorder, the risk of side effects from ectopic expression of the therapeutic gene would be avoided. Gene therapy, if targeted to the appropriate somatic cells, could thus promise lifelong/permanent correction of the genetic defect following only a single treatment. It is this promise that drives the numerous ongoing preclinical and clinical gene therapy studies for a wide range of diseases/disorders.

Translating Regenerative Medicine to the Clinic. http://dx.doi.org/10.1016/B978-0-12-800548-4.00012-7

While gene therapy approaches are actively being developed for many of the monogenic disorders, we will focus in this chapter on hemophilia, discussing its basic pathophysiology, explaining the fairly unique characteristics of this disease that make it an ideal disease to attempt to treat with gene therapy, and highlighting some of the key preclinical studies and clinical trials that have hinted at the ability of gene therapy to deliver on its promise of providing a permanent cure following a single treatment.

2. HEMOPHILIA: PATHOPHYSIOLOGY, HISTORY, AND CLINICAL MANAGEMENT

Hemophilia is one of the most frequent inheritable bleeding disorders, and is caused by either a mutation in the gene encoding coagulation factor FVIII (hemophilia A) or a mutation in the gene encoding coagulation factor IX (hemophilia B). Since factor VIII serves as a cofactor in the same step of the coagulation cascade that is catalyzed by factor IX, the absence (or lack of functionality) of either of these proteins is equally deleterious and produces a virtually identical clinical picture.[4] Hemophilia A is, however, far more prevalent, affecting roughly seven times as many individuals as hemophilia B.

Hemophilia was first described in the tenth century AD by Abulcasis, who astutely observed that males born into certain families bled excessively following circumcision, and often died of bleeding after only minor traumas. Despite this early recognition that hemophilia was somehow passed within families, roughly 900 years passed before Dr John Conrad Otto wrote a detailed account about "a hemorrhagic disposition existing in certain families" in which he called the affected males "bleeders." Otto recognized that the disorder was hereditary, that it affected mostly males, and that it could be passed down by healthy females.[5]

The genes encoding factors VIII and IX are both now known to reside on the X-chromosome, explaining why hemophilia almost always affects males, occurring when they inherit their single X-chromosome from a mother who carries a mutation in one copy of the factor VIII or factor IX gene. Since women possess two X-chromosomes, the one harboring the mutant coagulation factor gene is normally inactivated, and its loss is compensated by the normal allele. Carrier females are thus phenotypically normal, and are often unaware that they carry this mutation. Perhaps the most famous example of an individual who unknowingly harbored this mutation was Queen Victoria. Indeed, the appearance of hemophilia in Queen Victoria's eighth child, Leopold, upset and confused the Queen, who vehemently protested that the disease could not possibly have come from her side of the family, which was of "pure royal descent."[6,7] Nevertheless, the Queen's offspring went on to unwittingly spread a hemophilia-causing mutation through the royal families of Spain, Russia, and Prussia.[4,6–9] Despite its X-linkage, however, it is important to note that rare cases of hemophilia have been reported in women as well, as a result of inactivation of the X-chromosome encoding the normal/healthy allele for coagulation factor VIII or IX.[10–17]

The severity of hemophilia is traditionally based on plasma levels of factor VIII or IX, with persons exhibiting less than 1% normal factor being considered to have severe hemophilia, persons with 1–5% normal factor moderately severe, and persons with 5–40% of the normal factor levels mild.[18–20] While the clinical severity of hemophilia can vary from patient to patient, up to 70% of hemophilia patients are present with the severe form of the disease.[18–20] These patients suffer frequent spontaneous hemorrhaging, which leads to hematomas of subcutaneous connective tissue/muscle, recurrent joint bleeds (hemarthroses) which cause chronic painful and debilitating arthropathies, and potentially life-threatening internal bleeding.

Until the 1960s, there was no effectual treatment for these recurrent bleeds, and the life expectancy for patients with severe hemophilia did not normally extend beyond early adulthood. The discovery that coagulation factors VIII and IX were present in plasma and that the cryoprecipitated fraction of human plasma was highly enriched for these coagulation factors revolutionized the clinical treatment/management of hemophilia, and greatly increased the life expectancy of these patients.[21–23] Sadly, however, this breakthrough turned to tragedy in the 1980s, when it was realized that a high percentage of this young generation of hemophilia patients had inadvertently been infused with plasma-derived products that were contaminated with HIV and/or hepatitis C.[24–27] While the implementation of careful donor screening and the development of heat- and chemical-based methods for inactivating viral contaminants present in blood products greatly improved the safety of plasma-derived hemophilia therapeutics,[28–33] it was the cloning of the genes for factors VIII and IX[34–36] and the subsequent production of these proteins in recombinant form[37,38] that completely eliminated the risk of transmitting a known, or a new as-yet unidentified, pathogen when treating hemophilia.

With all these advances, current state-of-the-art treatment for hemophilia consists of frequent prophylactic infusions of plasma-derived or recombinant factor VIII or IX protein to maintain hemostasis. This treatment has dramatically improved the quality of life for many hemophilia patients, and has greatly increased the life expectancy. However, this approach to therapy is still far from ideal, since it requires lifelong infusions, without ever curing the underlying disease. In addition to being noncurative, such a treatment approach is also prohibitively expensive, costing $150,000 to >$600,000/year. Furthermore, even setting these shortcomings aside, factor replacement therapy is simply not available for ~75% of the world's hemophiliacs, placing these patients at great risk of severe, permanent disabilities and life-threatening bleeds.[39] Even for hemophilia patients who are fortunate enough to have access to factor VIII and the means to afford prophylactic infusions, there is no guarantee of a life that is free from treatment complications, since at least 30% of patients with severe hemophilia

A develop inhibitory antibodies to the infused factor VIII protein.[39–41] These antibodies/inhibitors greatly reduce treatment efficacy, increase morbidity/mortality, decrease quality of life, and can ultimately lead to treatment failure, placing the patient at risk of life-threatening hemorrhage.[42] While inhibitors are far less common in patients with hemophilia B,[43] their formation can trigger severe immune responses which can include anaphylaxis, placing patients in grave danger. Indeed, the development of antibodies that inhibit the function of infused coagulation factors is currently considered the most significant complication of hemophilia treatment.[41] There is thus a significant need to develop novel hemophilia therapies offering long-lasting benefit or permanent cure,[44] such as those based on gene therapy.

2.1 Hemophilia: An Archetype Genetic Disease for Correction by Gene Therapy

Although there are numerous candidate diseases being considered for correction with gene therapy, several aspects of the basic biology and pathophysiology of hemophilia makes it an ideal target disease.[44–53] The first of these is that neither factor VIII nor factor IX needs to be expressed in either a cell- or tissue-specific fashion to restore hemostasis. While the liver is thought to be the primary natural site of synthesis of these coagulation factors,[54] as long as these proteins are expressed in cells which have ready access to the circulation, they can be secreted into the bloodstream and exert their appropriate clotting activity. Equally important, expression of these coagulation factors in other tissues of the body exerts no observable deleterious effects. This is in stark contrast to many other genetic diseases in which the expression of the missing protein must be very tightly controlled with respect to tissue/cell type. In some circumstances, the expression must be further restricted to a specific subcellular locale for the protein to function correctly and to avoid deleterious effects.

A second feature of hemophilia A and B that sets them apart from many other diseases is the fact that only a very small quantity of the missing clotting factors are required to achieve a pronounced clinical improvement. Indeed, factor VIII or IX levels of only 3–5% of normal would convert severe hemophilia A or B, respectively, to a moderate/mild phenotype. Such a change would be expected to reduce or eliminate episodes of spontaneous bleeding, and greatly improve quality of life. Thus, even with the relatively low levels of gene transfer/transduction that can routinely be obtained with many of the current gene delivery systems, a marked clinical improvement would be anticipated in hemophilia patients. Conversely, even supraphysiologic levels of FVIII as high as 150% of normal are predicted to be well tolerated, making the therapeutic window extremely wide.[20] The remarkable progress that has taken place over the past three decades in the understanding of the molecular basis of the disease; the identification and characterization of genes for factors VIII and IX; and a detailed analysis of the structure, biochemistry, and basic biology of these proteins has furthered the interest and feasibility of treating hemophilia with gene therapy. Collectively, these factors impelled the American Society of Gene and Cell Therapy (www.ASGCT.org) to place hemophilia at the top of their list when they provided a road map to NIH director, Dr Francis Collins, in 2012, in which they detailed the most promising, "Target 10," group of diseases they felt would be viable with gene therapy products within the next 5–7 years.

3. PRECLINICAL TESTING OF GENE THERAPY FOR HEMOPHILIA

A number of animal models have been identified and/or developed to evaluate new methods of not only treatment of coagulation disorders, but also the prevention and treatment of inhibitor formation. Transient hemophilic rabbit models induced by infusion of plasma containing inhibitors have been used to evaluate the effect of different bypass products to factor VIII.[55] However, these models, while valuable for inhibitor studies, do not accurately recapitulate the human disease, precluding their use for gene therapy studies. Naturally occurring congenital deficiency of factors VIII and IX has been described in numerous breeds of dogs[56–74] and mouse models have been created using gene targeting and knockout technology.[74–80] These models have made it possible to study biology and function of factors VIII and IX, and have greatly facilitated the development and testing of various gene therapy-based approaches for treating hemophilia.

3.1 Adenoviral Vectors

A variety of viral-based vector systems and delivery strategies have been tested as means of expressing factor VIII or IX gene and thereby correcting hemophilia A or B, respectively. Based on its ease of production, high transduction efficiency, and lack of genomic integration (leading to a presumption that this would translate into increased safety), several early studies tested the ability of adenoviral-based vectors to deliver factors VIII and IX to treat hemophilia.[81–93] These studies revealed that, although therapeutic levels of these coagulation factors could be achieved following direct injection of adenoviral vectors encoding their respective genes, these levels were only maintained for a short period of time, likely due to the rapid loss of the nonintegrated episomal adenoviral genomes and, perhaps, the transient expression driven by some of the promoters used in these early studies. These and other studies employing adenoviral vectors also revealed that these early generation adenoviral vectors could trigger a rather robust inflammatory/immune response that could prove fatal, if unchecked.[94–102] These collective findings suggested that alternate vectors would likely be better suited for delivering the genes for factor VIII or IX and achieving long-term therapeutic effects, especially given the inherent immunogenicity of these proteins.

3.2 Retroviral Vectors

To overcome the issues with transient expression of the exogenous coagulation factors and address concerns over adenovirus' immunogenicity, several preclinical efforts to treat hemophilia were undertaken using vectors based on murine retroviruses,[103–110] based on the rationale that these vectors contain no residual viral genes (thereby minimizing the possibility of triggering anti-vector immunity) and the ability of these viruses to permanently integrate into the genome of cells following transduction. These studies collectively confirmed the superiority of integrating vectors for obtaining durable expression of the transferred coagulation factors, and showed that long-term phenotypic improvement could be achieved in both murine and canine models of hemophilia A and B, following direct injection of retroviral vectors to transduce liver or skeletal muscle. However, these studies also highlighted the therapeutic consequences of one of the limitations of vectors based on murine retroviruses; their inability to transduce quiescent/postmitotic mammalian cells. Since, in a mature animal, skeletal muscle is postmitotic and it is estimated that only 1 in 10,000 to 1 in 20,000 hepatocytes are actively cycling at any given time,[111] it proved very difficult to obtain sufficient levels of gene transfer to these tissues unless the gene delivery was performed during fetal or early neonatal life, or growth factors/injury were used to stimulate hepatocyte division.[109,110,112–117] The development of vectors based upon lentiviruses,[118] which encode a protein that facilitates active transport of the viral genome across the intact nuclear membrane, has overcome this requirement for cell division. Nevertheless, hepatocyte transduction in vivo has still been found to be relatively inefficient unless the hepatocytes are first "primed" by physical or chemical means.[119] In addition, their lack of cell specificity and the possibility that insertional mutagenesis can occur following the semirandom genomic integration of murine retroviral and lentiviral vectors have thus far precluded the clinical translation of gene therapy approaches based upon directly injecting these vectors in vivo.

3.3 Adeno-Associated Virus Vectors

Adeno-associated viruses (AAVs) possess many characteristics that have caused them to attract considerable interest as vectors for in vivo gene delivery/therapy. Their genome is easily manipulated, all viral genes can be deleted and replaced with the desired exogenous genetic material, AAV particles can be purified at very high titers, and they can then be lyophilized for easy handling/storage.[120–122] AAV efficiently transduces both proliferating and quiescent cells, and numerous serotypes exist in nature with differing tropisms, permitting some degree of tissue targeting.[123–126] Also, since AAVs are nonpathogenic, possess relatively low innate immunogenicity, and genomic integration of AAV vectors is rare, the general consensus has been that they may be inherently safer than other viral-based vectors.[127–130] Indeed, from fairly early on, AAV has been viewed as one of the most promising vector systems for treating hemophilia,[19,20,48,131,132] and clinical improvement has been seen in animal models of both hemophilia A and B following administration of AAV vectors encoding the appropriate coagulation factor.[133–139] Although AAV does not integrate into the genome of cells it transduces, by targeting the liver and skeletal muscle, investigators showed that it was possible to achieve stable, long-term expression of AAV-driven factors VIII and IX, and to obtain clinical improvement in both murine and canine models of hemophilia, even when working with mature/adult recipients. These studies also revealed, however, that achieving phenotypic correction in the dog is far more difficult than in mice, underscoring the value of large animal preclinical models for accurately predicting outcome in human patients.

3.4 Hemophilia A in Sheep

We recently used a variety of reproductive technologies to successfully reestablish an extinct line of hemophilia A sheep[140–142] and fully characterized both the clinical parameters and the precise molecular basis for their disease.[143] The affected animals showed extensive subcutaneous and intramuscular hematomas, and spontaneous hemarthroses were also frequent, leading to reduced locomotion and symptoms of pain in standing up, restricting nursing activity. Laboratory tests showed significantly increased partial thromboplastin time (PTT), and FVIII levels (as assessed by chromogenic assay) were undetectable in the plasma, explaining their severe phenotype. In addition, we identified a frameshift-induced premature stop codon as the molecular cause of the disease, just as that occurs in a percentage of human patients with hemophilia A, making this line of sheep unique among animal hemophilia A models, since hemophilia A mice were generated through targeted gene deletion and the hemophilia A dog colonies exhibit a gene inversion. Replacement therapy with human FVIII concentrate or fresh sheep plasma resulted in remission of disease and rapid clinical improvement. As has been seen in the hemophilia dogs, however, these sheep often made inhibitors to human factor VIII. We have now cloned sheep factor VIII and expressed recombinant protein,[144] which will greatly facilitate the maintenance of this line of animals, and allow studies to determine whether these sheep will also develop inhibitors to this same-species factor, an issue that has been stated to be tantamount to deriving clinically predictive data.[145] As will be detailed in the subsequent section, we have begun using this new model to test cell-based gene therapy treatments for hemophilia A.

4. USING CELLS AS VEHICLES TO DELIVER FACTORS VIII AND IX TO TREAT HEMOPHILIA

All of the preclinical studies discussed in the preceding sections utilized an approach in which vectors based upon various viruses were injected directly into the muscle, liver, or circulation of the recipients as a means of delivering an exogenous gene for factor VIII or IX. As these sections highlighted, although this approach can be successful, the in vivo administration of viral-based vectors is not without risks/shortcomings. An alternate approach that has received a great deal of attention over the past two plus decades has been to modify cells in vitro and transplant those cells into animal models of hemophilia A and B. This approach allows one to choose a vector system that is optimized for transducing the specific cell type in question, removes the risks associated with off-target gene transfer in vivo, should eliminate the possibility of triggering an immune response to the viral vector (since the recipient's immune system never actually "sees" the vector itself), and allows one to select and transplant only cells that have been successfully gene-modified and express high levels of the desired coagulation factor.

4.1 Fibroblasts

The first cell type that was explored as a possible vehicle for delivering factor VIII or FIX was the fibroblast, with the thought that dermal-derived fibroblasts could be obtained from the patient by punch biopsy, modified in vitro to express high levels of the needed coagulation factor, and these gene-modified cells could then be expanded and ultimately reimplanted in the skin, to serve as a sort of cellular "mini pump," continually releasing FVIII or FIX into the circulation to maintain hemostasis.[146–152] These studies showed that this approach could lead to detectable circulating levels of human coagulation factors in mice, rabbits, and dogs, and could produce at least partial phenotypic correction of bleeding phenotype, albeit transiently. These fairly preliminary findings were sufficiently promising to prompt the initiation of clinical trials in China testing this approach in human patients with hemophilia B.[153–155] Initial findings in these patients strongly suggested clinical benefit, with respect to both circulating levels of factor IX and clinical presentation/bleeding phenotype. However, no follow-up data were ever published on these patients, making it unclear how safe and effective this approach was beyond the initial period of observation.

4.2 Hematopoietic Stem Cells

Another cell type that has received a great deal of attention as a delivery vehicle for coagulation factor genes is the hematopoietic stem cell (HSC). Several unique properties of the HSC suggest that it may represent an ideal gene therapy target cell.[156–159] Despite being present only at very low frequency within the bone marrow under steady state, HSCs are responsible for producing all of the various blood cells present in the circulation throughout the lifetime of an organism. This is a remarkable feat when one considers that the turnover of cells in the hematopoietic system in a man weighing 70 kg is close to one trillion cells per day.[160] HSC possesses two key properties that are inherent to true stem cells and enable them to accomplish this task. The first of these is the ability to self-renew, or undergo cell division and give rise to progeny that are functionally identical to the original HSC. This property is essential to prevent rapid exhaustion of the HSC pool. The ability of the HSC to self-renew would make it possible for even a few gene-corrected cells to persist throughout the lifetime of a patient, enabling a single successful gene therapy treatment to provide lifelong correction of a genetic disease such as hemophilia. Indeed, the clinical success of bone marrow, cord blood, and mobilized peripheral blood transplantation unequivocally demonstrates the ability of genetically normal HSC to provide long-term correction for a variety of gene defects.[161]

In addition to their ability to self-renew, HSCs are truly multipotential, and can thus give rise to all of the various lineages of blood cells present in the circulation. This property makes it possible to not only achieve genetic correction in a wide variety of hematopoietic lineages following insertion of an exogenous gene into HSC, but also provides a means of delivering secreted gene products into the general circulation of the recipient. The ability to introduce a corrective gene product into the general circulation should make it possible for genetically modified HSC to deliver a coagulation factor throughout the body and thus treat hemophilia.

While HSCs and their myriad hematopoietic progeny do not naturally produce coagulation factors, early studies showed that they could be modified with viral vectors to express significant amounts of biologically active coagulation FVIII, and thus might be good vehicles for delivering factor VIII to treat hemophilia A.[162,163] Interestingly, to date, no studies have been undertaken to ascertain whether HSC may be equally well suited for delivering coagulation factor IX to treat hemophilia B. These early studies establishing the ability of HSC to be modified and generate progeny that produce functional factor VIII have led to intense research over the past several years, to optimize the efficiency with which the HSCs are transduced, the levels of factor VIII expression/secretion following transduction, and the conditioning regimen required to get sufficient levels of HSC engraftment to mediate a therapeutic effect in hemophilia A.[47,49,164–167] These studies have jointly demonstrated that transplanting HSCs that have been engineered with retroviral or lentiviral vectors to express human, mouse, porcine, or engineered high-expression human-porcine hybrid/chimeric factor FVIII molecules can result in sustained phenotypic correction of hemophilia A in mice. Moreover, by including T cell-depleting agents in the conditioning regimen, this approach makes it possible to achieve phenotypic correction of hemophilia A in mice that have high titer inhibitors to factor VIII at the time of treatment,[47] thus overcoming the most significant/serious problem that currently plagues the clinical treatment/management of patients with hemophilia A.[41] Based on these highly encouraging preclinical studies in mice, the use of gene-modified HSC to treat hemophilia A is rapidly advancing toward the clinic.

Another variation on the use of HSC as gene delivery vehicles for treating hemophilia A, which is also receiving a good deal of attention of late, is the modification of HSC with lentiviral vectors that restrict expression of factor VIII to the granules of platelets by placing the vector-encoded factor VIII gene under the transcriptional control of the ITGA2B gene promoter.[53,168–170] Investigators working on this approach hypothesize that storage of vector-encoded factor VIII within platelets may allow the factor VIII to be delivered directly to the site of active bleeds and thereby provide a highly efficient means of maintaining hemostasis in patients with hemophilia A. Moreover, by "hiding" the factor VIII within the platelets, it may be possible to achieve phenotypic correction in the presence of inhibitory antibodies. To date, studies in mice and one study in hemophilia dogs have supported these hypotheses and validated the benefits of this platelet-targeted approach to treating hemophilia A. However, concerns have recently been raised as to whether factor VIII delivered via platelets is biologically/functionally equivalent to that present within the plasma.[171] Further studies are clearly needed to address this clinically important issue.

Despite the immense potential of this approach, the use of HSCs as cellular vehicles for delivering a factor VIII gene is not, however, without shortcomings/risks. The first potential risk comes as a result of the very properties that make HSC an attractive target cell for gene delivery; namely, their massive proliferative capacity and ability to self-renew and persist for the life of the patient. These properties make it necessary to use vectors that integrate into the genome of the HSC, to avoid rapid dilution/loss of the therapeutic gene as the transplanted HSC repopulate the recipient's hematopoietic system. While insertional mutagenesis as a result of retroviral-mediated gene transfer was largely thought to be a theoretical risk, insertional mutagenesis-mediated leukemogenesis occurred in multiple children who received murine retroviral vector-transduced HSC as treatment for X-SCID,[172–174] despite never observing an adverse event of this nature in the decades of animal gene therapy experiments that led up to these clinical trials. Subsequent studies have provided compelling evidence that the specific disease (X-SCID) and the therapeutic γc transgene itself played a key role in the observed leukemogenesis, since overexpressing γc may confer enhanced proliferative potential upon transduced cells.[175] Because factor VIII harbors no oncogenic potential and cannot confer a proliferative/survival advantage on transduced cells, it seems highly unlikely that the use of HSC engineered to express factor FVIII will result in leukemogenesis. The use of lentiviral vectors, which lack the murine retrovirus' predilection for integrating near the transcriptional start sites of cellular genes,[176,177] and the inclusion of a 3'-modified long terminal repeat to produce self-inactivating vectors in which factor VIII is driven by an internal mammalian promoter that lacks an enhancer and cannot transactivate cellular genes,[178] should further safeguard against the risk of insertional mutagenesis.

Another possible concern with the use of gene-modified HSC for the treatment of hemophilia is the need to condition/myeloablate the recipient in order to achieve meaningful levels of engraftment. In contrast to disease like X-SCID, in which the transplanted gene-corrected HSC will have a survival advantage, HSC do not normally synthesize/produce factor VIII. As such, its deficit produces no deleterious effects on HSC, so the transplanted FVIII-expressing HSC will have no advantage over the recipient's endogenous FVIII-deficient HSC. As a result, radiation and/or chemotherapy is needed to knock down the host's endogenous hematopoiesis in order for the transplanted FVIII-expressing HSC to engraft at high enough levels to mediate a therapeutic effect. While this has not proven problematic in preclinical murine models, the phenotype of hemophilia A mice (and the hemophilia A dogs, for that matter) is markedly less severe than that seen in human patients with hemophilia A.[74] It is thus not clear how well tolerated the requisite conditioning regimens will be when applied to the far more fragile hemophilia A patient population. These data will, however, only be available once Phase 1 trials have begun or, could perhaps be obtained in the sheep model, given the greater clinical severity these animals exhibit when compared to the mice and dogs.[143]

4.3 Mesenchymal Stem Cells/Marrow Stromal Cells

More than 30 years after Friedenstein pioneered the concept that the marrow microenvironment resided within the so-called stromal cells of the marrow,[179,180] scientists have finally begun to realize the full potential of these microenvironmental cells, which are alternately referred to as mesenchymal stem cells (MSCs) or marrow stromal cells.[181] MSCs are a key part of the microenvironment/niche that supports HSC and drives the process of hematopoiesis, yet despite serving this vital function, MSC only comprise ~0.001–0.01% of cells within the marrow,[182] making methods for enriching and for expanding these cells essential. The most straightforward method for obtaining MSC is to exploit their propensity to adhere to plastic and their ability to be passaged with trypsin (contaminating hematopoietic cells do not passage with trypsin) to rapidly obtain a relatively morphologically homogeneous population of fibroblastic cells from a bulk mononuclear cell preparation.[183–185] Unfortunately, true MSC (defined by function) accounts for only a small percentage of this "plastic-adherent" population. While no marker has been identified to date that uniquely identifies MSC, we and others have found that by tri-labeling human bone marrow cells with Stro-1,[186] anti-CD45, and anti-GlyA, and selecting the Stro-1+CD45-GlyA cells, it is possible to consistently obtain a homogeneous population that is highly enriched for MSC.[187–193]

We and others have shown that MSCs possess a battery of unique properties which make them ideally suited for cellular therapies and as vehicles for gene delivery,[183,189,192,194–201] including: (1) they exhibit widespread distribution throughout the body[202–209]; (2) they are easy to isolate and have the ability to be extensively expanded in culture without loss of potential; (3) they have the ability to differentiate into a wide array of functional cell types in vitro and in vivo and release trophic factors that trigger the tissue's own endogenous repair pathways[196,210–227]; (4) they exert pronounced anti-inflammatory and immunomodulatory effects upon transplantation[194,195,228–249]; and (5) they have the ability to home to damaged tissues and sites of inflammation/injury following in vivo administration.[250–252]

While MSCs possess tremendous therapeutic potential by virtue of their ability to lodge/engraft within multiple tissues in the body and both give rise to tissue-specific cells and release trophic factors that trigger the tissue's own endogenous repair pathways,[196,210–227] gene therapists have realized that these properties are just the beginning of the therapeutic applications for MSC.[207,253,254] By using gene therapy to engineer MSC to either augment their own natural production of specific desired proteins or to enable them to express proteins outside of their native repertoire, it is possible to greatly broaden the spectrum of diseases for which MSC could provide therapeutic benefit. Unlike HSCs, which are notoriously difficult to modify with most viral vectors while preserving their in vivo potential, MSC can be readily transduced with all of the major clinically prevalent viral vector systems including adenoviruses,[255–257] murine retroviruses,[257–261] lentiviruses,[262–267] and AAV,[268,269] and efficiently produce a wide range of cytoplasmic, membrane-bound, and secreted protein products. This ease of transduction coupled with the ability to subsequently select and expand only the gene-modified cells in vitro to generate adequate numbers for transplantation combine to make MSC one of the most promising stem cell populations for use in gene therapy studies and trials.

Importantly from a safety standpoint, there is no evidence that MSC transforms or progresses to clonal dominance following transduction. This is in marked contrast to what has been observed following genetic modification of HSC.[172,174,270,271] Human MSCs, unlike their murine counterparts, appear to be inherently nontumorigenic.[272] Sheep transplanted in utero with human MSC have been followed for up to 2 years after transplant with no evidence of tumorigenesis. Moreover, even if genomic instability is intentionally induced in vitro, human MSCs undergo terminal differentiation rather than transformation.[273] MSC should thus represent safe cellular vehicles for delivering a therapeutic gene to treat hemophilia.

Looking specifically at gene therapy for hemophilia, in ongoing studies, we have shown that MSCs produce factor VIII endogenously,[274] albeit at relatively low levels. Although these low levels of endogenous production would not likely make MSC therapeutic on their own, it does establish that MSCs (in contrast to many other cell types) possess the necessary cellular machinery to efficiently express/secrete factor VIII. As such, they may be ideally suited as cellular vehicles for delivering a factor VIII transgene, and thereby treating hemophilia A. In support of this contention, we and others have shown that MSC can be modified with viral vectors to secrete high levels of FVIII that is indistinguishable from native FVIII.[275–280]

To begin experimentally testing the ability of MSC to serve as factor VIII delivery vehicles and thus treat hemophilia A, we recently tested a novel, nonablative transplant-based gene therapy in two pediatric hemophilia A lambs.[278] During the first 3–5 months of life, both these animals had received frequent, on-demand infusions of human FVIII for multiple hematomas and chronic, progressive, debilitating hemarthroses of the leg joints which had resulted in severe defects in posture and gait, rendering them nearly immobile. In an ideal situation, one would use autologous cells to deliver an FVIII transgene, and thus avoid any complications due to MHC mismatching. Unfortunately, the severe life-threatening phenotype of the hemophilia A sheep prevented us from collecting bone marrow aspirates to isolate autologous cells. We therefore elected to utilize cells from the ram that had sired the two hemophiliac lambs, hoping that, by using paternal (haploidentical) MSC, immunologic incompatibility between the donor and recipient should be minimized sufficiently to allow engraftment, especially given the large body of evidence now accumulating that MSC can be transplanted across allogeneic barriers without eliciting an immune response.[281,282]

Based on our prior work in the fetal sheep model, we knew that the intraperitoneal (IP) transplantation of MSC results in widespread engraftment throughout all of the major organs[189,283–287] and durable expression of vector-encoded genes.[285,286,288] Importantly, the use of the IP route would avoid the lung-trapping which hinders the efficient trafficking of MSC to desired target organs following intravenous administration, and also poses clinical risks due to emboli formation.[289,290]

Following isolation, MSCs were simultaneously transduced with two lentiviral vectors, the first of which encoded an expression-/secretion-optimized porcine factor VIII (pFVIII) transgene.[291] We selected a pFVIII transgene for two reasons. First, we had not yet cloned the ovine factor VIII cDNA and constructed a B domain-deleted cassette that would fit in a lentiviral vector. Second, the pFVIII transgene had previously been shown, in human cells, to be expressed/secreted at 10–100 times higher levels than human FVIII.[46,47,292] We thus felt that these very high levels of expression/secretion might enable us to achieve a therapeutic benefit, even if we obtained very low levels of engraftment of the transduced paternal MSC. The second lentivector-encoded eGFP facilitates tracking and identification of donor cells in vivo. Once the transduced MSC had been sufficiently expanded, the first animal to be transplanted was treated with human factor VIII to ensure no procedure-related bleeding occurred, sedated and 30×10^6 transduced MSCs were transplanted into the peritoneal cavity under ultrasound guidance in the absence of any preconditioning.

Following transplantation, chromogenic assay revealed no detectable factor VIII activity in the circulation, but this animal's clinical picture improved dramatically. All spontaneous bleeding events ceased, and he enjoyed an event-free clinical course, devoid of spontaneous bleeds, enabling us to cease human factor VIII infusions. Existing hemarthroses resolved, the animal's joints recovered fully and resumed normal appearance, and he regained normal posture and gait, resuming a normal activity level. This study represented the first report of phenotypic correction of severe hemophilia A in a large animal model following transplantation of cells engineered to produce FVIII, and the first time that reversal of chronic debilitating hemarthroses had been achieved.

Based on the remarkable clinical improvement we had achieved in this first animal, we transplanted a second animal with 120×10^6 transduced paternal MSC. In similarity to the first animal, hemarthroses present in this second animal at the time of transplant resolved, and he resumed normal activity shortly after transplantation. This second animal also became factor-independent following the transplant. These results thus confirm the ability of this MSC-based approach to provide phenotypic correction in this large animal model of hemophilia A. However, just as we had observed in the first animal, no factor VIII activity was detectable in the circulation of this animal, making the mechanism by which this procedure mediated such pronounced clinical improvement uncertain.

Despite the pronounced clinical improvement we observed in these animals, both mounted a rapid and fairly robust immune response to factor VIII, in similarity to prior studies performed with hemophilia A mice.[292] Before transplant, these animals had very low (animal 1) and undetectable (animal 2) Bethesda titers against human FVIII, yet this lifesaving procedure resulted in a marked rise in Bethesda titer against the vector-encoded pFVIII, and these antibodies exhibited pronounced cross-reactivity to the human protein as well. This cross-reactivity was surprising, given the well-established ability to successfully use porcine FVIII products in human patients to bypass existing antihuman FVIII inhibitors.[107,293–295]

Following euthanasia of these animals, PCR analysis demonstrated readily detectable levels of MSC engraftment in nearly all tissues analyzed, including liver, lymph nodes, intestine, lung, kidney, omentum, and thymus, thereby proving it is possible to achieve widespread durable engraftment of MSC following transplantation in a postnatal setting in a large animal model without the need for preconditioning/ablation, and in the absence of any selective advantage for the donor cells.

Confocal immunofluorescence analysis revealed large numbers of FVIII-expressing MSC within the synovium of the joints which exhibited hemarthrosis at the time of transplant, demonstrating that transplanted MSCs possess the intrinsic ability to home and persist within sites of ongoing injury/inflammation in hemophiliac recipients, and then release FVIII locally within the joint. This finding thus provides an explanation for the dramatic improvement we observed in the animals' joints. This marked improvement in the joints without detectable circulating factor also agrees with prior studies,[296] showing that local delivery of FIX-AAV to the joints of mice with injury-induced hemarthroses (hemophilia A and B mice, unlike human patients, do not exhibit spontaneous bleeding) led to resolution of the hemarthroses in the absence of any detectable circulating FIX. While this finding provides an explanation for the reversal of the joint pathology present in these animals at transplant, it does not explain the observed systemic benefits, such as the cessation of spontaneous bleeding events.

Confocal analysis also revealed engrafted cells within the small intestine, demonstrating that MSC can still engraft within the intestine following postnatal transplantation, just as had observed in prior studies in fetal recipients.[285] Given the ease with which proteins secreted from cells within the intestine can enter the circulation, future studies aimed at improving the levels of engraftment within the intestine have the potential to greatly improve the systemic release of FVIII.

The marked phenotypic improvement and improvement in quality of life observed in the sheep model thus support the further development of therapeutic strategies for hemophilia A and, perhaps, other coagulation disorders, employing MSC as cellular vehicles to deliver the required transgene.

5. HUMAN CLINICAL GENE THERAPY TRIALS FOR HEMOPHILIA

Based on the promise of lifelong cure following a single treatment, and the encouraging preclinical results in murine and canine models, several clinical gene therapy trials in both hemophilia A and B have been undertaken in the US. In each case, the treatment was well tolerated with the vectors and doses used. Although preliminary data from these studies[19,20,48,132,145,297–312] showing gene transfer, and persistence of both the transgene and the expression of the vector-encoded coagulation factor for up to 10 years post-gene transfer[306] have been encouraging, the plasma levels of factor VIII and factor IX achieved in these trials have been insufficient to free the patients from the need of exogenous factor. In addition, expression of the factors VIII and IX transgenes was often transient, and none of the trials exhibited a clear relationship between vector dose and resultant circulating factor levels. Thus, despite the decades of promising results using gene therapy to correct hemophilia A in animal models, to date, no clinical gene therapy trial has shown phenotypic/clinical improvement of hemophilia A in human patients. Based on the disappointing results to date, there are currently no active hemophilia A clinical gene therapy trials, even though hemophilia A accounts for roughly 80% of all cases of hemophilia. These clinical trials also raised the troubling possibility of inadvertent germ line alteration when semen samples were found to be transiently positive for vector DNA.[313,314] This is in contrast to prior studies conducted in experimental animals,[315,316] reinforcing the importance of the choice of animal model system employed when conducting preclinical studies, if one wishes to extrapolate results obtained in their model to what would likely happen in the clinic.

Despite these rather discouraging initial results from clinical trials exploring gene therapy for hemophilia, recently, a highly successful ongoing trial being conducted jointly by St. Jude Children's Research Hospital and University College London has finally highlighted the tremendous potential of gene therapy for the treatment of human hemophilia B.[132,317,318] This trial has shown it is possible, by carefully selecting a serotype of AAV that exhibits a natural tropism for the liver, to attain expression of therapeutic levels of factor IX for over 3 years, to date, in six adults with severe hemophilia B, following a single systemic dose of a factor IX-encoding AAV-based vector. While not tremendously high, the levels of circulating factor IX achieved with this approach have enabled four of these six subjects to discontinue routine factor prophylaxis. While the other two patients have not achieved complete factor independence, this gene therapy-based treatment has allowed them to significantly reduce the frequency with which they need to administer their prophylactic infusions.[132] Moreover, by choosing a hepatotropic serotype of AAV, it appears it may be possible to avoid induction of antibodies to exogenous factor IX delivered by gene therapy, which is in agreement with a large body of preclinical evidence that limiting expression of transgenes to hepatic cells can result in induction of immune tolerance.[137,319–336] However, this critical point is not yet entirely clear, as several of the patients developed elevated liver enzymes at roughly 2 months post-vector infusion, and in one patient, this led to a rapid decrease in circulating factor IX levels, which the investigators concluded was due to a cellular immune response against the transduced hepatocytes.[337] Despite these issues, however, these results represent a leap forward in the treatment/management of hemophilia B, and the gene therapy community will eagerly await news on whether these groups succeed with their plans to adapt this strategy to the treatment of hemophilia A.[338]

6. FUTURE DIRECTIONS IN GENE THERAPY FOR HEMOPHILIA

As discussed earlier, one of the major hurdles that have plagued the successful phenotypic correction of the hemophilias by factor replacement therapy is inhibitory antibody formation, which occurs in at least 30% of patients with severe hemophilia A. The formation of these inhibitors greatly reduces the efficacy of subsequent FVIII infusions, and can ultimately lead to treatment failure, placing the patient at risk of life-threatening hemorrhage. Not surprisingly, inhibitor formation has also observed when gene therapy has been used in an attempt to treat hemophilia A and B preclinically,[83,93,110,311,339] since the recipient's immune system still "sees" the vector-encoded factor VIII or IX as a foreign protein. Given that patients with hemophilia A are more than 10 times as likely to develop inhibitory antibodies to the exogenous coagulation factor than patients with hemophilia B,[340,341] it appears that factor VIII may represent a far more immunogenic protein than factor IX. As such, it is not yet clear whether the use of a liver-tropic AAV vector (as in the ongoing successful hemophilia B trial) will be sufficient to avoid inhibitor induction in the context of hemophilia A. Similarly, strategies using cells as vehicles to deliver factor VIII, rather than directly administering the vector, will be faced with this same immune hurdle. Moreover, in the case of cell-based treatments, adequate numbers of factor VIII-expressing cells will need to be generated, and some conditioning or cytoreduction/ablation may be necessary to ensure adequate levels of engraftment of these factor VIII-expressing cells are obtained after infusion.

Importantly, many of the hurdles that have thus far prevented gene therapy from curing patients with hemophilia, or many other genetic diseases that have been investigated, could likely be circumvented by performing gene therapy prior to birth. Despite still being viewed as "experimental," in utero stem cell transplantation (IUTx) has been performed in 46 human patients, for 14 different genetic disorders,[342–345] with many patients receiving three injections spaced 1 week apart, proving that accessing the early human fetus multiple times, using a minimally invasive, ultrasound-guided approach, poses minimal risk.[342–349]

For individuals with a family history of hemophilia A (~75% of cases), prenatal diagnosis is feasible, available, and is encouraged in most Western and developing countries,[350–360] giving the opportunity to intervene therapeutically during the fetal period. Digital PCR has now made it possible to analyze free fetal DNA present within maternal plasma, and thereby diagnose hemophilia A in utero noninvasively.[359]

Interestingly, although ~75% of hemophilia A patients are born into families with a history of hemophilia, and prenatal screening is available and encouraged in these families, in utero attempts to treat this disease have thus far not been explored (with the exception of a 1 patient pilot study (J.L. Touraine, personal communication and Ref. 361)). While concerns could be raised over potential bleeding risks arising from in utero intervention in the context of hemophilia, during early fetal life, activation of factor X occurs predominantly via tissue factor activity, making it largely independent of the factor IXa/factor VIIIa phospholipid complex.[362] As a result, the fetus develops without hemorrhage, despite having little or no expression of factors VIII and IX.[362–364] The unique hemostasis of the fetus should thus allow IUTx to be performed safely for hemophilia, as was demonstrated when IUTx was successfully performed on a human fetus with severe hemophilia A without causing bleeding.[361]

While only one hemophilia A patient has thus far been treated by IUTx, this patient has thus far exhibited reduced severity of disease compared to his siblings, even though this study relied on the ability of unpurified fetal liver cells to endogenously produce sufficient levels of factor VIII, after transplant, to mediate correction. The only partial correction observed in this patient supports the approach of genetically modifying the donor cells, to ensure they express high levels of factor VIII upon engraftment, and can thus mediate the required therapeutic effect.

One of the greatest advantages of performing IUTx for hemophilia A is that it could induce stable tolerance to factor VIII, thus allowing subsequent postnatal treatment (either factor replacement or gene therapy) without the risk of inhibitors. We previously showed that performing IUTx or in utero gene therapy takes advantage of multiple tolerogenic avenues present early in gestation and induces durable tolerance to both the transplanted donor cells and the vector-encoded proteins.[365,366] In addition, others have proven that tolerance can be induced to coagulation factors (factors VIII and IX) by early in utero gene therapy.[367–369] Human clinical trials have also shown that antigen-/donor-specific immune tolerance is induced following IUTx.[348,370–373] Given these unique immunological advantages presented by the early fetus, one can envision that in utero gene therapy would be an ideal approach for treating hemophilia A, since lifelong tolerance could be induced to FVIII. This would thus ensure that, even if in utero gene therapy was not curative, postnatal gene therapy or protein replacement could proceed safely without the risk of inhibitor formation. Indeed, the one hemophilia A patient who received IUTx has not developed inhibitors with factor VIII treatment, in contrast to all of his affected siblings (J.L. Touraine, personal communication and Ref. 361), validating IUTx's ability to induce tolerance to factor VIII.

Nevertheless, despite the availability of prenatal screening and the medical ability to intervene therapeutically prior to birth, roughly 1 in 5000 boys born each year worldwide are affected with hemophilia A.[374] In the US alone, correcting this disease prior to birth could thus benefit the ~240 patients/year born into families with a history of hemophilia A. In utero treatment could promise the birth of a healthy infant who required no further treatments, removing the heavy physical, psychological, and monetary burden on the patients, their families, and the health-care system. The current estimate for the lifetime cost of prophylactic treatment for one patient with hemophilia A is $20 million. Curative in utero treatment would thus save roughly $48 billion over the lifetime of the hemophilia A patients born just this year in the US.

Given the numerous therapeutic and socioeconomic advantages of treating hemophilia prior to birth, we envision fetal intervention receiving a great deal more attention in the near future to treat hemophilia. Indeed, when a panel of stem cell, gene therapy, and fetal surgery experts from around the world recently met at UCSF in a CIRM-sponsored conference, they reached a consensus that sufficient animal data exist to begin moving forward with the use of gene-modified cells to treat diseases prior to birth, and identified hemophilia as one of the prime target diseases.[375]

7. CONCLUSIONS

In conclusion, gene therapy offers many advantages over existing treatment options for genetic diseases, both from a scientific standpoint and from a socioeconomic point of view, since it is one of the only therapies that could promise a complete cure following a single treatment. Among the diseases being considered as candidates, hemophilia has many unique characteristics that make it ideally suited for correction by gene therapy. Decades of preclinical studies in hemophiliac mice, dogs, and more recently in hemophilia sheep, have provided compelling evidence that a variety of vectors and delivery strategies can result in pronounced phenotypic/clinical improvement in these animal models. However, translating these preclinical findings into success in human patients has been difficult. The recent success in an ongoing clinical trial for hemophilia B has highlighted the tremendous therapeutic potential of gene therapy, but has also identified additional hurdles that will need to be overcome to make gene therapy fully curative for hemophilia B and its more prevalent and technically more challenging sibling, hemophilia A. Many of these hurdles could likely be overcome by intervening early and initiating treatment prior to birth. Performing gene therapy before birth should also induce immune tolerance to factors VIII and IX. As such, even if noncurative, this approach would allow patients to receive postnatal gene therapy or factor replacement without the risk of inhibitor formation.

REFERENCES

1. Nathwani AC, Benjamin R, Nienhuis AW, Davidoff AM. Current status and prospects for gene therapy. *Vox Sang* 2004;**87**(2):73–81. Epub 2004/09/10. PubMed PMID: 15355497.
2. Nathwani AC, Nienhuis AW, Davidoff AM. Current status of gene therapy for hemophilia. *Curr Hematol Rep* 2003;**2**(4):319–27. Epub 2003/08/07. PubMed PMID: 12901329.
3. Podsakoff GM, Engel BC, Kohn DB. Perspectives on gene therapy for immune deficiencies. *Biol Blood Marrow Transplant* 2005;**11**(12):972–6. Epub 2005/12/13. PubMed PMID: 16338618.
4. High KA. Gene therapy for hemophilia: the clot thickens. *Hum Gene Ther* 2014;**25**(11):915–22. PubMed PMID: 25397928 PMCID: PMC4236063.
5. Hays J, Mitchel J. *History of medicine: hemophilia*. New York: Target Health, Inc. Available from: http://blog.targethealth.com/?p=28118; [cited 26.12.14].
6. Aronova-Tiuntseva Y, Herreid CF. *Hemophilia "The royal disease"*. 2003. Available from: http://sciencecases.lib.buffalo.edu/cs/files/hemo.pdf. [cited 25.12.14].
7. Potts DM, Potts WTW. *Queen Victoria's gene: haemophilia and the royal family*. (Stroud, UK): Sutton Publishing; 1999.
8. Rogaev EI, Grigorenko AP, Faskhutdinova G, Kittler EL, Moliaka YK. Genotype analysis identifies the cause of the "royal disease". *Science* 2009;**326**(5954):817. PubMed PMID: 19815722.
9. Massie RK. *The Romanovs: the final chapter*. New York (NY): Random House; 1996.
10. Bennett CM, Boye E, Neufeld EJ. Female monozygotic twins discordant for hemophilia A due to nonrandom X-chromosome inactivation. *Am J Hematol* 2008;**83**(10):778–80. PubMed PMID: 18645989.
11. Di Michele DM, Gibb C, Lefkowitz JM, Ni Q, Gerber LM, Ganguly A. Severe and moderate haemophilia A and B in US females. *Haemophilia* 2014;**20**(2):e136–43. PubMed PMID: 24533955.
12. Esquilin JM, Takemoto CM, Green NS. Female factor IX deficiency due to maternally inherited X-inactivation. *Clin Genet* 2012;**82**(6):583–6. PubMed PMID: 22233509.
13. Ingerslev J, Schwartz M, Lamm LU, Kruse TA, Bukh A, Stenbjerg S. Female haemophilia A in a family with seeming extreme bidirectional lyonization tendency: abnormal premature X-chromosome inactivation?. *Clin Genet* 1989;**35**(1):41–8. PubMed PMID: 2564325.
14. Miyawaki Y, Suzuki A, Fujimori Y, Takagi A, Murate T, Suzuki N, et al. Severe hemophilia A in a Japanese female caused by an F8-intron 22 inversion associated with skewed X chromosome inactivation. *Int J Hematol* 2010;**92**(2):405–8. PubMed PMID: 20700669.
15. Pavlova A, Brondke H, Musebeck J, Pollmann H, Srivastava A, Oldenburg J. Molecular mechanisms underlying hemophilia A phenotype in seven females. *J Thromb Haemost* 2009;**7**(6):976–82. PubMed PMID: 19302446.
16. Seeler RA, Vnencak-Jones CL, Bassett LM, Gilbert JB, Michaelis RC. Severe haemophilia A in a female: a compound heterozygote with nonrandom X-inactivation. *Haemophilia* 1999;**5**(6):445–9. PubMed PMID: 10583534.
17. Wang X, Lu Y, Ding Q, Dai J, Xi X, Wang H. Haemophilia A in two unrelated females due to F8 gene inversions combined with skewed inactivation of X chromosome. *Thromb Haemost* 2009;**101**(4):775–8. PubMed PMID: 19350126.
18. Agaliotis D. *Hemophilia, overview*. 2006.
19. High KA. Gene transfer as an approach to treating hemophilia. *Semin Thromb Hemost* 2003;**29**(1):107–20. PubMed PMID: 12640573.
20. Kay MA, High K. Gene therapy for the hemophilias. *Proc Natl Acad Sci USA* 1999;**96**(18):9973–5. PubMed PMID: 10468539.
21. Pool JG. Cryoprecipitated factor VIII concentrate. *Thromb Diath Haemorrh Suppl* 1969;**35**:35–40.
22. Osterud B, Flengsrud R. Purification and some characteristics of the coagulation factor IX from human plasma. *Biochem J* 1975;**145**(3):469–74. PubMed PMID: 1171684 PMCID: PMC1165246.
23. Wagner BH, McLester WD, Smith M, Brinkhous KM. Purification of antihemophilic factor (factor Viii) by amino acid precipitation. *Thromb Diath Haemorrh* 1964;**11**:64–74. PubMed PMID: 14167116.
24. Ragni MV, Tegtmeier GE, Levy JA, Kaminsky LS, Lewis JH, Spero JA, et al. AIDS retrovirus antibodies in hemophiliacs treated with factor VIII or factor IX concentrates, cryoprecipitate, or fresh frozen plasma: prevalence, seroconversion rate, and clinical correlations. *Blood* 1986;**67**(3):592–5. PubMed PMID: 3081062.
25. Ragni MV, Winkelstein A, Kingsley L, Spero JA, Lewis JH. 1986 update of HIV seroprevalence, seroconversion, AIDS incidence, and immunologic correlates of HIV infection in patients with hemophilia A and B. *Blood* 1987;**70**(3):786–90. PubMed PMID: 2887224.
26. Norkrans G, Widell A, Teger-Nilsson AC, Kjellman H, Frosner G, Iwarson S. Acute hepatitis non-A, non-B following administration of factor VIII concentrates. *Vox Sang* 1981;**41**(3):129–33. PubMed PMID: 6800131.
27. Schimpf K. Post-transfusion hepatitis and its sequelae in the treatment of hemophilia. *Behring Inst Mitt* 1983;**73**:111–7. PubMed PMID: 6433878.
28. Ways to reduce the risk of transmission of viral infections by plasma and plasma products. A comparison of methods, their advantages and disadvantages. *Vox Sang* 1988;**54**(4):228–45. PubMed PMID: 3388821.

29. Cuthbert RJ, Ludlam CA, Brookes E, McClelland DB. Efficacy of heat treatment of factor VIII concentrate. *Vox Sang* 1988;**54**(4):199–200. PubMed PMID: 3133880.

30. Griffith M. Ultrapure plasma factor VIII produced by anti-F VIII c immunoaffinity chromatography and solvent/detergent viral inactivation. Characterization of the Method M process and Hemofil M antihemophilic factor (human). *Ann Hematol* 1991;**63**(3):131–7. PubMed PMID: 1932287.

31. Mannucci PM. Clinical evaluation of viral safety of coagulation factor VIII and IX concentrates. *Vox Sang* 1993;**64**(4):197–203. PubMed PMID: 8517048.

32. McGrath KM, Thomas KB, Herrington RW, Turner PJ, Taylor L, Ekert H, et al. Use of heat-treated clotting-factor concentrates in patients with haemophilia and a high exposure to HTLV-III. *Med J Aust* 1985;**143**(1):11–3. PubMed PMID: 2989666.

33. Ramsey RB, Evatt BL, McDougal JS, Feorino P, Jackson D. Antibody to human immunodeficiency virus in factor-deficient plasma. Effect of heat treatment on lyophilized plasma. *Am J Clin Pathol* 1987;**87**(2):263–6. PubMed PMID: 3643746.

34. Choo KH, Gould KG, Rees DJ, Brownlee GG. Molecular cloning of the gene for human anti-haemophilic factor IX. *Nature* 1982;**299**(5879):178–80. PubMed PMID: 6287289.

35. Gitschier J, Wood WI, Goralka TM, Wion KL, Chen EY, Eaton DH, et al. Characterization of the human factor VIII gene. *Nature* 1984;**312**(5992):326–30. PubMed PMID: 6438525.

36. Kurachi K, Davie EW. Isolation and characterization of a cDNA coding for human factor IX. *Proc Natl Acad Sci USA* 1982;**79**(21):6461–4. PubMed PMID: 6959130 PMCID: PMC347146.

37. Anson DS, Austen DE, Brownlee GG. Expression of active human clotting factor IX from recombinant DNA clones in mammalian cells. *Nature* 1985;**315**(6021):683–5. PubMed PMID: 2989700.

38. Wood WI, Capon DJ, Simonsen CC, Eaton DL, Gitschier J, Keyt B, et al. Expression of active human factor VIII from recombinant DNA clones. *Nature* 1984;**312**(5992):330–7. PubMed PMID: 6438526.

39. Kaveri SV, Dasgupta S, Andre S, Navarrete AM, Repesse Y, Wootla B, et al. Factor VIII inhibitors: role of von Willebrand factor on the uptake of factor VIII by dendritic cells. *Haemophilia* 2007;**13**(Suppl. 5):61–4. Epub 2008/01/31. PubMed PMID: 18078399.

40. Green D. Factor VIII inhibitors: a 50-year perspective. *Haemophilia* 2011;**17**(6):831–8. Epub 2011/05/20. PubMed PMID: 21592257.

41. Kempton CL, Meeks SL. Toward optimal therapy for inhibitors in hemophilia. *Blood* 2014;**124**(23):3365–72. PubMed PMID: 25428222.

42. Ananyeva NM, Lacroix-Desmazes S, Hauser CA, Shima M, Ovanesov MV, Khrenov AV, et al. Inhibitors in hemophilia A: mechanisms of inhibition, management and perspectives. *Blood Coagul Fibrinolysis* 2004;**15**(2):109–24. PubMed PMID: 15090997.

43. Lusher JM. Is the incidence and prevalence of inhibitors greater with recombinant products? *No J Thromb Haemost* 2004;**2**(6):863–5. PubMed PMID: 15140116.

44. Tellez J, Finn JD, Tschernia N, Almeida-Porada G, Arruda VR, Porada CD. Sheep harbor naturally-occurring antibodies to human AAV: a new large animal model for AAV immunology. *Mol Ther* 2010;**18**(Suppl. 1):S213.

45. Arruda VR. Toward gene therapy for hemophilia A with novel adenoviral vectors: successes and limitations in canine models. *J Thromb Haemost* 2006;**4**(6):1215–7. Epub 2006/05/19. PubMed PMID: 16706962.

46. Doering CB, Denning G, Dooriss K, Gangadharan B, Johnston JM, Kerstann KW, et al. Directed engineering of a high-expression chimeric transgene as a strategy for gene therapy of hemophilia A. *Mol Ther* 2009;**17**(7):1145–54. Epub 2009/03/05. PubMed PMID: 19259064.

47. Doering CB, Gangadharan B, Dukart HZ, Spencer HT. Hematopoietic stem cells encoding porcine factor VIII induce pro-coagulant activity in hemophilia A mice with pre-existing factor VIII immunity. *Mol Ther* 2007;**15**(6):1093–9. PubMed PMID: 17387335.

48. High KA. Gene therapy for haemophilia: a long and winding road. *J Thromb Haemost* 2011;**9**(Suppl. 1):2–11. Epub 2011/08/04. PubMed PMID: 21781236.

49. Ide LM, Gangadharan B, Chiang KY, Doering CB, Spencer HT. Hematopoietic stem-cell gene therapy of hemophilia A incorporating a porcine factor VIII transgene and nonmyeloablative conditioning regimens. *Blood* 2007;**110**(8):2855–63. PubMed PMID: 17569821.

50. Lipshutz GS, Sarkar R, Flebbe-Rehwaldt L, Kazazian H, Gaensler KM. Short-term correction of factor VIII deficiency in a murine model of hemophilia A after delivery of adenovirus murine factor VIII in utero. *Proc Natl Acad Sci USA* 1999;**96**(23):13324–9. PubMed PMID: 10557319.

51. Nichols TC, Dillow AM, Franck HW, Merricks EP, Raymer RA, Bellinger DA, et al. Protein replacement therapy and gene transfer in canine models of hemophilia A, hemophilia B, Von Willebrand disease, and factor VII deficiency. *ILAR J* 2009;**50**(2):144–67. Epub 2009/03/19. PubMed PMID: 19293459.

52. Ponder KP. Gene therapy for hemophilia. *Curr Opin Hematol* 2006;**13**(5):301–7. PubMed PMID: 16888433.

53. Shi Q, Fahs SA, Wilcox DA, Kuether EL, Morateck PA, Mareno N, et al. Syngeneic transplantation of hematopoietic stem cells that are genetically modified to express factor VIII in platelets restores hemostasis to hemophilia A mice with preexisting FVIII immunity. *Blood* 2008;**112**(7):2713–21. Epub 2008/05/23. PubMed PMID: 18495954.

54. Hollestelle MJ, Thinnes T, Crain K, Stiko A, Kruijt JK, van Berkel TJ, et al. Tissue distribution of factor VIII gene expression in vivo – a closer look. *Thromb Haemost* 2001;**86**(3):855–61. PubMed PMID: 11583319.

55. Turecek PL, Gritsch H, Richter G, Auer W, Pichler L, Schwarz HP. Assessment of bleeding for the evaluation of therapeutic preparations in small animal models of antibody-induced hemophilia and Von Willebrand disease. *Thromb Haemost* 1997;**77**(3):591–9. PubMed PMID: 9066015.

56. Hough C, Kamisue S, Cameron C, Notley C, Tinlin S, Giles A, et al. Aberrant splicing and premature termination of transcription of the FVIII gene as a cause of severe canine hemophilia A: similarities with the intron 22 inversion mutation in human hemophilia. *Thromb Haemost* 2002;**87**(4):659–65. PubMed PMID: 12008949.

57. Lozier JN, Dutra A, Pak E, Zhou N, Zheng Z, Nichols TC, et al. The Chapel Hill hemophilia A dog colony exhibits a factor VIII gene inversion. *Proc Natl Acad Sci USA* 2002;**99**(20):12991–6. PubMed PMID: 12242334.

58. Campbell KL, Greene CE, Dodds WJ. Factor IX deficiency (hemophilia B) in a Scottish terrier. *J Am Veterinary Med Assoc* 1983;**182**(2):170–1. PubMed PMID: 6826437.

59. Fogh JM, Nygaard L, Andresen E, Nilsson IM. Hemophilia in dogs, with special reference to hemophilia A among German shepherd dogs in Denmark. I: pathophysiology, laboratory tests and genetics. *Nord Veterinaermed* 1984;**36**(7–8):235–40. PubMed PMID: 6436787.

60. Littlewood JD, Matic SE, Smith N. Factor IX deficiency (haemophilia B, Christmas disease) in a crossbred dog. *Vet Rec* 1986;**118**(14):400–1. PubMed PMID: 3716095.

61. Mustard JF, Rowsell HC, Robinson GA, Hoeksema TD, Downie HG. Canine haemophilia B (Christmas disease). *Br J Haematol* 1960;**6**:259–66. PubMed PMID: 13727144.

62. Peterson ME, Dodds WJ. Factor IX deficiency in an Alaskan Malamute. *J Am Vet Med Assoc* 1979;**174**(12):1326–7. PubMed PMID: 511734.

63. Rowsell HC, Downie HG, Mustard JF, Leeson JE, Archibald JA. A disorder resembling hemophilia B (Christmas disease) in dogs. *J Am Vet Med Assoc* 1960;**137**:247–50. PubMed PMID: 14439728.

64. Sherding RG, DiBartola SP. Hemophilia B (factor IX deficiency) in an Old English Sheepdog. *J Am Vet Med Assoc* 1980;**176**(2):141–2. PubMed PMID: 7353989.

65. Slappendel RJ. Hemophilia A and hemophilia B in a family of French bulldogs. *Tijdschr Diergeneeskd* 1975;**100**(20):1075–88. PubMed PMID: 1209580.

66. Verlander JW, Gorman NT, Dodds WJ. Factor IX deficiency (hemophilia B) in a litter of Labrador retrievers. *J Am Vet Med Assoc* 1984;**185**(1):83–4. PubMed PMID: 6746380.

67. Brinkhous KM, Graham JB. Hemophilia in the female dog. *Science* 1950;**111**(2896):723–4. PubMed PMID: 15431070.

68. Brock WE, Buckner RG, Hampton JW, Bird RM, Wulz CE. Canine hemophilia. Establishment of a new colony. *Arch Pathol* 1963;**76**:464–9. PubMed PMID: 14054168.

69. Didisheim P, Bunting DL. Canine hemophilia. *Thromb Diath Haemorrh* 1964;**12**:377–81. PubMed PMID: 14254696.

70. Graham JB, Buckwalter JA, et al. Canine hemophilia; observations on the course, the clotting anomaly, and the effect of blood transfusions. *J Exp Med* 1949;**90**(2):97–111. PubMed PMID: 18136192 PMCID: PMC2135899.

71. Parks BJ, Brinkhous KM, Harris PF, Penick GD. Laboratory detection of female carriers of canine hemophilia. *Thromb Diath Haemorrh* 1964;**12**:368–76. PubMed PMID: 14254695.

72. Pick JR, Goyer RA, Graham JB, Renwick JH. Subluxation of the carpus in dogs. An X chromosomal defect closely linked with the locus for hemophilia A. *Lab Invest* 1967;**17**(3):243–8. PubMed PMID: 6069272.

73. Wagner RH, Langdell RD, Richardson BA, Farrell RA, Brinkhous KM. Antihemophilic factor (AHF): plasma levels after administration of AHF preparations to hemophilic dogs. *Proc Soc Exp Biol Med* 1957;**96**(1):152–5. PubMed PMID: 13485042.

74. Sabatino DE, Nichols TC, Merricks E, Bellinger DA, Herzog RW, Monahan PE. Animal models of hemophilia. *Prog Mol Biol Transl Sci* 2012;**105**:151–209. Epub 2011/12/06. PubMed PMID: 22137432 PMCID: PMC3713797.

75. Bi L, Lawler AM, Antonarakis SE, High KA, Gearhart JD, Kazazian Jr HH. Targeted disruption of the mouse factor VIII gene produces a model of haemophilia A. *Nat Genet* 1995;**10**(1):119–21. PubMed PMID: 7647782.

76. Bi L, Sarkar R, Naas T, Lawler AM, Pain J, Shumaker SL, et al. Further characterization of factor VIII-deficient mice created by gene targeting: RNA and protein studies. *Blood* 1996;**88**(9):3446–50. PubMed PMID: 8896409.

77. Kundu RK, Sangiorgi F, Wu LY, Kurachi K, Anderson WF, Maxson R, et al. Targeted inactivation of the coagulation factor IX gene causes hemophilia B in mice. *Blood* 1998;**92**(1):168–74. PubMed PMID: 9639513.

78. Lin HF, Maeda N, Smithies O, Straight DL, Stafford DW. A coagulation factor IX-deficient mouse model for human hemophilia B. *Blood* 1997;**90**(10):3962–6. PubMed PMID: 9354664.

79. Muchitsch EM, Turecek PL, Zimmermann K, Pichler L, Auer W, Richter G, et al. Phenotypic expression of murine hemophilia. *Thromb Haemost* 1999;**82**(4):1371–3. PubMed PMID: 10544939.

80. Wang L, Zoppe M, Hackeng TM, Griffin JH, Lee KF, Verma IM. A factor IX-deficient mouse model for hemophilia B gene therapy. *Proc Natl Acad Sci USA* 1997;**94**(21):11563–6. PubMed PMID: 9326649 PMCID: PMC23538.

81. Connelly S, Gardner JM, Lyons RM, McClelland A, Kaleko M. Sustained expression of therapeutic levels of human factor VIII in mice. *Blood* 1996;**87**(11):4671–7. PubMed PMID: 8639836.

82. Connelly S, Gardner JM, McClelland A, Kaleko M. High-level tissue-specific expression of functional human factor VIII in mice. *Hum Gene Ther* 1996;**7**(2):183–95. PubMed PMID: 8788169.

83. Connelly S, Mount J, Mauser A, Gardner JM, Kaleko M, McClelland A, et al. Complete short-term correction of canine hemophilia A by in vivo gene therapy. *Blood* 1996;**88**(10):3846–53. PubMed PMID: 8916949.

84. Fang B, Eisensmith RC, Wang H, Kay MA, Cross RE, Landen CN, et al. Gene therapy for hemophilia B: host immunosuppression prolongs the therapeutic effect of adenovirus-mediated factor IX expression. *Hum Gene Ther* 1995;**6**(8):1039–44. PubMed PMID: 7578416.

85. Fang B, Wang H, Gordon G, Bellinger DA, Read MS, Brinkhous KM, et al. Lack of persistence of E1- recombinant adenoviral vectors containing a temperature-sensitive E2A mutation in immunocompetent mice and hemophilia B dogs. *Gene Ther* 1996;**3**(3):217–22. PubMed PMID: 8646552.

86. Kay MA, Landen CN, Rothenberg SR, Taylor LA, Leland F, Wiehle S, et al. In vivo hepatic gene therapy: complete albeit transient correction of factor IX deficiency in hemophilia B dogs. *Proc Natl Acad Sci USA* 1994;**91**(6):2353–7. PubMed PMID: 8134398.

87. Lozier JN, Thompson AR, Hu PC, Read M, Brinkhous KM, High KA, et al. Efficient transfection of primary cells in a canine hemophilia B model using adenovirus-polylysine-DNA complexes. *Hum Gene Ther* 1994;**5**(3):313–22. PubMed PMID: 8018746.

88. Smith TA, Mehaffey MG, Kayda DB, Saunders JM, Yei S, Trapnell BC, et al. Adenovirus mediated expression of therapeutic plasma levels of human factor IX in mice. *Nat Genet* 1993;**5**(4):397–402. PubMed PMID: 8298650.

89. Yao SN, Farjo A, Roessler BJ, Davidson BL, Kurachi K. Adenovirus-mediated transfer of human factor IX gene in immunodeficient and normal mice: evidence for prolonged stability and activity of the transgene in liver. *Viral Immunol* 1996;**9**(3):141–53. PubMed PMID: 8890472.

90. Gallo-Penn AM, Shirley PS, Andrews JL, Kayda DB, Pinkstaff AM, Kaloss M, et al. In vivo evaluation of an adenoviral vector encoding canine factor VIII: high-level, sustained expression in hemophiliac mice. *Hum Gene Ther* 1999;**10**(11):1791–802. PubMed PMID: 10446919.

91. Reddy PS, Sakhuja K, Ganesh S, Yang L, Kayda D, Brann T, et al. Sustained human factor VIII expression in hemophilia A mice following systemic delivery of a gutless adenoviral vector. *Mol Ther* 2002;**5**(1):63–73. PubMed PMID: 11786047.

92. Brunetti-Pierri N, Nichols TC, McCorquodale S, Merricks E, Palmer DJ, Beaudet AL, et al. Sustained phenotypic correction of canine hemophilia B after systemic administration of helper-dependent adenoviral vector. *Hum Gene Ther* 2005;**16**(7):811–20. Epub 2005/07/08. PubMed PMID: 16000063.

93. Gallo-Penn AM, Shirley PS, Andrews JL, Tinlin S, Webster S, Cameron C, et al. Systemic delivery of an adenoviral vector encoding canine factor VIII results in short-term phenotypic correction, inhibitor development, and biphasic liver toxicity in hemophilia A dogs. *Blood* 2001;**97**(1):107–13. PubMed PMID: 11133749.

94. Carmen IH. A death in the laboratory: the politics of the Gelsinger aftermath. *Mol Ther* 2001;**3**(4):425–8. PubMed PMID: 11319902.

95. DeMatteo RP, Chu G, Ahn M, Chang E, Burke C, Raper SE, et al. Immunologic barriers to hepatic adenoviral gene therapy for transplantation. *Transplantation* 1997;**63**(2):315–9. PubMed PMID: 9020337.

96. DeMatteo RP, Markmann JF, Kozarsky KF, Barker CF, Raper SE. Prolongation of adenoviral transgene expression in mouse liver by T lymphocyte subset depletion. *Gene Ther* 1996;**3**(1):4–12. PubMed PMID: 8929906.

97. DeMatteo RP, Raper SE, Ahn M, Fisher KJ, Burke C, Radu A, et al. Gene transfer to the thymus. A means of abrogating the immune response to recombinant adenovirus. *Ann Surg* 1995;**222**(3):229–39. Discussion 39–42. PubMed PMID: 7677454.

98. DeMatteo RP, Yeh H, Friscia M, Caparrelli D, Burke C, Desai N, et al. Cellular immunity delimits adenoviral gene therapy strategies for the treatment of neoplastic diseases. *Ann Surg Oncol* 1999;**6**(1):88–94. PubMed PMID: 10030420.

99. Lehrman S. Virus treatment questioned after gene therapy death. *Nature* 1999;**401**(6753):517–8. PubMed PMID: 10524611.

100. Marshall E. Gene therapy death prompts review of adenovirus vector. *Science* 1999;**286**(5448):2244–5. PubMed PMID: 10636774.

101. Morsy MA, Gu M, Motzel S, Zhao J, Lin J, Su Q, et al. An adenoviral vector deleted for all viral coding sequences results in enhanced safety and extended expression of a leptin transgene. *Proc Natl Acad Sci USA* 1998;**95**(14):7866–71. PubMed PMID: 9653106 PMCID: PMC20895.

102. Smaglik P. Tighter watch urged on adenoviral vectors…with proposal to report all 'adverse events'. *Nature* 1999;**402**(6763):707. PubMed PMID: 10617181.

103. Baru M, Sha'anani J, Nur I. Retroviral-mediated in vivo gene transfer into muscle cells and synthesis of human factor IX in mice. *Intervirology* 1995;**38**(6):356–60. PubMed PMID: 8880387.

104. Chuah MK, Vandendriessche T, Morgan RA. Development and analysis of retroviral vectors expressing human factor VIII as a potential gene therapy for hemophilia A. *Hum Gene Ther* 1995;**6**(11):1363–77. PubMed PMID: 8573610.

105. Kay MA, Rothenberg S, Landen CN, Bellinger DA, Leland F, Toman C, et al. In vivo gene therapy of hemophilia B: sustained partial correction in factor IX-deficient dogs. *Science* 1993;**262**(5130):117–9. PubMed PMID: 8211118.

106. Park F, Ohashi K, Kay MA. Therapeutic levels of human factor VIII and IX using HIV-1-based lentiviral vectors in mouse liver. *Blood* 2000;**96**(3):1173–6. PubMed PMID: 10910939.

107. VandenDriessche T, Vanslembrouck V, Goovaerts I, Zwinnen H, Vanderhaeghen ML, Collen D, et al. Long-term expression of human coagulation factor VIII and correction of hemophilia A after in vivo retroviral gene transfer in factor VIII-deficient mice. *Proc Natl Acad Sci USA* 1999;**96**(18):10379–84. Epub 1999/09/01. PubMed PMID: 10468616.

108. Wang JM, Zheng H, Sugahara Y, Tan J, Yao SN, Olson E, et al. Construction of human factor IX expression vectors in retroviral vector frames optimized for muscle cells. *Hum Gene Ther* 1996;**7**(14):1743–56. PubMed PMID: 8886845.

109. Xu L, Gao C, Sands MS, Cai SR, Nichols TC, Bellinger DA, et al. Neonatal or hepatocyte growth factor-potentiated adult gene therapy with a retroviral vector results in therapeutic levels of canine factor IX for hemophilia B. *Blood* 2003;**101**(10):3924–32. PubMed PMID: 12531787.

110. Zhang J, Xu L, Haskins ME, Parker Ponder K. Neonatal gene transfer with a retroviral vector results in tolerance to human factor IX in mice and dogs. *Blood* 2004;**103**(1):143–51. PubMed PMID: 12969967.

111. Fausto N, Webber EM. In: Arias AM, Boyer JL, Fausto N, Jacoby WB, Schachter D, Shafritz DA, editors. *The liver: biology and pathobiology*. New York: Raven; 1994. p. 1059–84.

112. Porada CD, Almeida-Porada G. Treatment of hemophilia A in utero and postnatally using sheep as a model for cell and gene delivery. *J Genet Syndr Gene Ther* 2012:S1. PubMed PMID: 23264887 PMCID: PMC3526064.

113. Porada CD, Park PJ, Almeida-Porada G, Liu W, Ozturk F, Glimp HA, et al. Gestational age of recipient determines pattern and level of transgene expression following in utero retroviral gene transfer. *Mol Ther* 2005;**11**(2):284–93. PubMed PMID: 15668140.

114. Porada CD, Rodman C, Ignacio G, Atala A, Almeida-Porada G, Hemophilia A: an ideal disease to correct in utero. *Front Pharmacol* 2014;**5**:276. http://dx.doi.org/10.3389/fphar.2014.00276. eCollection 2014.

115. Ferry N, Duplessis O, Houssin D, Danos O, Heard JM. Retroviral-mediated gene transfer into hepatocytes in vivo. *Proc Natl Acad Sci USA* 1991;**88**(19):8377–81. PubMed PMID: 1656443 PMCID: PMC52511.

116. Kolodka TM, Finegold M, Kay MA, Woo SL. Hepatic gene therapy: efficient retroviral-mediated gene transfer into rat hepatocytes in vivo. *Somat Cell Mol Genet* 1993;**19**(5):491–7. PubMed PMID: 7980740.

117. Moscioni AD, Rozga J, Neuzil DF, Overell RW, Holt JT, Demetriou AA. In vivo regional delivery of retrovirally mediated foreign genes to rat liver cells: need for partial hepatectomy for successful foreign gene expression. *Surgery* 1993;**113**(3):304–11. PubMed PMID: 8382843.

118. Naldini L, Blomer U, Gallay P, Ory D, Mulligan R, Gage FH, et al. In vivo gene delivery and stable transduction of nondividing cells by a lentiviral vector. *Science* 1996;**272**(5259):263–7. PubMed PMID: 8602510.

119. Pichard V, Boni S, Baron W, Nguyen TH, Ferry N. Priming of hepatocytes enhances in vivo liver transduction with lentiviral vectors in adult mice. *Hum Gene Ther methods* 2012;**23**(1):8–17. PubMed PMID: 22428976.

120. Binny CJ, Nathwani AC. Vector systems for prenatal gene therapy: principles of adeno-associated virus vector design and production. *Methods Mol Biol* 2012;**891**:109–31. Epub 2012/06/01. PubMed PMID: 22648770.

121. Grieger JC, Samulski RJ. Adeno-associated virus as a gene therapy vector: vector development, production and clinical applications. *Adv Biochem Eng Biotechnol* 2005;**99**:119–45. Epub 2006/03/30. PubMed PMID: 16568890.

122. Ortolano S, Spuch C, Navarro C. Present and future of adeno associated virus based gene therapy approaches. *Recent Pat Endocr Metab Immune Drug Discov* 2012;**6**(1):47–66. Epub 2012/01/24. PubMed PMID: 22264214.

123. Agbandje-McKenna M, Kleinschmidt J. AAV capsid structure and cell interactions. *Methods Mol Biol* 2011;**807**:47–92. Epub 2011/10/29. PubMed PMID: 22034026.

124. Mitchell AM, Nicolson SC, Warischalk JK, Samulski RJ. AAV's anatomy: roadmap for optimizing vectors for translational success. *Curr Gene Ther* 2010;**10**(5):319–40. Epub 2010/08/18. PubMed PMID: 20712583.

125. Sharma A, Tovey JC, Ghosh A, Mohan RR. AAV serotype influences gene transfer in corneal stroma in vivo. *Exp Eye Res* 2010;**91**(3):440–8. Epub 2010/07/06. PubMed PMID: 20599959.

126. Wang J, Faust SM, Rabinowitz JE. The next step in gene delivery: molecular engineering of adeno-associated virus serotypes. *J Mol Cell Cardiol* 2011;**50**(5):793–802. Epub 2010/10/30. PubMed PMID: 21029739.

127. Bueler H. Adeno-associated viral vectors for gene transfer and gene therapy. *Biol Chem* 1999;**380**(6):613–22. Epub 1999/08/03. PubMed PMID: 10430026.

128. Carter PJ, Samulski RJ. Adeno-associated viral vectors as gene delivery vehicles. *Int J Mol Med* 2000;**6**(1):17–27. Epub 2000/06/14. PubMed PMID: 10851261.

129. Goncalves MA. Adeno-associated virus: from defective virus to effective vector. *Virol J* 2005;**2**:43. Epub 2005/05/10. PubMed PMID: 15877812.

130. Warrington Jr KH, Herzog RW. Treatment of human disease by adeno-associated viral gene transfer. *Hum Genet* 2006;**119**(6):571–603. Epub 2006/04/14. PubMed PMID: 16612615.

131. High KA. Gene therapy: a 2001 perspective. *Haemophilia* 2001;**7**(Suppl. 1):23–7. PubMed PMID: 11240615.

132. High KA. The gene therapy journey for hemophilia: are we there yet?. *Hematol Am Soc Hematol Educ Program* 2012;**2012**:375–81. Epub 2012/12/13. PubMed PMID: 23233607.

133. Chao H, Samulski R, Bellinger D, Monahan P, Nichols T, Walsh C. Persistent expression of canine factor IX in hemophilia B canines. *Gene Ther* 1999;**6**(10):1695–704. PubMed PMID: 10516718.

134. Finn JD, Nichols TC, Svoronos N, Merricks EP, Bellenger DA, Zhou S, et al. The efficacy and the risk of immunogenicity of FIX Padua (R338L) in hemophilia B dogs treated by AAV muscle gene therapy. *Blood* 2012;**120**(23):4521–3. Epub 2012/08/25. PubMed PMID: 22919027 PMCID: PMC3512231.

135. Herzog RW, Hagstrom JN, Kung SH, Tai SJ, Wilson JM, Fisher KJ, et al. Stable gene transfer and expression of human blood coagulation factor IX after intramuscular injection of recombinant adeno-associated virus. *Proc Natl Acad Sci USA* 1997;**94**(11):5804–9. PubMed PMID: 9159155 PMCID: PMC20861.

136. McIntosh J, Lenting PJ, Rosales C, Lee D, Rabbanian S, Raj D, et al. Therapeutic levels of FVIII following a single peripheral vein administration of rAAV vector encoding a novel human factor VIII variant. *Blood* 2013;**121**(17):3335–44. PubMed PMID: 23426947 PMCID: PMC3637010.

137. Mount JD, Herzog RW, Tillson DM, Goodman SA, Robinson N, McCleland ML, et al. Sustained phenotypic correction of hemophilia B dogs with a factor IX null mutation by liver-directed gene therapy. *Blood* 2002;**99**(8):2670–6. PubMed PMID: 11929752.

138. Snyder RO, Miao CH, Patijn GA, Spratt SK, Danos O, Nagy D, et al. Persistent and therapeutic concentrations of human factor IX in mice after hepatic gene transfer of recombinant AAV vectors. *Nat Genet* 1997;**16**(3):270–6. PubMed PMID: 9207793.

139. Monahan PE, Samulski RJ, Tazelaar J, Xiao X, Nichols TC, Bellinger DA, et al. Direct intramuscular injection with recombinant AAV vectors results in sustained expression in a dog model of hemophilia. *Gene Ther* 1998;**5**(1):40–9. PubMed PMID: 9536263.

140. Neuenschwander S, Kissling-Albrecht L, Heiniger J, Backfisch W, Stranzinger G, Pliska V. Inherited defect of blood clotting factor VIII (haemophilia A) in sheep. *Thromb haemost* 1992;**68**(5):618–20. PubMed PMID: 1455410.

141. Backfisch W, Neuenschwander S, Giger U, Stranzinger G, Pliska V. Carrier detection of ovine hemophilia A using an RFLP marker, and mapping of the factor VIII gene on the ovine X-chromosome. *J Hered* 1994;**85**(6):474–8. PubMed PMID: 7995928.

142. Neuenschwander S, Pliska V. Factor VIII in blood plasma of haemophilic sheep: analysis of clotting time-plasma dilution curves. *Haemostasis* 1994;**24**(1):27–35. PubMed PMID: 7959353.

143. Porada CD, Sanada C, Long CR, Wood JA, Desai J, Frederick N, et al. Clinical and molecular characterization of a re-established line of sheep exhibiting hemophilia A. *J Thromb Haemost* 2010;**8**(2):276–85. Epub 2009/12/01. PubMed PMID: 19943872.

144. Zakas PM, Gangadharan B, Almeida-Porada G, Porada CD, Spencer HT, Doering CB. Development and characterization of recombinant ovine coagulation factor VIII. *PLoS One* 2012;**7**(11):e49481. PubMed PMID: 23152911 PMCID: PMC3494657.

145. High K. Gene transfer for hemophilia: can therapeutic efficacy in large animals be safely translated to patients?. *J Thromb Haemost* 2005;**3**(8): 1682–91. PubMed PMID: 16102034.

146. Axelrod JH, Read MS, Brinkhous KM, Verma IM. Phenotypic correction of factor IX deficiency in skin fibroblasts of hemophilic dogs. *Proc Natl Acad Sci USA* 1990;**87**(13):5173–7. PubMed PMID: 2367529 PMCID: PMC54284.

147. Dai YF, Qiu XF, Xue JL, Liu ZD. High efficient transfer and expression of human clotting factor IX cDNA in cultured human primary skin fibroblasts from hemophilia B patient by retroviral vectors. *Sci China Ser B Chem Life Sci Earth Sci* 1992;**35**(2):183–93. PubMed PMID: 1581003.

148. Dwarki VJ, Belloni P, Nijjar T, Smith J, Couto L, Rabier M, et al. Gene therapy for hemophilia A: production of therapeutic levels of human factor VIII in vivo in mice. *Proc Natl Acad Sci USA* 1995;**92**(4):1023–7. PubMed PMID: 7862626 PMCID: PMC42629.

149. Hoeben RC, Fallaux FJ, Van Tilburg NH, Cramer SJ, Van Ormondt H, Briet E, et al. Toward gene therapy for hemophilia A: long-term persistence of factor VIII-secreting fibroblasts after transplantation into immunodeficient mice. *Hum Gene Ther* 1993;**4**(2):179–86. PubMed PMID: 8494927.

150. Palmer TD, Thompson AR, Miller AD. Production of human factor IX in animals by genetically modified skin fibroblasts: potential therapy for hemophilia B. *Blood* 1989;**73**(2):438–45. PubMed PMID: 2917183.

151. Zhou JM, Dai YF, Qiu XF, Hou GY, Akira Y, Xue JL. Expression of human factor IX cDNA in mice by implants of genetically modified skin fibroblasts from a hemophilia B patient. *Sci China Ser B Chem Life Sci Earth Sci* 1993;**36**(9):1082–92. PubMed PMID: 8274202.

152. Zhou JM, Qiu XF, Lu DR, Lu JY, Xue JL. Long-term expression of human factor IX cDNA in rabbits. *Sci China Ser B Chem Life Sci Earth Sci* 1993;**36**(11):1333–41. PubMed PMID: 8142022.

153. Hsueh JL. Clinical protocol of human gene transfer for hemophilia B. *Hum Gene Ther* 1992;**3**(5):543–52. PubMed PMID: 1420453.

154. Lu DR, Zhou JM, Zheng B, Qiu XF, Xue JL, Wang JM, et al. Stage I clinical trial of gene therapy for hemophilia B. *Sci China Ser B Chem Life Sci Earth Sci* 1993;**36**(11):1342–51. PubMed PMID: 8142023.

155. Qiu X, Lu D, Zhou J, Wang J, Yang J, Meng P, et al. Implantation of autologous skin fibroblast genetically modified to secrete clotting factor IX partially corrects the hemorrhagic tendencies in two hemophilia B patients. *Chin Med J (Engl)* 1996;**109**(11):832–9. PubMed PMID: 9275366.

156. Bank A. Human somatic cell gene therapy. *BioEssays* 1996;**18**(12):999–1007. PubMed PMID: 8976157.

157. Barranger JA. Hematopoietic stem cell gene transfer. *Gene Ther* 1996;**3**(5):379–80. PubMed PMID: 9156797.

158. Kohn DB. Gene therapy for haematopoietic and lymphoid disorders. *Clin Exp Immunol* 1997;**107**(Suppl. 1):54–7. PubMed PMID: 9020937.

159. Swaney WP, Novelli EM, Bahnson AB, Barranger JA. Retrovirus-mediated gene transfer to human hematopoietic stem cells. *Methods Mol Med* 2002;**69**:187–202. PubMed PMID: 11987778.

160. Ogawa M. Differentiation and proliferation of hematopoietic stem cells. *Blood* 1993;**81**(11):2844–53. PubMed PMID: 8499622.

161. Gyurkocza B, Rezvani A, Storb RF. Allogeneic hematopoietic cell transplantation: the state of the art. *Expert Rev Hematol* 2010;**3**(3):285–99. PubMed PMID: 20871781 PMCID: PMC2943393.

162. Hao QL, Malik P, Salazar R, Tang H, Gordon EM, Kohn DB. Expression of biologically active human factor IX in human hematopoietic cells after retroviral vector-mediated gene transduction. *Hum Gene Ther* 1995;**6**(7):873–80. PubMed PMID: 7578406.

163. Hoeben RC, Einerhand MP, Briet E, van Ormondt H, Valerio D, van der Eb AJ. Toward gene therapy in haemophilia A: retrovirus-mediated transfer of a factor VIII gene into murine haematopoietic progenitor cells. *Thromb Haemost* 1992;**67**(3):341–5. PubMed PMID: 1641825.

164. Ide LM, Iwakoshi NN, Gangadharan B, Jobe S, Moot R, McCarty D, et al. Functional aspects of factor VIII expression after transplantation of genetically-modified hematopoietic stem cells for hemophilia A. *J Gene Med* 2010;**12**(4):333–44. Epub 2010/03/09. PubMed PMID: 20209485.

165. Moayeri M, Ramezani A, Morgan RA, Hawley TS, Hawley RG. Sustained phenotypic correction of hemophilia a mice following oncoretroviral-mediated expression of a bioengineered human factor VIII gene in long-term hematopoietic repopulating cells. *Mol Ther* 2004;**10**(5):892–902. PubMed PMID: 15509507.

166. Ramezani A, Hawley RG. Correction of murine hemophilia A following nonmyeloablative transplantation of hematopoietic stem cells engineered to encode an enhanced human factor VIII variant using a safety-augmented retroviral vector. *Blood* 2009;**114**(3):526–34. PubMed PMID: 19470695 PMCID: PMC2713478.

167. Zakas PM, Spencer HT, Doering CB. Engineered hematopoietic stem cells as therapeutics for hemophilia a. *J Genet Syndr Gene Ther* 2011;**1**(3). PubMed PMID: 25383239 PMCID: PMC4220243.

168. Du LM, Nurden P, Nurden AT, Nichols TC, Bellinger DA, Jensen ES, et al. Platelet-targeted gene therapy with human factor VIII establishes haemostasis in dogs with haemophilia A. *Nat Commun* 2013;**4**:2773. PubMed PMID: 24253479 PMCID: PMC3868233.

169. Kuether EL, Schroeder JA, Fahs SA, Cooley BC, Chen Y, Montgomery RR, et al. Lentivirus-mediated platelet gene therapy of murine hemophilia A with pre-existing anti-factor VIII immunity. *J Thromb Haemost* 2012;**10**(8):1570–80. PubMed PMID: 22632092 PMCID: PMC3419807.

170. Shi Q, Kuether EL, Chen Y, Schroeder JA, Fahs SA, Montgomery RR. Platelet gene therapy corrects the hemophilic phenotype in immunocompromised hemophilia A mice transplanted with genetically manipulated human cord blood stem cells. *Blood* 2014;**123**(3):395–403. PubMed PMID: 24269957 PMCID: PMC3894495.

171. Greene TK, Lambert MP, Poncz M. Ectopic platelet-delivered factor (F) VIII for the treatment of hemophilia A: plasma and platelet FVIII, is it all the same? *J Genet Syndr Gene Ther* 2011;(Suppl. 1(1)). PubMed PMID: 24319630 PMCID: PMC3852407.

172. Hacein-Bey-Abina S, Von Kalle C, Schmidt M, McCormack MP, Wulffraat N, Leboulch P, et al. LMO2-associated clonal T cell proliferation in two patients after gene therapy for SCID-X1. *Science* 2003;**302**(5644):415–9. PubMed PMID: 14564000.

173. Howe SJ, Mansour MR, Schwarzwaelder K, Bartholomae C, Hubank M, Kempski H, et al. Insertional mutagenesis combined with acquired somatic mutations causes leukemogenesis following gene therapy of SCID-X1 patients. *J Clin Invest* 2008;**118**(9):3143–50. Epub 2008/08/09. PubMed PMID: 18688286.

174. Thrasher AJ, Gaspar HB, Baum C, Modlich U, Schambach A, Candotti F, et al. Gene therapy: X-SCID transgene leukaemogenicity. *Nature* 2006;**443**(7109):E5–6. discussion E-7. Epub 2006/09/22. PubMed PMID: 16988659.

175. Pike-Overzet K, van der Burg M, Wagemaker G, van Dongen JJ, Staal FJ. New insights and unresolved issues regarding insertional mutagenesis in X-linked SCID gene therapy. *Mol Ther* 2007;**15**(11):1910–6. PubMed PMID: 17726455.

176. LaFave MC, Varshney GK, Gildea DE, Wolfsberg TG, Baxevanis AD, Burgess SM. MLV integration site selection is driven by strong enhancers and active promoters. *Nucleic Acids Res* 2014;**42**(7):4257–69. PubMed PMID: 24464997 PMCID: PMC3985626.

177. Wu X, Li Y, Crise B, Burgess SM. Transcription start regions in the human genome are favored targets for MLV integration. *Science* 2003;**300**(5626):1749–51. PubMed PMID: 12805549.

178. Sorrentino B. Assessing the risk of T-cell malignancies in mouse models of SCID-X1. *Mol Ther* 2010;**18**(5):868–70. PubMed PMID: 20436493.

179. Friedenstein AJ. Osteogenic stem cells in the bone marrow. *Bone Mineral* 1991;**7**:243–72.

180. Friedenstein AJ, Chailakhyan RK, Latsinik NV, Panasyuk AF, Keiliss-Borok IV. Stromal cells responsible for transferring the microenvironment of the hemopoietic tissues. Cloning in vitro and retransplantation in vivo. *Transplantation* 1974;**17**(4):331–40. PubMed PMID: 4150881.

181. Caplan AI. Mesenchymal stem cells. *J Orthop Res* 1991;**9**(5):641–50. Epub 1991/09/01. PubMed PMID: 1870029.

182. Galotto M, Berisso G, Delfino L, Podesta M, Ottaggio L, Dallorso S, et al. Stromal damage as consequence of high-dose chemo/radiotherapy in bone marrow transplant recipients. *Exp Hematol* 1999;**27**(9):1460–6. Epub 1999/09/10. PubMed PMID: 10480437.

183. Kassem M. Mesenchymal stem cells: biological characteristics and potential clinical applications. *Cloning Stem Cells* 2004;**6**(4):369–74. Epub 2005/01/27. PubMed PMID: 15671665.

184. Luria EA, Panasyuk AF, Friedenstein AY. Fibroblast colony formation from monolayer cultures of blood cells. *Transfusion* 1971;**11**(6):345–9. Epub 1971/11/01. PubMed PMID: 5136066.

185. Pittenger MF, Mackay AM, Beck SC, Jaiswal RK, Douglas R, Mosca JD, et al. Multilineage potential of adult human mesenchymal stem cells. *Science* 1999;**284**(5411):143–7. PubMed PMID: 10102814.

186. Simmons PJ, Torok-Storb B. Identification of stromal cell precursors in human bone marrow by a novel monoclonal antibody, STRO-1. *Blood* 1991;**78**(1):55–62. Epub 1991/07/01. PubMed PMID: 2070060.

187. Airey JA, Almeida-Porada G, Colletti EJ, Porada CD, Chamberlain J, Movsesian M, et al. Human mesenchymal stem cells form Purkinje fibers in fetal sheep heart. *Circulation* 2004;**109**(11):1401–7. PubMed PMID: 15023887.

188. Almeida-Porada MG, Porada C, ElShabrawy D, Simmons PJ, Zanjani ED. Human marrow stromal cells (MSC) represent a latent pool of stem cells capable of generating long-term hematopoietic cells. *Blood* 2001;**98**(Part 1):713a.

189. Chamberlain J, Yamagami T, Colletti E, Theise ND, Desai J, Frias A, et al. Efficient generation of human hepatocytes by the intrahepatic delivery of clonal human mesenchymal stem cells in fetal sheep. *Hepatology (Baltimore, Md.)* 2007;**46**(6):1935–45. PubMed PMID: 17705296.

190. Colletti E, Zanjani ED, Porada CD, Almeida-Porada MG. Tales from the crypt: mesenchymal stem cells for replenishing the intestinal stem cell pool. *Blood* 2008;**112**. Abstract 390.

191. Colletti EJAJA, Zanjani ED, Porada CD, Almeida-Porada G. Human mesenchymal stem cells differentiate promptly into tissue-specific cell types without cell fusion, mitochondrial or membrane vesicular transfer in fetal sheep. *Blood* 2007;**110**(11):135a.

192. Colletti EJ, Airey JA, Liu W, Simmons PJ, Zanjani ED, Porada CD, et al. Generation of tissue-specific cells from MSC does not require fusion or donor-to-host mitochondrial/membrane transfer. *Stem Cell Res* 2009;**2**(2):125–38. Epub 2009/04/23. PubMed PMID: 19383418.

193. Colletti EJ, Almeida-Porada G, Chamberlain J, Zanjani ED, Airey JA. The time course of engraftment of human mesenchymal stem cells in fetal heart demonstrates that Purkinje fiber aggregates derive from a single cell and not multi-cell homing. *Exp Hematol* 2006;**34**(7):926–33. PubMed PMID: 16797420.

194. Aggarwal S, Pittenger MF. Human mesenchymal stem cells modulate allogeneic immune cell responses. *Blood* 2005;**105**(4):1815–22. PubMed PMID: 15494428.

195. Batten P, Sarathchandra P, Antoniw JW, Tay SS, Lowdell MW, Taylor PM, et al. Human mesenchymal stem cells induce T cell anergy and down-regulate T cell allo-responses via the TH2 pathway: relevance to tissue engineering human heart valves. *Tissue Eng* 2006;**12**(8):2263–73. Epub 2006/09/14. PubMed PMID: 16968166.

196. Caplan AI, Dennis JE. Mesenchymal stem cells as trophic mediators. *J Cell Biochem* 2006;**98**(5):1076–84. PubMed PMID: 16619257.

197. Chapel A, Bertho JM, Bensidhoum M, Fouillard L, Young RG, Frick J, et al. Mesenchymal stem cells home to injured tissues when co-infused with hematopoietic cells to treat a radiation-induced multi-organ failure syndrome. *J Gene Med* 2003;**5**(12):1028–38. Epub 2003/12/09. PubMed PMID: 14661178.

198. Hamada H, Kobune M, Nakamura K, Kawano Y, Kato K, Honmou O, et al. Mesenchymal stem cells (MSC) as therapeutic cytoreagents for gene therapy. *Cancer Sci* 2005;**96**(3):149–56. Epub 2005/03/18. PubMed PMID: 15771617.

199. Liechty KW, MacKenzie TC, Shaaban AF, Radu A, Moseley AM, Deans R, et al. Human mesenchymal stem cells engraft and demonstrate site-specific differentiation after in utero transplantation in sheep. *Nat Med* 2000;**6**(11):1282–6. PubMed PMID: 11062543.

200. Mackenzie TC, Flake AW. Multilineage differentiation of human MSC after in utero transplantation. *Cytotherapy* 2001;**3**(5):403–5. PubMed PMID: 11953022.

201. Porada CD, Almeida-Porada G. Mesenchymal stem cells as therapeutics and vehicles for gene and drug delivery. *Adv Drug Deliv Rev* 2010;**62**(12):1156–66. Epub 2010/09/11. PubMed PMID: 20828588.

202. Almeida-Porada G, El Shabrawy D, Porada C, Zanjani ED. Differentiative potential of human metanephric mesenchymal cells. *Exp Hematol* 2002;**30**(12):1454–62. PubMed PMID: 12482508.

203. Fan CG, Tang FW, Zhang QJ, Lu SH, Liu HY, Zhao ZM, et al. Characterization and neural differentiation of fetal lung mesenchymal stem cells. *Cell Transplant* 2005;**14**(5):311–21. Epub 2005/08/02. PubMed PMID: 16052912.

204. Gotherstrom C, West A, Liden J, Uzunel M, Lahesmaa R, Le Blanc K. Difference in gene expression between human fetal liver and adult bone marrow mesenchymal stem cells. *Haematologica* 2005;**90**(8):1017–26. Epub 2005/08/05. PubMed PMID: 16079100.

205. in 't Anker PS, Noort WA, Scherjon SA, Kleijburg-van der Keur C, Kruisselbrink AB, van Bezooijen RL, et al. Mesenchymal stem cells in human second-trimester bone marrow, liver, lung, and spleen exhibit a similar immunophenotype but a heterogeneous multilineage differentiation potential. *Haematologica* 2003;**88**(8):845–52. Epub 2003/08/26. PubMed PMID: 12935972.

206. Lee OK, Kuo TK, Chen WM, Lee KD, Hsieh SL, Chen TH. Isolation of multipotent mesenchymal stem cells from umbilical cord blood. *Blood* 2004;**103**(5):1669–75. Epub 2003/10/25. PubMed PMID: 14576065.

207. Morizono K, De Ugarte DA, Zhu M, Zuk P, Elbarbary A, Ashjian P, et al. Multilineage cells from adipose tissue as gene delivery vehicles. *Hum Gene Ther* 2003;**14**(1):59–66. Epub 2003/02/08. PubMed PMID: 12573059.

208. Zuk PA, Zhu M, Ashjian P, De Ugarte DA, Huang JI, Mizuno H, et al. Human adipose tissue is a source of multipotent stem cells. *Mol Biol Cell* 2002;**13**(12):4279–95. Epub 2002/12/12. PubMed PMID: 12475952.

209. Zuk PA, Zhu M, Mizuno H, Huang J, Futrell JW, Katz AJ, et al. Multilineage cells from human adipose tissue: implications for cell-based therapies. *Tissue Eng* 2001;**7**(2):211–28. Epub 2001/04/17. PubMed PMID: 11304456.

210. Chen TS, Lai RC, Lee MM, Choo AB, Lee CN, Lim SK. Mesenchymal stem cell secretes microparticles enriched in pre-microRNAs. *Nucleic Acids Res* 2010;**38**(1):215–24. Epub 2009/10/24. PubMed PMID 19850715.

211. Dai W, Hale SL, Kloner RA. Role of a paracrine action of mesenchymal stem cells in the improvement of left ventricular function after coronary artery occlusion in rats. *Regen Med* 2007;**2**(1):63–8. Epub 2007/05/01. PubMed PMID: 17465776.

212. Gnecchi M, He H, Liang OD, Melo LG, Morello F, Mu H, et al. Paracrine action accounts for marked protection of ischemic heart by Akt-modified mesenchymal stem cells. *Nat Med* 2005;**11**(4):367–8. Epub 2005/04/07. PubMed PMID: 15812508.

213. Haynesworth SE, Baber MA, Caplan AI. Cytokine expression by human marrow-derived mesenchymal progenitor cells in vitro: effects of dexamethasone and IL-1 alpha. *J Cell Physiol* 1996;**166**(3):585–92. PubMed PMID: 8600162.

214. Huang NF, Lam A, Fang Q, Sievers RE, Li S, Lee RJ. Bone marrow-derived mesenchymal stem cells in fibrin augment angiogenesis in the chronically infarcted myocardium. *Regen Med* 2009;**4**(4):527–38. Epub 2009/07/08. PubMed PMID: 19580402.

215. Kuo TK, Hung SP, Chuang CH, Chen CT, Shih YR, Fang SC, et al. Stem cell therapy for liver disease: parameters governing the success of using bone marrow mesenchymal stem cells. *Gastroenterology* 2008;**134**(7):2111–21. 2121.e1–3. PubMed PMID: 18455168.

216. Ladage D, Brixius K, Steingen C, Mehlhorn U, Schwinger RH, Bloch W, et al. Mesenchymal stem cells induce endothelial activation via paracine mechanisms. *Endothelium* 2007;**14**(2):53–63. Epub 2007/05/15. PubMed PMID: 17497361.

217. Lai RC, Arslan F, Lee MM, Sze NS, Choo A, Chen TS, et al. Exosome secreted by MSC reduces myocardial ischemia/reperfusion injury. *Stem Cell Res* 2010;**4**(3):214–22. Epub 2010/02/09. PubMed PMID: 20138817.

218. Li Z, Guo J, Chang Q, Zhang A. Paracrine role for mesenchymal stem cells in acute myocardial infarction. *Biol Pharm Bull* 2009;**32**(8):1343–6. Epub 2009/08/05. PubMed PMID: 19652371.

219. Mias C, Lairez O, Trouche E, Roncalli J, Calise D, Seguelas MH, et al. Mesenchymal stem cells promote matrix metalloproteinase secretion by cardiac fibroblasts and reduce cardiac ventricular fibrosis after myocardial infarction. *Stem Cells* 2009;**27**(11):2734–43. Epub 2009/07/11. PubMed PMID: 19591227.

220. Ohnishi S, Sumiyoshi H, Kitamura S, Nagaya N. Mesenchymal stem cells attenuate cardiac fibroblast proliferation and collagen synthesis through paracrine actions. *FEBS Lett* 2007;**581**(21):3961–6. Epub 2007/07/31. PubMed PMID: 17662720.

221. Parekkadan B, van Poll D, Megeed Z, Kobayashi N, Tilles AW, Berthiaume F, et al. Immunomodulation of activated hepatic stellate cells by mesenchymal stem cells. *Biochem Biophys Res Commun* 2007;**363**(2):247–52. PubMed PMID: 17869217.

222. Parekkadan B, van Poll D, Suganuma K, Carter EA, Berthiaume F, Tilles AW, et al. Mesenchymal stem cell-derived molecules reverse fulminant hepatic failure. *PLoS One* 2007;**2**(9):e941. Epub 2007/09/27. PubMed PMID: 17895982.

223. Shabbir A, Zisa D, Suzuki G, Lee T. Heart failure therapy mediated by the trophic activities of bone marrow mesenchymal stem cells: a noninvasive therapeutic regimen. *Am J Physiol Heart Circ Physiol* 2009;**296**(6):H1888–97. Epub 2009/04/28. PubMed PMID: 19395555.

224. Timmers L, Lim SK, Arslan F, Armstrong JS, Hoefer IE, Doevendans PA, et al. Reduction of myocardial infarct size by human mesenchymal stem cell conditioned medium. *Stem Cell Res* 2007;**1**(2):129–37. Epub 2007/11/01. PubMed PMID: 19383393.

225. van Poll D, Parekkadan B, Cho CH, Berthiaume F, Nahmias Y, Tilles AW, et al. Mesenchymal stem cell-derived molecules directly modulate hepatocellular death and regeneration in vitro and in vivo. *Hepatology (Baltimore, Md.)* 2008;**47**(5):1634–43. Epub 2008/04/09. PubMed PMID: 18395843.

226. Xiang MX, He AN, Wang JA, Gui C. Protective paracrine effect of mesenchymal stem cells on cardiomyocytes. *J Zhejiang Univ Sci B* 2009;**10**(8):619–24. Epub 2009/08/04. PubMed PMID: 19650201.

227. Yu XY, Geng YJ, Li XH, Lin QX, Shan ZX, Lin SG, et al. The effects of mesenchymal stem cells on c-kit up-regulation and cell-cycle re-entry of neonatal cardiomyocytes are mediated by activation of insulin-like growth factor 1 receptor. *Mol Cell Biochem* 2009;**332**(1–2):25–32. Epub 2009/06/10. PubMed PMID: 19507001.

228. Le Blanc K, Ringden O. Immunobiology of human mesenchymal stem cells and future use in hematopoietic stem cell transplantation. *Biol Blood Marrow Transplant* 2005;**11**(5):321–34. PubMed PMID: 15846285.

229. Le Blanc K, Tammik L, Sundberg B, Haynesworth SE, Ringden O. Mesenchymal stem cells inhibit and stimulate mixed lymphocyte cultures and mitogenic responses independently of the major histocompatibility complex. *Scand J Immunol* 2003;**57**(1):11–20. PubMed PMID: 12542793.

230. Puissant B, Barreau C, Bourin P, Clavel C, Corre J, Bousquet C, et al. Immunomodulatory effect of human adipose tissue-derived adult stem cells: comparison with bone marrow mesenchymal stem cells. *Br J Haematol* 2005;**129**(1):118–29. PubMed PMID: 15801964.

231. Di Nicola M, Carlo-Stella C, Magni M, Milanesi M, Longoni PD, Matteucci P, et al. Human bone marrow stromal cells suppress T-lymphocyte proliferation induced by cellular or nonspecific mitogenic stimuli. *Blood* 2002;**99**(10):3838–43. Epub 2002/05/03. PubMed PMID: 11986244.

232. Groh ME, Maitra B, Szekely E, Koc ON. Human mesenchymal stem cells require monocyte-mediated activation to suppress alloreactive T cells. *Exp Hematol* 2005;**33**(8):928–34. Epub 2005/07/26. PubMed PMID: 16038786.

233. Krampera M, Glennie S, Dyson J, Scott D, Laylor R, Simpson E, et al. Bone marrow mesenchymal stem cells inhibit the response of naive and memory antigen-specific T cells to their cognate peptide. *Blood* 2003;**101**(9):3722–9. Epub 2002/12/31. PubMed PMID: 12506037.

234. Jones S, Horwood N, Cope A, Dazzi F. The antiproliferative effect of mesenchymal stem cells is a fundamental property shared by all stromal cells. *J Immunol* 2007;**179**(5):2824–31. Epub 2007/08/22. PubMed PMID: 17709496.

235. Klyushnenkova E, Mosca JD, Zernetkina V, Majumdar MK, Beggs KJ, Simonetti DW, et al. T cell responses to allogeneic human mesenchymal stem cells: immunogenicity, tolerance, and suppression. *J Biomed Sci* 2005;**12**(1):47–57. PubMed PMID: 15864738.

236. Tse WT, Pendleton JD, Beyer WM, Egalka MC, Guinan EC. Suppression of allogeneic T-cell proliferation by human marrow stromal cells: implications in transplantation. *Transplantation* 2003;**75**(3):389–97. Epub 2003/02/18. PubMed PMID: 12589164.

237. Djouad F, Charbonnier LM, Bouffi C, Louis-Plence P, Bony C, Apparailly F, et al. Mesenchymal stem cells inhibit the differentiation of dendritic cells through an interleukin-6-dependent mechanism. *Stem Cells* 2007;**25**(8):2025–32. Epub 2007/05/19. PubMed PMID: 17510220.

238. Jiang XX, Zhang Y, Liu B, Zhang SX, Wu Y, Yu XD, et al. Human mesenchymal stem cells inhibit differentiation and function of monocyte-derived dendritic cells. *Blood* 2005;**105**(10):4120–6. Epub 2005/02/05. PubMed PMID: 15692068.

239. Nauta AJ, Kruisselbrink AB, Lurvink E, Willemze R, Fibbe WE. Mesenchymal stem cells inhibit generation and function of both CD34+-derived and monocyte-derived dendritic cells. *J Immunol* 2006;**177**(4):2080–7. Epub 2006/08/05. PubMed PMID: 16887966.

240. Zhang W, Ge W, Li C, You S, Liao L, Han Q, et al. Effects of mesenchymal stem cells on differentiation, maturation, and function of human monocyte-derived dendritic cells. *Stem Cells Dev* 2004;**13**(3):263–71. Epub 2004/06/10. PubMed PMID: 15186722.

241. Corcione A, Benvenuto F, Ferretti E, Giunti D, Cappiello V, Cazzanti F, et al. Human mesenchymal stem cells modulate B-cell functions. *Blood* 2006;**107**(1):367–72. Epub 2005/09/06. PubMed PMID: 16141348.

242. Crop M, Baan C, Weimar W, Hoogduijn M. Potential of mesenchymal stem cells as immune therapy in solid-organ transplantation. *Transpl Int* 2009;**22**(4):365–76. Epub 2008/11/13. PubMed PMID: 19000235.

243. Maccario R, Podesta M, Moretta A, Cometa A, Comoli P, Montagna D, et al. Interaction of human mesenchymal stem cells with cells involved in alloantigen-specific immune response favors the differentiation of CD4+ T-cell subsets expressing a regulatory/suppressive phenotype. *Haematologica* 2005;**90**(4):516–25. Epub 2005/04/12. PubMed PMID: 15820948.

244. Prevosto C, Zancolli M, Canevali P, Zocchi MR, Poggi A. Generation of CD4+ or CD8+ regulatory T cells upon mesenchymal stem cell-lymphocyte interaction. *Haematologica* 2007;**92**(7):881–8. Epub 2007/07/04. PubMed PMID: 17606437.

245. Sato K, Ozaki K, Oh I, Meguro A, Hatanaka K, Nagai T, et al. Nitric oxide plays a critical role in suppression of T-cell proliferation by mesenchymal stem cells. *Blood* 2007;**109**(1):228–34. Epub 2006/09/21. PubMed PMID: 16985180.

246. Nasef A, Chapel A, Mazurier C, Bouchet S, Lopez M, Mathieu N, et al. Identification of IL-10 and TGF-β transcripts involved in the inhibition of T-lymphocyte proliferation during cell contact with human mesenchymal stem cells. *Gene Expr* 2007;**13**(4–5):217–26. Epub 2007/07/04. PubMed PMID: 17605296.

247. Nasef A, Mathieu N, Chapel A, Frick J, Francois S, Mazurier C, et al. Immunosuppressive effects of mesenchymal stem cells: involvement of HLA-G. *Transplantation* 2007;**84**(2):231–7. Epub 2007/08/02. PubMed PMID: 17667815.

248. Hainz U, Jurgens B, Heitger A. The role of indoleamine 2,3-dioxygenase in transplantation. *Transpl Int* 2007;**20**(2):118–27. Epub 2007/01/24. PubMed PMID: 17239019.

249. Meisel R, Zibert A, Laryea M, Gobel U, Daubener W, Dilloo D. Human bone marrow stromal cells inhibit allogeneic T-cell responses by indoleamine 2,3-dioxygenase-mediated tryptophan degradation. *Blood* 2004;**103**(12):4619–21. Epub 2004/03/06. PubMed PMID: 15001472.

250. Kidd S, Spaeth E, Dembinski JL, Dietrich M, Watson K, Klopp A, et al. Direct evidence of mesenchymal stem cell tropism for tumor and wounding microenvironments using in vivo bioluminescent imaging. *Stem Cells* 2009;**27**(10):2614–23. Epub 2009/08/04. PubMed PMID: 19650040.

251. Spaeth EL, Kidd S, Marini FC. Tracking inflammation-induced mobilization of mesenchymal stem cells. *Methods Mol Biol* 2012;**904**:173–90. Epub 2012/08/15. PubMed PMID: 22890932.

252. Spaeth EL, Marini FC. Dissecting mesenchymal stem cell movement: migration assays for tracing and deducing cell migration. *Methods Mol Biol* 2011;**750**:241–59. Epub 2011/05/28. PubMed PMID: 21618096.

253. Ozawa K, Sato K, Oh I, Ozaki K, Uchibori R, Obara Y, et al. Cell and gene therapy using mesenchymal stem cells (MSCs). *J Autoimmun* 2008;**30**(3):121–7. Epub 2008/02/06. PubMed PMID: 18249090.

254. Reiser J, Zhang XY, Hemenway CS, Mondal D, Pradhan L, La Russa VF. Potential of mesenchymal stem cells in gene therapy approaches for inherited and acquired diseases. *Expert Opin Biol Ther* 2005;**5**(12):1571–84. Epub 2005/12/02. PubMed PMID: 16318421.

255. Bosch P, Fouletier-Dilling C, Olmsted-Davis EA, Davis AR, Stice SL. Efficient adenoviral-mediated gene delivery into porcine mesenchymal stem cells. *Mol Reprod Dev* 2006;**73**(11):1393–403. Epub 2006/08/10. PubMed PMID: 16897738.

256. Bosch P, Stice SL. Adenoviral transduction of mesenchymal stem cells. *Methods Mol Biol* 2007;**407**:265–74. Epub 2008/05/06. PubMed PMID: 18453261.

257. Roelants V, Labar D, de Meester C, Havaux X, Tabilio A, Gambhir SS, et al. Comparison between adenoviral and retroviral vectors for the transduction of the thymidine kinase PET reporter gene in rat mesenchymal stem cells. *J Nucl Med* 2008;**49**(11):1836–44. Epub 2008/11/06. PubMed PMID: 18984872.

258. Gnecchi M, Melo LG. Bone marrow-derived mesenchymal stem cells: isolation, expansion, characterization, viral transduction, and production of conditioned medium. *Methods Mol Biol* 2009;**482**:281–94. Epub 2008/12/18. PubMed PMID: 19089363.

259. Meyerrose TE, De Ugarte DA, Hofling AA, Herrbrich PE, Cordonnier TD, Shultz LD, et al. In vivo distribution of human adipose-derived mesenchymal stem cells in novel xenotransplantation models. *Stem Cells* 2007;**25**(1):220–7. Epub 2006/09/09. PubMed PMID: 16960135.

260. Piccoli C, Scrima R, Ripoli M, Di Ianni M, Del Papa B, D'Aprile A, et al. Transformation by retroviral vectors of bone marrow-derived mesenchymal cells induces mitochondria-dependent cAMP-sensitive reactive oxygen species production. *Stem Cells* 2008;**26**(11):2843–54. Epub 2008/09/13. PubMed PMID: 18787213.

261. Sales VL, Mettler BA, Lopez-Ilasaca M, Johnson Jr JA, Mayer Jr JE. Endothelial progenitor and mesenchymal stem cell-derived cells persist in tissue-engineered patch in vivo: application of green and red fluorescent protein-expressing retroviral vector. *Tissue Eng* 2007;**13**(3):525–35. Epub 2007/05/24. PubMed PMID: 17518601.

262. Fan L, Lin C, Zhuo S, Chen L, Liu N, Luo Y, et al. Transplantation with survivin-engineered mesenchymal stem cells results in better prognosis in a rat model of myocardial infarction. *Eur J Heart Fail* 2009;**11**(11):1023–30. Epub 2009/10/31. PubMed PMID: 19875403.

263. Meyerrose TE, Roberts M, Ohlemiller KK, Vogler CA, Wirthlin L, Nolta JA, et al. Lentiviral-transduced human mesenchymal stem cells persistently express therapeutic levels of enzyme in a xenotransplantation model of human disease. *Stem Cells* 2008;**26**(7):1713–22. Epub 2008/04/26. PubMed PMID: 18436861.

264. Wang F, Dennis JE, Awadallah A, Solchaga LA, Molter J, Kuang Y, et al. Transcriptional profiling of human mesenchymal stem cells transduced with reporter genes for imaging. *Physiol Genomics* 2009;**37**(1):23–34. Epub 2009/01/01. PubMed PMID: 19116247.

265. Xiang J, Tang J, Song C, Yang Z, Hirst DG, Zheng QJ, et al. Mesenchymal stem cells as a gene therapy carrier for treatment of fibrosarcoma. *Cytotherapy* 2009;**11**(5):516–26. Epub 2009/06/30. PubMed PMID: 19562576.

266. Zhang XY, La Russa VF, Reiser J. Transduction of bone-marrow-derived mesenchymal stem cells by using lentivirus vectors pseudotyped with modified RD114 envelope glycoproteins. *J Virol* 2004;**78**(3):1219–29. Epub 2004/01/15. PubMed PMID: 14722277.

267. Zhang XY, La Russa VF, Bao L, Kolls J, Schwarzenberger P, Reiser J. Lentiviral vectors for sustained transgene expression in human bone marrow-derived stromal cells. *Mol Ther* 2002;**5**(5 Pt 1):555–65. Epub 2002/05/07. PubMed PMID: 11991746.

268. Kumar S, Mahendra G, Nagy TR, Ponnazhagan S. Osteogenic differentiation of recombinant adeno-associated virus 2-transduced murine mesenchymal stem cells and development of an immunocompetent mouse model for ex vivo osteoporosis gene therapy. *Hum Gene Ther* 2004;**15**(12):1197–206. Epub 2005/02/03. PubMed PMID: 15684696.

269. Stender S, Murphy M, O'Brien T, Stengaard C, Ulrich-Vinther M, Soballe K, et al. Adeno-associated viral vector transduction of human mesenchymal stem cells. *Eur Cell Mater* 2007;**13**:93–9. Discussion 9. Epub 2007/06/01. PubMed PMID: 17538898.

270. Hacein-Bey-Abina S, Garrigue A, Wang GP, Soulier J, Lim A, Morillon E, et al. Insertional oncogenesis in 4 patients after retrovirus-mediated gene therapy of SCID-X1. *J Clin Invest* 2008;**118**(9):3132–42. Epub 2008/08/09. PubMed PMID: 18688285 PMCID: PMC2496963.

271. Hacein-Bey-Abina SvKC, Schmidt M, Le Deist F, Wulffraat N, McIntyre E, Radford I, et al. A serious adverse event after successful gene therapy for X-linked severe combined immunodeficiency. *N Engl J Med* 2003;**348**(3):255–6.

272. Bernardo ME, Zaffaroni N, Novara F, Cometa AM, Avanzini MA, Moretta A, et al. Human bone marrow derived mesenchymal stem cells do not undergo transformation after long-term in vitro culture and do not exhibit telomere maintenance mechanisms. *Cancer Res* 2007;**67**(19):9142–9. Epub 2007/10/03. PubMed PMID: 17909019.

273. Altanerova V, Horvathova E, Matuskova M, Kucerova L, Altaner C. Genotoxic damage of human adipose-tissue derived mesenchymal stem cells triggers their terminal differentiation. *Neoplasma* 2009;**56**(6):542–7. Epub 2009/09/05. PubMed PMID: 19728764.

274. Sanada C, Kuo CJ, Colletti EJ, Soland M, Mokhtari S, Knovich MA, et al. Mesenchymal stem cells contribute to endogenous FVIII:c production. *J Cell Physiol* 2013;**228**(5):1010–6. Epub 2012/10/09. PubMed PMID: 23042590.

275. Chuah MK, Brems H, Vanslembrouck V, Collen D, VandenDriessche T. Bone marrow stromal cells as targets for gene therapy of hemophilia A. *Hum Gene Ther* 1998;**9**(3):353–65. Epub 1998/03/21. PubMed PMID: 9508053.

276. Chuah MK, Van Damme A, Zwinnen H, Goovaerts I, Vanslembrouck V, Collen D, et al. Long-term persistence of human bone marrow stromal cells transduced with factor VIII-retroviral vectors and transient production of therapeutic levels of human factor VIII in nonmyeloablated immunodeficient mice. *Hum Gene Ther* 2000;**11**(5):729–38. Epub 2000/04/11. PubMed PMID: 10757352.

277. Doering CB. Retroviral modification of mesenchymal stem cells for gene therapy of hemophilia. *Methods Mol Biol* 2008;**433**:203–12. Epub 2008/08/06. PubMed PMID: 18679625.

278. Porada CD, Sanada C, Kuo CJ, Colletti E, Mandeville W, Hasenau J, et al. Phenotypic correction of hemophilia A in sheep by postnatal intraperitoneal transplantation of FVIII-expressing MSC. *Exp Hematol* 2011;**39**(12):1124–35. Epub 2011/09/13. PubMed PMID: 21906573.

279. Van Damme A, Chuah MK, Dell'accio F, De Bari C, Luyten F, Collen D, et al. Bone marrow mesenchymal cells for haemophilia A gene therapy using retroviral vectors with modified long-terminal repeats. *Haemophilia* 2003;**9**(1):94–103. Epub 2003/02/01. PubMed PMID: 12558785.

280. Chiang GG, Rubin HL, Cherington V, Wang T, Sobolewski J, McGrath CA, et al. Bone marrow stromal cell-mediated gene therapy for hemophilia A: in vitro expression of human factor VIII with high biological activity requires the inclusion of the proteolytic site at amino acid 1648. *Hum Gene Ther* 1999;**10**(1):61–76. PubMed PMID: 10022531.

281. Bartholomew A, Patil S, Mackay A, Nelson M, Buyaner D, Hardy W, et al. Baboon mesenchymal stem cells can be genetically modified to secrete human erythropoietin in vivo. *Hum Gene Ther* 2001;**12**(12):1527–41. PubMed PMID: 11506695.

282. Devine SM, Bartholomew AM, Mahmud N, Nelson M, Patil S, Hardy W, et al. Mesenchymal stem cells are capable of homing to the bone marrow of non-human primates following systemic infusion. *Exp Hematol* 2001;**29**(2):244–55. PubMed PMID: 11166464.

283. Almeida-Porada G, Porada C, Zanjani ED. Plasticity of human stem cells in the fetal sheep model of human stem cell transplantation. *Int J Hematol* 2004;**79**(1):1–6. PubMed PMID: 14979471.

284. Russo FP, Alison MR, Bigger BW, Amofah E, Florou A, Amin F, et al. The bone marrow functionally contributes to liver fibrosis. *Gastroenterology* 2006;**130**(6):1807–21. PubMed PMID: 16697743.

285. Feldmann G, Scoazec JY, Racine L, Bernuau D. Functional hepatocellular heterogeneity for the production of plasma proteins. *Enzyme* 1992;**46**(1–3):139–54. PubMed PMID: 1289079.

286. Colletti E, Airey JA, Liu W, Simmons PJ, Zanjani ED, Porada CD, et al. Generation of tissue-specific cells by MSC does not require fusion or donor to host mitochondrial/membrane transfer. *Stem Cell Res* 2009;**2**(2):125–38. http://dx.doi.org/10.1016/j.scr.2008.08.002. Epub 2008/09/16.

287. Aurich I, Mueller LP, Aurich H, Luetzkendorf J, Tisljar K, Dollinger M, et al. Functional integration of human mesenchymal stem cell-derived hepatocytes into mouse livers. *Gut* 2006;**56**(3):405–15. PubMed PMID: 16928726.

288. Zanjani ED, Ascensao JL, Tavassoli M. Liver-derived fetal hematopoietic stem cells selectively and preferentially home to the fetal bone marrow. *Blood* 1993;**81**(2):399–404. PubMed PMID: 8093667.

289. Traas AM, Wang P, Ma X, Tittiger M, Schaller L, O'Donnell P, et al. Correction of clinical manifestations of canine mucopolysaccharidosis I with neonatal retroviral vector gene therapy. *Mol Ther* 2007;**15**(8):1423–31. Epub 2007/05/24. PubMed PMID: 17519893.

290. Mancuso ME, Graca L, Auerswald G, Santagostino E. Haemophilia care in children–benefits of early prophylaxis for inhibitor prevention. *Haemophilia* 2009;**15**(Suppl. 1):8–14. Epub 2009/01/15. PubMed PMID: 19125935.

291. Yamagami T, Porada C, Chamberlain J, Zanjani E, Almeida-Porada G. Alterations in host immunity following in utero transplantation of human mesenchymal stem cells (MSC). *Exp Hematol* 2006;**34**(9 Suppl. 1):39.

292. Gangadharan B, Parker ET, Ide LM, Spencer HT, Doering CB. High-level expression of porcine factor VIII from genetically modified bone marrow-derived stem cells. *Blood* 2006;**107**(10):3859–64. PubMed PMID: 16449528.

293. Bhakta S, Hong P, Koc O. The surface adhesion molecule CXCR4 stimulates mesenchymal stem cell migration to stromal cell-derived factor-1 in vitro but does not decrease apoptosis under serum deprivation. *Cardiovasc Revasc Med* 2006;**7**(1):19–24. PubMed PMID: 16513519.

294. Brown BD, Lillicrap D. Dangerous liaisons: the role of "danger" signals in the immune response to gene therapy. *Blood* 2002;**100**(4):1133–40. Epub 2002/08/01. PubMed PMID: 12149189.

295. Son BR, Marquez-Curtis LA, Kucia M, Wysoczynski M, Turner AR, Ratajczak J, et al. Migration of bone marrow and cord blood mesenchymal stem cells in vitro is regulated by stromal-derived factor-1-CXCR4 and hepatocyte growth factor-c-met axes and involves matrix metalloproteinases. *Stem Cells* 2006;**24**(5):1254–64. PubMed PMID: 16410389.

296. Sun J, Hakobyan N, Valentino LA, Feldman BL, Samulski RJ, Monahan PE. Intraarticular factor IX protein or gene replacement protects against development of hemophilic synovitis in the absence of circulating factor IX. *Blood* 2008;**112**(12):4532–41. Epub 2008/08/22. PubMed PMID: 18716130.

297. Chuah MK, Collen D, VandenDriessche T. Clinical gene transfer studies for hemophilia A. *Semin Thromb Hemost* 2004;**30**(2):249–56. PubMed PMID: 15118936.

298. White 2nd GC. Gene therapy in hemophilia: clinical trials update. *Thromb Haemost* 2001;**86**(1):172–7. PubMed PMID: 11487005.

299. Chuah MK, Collen D, Vandendriessche T. Preclinical and clinical gene therapy for haemophilia. *Haemophilia* 2004;**10**(Suppl. 4):119–25. PubMed PMID: 15479384.

300. Graw J, Brackmann HH, Oldenburg J, Schneppenheim R, Spannagl M, Schwaab R, et al. from mutation analysis to new therapies. *Nat Rev Genet* 2005;**6**(6):488–501. PubMed PMID: 15931172.

301. Herzog RW, Arruda VR. Update on gene therapy for hereditary hematological disorders. *Expert Rev Cardiovasc Ther* 2003;**1**(2):215–32. PubMed PMID: 15030282.

302. Hough C, Lillicrap D. Gene therapy for hemophilia: an imperative to succeed. *J Thromb Haemost* 2005;**3**(6):1195–205. PubMed PMID: 15946210.

303. Nathwani AC, Davidoff AM, Tuddenham EG. Prospects for gene therapy of haemophilia. *Haemophilia* 2004;**10**(4):309–18. PubMed PMID: 15230943.

304. Roth DA, Tawa Jr NE, O'Brien JM, Treco DA, Selden RF. Nonviral transfer of the gene encoding coagulation factor VIII in patients with severe hemophilia A. *N Engl J Med* 2001;**344**(23):1735–42. PubMed PMID: 11396439.

305. VandenDriessche T, Collen D, Chuah MK. Gene therapy for the hemophilias. *J Thromb Haemost* 2003;**1**(7):1550–8. PubMed PMID: 12871290.

306. Buchlis G, Podsakoff GM, Radu A, Hawk SM, Flake AW, Mingozzi F, et al. Factor IX expression in skeletal muscle of a severe hemophilia B patient 10 years after AAV-mediated gene transfer. *Blood* 2012;**119**(13):3038–41. Epub 2012/01/25. PubMed PMID: 22271447 PMCID: PMC3321866.

307. Kay MA, Manno CS, Ragni MV, Larson PJ, Couto LB, McClelland A, et al. Evidence for gene transfer and expression of factor IX in haemophilia B patients treated with an AAV vector. *Nat Genet* 2000;**24**(3):257–61. PubMed PMID: 10700178.

308. Margaritis P, High KA. Gene therapy in haemophilia–going for cure? *Haemophilia* 2010;**16**(Suppl. 3):24–8. Epub 2010/07/09. PubMed PMID: 20586798.

309. Mingozzi F, High KA. Therapeutic in vivo gene transfer for genetic disease using AAV: progress and challenges. *Nat Rev Genet* 2011;**12**(5):341–55. Epub 2011/04/19. PubMed PMID: 21499295.

310. Petrus I, Chuah M, VandenDriessche T. Gene therapy strategies for hemophilia: benefits versus risks. *J Gene Med* 2010;**12**(10):797–809. Epub 2010/09/18. PubMed PMID: 20848668.

311. Scott DW, Lozier JN. Gene therapy for haemophilia: prospects and challenges to prevent or reverse inhibitor formation. *Br J Haematol* 2012;**156**(3):295–302. Epub 2011/11/08. PubMed PMID: 22055221 PMCID: PMC3257353.

312. Viiala NO, Larsen SR, Rasko JE. Gene therapy for hemophilia: clinical trials and technical tribulations. *Semin Thromb Hemost* 2009;**35**(1):81–92. Epub 2009/03/25. PubMed PMID: 19308896.

313. Boyce N. Trial halted after gene shows up in semen. *Nature* 2001;**414**(6865):677. PubMed PMID: 11742355.

314. Marshall E. Gene therapy. Panel reviews risks of germ line changes. *Science* 2001;**294**(5550):2268–9. PubMed PMID: 11743173.

315. Arruda VR, Fields PA, Milner R, Wainwright L, De Miguel MP, Donovan PJ, et al. Lack of germline transmission of vector sequences following systemic administration of recombinant AAV-2 vector in males. *Mol Ther* 2001;**4**(6):586–92. PubMed PMID: 11735343.

316. Roehl HH, Leibbrandt ME, Greengard JS, Kamantigue E, Glass WG, Giedlin M, et al. Analysis of testes and semen from rabbits treated by intravenous injection with a retroviral vector encoding the human factor VIII gene: no evidence of germ line transduction. *Hum Gene Ther* 2000;**11**(18):2529–40. PubMed PMID: 11119423.

317. Nathwani AC, Tuddenham EG, Rangarajan S, Rosales C, McIntosh J, Linch DC, et al. Adenovirus-associated virus vector-mediated gene transfer in hemophilia B. *N Engl J Med* 2011;**365**(25):2357–65. Epub 2011/12/14. PubMed PMID: 22149959.

318. Nathwani A, Tuddenham E, Rosales C, McIntosh J, Riddell A, Rustagi P, et al. Early clinical trial results following administration of a low dose of a novel self complementary adeno-associated viral vector encoding human factor IX in two subjects with severe hemophilia B. *Blood* 2010;**116**:248a.

319. Dobrzynski E, Fitzgerald JC, Cao O, Mingozzi F, Wang L, Herzog RW. Prevention of cytotoxic T lymphocyte responses to factor IX-expressing hepatocytes by gene transfer-induced regulatory T cells. *Proc Natl Acad Sci USA* 2006;**103**(12):4592–7. PubMed PMID: 16537361.

320. Herzog RW, Dobrzynski E. Immune implications of gene therapy for hemophilia. *Semin Thromb Hemost* 2004;**30**(2):215–26. PubMed PMID: 15118933.

321. Knolle PA, Gerken G. Local control of the immune response in the liver. *Immunol Rev* 2000;**174**:21–34. PubMed PMID: 10807504.

322. Mingozzi F, Liu YL, Dobrzynski E, Kaufhold A, Liu JH, Wang Y, et al. Induction of immune tolerance to coagulation factor IX antigen by in vivo hepatic gene transfer. *J Clin Invest* 2003;**111**(9):1347–56. PubMed PMID: 12727926.

323. Arnold B. Parenchymal cells in immune and tolerance induction. *Immunol Lett* 2003;**89**(2–3):225–8. PubMed PMID: 14556982.

324. De Geest BR, Van Linthout SA, Collen D. Humoral immune response in mice against a circulating antigen induced by adenoviral transfer is strictly dependent on expression in antigen-presenting cells. *Blood* 2003;**101**(7):2551–6. PubMed PMID: 12446451.

325. Cao O, Dobrzynski E, Wang L, Nayak S, Mingle B, Terhorst C, et al. Induction and role of regulatory CD4+CD25+ T cells in tolerance to the trans-gene product following hepatic in vivo gene transfer. *Blood* 2007;**110**(4):1132–40. PubMed PMID: 17438084.

326. Dobrzynski E, Mingozzi F, Liu YL, Bendo E, Cao O, Wang L, et al. Induction of antigen-specific CD4+ T-cell anergy and deletion by in vivo viral gene transfer. *Blood* 2004;**104**(4):969–77. PubMed PMID: 15105293.

327. Markusic DM, Herzog RW. Liver-Directed adeno-associated viral gene therapy for hemophilia. *J Genet Syndr Gene Ther* 2012;**1**:1–9. Epub 2012/01/18. PubMed PMID: 23565343 PMCID: PMC3615444.

328. Markusic DM, Herzog RW, Aslanidi GV, Hoffman BE, Li B, Li M, et al. High-efficiency transduction and correction of murine hemophilia B using AAV2 vectors devoid of multiple surface-exposed tyrosines. *Mol Ther* 2010;**18**(12):2048–56. Epub 2010/08/26. PubMed PMID: 20736929.

329. Niemeyer GP, Herzog RW, Mount J, Arruda VR, Tillson DM, Hathcock J, et al. Long-term correction of inhibitor-prone hemophilia B dogs treated with liver-directed AAV2-mediated factor IX gene therapy. *Blood* 2009;**113**(4):797–806. Epub 2008/10/30. PubMed PMID: 18957684.

330. Bowen DG, McCaughan GW, Bertolino P. Intrahepatic immunity: a tale of two sites?. *Trends Immunol* 2005;**26**(10):512–7. PubMed PMID: 16109501.

331. Brown BD, Cantore A, Annoni A, Sergi LS, Lombardo A, Della Valle P, et al. A microRNA-regulated lentiviral vector mediates stable correction of hemophilia B mice. *Blood* 2007;**110**(13):4144–52. Epub 2007/08/30. PubMed PMID: 17726165.

332. Brown BD, Venneri MA, Zingale A, Sergi Sergi L, Naldini L. Endogenous microRNA regulation suppresses transgene expression in hematopoietic lineages and enables stable gene transfer. *Nat Med* 2006;**12**(5):585–91. PubMed PMID: 16633348.

333. Ferber I, Schonrich G, Schenkel J, Mellor AL, Hammerling GJ, Arnold B. Levels of peripheral T cell tolerance induced by different doses of tolero-gen. *Science* 1994;**263**(5147):674–6. PubMed PMID: 8303275.

334. Follenzi A, Battaglia M, Lombardo A, Annoni A, Roncarolo MG, Naldini L. Targeting lentiviral vector expression to hepatocytes limits transgene-specific immune response and establishes long-term expression of human antihemophilic factor IX in mice. *Blood* 2004;**103**(10):3700–9. PubMed PMID: 14701690.

335. Follenzi A, Sabatino G, Lombardo A, Boccaccio C, Naldini L. Efficient gene delivery and targeted expression to hepatocytes in vivo by improved lentiviral vectors. *Hum Gene Ther* 2002;**13**(2):243–60. PubMed PMID: 11812281.

336. Franco LM, Sun B, Yang X, Bird A, Zhang H, Schneider A, et al. Evasion of immune responses to introduced human acid α-glucosidase by liver-restricted expression in glycogen storage disease type II. *Mol Ther* 2005;**12**(5):876–84. PubMed PMID: 16005263.

337. Nathwani AC, Nienhuis AW, Davidoff AM. Our journey to successful gene therapy for hemophilia B. *Hum Gene Ther* 2014;**25**(11):923–6. PubMed PMID: 25397929 PMCID: PMC4236090.

338. Charlesworth D. *BioMarin licenses factor VIII gene therapy program for hemophilia A from University College London and St. Jude Children's Research Hospital*. San Rafael: GLOBE NEWSWIRE; 2013. Available from: http://investors.bmrn.com/releasedetail.cfm?ReleaseID=742285. [cited 22.08.13].

339. Ye P, Thompson AR, Sarkar R, Shen Z, Lillicrap DP, Kaufman RJ, et al. Naked DNA transfer of factor VIII induced transgene-specific, species-independent immune response in hemophilia A mice. *Mol Ther* 2004;**10**(1):117–26. PubMed PMID: 15233948.

340. Chitlur M, Warrier I, Rajpurkar M, Lusher JM. Inhibitors in factor IX deficiency a report of the ISTH-SSC international FIX inhibitor registry (1997–2006). *Haemophilia* 2009;**15**(5):1027–31. Epub 2009/06/12. PubMed PMID: 19515028.

341. Ehrenforth S, Kreuz W, Scharrer I, Linde R, Funk M, Gungor T, et al. Incidence of development of factor VIII and factor IX inhibitors in haemo-philiacs. *Lancet* 1992;**339**(8793):594–8. Epub 1992/03/07. PubMed PMID: 1347102.

342. Pearson EG, Flake AW. Stem cell and genetic therapies for the fetus. *Semin Pediatr Surg* 2013;**22**(1):56–61. Epub 2013/02/12. PubMed PMID: 23395147.

343. Troeger C, Surbek D, Schoberlein A, Schatt S, Dudler L, Hahn S, et al. In utero haematopoietic stem cell transplantation. Experiences in mice, sheep and humans. *Swiss Med Wkly* 2006;**136**(31–32):498–503. PubMed PMID: 16947088.

344. Vrecenak JD, Flake AW. In utero hematopoietic cell transplantation-recent progress and the potential for clinical application. *Cytotherapy* 2013;**15**(5):525–35. Epub 2013/02/19. PubMed PMID: 23415921.

345. Le Blanc K, Gotherstrom C, Ringden O, Hassan M, McMahon R, Horwitz E, et al. Fetal mesenchymal stem-cell engraftment in bone after in utero transplantation in a patient with severe osteogenesis imperfecta. *Transplantation* 2005;**79**(11):1607–14. PubMed PMID: 15940052.

346. Flake AW. In utero stem cell transplantation. *Best Pract Res Clin Obstet Gynaecol* 2004;**18**(6):941–58. PubMed PMID: 15582548.

347. Flake AW, Zanjani ED. In utero hematopoietic stem cell transplantation: ontogenic opportunities and biologic barriers. *Blood* 1999;**94**(7):2179–91. PubMed PMID: 10498587.

348. Merianos D, Heaton T, Flake AW. In utero hematopoietic stem cell transplantation: progress toward clinical application. *Biol Blood Marrow Trans-plant* 2008;**14**(7):729–40. PubMed PMID: 18541191.

349. Roybal JL, Santore MT, Flake AW. Stem cell and genetic therapies for the fetus. *Semin Fetal Neonatal Med* 2010;**15**(1):46–51. Epub 2009/06/23. PubMed PMID: 19540822.

350. Balak DM, Gouw SC, Plug I, Mauser-Bunschoten EP, Vriends AH, Van Diemen-Homan JE, et al. Prenatal diagnosis for haemophilia: a nationwide survey among female carriers in the Netherlands. *Haemophilia* 2012;**18**(4):584–92. Epub 2012/01/19. PubMed PMID: 22250892.

351. Chalmers E, Williams M, Brennand J, Liesner R, Collins P, Richards M. Guideline on the management of haemophilia in the fetus and neonate. *Br J Haematol* 2011;**154**(2):208–15. Epub 2011/05/11. PubMed PMID: 21554256.

352. Dai J, Lu Y, Ding Q, Wang H, Xi X, Wang X. The status of carrier and prenatal diagnosis of haemophilia in China. *Haemophilia* 2012;**18**(2):235–40. Epub 2011/09/14. PubMed PMID: 21910785.

353. Deka D, Dadhwal V, Roy KK, Malhotra N, Vaid A, Mittal S. Indications of 1342 fetal cord blood sampling procedures performed as an integral part of high risk pregnancy care. *J Obstet Gynaecol India* 2012;**62**(1):20–4. Epub 2013/02/02. PubMed PMID: 23372284 PMCID: PMC3366572.

354. Massaro JD, Wiezel CE, Muniz YC, Rego EM, de Oliveira LC, Mendes-Junior CT, et al. Analysis of five polymorphic DNA markers for indirect genetic diagnosis of haemophilia A in the Brazilian population. *Haemophilia* 2011;**17**(5):e936–43. Epub 2011/06/09. PubMed PMID: 21649803.

355. Peyvandi F. Carrier detection and prenatal diagnosis of hemophilia in developing countries. *Semin Thromb Hemost* 2005;**31**(5):544–54. Epub 2005/11/09. PubMed PMID: 16276463.

356. Shetty S, Ghosh K, Jijina F. First-trimester prenatal diagnosis in haemophilia A and B families–10 years experience from a centre in India. *Prenat Diagn* 2006;**26**(11):1015–7. Epub 2006/08/31. PubMed PMID: 16941728.

357. Silva Pinto C, Fidalgo T, Salvado R, Marques D, Goncalves E, Martinho P, et al. Molecular diagnosis of haemophilia A at Centro Hospitalar de Coimbra in Portugal: study of 103 families–15 new mutations. *Haemophilia* 2012;**18**(1):129–38. Epub 2011/06/08. PubMed PMID: 21645180.

358. Sasanakul W, Chuansumrit A, Ajjimakorn S, Krasaesub S, Sirachainan N, Chotsupakarn S, et al. Cost-effectiveness in establishing hemophilia carrier detection and prenatal diagnosis services in a developing country with limited health resources. *Southeast Asian J Trop Med Public Health* 2003;**34**(4):891–8. Epub 2004/04/30. PubMed PMID: 15115107.

359. Tsui NB, Kadir RA, Chan KC, Chi C, Mellars G, Tuddenham EG, et al. Noninvasive prenatal diagnosis of hemophilia by microfluidics digital PCR analysis of maternal plasma DNA. *Blood* 2011;**117**(13):3684–91. Epub 2011/01/26. PubMed PMID: 21263151.

360. Hussein IR, El-Beshlawy A, Salem A, Mosaad R, Zaghloul N, Ragab L, et al. The use of DNA markers for carrier detection and prenatal diagnosis of haemophilia A in Egyptian families. *Haemophilia* 2008;**14**(5):1082–7. Epub 2008/06/13. PubMed PMID: 18547262.

361. Touraine JL. Transplantation of human fetal liver cells into children or human fetuses. In: Bhattacharya N, editor. *Human fetal tissue transplantation*. Springer; 2013.

362. Hassan HJ, Leonardi A, Chelucci C, Mattia G, Macioce G, Guerriero R, et al. Blood coagulation factors in human embryonic-fetal development: preferential expression of the FVII/tissue factor pathway. *Blood* 1990;**76**(6):1158–64. Epub 1990/09/15. PubMed PMID: 1698100.

363. Ong K, Horsfall W, Conway EM, Schuh AC. Early embryonic expression of murine coagulation system components. *Thromb Haemost* 2000;**84**(6):1023–30. Epub 2001/01/12. PubMed PMID: 11154109.

364. Manco-Johnson MJ. Development of hemostasis in the fetus. *Thromb Res* 2005;**115**(Suppl. 1):55–63. Epub 2005/03/26. PubMed PMID: 15790157.

365. Colletti E, Lindstedt S, Park P, Almeida-Porada G, Porada C. Early fetal gene delivery utilizes both central and peripheral mechanisms of tolerance induction. *Exp Hematol* 2008;**36**(7):816–22.

366. Tran ND, Porada CD, Almeida-Porada G, Glimp HA, Anderson WF, Zanjani ED. Induction of stable prenatal tolerance to β-galactosidase by in utero gene transfer into preimmune sheep fetuses. *Blood* 2001;**97**(11):3417–23. PubMed PMID: 11369632.

367. Lipshutz GS, Flebbe-Rehwaldt L, Gaensler KM. Reexpression following readministration of an adenoviral vector in adult mice after initial in utero adenoviral administration. *Mol Ther* 2000;**2**(4):374–80. PubMed PMID: 11020353.

368. Waddington SN, Buckley SM, Nivsarkar M, Jezzard S, Schneider H, Dahse T, et al. In utero gene transfer of human factor IX to fetal mice can induce postnatal tolerance of the exogenous clotting factor. *Blood* 2003;**101**(4):1359–66. PubMed PMID: 12393743.

369. Waddington SN, Nivsarkar MS, Mistry AR, Buckley SM, Kemball-Cook G, Mosley KL, et al. Permanent phenotypic correction of hemophilia B in immunocompetent mice by prenatal gene therapy. *Blood* 2004;**104**(9):2714–21. PubMed PMID: 15231566.

370. Roncarolo MG, Bacchetta R. T cell repertoire and tolerance after fetal stem cell transplantation. *Bone Marrow Transplant* 1992;**9**(Suppl. 1):127–8.

371. Santore MT, Roybal JL, Flake AW. Prenatal stem cell transplantation and gene therapy. *Clin Perinatol* 2009;**36**(2):451–71. xi. Epub 2009/06/30. PubMed PMID: 19559331.

372. Touraine JL, Roncarolo MG, Raudrant D, Bacchetta R, Golfier F, Sembeil R, et al. Induction of transplantation tolerance in humans using fetal cell transplants. *Transplant Proc* 2005;**37**(1):65–6. Epub 2005/04/06. PubMed PMID: 15808548.

373. Touraine JL, Sanhadji K. Transplantation tolerance induced in humans at the fetal or the neonatal stage. *J Transplant* 2011;**2011**:760319. Epub 2011/08/31. PubMed PMID: 21876781.

374. Mannucci PM, Tuddenham EG. The hemophilias–from royal genes to gene therapy. *N Engl J Med* 2001;**344**(23):1773–9. PubMed PMID: 11396445.

375. MacKenzie TC, David A, Flake AW, Almeida-Porada G. Consensus statement from the first international conference for in utero stem cell transplantation and gene therapy. *Front Pharmacol Integr Regen Pharmacol* 2014;**6**:15.

Chapter 13

Gene Therapy in Monogenic Congenital Myopathies

Xuan Guan*, Melissa A. Goddard*, David L. Mack, Martin K. Childers

Key Concepts

1. Gene therapy: introduction of nucleic acids, including DNA, RNA, and their analogs into cells of living organism to treat diseases.
2. Monogenic congenital myopathy: Genetic muscular diseases resulting from the alteration of single gene.
3. Gene therapy vector: the vehicles shuttle therapeutic genetic material into cells.
4. Preclinical disease models: biological model systems, which recapitulates certain aspects of human diseases, can be utilized as testbed to test novel therapies before entering clinical trials.

1. INTRODUCTION TO MONOGENIC CONGENITAL MYOPATHIES

Current treatment options for patients with monogenic congenital myopathies (MCM) ameliorate the symptoms of the disorder without resolving the underlying cause. However, therapies are being developed where the mutated or deficient gene target is replaced. Thousands of clinical trials have been undertaken relating to gene therapy, with around 9% focused on monogenic diseases such as Duchenne muscular dystrophy (DMD) and limb girdle muscular dystrophy (LGMD).[1] Preclinical findings in animal models have been promising, as illustrated by studies of a potential treatment for X-linked myotubular myopathy (XLMTM) in canine and murine models.[2] We will therefore discuss the prospective applications and approaches of gene replacement therapy, using these disorders as examples. Both limb girdle muscular dystrophy type 2C and Duchenne muscular dystrophy are part of a subclass of myopathies known as dystrophies, diseases where muscle degeneration is accompanied by replacement with fatty or connective tissue. DMD is caused by X chromosome-linked genetic mutations leading to the absence of membrane-anchored dystrophin protein, the centerpiece of the large dystroglycan complex that plays a pivotal role in sarcolemma stability during muscle contraction[3] (Table 1). The symptoms are visible as early as 2–3 years of age, a progressive decrease in striated muscle function, starting from proximal muscle such as legs and pelvis and eventually involve the whole body. Most patients are wheelchair-dependent starting from early teen. The average life expectancy is around 25 years (http://www.nlm.nih.gov/medlineplus/ency/article/000705.htm), with respiratory failure and cardiac complications the highest causes of mortality. Congenital centronuclear myopathies are inherited muscle diseases where the nucleus is located in the center of the muscle fiber instead of the periphery. X-linked myotubular myopathy (XLMTM) is the most common centronuclear myopathy, affecting an estimated 1 in 50,000 male births (Table 1).[4,5] The disease is due to a mutation on the long-arm of the X chromosome, usually inherited by hemizygous boys from an asymptomatic carrier mother.[6] This mutation causes a deficiency of the protein myotubularin.[7] Myotubularin has been identified as a phosphoinositol phosphatase and may be critical to normal excitation–contraction coupling and remodeling of the sarcoplasmic reticulum in muscle.[8] When XLMTM patients are first born, they typically exhibit hypotonia and may be blue due to respiratory insufficiency.[6] The disease is often fatal in the first year of life and long-term survivors may require ventilatory support.[9] Affected boys are particularly susceptible to infection and respiratory dysfunction is the leading cause of death.[10]

Although there are differences between the symptomatic presentations of these diseases, there are some shared difficulties to consider when designgene therapies. High vector titers may be required to reach an effective dose,[11] increasing the chance of adverse effects in patients. In addition, the need to treat respiratory muscles as well as the heart in DMD

---

* Equal contribution.

Translating Regenerative Medicine to the Clinic. http://dx.doi.org/10.1016/B978-0-12-800548-4.00013-9

TABLE 1 A comparison of Some of the Monogenic Congenital Myopathies Presently under Study

	Duchenne muscular dystrophy (DMD)	X-linked myotubular myopathy (XLMTM)	Facioscapulohumeral muscular dystrophy (FSHD)	Myotonic Dystrophy (DM)	Limb-Girdle muscular dystrophy (LGMD) 2C and 2D
Inheritance	Single gene mutation on the X chromosome	Single gene mutation at q28 on the X chromosome[17]	Autosomal dominant, contraction of D4Z4 repeat on chromosome 4q35 and toxic gain of function of the DUX4 gene[18]	Autosomal dominant. DM1 CTG triplet repeats expansion of DMPK gene locates on chromosome 19.[19] DM2 CCTG tetranucleiotide repeat expansion of ZNF9 gene on chromosome 3[20]	Autosomal recessive. Single gene mutation on chromosome 13 and 17 (2C and 2D, respectively)
Molecular biology	Deficiency of the protein dystrophin	Deficiency of the phosphatase myotubularin[17]	Toxic protein product of DUX4[18]	Malfunctioned DMPK and ZNF9 proteins	Deficiency of gamma and alpha Sarcoglycan (2C and 2D, respectively).
Clinical symptoms	Weakness of the skeletal muscles; Respiratory insufficiency in teens; Cardiac dysfunction; Leading cause of death is cardiorespiratory failure	Weakness of the skeletal muscles[10]; Wheelchair dependence[10]; Respiratory insufficiency at birth[21]; No cardiac phenotype; Leading cause of death is respiratory dysfunction[22]	Initial weakness of facial, scapula and humeral muscle, progressively involving other muscles; Sparing respiratory muscle.	Muscle wasting and myotonic; Heart conduction block; Cataract; Infertility.	Muscle wasting primarily involve proximal muscle such as hip and shoulder
Demographics	Presentation around 2-3 years of age; Average life expectancy of 25 years.	Affects 1 in 5000 male births[4]; Presentation typically at birth[23]; Average life expectancy of 29 months[9].	Affects 12/100,000[24]	Affects 1/8000 people worldwide. Type 1 most common in most countries.[25]	Up to 68% of individuals with childhood onset and ~10% with adult onset[26]
Histology	Increased fibre size variability; Cycling of fibre regeneration and degeneration.	Centrally located nucleus[27]; Variably-sized myofibers with an abnormally large number of small fibers; Organelle abnormality and "necklace fibers"[28].	Non-specific fiber necrosis, increased variation in fiber size, internal nuclei, fiber type variability, connective tissue and fat proliferation. Mononuclear cell filtration.[29]	Fiber atrophy, internal nuclei, pyknotic nuclear clumps, lipofuscin accumulation, increased fiber size variation[30]	Variation of fiber diameter, fiber degeneration and regeneration, split fibers, ring fibers

and XLMTM may complicate delivery. Improvements in delivery methods[12] and in vector characterization to increase efficiency may address this problem.[13] Vector modification may also ensure more efficient delivery to the muscle and improve safety by reducing off-target delivery to organs like the liver.[11,14] Tissue-specific promoters is another strategy to secure tissue-specific transgene expression. Immune response is a major concern, particularly in genetically-null patients who may have antibodies against the gene product produced by the treatment[15] and immunosuppression before and during treatment may have to be considered. There are also challenges specific to each disease. For example, the large size of the dystrophin gene limits the choice of vector to be used in treatment. Overexpression of γ-sarcoglycan in LGMD patients may exacerbate the condition.[16] Significant wasting in XLMTM patients leaves very little muscle to treat and, due to the young age of the patients, selecting an appropriate and reproducible outcome measure may prove difficult. We will be discussing new developments that address these concerns, including modifications of the vector and the combination of gene therapy with other approaches.

2. GENE THERAPY

2.1 What Is Gene Therapy

Gene therapy is defined as the introduction of nucleic acids, including DNA, RNA, and their analogs into cells of living organism to treat diseases.[31] This occurs through modified expression of genes of interest to trigger alterations of certain biological functions. Gene therapy targets living cells, primarily because cell's intrinsic gene expression machinery is indispensable to mediate the production of therapeutic molecules, including protein, shRNA, and microRNA (miRNA).

Since classical gene therapy acts on native tissues, the abundance of target cells largely determines the effect of the gene therapy. This is especially true in congenital myopathies. In the advanced stage of diseases, such as DMD and XLMTM, surviving myocytes are so limited that even if the function of individual myofibers were fully restored, there would be no appreciable functional improvement on tissue level. The advent of stem cell technology, especially the discovery of induced pluripotent stem cells (iPSCs),[32,33] has the potential to overcome this hurdle. Pluripotent stem cells may be able to replenish tissue loss through their indefinite self-replicating potential and capacity to be converted into nearly all cell types within the body. The advantages of combining stem cell therapy with gene therapy have been demonstrated in several animal studies, in which vectors were administered ex vivo and modified donor cells were later engrafted into native tissue.[34–36]

Various gene therapy strategies target gene expression and regulatory network at different levels. For example, genetic sequence can be permanently inserted into genome for long-term expression, using retrovirus or lentivirus.[31] With the development of genome modification tools[37] such as clustered regularly interspersed short palindromic repeats (CRISPR) enzymes and Transcription activator-like effector nuclease (TALEN), the technical barrier of *in situ* editing eukaryotic genomic DNA has been substantially lowered. These techniques hold the potential to seamlessly restore the genome to a disease-free state without the introduction of foreign genetic elements, eliminating the widely shared concern of increased tumorigenic risk. Alternatively, the therapeutic genetic sequence can be designed to persist within cells as a stable episome, allowing maintenance of long-term expression without genomic integration.[31]

Introduced nucleic acids can be further divided into protein-coding and non-coding sequence. While protein-coding sequencesserve as templates for protein production, non-coding nucleic acidsfunction to modulateepigenetic processes controlling gene expression. A good example is RNA interference, in which microRNA (miRNA)[38] or small interfering RNA (siRNA)[38] bind to mRNA molecules to halt protein translation and mediate mRNA degradation.

2.2 What Will Be Administered

Based on the mechanism of action, gene therapy can be categorized into gene addition, gene correction, or gene subtraction. In gene addition, exogenous genetic sequence is introduced into cells. Gene correction alters the diseased loci. Attenuating or silencing the expression of single or a network of genes is known as gene subtraction.

The root causes of diseases dictate the gene therapy strategy, which differs dramatically among diseases (Table 2). Genomic mutations sometimes interrupt the reading frame of the protein-coding genes, leading to the absence of proteins, as occurs with dystrophin[39] and myotubularin[40] in DMD and XLMTM. Consequently the focus of gene therapy has been devoted to supplement a protein-coding sequence to replace the defective gene. Alternately, the mRNA splicing event can be modified to bypass the mutated region, restoring the reading frame with a truncated but functional protein product.[41] In other cases, where the disease is caused by pathogenic overexpression, it is imperative to silence the gene expression.[42] One such example is facioscapulohumeral muscular dystrophy (FSHD), which is caused by the overexpression of the myopathic DUX4 gene.[18]

TABLE 2 The Different Strategies for Various Monogenic Myopathies

Gene Therapy Strategy	Diseases
Gene addition	DMD,[39] XLMTM,[2] LGMD2D[43]
Gene correction	DMD[3]
Gene subtraction	FSHD,[42] myotonic dystrophy[44]

Abbreviations: DMD, Duchenne muscular dystrophy; XLMTM, X-linked myotubular myopathy; LFMD2D, limb girdle muscular dystrophy type 2D; FSHD, facioscapulohumeral muscular dystrophy.

2.3 Advantages of Gene Therapy in Treating Congenital Myopathies

The goal of gene therapy in treating congenital myopathies is twofold. For some monogenic diseases, the current technology is adequate to completely restore the genetic defect in somatic cells, representing a cure that is unachievable by any other methods. Recent findings in canine and mouse models of XLMTM are good examples, where a single treatment recovered animals with a severe monogenic disease to normal function.[2] For myopathies with complex genetic makeup, a more realistic goal is to delay disease progression and preserve muscle function in order to maintain life quality.

There are several advantages associated with gene therapies in comparison to conventional pharmacotherapies, which are as follows:

- Complete rectification of the abnormal genetic code
  There are currently very limited therapeutic options that effectively target congenital myopathies. Most available drugs are largely symptom-alleviating agents with transient effect. In contrast, gene therapy is designed to target the root genetic cause and thus represent a potential cure for some monogenic diseases.
- High specificity
  Conventional pharmacotherapy utilizes natural or synthetic small molecules aiming to alter certain biological functions. However, it is difficult to identify high-specificity molecules that only interact with the molecule of interest. As a result, pleiotropic effects of these agents are the main source of undesired adverse effects. Gene therapy functions to modulate the production of native proteins and is therefore more specific and effective.
- Long-term duration of efficacy
  Certain vectors, such as lentivirus, retrovirus or AAV,[31] demonstrate persistent effects leading to extended phenotypic correction/disease remission. This is in stark contrast to conventional drugs, requiring repetitive dosing to reach a steady-state drug concentration for stable effects.

2.4 Gene Therapy Clinical Trails for Congenital Myopathies

Ongoing clinical trails are mainly targeting diseases with defective dystrophoglycan complex, including DMD, BMD and LGMD. Local delivery of full-length dystrophin plasmid to patients' radialis muscle results in detection of dystrophin mRNA and protein in 6 out of 9 patients.[45] Encouraged by this study, various gene therapy strategies have been employed in clinical trails aiming to restore dystrophin expression, including read-through agents (Ataluren/PTC124, Arbekacin Sulfate/NPC14 and Gentamicin), exon skipping oligonucleotides (AVI-4658, Drisapersen, Pro044/045/053, SRP4045/4043 and NS-065/NCNP-01) and virus mediated delivery of protein encoding genes, such as minidystrophin and Follistin.Mendell *et al.* reported delivery of truncated minidystrophin gene (NCT00428935)elicited dystrophin specific T cells, without direct visualization of the protein in muscle, suggesting immune responseto be a major hurdlefor successful gene therapy.[15] The Phase 2a study of the read-through agent ataluren reported 61% patients demonstrated increase of dystrophin expression after a course of 28 days treatment (NCT00264888).[46] Intramuscular injection of the exon skipping agent AVI-4658(NCT00159250)resulted in 17% increase of the mean dystrophin signal, reaching 22% to 32% of the healthy control.[47] For LGMD, AAV packaged alpha and gamma sarcoglycan are now being tested in clinical trails. Local injection of AAV1.tMCK.hSGCA (NCT00494195, NCT01976091) led to persistent α-sarcoglycan expression up to 6 months and augment muscle fiber size.[43,48] A trial with AAV1-γ-sarcoglycan vector in LGMD 2C patients has completed though result has not yet been disclosed (NCT01344798). The ongoing clinical trials are summarized in the Table 3.

TABLE 3 Ongoing Gene Therapy Clinical Trails for Congenital Myopathy

NCT Number	Conditions	Interventions	Phases
NCT02354781	DMD	rAAV1.CMV.huFollistin344	Phase 1/Phase 2
NCT01519349	BMD	rAAV1.CMV.huFollistatin344	Phase 1
NCT02255552	DMD	eteplirsen injection	Phase 3
NCT01918384	DMD	NPC-14	Phase 2
NCT02376816	DMD	rAAVrh74.MCK.micro-Dystrophin	Phase 1
NCT01247207	DMD/BMD	Ataluren	Phase 3
NCT01557400	DMD/BMD	Ataluren	Phase 3
NCT01826487	DMD	Ataluren	Phase 3
NCT02090959	DMD	Ataluren	Phase 3
NCT02329769	DMD	PRO044	Phase 2
NCT01826474	DMD	PRO045	Phase 1/Phase 2
NCT01910649	DMD/BMD	Drisapersen	Phase 2
NCT01957059	DMD	PRO053	Phase 1/Phase 2
NCT02310906	DMD	SRP-4053	Phase 1/Phase 2
NCT02500381	DMD	SRP-4045/SRP-4053	Phase 3
NCT02081625	DMD	NS-065/NCNP-01	Phase 1
NCT01976091	LGMD2D	scAAVrh74.tMCK.hSGCA	Phase 1/Phase 2

3. VECTOR TOOLBOX

A critical element of gene therapy is the vector, the vehicle that facilitates the transfer of genetic material. Due to the size and negative charge of ribonucleic acid, shuttles are needed to carry cargo across multiple biological barriers to reach target cells. The vector toolbox is composed of viral vectors and nonviral vectors. Viruses have naturally evolved sophisticated machinery to target cells with high efficiency making them the ideal tool for gene transfer. A summary of the advantages and disadvantages of commonly used viral vectors is outlined in Table 4.

AAV is the preferred viral vector for many ongoing clinical trials, largely due to the fact that AAV efficiently targets post-mitotic parenchymal cells, such as neurons and skeletal myofibers, which are usually impermissive to other vectors.[50] Unlike integrating viral vectors, AAV exists as epichromosome, reducing the likelihood of insertional mutagenesis while maintaining long-term transgene expression. Another advantage of AAV is that the vector genome can be pseudotyped with alternative capsids. The viral capsid largely determines tissue tropism, gene transfer efficacy, and the vector's dose-dependent toxicity. For example, AAV 1, 6, 9 transduce muscle with high efficiency.[51] This array of available serotypes allow the vector to be tailored to different applications. Though many natural AAV variants have been identified, more efforts are being devoted to engineer synthetic viral capsids to address specific clinical challenges. Mutating tyrosine residues (Y445F and Y731F) in the AAV6 capsid can improve the skeletal muscle gene transfer.[52] Moreover, retention of AAV vector in liver has been a problem of systemic infusion. Vectors that "detarget" liver have been created through randomly mutating the surface-exposing region of AAV9 capsid (AAV9.45 and AAV9.61)[53] or by engineering a chimeric capsid with AAV2 and AAV8 (AAV2i8).[54] Both strategies redirected the vector away from liver while maintaining high transduction efficiency to skeletal muscle.

Viral vectors have been widely used in ongoing gene therapy clinical trials 1. Despite their differences, however, the use of any viral-based vector carries with it certain risks including tumorigenicity, immunogenicity, and limited cargo space. Most of these risks can be minimized by nonviral vectors. On the other hand, the main drawbacks of nonviral vectors, including the capacity to cross various biological barriers and stability, have been largely addressed by late breakthroughs. For example, material science has provided various lipid-based and polymer-based DNA vectors, many of which have been tested in clinical trials.[55] Nucleic acid chemistry evolution has resulted in the development of nucleic acid analogs, such as

TABLE 4 Advantages and Disadvantages of the Various Types of Viral Vector

Viral Vector	Advantage	Disadvantages
Retrovirus	• Stable integration • Allowing to be pseudotyped	• Only infect dividing cells. • Size limit of 8kb • Insertional mutagenesis associated tumor risk • Low titers[49]
Lentivirus	• Stable integration for persistent transgene expression; • Infect both dividing and non-dividing cells; • Allowing to be pseudotyped with the VSVG (vesicular stomatitis virus G), making concentration easier.	• Possibility of insertional mutagenesis;
Adenovirus	• Large cargo space (36kb adenoviral genome) • High titer; • Permissiveness to non-dividing cells; • Persists as non-integrating episome.[49]	• Strong immune response associated toxicity; • Transient expression due to immune response;
Adeno-associated virus	• Easy to concentrate; • Availability of numerous serotypes makes it possible to choose a vector with a degree of selectivity for the desired target cell type; • Transducing non-dividing cells; • Mostly remain non-integrating episome • Naturally non-pathogenic	• Limited cargo space around 4kb;[31]

TABLE 5 Some of the Treatments Currently under Clinical Trial for Duchenne Muscular Dystrophy

Drug Name	Description	Clinical Trial Number	Delivery Route
GSK2402968 (Pro051)/Drisapersen	2′OMePS AON to exon 51	NCT01803412	SC
Pro044	2′OMePS AON to exon 44	NCT01037309	SC/IV
Pro045	2′OMePS AON to exon 45	NCT01826474	SC
AVI-4658 (eteplirsen)	PMO morpholino to exon 51	NCT02255552	IV

Abbreviations: SC, Subcutaneous; IV, Intravenous; AON, antisense oligonucleotide; 2′OMePS, 2′-O-methyl-phosphorothioate phosphodiester; PMO, phosphorodiamidate oligonucleotide.

2′-O-methyl-phosphorothioate (2′OMe) and morpholino phosphorodiamidate oligonucleotide (PMO). While retaining the same nucleobase to enable Watson-Crick base pair with natural nucleotides, ribonuclease-resistant moieties have replaced natural ribose ring and backbone phosphodiester linkage. These modifications lead to increased molecular stability against enzyme degradation. Direct infusion of analog oligonucleotides is associated with efficient targeting.[56–59] 2′OMe and PMO have been employed as antisense oligonucleotide to mediate exon skipping in DMD treatment (Table 5). Both Drisapersen (2′OMePS AON to exon 51) and Eteplirsen (PMO morpholino to exon 51) have also demonstrated restored dystrophin expression and even mild clinical improvement as measured by 6-min walk.[60,61]

4. ROUTES OF DELIVERY

The success of gene therapy is largely determined by the efficient delivery of vectors to target tissues directly determines the success. Effective delivery approaches include direct injection, locoregional perfusion, and systemic delivery.

4.1 Direct Injection

Direct injection into the target tissue is commonly used to determine the efficacy of a potential new therapy.[2,62–64] This method ensures that the desired organ receives the necessary therapeutic dose and restricts the treatment to that organ, reducing off-target effects. However, distribution may be limited, even within the injected muscle.[65] For example, XLMTM dogs treated by AAV8-*MTM*1 injection into the cranial tibialis of the hindlimb show improvements in the strength, size, and histopathology of the treated limb[2] but show a continued progression of the disease including muscle weakness, impaired ambulation, and early death.[2,66] This limited distribution is therefore disadvantageous when the disease affects multiple systems or the entire body, as is often the case with congenital myopathies. Also, direct treatment of organs like the heart or diaphragm may require invasive surgery or complicated techniques,[67,68] which can be difficult in chronically ill patients.

4.2 Locoregional Perfusion

Locoregional perfusion, where a limb is isolated before intravascular infusion under high pressure, is another approach to gene therapy delivery. Despite the use of high pressure, it is a safe, relatively painless option for human patients[12] and has been used successfully in animal models, including the XLMTM dogs, where there was widespread clinical improvement.[2,69,70] However, vector titers may be reduced due to the low permeability of the vascular endothelium[71] and, like direct injection, isolated locoregional perfusion does not address treatment of the cardiorespiratory system.

4.3 Systemic Delivery

In systemic delivery, a potential gene therapy vector is introduced to the entire body. This is of particular importance in congenital myopathies where cardiorespiratory failure is the leading cause of death but there are effects of the disease throughout the body.[72,73] As with locoregional perfusion, the vascular endothelium could hinder distribution to the skeletal muscle and cardiorespiratory systems most affected by disease.[71,74] Systemic dosing also increases the likelihood of off-target gene delivery, which may require the use of additional safety measures like tissue-specific promoters. Finally, higher doses may be required for systemic delivery, due to circulating antibodies and filtration by the liver. However, modifications of the vector or immunosuppression can address these problems.[75–77]

5. PRECLINICAL DISEASE MODEL SYSTEMS

Reliable model systems are indispensable for the critical transition from bench to bedside. It is required by the FDA and the potency and toxicity of vectors be validated on several levels of preclinical models, before entering clinical trials. The following section summarizes model systems at different level from cell culture to animal models.

5.1 Cell Culture Models

Cell culture models can be used to test the potency and cellular toxicity of a biological therapy. There are several advantages, such as intricate manipulation of experiment conditions and considerably lower costs. Early phase vector validations are usually carried out in relevant cell cultures. Immortalized mammalian human embryonic kidney (HEK) 293 cells and C2 myoblasts were employed to validate the vector expressing a truncated dystrophin protein.[78] Engineered HEK293 cells overexpressing pathogenic DUX4 gene have been utilized to confirm the efficacy of RNAi vector.[42] Cells derived from diseased tissues, such as myoblasts isolated from animal models or human patients, have been utliized to test potency for vectors developed for DMD,[79] Pompe disease,[80] and myotonic dystrophy.[44]

Due to physiological differences between animal models and human patients, it is not rare therapies that are effective in diseased model animals failed to show benefits in clinical trials,[81] including DMD.[82] Moreover, mutations may vary among patients, which requires vector personalization, such as oligonucleotides for exon skipping therapy.[83] Consequently, a personalized human cell culture system is highly desirable for potency testing.

The emergence of human iPSC offers a novel cell culture platform to meet both these criteria. Unlimited disease relevant cells with specific patient mutations could be generated. More importantly, these cells demonstrate disease-associated phenotypes reflecting disease severity.[84] As a result, correction of these disease-associated phenotypes can be measured as efficacy. Moreover, testing vectors on these personalized cells will not only demonstrate the presence of the transgene product,[85] but also enable quantified measurement of biological activities, as required by the FDA for gene therapy products.

5.2 Animal Models

While in vitro models allow for the testing of potential therapies in human cells, the ability to adequately recreate the complexities of the human body is limited. As such, preclinical testing in an animal model remains the gold standard for investigating the efficacy and potential toxicity of a putative treatment.

Nonmammalian models such as zebrafish have been invaluable to the understanding of disease mechanisms within a complex organism, particularly in monogenetic neuromuscular disorders.[86,87] However, physiological and phenotypic dissimilarities with humans limit their translational power[88,89] (Tables 6-8).

TABLE 6 Available Mouse Models for Monogenic Congenital Myopathies

DMD mouse models		
	Mutation	**Phenotype**
Mdx mouse[95]	• Spontaneous point mutation in exon 23	• Similar but milder muscle phenotype compared to human patients
Mdx52 mouse[96]	• Exon 52 deletion	• Absence of shorter isoforms Dp260 and Dp140
Mdx/Utrphin double knockout mouse[97]	• Utrophin knockout on the basis of *Mdx*	• Severe phenotype with early deterioration

LGMD 2C Mouse Models

	Mutation	**Phenotype**
γ-Sarcoglycan null mouse[98]	• Exon 2 deletion	• Muscular dystropy • Progressive but mild • Gait abnormalities • Reduced activity • Increased serum CK • Pseudohypertrophy of the diaphragm

XLMTM Mouse Models

	Mutation	**Phenotype**
XLMTM knockout mouse[99]	• Deletion of exon 4 causes an absence of myotubularin	• Centronuclear myopathy • Severely affected • Skeletal muscle wasting, resulting in pronounced and progressive weakness • Shortened lifespan
MTM1 p.R69C mutant mouse[100]	• C to T point mutation in exon 4	• Milder phenotype

TABLE 7 Available Dog Models for Monogenic Congenital Myopathies

DMD dog models		
	Mutation	**Phenotype**
Golden Retriever muscular dystrophy[101]	• X-linked mutation	• Dystrophin deficient • Muscle hypertrophy and weakness • Contracture • Esophageal dysfunction due to enlargement of the tongue • Cardiac phenotype

LGMD 2C Dog Models

	Mutation	**Phenotype**
SG-deficient Boston terrier[102]	• Mutation unknown	• Reduced α-sarcoglycan; β- and γ-sarcoglycan absent • Muscular dystrophy
SG-deficient Chihuahua[92]	• Mutation unknown	• α-, β- and γ-sarcoglycan absent • Muscular dystrophy
SG-deficient Cocker Spaniel[92]	• Mutation unknown	• Reduced α- and β-sarcoglycan; γ-sarcoglycan absent • Muscular dystrophy

XLMTM Dog Models

	Mutation	**Phenotype**
XLMTM dog[103-105]	• Missense mutation in exon	• Myotubularin deficient • Centronuclear myopathy • Severely affected • Skeletal muscle wasting, resulting in pronounced and progressive weakness • Shortened lifespan

TABLE 8 Comparison of Cell Culture, Small Animal and Large Animal Models

	Cell culture	Small animal	Large animal
Cost	• Cheap	• Relatively inexpensive	• Very expensive
Mutations	• Harbors patient specific mutations • Easy to engineer specific mutations	• Created through the knockdown or knock out of related genes	• Often naturally-occurring mutations that mirror human disease
Phenotype	• Cellular phenotype	• May differ from human phenotype	• Often similar to human disease
Size	• N/A	• Small size	• Larger size • Organ size closer to that of humans • Able to adapt test originally developed for use in the human clinic, increasing translational power
Lifespan	• N/A	• Shorter lifespan may make assessing long term effects difficult	• Longer lifetime allows for extended study of a single organism
Reproduction	• Primary cells have limited culture time • iPS cells theoretically can be propagated indefinitely	• Reproductive fecundity allows for multiple replicates for more fully powered experiments	• Relatively slow reproduction with smaller litter size can limit study
Expressivity	• Primary cells from patients with distinct genetic background	• Genetically very similar to wild-type outside of loci of interest	• Some variation in genetic background, even when using true littermate controls
Immunological	• N/A	• Immune response may differ from that seen in humans	• Immune response may differ from that seen in humans

Small mammalian models like rodents are often used in preclinical assessments of efficacy and toxicity. With extensive research into study of the murine genome,[90] mice in particular are a powerful tool in the study of monogenic disorders and knockdown or knockout mouse models have been developed to study musculoskeletal disease.

However, the small size of the mouse as well as anatomical and phenotypic differences make them less than ideal candidates for the preclinical assessment of gene therapies.[91] Therefore, a larger animal model like a dog, where organ size more closely approximates that of humans, may be more suitable. Methods of functional assessment developed for use in the clinic have also been successfully adapted for use in dogs.[22] Many naturally occurring musculoskeletal diseases, similar to those seen in human patients, have been identified and characterized[92] (Tables 6-8) facilitating the use of that dog model for the preclinical assessment of potential treatments.[2,93,94]

AAV8-mediated delivery of myotubularin in the XLMTM dog provides a successful example of gene therapy in preclinical practice. Due to the gene's small size, a full-length canine *MTM1* cDNA was carried by the muscle-tropic AAV8. This was packaged with a human desmin promoter to ensure muscle-specific expression. Dogs were treated by direct injection and by locoregional hindlimb perfusion, where systemic effects were observed likely due to leakage. Treated dogs have near normal muscle strength, with the marked improvements in respiratory function and increased survival for dogs treated intravascularly. Indeed, these animals continue to survive more than 2 years after treatment and continue to thrive and breed. The size of the dog has allowed for the inclusion of many other relevant assessments at multiple timepoints, including neurological scoring, MRI, EMG, and gait testing as we investigate potential outcome measures for translation of this therapy to human patients.

ACKNOWLEDGMENTS

Funding: American Heart Association fellowship to X.G., Muscular Dystrophy Association to M.K.C, Association Française contre les Myopathies to M.K.C.; NIH grants R21 AR064503 and R01 HL115001 to M.K.C.; Joshua Frase Foundation to M.K.C.; Where There's a Will There's a Cure to DM and MKC; Peter Khuri Myopathy Research Foundation to M.K.C.

REFERENCES

1. Ginn SL, Alexander IE, Edelstein ML, Abedi MR, Wixon J. Gene therapy clinical trials worldwide to 2012-an update. *J Gene Med* 2013;**15**(2):65–77.
2. Childers MK, Joubert R, Poulard K, et al. Gene therapy prolongs survival and restores function in murine and canine models of myotubular myopathy. *Sci Transl Med* 2014;**6**(220):220ra210.
3. Chamberlain JS. Gene therapy of muscular dystrophy. *Hum Mol Genet* 2002;**11**(20):2355–62.

4. Biancalana V, Beggs AH, Das S, et al. Clinical utility gene card for: centronuclear and myotubular myopathies. *Eur J Hum Genet* 2012;**20**(10).

5. Amburgey K, McNamara N, Bennett LR, McCormick ME, Acsadi G, Dowling JJ. Prevalence of congenital myopathies in a representative pediatric united states population. *Ann Neurol* 2011;**70**(4):662–5.

6. Heckmatt JZ, Sewry CA, Hodes D, Dubowitz V. Congenital centronuclear (myotubular) myopathy. A clinical, pathological and genetic study in eight children. *Brain* 1985;**108**(Pt 4):941–64.

7. Blondeau F, Laporte J, Bodin S, Superti-Furga G, Payrastre B, Mandel JL. Myotubularin, a phosphatase deficient in myotubular myopathy, acts on phosphatidylinositol 3-kinase and phosphatidylinositol 3-phosphate pathway. *Hum Mol Genet* 2000;**9**(15):2223–9.

8. Amoasii L, Hnia K, Chicanne G, et al. Myotubularin and PtdIns3P remodel the sarcoplasmic reticulum in muscle in vivo. *J Cell Sci* 2013;**126**(Pt 8): 1806–19.

9. McEntagart M, Parsons G, Buj-Bello A, et al. Genotype-phenotype correlations in X-linked myotubular myopathy. *Neuromuscul Disord* 2002;**12**(10):939–46.

10. Herman GE, Finegold M, Zhao W, de Gouyon B, Metzenberg A. Medical complications in long-term survivors with X-linked myotubular myopathy. *J Pediatr* 1999;**134**(2):206–14.

11. Muntoni F, Wells D. Genetic treatments in muscular dystrophies. *Curr Opin Neurol* 2007;**20**(5):590–4.

12. Fan Z, Kocis K, Valley R, et al. Safety and feasibility of high-pressure transvenous limb perfusion with 0.9% saline in human muscular dystrophy. *Mol Ther* 2012;**20**(2):456–61.

13. Burger C, Gorbatyuk OS, Velardo MJ, et al. Recombinant AAV viral vectors pseudotyped with viral capsids from serotypes 1, 2, and 5 display differential efficiency and cell tropism after delivery to different regions of the central nervous system. *Mol Ther* 2004;**10**(2):302–17.

14. Gigout L, Rebollo P, Clement N, et al. Altering AAV tropism with mosaic viral capsids. *Mol Ther* 2005;**11**(6):856–65.

15. Mendell JR, Campbell K, Rodino-Klapac L, et al. Dystrophin immunity in Duchenne's muscular dystrophy. *N Engl J Med* 2010;**363**(15):1429–37.

16. Zhu X, Hadhazy M, Groh ME, Wheeler MT, Wollmann R, McNally EM. Overexpression of gamma-sarcoglycan induces severe muscular dystrophy. Implications for the regulation of Sarcoglycan assembly. *J Biol Chem* 2001;**276**(24):21785–90.

17. Laporte J, Hu LJ, Kretz C, et al. A gene mutated in X-linked myotubular myopathy defines a new putative tyrosine phosphatase family conserved in yeast. *Nat Genet* 1996;**13**(2):175–82.

18. Lemmers RJ, van der Vliet PJ, Klooster R, et al. A unifying genetic model for facioscapulohumeral muscular dystrophy. *Science* 2010;**329**(5999):1650–3.

19. Turner C, Hilton-Jones D. The myotonic dystrophies: diagnosis and management. *J Neurol Neurosurg Psychiatry* 2010;**81**(4):358–67.

20. Day JW, Ricker K, Jacobsen JF, et al. Myotonic dystrophy type 2: molecular, diagnostic and clinical spectrum. *Neurology* 2003;**60**(4):657–64.

21. Shahrizaila N, Kinnear WJ, Wills AJ. Respiratory involvement in inherited primary muscle conditions. *J Neurol Neurosurg Psychiatry* 2006;**77**(10):1108–15.

22. Smith BK, Goddard M, Childers MK. Respiratory assessment in centronuclear myopathies. *Muscle Nerve* 2014;**50**(3):315–26.

23. Laporte J, Biancalana V, Tanner SM, et al. MTM1 mutations in X-linked myotubular myopathy. *Hum Mutat* 2000;**15**(5):393–409.

24. Deenen JC, Arnts H, van der Maarel SM, et al. Population-based incidence and prevalence of facioscapulohumeral dystrophy. *Neurology* 2014;**83**(12):1056–9.

25. Meola G. Myotonic dystrophies. *Curr Opin Neurol* 2000;**13**(5):519–25.

26. Vainzof M, Passos-Bueno MR, Pavanello RC, Marie SK, Oliveira AS, Zatz M. Sarcoglycanopathies are responsible for 68% of severe autosomal recessive limb-girdle muscular dystrophy in the Brazilian population. *J Neurol Sci* 1999;**164**(1):44–9.

27. Spiro AJ, Shy GM, Gonatas NK. Myotubular myopathy. Persistence of fetal muscle in an adolescent boy. *Arch Neurol* 1966;**14**(1):1–14.

28. Bevilacqua JA, Bitoun M, Biancalana V, et al. "Necklace" fibers, a new histological marker of late-onset MTM1-related centronuclear myopathy. *Acta Neuropathol* 2009;**117**(3):283–91.

29. Cooper D, Upadhhyaya M. *Facioscapulohumeral muscular dystrophy (FSHD): clinical medicine and molecular cell biology.* Garland Science; 2004.

30. Nadaj-Pakleza A, Lusakowska A, Sulek-Piatkowska A, et al. Muscle pathology in myotonic dystrophy: light and electron microscopic investigation in eighteen patients. *Folia Morphol (Warsz)* 2011;**70**(2):121–9.

31. Thomas CE, Ehrhardt A, Kay MA. Progress and problems with the use of viral vectors for gene therapy. *Nat Rev Genet* 2003;**4**(5):346–58.

32. Takahashi K, Yamanaka S. Induction of pluripotent stem cells from mouse embryonic and adult fibroblast cultures by defined factors. *Cell* 2006;**126**(4):663–76.

33. Yu J, Vodyanik MA, Smuga-Otto K, et al. Induced pluripotent stem cell lines derived from human somatic cells. *Science* 2007;**318**(5858):1917–20.

34. Filareto A, Parker S, Darabi R, et al. An ex vivo gene therapy approach to treat muscular dystrophy using inducible pluripotent stem cells. *Nat Commun* 2013;**4**:1549.

35. Corti S, Nizzardo M, Simone C, et al. Genetic correction of human induced pluripotent stem cells from patients with spinal muscular atrophy. *Sci Transl Med* 2012;**4**(165):165ra162.

36. Tedesco FS, Gerli MF, Perani L, et al. Transplantation of genetically corrected human iPSC-derived progenitors in mice with limb-girdle muscular dystrophy. *Sci Transl Med* 2012;**4**(140):140ra189.

37. Peters DT, Cowan CA, Musunuru K. *Genome editing in human pluripotent stem cells.* Cambridge (MA): StemBook, 2008.

38. Ambros V. The functions of animal microRNAs. *Nature* 2004;**431**(7006):350–5.

39. Muntoni F, Torelli S, Ferlini A. Dystrophin and mutations: one gene, several proteins, multiple phenotypes. *Lancet Neurol* 2003;**2**(12):731–40.

40. Kioschis P, Rogner UC, Pick E, et al. A 900-kb cosmid contig and 10 new transcripts within the candidate region for myotubular myopathy (MTM1). *Genomics* 1996;**33**(3):365–73.

41. Aartsma-Rus A, van Ommen GJ. Antisense-mediated exon skipping: a versatile tool with therapeutic and research applications. *RNA* 2007;**13**(10): 1609–24.

42. Wallace LM, Liu J, Domire JS, et al. RNA interference inhibits DUX4-induced muscle toxicity in vivo: implications for a targeted FSHD therapy. *Mol Ther* 2012;**20**(7):1417–23.

43. Mendell JR, Rodino-Klapac LR, Rosales XQ, et al. Sustained alpha-sarcoglycan gene expression after gene transfer in limb-girdle muscular dystrophy, type 2D. *Ann Neurol* 2010;**68**(5):629–38.

44. Furling D, Doucet G, Langlois MA, et al. Viral vector producing antisense RNA restores myotonic dystrophy myoblast functions. *Gene Ther* 2003;**10**(9):795–802.

45. Romero NB, Braun S, Benveniste O, et al. Phase I study of Gene Therapy in Duchenne/Becker muscular dystrophy. *Hum Gene Ther* 2004;**15**(11):1065–76.

46. Finkel RS, Flanigan KM, Wong B, et al. Phase 2a study of ataluren-mediated dystrophin production in patients with nonsense mutation Duchenne muscular dystrophy. *PLoS One* 2013;**8**(12):e81302.

47. Kinali M, Arechavala-Gomeza V, Feng L, et al. Local restoration of dystrophin expression with the morpholino oligomer AVI-4658 in Duchenne muscular dystrophy: a single-blind, placebo-controlled, dose-escalation, proof-of-concept study. *Lancet Neurol* 2009;**8**(10):918–28.

48. Mendell JR, Rodino-Klapac LR, Rosales-Quintero X, et al. Limb-girdle muscular dystrophy type 2D gene therapy restores alpha-sarcoglycan and associated proteins. *Annals of neurology* 2009;**66**(3):290–7.

49. Ponder KP. Vectors of gene therapy. In: Kresina TF, editor. *An introduction to molecular medicine and gene therapy*. Wiley-Liss, Inc; 2001.

50. Lovric J, Mano M, Zentilin L, Eulalio A, Zacchigna S. Terminal differentiation of cardiac and skeletal myocytes induces permissivity to AAV transduction by relieving inhibition imposed by DNA damage response proteins. *Mol Ther* 2012;**20**(11):2087–97.

51. Mingozzi F, High KA. Immune responses to AAV vectors: overcoming barriers to successful gene therapy. *Blood* 2013;**122**(1):23–36.

52. Qiao C, Zhang W, Yuan Z, et al. Adeno-associated virus serotype 6 capsid tyrosine-to-phenylalanine mutations improve gene transfer to skeletal muscle. *Hum Gene Ther* 2010;**21**(10):1343–8.

53. Pulicherla N, Shen S, Yadav S, et al. Engineering liver-detargeted AAV9 vectors for cardiac and musculoskeletal gene transfer. *Mol Ther* 2011;**19**(6):1070–8.

54. Asokan A, Conway JC, Phillips JL, et al. Reengineering a receptor footprint of adeno-associated virus enables selective and systemic gene transfer to muscle. *Nat Biotechnol* 2010;**28**(1):79–82.

55. Yin H, Kanasty RL, Eltoukhy AA, Vegas AJ, Dorkin JR, Anderson DG. Non-viral vectors for gene-based therapy. *Nat Rev Genet* 2014;**15**(8):541–55.

56. Inoue H, Hayase Y, Imura A, Iwai S, Miura K, Ohtsuka E. Synthesis and hybridization studies on two complementary nona(2′-O-methyl)ribonucleotides. *Nucleic Acids Res* 1987;**15**(15):6131–48.

57. Ecker DJ, Vickers TA, Bruice TW, et al. Pseudo–half-knot formation with RNA. *Science* 1992;**257**(5072):958–61.

58. Hudziak RM, Barofsky E, Barofsky DF, Weller DL, Huang SB, Weller DD. Resistance of morpholino phosphorodiamidate oligomers to enzymatic degradation. *Antisense Nucleic Acid Drug Dev* 1996;**6**(4):267–72.

59. Summerton J, Weller D. Morpholino antisense oligomers: design, preparation, and properties. *Antisense Nucleic Acid Drug Dev* 1997;**7**(3):187–95.

60. Ganea R, Jeannet PY, Paraschiv-Ionescu A, et al. Gait assessment in children with Duchenne muscular dystrophy during long-distance walking. *J Child Neurol* 2012;**27**(1):30–8.

61. Cirak S, Arechavala-Gomeza V, Guglieri M, et al. Exon skipping and dystrophin restoration in patients with Duchenne muscular dystrophy after systemic phosphorodiamidate morpholino oligomer treatment: an open-label, phase 2, dose-escalation study. *Lancet* 2011;**378**(9791):595–605.

62. Buj-Bello A, Fougerousse F, Schwab Y, et al. AAV-mediated intramuscular delivery of myotubularin corrects the myotubular myopathy phenotype in targeted murine muscle and suggests a function in plasma membrane homeostasis. *Hum Mol Genet* 2008;**17**(14):2132–43.

63. Musaro A, McCullagh K, Paul A, et al. Localized Igf-1 transgene expression sustains hypertrophy and regeneration in senescent skeletal muscle. *Nat Genet* 2001;**27**(2):195–200.

64. Vassalli G, Büeler H, Dudler J, von Segesser LK, Kappenberger L. Adeno-associated virus (AAV) vectors achieve prolonged transgene expression in mouse myocardium and arteries in vivo: a comparative study with adenovirus vectors. *Int J Cardiol* 2003;**90**(2–3):229–38.

65. Acsadi G, Dickson G, Love DR, et al. Human dystrophin expression in mdx mice after intramuscular injection of DNA constructs. *Nature* 1991;**352**(6338):815–8.

66. Goddard MA, Burlingame E, Beggs AH, et al. Gait characteristics in a canine model of X-linked myotubular myopathy. *J Neurol Sci* 2014;**346**(1–2):221–6.

67. Hoshijima M, Ikeda Y, Iwanaga Y, et al. Chronic suppression of heart-failure progression by a pseudophosphorylated mutant of phospholamban via in vivo cardiac rAAV gene delivery. *Nat Med* 2002;**8**(8):864–71.

68. Conlon TJ, Erger K, Porvasnik S, et al. Preclinical toxicology and biodistribution studies of recombinant adeno-associated virus 1 human acid alpha-glucosidase. *Hum Gene Ther Clin Dev* 2013;**24**(3):127–33.

69. Rodino-Klapac LR, Janssen PML, Montgomery CL, et al. A translational approach for limb vascular delivery of the micro-dystrophin gene without high volume or high pressure for treatment of Duchenne muscular dystrophy. *J Transl Med* 2007;**5**.

70. Le Guiner C, Montus M, Servais L, et al. Forelimb treatment in a large Cohort of dystrophic dogs supports delivery of a recombinant AAV for exon skipping in Duchenne patients. *Mol Ther* 2014;**22**(11):1923–35.

71. Gregorevic P, Blankinship MJ, Allen JM, et al. Systemic delivery of genes to striated muscles using adeno-associated viral vectors. *Nat Med* 2004;**10**(8):828–34.

72. Yue Y, Ghosh A, Long C, et al. A single intravenous injection of adeno-associated virus serotype-9 leads to whole body skeletal muscle transduction in dogs. *Mol Ther* 2008;**16**(12):1944–52.

73. Yue Y, Shin JH, Duan D. Whole body skeletal muscle transduction in neonatal dogs with AAV-9. *Methods Mol Biol* 2011;**709**:313–29.

74. Greelish JP, Su LT, Lankford EB, et al. Stable restoration of the sarcoglycan complex in dystrophic muscle perfused with histamine and a recombinant adeno-associated viral vector. *Nat Med* 1999;**5**(4):439–43.

75. Mingozzi F, High KA. Therapeutic in vivo gene transfer for genetic disease using AAV: progress and challenges. *Nat Rev Genet* 2011;**12**(5): 341–55.

76. Arruda VR. The role of immunosuppression in gene- and cell-based treatments for Duchenne muscular dystrophy. *Mol Ther* 2007;**15**(6):1040–1.

77. Arruda VR, Stedman HH, Haurigot V, et al. Peripheral transvenular delivery of adeno-associated viral vectors to skeletal muscle as a novel therapy for hemophilia B. *Blood* 2010;**115**(23):4678–88.

78. Ragot T, Vincent N, Chafey P, et al. Efficient adenovirus-mediated transfer of a human minidystrophin gene to skeletal muscle of mdx mice. *Nature* 1993;**361**(6413):647–50.

79. Mann CJ, Honeyman K, Cheng AJ, et al. Antisense-induced exon skipping and synthesis of dystrophin in the mdx mouse. *Proc Natl Acad Sci USA* 2001;**98**(1):42–7.

80. Pauly DF, Fraites TJ, Toma C, et al. Intercellular transfer of the virally derived precursor form of acid alpha-glucosidase corrects the enzyme deficiency in inherited cardioskeletal myopathy Pompe disease. *Hum Gene Ther* 2001;**12**(5):527–38.

81. Perel P, Roberts I, Sena E, et al. Comparison of treatment effects between animal experiments and clinical trials: systematic review. *BMJ* 2007;**334**(7586):197.

82. Leung DG, Herzka DA, Thompson WR, et al. Sildenafil does not improve cardiomyopathy in Duchenne/Becker muscular dystrophy. *Ann Neurol* 2014;**76**(4):541–9.

83. Aartsma-Rus A, Janson AA, Kaman WE, et al. Antisense-induced multiexon skipping for Duchenne muscular dystrophy makes more sense. *Am J Hum Genet* 2004;**74**(1):83–92.

84. Guan X, Mack DL, Moreno CM, et al. Dystrophin-deficient cardiomyocytes derived from human urine: new biologic reagents for drug discovery. *Stem Cell Res* 2014;**12**(2):467–80.

85. Dick E, Kalra S, Anderson D, et al. Exon skipping and gene transfer restore dystrophin expression in human induced pluripotent stem cells-cardiomyocytes harboring DMD mutations. *Stem Cells Dev* 2013;**22**(20):2714–24.

86. Berger J, Currie PD. Zebrafish models flex their muscles to shed light on muscular dystrophies. *Dis Model Mech* 2012;**5**(6):726–32.

87. Gibbs EM, Horstick EJ, Dowling JJ. Swimming into prominence: the zebrafish as a valuable tool for studying human myopathies and muscular dystrophies. *FEBS J* 2013;**280**(17):4187–97.

88. Lieschke GJ, Currie PD. Animal models of human disease: zebrafish swim into view. *Nat Rev Genet* 2007;**8**(5):353–67.

89. Goddard MA, Mitchell EL, Smith BK, Childers MK. Establishing clinical end points of respiratory function in large animals for clinical translation. *Phys Med Rehabil Clin N Am* 2012;**23**(1):75–94. xi.

90. Beckers J, Wurst W, de Angelis MH. Towards better mouse models: enhanced genotypes, systemic phenotyping and envirotype modelling. *Nat Rev Genet* 2009;**10**(6):371–80.

91. Nowend KL, Starr-Moss AN, Murphy KE. The function of dog models in developing gene therapy strategies for human health. *Mamm Genome* 2011;**22**(7–8):476–85.

92. Shelton GD, Engvall E. Canine and feline models of human inherited muscle diseases. *Neuromuscul Disord* 2005;**15**(2):127–38.

93. Vulin A, Barthélémy I, Goyenvalle A, et al. Muscle function recovery in golden retriever muscular dystrophy after AAV1-U7 exon skipping. *Mol Ther* 2012;**20**(11):2120–33.

94. Shin JH, Yue Y, Srivastava A, Smith B, Lai Y, Duan D. A simplified immune suppression scheme leads to persistent micro-dystrophin expression in Duchenne muscular dystrophy dogs. *Hum Gene Ther* 2012;**23**(2):202–9.

95. Bulfield G, Siller WG, Wight PA, Moore KJ. X chromosome-linked muscular dystrophy (mdx) in the mouse. *Proc Natl Acad Sci USA* 1984;**81**(4): 1189–92.

96. Araki E, Nakamura K, Nakao K, et al. Targeted disruption of exon 52 in the mouse dystrophin gene induced muscle degeneration similar to that observed in Duchenne muscular dystrophy. *Biochem Biophys Res Commun* 1997;**238**(2):492–7.

97. Deconinck AE, Rafael JA, Skinner JA, et al. Utrophin-dystrophin-deficient mice as a model for Duchenne muscular dystrophy. *Cell* 1997;**90**(4): 717–27.

98. Hack AA, Ly CT, Jiang F, et al. Gamma-sarcoglycan deficiency leads to muscle membrane defects and apoptosis independent of dystrophin. *J Cell Biol* 1998;**142**(5):1279–87.

99. Buj-Bello A, Laugel V, Messaddeq N, et al. The lipid phosphatase myotubularin is essential for skeletal muscle maintenance but not for myogenesis in mice. *Proc Natl Acad Sci USA* 2002;**99**(23):15060–5.

100. Pierson CR, Dulin-Smith AN, Durban AN, et al. Modeling the human MTM1 p.R69C mutation in murine Mtm1 results in exon 4 skipping and a less severe myotubular myopathy phenotype. *Hum Mol Genet* 2012;**21**(4):811–25.

101. Sharp NJ, Kornegay JN, Van Camp SD, et al. An error in dystrophin mRNA processing in golden retriever muscular dystrophy, an animal homologue of Duchenne muscular dystrophy. *Genomics* 1992;**13**(1):115–21.

102. Deitz K, Morrison JA, Kline K, Guo LT, Shelton GD. Sarcoglycan-deficient muscular dystrophy in a Boston Terrier. *J Vet Intern Med* 2008;**22**(2): 476–80.

103. Childers MK, Joubert R, Poulard K, et al. Gene Therapy Prolongs Survival and Restores Function in Murine and Canine Models of Myotubular Myopathy. *Science Translational Medicine* 2014;**6**(220):220ra210.

104. Grange RW, Doering J, Mitchell E, et al. Muscle function in a canine model of X-linked myotubular myopathy. *Muscle & Nerve* 2012;**46**(4): 588–91.

105. Beggs AH, Bohm J, Snead E, et al. MTM1 mutation associated with X-linked myotubular myopathy in Labrador Retrievers. *Proc Natl Acad Sci U S A* 2010;**107**(33):14697–702.

Chapter 14

Microvesicles as Mediators of Tissue Regeneration

Keith Sabin, Nobuaki Kikyo

Key Concepts

1. Microvesicles are small membranous particles secreted by many types of cells and ubiquitously distributed in body fluids.
2. They carry specific proteins, lipids, mRNAs, and miRNAs unique to secreting cells and serve as vehicles to transfer the contents to recipient cells.
3. Acute kidney injury induced by ischemia and chemicals is the most extensively used model to study the benefits of microvesicles for tissue regeneration. Microvesicles secreted by mesenchymal stem cells and others prevent apoptosis, increase cell proliferation, and preserve renal functions.
4. Microvesicles prepared from mesenchymal stem cells and progenitor/stem cells of various sources can also diminish the damage or promote regeneration of cardiac, hepatic, and neural tissues.
5. Although some key cargos of microvesicles have been identified, molecular mechanisms underlying tissue regeneration mediated by microvesicles remain elusive. Overall, microvesicle-mediated regenerative medicine is in its infancy.

List of Abbreviations

AKI Acute kidney injury
BrdU 5-bromo-2′-deoxyuridine
CPC Cardiomyocyte progenitor cell
EGFRvIII Epidermal growth factor receptor variant III
EMMPRIN Extracellular matrix metalloproteinase inducer
EPC Endothelial progenitor cell
ESC Embryonic stem cell
FasL Fas ligand
HGF Hepatocyte growth factor
HLSC Human liver stem cell
iPSC Induced pluripotent stem cell
MSC Mesenchymal stem cell
MV Microvesicle
MVB Multivesicular body
siRNA Short interfering RNA
TGF-β Transforming growth factor β
VEGF Vascular endothelial growth factor

1. INTRODUCTION

Tissue modeling and maintenance is a complex and dynamic process that requires constant input from many different cell types. Historically, the study of cell–cell communication has been primarily confined to the context of chemical and physical signals, in the form of soluble proteins, bioactive lipids, gases, and electrical impulses. More recently, a growing body of evidence has shown that membranous vesicles released from multiple cell types also contribute to cell–cell communication.[1–6] Two such populations of membranous vesicles released from cells are exosomes and microparticles.

Exosomes are small (50–100 nm) membranous vesicles that arise in the endocytic pathway and are released by numerous cell types, including neurons, immune cells, cancer cells, and stem cells.[3,5,6] As early endosomes mature, they become

Translating Regenerative Medicine to the Clinic. http://dx.doi.org/10.1016/B978-0-12-800548-4.00014-0

FIGURE 1 Synthesis and release of microvesicles. (A) Synthesis of exosomes during the process of endocytosis. Exosomes arise from early endosomes through the invagination of the endosomal membrane to form intraluminal vesicles within multivesicular bodies. The intraluminal vesicles are then released as exosomes. (B) Synthesis of microparticles independently of endocytosis. Increases in intracellular Ca^{2+} activate key enzymes, leading to the dynamic redistribution of phospholipids and membrane budding.

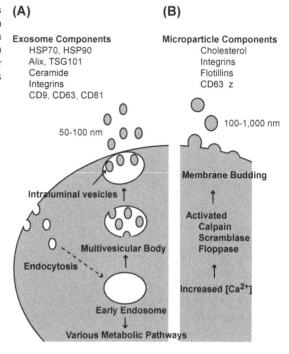

multivesicular bodies (MVBs) by accumulating intraluminal vesicles through the invagination of the limiting membrane of the endosome (Figure 1(A)). MVBs destined for degradation fuse with lysosomes, but a subset of MVBs fuse with the plasma membrane and release intraluminal vesicles into the extracellular environment as exosomes. Exosomes are enriched in heat shock proteins (HSP70 and HSP90) as well as endosome-specific proteins, such as Alix and TSG101. In addition, exosomes contain cholesterol, ceramide, integrins, and tetraspanins (CD9, CD63, and CD81), all of which are typical components of microdomains in the plasma membrane called lipid rafts. Lipid rafts are rigid membrane domains involved in sorting lipids and proteins during endocytosis.[7–9]

Microparticles (100–1000 nm) are shed vesicles that arise from the budding of the plasma membrane through the dynamic redistribution of phospholipids. Plasma membrane reorganization and microparticle budding are triggered by increased intracellular Ca^{2+} concentrations and subsequent activation of several key enzymes, notably calpain, scramblase, and floppase (Figure 1(B)).[1,3,10,11] Microparticle secretion has been best characterized in platelets, red blood cells, and endothelial cells. Microparticles lack proteins of the endocytic pathway but are enriched in cholesterol and lipid raft-associated proteins, such as integrins and flotillins. Although tetraspanins are commonly used as unique markers for exosomes, they can be detected in microparticles in some cases.[12] This overlap in molecular markers makes the distinction between exosomes and microparticles sometimes ambiguous and, therefore, both types of membranous vesicles will collectively be referred to as microvesicles (MVs) in the following discussion.

2. FUNCTIONS OF MVs

MVs have been isolated from many types of biological fluids, including serum, cerebrospinal fluid, and urine, by use of ultracentrifugation and size exclusion column chromatography.[6,13,14] The physiological functions of MVs in many tissues remain largely unknown; however, their potential roles in pathological settings have been widely studied in oncology and immunology. A subset of molecules, including proteins and RNAs, have been identified in association with MVs, and their effects on recipient cells have been intensively studied in these fields, primarily in vitro. The ExoCarta website (http://exocarta.org/#) provides a comprehensive list of MV-associated proteins, RNAs, and lipids reported from more than 140 studies. Several examples of MV-mediated cell–cell communication are briefly discussed below.

Several types of cancer cells secrete MVs containing the proapoptotic molecule Fas ligand (FasL) on the surface, which then binds to its receptor Fas on T cells, activating the apoptotic pathway and resulting in T cell death (Figure 2(A)).[15–17] When T cells are pretreated with a neutralizing antibody that inhibits Fas–FasL interactions, they become insensitive to MV-mediated apoptosis. In this way, the release of MVs by tumor cells creates an immunosuppressed microenvironment optimal for tumor progression. MVs can also transfer active cytokine receptors to recipient cells. For example,

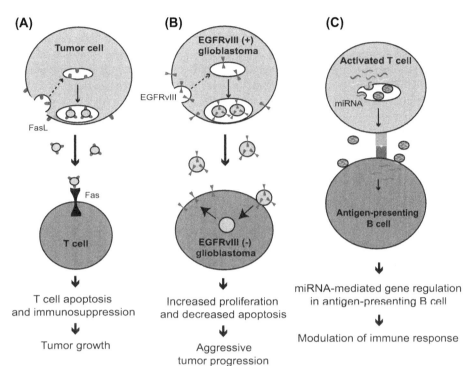

FIGURE 2 Examples of microvesicle-mediated transfer of protein and RNA. (**A**) FasL on microvesicles from tumor cells binds to Fas on T cells, causing T cell apoptosis and resulting in tumor progression. (**B**) Transfer of the oncogenic receptor EGFRvIII into glioblastoma cells causes more aggressive growth of the cells. (**C**) Activated T cells transfer miRNAs into antigen-presenting B cells, modulating their immune response.

tumor-derived MVs isolated from the serum of glioblastoma patients contain mRNA encoding the truncated and oncogenic receptor epidermal growth factor receptor variant III (EGFRvIII).[18] Cultured glioma cells that overexpress EGFRvIII secrete the oncogenic receptor protein in association with MVs. EGFRvIII-negative glioblastoma cells take up these MVs, which causes overexpression of antiapoptotic genes and increased proliferation, thus conferring a more aggressive phenotype to the recipient glioblastoma cells (Figure 2(B)).[19] The transfer of oncogenic molecules via MVs could be a potential mechanism leading to tumor progression.

In addition to the transportation of proteins, MVs also transfer miRNA between cells (Figure 2(C)).[20] It was recently demonstrated that T cells release miRNA-enriched MVs following stimulation by antigen-presenting B cells.[21] Some miRNAs are specifically enriched in MVs compared with the parent cells, while other miRNAs are more abundant in parent cells than in MVs.[21] This skewed miRNA content indicates a selective uptake of miRNAs into MVs, which has been reported in many other cases, although the mechanism is unknown.[22–25] Furthermore, MV-associated miRNAs are transferred from activated T cells to antigen-presenting B cells, where they effectively regulate target gene expression. This was demonstrated by the ability of transferred miR-335 to knockdown expression of a luciferase reporter gene containing the seed sequence of miR-335 at the 3′-UTR.[21] These data provide evidence for a novel form of communication between T cells and antigen-presenting cells.

3. MESENCHYMAL STEM CELLS AND REGENERATIVE MEDICINE

Recent progress in basic stem cell biology has greatly contributed to the realization of regenerative medicine. Differentiated cells derived from various types of stem cells are expected to facilitate regeneration of damaged tissues. In particular, the establishment of human embryonic stem cells (ESCs)[26] and more recent creation of induced pluripotent stem cells (iPSCs)[27] provided a major driving force for the medical and public enthusiasm toward regenerative medicine. While ESCs are pluripotent stem cells established from the inner cell mass of the blastocyst,[28] iPSCs are commonly established from human skin fibroblasts by overexpressing four transcription factor genes—Oct4, Sox2, Klf4, and c-Myc. Patient-derived iPSCs provided hitherto unimaginable opportunities for autologous transplantation, disease modeling, and drug screening. Despite their clinical appeal, however, no ongoing human trials have been undertaken with iPSCs primarily due to technological restrains. The most commonly used type of stem cells in regenerative medicine is autologous mesenchymal stem cells (MSCs), which will be discussed currently.

MSCs are generally defined as a heterogeneous population of nonhematopoietic stem cells with the capability to differentiate into cells of endoderm, mesoderm, and ectoderm origin. However, because of their heterogeneous nature, MSCs were defined by the International Society of Cellular Therapy more stringently for use in clinical trials, based on adhesion to plastic, the presence or absence of specific cell surface markers, and their ability to differentiate into mesodermal lineages.[29] MSCs can be readily isolated from many sources, including bone marrow and adipose tissue, and expanded ex vivo for transplantation. Transplanted MSCs stimulate angiogenesis, cell survival, and proliferation and regulate the inflammatory process in host tissues. As of August 2013, 340 clinical trials using MSCs to treat various diseases had been registered with the US National Institutes of Health (clinicaltrials.gov).

Although MSCs have been shown to facilitate regeneration of diseased tissue in various animal models, the exact mechanism underlying this benefit is not fully understood. MSCs are generally thought to achieve this function through paracrine mechanisms, rather than engraftment and subsequent repopulation of the tissues.[30–32] Although engraftment of transplanted MSCs has been documented in some cases,[33–35] only a small percentage of injected MSCs successfully engraft in various models of myocardial infarction and acute kidney injury (AKI).[36–39] Consistent with these findings, injection of mice with cell-free, MSC-conditioned medium is sufficient to stimulate the structural and functional regeneration of cardiac and renal tissues in these models.[36,40,41] This indicates an integral role for MSC-secreted signaling molecules in stimulating tissue repair. Indeed, MSCs are known to promote tissue regeneration through the secretion of cytokines, such as transforming growth factor-β (TGF-β), hepatocyte growth factor (HGF), and vascular endothelial growth factor (VEGF), to name a few.[30] Recent findings also indicate a proregenerative role for MVs released by MSCs (MSC-MVs) in several models of tissue regeneration, including regeneration of kidney, heart, liver, and nervous tissues, as discussed below.

4. MSC-MVs IN KIDNEY REGENERATION

Kidney regeneration mediated by MSC-MVs is commonly studied in AKI models induced by cisplatin, ischemia and subsequent reperfusion, or glycerol.[42–46] Specifically, cisplatin-induced injury of cultured renal tubular epithelial cells is mitigated after the addition of purified MSC-MVs. MSC-MVs increase the amount of cellular incorporation of 5-bromo-2′-deoxyuridine (BrdU), indicative of proliferation, while decreasing the number of TUNEL-positive apoptotic cells (Figure 3(A)).[46] Similarly, the use of purified MSC-MVs to treat AKI in mice recapitulates the in vitro data, leading to an increase of BrdU-positive cells and a decrease of TUNEL-positive cells. Consequently, functional impairment is similarly reduced as indicated by decreased serum levels of blood urea nitrogen, creatinine, and renal cellular necrosis.[46] Additionally, microarray analysis of the MV mRNA detected 239 species of mRNA molecules, with specific enrichment of mRNAs involved in cell proliferation, transcription, and immune regulation, providing another example of selective uptake of RNA species into MVs.

These processes appear to be mediated, at least in part, by the transfer of RNA by MSC-MVs, as indicated by the loss of regenerative effects after RNase treatment of the MVs.[46] However, these results are contradictory to other studies that show MV-associated RNA is resistant to RNase treatment.[18,47,48] Notably, these studies did not perform functional experiments with the RNase-treated MVs but instead isolated RNA from the MVs and either quantified total RNA or amplified specific

FIGURE 3   **Kidney regeneration by the microvesicles released by mesenchymal stem cells and endothelial progenitor cells (EPCs).** (A) Microvesicles derived from mesenchymal stem cells (MSCs) inhibit apoptosis and stimulate proliferation of renal tubular epithelial cells. Microarray analysis of microvesicle RNA detected 239 mRNA species, including species with the functions described at the right. (B) Microvesicles released from EPCs induce quiescent endothelial cells to reenter the cell cycle and form blood vessels. In addition to mRNA species, miRNAs are also transferred by the microvesicles.

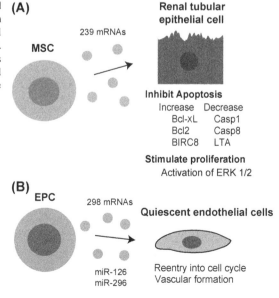

targets using RT-PCR to demonstrate RNase insensitivity. These results are consistent with recent observations that RNA can be isolated from biologic fluids with high intrinsic RNase activity, such as bovine milk or human plasma, indicating some protective mechanism conferring RNase resistance.[47,49] One possible explanation for these conflicting reports could stem from residual RNase that remained attached to the surface of the MVs and were subsequently incorporated into the cell with the MVs. The RNase could then degrade nearby MV-derived RNAs, leading to abrogation of the regenerative response. To address this possibility, MVs could be treated with RNase-coated beads and thereby eliminating the possibility of soluble RNase being incorporated into cells.

Recent reports have begun deciphering the molecular pathways modulated by MSC-MVs in the context of renal regeneration. Specifically, MSC-MVs induce the expression of several antiapoptotic genes, including Bcl-XL, Bcl2, and BIRC8, in renal tubular epithelial cells while simultaneously downregulating proapoptotic genes, such as Casp1, Casp8, and LTA.[50] In this manner, MSC-MVs confer an antiapoptotic phenotype necessary for tissue repair. Additionally, MSC-MVs stimulate renal cell proliferation by inducing the phosphorylation and subsequent activation of extracellular regulated kinase (ERK) 1/2.[51] Blockade of ERK activation with a chemical inhibitor significantly reduces cell proliferation after MSC-MV treatment.[51] Although the exact molecules in the MVs that mediate the antiapoptotic and proproliferative effects have not been identified, these data demonstrate the ability of MSC-MVs to simultaneously modulate several different pathways to stimulate renal regeneration.

While the ability of MSC-MVs to stimulate renal regeneration has been intensively studied, it is becoming clear that MVs from diverse cell types could potentially aid in the regenerative process. For example, endothelial progenitor cells (EPCs) may also contribute to tissue regeneration through released MVs.[52,53] EPCs are a subpopulation of nonhematopoietic bone marrow cell with a differentiation potential restricted to the endothelial lineage. EPC-derived MVs (EPC-MVs) activate an angiogenic program in quiescent endothelial cells by stimulating reentry into the cell cycle and promoting blood vessel formation (Figure 3(B)).[54] This process is dependent on the transfer of mRNAs by incorporation of EPC-MVs into endothelial cells. If MV incorporation is blocked using α4-or β1-integrin neutralizing antibodies, EPC-MVs do not activate quiescent endothelial cells. A microarray analysis of the mRNA in EPC-MVs detected 298 species of mRNA, including those involved in the phosphatidylinositol 3-kinase/AKT signaling pathway, which is critical for the activation of angiogenesis. Similarly, MVs released by injured renal epithelial cells promote proliferation of fibroblasts and secretion of type I collagen from fibroblasts, initiating a tissue repair process. This process is dependent on the presence and transfer of TGF-β1 mRNA by MVs.[55] When TGF-β1 is knocked down using short interfering RNA (siRNA), renal epithelial cell-derived MVs could no longer stimulate tissue repair.

In addition to mRNAs, the transfer of miRNAs by MVs is important in the context of kidney regeneration. In a rat model of AKI, intravenously injected EPC-MVs mainly localize in the peritubular capillaries and renal tubular cells, and subsequently promote tissue repair and reduce functional impairment.[22] Their protective effects can be inhibited by the knockdown of Dicer, which is an RNase essential for processing of pre-miRNA to miRNA, or by simultaneous inhibition of miR-126 and miR-296 in the parental EPCs. These two miRNAs are known to be essential for angiogenesis in many contexts.[56] These data indicate an important and potentially underappreciated role of miRNA transfer by MVs in stimulating tissue regeneration.

5. MSC-MVs IN CARDIAC REGENERATION

In addition to contributing to kidney regeneration, MSCs mediate functional and structural regeneration of cardiac tissue in a largely paracrine manner.[57,58] For instance, injection of rats with MSCs engineered to express the prosurvival gene Akt1 stimulates cardiac regeneration after ischemia.[40,59] Recovery of left ventricular function occurs as early as 72 h after the addition of MSCs, suggesting that the functional recovery is not dependent on cellular engraftment and subsequent differentiation.[40] Supporting this interpretation, injection of mice with MSC-Akt-conditioned medium significantly reduces the infarct size and increases left ventricular systolic pressure as compared with controls.[40,59] Similarly, intravenous injection of medium conditioned by wild-type MSCs also reduces the infarct size caused by ischemia and reperfusion.[41] In this model of myocardial infarction, the cardioprotective effect of MSC-conditioned medium could be observed as early as 4 h after reperfusion. Subsequent ultracentrifugation and biochemical characterization showed that MVs were the key components in the conditioned medium mediating myocardial protection.[60] Administration of purified MSC-MVs 5 min before reperfusion significantly reduces infarct size and drastically improves left ventricular function. Furthermore, MSCs derived from fetal tissue are highly proliferative, with the potential of producing up to 10^{19} cells, and are therefore capable of producing substantial quantities of cardioprotective MVs for therapeutic use.[61]

In addition to MSCs, MVs isolated from cardiac progenitor cells (CPCs) may also contribute to cardiac regeneration.[62] CPCs have been isolated from embryonic and adult hearts based on the expression of several specific marker proteins.[63,64] For example, the first purified CPCs from adult rat hearts were negative for the expression of blood lineage markers (Lin⁻) and positive for the expression of the stem cell-related surface antigen c-kit.[65] These cells are multipotent, being able to differentiate into cardiac muscle, smooth muscle, and endothelial cells, and promote cardiac regeneration after myocardial infarction. MVs isolated from the supernatant of the culture of CPCs stimulate endothelial cell migration through the extracellular matrix metalloproteinase inducer (EMMPRIN) expressed on the MV surface. EMMPRIN signaling subsequently activates the ERK1/2 pathway and induces expression of matrix metalloproteinases and secretion of VEGF, thus promoting angiogenesis.[66,67] The ability of MVs to activate an angiogenic program in endothelial cells could provide, in part, a mechanism for their ability to promote cardiac regeneration.

6. MVs IN REGENERATION OF OTHER TISSUES

Although the proregenerative role of MSC-MVs has been most intensely studied in the context of kidney and cardiac regeneration, recent reports indicate a protective role for MVs in other tissues as well. For example, MSC-MVs reduce tetrachloride-induced liver fibrosis, potentially by inhibiting the epithelial-to-mesenchymal transition of hepatocytes and collagen production.[68] Human liver stem cells (HLSCs) have also been used as a source for MVs to stimulate liver regeneration after physical and chemical injury. HLSCs were first established after culture of primary hepatocytes for 2 weeks in selective media as a population of actively proliferating cells.[69] These cells can differentiate into bone, adipose cells, and endothelial cells in addition to liver cells. MVs isolated from HLSCs can protect cultured hepatocytes from D-galactosamine-induced apoptosis when simultaneously added as determined by TUNEL assay. In parallel with the in vitro data, administration of HLSC-MVs to 70% hepatectomized rats significantly contributed to structural and functional hepatic recovery. Specifically, the amounts of the hepatic enzymes aspartate aminotransferase and alanine aminotransferase in the serum diminished and the production of albumin increased. At the histological level, more hepatic cells incorporated BrdU, indicating proliferation, and fewer cells were apoptotic than in control liver. HLSC-MVs were internalized into hepatocytes via a specific α4 integrin-dependent process and horizontally transferred a specific subset of mRNAs to the hepatocytes. A majority of the transcripts present in HLSC-MVs are involved in regulation of transcription, translation, and cell proliferation.

A role for MSC-MVs has also been proposed in an ischemic model of neural regeneration. Exposure of MSCs to ischemic rat brain extract for 72 h induces the expression of miR-133b in MSCs and its subsequent presence in MSC-MVs.[70] miR-133b is essential for the functional regeneration of motor neuron axons after spinal cord injury in zebrafish.[71] Treatment of healthy primary rat cortical neurons with MVs derived from the brain extract-treated MSCs increases the total number of neurites and neurite length after 48 h. Furthermore, neurons previously transfected with a miR-133b inhibitor do not exhibit changes in neurite morphology after MSC-MV treatment. The changes in neurite number and length seem to result from the targeting and inhibition of the small GTPase RhoA, a known inhibitor of neurite growth, by miR-133b.[70]

7. MVs AND EMBRYONIC STEM CELLS

Because of their potential use in medical applications, ESCs have been extensively studied in the past decade, although MVs released from ESCs (ESC-MVs) have not attracted wide interest. ESC-MVs contain mRNA encoding Oct4, Sox2, and Rex1, which are key genes for pluripotency, as well as the Oct4 protein.[72] After incorporation of ESC-MVs, hematopoietic progenitor cells translate mRNA encoding Oct4, the master transcription factor for pluripotency, although this has not been shown to result in pluripotency of the recipient cells. A subset of miRNAs can be transferred to irradiated fibroblasts via ESC-MVs (Figure 4); specifically, several miRNAs of the miR-290 cluster, which are the most abundant miRNAs in ESCs[73] and are involved in ESC-specific rapid cell cycle regulation, were found to have transferred to the fibroblasts 1 h after addition of the MVs.[25] The ability of MVs to transfer miRNA and mRNA has potential implications for iPSC formation from fibroblasts. For example, MVs from partially reprogrammed iPSCs might function in an autocrine manner to facilitate reprogramming nearby cells within a single colony. On the other hand, fibroblasts might secret inhibitory MVs for iPSC formation and therefore, inhibition of fibroblast-derived MVs might promote iPSC formation. Given the recent report on the creation of iPSCs from fibroblasts with miRNAs alone,[74] albeit different miRNAs than those found in ESC-MVs, MV-mediated delivery of miRNAs warrants further investigation as a possible unrecognized communication process during iPSC development.

Furthermore, ESC-MVs can reprogram Müller glia cells to less-differentiated, multipotent retinal progenitor cells.[75] Through the selective transfer of members of the miR-290 cluster and mRNA encoding Oct4 and Sox2 to Müller cells, ESC-MVs induce drastic morphological changes and upregulation of early retinal genes and pluripotency genes, indicative of dedifferentiation. Although the long-term effects of the transfer and the stability of the newly acquired phenotype were not reported, this work indicates a potential application for the use of ESC-MVs to reprogram differentiated cells. Specifically, the reprogramming of Müller glia cells into retinal progenitor cells by ESC-MVs could prove to be a novel treatment for retinal diseases such as macular degeneration or cone–rod dystrophy. Moreover, in the context of tissue regeneration, this strategy of reprogramming quiescent tissues with ESC-MVs could be broadly applied to neurodegenerative, heart, lung, and kidney disease.

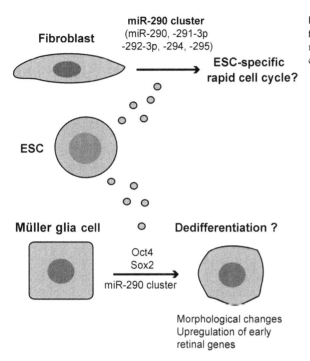

FIGURE 4 Potential dedifferentiation induced by microvesicles derived from embryonic stem cells. Embryonic stem cell (ESC)-specific components of microvesicles, including the abundant miR-290 cluster, potentially affect the cell cycle regulation of fibroblasts and induce dedifferentiation of Müller glia cells.

8. FUTURE PERSPECTIVES

The functional significance of MVs in regenerative medicine has yet to be firmly established. Although many studies have demonstrated that MVs isolated from various cells are capable of supporting tissue regeneration, it is not clear whether MVs are indeed necessary components of the regeneration process. To prove this, release or incorporation of MVs needs to be specifically blocked in vivo, which requires deeper understanding of their synthetic pathways, selective uptake of their contents, secretion, and incorporation by target cells. Fortunately, possible molecular targets involved in MV biogenesis and secretion have recently been elucidated in vitro. For instance, the GTPases Rab27a and Rab27b are necessary for the docking of MVBs to the plasma membrane before release to the extracellular space in several types of cancer cells.[76,77] Similarly, a role for the guanine nucleotide exchange factor BIG2 has been demonstrated for the constitutive release of MVs from human vascular endothelial cells.[78] However, the necessity of these molecules for MV synthesis and secretion in vivo has not yet been demonstrated.

Regardless of our understanding of MV biogenesis, the clinical benefits of MVs are already being exploited. As of August 2013, there were eight registered clinical trials using exosomes, primarily in the field of cancer diagnostics and therapeutics (clinicaltrials.gov). Given the promising results of MSC-MVs in animal models of tissue regeneration described above, human clinical trials with MVs are also likely to be conducted in the near future. However, several factors must be considered before the use of MVs in human clinical trials can be realized.[79,80]

First, the type of effector cargo molecule (i.e., protein, mRNA, or miRNA) and the ability of that molecule to localize in MVs need to be determined. Proteins can be specifically targeted to MVs through fusion to MV-enriched molecules, such as lactadherin, Lamp2b, and the transmembrane domain of platelet-derived growth factor receptor.[81–83] Additionally, exogenous molecules, such as siRNA, can be inserted into MVs via electroporation.[81]

Second, the source of membranous particles needs to be considered.[75] Injection of MSCs would ensure the long-term, continuous release of pro-regenerative MVs. However, the large-scale production of a clinically acceptable population of MSCs could prove expensive. Therefore, the use of purified MVs or synthetic liposomes could be preferable in a clinical setting. One benefit of using purified MVs is that they are empirically known to contain the necessary components to dock with and be incorporated into injured tissues to promote regeneration. The specific molecules responsible for these effects are largely unknown; therefore, at this time, using cell-derived MVs would be superior to synthetic liposomes. However, once these molecules are uncovered, large-scale production of synthetic liposomes enriched with these molecules would be more cost-effective and safer than injection of living cells.

Third, targeting of MVs to specific tissues is a critical determinant for therapeutic efficacy. Tissue-specific differences in MV incorporation have been reported. For example, MSC-MVs are incorporated into renal epithelial cells via a CD44-and CD29-dependent pathway, whereas incorporation of EPC-MVs into human microvascular endothelial cells is dependent on the presence of α4 integrin and CD29.[46,54] Target-specific molecules have been used in delivering MVs to neural tissues for therapeutic use. For example, dendritic cells were genetically engineered to express the rabies virus glycoprotein, which specifically targets neural tissues, fused to the MV-enriched molecule Lamp2b.[81] MVs were isolated from these genetically manipulated dendritic cells and loaded with an siRNA targeting BACE2, a common therapeutic target in Alzheimer's disease. Subsequent injection of these targeted MVs into mice facilitated the knockdown of BACE2 expression only in neural tissues. Depending on the target tissue, parent cells would need to be chosen based on the endogenous levels of specific targeting molecules or would need to be genetically manipulated to express the targeting molecules.

Last, genetic manipulation of the parent cells could be used to enrich MVs with therapeutic cargo. A potential novel application of this technology would be the MV-mediated delivery of a defined set of mRNAs to damaged tissues, with the aim of regenerating the tissues by reprogramming resident cells, such as fibroblasts. A cocktail of several transcription factor genes can reprogram fibroblasts into various types of differentiated cells in vivo.[84] For example, injection of a specific combination of three or four cardiac transcription factor genes into an infarct area can convert cardiac fibroblasts into cardiomyocytes in situ, improving cardiac function after infarction.[85–87] Similarly, pancreatic exocrine cells can be converted to β cells with three β cell-specific transcription factor genes in only 3 days, ameliorating hyperglycemia in a mouse model of diabetes.[88] Therefore, MVs purified from the culture medium of MSCs that overexpress these genes could act as a novel delivery vector for fragile mRNAs protected from physical and chemical damage. Another application would be to inject the engineered MSCs, expecting that these cells would deliver the MVs to the diseased tissues. The in situ reprogramming approach could be used to treat a variety of diseases including cardiovascular, neurodegenerative, and musculoskeletal disease and diabetes. Although the role of MVs in regenerative medicine is still in its infancy, the use of MVs to stimulate tissue repair provides novel and exciting possibilities.

ACKNOWLEDGMENT

We are grateful to the US National Institutes of Health (R01 GM098294), the Engdahl Family Foundation, and the Schulze Family Foundation for their support of our work related to this topic. KS would like to thank Drs Robert Winn, Robert Belton, Richard Rovin, and Johnathan Lawrence, and the Upper Michigan Brain Tumor Center for supporting his previous work in the field of microvesicles. All authors have read the journal's policy on conflicts of interest and have none to declare.

REFERENCES

1. Mathivanan S, Ji H, Simpson RJ. Exosomes: extracellular organelles important in intercellular communication. *J Proteomics* 2010;**73**:1907–20.
2. Simons M, Raposo G. Exosomes–vesicular carriers for intercellular communication. *Curr Opin Cell Biol* 2009;**21**:575–81.
3. Gyorgy B, Szabo TG, Pasztoi M, et al. Membrane vesicles, current state-of-the-art: emerging role of extracellular vesicles. *Cell Mol Life Sci* 2011;**68**:2667–88.
4. Vlassov AV, Magdaleno S, Setterquist R, Conrad R. Exosomes: current knowledge of their composition, biological functions, and diagnostic and therapeutic potentials. *Biochim Biophys Acta* 2012;**1820**:940–8.
5. Record M, Subra C, Silvente-Poirot S, Poirot M. Exosomes as intercellular signalosomes and pharmacological effectors. *Biochem Pharmacol* 2011;**81**:1171–82.
6. Pant S, Hilton H, Burczynski ME. The multifaceted exosome: biogenesis, role in normal and aberrant cellular function, and frontiers for pharmacological and biomarker opportunities. *Biochem Pharmacol* 2012;**83**:1484–94.
7. Gagescu R, Demaurex N, Parton RG, Hunziker W, Huber LA, Gruenberg J. The recycling endosome of madin-darby canine kidney cells is a mildly acidic compartment rich in raft components. *Mol Biol Cell* 2000;**11**:2775–91.
8. Fullekrug J, Simons K. Lipid rafts and apical membrane traffic. *Ann N Y Acad Sci* 2004;**1014**:164–9.
9. Leitinger B, Hogg N. The involvement of lipid rafts in the regulation of integrin function. *J Cell Sci* 2002;**115**:963–72.
10. Del Conde I, Cruz MA, Zhang H, Lopez JA, Afshar-Kharghan V. Platelet activation leads to activation and propagation of the complement system. *J Exp Med* 2005;**201**:871–9.
11. Piccin A, Murphy WG, Smith OP. Circulating microparticles: pathophysiology and clinical implications. *Blood Rev* 2007;**21**:157–71.
12. Dale GL, Remenyi G, Friese P. Tetraspanin CD9 is required for microparticle release from coated-platelets. *Platelets* 2009;**20**:361–6.
13. Caby MP, Lankar D, Vincendeau-Scherrer C, Raposo G, Bonnerot C. Exosomal-like vesicles are present in human blood plasma. *Int Immunol* 2005;**17**:879–87.
14. Simpson RJ, Lim JW, Moritz RL, Mathivanan S. Exosomes: proteomic insights and diagnostic potential. *Expert Rev Proteomics* 2009;**6**:267–83.
15. Abusamra AJ, Zhong Z, Zheng X, et al. Tumor exosomes expressing Fas ligand mediate CD8+ T-cell apoptosis. *Blood Cells Mol Dis* 2005;**35**:169–73.
16. Andreola G, Rivoltini L, Castelli C, et al. Induction of lymphocyte apoptosis by tumor cell secretion of FasL-bearing microvesicles. *J Exp Med* 2002;**195**:1303–16.
17. Huber V, Fais S, Iero M, et al. Human colorectal cancer cells induce T-cell death through release of proapoptotic microvesicles: role in immune escape. *Gastroenterology* 2005;**128**:1796–804.
18. Skog J, Wurdinger T, van Rijn S, et al. Glioblastoma microvesicles transport RNA and proteins that promote tumour growth and provide diagnostic biomarkers. *Nat Cell Biol* 2008;**10**:1470–6.

19. Al-Nedawi K, Meehan B, Micallef J, et al. Intercellular transfer of the oncogenic receptor EGFRvIII by microvesicles derived from tumour cells. *Nat Cell Biol* 2008;**10**:619–24.

20. Ramachandran S, Palanisamy V. Horizontal transfer of RNAs: exosomes as mediators of intercellular communication. *Wiley Interdiscip Rev RNA* 2012;**3**:286–93.

21. Mittelbrunn M, Gutierrez-Vazquez C, Villarroya-Beltri C, et al. Unidirectional transfer of microRNA-loaded exosomes from T cells to antigen-presenting cells. *Nat Commun* 2011;**2**:282.

22. Cantaluppi V, Gatti S, Medica D, et al. Microvesicles derived from endothelial progenitor cells protect the kidney from ischemia-reperfusion injury by microRNA-dependent reprogramming of resident renal cells. *Kidney Int* 2012;**82**:412–27.

23. Collino F, Deregibus MC, Bruno S, et al. Microvesicles derived from adult human bone marrow and tissue specific mesenchymal stem cells shuttle selected pattern of miRNAs. *PLoS One* 2010;**5**:e11803.

24. Chen TS, Lai RC, Lee MM, Choo AB, Lee CN, Lim SK. Mesenchymal stem cell secretes microparticles enriched in pre-microRNAs. *Nucleic Acids Res* 2010;**38**:215–24.

25. Yuan A, Farber EL, Rapoport AL, et al. Transfer of microRNAs by embryonic stem cell microvesicles. *PLoS One* 2009;**4**:e4722.

26. Thomson JA, Itskovitz-Eldor J, Shapiro SS, et al. Embryonic stem cell lines derived from human blastocysts. *Science* 1998;**282**:1145–7.

27. Takahashi K, Yamanaka S. Induction of pluripotent stem cells from mouse embryonic and adult fibroblast cultures by defined factors. *Cell* 2006;**126**:663–76.

28. Smith AG. Embryo-derived stem cells: of mice and men. *Annu Rev Cell Dev Biol* 2001;**17**:435–62.

29. Dominici M, Le Blanc K, Mueller I, et al. Minimal criteria for defining multipotent mesenchymal stromal cells. The International Society for Cellular Therapy position statement. *Cytotherapy* 2006;**8**:315–7.

30. Wang S, Qu X, Zhao RC. Clinical applications of mesenchymal stem cells. *J Hematol Oncol* 2012;**5**:19.

31. Lavoie JR, Rosu-Myles M. Uncovering the secretes of mesenchymal stem cells. *Biochimie* 2012;**95**:2212–21. [Epub ahead of print].

32. Quesenberry PJ, Aliotta JM. The paradoxical dynamism of marrow stem cells: considerations of stem cells, niches, and microvesicles. *Stem Cell Rev* 2008;**4**:137–47.

33. Kotton DN, Ma BY, Cardoso WV, et al. Bone marrow-derived cells as progenitors of lung alveolar epithelium. *Development* 2001;**128**:5181–8.

34. Ortiz LA, Gambelli F, McBride C, et al. Mesenchymal stem cell engraftment in lung is enhanced in response to bleomycin exposure and ameliorates its fibrotic effects. *Proc Natl Acad Sci USA* 2003;**100**:8407–11.

35. Liu Y, Yan X, Sun Z, et al. Flk-1+ adipose-derived mesenchymal stem cells differentiate into skeletal muscle satellite cells and ameliorate muscular dystrophy in mdx mice. *Stem Cells Dev* 2007;**16**:695–706.

36. Bi B, Schmitt R, Israilova M, Nishio H, Cantley LG. Stromal cells protect against acute tubular injury via an endocrine effect. *J Am Soc Nephrol* 2007;**18**:2486–96.

37. Duffield JS, Park KM, Hsiao LL, et al. Restoration of tubular epithelial cells during repair of the postischemic kidney occurs independently of bone marrow-derived stem cells. *J Clin Invest* 2005;**115**:1743–55.

38. Togel F, Hu Z, Weiss K, Isaac J, Lange C, Westenfelder C. Administered mesenchymal stem cells protect against ischemic acute renal failure through differentiation-independent mechanisms. *Am J Physiol Ren Physiol* 2005;**289**:F31–42.

39. Chimenti I, Smith RR, Li TS, et al. Relative roles of direct regeneration versus paracrine effects of human cardiosphere-derived cells transplanted into infarcted mice. *Circ Res* 2010;**106**:971–80.

40. Gnecchi M, He H, Noiseux N, et al. Evidence supporting paracrine hypothesis for AKT-modified mesenchymal stem cell-mediated cardiac protection and functional improvement. *FASEB J* 2006;**20**:661–9.

41. Timmers L, Lim SK, Arslan F, et al. Reduction of myocardial infarct size by human mesenchymal stem cell conditioned medium. *Stem Cell Res* 2007;**1**:129–37.

42. Bruno S, Camussi G. Role of mesenchymal stem cell-derived microvesicles in tissue repair. *Pediatr Nephrol* 2013;**28**:2249–54.

43. Camussi G, Deregibus MC, Tetta C. Paracrine/endocrine mechanism of stem cells on kidney repair: role of microvesicle-mediated transfer of genetic information. *Curr Opin Nephrol Hypertens* 2010;**19**:7–12.

44. Dorronsoro A, Robbins PD. Regenerating the injured kidney with human umbilical cord mesenchymal stem cell-derived exosomes. *Stem Cell Res Ther* 2013;**4**:39.

45. Ratajczak MZ. The emerging role of microvesicles in cellular therapies for organ/tissue regeneration. *Nephrol Dial Transpl* 2011;**26**:1453–6.

46. Bruno S, Grange C, Deregibus MC, et al. Mesenchymal stem cell-derived microvesicles protect against acute tubular injury. *J Am Soc Nephrol* 2009;**20**:1053–67.

47. Hata T, Murakami K, Nakatani H, Yamamoto Y, Matsuda T, Aoki N. Isolation of bovine milk-derived microvesicles carrying mRNAs and microR-NAs. *Biochem Biophys Res Commun* 2010;**396**:528–33.

48. Li L, Zhu D, Huang L, et al. Argonaute 2 complexes selectively protect the circulating microRNAs in cell-secreted microvesicles. *PLoS One* 2012;**7**:e46957.

49. Kosaka N, Iguchi H, Yoshioka Y, Takeshita F, Matsuki Y, Ochiya T. Secretory mechanisms and intercellular transfer of microRNAs in living cells. *J Biol Chem* 2010;**285**:17442–52.

50. Bruno S, Grange C, Collino F, et al. Microvesicles derived from mesenchymal stem cells enhance survival in a lethal model of acute kidney injury. *PLoS One* 2012;**7**:e33115.

51. Zhou Y, Xu H, Xu W, et al. Exosomes released by human umbilical cord mesenchymal stem cells protect against cisplatin-induced renal oxidative stress and apoptosis in vivo and in vitro. *Stem Cell Res Ther* 2013;**4**:34.

52. Asahara T, Murohara T, Sullivan A, et al. Isolation of putative progenitor endothelial cells for angiogenesis. *Science* 1997;**275**:964–7.

53. Zampetaki A, Kirton JP, Xu Q. Vascular repair by endothelial progenitor cells. *Cardiovasc Res* 2008;**78**:413–21.

54. Deregibus MC, Cantaluppi V, Calogero R, et al. Endothelial progenitor cell derived microvesicles activate an angiogenic program in endothelial cells by a horizontal transfer of mRNA. *Blood* 2007;**110**:2440–8.

55. Borges FT, Melo SA, Ozdemir BC, et al. TGF-beta1-containing exosomes from injured epithelial cells activate fibroblasts to initiate tissue regenerative responses and fibrosis. *J Am Soc Nephrol* 2013;**24**:385–92.

56. Wang S, Olson EN. AngiomiRs–key regulators of angiogenesis. *Curr Opin Genet Dev* 2009;**19**:205–11.

57. Lai RC, Chen TS, Lim SK. Mesenchymal stem cell exosome: a novel stem cell-based therapy for cardiovascular disease. *Regen Med* 2011;**6**:481–92.

58. Mummery CL, Davis RP, Krieger JE. Challenges in using stem cells for cardiac repair. *Sci Transl Med* 2010;**2**:27ps17.

59. Gnecchi M, He H, Liang OD, et al. Paracrine action accounts for marked protection of ischemic heart by AKT-modified mesenchymal stem cells. *Nat Med* 2005;**11**:367–8.

60. Lai RC, Arslan F, Lee MM, et al. Exosome secreted by MSC reduces myocardial ischemia/reperfusion injury. *Stem Cell Res* 2010;**4**:214–22.

61. Lai RC, Choo A, Lim SK. Derivation and characterization of human ESC-derived mesenchymal stem cells. *Methods Mol Biol* 2011;**698**:141–50.

62. Vrijsen KR, Sluijter JP, Schuchardt MW, et al. Cardiomyocyte progenitor cell-derived exosomes stimulate migration of endothelial cells. *J Cell Mol Med* 2010;**14**:1064–70.

63. Guan K, Hasenfuss G. Do stem cells in the heart truly differentiate into cardiomyocytes? *J Mol Cell Cardiol* 2007;**43**:377–87.

64. Aguirre A, Sancho-Martinez I. Izpisua Belmonte JC. Reprogramming toward heart regeneration: stem cells and beyond. *Cell Stem Cell* 2013;**12**:275–84.

65. Beltrami AP, Barlucchi L, Torella D, et al. Adult cardiac stem cells are multipotent and support myocardial regeneration. *Cell* 2003;**114**:763–76.

66. Sidhu SS, Mengistab AT, Tauscher AN, LaVail J, Basbaum C. The microvesicle as a vehicle for emmprin in tumor-stromal interactions. *Oncogene* 2004;**23**:956–63.

67. Belton Jr RJ, Chen L, Mesquita FS, Nowak RA. Basigin-2 is a cell surface receptor for soluble basigin ligand. *J Biol Chem* 2008;**283**:17805–14.

68. Li T, Yan Y, Wang B, et al. Exosomes derived from human umbilical cord mesenchymal stem cells alleviate liver fibrosis. *Stem Cells Dev* 2013;**22**:845–54.

69. Herrera MB, Fonsato V, Gatti S, et al. Human liver stem cell-derived microvesicles accelerate hepatic regeneration in hepatectomized rats. *J Cell Mol Med* 2010;**14**:1605–18.

70. Xin H, Li Y, Buller B, et al. Exosome-mediated transfer of mir-133b from multipotent mesenchymal stromal cells to neural cells contributes to neurite outgrowth. *Stem Cells* 2012;**30**:1556–64.

71. Yu YM, Gibbs KM, Davila J, et al. MicroRNA mir-133b is essential for functional recovery after spinal cord injury in adult zebrafish. *Eur J Neurosci* 2011;**33**:1587–97.

72. Ratajczak J, Miekus K, Kucia M, et al. Embryonic stem cell-derived microvesicles reprogram hematopoietic progenitors: evidence for horizontal transfer of mRNA and protein delivery. *Leukemia* 2006;**20**:847–56.

73. Calabrese JM, Seila AC, Yeo GW, Sharp PA. RNA sequence analysis defines Dicer's role in mouse embryonic stem cells. *Proc Natl Acad Sci USA* 2007;**104**:18097–102.

74. Miyoshi N, Ishii H, Nagano H, et al. Reprogramming of mouse and human cells to pluripotency using mature microRNAs. *Cell Stem Cell* 2011;**8**:633–8.

75. Katsman D, Stackpole EJ, Domin DR, Farber DB. Embryonic stem cell-derived microvesicles induce gene expression changes in Müller cells of the retina. *PLoS One* 2012;**7**:e50417.

76. Bobrie A, Krumeich S, Reyal F, et al. Rab27a supports exosome-dependent and -independent mechanisms that modify the tumor microenvironment and can promote tumor progression. *Cancer Res* 2012;**72**:4920–30.

77. Ostrowski M, Carmo NB, Krumeich S, et al. Rab27a and Rab27b control different steps of the exosome secretion pathway. *Nat Cell Biol* 2010;**12**:19–30. sup pp. 1–13.

78. Islam A, Shen X, Hiroi T, Moss J, Vaughan M, Levine SJ. The brefeldin A-inhibited guanine nucleotide-exchange protein, BIG2, regulates the constitutive release of TNFR1 exosome-like vesicles. *J Biol Chem* 2007;**282**:9591–9.

79. Kooijmans SA, Vader P, van Dommelen SM, van Solinge WW, Schiffelers RM. Exosome mimetics: a novel class of drug delivery systems. *Int J Nanomedicine* 2012;**7**:1525–41.

80. Marcus ME, Leonard JN. Fedexosomes: engineering therapeutic biological nanoparticles that truly deliver. *Pharm (Basel)* 2013;**6**:659–80.

81. Alvarez-Erviti L, Seow Y, Yin H, Betts C, Lakhal S, Wood MJ. Delivery of siRNA to the mouse brain by systemic injection of targeted exosomes. *Nat Biotechnol* 2011;**29**:341–5.

82. Hartman ZC, Wei J, Glass OK, et al. Increasing vaccine potency through exosome antigen targeting. *Vaccine* 2011;**29**:9361–7.

83. Ohno S, Takanashi M, Sudo K, et al. Systemically injected exosomes targeted to EGFR deliver antitumor microRNA to breast cancer cells. *Mol Ther* 2013;**21**:185–91.

84. Sancho-Martinez I, Baek SH, Izpisua Belmonte JC. Lineage conversion methodologies meet the reprogramming toolbox. *Nat Cell Biol* 2012;**14**:892–9.

85. Inagawa K, Miyamoto K, Yamakawa H, et al. Induction of cardiomyocyte-like cells in infarct hearts by gene transfer of Gata4, Mef2c, and Tbx5. *Circ Res* 2012;**111**:1147–56.

86. Qian L, Huang Y, Spencer CI, et al. In vivo reprogramming of murine cardiac fibroblasts into induced cardiomyocytes. *Nature* 2012;**485**:593–8.

87. Song K, Nam YJ, Luo X, et al. Heart repair by reprogramming non-myocytes with cardiac transcription factors. *Nature* 2012;**485**:599–604.

88. Zhou Q, Brown J, Kanarek A, Rajagopal J, Melton DA. In vivo reprogramming of adult pancreatic exocrine cells to beta-cells. *Nature* 2008;**455**:627–32.

Part IV

Cell Therapies and Other Applications

Chapter 15

Nature or Nurture: Innate versus Cultured Mesenchymal Stem Cells for Tissue Regeneration

Alvaro Santamaria, Greg Asatrian, William C.W. Chen, Aaron W. James, Winters Hardy, Kang Ting, Arnold I. Caplan, Chia Soo, Bruno Péault

Key Concepts

1. Mesenchymal stem cells (MSCs) are potent regenerative cells selected retrospectively by adherence and proliferation during long-term culture, and already used in a large number of clinical trials to treat multiple, diverse conditions.
2. Native MSC precursors have been prospectively identified as perivascular cells. However, whether perivascular MSC ascendants display naturally in vivo all the potentials exhibited by conventional culture-selected MSCs is unknown.
3. Prospectively purified perivascular cells show, experimentally, superior ability to generate/repair bone Potential benefits of using in patients prospectively purified PSCs in place of conventional, heterogeneous, culture-selected MSCs include: a- compliance with FDA criteria in terms of cell identity, purity, and potency; b- absence of contaminating endothelial cells, known to interfere with MSC differentiation; c- absence of in vitro expansion, avoiding xenoimmunization, and limiting risks of genetic instability and malignant transformation.

List of Abbreviations

AP Adventitial pericyte
BMC Bone marrow mononuclear cell
CDAI Crohn's disease activity index
CFR Code of Federal Regulations
cMSC Cardiac mesenchymal stem/stromal cell
CSC Cardiac stem/progenitor cell
DBX Demineralized bone matrix
DLI Donor lymphocyte infusion
ESC Embryonic stem cell
FDA Food and Drug Administration
FGF Fibroblast growth factor
GMP Good manufacturing practices
GVHD Graft-versus-host disease
HB-EGF Heparin-binding epidermal growth factor
hHP Human heart pericyte
HLA Human leukocyte antigen
hSkMP Human skeletal muscle-derived pericyte
hSVF Human stromal vascular fraction
HUVEC Human umbilical cord vein endothelial cell
IBD Inflammatory bowel disease
ICM Induced cardiac myocyte
IFN-γ Interferon γ
IL Interleukin
IND investigational new drug

Translating Regenerative Medicine to the Clinic. http://dx.doi.org/10.1016/B978-0-12-800548-4.00015-2

IPSC Induced pluripotent stem cell
LIF Leukemia inhibiting factor
MI Myocardial infarction
miR MicroRNA
MSC Mesenchymal stem cell
NK cell Natural killer cell
OCN Osteocalcin
OI Osteogenesis imperfecta
OPN Osteopontin
PDAI Perianal disease activity index
PDGFR Platelet-derived growth factor receptor
PLGA Poly(lactic-co-glycolic acid)
PSC Perivascular stem cell
SCID Severe combined immunodeficiency
TGF Transforming growth factor
TNF-α Tumor necrosis factor α
VEGF Vascular endothelial growth factor

1. INTRODUCTION

Mesenchymal stem cells (MSCs) occupy a special niche in stem cell science. These cells have been and remain very popular among cell therapists and tissue engineers, as shown by the use of MSCs in over 480 clinical trials (see clinical-trials.gov). This success is justified by the diverse positive contributions exerted by MSCs toward tissue (re)generation and repair as tissue progenitors, proangiogenic cells, immunosuppressors, and supportive cells for lineage-committed stem cells. MSCs are, in addition, remarkably easy to derive and expand since MSC extraction is a mere primary culture of unselected dissociated cells. Moreover, MSCs can be grown from virtually any vascularized human organ, leaving a choice of convenient, abundant, and dispensable sources of these cells such as adult abdominal adipose tissue and fetal appendages at birth.

On the other hand, exclusively indirect selection by adherence and proliferation on culture plastic has long obscured the biologic characteristics of innate MSCs. *Bona fide* MSCs being by essence long-term cultured cells, the native embryonic origin, identity, lineage affiliation, tissue distribution, frequency, and—importantly—actual role of these cells in normal tissue homeostasis and repair remained unknown decades after their initial discovery.

In the past 10 years, the very identity of native MSCs has been gradually uncovered, revealing a perivascular origin for these elusive regenerative cells. The prospective identification of innate MSCs now opens the possibility of using highly purified—and, in some instances, uncultured—precisely characterized perivascular cells for cell therapies, in place of their heterogeneous, culture-selected conventional progeny. In this chapter, we have reviewed the clinical use of customary, in vitro-derived MSCs, and put in perspective recent attempts to achieve tissue regeneration using their perivascular native ancestors.

2. THE CONVENTIONAL CULTURED MSC: A BRIEF HISTORIC PERSPECTIVE

Alexander Friedenstein (1924–1998) is usually credited for the discovery of MSCs. A scientist at the Academy of Sciences of USSR, Friedenstein was studying transplantation of the cellular environment of hematopoiesis and observed that stromal cells in the bone marrow, spleen and thymus yield clonal fibroblast colonies in culture[1] and contain inducible osteogenic precursor cells.[2] Relying on expertise in the embryonic development of mesodermal tissues, Arnold Caplan later proposed that *MSCs*, able to form bone and cartilage, persist in the adult and documented the existence of such progenitor cells in chick and rat bone marrow, predicting the potential considerable therapeutic value of equivalent human cells.[3] Culture-adherent osteogenic cells were identified in human bone marrow and typified by expression of the SH2, SH3, and SH4 surface markers.[4] Definitive characterization of human MSCs was published in 1997, as bone marrow-derived culture-adherent cells expressing SH2, SH3, CD44, CD71, and CD90, among other markers, were capable of differentiation into fat, bone, and cartilage.[5] Importantly, clonal derivation in vitro confirmed culture-selected MSCs to be truly multipotent cells and not a combination of diversely restricted progenitors.[5] Criteria defining culture-isolated MSCs have remained essentially the same since then, as formalized in 2006 by the International Society for Cell Therapy.[6] Even though MSCs were initially isolated from bone marrow, multiple investigators have showed since the early 2000s that related or similar cells can be isolated in culture from virtually every tissue and organ, including dental pulp,[7] dermis,[8] white adipose tissue,[9] synovial membrane,[10] perichondrium,[11] pancreas,[12] fetal blood-forming organs,[13] umbilical cord,[14] umbilical cord blood,[15] cornea,[16] lung,[17] skeletal muscle,[18] and endometrium.[19] Last, a recent publication documents that clonally derived MSCs can be serially transplanted into a murine marrow injury model. This could be interpreted to support their stem cell designation.[20]

3. MEDICAL USE OF MSCs

The advent of MSCs for therapeutic purposes has boosted optimism for the treatment of diseases that are currently incurable, or are in dire need of a superior therapeutic alternative. MSC applications have a wide variety of uses that are primarily being developed for treating diseases at the bedside, either by direct administration to patients or indirectly by developing therapies through tissue engineering. Furthermore, their use is not limited to direct disease intervention. MSCs are also being industrialized for stem cell-based testing and diagnostic techniques, which will see a rise in use as MSC therapies are developed. Other applications show great promise in pharmaceuticals, where MSCs are used for drug discovery and pharmacologic testing.

Research and development using MSCs is essentially focused on addressing unmet health-care needs in a range of diseases. Investigations at both preclinical and clinical stages have shown that MSC potential lies in the treatment of both immunologic and nonimmunologic diseases, further validating this mode of therapy as a mainstay in regenerative medicine. The main biological properties of MSCs being exploited focus on controlling inflammation and promoting tissue regeneration. This is achieved by nonmutually exclusive mechanisms that include, but are not limited to, immunomodulation, differentiation into a variety of cell types, secretion of trophic factors, and homing to sites of inflammation.[21,22] Because of the profound paracrine effects of MSCs, Caplan has suggested that they could be referred to as Medicinal Signaling Cells.[23]

This section will explore how MSCs are being used to target diseases and showcase their efficacy in treating those diseases. The postulated mechanism of action will be specified and linked to preclinical studies where applicable. A summary of diseases where MSCs are being used in clinical trials will be provided to allow for a comprehensive understanding of their use. Finally, the status of MSC therapy as a whole will be explored with regard to their transition into clinical practice as an Food and Drug Administration (FDA)-approved therapeutic.

3.1 Diseases of the Immune System

Immune modulation by MSCs appears to play an important role in disease intervention. Some available therapies that postulate immunomodulation as their main mechanism of action target diseases such as graft-versus-host disease (GVHD), diabetes, liver cirrhosis, Crohn's disease, and rheumatoid arthritis.[22] Key mechanisms by which MSC immunomodulation is achieved involve suppressing lymphocyte differentiation, the induction of regulatory T cells, and the production of anti-inflammatory cytokines.[24,25]

3.1.1 Graft-versus-Host Disease

Acute GVHD is a life-threatening and frequent complication of allogeneic hematopoietic stem cell transplantation and donor lymphocyte infusion. Currently, steroids are the first-line treatment for this complication; however, an overall response is only seen in 50% of patients.[26] Furthermore, nonresponders provided second-line treatment with other immunosuppressive agents exhibit a 1-year survival rate of 30%.[27]

Significant improvements to GVHD management are already being seen through MSC therapies in clinical trials, with the most advanced study being in Phase III. In a Phase-II study using bone marrow-derived MSCs for steroid-resistant, acute GVHD, patients who experienced complete resolution of the disease had a lower mortality rate after 1 year compared to those who had a partial or no response (37% vs 72%; p=0.002).[28] Similarly, in a Phase-II study using premanufactured bone marrow-derived MSCs along with corticosteroids, patients who had a complete response, versus those who did not, had a significantly higher survival rate at day 90 (88% vs 14%; p=0.00008).[29] Finally, in a Phase-I/II study using premanufactured bone marrow-derived MSCs by Muroi et al., patients with a complete response had an improved survival rate at 2 years versus noncomplete responders, with an overall survival of 57%.[30] Overall, clinical studies consistently show improved symptoms and overall survival in comparison to existing therapies for GVHD. However, the mechanism by which this is achieved is largely unknown and is principally explained by data from preclinical studies.

In vitro and in vivo studies suggest that the immunosuppressive properties of MSCs account for their efficacy in treating GVHD. MSCs act on B and T cells, natural killer (NK) cells, and antigen-presenting cells through a variety of mechanisms.[28]

MSCs suppress the activation and proliferation of T cells through both contact-dependent and contact-independent ways.[24] The cytokine secretion profile of T cells (T_H1, T_H2), among other immune cells, is altered in such a way that it becomes anti-inflammatory. IFN-γ and IL-2 secretion are decreased, while IL-4 secretion is increased.[24,31] Differential distribution of T cells is also altered under the presence of MSCs, leading to an increased representation of regulatory T cells (T_{regs}).[24,31] Dendritic cells are directed to decrease TNF-α and increase IL-10 secretion, which further suppresses T-cell proliferation among other anti-inflammatory properties.[31] Interactions between MSCs and NK cells lead to decreased IFN-γ and further suppress the cytokine-induced pathology.[31] The immunologic tolerance created by a shift from proinflammatory to anti-inflammatory cytokine secretion and the diminished activation of alloreactive T cells is the mechanism by which MSCs may be operating to attenuate the pathology in GVHD. Observing these effects in clinical studies may be difficult to assess given the number of variables that may be at play in these studies, such as cointerventions and the tissue source of MSCs.

Determining the efficacy of MSC therapy in GVHD is confounded by multiple variables that form the basis of each individual clinical study. First and foremost, sources of MSCs and preparation thereof prior to infusion vary greatly between studies. In the aforementioned studies, MSCs produced no adverse reactions and generally relieved patient symptoms, despite the fact that some were obtained from HLA-identical sibling donors and haploidentical donors, while others derived from third-party HLA mismatched donors. While this showed that huge amounts of MSCs can be available as off-the-shelf items, the optimal type and source of MSCs was yet unknown. Additionally, many studies included coadministration of immunosuppressive agents that could act in synergy with MSCs, a fact which is not well documented in the absence of proper controls. Finally, patient profiles in terms of age and GVHD stage were variable, which could limit showing a clear effect in one subpopulation and prolong the FDA approval process for MSC use in GVHD.

3.1.2 Crohn's Disease

Crohn's disease is a systemic inflammatory disorder that belongs to a group of conditions known as inflammatory bowel diseases. Crohn's disease mainly affects the gastrointestinal tract and often presents with abdominal pain, cramping, constipation, or diarrhea with blood or mucus.[32] Primary goals in Crohn's disease therapy aim at halting gastrointestinal mucosal destruction and managing symptoms. Therapeutic goals are primarily achieved through lifestyle changes, anti-inflammatory and antineoplastic drugs such as anti-TNF biologics and thiopurines.[32] Despite the use of biologic therapy, the annual hospital admission rate for Crohn's disease is 20%, with 25% of all patients requiring surgery within 5 years after diagnosis, and 50% after 10 years.[32,33]

MSC use in Crohn's disease has shown promising results in clinical studies that have propelled their use in Phase-III trials. In a Phase-II study by Forbes et al., biologic therapy refractory patients underwent intravenous infusions of allogeneic bone marrow-derived MSCs for 4 weeks.[33] Study outcomes included a clinical response measured by a decreased value in the Crohn's disease activity index (CDAI), clinical remission, endoscopic improvement, normalization of C-reactive protein, and improved quality of life. Study results showed 80% of patients had a clinical response, 53% had clinical remission, and 47% had endoscopic improvement.[33] In a similar study by Ciccocioppo et al., autologous bone marrow-derived MSCs were locally injected to treat Crohn's disease, with the specific aim of healing external Crohn's disease-derived fistulas.[34] Of the ten patients treated with MSCs, seven experienced full closure of fistula tracks and the remaining three saw partial closure. Furthermore, all patients had a reduction of Crohn's disease as measured by CDAI ($p < 0.001$) and reduced perianal disease activity as measured by the perianal disease activity index (PDAI) ($p < 0.001$) (see Figure 1).[34]

Crohn's disease exhibits an immune dysregulation to intestinal bacteria causing severe intestinal tissue damage.[34] As a result, populations of regulatory T cells, T_H1, and T_H17 cells are altered in a proinflammatory fashion.[35] Aside from using MSCs for immune regulation as with GVHD, MSCs are thought to specifically target this T-cell subpopulation imbalance by enhancing regulatory T-cell differentiation and suppressing proinflammatory T_H1 and T_H17 cells.[35,36] The ability of MSCs to mediate this effect is in part due to their ability to localize, or home, to the site of inflammation.[37]

Clinical amelioration in Crohn's disease is measured by the CDAI. In the above mentioned studies, an improvement in CDAI was seen with statistical significance.[33,34] However, similar to studies in GVHD, the coadministration of immunosuppressive agents along with MSCs makes it difficult to observe which MSC population and preparation is most effective in achieving treatment outcomes. Understanding which MSC formulation is most successful and potent is essential for developing a therapy, facilitating FDA approval, and allowing for aspects of this technology to be used in other avenues (Table 1).

3.2 Nonimmune Diseases

MSCs have unique biological properties that modulate disease through direct differentiation and regeneration capabilities, in addition to immunomodulation. Therapies that make use of MSC differentiation currently target diseases such as

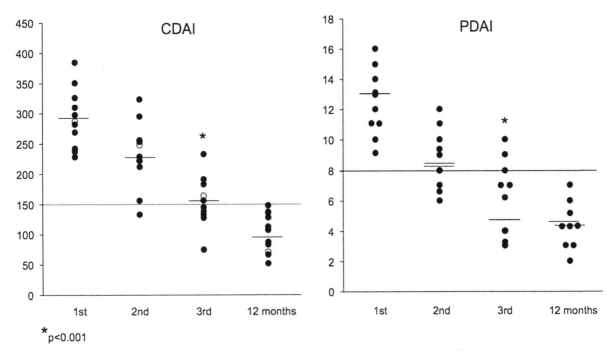

*p<0.001

FIGURE 1 Representation of Crohn's disease activity index (CDAI) and perianal disease activity index (PDAI) score values throughout the study. The CDAI and PDAI values of each patient at the time of the first (1st), second (2nd), and third (3rd) injection of autologous bone marrow-derived mesenchymal stromal cells, and after 12 months follow-up (12 months) are shown in the left and right panel, respectively. A significant decrease of activity index was observed at the time of the third procedure with induction of remission (CDAI·150 and PDAI·8), which persisted after the 12-month follow-up. The values indicate mean±SD, the horizontal lines indicate the threshold values between quiescent and active disease.

TABLE 1 Summary of MSC Use in Some Clinical Applications

Target	MSC Source	Mode of Delivery	Proposed Mechanism	Outcome
Liver cirrhosis[38]	Autologous bone marrow-derived MSC	Injected through one of the main branches of the portal vein	Migration, differentiation	Improved model for end-stage liver disease score (MELD) and INR, decreased serum creatinine, improved quality of life
Multiple sclerosis[39]	Autologous bone marrow-derived MSC	Intravenous infusion	Neuroprotection through paracrine signaling, immunomodulation	Improved visual acuity and visual evoked response latency
Systemic lupus[40]	Allogeneic non-human leukocyte antigen matched bone marrow-derived MSC	Intravenous infusion	Immunomodulation, expansion of regulatory T cells, differentiation of MSC to endothelial cells in nephron	Decreased SLE activity index, decreased anti-dsDNA levels, decreased proteinuria
Osteoarthritis[41]	Autologous bone marrow-derived MSC	MSCs embedded in collagen gels transplanted into articular cartilage defect	Differentiation, secretion of soluble factors that promote tissue regeneration	Arthroscopic and histological grading score of cell transplanted group was better than that of the control group

Note: Many of these studies were feasibility or safety studies that provided information about efficacy as preliminary results. MSC, mesenchymal stem cell; INR, international normalized ratio; SLE, systemic lupus erythematosus.

osteoarthritis, osteogenesis imperfecta (OI), and while still a subject of debate, myocardial infarction. Mechanisms behind disease modulation for these and related conditions have been approached in vitro and in animals but still remain elusive, and largely unknown in humans.

3.2.1 Myocardial Infarction

Several clinical trials are underway to document the safety and efficacy of MSCs in treating cardiovascular diseases and some of the results are promising. In a Phase-1 randomized, double-blinded, placebo-controlled trial by Hare et al., patients underwent intravenous infusion of non-HLA-matched allogeneic bone marrow-derived MSCs (Prochymal) after acute myocardial infarction (AMI).[42] The main study end point of evaluating safety in humans was established, but it is the provisional efficacy data that further support its use in treating AMI. Experimental efficacy end points evaluated were ejection fraction, tachycardy episodes, pulmonary function, and overall health. All efficacy end points were significantly improved and reverse remodeling of heart tissue that prevented chamber enlargement was thought to contribute to these findings.[42]

Other clinical studies have similarly shown promising results using different sources of MSCs. In a study by Chen et al., patients received autologous bone marrow-derived MSCs that had been expanded in culture for 10 days prior to direct injection into the occluded coronary artery, 18 days after percutaneous coronary intervention. In this study, left ventricular function improved significantly when compared to its state before infusion and to the control group.[43]

These two studies serve as examples of the diverse sources and formulations of MSCs that can be used to regenerate a damaged myocardium. Bone marrow mononuclear cells are another MSC source that has been demonstrated to improve left ventricular ejection fraction.[44,45] In comparing MSC therapies to treat AMI, several factors come into play. As previously explored, the source of MSCs can have a significant impact on therapy due to the time of MSC availability and available dose. With regard to an AMI, the timing of treatment delivery after an acute event has been shown to alter treatment outcomes.[43]

The mechanisms by which MSCs are thought to improve cardiac function after AMI are several: secretion of paracrine factors, localization and engraftment to damaged cardiac tissue, and differentiation into various cells.[46] In animal studies MSCs improve cardiac function by differentiating to cardiomyocytes and vascular endothelial cells.[47,48] In addition, they decrease collagen volume fraction and stimulate neovascularization.[48]

3.2.2 Osteogenesis Imperfecta

OI is a complex congenital bone disorder that is mainly caused by genetic defects of Type-I collagen synthesis.[49] Clinical manifestations from OI are frequent bone fractures, osteopenia, and short stature.[49] While there is no cure for this disorder, therapy in the form of bisphosphonates and orthopedic surgery are frequently used to manage OI, with minimal relief.[50] Gene therapy and cell-based therapies have been postulated as potential future treatments for OI, and MSC use has already shown positive results in managing this disorder.

In a study by Horwitz et al., three children underwent allogeneic bone marrow-derived MSC transplantation from an HLA matched or single antigen mismatch sibling.[50] Recipients saw engraftment of cells that differentiated into osteoblasts. 1.5–2% of the osteoblasts were of donor origin and significantly contributed to bone remodeling. Histology of bone showed improved bone formation when compared to samples taken prior to infusion. Additionally, the bone mineral content of patients increased, further promoting proper bone deposition and mineralization that led to fewer fractures and improved bone growth.[50]

In a similar study, six children underwent isolated MSC transplantation after a bone marrow transplant. This is different from the previous study[50] in that MSCs were not isolated prior to infusion, but instead were infused as unmanipulated bone marrow cells. Similarly, five out of six patients saw engraftment of MSCs and accelerated bone growth up to 6 months after infusion.[51] Results of these studies undeniably point toward improved effects due to either whole bone marrow-derived MSCs or isolated MSCs. However, in the former, it is difficult to pinpoint the effects solely to MSCs, as other bone marrow-derived cells were included in the infusion.

3.3 Regulatory Challenges Faced by MSC-Based Therapies

Data generated from research over the past few decades have shown that MSCs are safe and often efficacious at treating targeted diseases. However, despite evidence promoting their therapeutic use, MSC approval by the FDA has been slow to materialize. The challenge this product has faced in gaining approval calls into question the magnitude of efficacy of MSCs, their manufacturing process, and begs for a thorough evaluation of the requirements outlined in regulatory guidelines used in the approval process.

The FDA requires that the safety, identity, purity, and potency of MSCs be clearly established. This requires compliance with Title 21 of the Code of Federal Regulations (CFR) part 1271, which is applicable to human cells and tissues intended for transplantation into a human recipient.[52] An additional level of complexity is added to the MSC approval process when the cells are processed ex vivo, requiring compliance with good manufacturing practices.[53] There are several ways to overcome these difficulties and make MSC technology widely available in medical practice. These include standardization and better documentation of the MSC isolation process, proper characterization of MSCs with well-controlled assays, and quantitation of product efficacy permitting comparison of MSCs produced in different laboratories.[54]

MSCs produced by Osiris Therapeutics under the name Prochymal appear to be one of a few products in the advanced stages of the approval process. These MSCs are currently in Phase-III clinical trials for acute GVHD, in Phase III for Crohn's disease, and in Phase II for type-I diabetes and AMI. Additionally, Osiris has FDA clearance for Grafix®, an MSC-based therapy for use in acute and chronic skin wounds. What appears to allow for continued development of this therapy, aside from MSCs' ability to treat disease, is the consistency by which MSCs are manufactured and precisely characterized with fluorescence-activated cell sorting (FACS). Other laboratories have been able to produce MSCs according to the Prochymal protocol and seen comparable results in treating diseases.

In order to move forward and integrate MSCs into clinical practice and patient care, large, randomized controlled trials with high statistical power must precede any mainstream therapeutic use. Studies administering other treatments alongside MSCs must account for and analyze the combined effects of those treatments thoroughly. Better characterization of safety, identity, purity, and potency needs to be developed in order to quantitatively confirm the efficacy of MSCs and be able to compare their therapeutic effects between different laboratories.

4. THE ORIGINAL TISSUE RESIDENT MSC: A BETTER THERAPEUTIC ALTERNATIVE?

4.1 Prospective Identification of Native MSCs

Our search for innate MSCs within human tissues was guided by similarities described between MSCs and perivascular "mural" cells—aka pericytes[55]—in terms of phenotype and gene expression,[56–61] suggesting an affiliation between these two cell populations.[62] Human pericytes from multiple tissues and organs (skeletal muscle, myocardium, pancreas, adipose tissue, skin, placenta, umbilical cord, bone marrow, kidney, lung, brain, dental pulp) can be identified by surface expression of NG2, CD146, and PDGFR-β as well as the MSC markers CD44, CD73, CD90, and CD105, and the absence of known hematopoietic, endothelial, and myogenic cell markers.[63] Pericytes purified to homogeneity from diverse human organs by flow cytometry as CD146[+] CD34[−] CD45[−] CD56[−] cells and seeded in culture were indistinguishable from conventionally derived MSCs in terms of adherence, morphology, proliferation, surface antigen expression and, most importantly, developmental potential. At the clonal level, cultured pericytes differentiate into bone, cartilage, and fat cells in culture, and are strongly myogenic and osteogenic in vivo in immunodeficient mice.[63–65] We further explored whether *all* MSCs descend from pericytes[66] by dissecting by flow cytometry, using multiple cell markers, the stromal cell compartments of human white adipose tissue, lung, and skeletal muscle. This led to the identification of a population of CD146[−] CD34[hi] CD31[−] CD45[−] perivascular cells, which, although phenotypically and anatomically distinct from pericytes—these cells reside in the *tunica adventitia* of arteries and veins—exhibit the same ability to give rise to MSCs in culture.[67] Of note, related progenitors exhibiting MSC phenotypic and developmental properties have been described in the bovine artery wall[68] and isolated from the tunica adventitia of the human pulmonary artery.[69]

In summary, we have identified and characterized two separate perivascular presumptive MSCs: pericytes lining microvessels and adventitial cells ensheathing larger vessels, designated collectively as perivascular stem cells (PSCs).[70–73] As an important remark, the documented MSC potential of perivascular cells does not reflect a mere culture artifact resulting in the reprogramming of these cells in vitro. Leydig cells in the rat testis can be regenerated by pericytes following chemical injury[74] and mural cells act as white adipocytes progenitors in murine fat tissue.[75]

Pericytes residing in skeletal muscle differentiate into muscle fibers and generate satellite cells following chemical damage,[76] and direct differentiation of pericytes into follicular dendritic cells was demonstrated in the mouse.[77] During development, pericytes give rise to diverse mesodermal derivatives[78] in response to PDGFRβ signaling.[79] These recent data point to the existence of professional tissue (re)generative cells at the periphery of blood vessels throughout embryonic, fetal, and postnatal life.

4.2 PSCs and Cardiac Regeneration

Stem cell-based cardiac regeneration has been extensively investigated in the past decade. A wide selection of precursor cell sources have been examined for their therapeutic efficacy, including endogenous c-kit[+] cardiac stem/progenitor cells (CSCs)[80] and Wt1[+] epicardium-derived progenitor cells,[81] embryonic stem cells[82] and induced pluripotent stem cells,[83]

FIGURE 2 Putative repair mechanisms by donor human skeletal muscle-derived pericyte (hSkMPs) within the ischemic myocardium. The size of arrows indicates the presumable contribution of each mechanism.

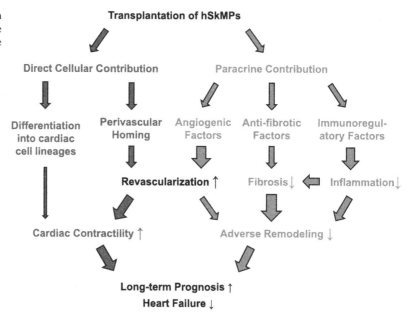

induced cardiac myocytes,[84] circulating CD34+ cells,[85] and exogenous adult stem/progenitor cells derived from different tissues,[86–88] including conventionally derived MSCs (see above). In recent years, the role of perivascular stromal cells, particularly microvascular pericytes, in cardiac repair and regeneration has gained attention due to their inherent multipotency and robust paracrine function.[63,89]

Our group previously examined the therapeutic potential of human skeletal muscle-derived pericytes (hSkMPs) for cardiac regeneration in an immunodeficient mouse model of AMI.[90] Functionally, intramyocardial transplantation of hSkMPs significantly improved cardiac contractility and attenuated adverse ventricular remodeling. Histologically, hSkMP treatment resulted in notable increase in host angiogenesis and substantial reduction of chronic inflammation and myocardial fibrosis at the infarct site. Although we have detected direct cellular involvement of donor cells in the regenerative process, including homing of hSkMPs to myocardial microvasculature, formation of gap junctions between hSkMPs and host cells, and differentiation of a minor fraction of engrafted hSkMPs into cardiac cell lineages, the benefits were largely attributable to the active secretion of trophic factors and cytokines including VEGF-A, PDGF-β, TGF-β1, IL-6, LIF, COX-2, and HMOX-1 by hSkMPs, even under stress conditions.[90] A schematic depiction of putative mechanisms of cardiac repair by hSkMP transplantation is shown in Figure 2.

Similarly, the potency of human CD34+/31− adventitial pericytes (APs), derived from the adventitial *vasa vasorum* of large blood vessels, for treating ischemic heart disease has been examined.[91] AP treatment led to improved cardiac function, ameliorated ventricular dilation and wall thinning, increased neovascularization, augmented myocardial blood flow, and reduced vascular permeability, cardiomyocyte apoptosis, and myocardial fibrosis.[92] The paracrine secretion of proangiogenic factors and chemokines, including microRNA-132 (miR-132), by APs presumably activated the proangiogenic and prosurvival Akt/eNOS/Bcl-2 signaling pathway. Blocking miR-132 function in APs significantly reduced their vascular supportive capacity in vitro, revascularization in the ischemic myocardium, and cardiac protective functions, suggesting the importance of miR-132 in AP-mediated cardiac regeneration.[92] Lately, a combinatory therapy enlisting both APs and CSCs exhibited equal or better effects functionally and histologically than administration of either APs or CSCs alone, suggesting that a complementary benefit derives from the use of two or more therapeutic stem/progenitor cell populations.[93]

On the other hand, the origin and role of endogenous cardiac mesenchymal stem/stromal cells (cMSCs) has not been clarified until recently. Murine cMSCs were not only able to differentiate into major cardiac cell types, including cardiomyocytes, smooth muscle cells, and endothelial cells, but also capable of broadly developing into cell lineages of all three germ layers.[94] However, whether cMSCs reside in perivascular locations in the human heart is not known. Lately we have successfully identified and purified human heart pericytes (hHPs) by flow cytometry based on their surface antigen expression[95] (Figure 3). A fraction of hHPs (CD146+/CD34−/CD45−/CD56−/CD117−) demonstrated immature cardiomyogenic potential in vitro and yielded mature cardiomyocyte phenotypes after intramyocardial transplantation into healthy or infarcted immunodeficient mouse hearts.[95] hHPs shared many phenotypic and developmental similarities with skeletal muscle homologs such as classic pericyte and MSC marker expression and osteo-, chondro-, and adipogenic potentials; yet hHPs exhibited very different myogenic and

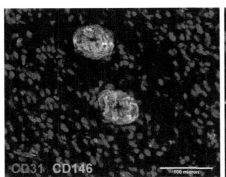
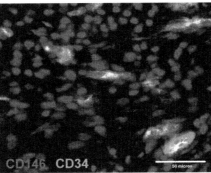

FIGURE 3 CD146⁺ perivascular stromal cells tightly surround CD31⁺ or CD34⁺ vascular endothelial cells at the microvascular level (<100 μm) in the human myocardium.

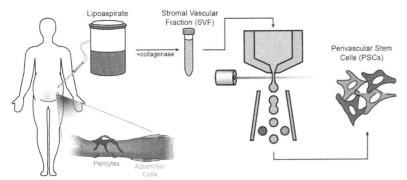

FIGURE 4 Perivascular stem cell purification from human adipose tissue.

angiogenic properties from hSkMPs. Cultured hHPs did not form skeletal myotubes or express myogenic markers under inductive conditions while isogenic hSkMPs undertook skeletal myogenesis as did pericytes from other organs. In three-dimensional Matrigel coculture with human umbilical cord vein endothelial cells, hHPs showed prevailing angiogenic and microvascular supportive responses under hypoxic conditions when compared with isogenic hSkMPs. Our data illustrate, for the first time, the likely developmental and functional divergence of pericytes in different human organs due to tissue specification.[95] Further investigation is needed to decipher such specifying signal(s) and/or mechanism(s) at distinct anatomical locations.

4.3 PSCs for Bone Repair and Regeneration

The utility of human perivascular stem cells (hPSCs) for bone tissue engineering/regeneration, and the advantage of hPSCs over unpurified cell populations, has been extensively documented across three small animal models: (1) a mouse intramuscular implantation model,[96,97] (2) a mouse critical-size calvarial defect model,[98] and (3) a rat lumbar spinal fusion model.[99] In each case, FACS purified adipose-derived hPSCs were compared to an unsorted population (hSVF—human stromal vascular fraction) obtained from the same patient (Figure 4). Of note, multiple different biomaterial scaffolds have been used as successful carriers for PSCs, including demineralized bone matrix (DBX)[96,97,100] as well as poly(lactic-co-glycolic acid) (PLGA).[98] In each model, we have shown promising preliminary evidence of PSC-mediated improved bone-formation in comparison to the unsorted stem cell population, presented sequentially below (Figure 5).

4.3.1 Improved PSC-Mediated Bone Formation in a Rodent Intramuscular Model

For intramuscular bone formation, equal numbers of unsorted hSVF or purified hPSCs were taken from the same patient and implanted into opposite limbs of the same severe combined immunodeficiency mouse using a DBX scaffold for delivery (Figure 2).[96] Results indicated that hPSC-induced robust bone formation, confirmed by immunohistochemical stainings for osteopontin and osteocalcin, as well as greater vascular endothelial growth factor (VEGF) expression and capillary density in the interstices between bone particles.[100] This indicates greater bone formation and vasculogenesis with PSC implantation. Importantly and in all 165 implants performed, neither cellular atypia nor tumor formation has been observed with any hPSC muscle pouch implantation. These results suggested the superiority of hPSCs to unsorted MSC populations in bone formation, and the high level of safety of hPSC for in vivo use.

FIGURE 5 Assessing the bone-forming potential of human adipose tissue-derived perivascular stem cells in immunodeficient rodents. Total stromal vascular fraction cells or fluorescence-activated cell-sorted pericytes and adventitial cells were used for mouse intramuscular implantation, mouse critical-size calvarial defect regeneration, and rat lumbar spinal fusion.

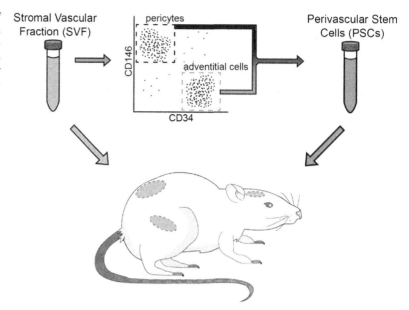

4.3.2 Improved PSC-Mediated Bone Formation in a Rodent Calvarial Defect Model

To further test the advantages of hPSCs over hSVF, a mouse calvarial defect was next employed.[97] Here, equal numbers of hPSCs or hSVF cells derived from the same patient were implanted in a critical-size 3-mm calvarial defect, using a custom-fabricated PLGA scaffold for cell delivery. Results demonstrated that hPSCs led to significantly greater bone healing than did hSVF, shown both in microCT reconstructions and relative defect healing. hPSC-mediated defect healing was accompanied by an increased presence of bone markers, the persistence of human-derived cells, and an increased elaboration of provasculogenic growth factors such as VEGF.[97] These results extended our findings to a bone defect model, again finding a superior bone-forming effect with PSC implantation over unpurified MSCs.

4.3.3 Improved PSC-Mediated Bone Formation in a Rodent Spinal Fusion Model

Next, we investigated whether hPSCs hold promise for spinal fusion. A rat lumbar intertransverse spinal fusion model was chosen, which we successfully employed in previous bone regeneration studies.[101,102] A DBX scaffold was utilized for cell delivery. Four weeks after implantation, hPSCs induced complete bone bridging between the L4:L5 levels, confirmed by high-resolution microCT reconstructions, while the scaffold alone or unpurified hSVF did not lead to reliable fusion. Histologic analyses showed that hPSC treatment resulted in robust endochondral ossification and the persistence of human antigens in bone-lining osteoblasts and lacunar osteocytes. Further, manual palpation scoring confirmed complete fusion among all hPSC-treated samples, in marked contrast to control groups. Of note, across both calvarial and spinal fusion models, no significant adverse events have occurred, no incidence of ectopic bone formation, nor observed morbidity or mortality. In aggregate, these results showed that hPSCs exhibit a high degree of promise in a clinically relevant rat spinal fusion model.

4.3.4 Robust Paracrine Effects of hPSCs

Despite accumulating data to suggest the preclinical efficacy of hPSC-mediated bone tissue engineering—multiple lines of evidence suggest that hPSCs do not themselves directly produce the majority of new bone. Indeed, focal evidence for direct participation of implanted PSCs in ossification has been found in vivo. Here, colocalization of labels of mineralization with human-specific antigens identified foci of direct PSC ossification. Nevertheless, when expression of human- and rat-specific antigens was examined in bone tissue regenerates, it became clear that osteoblasts and osteocytes within new bone were predominately of rat origin (up to 30 times more rat-specific immunoreactive cells). Thus, the majority of newly formed bone was clearly host rather than donor derived, and an underlying paracrine mechanism was hypothesized for PSC-induced bone formation. Support for this hypothesis is evident; even in our original description of pericytes and PSCs where it was clear that these cells were robust cytokine sources.[89] For example, human pericytes secrete 5–20 times more heparin-binding epidermal growth factor, fibroblast growth factor-2 (FGF-2), VEGF, and keratinocyte growth factor than classically derived adipose tissue or cord blood MSCs.[89] Thus, the robust osteogenic/vasculogenic effects of hPSCs may be predominantly paracrine in nature.

5. PERSPECTIVES

MSCs represent a paradoxical example of clinically prevalent adult regenerative cells endowed with multifaceted tissue development/repair potential but still crudely characterized in terms of biologic identity and natural history. Research developed in the past decade has attributed to MSCs an original perivascular individuality,[73] although the existence of MSC forerunners away from blood vessels cannot yet be definitively ruled out. Innate physical association with blood vessels may explain why MSCs have been grown from virtually all organs[7–19] and allow for the prospective purification of these cells to homogeneity, by flow cytometry, as pericytes and adventitial progenitors.[73] We have illustrated in this chapter the use of purified PSCs for cardiac repair and bone regeneration, as modeled in human/rodent xenochimeras. Many other clinical applications of these cells can be, however, envisioned as PSCs have been also used experimentally for lung repair,[103] vascular graft engineering,[104] wound healing,[105] renin-producing cell generation,[106] and long-term maintenance of hematopoietic stem cells.[107] What are the benefits of using in the clinic prospectively purified PSCs in place of conventional, heterogeneous, culture-selected MSCs? Compliance with criteria edicted by the FDA in terms of cell identity, purity, and potency is a prevailing argument and will undoubtedly facilitate FDA approval.[52] In addition, the use of PSCs will guarantee the absence of contaminating endothelial cells which have been shown to interfere with MSC differentiation.[108] Last, but not least, PSCs can be used in some indications, such as bone repair and spine fusion, in the absence of in vitro expansion (see above). This will allow to avoid therapeutic cell contact with animal products and resulting xenoimmunization, and limit the risks of genetic instability and malignant transformation associated with cell culture over the long term. Our group is presently pursuing the validation of a composite product including freshly sorted autologous PSCs as an investigational new drug for bone regeneration. Clinical experience will test the validity of the proposed use of a highly pure population of regenerative cells.

ACKNOWLEDGMENTS

The work described in the present chapter has been made possible by grants from the National Institute of Health, California Institute for Regenerative Medicine, British Heart Foundation, and Medical Research Council.

REFERENCES

1. Friedenstein AJ, Chailakhjan RK, Lalykina KS. The development of fibroblast colonies in monolayer cultures of guinea-pig bone marrow and spleen cells. *Cell Tissue Kinet* 1970;**3**:393–403.
2. Friedenstein AJ, Lalykina KS. Thymus cells are inducible to osteogenesis. *Eur J Immunol* 1972;**2**:602–3.
3. Caplan AI. Mesenchymal stem cells. *J Orthop Res* 1991;**9**:641–50.
4. Haynesworth SE, Barer MA, Caplan AI. Cell surface antigens on human marrow-derived mesenchymal cells are detected by monoclonal antibodies. *Bone* 1992;**13**:69–80.
5. Pittenger MF, Mackay AM, Beck SC, et al. Multilineage potential of adult human mesenchymal stem cells. *Science* 1997;**284**:143–7.
6. Dominici M, Le Blanc K, Mueller I, et al. Minimal criteria for defining multipotent mesenchymal stromal cells. The International Society for Cellular Therapy position statement. *Cytotherapy* 2006;**8**(4):315–7.
7. Gronthos S, Mankani M, Brahim J, et al. Postnatal human dental pulp stem cells (DPSCs) in vitro and in vivo. *Proc Natl Acad Sci USA* 2000;**97**(25):13625–30.
8. Young HE, Steele TA, Bray RA, et al. Human reserve pluripotent mesenchymal stem cells are present in the connective tissues of skeletal muscle and dermis derived from fetal, adult, and geriatric donors. *Anat Rec* 2001;**264**(1):51–62.
9. Zuk PA, Zhu M, Mizuno H, et al. Multilineage cells from human adipose tissue: implications for cell-based therapies. *Tissue Eng* 2001;**7**(2):211–28.
10. De Bari C, Dell'Accio F, Tylzanowski P, et al. Multipotent mesenchymal stem cells from adult human synovial membrane. *Arthritis Rheum* 2001;**44**(8):1928–42.
11. Arai F, Ohneda O, Miyamoto T, et al. Mesenchymal stem cells in perichondrium express activated leukocyte cell adhesion molecule and participate in bone marrow formation. *J Exp Med* 2002;**195**(12):1549–63.
12. Hua Y, Liaoa L, Wanga Q, et al. Isolation and identification of mesenchymal stem cells from human fetal pancreas. *J Lab Clin Med* 2003;**141**(5):342–9.
13. in't Anker PS, Noort WA, Scherjon SA, et al. Mesenchymal stem cells in human second-trimester bone marrow, liver, lung, and spleen exhibit a similar immunophenotype but a heterogeneous multilineage differentiation potential. *Haematologica* 2003;**88**(8):845–52.
14. Romanov YA, Svintsitskaya VA, Smirnov VN. Searching for alternative sources of postnatal human mesenchymal stem cells: candidate MSC-like cells from umbilical cord. *Stem Cells* 2003;**21**:105–10.

15. Wagner W, Wein F, Seckinger A, et al. Comparative characteristics of mesenchymal stem cells from human bone marrow, adipose tissue, and umbilical cord blood. *Exp Hematol* 2005;**33**(11):1402–16.

16. Choong PF, Mok PL, Cheong SK, et al. Mesenchymal stromal cell-like characteristics of corneal keratocytes. *Cytotherapy* 2007;**9**(3):252–8.

17. Lama VN, Smith L, Badri L, et al. Evidence for tissue-resident mesenchymal stem cells in human adult lung from studies of transplanted allografts. *J Clin Invest* 2007;**117**(4):989–96.

18. Zheng B, Cao B, Crisan M, et al. Prospective identification of myogenic endothelial cells in human skeletal muscle. *Nat Biotech* 2007;**25**(9): 1025–34.

19. Gargett CE, Schwab KE, Zillwood RM, et al. Isolation and culture of epithelial progenitors and mesenchymal stem cells from human endometrium. *Biol Reprod* 2009;**80**:1136–45.

20. Lin P, Corea D, Kean TJ, Awadallah A, Dennis JE, Caplan AI. Serial transplantation and long-term engraftment of intra-arterially delivered clonally derived mesenchymal stem cells to injured bone marrow. *Mol Ther* 2014;**22**(1):160–8.

21. Barry FP, Murphy JM. Mesenchymal stem cells: clinical applications and biological characterization. *Int J Biochem Cell Biol* 2004;**36**(4):568–84.

22. Wei X, et al. Mesenchymal stem cells: a new trend for cell therapy. *Acta Pharmacol Sin* 2013;**34**(6):747–54.

23. Caplan AI. What's in a name? *Tissue Eng Part A* 2010;**16**(8):2415–7.

24. Singer NG, Caplan AI. Mesenchymal stem cells: mechanisms of inflammation. *Ann Rev Pathol* 2011;**6**:457–78.

25. Caplan AI, Dennis JE. Mesenchymal stem cells as trophic mediators. *J Cell Biochem* 2006;**98**(5):1076–84.

26. Ringden O, Keating A. Mesenchymal stromal cells as treatment for chronic GVHD. *Bone Marrow Transpl* 2011;**46**(2):163–4.

27. Bacigalupo A. Management of acute graft-versus-host disease. *Br J Haematol* 2007;**137**(2):87–98.

28. Le Blanc K, Frassoni F, Ball L, Locatelli F, Roelofs H, Lewis I, et al. Mesenchymal stem cells for treatment of steroid-resistant, severe, acute graft-versus-host disease: a phase II study. *Lancet* 2008;**371**(9624):1579–86.

29. Kebriaei P, Isola L, Bahceci E, et al. Adult human mesenchymal stem cells added to corticosteroid therapy for the treatment of acute graft-versus-host disease. *Biol Blood Marrow Transplant* 2009;**15**(7):804–11.

30. Muroi K, Miyamura K, Ohashi K, et al. Unrelated allogeneic bone marrow-derived mesenchymal stem cells for steroid-refractory acute graft-versus-host disease: a phase I/II study. *Int J Hematol* 2013;**98**(2):206–13.

31. Aggarwal S, Pittenger MF. Human mesenchymal stem cells modulate allogeneic immune cell responses. *Blood* 2005;**105**(4):1815–22.

32. Baumgart DC, Sandborn WJ. Crohn's disease. *Lancet* 2012;**380**(9853):1590–605.

33. Forbes GM, Sturm MJ, Leong RW, et al. A phase 2 study of allogeneic mesenchymal stromal cells for luminal Crohn's disease refractory to biologic therapy. *Clin Gastroenterol Hepatol* 2014;**12**(1):64–71.

34. Ciccocioppo R, Bernardo ME, Sgarella A, et al. Autologous bone marrow-derived mesenchymal stromal cells in the treatment of fistulising Crohn's disease. *Gut* 2011;**60**(6):788–98.

35. Brand S. Crohn's disease: Th1, Th17 or both? The change of a paradigm: new immunological and genetic insights implicate Th17 cells in the pathogenesis of Crohn's disease. *Gut* 2009;**58**(8):1152–67.

36. Bai L, Lennon DP, Eaton V, et al. Human bone marrow-derived mesenchymal stem cells induce Th2-polarized immune response and promote endogenous repair in animal models of multiple sclerosis. *Glia* 2009;**57**(11):1192–203.

37. Karp JM, Leng Teo GS. Mesenchymal stem cell homing: the devil is in the details. *Cell Stem Cell* 2009;**4**(3):206–16.

38. Kharaziha P, Hellström PM, Noorinayer B, et al. Improvement of liver function in liver cirrhosis patients after autologous mesenchymal stem cell injection: a phase I–II clinical trial. *Eur J Gastroenterol Hepatol* 2009;**21**(10):1199–205.

39. Connick P, Kolappan M, Crawley C, et al. Autologous mesenchymal stem cells for the treatment of secondary progressive multiple sclerosis: an open-label phase 2 proof-of-concept study. *Lancet Neurol* 2012;**11**(2):150–6.

40. Liang J, Zhang H, Hua B, et al. Allogenic mesenchymal stem cells transplantation in refractory systemic lupus erythematosus: a pilot clinical study. *Ann Rheum Dis* 2010;**69**(8):1423–9.

41. Wakitani S, Nawata M, Tensho K, et al. Repair of articular cartilage defects in the patello-femoral joint with autologous bone marrow mesenchymal cell transplantation: three case reports involving nine defects in five knees. *J Tissue Eng Regen Med* 2007;**1**(1):74–9.

42. Hare JM, Traverse JH, Henry TD, et al. A randomized, double-blind, placebo-controlled, dose-escalation study of intravenous adult human mesenchymal stem cells (prochymal) after acute myocardial infarction. *J Am Coll Cardiol* 2009;**54**(24):2277–86.

43. Chen SL, Fang WW, Ye F, et al. Effect on left ventricular function of intracoronary transplantation of autologous bone marrow mesenchymal stem cell in patients with acute myocardial infarction. *Am J Cardiol* 2004;**94**(1):92–5.

44. Wollert KC, Meyer GP, Lotz J, et al. Intracoronary autologous bone-marrow cell transfer after myocardial infarction: the BOOST randomised controlled clinical trial. *Lancet* 2004;**364**(9429):141–8.

45. Schaefer A, Meyer GP, Fuchs M, et al. Impact of intracoronary bone marrow cell transfer on diastolic function in patients after acute myocardial infarction: results from the BOOST trial. *Eur Heart J* 2006;**27**:929–35.

46. Mirotsou M, Zhang Z, Deb A, et al. Secreted frizzled related protein 2 (Sfrp2) is the key Akt-mesenchymal stem cell-released paracrine factor mediating myocardial survival and repair. *Proc Natl Acad Sci USA* 2007;**104**(5):1643–8.

47. Mazhari R, Hare JM. Mechanisms of action of mesenchymal stem cells in cardiac repair: potential influences on the cardiac stem cell niche. *Nat Clin Pract Cardiovasc Med* 2007;**4**:S21–6.

48. Nagaya N, et al. Transplantation of mesenchymal stem cells improves cardiac function in a rat model of dilated cardiomyopathy. *Circulation* 2005;**112**(8):1128–35.

49. van Dijk FS, Cobben JM, Kariminejad A, et al. Osteogenesis imperfecta: a review with clinical examples. *Mol Syndr* 2011;**2**(1):1–20.

50. Horwitz EM, Prockop DJ, Fitzpatrick LA, et al. Transplantability and therapeutic effects of bone marrow-derived mesenchymal cells in children with osteogenesis imperfecta. *Nat Med* 1999;**5**(3):309–13.

51. Horwitz EM, Gordon PL, Koo WK, et al. Isolated allogeneic bone marrow-derived mesenchymal cells engraft and stimulate growth in children with osteogenesis imperfecta: implications for cell therapy of bone. *Proc Natl Acad Sci USA* 2002;**99**(13):8932–7.

52. *CFR - Code of Federal Regulations.* Title 21. U.S. Food and Drug Administration. U.S. Department of Health and Human Services. http://www.accessdata.fda.gov/scripts/cdrh/cfdocs/cfcfr/CFRSearch.cfm?CFRPart=1271; September 1, 2014 [Web 28.01.15].

53. Tyndall A. Mesenchymal stem cell treatments in rheumatology—a glass half full? *Nat Rev Rheumatol* 2014;**10**:117–24.

54. Prockop DJ, Prockop SE, Bertoncello I. Are clinical trials with mesenchymal stem/progenitor cells too far ahead of the science? Lessons from experimental hematology. *Stem Cells* 2014;**32**(12):3055–61.

55. Betsholtz C, Lindblom P, Gerhardt H. Role of pericytes in vascular morphogenesis. *EXS* 2005:115–25.

56. Shi S, Gronthos S. Perivascular niche of postnatal mesenchymal stem cells in human bone marrow and dental pulp. *J Bone Miner Res* 2003;**18**: 696–704.

57. Schwab KE, Gargett CE. Co-expression of two perivascular cell markers isolates mesenchymal stem-like cells from human endometrium. *Hum Reprod* 2007;**22**:2903–11.

58. Zannettino AC, Paton S, Arthur A, et al. Multipotential human adipose-derived stromal stem cells exhibit a perivascular phenotype in vitro and in vivo. *J Cell Physiol* 2008;**214**:413–21.

59. Covas DT, Panepucci RA, Fontes AM, et al. Multipotent mesenchymal stromal cells obtained from diverse human tissues share functional properties and gene-expression profile with CD146⁺ perivascular cells and fibroblasts. *Exp Hematol* 2008;**36**:642–54.

60. Traktuev DO, Merfeld-Clauss S, Li J, et al. A population of multipotent CD34-positive adipose stromal cells share pericyte and mesenchymal surface markers, reside in a periendothelial location, and stabilize endothelial networks. *Circ Res* 2008;**102**:77–85.

61. Sarugaser R, Lickorish D, Baksh D, Hosseini MM, Davies JE. Human umbilical cord perivascular (HUCPV) cells: a source of mesenchymal progenitors. *Stem Cells* 2005;**23**:220–9.

62. da Silva Meirelles L, Caplan AI, Nardi MB. In search of the in vivo identity of mesenchymal stem cells. *Stem Cells* 2008;**26**:2287–99.

63. Crisan M, Yap S, Casteilla L, et al. A perivascular origin for mesenchymal stem cells in multiple human organs. *Cell Stem Cell* 2008;**3**:301–13.

64. Dellavalle A, Sampaolesi M, Tonlorenzi R, et al. Pericytes of human skeletal muscle are myogenic precursors distinct from satellite cells. *Nat Cell Biol* 2007;**9**:255–67.

65. Park TS, Gavina M, Chen W, et al. Placental perivascular cells for human muscle regeneration. *Stem Cells Dev* 2011;**20**:451–63.

66. Caplan AI. All MSCs are pericytes? *Cell Stem Cell* 2008;**3**:229–30.

67. Corselli M, Chen CW, Sun B, et al. The tunica adventitia of human arteries and veins as a source of mesenchymal stem cells. *Stem Cells Dev* 2012;**21**(8):1299–308.

68. Tintut Y, Alfonso Z, Saini T, et al. Multilineage potential of cells from the artery wall. *Circulation* 2003;**108**(20):2505–10.

69. Hoshino A, Chiba H, Nagai K, et al. Human vascular adventitial fibroblasts contain mesenchymal stem/progenitor cells. *Biochem Biophys Res Commun* 2008;**368**(2):305–10.

70. Corselli M, Chen CW, Crisan M, Lazzari L, Péault B. Stem cell storage in blood vessel walls. *Arterioscler Thromb Vasc Biol* 2010;**30**:1104–9.

71. Crisan M, Corselli M, Chen CW, Péault B. Multilineage stem cells in the adult: a perivascular legacy? *Organogenesis* 2011;**7**(2):101–4.

72. Murray IR, West C, Hardy WR, James AW, Park TS, Nguyen A, et al. Natural history of mesenchymal stem cells, from vessel walls to culture vessels. *Cell Mol Life Sci* 2013. http://dx.doi.org/10.1007/s00018-013-1462-6. [Epub ahead of print].

73. Corselli M, Crisan M, Murray IR, West CC, Scholes J, Codrea F, et al. Identification of perivascular mesenchymal stromal/stem cells by flow cytometry. *Cytometry* 2013;**83**(8):714–20.

74. Davidoff MS, Middendorff R, Enikolopov G, et al. Progenitor cells of the testosterone-producing Leydig cells revealed. *J Cell Biol* 2004;**167**(5): 935–44.

75. Tang W, Zeve D, Suh JM, et al. White fat progenitor cells reside in the adipose vasculature. *Science* 2008;**322**(5901):583–6.

76. Dellavalle A, Maroli G, Covarello D, et al. Pericytes resident in postnatal skeletal muscle differentiate into muscle fibres and generate satellite cells. *Nat Commun* 2011;**2**:499.

77. Krautler NJ, Kana V, Kranich J, et al. Follicular dendritic cells emerge from ubiquitous perivascular precursors. *Cell* 2012;**150**(1):194–206.

78. Bouacida A, Rosset P, Trichet V, et al. Pericyte-like progenitors show high immaturity and engraftment potential as compared with mesenchymal stem cells. *PLoS One* 2012;**7**(11):e48648.

79. Olson LE, Soriano P. PDGFRbeta signaling regulates mural cell plasticity and inhibits fat development. *Dev Cell* 2011;**20**(6):815–26.

80. Urbanek K, Torella D, Sheikh F, et al. Myocardial regeneration by activation of multipotent cardiac stem cells in ischemic heart failure. *Proc Natl Acad Sci USA* 2005;**102**:8692–7.

81. Smart N, Bollini S, Dube KN, et al. De novo cardiomyocytes from within the activated adult heart after injury. *Nature* 2011;**474**:640–4.

82. Chong JJH, Yang X, Don CW, et al. Human embryonic-stem-cell-derived cardiomyocytes regenerate non-human primate hearts. *Nature* 2014;**510**:273–7.

83. Mauritz C, Martens A, Rojas SV, et al. *Induced pluripotent stem cell (iPSC)-derived Flk-1 progenitor cells engraft, differentiate, and improve heart function in a mouse model of acute myocardial infarction,* vol. 32. 2011.

84. Qian L, Huang Y, Spencer CI, et al. In vivo reprogramming of murine cardiac fibroblasts into induced cardiomyocytes. *Nature* 2012;**485**:593–8.

85. Kawamoto A, Iwasaki H, Kusano K, et al. CD34-Positive cells exhibit increased potency and safety for therapeutic neovascularization after myocardial infarction compared with total mononuclear cells. *Circulation* 2006;**114**:2163–9.

86. Hatzistergos KE, Quevedo H, Oskouei BN, et al. Bone marrow mesenchymal stem cells stimulate cardiac stem cell proliferation and differentiation. *Circ Res* 2010;**107**:913–22.

87. Mazo M, Planat-Bénard V, Abizanda G, et al. Transplantation of adipose derived stromal cells is associated with functional improvement in a rat model of chronic myocardial infarction. *Eur J Heart Fail* 2008;**10**:454–62.

88. Okada M, Payne TR, Drowley L, et al. Human skeletal muscle cells with a slow adhesion rate after isolation and an enhanced stress resistance improve function of ischemic hearts. *Mol Ther* 2012;**20**:138–45.

89. Chen CW, Montelatici E, Crisan M, et al. Perivascular multi-lineage progenitor cells in human organs: regenerative units, cytokine sources or both? *Cytokine Growth Factor Rev* 2009;**20**:429–34.

90. Chen CW, Okada M, Proto JD, et al. Human pericytes for ischemic heart repair. *Stem Cells* 2013;**31**:305–16.

91. Katare RG, Madeddu P. Pericytes from human veins for treatment of myocardial ischemia. *Trends Cardiovasc Med* 2013;**23**(3).

92. Katare R, Riu F, Mitchell K, et al. Transplantation of human pericyte progenitor cells improves the repair of infarcted heart through activation of an angiogenic program involving micro-RNA-132/Novelty and significance. *Circ Res* 2011;**109**:894–906.

93. Avolio E, Meloni M, Spencer HL, et al. Combined intramyocardial delivery of human pericytes and cardiac stem cells additively improves the healing of mouse infarcted hearts through stimulation of vascular and muscular repair. *Circ Res* 2015. http://dx.doi.org/10.1161/CIRCRESAHA.115.306146.

94. Chong JJH, Chandrakanthan V, Xaymardan M, et al. Adult cardiac-resident MSC-like stem cells with a proepicardial origin. *Cell Stem Cell* 2011;**9**:527–40.

95. Chen WCW, Baily JE, Corselli M, et al. Human myocardial pericytes: multipotent mesodermal precursors exhibiting cardiac specificity. *Stem Cells* 2015;**33**:557–73.

96. James AW, Zara J, Zhang X, et al. Perivascular stem cells: a prospectively purified mesenchymal stem cell population for bone tissue engineering. *Stem Cells Transl Med* 2012;**1**:510–9.

97. James AW, Zara JN, Corselli M, et al. Use of human perivascular stem cells for bone regeneration. *J Vis Exp* 2012:e2952.

98. James AW, Zara J, Corselli M, et al. An abundant perivascular source of stem cells for bone tissue engineering. *Stem Cells Transl Med* 2012;**1**(9). [Epub ahead of print].

99. Chung CG, James AW, Asatrian G, et al. Human perivascular stem cell-based bone graft substitute induces rat spinal fusion. *Stem Cells Transl Med* 2014;**3**:1231–41.

100. Askarinam A, James AW, Zara JN, et al. Human perivascular stem cells show enhanced osteogenesis and vasculogenesis with Nel-like molecule I protein. *Tissue Eng Part A* 2013;**19**:1386–97.

101. Li W, Lee M, Whang J, et al. Delivery of lyophilized Nell-1 in a rat spinal fusion model. *Tissue Eng Part A* 2010;**16**:2861–70.

102. Siu RK, Lu SS, Li W, et al. Nell-1 protein promotes bone formation in a sheep spinal fusion model. *Tissue Eng Part A* 2011;**17**:1123–35.

103. Montemurro T, Andriolo G, Montelatici E, et al. Differentiation and migration properties of human fetal umbilical cord perivascular cells: potential for lung repair. *J Cell Mol Med* 2011;**15**(4):796–808.

104. He W, Nieponice A, Soletti L, et al. Pericyte-based human tissue engineered vascular grafts: fabrication, characterization and in vivo assessment. *Biomaterials* 2010;**31**:8235–44.

105. Tottey S, Corselli M, Jeffries E, et al. Extracellular matrix degradation products and low oxygen conditions enhance the regenerative potential of perivascular stem cells. *Tissue Eng Part A* 2011;**17**:37–44.

106. Stefańska AM, Péault B, Mullins JJ. Renal pericytes: multifunctional cells of the kidneys. *Pflügers Arch* 2013;**465**(6):767–73.

107. Corselli M, Chin C, Parekh C, et al. Perivascular support of human hematopoietic stem/progenitor cells. *Blood* 2013;**121**(15):2891–901.

108. Rajashekhar G, Traktuev DO, Roell WC, et al. IFATS collection: adipose stromal cell differentiation is reduced by endothelial cell contact and paracrine communication: role of canonical Wnt signaling. *Stem Cells* 2008;**26**(10):2674–81.

Chapter 16

Adipose Tissue as a Plentiful Source of Stem Cells for Regenerative Medicine Therapies

Jolene E. Valentin, Albert Donnenberg, Kacey G. Marra, J. Peter Rubin

Key Concepts

1. Adipose tissue is an abundant and easily obtainable source of adult mesenchymal stem cells (MSCs), which share similar characteristics to bone marrow MSCs, both in phenotypic marker identification and in functionality. These adipose stromal stem/progenitor cells (ASCs) are highly proliferative and possess the ability to differentiate into adipocytes, chondrocytes, osteoblasts, and myocytes. ASCs also participate in paracrine signaling by the secretion of cytokines and growth factors vital to angiogenesis and differentiation.
2. Liposuction is the most common method to harvest ASCs, and can be performed either as an inpatient procedure or as a minimally invasive outpatient procedure using a tumescent technique for local anesthesia, which allows for the safe harvest of adipose tissue with low risk.
3. Stromal vascular fraction (SVF) can be isolated manually or using an automated device. The adipose tissue is digested with one or more enzymatic solutions and then centrifuged to separate the digested tissue into three fractions: a lipid fraction at the top; a clear amber liquid in the middle; and a cell pellet at the bottom. The cell pellet, which contains the SVF, is then resuspended until ready for use. The cell populations in the SVF include ASCs, pericytes, endothelial progenitor and mature cells, and hematopoietic cells.
4. The regulatory guidelines for adipose tissue administration encompass two general categories in the United States. One category, 21CFR 1271, is designated for human cells and tissue products. The manufacturer must adhere to rigorous guidelines for donor selection criteria, specimen handling, and testing protocols. An exception to 21CFR 1271 would be for procedures in which cells and tissues are removed from the patient and reimplanted within the same operative procedure. A second category would classify the cell product as a "biologic drug," with a requirement for phase 1, 2, and 3 clinical studies and a biologic license.
5. ASCs can be locally delivered to areas of injury or intravenously. ASCs can be used in tissue engineering using plastic surgery procedures that repair soft tissue defects by lipofilling with an SVF-enriched fat graft, also known as cell-assisted lipotransfer (CAL). CAL can be used to correct craniofacial asymmetry or to repair craniofacial defects after blast injury by transforming the lipoaspirate from progenitor-poor into progenitor-rich fat grafting material.

1. THERAPEUTIC POTENTIAL OF ADIPOSE-DERIVED STEM CELLS

Human adipose tissue is often discarded as medical waste during cosmetic surgery procedures. As fate would have it, adipose tissue is now known to be a rich and easily obtainable source of adult mesenchymal stem cells (MSCs). Adipocytes, which make up the bulk of adipose tissue, are organized into lobules. A heterogeneous cell population, called the stromal vascular fraction (SVF), surrounds the mature adipocytes. SVF includes adipose stromal stem/progenitor cells (ASCs), pericytes, mature and immature (capillary) vascular endothelial cells, fibroblasts, and hematopoietic-lineage cells. The removal of the SVF from the adipose tissue can be completed in less than 2 h. The isolated SVF can then be admixed directly into processed lipoaspirate for cell-enhanced fat grafting or directly injected into injury or disease sites during the same OR procedure.

ASCs and bone marrow MSCs (BM-MSCs) have similar characteristics. A joint statement made by leading researchers from the International Federation for Adipose Therapeutics and Science (IFATS) and the International Society for Cellular Therapy (ISCT) described the known differences between freshly isolated stromal (connective tissue) cells from adipose and culture-expanded plastic-adherent ASCs through functional characterization and phenotypic identification.[1] Using flow cytometry, cells from SVF have been identified as having the phenotype CD45−/CD235a−/CD31−/CD34+. Other markers of significance may also include CD13, CD73, CD90, and CD105. As with BM-MSCs, ASCs adhere to plastic tissue culture surfaces and preserve these properties after serial passage. When plated and expanded, ASCs have an elongated or fibroblast-like morphology, and continue to express stromal markers commonly associated with other MSCs including CD90, CD73, CD105, and CD44, and remain negative for CD45. A typical yield from an isolation of SVF from adipose tissue is 3×10^5 cells/g of adipose tissue, and adipose progenitor cells comprise approximately 30–80% of the nonheme portion of SVF.[2,3] Adipose tissue may be a better candidate for harvesting compared to bone marrow, which yields only 0.0001–0.01% of all nucleated cells.

As with other MSC types, ASCs are highly proliferative and can differentiate into cell types originating in the mesodermal layer, including adipocytes, chondrocytes, osteoblasts, and myocytes, when exposed to inductive factors that can be included in differentiation culture media.[3] Histologic stains or biomarker assays such as RT-PCR, Western blotting, or enzyme-linked immunosorbant assay (ELISA) are used to confirm differentiation.[1] ASCs differentiated into adipocytes will stain with oil red O and express adiponectin, fatty acid binding protein 4, leptin, and peroxisome proliferator-activated receptor γ. Chondrogenic differentiation can be confirmed with alcian blue or safranin O staining, and express the markers aggrecan, collagen type II, and Sox9. ASCs differentiated into osteoblasts can be confirmed by alizarin red or von Kossa staining, and will demonstrate increased expression of alkaline phosphatase, osteocalcin, and osterix. ASCs, like other MSCs, are involved in paracrine signaling by way of the secretion of many cytokines and growth factors important for differentiation and angiogenic induction, including vascular endothelial growth factor (VEGF), insulin-like growth factor (IGF), transforming growth factor (TGF), and hepatocyte growth factor (HGF). ASCs have also shown to possess immuno-modulatory properties. Using mixed lymphocyte reaction (MLR) assays, Puissant et al. discovered that ASCs do not elicit a response from allogeneic lymphocytes. Also, ASCs inhibit a proliferative response from lymphocytes induced by mitogens or by MHC mismatched peripheral blood mononuclear cells. This inhibitory effect is a function of cell concentration and of exposure time. However, the supernatant alone of ASCs is not effective in inhibiting proliferation in the MLR. Considering these characteristics, ASCs may be a promising allogeneic source of therapeutic cellular products. Indeed, the prospect of an off-the-shelf stem cell injection could potentially be a ground-breaking contribution to regenerative medicine therapies.

It is well documented that human adipose tissue is an abundant and easily obtainable source of MSCs. However, the health status of the donor must be taken into account when therapies employ human-origin cells. The age, gender, diabetic status, and body mass index are the most often cited factors that indicate the general health status of the donor. It has been hypothesized that the variability in fat graft retention observed clinically may actually be due to inherent biological properties of the adipose tissue that supplies the MSCs. Philips et al conducted a comprehensive study on lipoaspirate obtained from human subjects that investigated the effect of fundamental biological differences in the freshly isolated SVF. In vitro analysis included differences in proliferation, differentiation, and growth factor secretion in normoxic or hypoxic conditions. Fat graft retention and blood vessel quantification was also studied in a nude mouse xenograft model.[4] A strong correlation was found between the concentration of CD34+ progenitor cells identified within the SVF and high graft retention, which indicates that CD34+ cells may be a useful and reliable marker in predicting the outcome of fat grafting in the clinical setting.[4] More confirming evidence is needed, but patient-to-patient variability may in part be due to the inherent biological characteristics of the SVF that comes from the adipose tissue of the donor.

2. LIPOHARVEST METHODS

Liposuction has progressed from an in-patient procedure requiring general anesthesia into a minimally invasive outpatient procedure, due in part to the development of the tumescent technique for anesthesia. Tumescent local anesthesia (TLA) was developed by Dr Jeffrey Klein,[5] and consists of a dilute wetting solution of saline or lactated Ringers, lidocaine or another suitable short-acting analgesic, epinephrine, and sodium bicarbonate. The tumescent technique includes the infiltration of 3–4 mL of wetting solution per mL of lipoaspirate, which results in significant tissue turgor.[6] The use of TLA allows safe harvest of small amounts of adipose tissue (e.g., 25–100 cc), even when medical comorbidities are present. Lipoharvesting using TLA reduces blood loss from 20–45% using dry liposuction to 1% of volume aspirated, thereby, reducing or eliminating the risks associated with general anesthesia, which may include hemorrhage, hematoma formation, large shifts of fluid into tissues, and prolonged recovery times.[7,8] The vasoconstrictor epinephrine in the tumescent solution also reduces the absorption of lidocaine, thus enabling higher doses of anesthetic without systemic toxicity.[9] The solution is delivered locally in the subcutaneous space through an infusion cannula, and the flow of fluid can be controlled manually or by using a peristaltic pump. For fat harvesting in larger areas of the body, such as the thighs or abdomen, a longer cannula manufactured with multiple openings is appropriate for delivering the tumescent solution.

The adipose tissue can be aspirated using syringe suctioning, a technique that applies negative pressure to draw out fat particles into various sized syringes or mechanical aspirators that are vane- or piston-driven pumps.[7] Cannulas that extract the fat tissue are usually manufactured out of stainless steel. Blunt-tipped cannulas are used to avoid damaging the underlying tissues and blood vessels, while V-shaped cannulas are primarily used to break apart subcutaneous fibrous bands. Cannulas with multiple holes are used mostly for rapid fat tissue harvesting, and the location, number, and site of the holes vary considerably. Powered reciprocal cannulas are also available to ease the movement through the subcutaneous space, thereby reducing surgeon fatigue and increasing fat-removal efficiency. The complication rate with powered cannulas is similar to traditional liposuction with tumescent anesthesia.[7]

3. METHODS OF SVF ISOLATION: AUTOMATED VERSUS MANUAL

Once the processed lipoaspirate is prepared, the SVF can be isolated manually or by using automated devices. For manual isolation of SVF from human adipose tissue, the investigators or technicians must be specifically trained on the safe handling of human tissue and cell culture techniques. All work should be performed using proper aseptic technique in a clean room environment, such as a clinical Good Manufacturing Practice (cGMP) laboratory, and universal precautions should be taken during the handling of any cellular therapy product. Automated isolation devices can be used to isolate therapeutic doses of SVF in an aseptic room near the OR, thus eliminating the requirement for a cGMP facility for clinical cell isolations (Table 1). Whether the clinician decides to use an automated system or a manual protocol, strict control of all procedures is required in a clinical setting.

The adipose tissue can be digested using one or a combination of the following enzymes: collagenase, dispase, trypsin, or other related enzymes. In most laboratory research protocols, isolation reagents contain animal-sourced components, which may pose risks for immune reactions or exposure to infectious agents. A comparative study performed by Carvalho et al determined that commercially available, clinical grade enzymatic products such as "animal origin free" collagenase—CLSAFA (Worthington) and Liberase TM, a xenofree lyophilisates consisting of highly purified collagenase and neutral protease enzymes (Roche), were suitable replacements for traditional research-quality reagents, such as the crude collagenase type I that may be animal based (CLS1 (Worthington)) or Collagenase NB 4, a blend of crude collagenase and other proteases (SERVA).[10] This study reported no differences in SVF yield and viability, or ASC immunophenotype and differentiation potential, which suggests that clinical grade enzymatic solutions that are xenofree are the preferred choice for clinical use.

Major steps in a manual isolation protocol are briefly described. After the processed lipoaspirate is collected from the OR, the aqueous phase is decanted following centrifugation, reserving a sample of the aqueous phase for sterility analysis. A prewarmed 37 °C collagenase working solution, containing 2.5% collagenase in 5% human serum albumin/saline solution, is added in equal volume to the fat tissue. The digesting adipose tissue is then placed in a shaking water bath until tissue digestion is observed. After 15 min of digestion, the tissue should be visually inspected to determine degree of degradation. If there are no large tissue fragments, digestion is complete. If there is a clear (acellular) lipid layer on top, similar in appearance to melted butter, the sample may be overdigested. If large tissue fragments (but no clear lipid layer)

TABLE 1 Automated Cell Isolation Systems Developed to Obtain SVF from Adipose Tissue

Device (Reference)	Company (Location)	Details	Stage of Development
Tissue Genesis Icellator® (TGI 1200™)	Tissue Genesis (Honolulu, HI) www.tissuegenesis.com	• Aspirated fat tissue (20–60 mL) placed into processing chamber and enzymatically digested using Adipase® collagenase. • Final product is 35 mL SVF cell suspension. • Processing time is 1 hour	CE Mark certified, not sold for clinical use in US
Stempeutron™	Stempeutics (Bangalore, India) www.stempeutics.com	• Aspirated fat tissue placed into processing chamber and enzymatically digested. • Processing time is 120 minutes	In beta phase of development
Celution® 27,28	Cytori Therapeutics, Inc. (San Diego, CA)	• Aspirated fat tissue (up to 400 mL) is injected into port and enzymatically digested using Celase™ solution. • Final product is 5 mL SVF cells in lactated Ringers. • Processing time is within 2.5 h	CE Mark certified, not sold for clinical use in US
GID SVF-1™	GID Group, Inc (Louisville, CO) www.thegidgroup.com	• Sterile, single use disposable tissue canister for separation, and concentration of SVF	CE Mark certified, not sold for clinical use in US

are present, vigorously shake the mixture and return the tubes to the shaking water bath and agitate for an additional 5 min. This step may need to be repeated until digestion is complete. An EDTA-containing buffer is then added to stop enzymatic digestion. After centrifugation, the digested tissue will contain three layers: the fat layer at the top, the buffer/collagenase solution in the middle layer, and the cellular pellet at the bottom layer (Figure 1(A)). The digested fatty material is decanted into a sterile container with a lid (Figure 1(B)). This portion of the digest will be sieved last. Sieves, such as those with a progressively smaller mesh size, filter out the fibrous tissue and larger fragments from the digested tissue, and the filtrate in the catch basin is then collected for further washing and centrifugation steps (Figure 1(C–E)).

Upon removal from the centrifuge, the contents of the conical tubes will have separated into three fractions: a lipid fraction at the top, which will appear bright to pale yellow; a clear pale amber liquid in the middle; and a cell pellet that may appear red with a white layer. The lipid and liquid portions are decanted, and the cell pellet is washed. The supernatant is

FIGURE 1 Photograph montage of the sieving process. Sieving of the digested adipose stroma is performed using OR type aseptic technique by an operator dressed in sterile gown and gloves. (A) After the adipose tissue is enzymatically digested and centrifuged, three layers are distinguishable. The top pale yellow layer is the adipocytes, the middle aqueous layer consists of buffer and collagenase, and the bottom layer is the stromal vascular fraction cellular pellet (denoted by white arrow). (B) The fatty portion of the supernatant is then decanted into a sterile Nalgene jar with lid. Because of the risk of clogging the sieve, this portion of the digest is sieved last. (C) The catch basin of the sieve set is placed on the bottom, the 180 μm sieve (No. 80) is placed directly on top of the catch basin, and the 425 μm sieve (No. 40) is placed directly on top of the 180 μm sieve. The pellet is resuspended in the remaining aqueous fluid, and saline with human serum albumin is used to rinse the conical tubes. All of the liquid is passed over the sieve set, followed by the fatty portion. (D) A sterile glass pestle is used if tissue fragments are present. (E) The filtrate is then collected in the catch basin for the wash and centrifugation steps.

decanted and saved for sterility, endotoxin, and gram stain testing. The cell pellet is resuspended in human serum albumin-containing saline and stored in a 4 °C refrigerator until released for use. The expiration of the fresh cell suspension is 6 h from the time of resuspension. A small portion of resuspended cell product is used to count the cells, determine viability, and analyze with flow cytometry.

Regardless of the processing method, the cell product should meet predetermined release criteria before continuing the cell enrichment process. Example criteria include:

- A cell yield greater than 1×10^5 total SVF cells/mL of adipose
- Viability greater than 50% (can be measured by acridine orange staining)
- Total endotoxin <5EU/kg recipient ideal body weight (withhold product if greater than 5EU/kg)
- Gram stain negative (withhold product if positive)

Additional quality control measures include an expected average yield of $3 \pm 0.5 \times 10^5$ viable nucleated cells/mL adipose stroma and an expected average viability of at least $60.1 \pm 2.0\%$. Additionally, aerobic, anaerobic, and fungal cultures should be negative at 14 days (results are known after product administration).

4. FLOW CYTOMETRY ANALYSIS

Cell populations in SVF can be characterized by multiparameter flow cytometry to identify the cells and their relative frequencies, including ASCs, pericytes, endothelial progenitor and mature cells, and hematopoietic cells.[2] Because there is no single cell surface marker that identifies these subpopulations, it is recommended that multiple markers are used to distinguish the population profile, along with a viability marker such as DAPI to eliminate dead or apoptotic cells from the analysis. Table 2 shows the commonly used markers to characterize the cell populations residing in SVF, and the estimated frequency. To identify cells of hematopoietic origin, cells expressing CD45 (common leukocyte antigen) or CD235a (glycophorin A, a marker for erythrocytes) should be used. Endothelial cells and their progenitors express CD31 (PECAM-1). Endothelial progenitors and adipose stromal cells express CD34, so to identify the stromal cell population a combination of CD31 and CD34 should be used.

5. REGULATORY PROCESS

ASCs fall into two general categories of regulation in the United States.[11] One category, which is a moderate-risk category, is designated for human cells and tissue products and is tied to a specific code of regulations delineated within 21CFR 1271. In this regulatory framework, the cellular product would be governed under the same rules pertaining to grafted tissue materials, whether autogenous or allogenous. Examples of products commonly regulated under this framework would include acellular human dermal matrix and demineralized bone matrix. The major criteria underpinning this framework are that the tissue product is removed from a human donor (which could be the same patient being treated with the product) and intended for implantation. Additionally, the tissue product must be "minimally manipulated" and for homologous use. The descriptor in the federal code for minimal manipulation is that the basic biologic characteristics have not been changed. Examples of situations in which cell products are clearly more than minimally manipulated include culture expansion of the cells, exposure to growth factors, or genetic transfection. Where there has been debate on the topic of minimal manipulation is the use of enzymes to separate the adipose stem cells from the surrounding tissue. If a cell product is deemed to fall under this category of human cells and tissue, the cell manufacturer or tissue manufacturer must adhere to a rigorous

TABLE 2 Phenotype of Cell Populations Identified in SVF[3]

Cell Type	Phenotype	Proportion of Nonheme (CD45⁻) Nucleated Cells
Stromal/preadipocytes	CD31⁻, CD34⁺, CD146⁻/⁺, CD90⁺	67.6±29.7%
Endothelial progenitor	CD31⁺,CD34⁺, CD146⁺, CD90⁺	5.2±6.1%
Endothelial mature	CD31⁺,CD34⁻, CD146⁻, CD90⁻	Variable with harvest technique
Pericytes	CD31⁻, CD34⁻, CD146⁺, CD90⁺	0.8±0.7%

set of guidelines concerning donor selection criteria and testing, specimen labeling practices, processing and storage rules, and testing of the product. Additionally, the manufacturer must register their products and facility with the FDA. Indeed, 21CFR 1271was established as specific code based on the authority of the FDA to control the spread of communicable diseases as authorized by Section 361 of the Public Health Services Act. Human cell and tissue products are often required to register as "361 products." The fact that adipose stem cells can be prepared in an autologous fashion directly at the point of care can potentially exempt the cell manufacturer from having to register their facility with the FDA. An exception to 21CFR 1271 designation would be if a cell product (that would otherwise fall into this category) is removed from a patient and reimplanted in the same operative procedure. Therefore, clinicians could practice adipose-derived stem cell therapy in a clinic setting if the processing was performed immediately upon extraction of the fat tissue and used during the same operative procedure. Commercial devices that have been designed for separating ASCs from fat tissue could be used in this manner. However, the devices themselves are regulated by the FDA as Class III medical devices and are not yet approved for general clinical use. In the US, they are still considered to be research tools, and, therefore, they must be used under an investigational device exemption IDE issued by the FDA.

The second category of regulation would designate the cell product as a "biologic drug." In this framework, the FDA regulates the cell product under the authority granted to them by the Public Health Services Act to grant a biologic license. Cell products falling under this category must complete specific phase 1, 2, and 3 clinical studies, with oversight and approval of these studies by the FDA, and then apply for a biologic license. Cell products that cannot be regulated under the human cells and tissue category would fall into this category. Cell products regulated under this category are commonly referred to "351 products." If an investigator or cell manufacturer would like to get guidance from the FDA as to which category the cell product would fall under, they should submit a protocol and request an opinion to the FDA Tissue Reference Group. This interdisciplinary body within the FDA framework will then render a nonbinding opinion. A binding ruling can be obtained by submitting an official request for determination to the FDA.

Regulatory frameworks regarding cell therapy outside of the US vary widely. There are different prospectives internationally on the concepts of minimal manipulation, as well as differences in regulatory structure between autologous therapies and allogeneic therapies. With patient safety as a primary goal, it is likely that there will be continued evolution of cell therapy regulations as these fields progress.

6. CURRENT CLINICAL TRIALS AND GROWING POSSIBILITIES

At the time of publication, over 130 ongoing clinical trials are being conducted globally that involves the clinical use of ASCs in the treatment of a variety of diseases or injuries, mostly in Phase I or I/II for safety (See Table 3 for selected list of clinical studies). After harvesting and processing, the lipoaspirate from the patient, the isolated SVF can be admixed with additional lipoaspirate for soft tissue reconstruction procedures, or the SVF can be cultured and expanded, thereby, further concentrating the adipose product to contain a greater quantity of ASCs.

6.1 Delivery of ASCs to Areas of Injury

ASCs may participate in wound repair and tissue regeneration when administered locally into areas of injury. Because of their ability to mobilize to the region of tissue damage and populate the zone of injury, ASCs can be delivered by intravenous injection as well. This minimally invasive methodology takes advantage of the capability of ASCs to home to sites of injury and release growth factors, such as VEGF, that are important for the repair of damaged tissue. There is an area of concern with the systemic infusion of autologous ASCs because theoretically infused stem cells with high proliferative capacity could possibly lodge and form benign or malignant growths. To help alleviate this concern, a Phase I clinical trial (clinicaltrials.gov ID NCT01274975) was conducted in which freshly isolated ASCs were injected at a dose of 4×10^8 cells in traumatic spinal cord injury patients who were otherwise healthy; this study showed that there was no evidence of any serious adverse events related to the ASC administration up to 12 weeks posttransplantation.[12] Other laboratories have investigated ASC localization therapy following an acute myocardial infarction (AMI). A porcine AMI/reperfusion model was used by Valina et al. to study the effect of intracoronary administration of freshly isolated ASCs following AMI. The ASC administration resulted in improved left ventricular function and myocardial perfusion, and ASCs were shown to engraft within the infarct region post-therapy, thereby, improving neoangiogenesis to the injury zone.[13] Cytori Therapeutics Inc. recently conducted a Phase I/IIa human clinical trial (NCT00442806) using freshly isolated SVF for the treatment of AMI, and the results have demonstrated that ASC therapy is safe and feasible within hours after the primary percutaneous coronary intervention.[14] Also, ASC intervention showed a trend toward improved cardiac function and reduction of myocardial scar formation.

TABLE 3 Selected Active Clinical Trials Using Adipose Stem Cells

Subject	Clinical Trial (Identifier, Study Start Date)	Intervention/Delivery Method	Primary Outcome Measures (Timeframe)
Cardiovascular	ACELLDream for Adipose Cell Derived Regenerative Endothelial Angiogenic Medicine (NCT01211028, January 2009)	One dose of 100 million expanded cells delivered via intramuscular injection in critical leg ischemia	Clinical evaluation (15 days and every month for 6 months)
	Safety and Efficacy of ASCs for Non-Ischemic Congestive Heart Failure (NCT01502501, May 2011)	IV, intramyocardial via percutaneous transluminal endomyocardial catheter	Improvement in 6 min walk test (3, 6 months); Minnesota living with HF quality of life (3, 6 months); AE (6 months)
	Treatment CLI Nonrevascularizable Lower Limb with Cell Therapy (NCT01824069, April 2013)	One intramuscular injection of expanded adult ASCs at 1 million cells per kilo of body weight	Clinical evaluation with MRI, tissue oximetry, and ankle-brachial index (1 year)
Gastroenterologic	Allogeneic ASCs for the Induction of Remission in Ulcerative Colitis (NCT01914887, June 2013)	Total of 60 million ASCs injected endoscopically in sites within affected colonic submucosa	Physical exam, vital signs, lab tests (0, 9–10 days, and 4, 8, 12 weeks)
	ASCs for Induction of Remission in Perianal Fistulizing Crohn's Disease (NCT01541579, July 2012)	Intralesional injection with Cx601 (allogeneic expanded ASCs) at a dose of 120 million cells (5 million cells/mL)	MRI to determine remission of perianal fistulizing Crohn's disease (24 weeks)
Neurologic	Treatment of Sequelae Caused by Severe Brain Injury with Autologous ASCs (NCT01649700, October 2011)	5 IV infusions, 1 month apart, of 50–70 million cells	Safety evaluation through vital signs, clinical lab tests, AE
	Reparative Therapy in Acute Ischemic Stroke with Allogeneic MSCs from Adipose Tissue (NCT01678534, October 2012)	IV dose of 1 million cells/kg within first 2 weeks after onset of stroke symptoms, compare to placebo (vehicle)	Follow-up to assess safety. AE includes neurological and systemic complications, or development of tumors (24 months)
	Safety and Effect of ASC Implantation in Patients with Spinal Cord Injury (NCT01769872, January 2013)	IV: 200 million cells in 20 mL; Intrathecal: 50 million cells in 2 mL; Spinal cord: 20 million cells in 1 mL	American Spinal Injury Association scale (32 weeks) to evaluate change of treated spinal cord (before implantation and 3, 6 months)
	Study to Assess the Safety and Effects of Autologous Adipose-Derived Stromal in Patients with Parkinson's Disease (NCT01453803, May 2011)	Autologous SVF from lipoaspirate delivered IV or in vertebral artery using catheter system	Improvement in UPDRS scale and modified Hoehn and Yahr staging, including mentation, ADLs, motor examination (3, 6 months)
Orthopedic	Treatment of Tendon Injury Using MSCs (NCT01856140, May 2013)	Injection into lesion under ultrasound guidance. first group: 1 million cells/mL for safety; second group: 10 million cells/mL for efficacy	Change from baseline in visual analog scale (6, 12 weeks)
	Effectiveness of ASCs as Osteogenic Component in Composite Grafts (NCT01532076, June 2012)	ASC isolation from lipoaspirate using Cellution 800/CRS and single use kits (Cytori Therapeutics Inc.) and wrapped around hydroxyapatite micro-granules after embedding in fibrin gel. Scaffold for augmentation of bone-void during open reduction and internal fixation	Development of secondary dislocation within 12 months post-op on plain radiographs (more than 20° varus collapse of humeral head fragment in relation to humeral shaft or screw penetration through humeral head). Clinical/radiological follow-up (6, 12 weeks and after 6, 9, 12 months).
	ADIPOA (Evaluate Safety of a Single Injection of Autologous ASCs in the Treatment of Severe Osteoarthritis of the Knee Joint (NCT01585857, April 2012)	Single intraarticular injection of ASCs at doses of 2, 10, or 50 million cells in 5 mL	Record AE, functional status of knee, WOMAC index, SAS, and range of motion. Imaging using MRI, dGEMRIC and T1rhoMRI (1 year)

Continued

TABLE 3 Selected Active Clinical Trials Using Adipose Stem Cells—cont'd

Subject	Clinical Trial (Identifier, Study Start Date)	Intervention/Delivery Method	Primary Outcome Measures (Timeframe)
Plastic	Safety Study of Antria Cell Preparation Process to Enhance Facial Fat Grafting with ASCs (NCT01828723, April 2013)	SVF enriched lipoinjections in regions of face that require enhancement	Physical exam, vital signs, 12-lead ECGs, lab tests (CBC/LFT/BMP), urinalysis (6 months)
	Pilot Study of Skin Quality Improvement after ASCs Transfer in Irradiated Breasts (NCT01801878, May 2013)	SVF cell transfer to half of irradiated breast	Change in breast skin thickness of pre- and post-SVF graft from baseline, as measured by breast ultrasonography (12 weeks)
	Enriched Autologous Fat Grafting for Treating Pain at Amputation Sites (AMP-5) (NCT01645722, July 2012)	SVF enriched lipoinjections	Limb anatomy, healing over time, and stability of new tissue assessed by high resolution CT scanning with 3D reconstruction (24 months)
	Assessment of the Subcutaneous Reinjection of Human Autologous Adipose-derived SVF in the Hands of Patients Suffering from Systemic Sclerosis (NCT01813279, November 2012)	Autologous SVF isolated using Celution (Cytori Therapeutics, Inc.) Local injections into fingers of patients with scleroderma	Improvement of functional index of Cochin (2 years)
	Effect of Concentrating Endogenous Stromal Cells in the Fat Graft Using TGI Device (NCT01924364, June 2012)	Autologous SVF isolated using TGI Cell Isolation System™ (CIS) device. Cell enrichment of fat grafts.	Soft tissue volume and contour changes using esthetic grading scales, 3D photography, and high-resolution CT scanning (Surgical visit and 1, 3, 6, 9, 12, 24 months)
Other	MSC for Occusive Disease of the Kidney (NCT01840540, April 2013)	Autologous expanded ASCs, single dose into affected kidney by intraarterial infusion guided by renal angiography	Renal blood flow and function before/ after MSC infusion (3 months). Evaluation for AE (24 h, 1, 4, 8 weeks and 6 months). Health assessment and blood draws (12, 24 months) with urinary cytology and MRI
	Stem Cells Treatment for Bilateral Limbic Associated Keratopathy (NCT01808378, February 2012)	Autologous expanded ASCs delivered by intralesional injection	Micro-ocular photography visual acuity (16 weeks)

ASC, adipose tissue-derived mesenchymal stem cells; SVF, stromal vascular fraction; MSC, mesenchymal stem/stromal cells; AE, adverse event.
Source: clinicaltrials.gov

Because of their immunosuppressive properties, ASCs may be culture-expanded ex vivo and then administered as an *allogeneic* cell product.[15] Before commercial cell banking can be available to the clinician as an off-the-shelf product, certain guidelines and quality standards must be met during the collection, isolation, testing, banking, packaging, and shipping, in order to reduce the potential risk of blood-borne pathogens or cross-contamination. Quality measures may include the use of animal-free serums and solutions, the elimination of toxic cryopreservation additives, tightly controlled freezing and thawing protocols, and viability assays, to name a few.[16] Such a product may be ideal for taking advantage of the immunosuppressive effect of ASCs on allograft transplantation procedures that is currently under investigation as a promising treatment for organ transplantation (graft vs host disease) and for the treatment of complex perianal fistula in Crohn's disease.[17]

6.2 ASCs in Tissue Engineering

Since the technique was pioneered by Illouz in the early 1980s, liposuction has become one of the most commonly performed plastic surgical procedures. Lipofilling, which uses subcutaneous adipose tissue isolated by liposuction from a different body area, is an increasingly popular intervention for plastic surgery procedures as a way to obtain a natural appearance of soft tissue. Although intended to reduce the need for more invasive interventions, such as microvascular free tissue transfer, the use of prosthetic materials, or regional flap procedures, lipofilling has shown to be grossly inconsistent in terms of resorption rate and fat graft survival.[18] This effect is thought to be due in part to an initial lack of an intact vascular network within the fat graft, which can lead to ischemia and downstream fat necrosis, including oil cyst formation, calcification, and connective tissue formation. In the last decade, the technique of supplementing the fat graft with adipose tissue-derived SVF has gained increasing attention. Adipocytes are especially sensitive to severe ischemia; in contrast, ASCs are able to survive prolonged ischemia (up to 72 h) and become activated. This in turn begins the reparative process through proliferation, migration, and differentiation. It is currently hypothesized that enriching the fat graft with SVF may promote angiogenesis through paracrine signaling such as the secretion VEGF. The enhanced angiogenic response of cells found in the SVF may therefore promote graft viability and volume retention. Using minimally invasive liposuction techniques, human adipose tissue can be safely harvested in significant quantities (several 100 cc of adipose tissue using local anesthesia) with low morbidity and limited time in the OR, making this technological advancement a win–win situation for patients that would otherwise undergo extensive reconstructive surgery or be at risk for donor site morbidity from flap or tissue transfer procedures.

A commonly used technique for processing the harvested lipoaspirate and considered to be the gold standard is the Coleman technique.[19,20] Briefly, a "wetting" solution of saline solution with dilute epinephrine is infiltrated in the adipose tissue to be harvested. The liposuction cannula is connected to a 10-cc Luer-Lok syringe to collect the adipose tissue. When the syringe is filled with tissue, the plunger is removed and the syringe is fitted with a plug for the centrifugation step. Centrifuging the fat at 1200 g for 3 min is ideal for the separation of the three phases of the harvested fat. The oily upper level is composed primarily of ruptured fat cells and is decanted and discarded. Sterile cotton surgical strips are also used to wick away the oil from the interface of the upper and middle layers. The lower level is the densest layer and is composed primarily of blood, water, and wetting solution, and is also decanted by uncapping the syringe and allowing gravity to drain the fluid out of the syringe. The middle portion consists of parcels of fat tissue, and can be used for fat grafting and/or for the isolation of SVF.

Cell-assisted lipotransfer (CAL), which is the process of adding freshly isolated SVF to liposuctioned fatty tissue, has been used for soft tissue reconstruction procedures as well as cosmetic breast augmentation[21] and facial contouring during a facelift procedure.[22] CAL can also be applied to correct craniofacial asymmetry due to hemifacial lipoatropy (Parry–Romberg Syndrome),[23–25] linear scleroderma "en coup de sabre"[26] or craniofacial defects after blast injury. For CAL procedures the SVF is added back into the processed lipoaspirate through gentle luer-to-luer mixing in order to transform the "progenitor-poor" lipoaspirate into "progenitor-rich" fat grafting material.[9] Although additional investigation is needed to confirm the use of enriched fat grafting, preliminary results from clinical trials have suggested that CAL is safe and effective for soft tissue reconstruction, and may result in improved fat retention when compared to standard lipofilling procedures.

7. CONCLUDING REMARKS

Regenerative medicine therapies that integrate the use of ASCs have experienced incredible advances in the fields of plastic and reconstructive surgery, transplantation medicine, and other areas of practice. The biology and mechanisms of action of ASCs, and how they interact with tissue grafting, biomaterials, and supplemental factors such as pharmacological-based therapies are topics of scientific analysis in order to better understand and improve upon regenerative medicine strategies. The promising potential of adipose-derived stem cell therapies will continue to be explored for numerous clinical applications.

REFERENCES

1. Bourin P, Bunnell BA, Casteilla L, Dominici M, Katz AJ, March KL, et al. Stromal cells from the adipose tissue-derived stromal vascular fraction and culture expanded adipose tissue-derived stromal/stem cells: a joint statement of the International Federation for Adipose Therapeutics and Science (IFATS) and the International Society for Cellular Therapy (ISCT). *Cytotherapy* 2013;**15**(6):641–8. Epub 2013/04/11.

2. Zimmerlin L, Donnenberg VS, Pfeifer ME, Meyer EM, Peault B, Rubin JP, et al. Stromal vascular progenitors in adult human adipose tissue. *Cytom Part A J Int Soc Anal Cytol* 2010;**77**(1):22–30. Epub 2009/10/24.

3. Li H, Zimmerlin L, Marra KG, Donnenberg VS, Donnenberg AD, Rubin JP. Adipogenic potential of adipose stem cell subpopulations. *Plast Reconstr Surg* 2011;**128**(3):663–72. Epub 2011/05/17.

4. Philips BJ, Grahovac TL, Valentin JE, Chung CW, Bliley JM, Pfeifer ME, et al. Prevalence of endogenous CD34+ adipose stem cells predicts human fat graft retention in a xenograft model. *Plast Reconstr Surg* 2013;**132**(4):845–58. Epub 2013/06/21.

5. Klein JA. Tumescent technique for local anesthesia improves safety in large-volume liposuction. *Plast Reconstr Surg* 1993;**92**(6):1085–98. discussion 99-100. Epub 1993/11/01.

6. Stephan PJ, Kenkel JM. Updates and advances in liposuction. *Aesthet Surg J* 2010;**30**(1):83–97. Quiz 8-100. Epub 2010/05/06.

7. Dover JS, Alam M. *Procedures in cosmetic dermatology series: liposuction*. Elsevier Inc; 2005.

8. Chia CT, Theodorou SJ. 1,000 consecutive cases of laser-assisted liposuction and suction-assisted lipectomy managed with local anesthesia. *Aesthetic Plast Surg* 2012;**36**(4):795–802. Epub 2012/03/27.

9. Rubin JP. *Body contouring and liposuction*, vol. xii. Edinburgh, New York: Elsevier Saunders; 2013. p. 614.

10. Carvalho PP, Gimble JM, Dias IR, Gomes ME, Reis RL. Xenofree enzymatic products for the isolation of human adipose-derived stromal/stem cells. *Tissue Eng Part C Methods* 2013;**19**(6):473–8. Epub 2012/11/07.

11. Naghsineh N, Brown S, Cederna PS, Levi B, Lisiecki J, D'Amico RA, et al. Demystifying the U.S. Food and Drug Administration: understanding regulatory pathways. *Plast Reconstr Surg* September 2014;**134**(3):559–69.

12. Ra JC, Shin IS, Kim SH, Kang SK, Kang BC, Lee HY, et al. Safety of intravenous infusion of human adipose tissue-derived mesenchymal stem cells in animals and humans. *Stem Cells Dev* 2011;**20**(8):1297–308. Epub 2011/02/10.

13. Valina C, Pinkernell K, Song YH, Bai X, Sadat S, Campeau RJ, et al. Intracoronary administration of autologous adipose tissue-derived stem cells improves left ventricular function, perfusion, and remodelling after acute myocardial infarction. *Eur Heart J* 2007;**28**(21):2667–77. Epub 2007/10/16.

14. Houtgraaf JH, den Dekker WK, van Dalen BM, Springeling T, de Jong R, van Geuns RJ, et al. First experience in humans using adipose tissue-derived regenerative cells in the treatment of patients with ST-segment elevation myocardial infarction. *J Am Coll Cardiol* 2012;**59**(5):539–40. Epub 2012/01/28.

15. Gimble JM, Bunnell BA, Frazier T, Rowan B, Shah F, Thomas-Porch C, et al. Adipose-derived stromal/stem cells: a primer. *Organogenesis* 2013;**9**(1):3–10. Epub 2013/03/30.

16. Thirumala S, Goebel WS, Woods EJ. Clinical grade adult stem cell banking. *Organogenesis* 2009;**5**(3):143–54. Epub 2010/01/05.

17. de la Portilla F, Alba F, Garcia-Olmo D, Herrerias JM, Gonzalez FX, Galindo A. Expanded allogeneic adipose-derived stem cells (eASCs) for the treatment of complex perianal fistula in Crohn's disease: results from a multicenter phase I/IIa clinical trial. *Int J Colorectal Dis* 2013;**28**(3):313–23. Epub 2012/10/12.

18. Trojahn Kolle SF, Oliveri RS, Glovinski PV, Elberg JJ, Fischer-Nielsen A, Drzewiecki KT. Importance of mesenchymal stem cells in autologous fat grafting: a systematic review of existing studies. *J Plastic Surg Hand Surg* 2012;**46**(2):59–68. Epub 2012/04/05.

19. Coleman SR. Hand rejuvenation with structural fat grafting. *Plast Reconstr Surg* 2002;**110**(7):1731–44. discussion 45-7. Epub 2002/11/26.

20. Coleman SR. Structural fat grafting: more than a permanent filler. *Plast Reconstr Surg* 2006;**118**(3 Suppl):108S–20S. Epub 2006/08/29.

21. Yoshimura K, Sato K, Aoi N, Kurita M, Hirohi T, Harii K. Cell-assisted lipotransfer for cosmetic breast augmentation: supportive use of adipose-derived stem/stromal cells. *Aesthetic Plast Surg* 2008;**32**(1):48–55. discussion 6-7. Epub 2007/09/04.

22. Lee SK, Kim DW, Dhong ES, Park SH, Yoon ES. Facial soft tissue augmentation using autologous fat mixed with stromal vascular fraction. *Arch Plastic Surg* 2012;**39**(5):534–9. Epub 2012/10/25.

23. Koh KS, Oh TS, Kim H, Chung IW, Lee KW, Lee HB, et al. Clinical application of human adipose tissue-derived mesenchymal stem cells in progressive hemifacial atrophy (Parry-Romberg disease) with microfat grafting techniques using 3-dimensional computed tomography and 3-dimensional camera. *Ann Plast Surg* 2012;**69**(3):331–7. Epub 2012/08/22.

24. Yoshimura K, Sato K, Aoi N, Kurita M, Inoue K, Suga H, et al. Cell-assisted lipotransfer for facial lipoatrophy: efficacy of clinical use of adipose-derived stem cells. *Dermatol Surg* 2008;**34**(9):1178–85. Epub 2008/06/03.

25. Castro-Govea Y, De La Garza-Pineda O, Lara-Arias J, Chacon-Martinez H, Mecott-Rivera G, Salazar-Lozano A, et al. Cell-assisted lipotransfer for the treatment of parry-romberg syndrome. *Arch Plastic Surg* 2012;**39**(6):659–62. Epub 2012/12/13.

26. Karaaltin MV, Akpinar AC, Baghaki S, Akpinar F. Treatment of "en coup de sabre" deformity with adipose-derived regenerative cell-enriched fat graft. *J Craniofac Surg* 2012;**23**(2):e103–5. Epub 2012/03/27.

27. Marino G, Moraci M, Armenia E, Orabona C, Sergio R, De Sena G, et al. Therapy with autologous adipose-derived regenerative cells for the care of chronic ulcer of lower limbs in patients with peripheral arterial disease. *J Surg Res* 2013;**185**(1):36–44. Epub 2013/06/19.

28. Hicok KC, Hedrick MH. Automated isolation and processing of adipose-derived stem and regenerative cells. *Methods Mol Biol* 2011;**702**:87–105. Epub 2010/11/18.

Chapter 17

Developing "Smart" Point-of-Care Diagnostic Tools for "Next-Generation" Wound Care

Anthony R. Sheets, Catalina K. Hwang, Ira M. Herman

Key Concepts

1. Nonhealing wounds occur in several chronic conditions, such as diabetes and vascular disease, and are a major clinical concern in the United States and worldwide. Based on the rising incidence and prevalence of these diseases, chronic ulcers will continue to be a source of morbidity and mortality worldwide.
2. Numerous pathologic changes underlie the progression of chronic wounds, stemming from bacterial infection and biofilm formation, inflammation, an imbalance in proteolytic enzymes and their inhibitors, and abnormalities in signal transduction related to angiogenesis, wound granulation, and re-epithelialization.
3. Current standards of wound care and assessment include debridement (the removal of necrotic tissue), controlling infection, and supportive measures for underlying comorbidities such as diabetes and vascular disease. However, there is no single, objective system for wound evaluation or prognostic factors. Moreover, current therapeutic interventions, including negative pressure wound therapy (NPWT), hyperbaric oxygen therapy (HBOT), becaplermin (recombinant platelet-derived growth factor (PDGF)-bb), and bioengineered skin substitutes fail to completely convert nonhealing ulcers into those capable of complete wound closure.
4. Ongoing studies focused on revealing mechanisms that drive wound chronicity aim to identify biomarkers of microbial burden and biofilm formation, inflammatory mediators, proteases and their inhibitors, and metabolic changes within the wound microenvironment. This can lead to a molecular "bar code" that can be used to standardize wound diagnosis, monitoring, and treatment.
5. The recent development of flexible, wearable sensors to monitor individual aspects of the chronic wound milieu, including tissue hydration, metabolic and ionic factors such as pH and oxygenation, and protease abundance and activity represent major advances toward developing "smart" point-of-care devices that might help to dynamically inform wound care treatment strategies. These prototypes are important steps in manufacturing diagnostic devices capable of surveying multiple parameters within a nonhealing wound. Recent advances in synthetic biology will surely enhance and significantly extend such "smart" wound dressings that integrate rapid, reliable molecular diagnostics coupled with the dynamic delivery of therapeutics possessing spatial and temporal control characteristics, thereby able to identify and treat wound chronicity in real time.

1. INTRODUCTION

Nonhealing wounds are a problem of increasing magnitude worldwide, and remain an area of vastly unmet clinical need in both developed and developing nations. In the United States alone, these injuries affect at least 2–3% of the adult population—nearly 6.5 million people—at an annual cost that may exceed $50 billion.[1–3] The estimated annual incidence of these injuries is greater than that of several major cancers, according to the latest projections by the American Cancer Society (Figure 1). Much of this expenditure is attributable to long hospital stays, frequent outpatient following, in-home nursing, and is perpetuated by a limited arsenal of efficacious, advanced molecular or cellular-based therapeutics. Furthermore, failure to heal confers a disabling psychosocial impact on the patient, and can lead to limb amputation or mortality.[4] Ischemic ulcers, a subset of chronic wounds, have a 5-year amputation rate of 29%, and a mortality rate of 55%—greater than that of some cancers.[5,6] With an increasing prevalence of diabetes and obesity, the incidence of chronic wounds is only expected to further increase.[1]

In spite of the several advances that have been made in our understanding the molecular and cellular factors that are vital for wound resolution, little has successfully been translated into the clinic. Between the high numbers of patients living

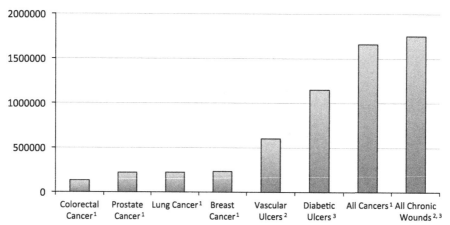

FIGURE 1 Estimated new cases in 2015: cancer versus chronic wounds. 1. *Cancer Facts & Figures 2015*. http://www.cancer.org/research/cancerfactsstatistics/cancerfactsfigures2015/index. 2. Abbade LP, Lastória S. Venous ulcer: epidemiology, physiopathology, diagnosis and treatment. *Int J Dermatol* June 2005;**44**(6):449–56. 3. Sen CK, et al. Human skin wounds: a major and snowballing threat to public health and the economy. *Wound Repair Regen* November–December 2009;**17**(6):763–71.

with these injuries, the morbidity and mortality associated with nonhealing ulcers, and the projected increases in conditions precipitating chronic wounds, it is of vital importance that new diagnostic and therapeutic modalities are developed and delivered for patient care. This chapter will offer a brief review of the pathogenesis of chronic wounds, current treatment options for these injuries, the need for biomarkers to identify and track healing progress, and the potential bases of advanced point-of-care (POC) devices for integrated real-time wound status monitoring and patient-specific care.

2. PATHOGENESIS OF CHRONIC WOUNDS

Numerous health complications may precipitate wounds that fail to heal. These include wounds resulting from venous hypertension or valvular incompetence, as well as ischemic peripheral artery disease. During diabetes mellitus, hyperglycemia drives neuropathy, microvascular disease, and altered foot architecture (i.e., Charcot arthropathy), which contribute to plantar ulcer formation. In immobilized individuals, the constant elevated pressure profiles existing over bony prominences give rise to decubitus ulcers, and, with aging, there are many comorbidities that give rise to complications and nonhealing wounds (Table 1).[1,7–9]

Acute, actively healing wounds progress through multiple and tightly regulated phases that foster tissue remodeling, regeneration, and wound closure. These include initial events linked to hemostatic control, and the ensuing inflammatory

TABLE 1 Chronic Wounds: Etiology, Epidemiology, Complications, and Costs of Care

Wound Type	Etiology[a,b,c,d]	Estimated Incidence[a,b,d]	Complications[a,d]	Cost of Care[d]
Vascular ulcers	*Venous insufficiency:* Varicosity deep vein thrombosis *Arterial disease:* Peripheral atherosclerosis ischemia	600,000 cases/year	Osteomyelitis Amputation Septicemia Death Neoplasms(vascular ulcers)	$9600 per case
Diabetic ulcer	Hyperglycemia Neuropathy Microvascular disease Macrovascular disease Charcot arthropathy	1,150,000 cases/year		$7439–20,622 per case $38,077 per amputation
Decubitus ulcer	Immobilization Elevated pressure over bony Prominences	15% of acute care patients 22% of critical care patients 29% of long-term care patients		$43,180 per case

[a]*Ref. 7.*
[b]*Ref. 8.*
[c]*Ref. 1.*
[d]*Ref. 9.*

responses initiating signaling cascades that give rise to recurrent rounds of cell proliferation and tissue remodeling. Importantly, these include dynamic and reciprocal signals that occur between the cells and their microenvironment, including the extracellular matrix (ECM).[10] In contrast, it has been postulated that over 100 different cellular pathways and microenvironmental factors are disrupted in chronic, nonhealing ulcers.[11,12] Nonhealing wounds typically remain arrested in a persistent inflammatory state that prevents the occurrence of the later repair phases. The elevated levels and activities of proinflammatory mediators and destructive proteases reduce growth factor bioavailability and negatively impact ECM synthesis, thus blunting angiogenesis, fibroblast activation, keratinocyte proliferation, and cellular migration.[13–15] The inability to successfully regrow blood vessels after injury leads to a lack of nutrients and oxygen needed to nourish the tissue and foster successful wound resolution.[16,17] In addition, chronic wound pH is known to be drastically different from that of healthy, healing injuries, which leads to increased susceptibility to bacterial infection; in turn, the complications from bacterial infection, especially biofilm-associated bacteria, impose a further constraint on wound closure.[18]

3. CHRONIC WOUND CARE

3.1 Wound Bed Preparation and Assessment

When patients present with nonhealing injuries such as diabetic foot ulcers and venous leg ulcers (DFU and VLU, respectively, Table 1), the wound must be adequately prepared in support of the anticipated healing. Therapeutic mainstays include infection control, removal of necrotic tissue (debridement), supportive care for comorbid conditions (including glycemic control and correcting nutritional deficiencies), and measures aimed at limiting patient discomfort.[19] The TIME principle, standing for Tissue, Infection/Inflammation, Moisture balance, and Edge of wound was set in place to direct treatment implementation as well as continuous monitoring and evaluation: combined, these components of wound management aim to minimize infection, maintain an optimal inflammatory state, and remove or agitate nonviable and/or infected tissue to maximize contraction and re-epithelialization of the wound edges.[20] However, there are few reliable indices to predict whether a wound will respond to these standards of treatment. In addition, debridement alone is insufficient to achieve healing, regardless of surgical, enzymatic, or biologic methods.[19]

Although infection control and removal of necrotic tissue via debridement paradigms represent commonly held practices across the chronic wound care community, best practices and technical approaches can and do vary across the care provider continuum. Furthermore, there is a compelling need to objectify standards of initial wound assessment or monitoring, such that each metric can reliably and reproducibly contribute to informed treatment and clinical outcomes, rather than merely offering mixed results that might equivocate either or both. As shown in Table 2, multiple wound scales and grading systems exist for healthcare providers to use as guides to the severity of a DFU or VLU, which may

TABLE 2 Wound Classification Systems

Wound Evaluation System	Metrics
Wagner[a]	"Grading": ulcer depth and presence of infection
University of Texas[b]	"Grading": ulcer depth and presence of infection
	"Staging": clinical signs of lower extremity ischemia
Wound bed "score"[c]	Healing edges
	Presence of eschar
	Greatest wound depth/granulation tissue
	Amount of exudate
	Edema
	Periwound dermatitis
	Periwound callus and/or fibrosis
	Pink/red wound bed

[a]Ref. 21.
[b]Ref. 22.
[c]Ref. 23.

rely on subjective factors such as wound color, depth, and overall area.[21–23] Although wound area measurements over time may be used with as much as 75% accuracy to predict wounds that are refractory to conventional treatments, this is operator-dependent and does not alter treatment decisions.[24–27] Clinical judgment supplemented with an objective vascular assessment, including skin blood flow, thermography and Doppler indices, has been used by surgeons and other wound care providers to assess the likelihood of wound healing following limb amputation, although with mixed results.[28,29] Still other prognostic approaches rely on the severity of underlying disease, such as glycemic control in diabetes, comorbidities such as congestive heart failure or evidence of hepatic injury, and risk factors such as immunosuppressive therapy, smoking, patient immobility, and age.[30,31] Psychological distress, which is estimated to occur in at least 25% of patients with chronic wounds, and nutritional deficiencies may also portend negative healing outcomes.[32,33] Hence, the subjectivity of current wound assessment presents an obstacle toward our achieving global standards and optimized best practices for the treatment of chronic wounds.

3.2 Current Modalities for Converting Chronic Wounds to Acute Wounds

Although there are several therapeutic options that a wound care specialist may consider in addition to debridement and supportive care, there is a paucity of evidence-based medicine to inform data-rich diagnostics, which would inform patient-specific therapeutic interventions or their hierarchical implementation. For instance, negative pressure wound therapy (NPWT), which is hypothesized to reduce inflammation and stimulate growth factor production and cellular activities through mechanical stimulation, and hyperbaric oxygen therapy (HBOT), which is thought to improve outcomes through increasing local oxygen tension, are popular modalities aimed at reversing chronic wounds, yet little high-quality information exists about their indications and overall success rates.[34,35] In efforts aimed at reaching consensus regarding their respective indications, meta-analyses and systematic reviews of the literature have revealed that the presence of ischemia may predict which wounds might best respond to either treatment.[34,36] Still, a number of patients failed to achieve complete healing in even the most successful trials, and these analyses noted a large number of unpublished and prematurely terminated trials that may obfuscate interpretations of what might be deemed successful outcomes.[37] Similarly, local application of recombinant growth factors and other survival factors or signaling entities, such as rhGM-CSF, PDGF (becaplermin), fibroblast growth factor, and vascular endothelial growth factor, have yielded little evidence demonstrating clinical efficacy compared with placebo, despite long-standing knowledge of their proangiogenic and re-epithelialization activities in animal models.[38,39] The inconsistent data regarding the utility of replacing single or small combinations of growth factors may not be surprising, given the complex spatiotemporal dynamics of growth factor and cytokine activities within the healing wound, as well as the persistent inflammation and elevated protease profiles present within the wound bed, which directly reduce growth factor bioavailability.[15] Moreover, recent studies suggest that cells within the diabetic microenvironment may not respond appropriately to growth factor stimulation, due to downregulation of growth factor receptors and intracellular inhibition of physiologically relevant signal transduction pathways that function to sustain acute wound closure.[40–42]

In addition, cell-based therapeutics have been developed, offering hope for those with nonhealing wounds. These include bioengineered skin equivalents that contain living cells, such as Apligraf, a bilayered skin substitute comprised of human neonatal foreskin-derived keratinocytes cultured upon a layer of human neonatal foreskin-derived fibroblasts embedded within a type I collagen matrix and Dermagraft, an artificial human dermis that contains neonatal foreskin-derived fibroblasts cultured onto a synthetic, bioabsorbable substrate.[43,44] Though the exact mechanisms of action of these skin equivalents is not agreed upon, it is hypothesized that the living cells contained within these modalities may be able to adapt to the wound microenvironment, where they may potentially reduce inflammation and contribute growth factors, cytokines, and ECM components to foster wound-healing angiogenesis, granulation, and re-epithelialization.[45] However, it is important to note that existing skin equivalents are not permanent solutions, as modest rates of successful engraftment and wound closure have been reported, and these treatment modalities have not been thoroughly evaluated through well-designed randomized-controlled trials.[45,46] Indeed, the absence of a true vascular cell component in Apligraf, Dermagraft, and other skin substitutes may underlie the reported engraftment failures; a well-perfused vascular network is necessary for oxygen and nutrient exchange to sustain all cells and tissues, and insufficient vascularization of skin equivalents or other engineered tissues may lead to improper cell integration or death.[47]

To this end, grafts of autologous, multipotent stem cells isolated from bone marrow and adipose tissues of adult patients have been suggested as a means of reversing wound chronicity, through local differentiation to achieve tissue revascularization, granulation, and wound closure.[48,49] But, there is no standard dosing or delivery regimen that has been established for stem cell grafts, and the inflamed, protease-rich chronic wound microenvironment may hinder stem cell engraftment.[50–52] Moreover, diabetes may impose a further limit on the utility of autologous stem cell grafts, as the hyperglycemia-driven "metabolic memory" within progenitor cell niches may directly contribute to low stem cell engraftment documented in preclinical animal models of chronic diabetic wounds.[53] In light of the current challenges regarding grafts of both stem cells and engineered tissues for nonhealing wounds, developing artificial skin substitutes that contain microfluidic networks lined with patient-derived microvascular endothelial cells might be a means by which to introduce a patent, functional vasculature into an angiogenically impaired chronic wound.[47] Engineered skin that includes preformed, cellularized vascular networks within biodegradable scaffolds could allow the gradual incorporation of graft microvessels with host tissues to achieve long-term engraftment and successful healing, and remains an area of active investigation.

Overall, there are neither algorithms to determine which wounds would benefit most from specific interventions, nor are there comprehensive diagnostics to logically inform the design of patient-specific treatment schedules. The limited successes of these current treatment modalities may lie in a failure to truly address the biological bases of nonhealing wounds. Without advanced diagnostics capable of informing and correcting the underlying roots of chronicity, therapeutics may remain inconsistent or ineffective. Thus, there is a great need for identification and characterization of chronic wound biomarker panels that will help to transform the manner in which care is offered to those in need of healing.

4. BIOMARKERS: MOLECULAR "BAR CODING" OF CHRONIC WOUNDS

The ability to unequivocally distinguish nonhealing wounds through molecular profiling represents a significant advance, just as it remains a comparable challenge. In cancer biology, efforts aimed at identifying molecular markers of specific malignancies have led to diagnostic, prognostic, and therapeutic innovations. For instance, next-generation sequencing and other complementary genomic analyses have allowed robust characterization of distinct subtypes of medulloblastoma, the most common malignant childhood brain tumor; in turn, these studies have garnered further insight into the genetic and epigenetic drivers of these neoplasms, and have uncovered novel targets with therapeutic implications.[54] Similarly, biomarkers such as HER2 and KRAS have successfully been used to direct treatment in breast and colorectal cancer, respectively.[55,56] Additionally, olaparib, a poly (ADP-ribose) polymerase (PARP) inhibitor, has been shown to have antitumor activity in cancers associated with *BRCA1/BRCA2* mutations.[57]

Indeed, such molecular "bar coding" could revolutionize diagnosis of nonhealing wounds, enhance our understanding of the molecular and cellular mechanisms of wound chronicity across etiologies, and herald the onset of therapeutic interventions for personalized wound care. Compared to acute wounds, several gene transcripts and proteins are differentially expressed in chronic ulcers, especially at the wound edge.[58,59] In addition, the heterogeneous responses to current wound-healing treatments are likely influenced by a combination of genetic and environmental factors, and may vary with wound etiology. This is evidenced by the recent identification of single nucleotide polymorphisms that may impact healing dynamics in chronic venous disease, and epigenetic, hyperglycemia-induced alterations in methylation-dependent promoter activation and gene transcription in several cells types within chronic diabetic wounds.[60–62] Hence, determining the molecular profiles of nonhealing ulcers would aid in diagnosing and stratifying injuries, and could effectively inform the treatment course, including debridement, where the molecular "signature" of healing tissue could be used to determine whether a wound has been sufficiently prepared for healing.[58] In turn, this could offer opportunities for changing the care administered within the treatment period, based on the dynamic molecular "bar code" of each patient's wound. However, a list of biomarkers that can be used with high sensitivity and specificity to predict wound-healing outcomes has not yet been established. Based on the factors that underlie the development and chronicity of recalcitrant wounds, an ideal set of diagnostic biomarkers would include molecules related to microbial colonization and infection, proinflammatory signaling cascades, ECM-degrading proteases, and metabolic and ionic parameters such as tissue oxygenation and pH.

4.1 Molecular Diagnostics of Microbial Burden

Managing infection is a major aspect of chronic wound care paradigms, yet successful identification and elimination of infectious microbial species remain considerable challenges. Traditional culture-based methods of bacterial identification take several days and only identify about 1% of the microbial flora within a chronic wound.[63] Chronic wound-associated infections are typically polymicrobial, and are often encased in dense polysaccharide capsules that are formed when a critical mass of microbes are present within a given environment, that is, within biofilms.[64,65] These diverse communities are often not amenable to culture-based identification, and possess distinct genomic and metabolic properties when compared to free-living, planktonic bacteria.[66] Metabolic alterations in biofilm-encapsulated bacteria may contribute to antibiotic failure, and biofilms themselves may have innate qualities that confer antibiotic tolerance or resistance.[67,68] Furthermore, these hard coatings may suppress host immune responses through anti-phagocytic properties, and as neutrophils and macrophages are unable to ingest the bacteria growing within these matrices, the immune cells release large amounts of proinflammatory cytokines and reactive oxygen species, which are destructive to nearby tissues and contribute to wound chronicity.[65,69]

The current treatment approaches for chronic wound biofilms have limited efficacy, and debridement is often the first line of therapy in order to remove diseased, biofilm-laden tissue.[70] NPWT may also be effective in reducing bacterial load, and HBOT may be toxic to anaerobic species within the wound and/or biofilm.[16,71,72] Still, much more attention has been directed at preventing biofilm formation, when possible, through antimicrobial strategies such as antibiotic-loaded sutures and silver-impregnated bandages, than to strategies tailored specifically to eliminating biofilms after their establishment.[70,73] This may be partly due to the lack of reliable diagnostics, as described above. Recently, culture-free, molecular diagnostic platforms have been described, to more effectively profile the microflora of nonhealing wounds. These include PCR-based tools, which may permit the identification of species growing within a chronic wound, especially fastidious anaerobic species and biofilm denizens that are unable to be cultured by current techniques, and NMR-based comparative metabolomics to identify markers of planktonic versus biofilm-associated bacteria.[74,75] Along with these methods, molecular diagnostics of the saccharide components of wound biofilms would be useful for the rational implementation of enzymatic debridement, wherein specific saccharidases could be applied in parallel or in sequence to disrupt the biofilm. This could allow for more reliable identification of microbial species encased within, enabling a determination of their antibiotic resistance profiles, and appropriate therapies could be initiated.

4.2 Inflammatory Mediators as Biomarkers: The "Omics" of Chronic Wound Fluid

Analysis of key pro- and anti-inflammatory mediators continues to provide valuable information regarding wound status, just as more sophistication is needed to significantly enrich what could become a robust database enabling treatment strategies that may not be possible through current diagnostics. For example, proteomic profiling of chronic wound fluid (CWF), or wound exudate, is a readily available, noninvasive substitute for biopsies that may be useful for identifying biomarkers indicative of healing status and treatment progression. It has long been known that CWF contains persistently elevated levels of inflammatory cytokines, proteolytic enzymes, and high levels of degraded cellular components that are inhibitory to wound healing, reflecting the harsh microenvironment of its source.[13,76] Recently, Edsberg et al. profiled proteomic changes within chronic pressure ulcers over 42 days, and identified several proteins whose expression changes over time, or that were differentially expressed between wounds of different degrees of healing.[77] They identified differential expression of chemokine (C-X-C motif) ligand 9 expression between healing and nonhealing wounds, and hypothesized that decreased expression levels may result in inhibited endothelial cell chemotaxis. This is but one example of how "omics" studies from a readily accessible clinical sample can further direct our understanding of chronic wound pathogenesis in a high-throughput manner.

However, there remains no "gold standard" or best practice that has been established as an optimized protocol for CWF collection, handling, storage, or assay. Furthermore, there is a lack of consensus on the inflammatory biomarkers most strongly associated with healing status.[78,79] Appreciable levels of relevant molecules contained within CWF may be difficult to detect, in light of the abundance of common proteins, such as albumin; indeed, recent work from Fernandez et al. employed an immunodepletion strategy to remove the highest abundant proteins from CWF, in order to better detect low-abundance moieties that could serve as diagnostic and/or prognostic markers.[80] This strategy allowed the investigators to uncover distinct proteomic profiles in CWF before and after immunodepletion, especially proinflammatory cytokines. These approaches may be hindered by the absorptive properties of different wound dressings, the amount of CWF required for analysis, or the potential for overdepletion, all of which may obfuscate detection.

4.3 Proteases as Chronic Wound Diagnostics

Nonhealing wounds are further distinguished by marked inductions in their expression of proteolytic enzymes, such as neutrophil elastase and several matrix metalloproteinases (MMPs). As such, these factors may also be useful as biomarkers to determine healing status, and have been suggested as targets of therapeutic interventions. In chronic wounds, MMPs are often expressed at levels 30- to100-fold higher than in healing skin, concomitant with diminished levels of the tissue inhibitors of metalloproteinases (TIMPs); together, these changes impede proper ECM remodeling, reduce the levels of growth factors within the chronic wound bed, and are associated with poor healing outcomes.[10,14,81] Further, this protease overload is thought to undermine the potential for successful treatment with numerous existing treatments, such as recombinant growth factors, as previously mentioned. While broad-spectrum MMP inhibition may seem an attractive target for therapeutic interventions, it is important to note that regulated proteolysis is an important part of cutaneous injury resolution: MMPs are necessary for several wound-healing processes, including keratinocyte migration at the wound edge, endothelial cell migration during angiogenesis, and fibroblast-driven remodeling of the granulation tissue, all of which are carefully metered by the activity of TIMPs.[14,82,83] Thus, the indiscriminate targeting or sequestering of MMPs by current therapeutics may not be an ideal strategy to convert a chronic wound, as a randomized-controlled trial comparing Promogran, a collagen/oxidized regenerated cellulose dressing that binds MMPs and allows release of growth factors into the microenvironment, versus moistened gauze showed no difference in wound closure.[84] Hence, therapeutic induction of TIMPs might offer greater benefits than indiscriminate MMP inhibition, given the importance of regulated ECM turnover in tissue repair and regeneration.

In light of these findings, there now exists a POC diagnostic for elevated protease activity in wounds—a metric that may have advantages over measuring overall abundance of specific proteases, given the many states and multimeric complexes in which these enzymes can exist.[81,85] In the future, advanced tools might not only aim to evaluate the baseline levels or activity of these enzymes at the onset of therapy, but also provide information regarding MMP/TIMP balance during the course of treatment, in order to inform additional interventions that may need to be applied to a nonhealing wound. In addition, based on the intriguing findings described by Fernandez et al., a future diagnostic approach to CWF proteomics may broaden the immunodepletion strategy to include MMPs and elastases, as these entities are often expressed many fold higher in chronic wounds than in successfully healing skin.[80] Antibody-mediated depletion of MMPs from CWF would still allow determination of protease expression levels, as well as potentially their activity, and perhaps provide further clarity into less abundant proteins that may have prognostic and therapeutic implications.

4.4 Metabolic Markers of Wound Chronicity

In addition to the critical roles of the aforementioned parameters for wound healing, metabolic and ionic factors, such as pH and oxygen tension, are known to impact healing dynamics. The acid–base balance within acute and chronic cutaneous ulcers affects the behaviors of microbial species within the wound, inflammation, activity of proteases and other enzymes, and migration and proliferation of endothelial cells, fibroblasts, and keratinocytes.[18] Further, numerous cellular and environmental processes modulate pH; it is dynamic and may vary throughout the chronic wound bed. Therefore, pH-driven diagnostics would require frequent sampling intervals and high spatial resolution to yield clinically applicable data for monitoring healing status.

Low levels of tissue oxygenation have been linked to poor healing outcomes, especially in the case of ischemic vascular disease.[16,86] Hence, transcutaneous oximetry has been suggested as a promising, noninvasive approach to identifying nonhealing wounds or to predict impediments to healing. Yet, as in the case of many other modalities discussed above, strong data regarding transcutaneous oxygen tension ($PtcO_2$) as an independent predictor of wound status are lacking. To this end, Arsenault et al. conducted a meta-analysis of studies regarding $PtcO_2$ measurements in wounds.[87] While over 2000 articles were considered as candidates for comparison, only a total of 4 studies met the authors' criteria for inclusion in their analyses. Despite the small number of studies analyzed, which enrolled a total of 901 patients with chronic lower extremity wounds, Arsenault et al. noted that $PtcO_2$ *might* be an independent prognostic factor for impaired healing, but based on the data presented therein, the cutoff level for significance may be 20 mmHg—the same as for critical limb ischemia. Thus, $PtcO_2$ remains an area of active investigation, but its utility as a stand-alone metric of healing status may be limited.

4.5 Therapeutic Implications of Chronic Wound Profiling

The identification of well-validated biomarkers will yield marked advancements in diagnosing, monitoring, and treating chronic wounds, as has been the case in breast, colorectal, and central nervous system cancers.[54–56] Based on these observations, it is apparent that applications of transcriptional, proteomic, and metabolomic analyses should not be limited to simply identifying chronic wounds and stratifying them based on their potential to benefit from current treatment protocols, but may also be used to uncover novel therapeutic targets for effective, individualized wound care. As stated previously, the microbial-derived proteins and metabolites identified through culture-free diagnostics would not only be informative about the populations present in a wound, but could also directly inform treatment strategies through an enhanced understanding of this microenvironment. In addition, molecular profiles of MMPs and other proteases—as well as their natural inhibitors—might allow targeted induction of specific TIMPs, to restore regulated proteolysis, and potentially identify combinations of small molecules for co-delivery to stimulate several pathways of tissue repair.

5. NOVEL DEVICES FOR WOUND ASSESSMENT

Based on the need for objective analyses of wound chronicity and healing progression, multiple groups have recently developed flexible, wearable sensors that can be placed into the wound at the time of occlusion and used to measure the physical and chemical components of the wound microenvironment, including moisture balance, protease activity,

TABLE 3 Wound Monitoring and Diagnostic Devices

Existing Tests and Sensors		Desired Tests and Sensors
Single Metrics	**Multiple Parameters**	**Currently Unavailable**
PH sensors[a,b]	Combined temperature, moisture, and pressure detector (prototype)[g]	Bacterial burden: Molecular diagnostic of biofilms Reliable drug susceptibility profiles
Transcutaneous O_2 meters[a,c] • Clark electrode • Galvanic probe		Inflammation: Pro/anti-inflammatory biomarkers
Bacterial burden: Swab and culture		Protease detection: MMP/TIMP balance
Moisture-sensing dressings[a,b,d]		Epidermal and dermal repair pathways: migration proliferation angiogenesis signal transduction abnormalities
Protease detection[b,e,f] • abundance • activity • sequestering-bandage		

MMP, matrix metalloproteinase.
[a]Ref. 89.
[b]Ref. 90.
[c]Ref. 87.
[d]Ref. 88.
[e]Ref. 85.
[f]Ref. 84.
[g]Ref. 91.

ionic composition, and metabolic activity.[88–90] While these "smart" devices offer distinctive advantages over current subjective means of wound assessment, the sensors are not contained within a single device (Table 3). To this end, Mehmood et al. recently developed a prototype wireless sensor that may allow temperature, moisture, and pressure data all to be monitored in real time on a computer, and can be worn by a patient.[91] Indeed, such sensors could potentially be coupled with advanced molecular diagnostics, in order to take stock of wound infection, inflammation, protease balance, and epidermal and dermal signal transduction pathways that regulate cell migration, proliferation, and wound-healing angiogenesis. Together, this information could guide treatment by identifying when a bandage needs to be changed, if pressure has been adequately applied to the site, if there is a fulminant infection, and whether the wound is resolving.

5.1 Opportunities for Advanced Diagnostic and Therapeutic Modalities: Lessons from Synthetic Biology?

With the many cellular and molecular aberrations that drive wound chronicity, it is difficult to assign primacy to any sole aspect or component of the wound microenvironment. As such, there have been considerable challenges in creating effective diagnostics that fully survey the chronic wound landscape while offering sufficient or any spatial–temporal analytics that might help to inform treatment strategies for one or another chronically inflamed or nonhealing wound. The prototypes described above are important first steps in achieving these goals of standardized diagnosis and therapy. Looking to the future, one promising means of developing multimodal, POC devices for wound care might lie in applying the tenets of synthetic biology. Broadly, synthetic biology is based on the notion that cells and biological systems are programmed with an intrinsic logic, similar to a computer circuit, which allows them to respond to environmental constraints in predictable ways; within this framework, as basic science continues to define the pathways necessary for tissue repair and regeneration, and identify those that are deficient in nonhealing wounds, novel synthetic networks can be engineered and potentially implemented to create advanced diagnostics and therapeutics targeting factors of chronicity including microbial infection, inflammation, proteolytic imbalance, and pathways driving wound-healing angiogenesis, granulation, and re-epithelialization.[92,93] Indeed, many therapeutic approaches already employ aspects of this discipline: the aforementioned moisture, pH- and metalloproteinase-sensing dressings can be thought of as simple, programmed circuits with defined

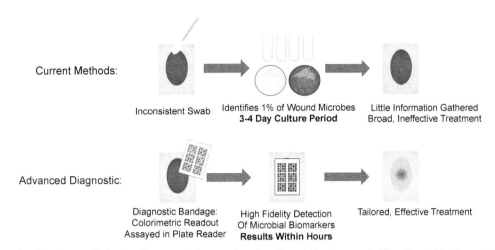

Current Methods:

Inconsistent Swab

Identifies 1% of Wound Microbes
3-4 Day Culture Period

Little Information Gathered
Broad, Ineffective Treatment

Advanced Diagnostic:

Diagnostic Bandage:
Colorimetric Readout
Assayed in Plate Reader

High Fidelity Detection
Of Microbial Biomarkers
Results Within Hours

Tailored, Effective Treatment

FIGURE 2 Assessing chronic wound infection. Current techniques are based on: (i) swabbing the wound, (ii) and bacterial culture, which takes 3–4 days and is able to identify only ~1% wound-associated microbes.

endpoints. The individual strengths and limitations of these first efforts, coupled with recent developments in synthetic gene-based diagnostics, may offer vital information for the creation of several tools for comprehensively evaluating and treating chronic wounds.

For example, concepts from synthetic biology have been harnessed to create inexpensive, paper-based gene networks for use as rapid diagnostics for infectious agents, including bacteria and viruses.[94] In 2014, Pardee et al. created synthetic, "toehold switches," platforms that can drive posttranscriptional activation of protein translation in the presence of specific RNA-based triggers, such that (1) bacterial genes coding for antibiotic resistance and (2) Ebola strain-specific RNAs would induce specific colorimetric changes that could be detected in a plate reader within minutes to hours, and successfully produce unique diagnostic outputs.[94,95] In addition, protein and small molecule ligand-sensing aptamers have been successfully fused to trans-acting noncoding RNAs, yielding systems that could similarly be adapted for diagnostic purposes.[96] Such sterile, cell-free systems could hold great promise for either transcriptional or proteomic analysis of nonhealing wounds, where the ability to rapidly and reliably detect molecular signatures could aid in diagnostics, predict successful treatment options, and chart therapeutic progress after initiating treatment.

Indeed, diagnostic tools driven by synthetic gene-based sensors would be a powerful means to quickly identify several factors of wound chronicity—and sensors of multiple parameters might be integrated into a single test or wearable diagnostic device. For instance, networks designed to produce colorimetric changes in response to the molecular "signatures" of biofilms, such as metabolites from polymicrobial communities, specific polysaccharide residues characteristic of such matrices, along with antibiotic resistance genes would enable rapid and reliable detection of bacterial infection.[74,75] The clinician(s) could then initiate treatments based on the composition of the biofilm and the antibiotic resistance profiles, within hours of sampling and testing—vast improvements over current microbial culture-based techniques, which take several days to perform and fail to identify many of the species present within an infected wound.[63] Figure 2 depicts a comparison between current and future methods of identifying bacterial infection and biofilm composition within nonhealing wounds. This technology could be further applied to detect oxygen abundance, ionic balance, and levels of pro- and anti-inflammatory cytokines across multiple areas of the wound; protease activity might be assayed using MMP-cleavable repressors of fluorescent genetic elements, wherein the fluorescence intensity detected would directly depend on the degree of protease activity. The information gained from these tools would be useful not only to inform therapeutic deliverables but also to chart healing dynamically, therein enabling alterations in treatment strategies.

Additionally, a synthetic network could be designed to diagnose dysfunctional signal transduction in the cells and tissue beds of chronic wounds. These could focus on, but not be limited to, angiogenic activation, granulation tissue formation/ remodeling, or re-epithelialization. As recent data indicate that reduced growth factor activity and angiogenic impairment in DFU may be caused by deviations in protein tyrosine phosphatase (PTP) expression or activity, PTP molecular sensors could be designed and implemented to determine whether chronic wounds might successfully be treated with specific PTP inhibitors, which have been suggested to restore growth factor sensitivity within DFU.[41,42,97] Together, these tools could quickly generate high-fidelity readouts of multiple parameters, and provide clinicians objective metrics of the chronic wound microenvironment.

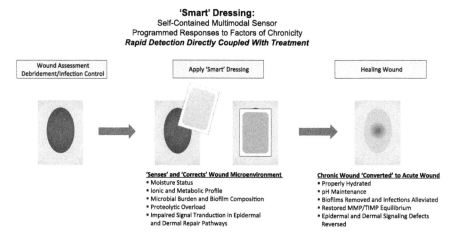

FIGURE 3 "Smart" Wound Dressing. Schematic depiction of a stylized "advanced" device for POC diagnostics coupled to dynamic, real-time wound treatment. Created through a combination of flexible biomaterials, micro- and nanoscale sensors, and synthetic gene networks programmed with conditional logic, a self-contained diagnostic and therapeutic bandage would offer rapid and accurate detection of multiple factors of chronicity, and dynamically deliver advanced therapies in a spatially and temporally controlled manner.

5.2 Programmed Synthetic Systems as "Smart" Wound Dressings: Combined Diagnostics and Therapeutics

More complex synthetic networks could be designed as combined diagnostic and therapeutic devices, reducing patient morbidity and mortality through coupling microenvironmental sensors and readouts with defined treatments. Based on the information gleaned from basic studies of acute and chronic wounds, "smart" dressings could be engineered to function in this capacity (Figure 3). Conditional signaling networks could be designed using either toehold switches or ligand-sensing aptamers, where specific inputs from the chronic wound milieu would activate defined transcriptional elements to yield controlled, therapeutic effects—thus creating a functional, self-contained combination of diagnostic–therapeutic platform. This could be a promising approach to better engineer autologous and/or syngeneic cell transplants and skin equivalents: in sensing the conditions of the wound, the cells contained within these modalities could adapt according to preprogrammed regimens to achieve better engraftment over cell-based therapeutics, and deploy their synthetically programmed healing mechanisms. As an example, through studies conducted by our lab and others, we appreciate the roles that both cellular traction-driven and environment-dependent mechanical forces play in regulating microvascular cell proliferation and angiogenic induction, and that extracellular matrices of different composition have unique mechanical properties that may foster or restrict vascular endothelial cell growth and remodeling through dynamic and reciprocal interactions with one another.[10,98–101] Based on these experiments, one potential cellular medicine could exploit the dynamic reciprocity of cell–matrix interactions to create programmed mechanosensory feedback loops: patient-derived fibroblasts could be engineered as ECM-sensors, and produce defined amounts of specific components in response to the matrix composition of the chronic wound, to manufacture a proangiogenic microenvironment in situ. Alternatively, cells might be programmed to alter their contractile properties in response to ECM stiffness, to achieve a similar end.

Matrix-sensing scaffolds could be created as a combined diagnostic–therapeutic modality for sensing and treating biofilm-associated microbes. Here, biofilm component-receptors, perhaps identified from profiling species contained within these polysaccharide capsules, might be fused to synthetic genetic elements that allow production of specific enzymes to degrade these protective coatings, thereby allowing the host defenses to penetrate. Cell-based systems of this variety could be engineered with optogenetically driven "kill switches," where cells might coexpress death mediators specifically induced following stimulation of bacterial or eukaryotic light-sensing opsins.[102] Upon confirming the desired therapeutic outcome, specific wavelengths of light could be introduced through existing technologies, triggering destruction of any and all cells expressing these constructs—including wound flora that may have acquired them through horizontal gene transfer. Together, these synthetic biology-based tools could potentially be integrated into a single "smart" device, capable of assaying and rectifying numerous biochemical and metabolic factors simultaneously, and eliminate the need for multiple diagnostics and therapeutics.

6. SUMMARY AND FUTURE OUTLOOK

There still exist significant gaps in our understanding of the molecular and cellular mechanisms that prevent closure in chronically inflamed, nonhealing wounds as well as in translating this basic knowledge into clinically relevant outcomes that would meet this enormous and unmet clinical need. Indeed, as the overall incidence and prevalence of chronic wounds continue to rise in the coming decades, narrowing the distance between basic science research and clinical implementation is of paramount importance. Several groups are undertaking transcriptional and proteomic approaches to identify biomarkers of chronic wounds and to track healing progress after specific interventions, as well toward developing advanced, POC devices to measure multiple parameters such as protease activity, tissue oxygenation, pH, and hydration. As is the case of the insulin pump for patients living with diabetes, the implementation of a "smart" wound-healing diagnostic device that could also administer bioactive or small molecules based on feedback and data derived from the wound bed could revolutionize the manner in which wound care will be delivered moving forward. However, further work must be performed in order to come closer to making this type of device a reality. There are still no validated biomarkers for chronic wounds, and additional efforts will be needed to not only comb through a growing wealth of information, but to also identify molecular entities and therapeutic tools that can be used to convert chronic ulcers into acute, healing wounds. There is also a need for standardized method of collection and analysis of wound samples. In the near future, synthetic biological systems might be adapted to create inexpensive tools to rapidly and reliably survey the chronic wound microenvironment to better determine treatment regimens. In turn, these insights may lead to development of advanced biomaterials and cellular medicines that can act as combined diagnostic therapeutic modalities to reduce microbial burden, restore the balance of proteases and their inhibitors, or reverse abnormalities in signal transduction pathways required to activate the cellular and tissue responses required for resolution and closure of nonhealing wounds.

REFERENCES

1. Sen CK, Gordillo GM, Roy S, Kirsner R, Lambert L, Hunt TK, et al. Human skin wounds: a major and snowballing threat to public health and the economy. *Wound Repair Regen* 2009;**17**:763–71.
2. Hess CT. Putting the squeeze on venous ulcers. *Nursing* 2004;**34**. Suppl Travel:8–13; quiz 13–4.
3. Driver VR, Fabbi M, Lavery LA, Gibbons G. The costs of diabetic foot: the economic case for the limb salvage team. *J Vasc Surg* 2010;**52**:17S–22S.
4. Pragnell J, Neilson J. The social and psychological impact of hard-to-heal wounds. *Br J Nurs* 2010;**19**:1248–52.
5. Moulik PK, Mtonga R, Gill GV. Amputation and mortality in new-onset diabetic foot ulcers stratified by etiology. *Diabetes Care* 2003;**26**:491–4.
6. Armstrong DG, Wrobel J, Robbins JM. Guest Editorial: are diabetes-related wounds and amputations worse than cancer? *Int Wound J* 2008:286–7.
7. Abbade LP, Lastória S. Venous ulcer: epidemiology, physiopathology, diagnosis and treatment. *Int J Dermatol* 2005;**44**:449–56.
8. Chiriano J, Bianchi C, Teruya TH, Mills B, Bishop V, Abou-Zamzam AM. Management of lower extremity wounds in patients with peripheral arterial disease: a stratified conservative approach. *Ann Vasc Surg* 2010;**24**:1110–6.
9. Gould L, Abadir P, Brem H, Carter M, Conner-Kerr T, Davidson J, et al. Chronic wound repair and healing in older adults: current status and future research. *Wound Repair Regen* 2015;**23**(1):1–13.
10. Schultz GS, Davidson JM, Kirsner RS, Bornstein P, Herman IM. Dynamic reciprocity in the wound microenvironment. *Wound Repair Regen* 2011;**19**:134–48.
11. Demidova-Rice TN, Hamblin MR, Herman IM. Acute and impaired wound healing: pathophysiology and current methods for drug delivery, part 1: normal and chronic wounds: biology, causes, and approaches to care. *Adv Skin Wound Care* 2012;**25**:304–14.
12. Brem H, Tomic-Canic M. Cellular and molecular basis of wound healing in diabetes. *J Clin Invest* 2007;**117**:1219–22.
13. Trengove NJ, Stacey MC, MacAuley S, Bennett N, Gibson J, Burslem F, et al. Analysis of the acute and chronic wound environments: the role of proteases and their inhibitors. *Wound Repair Regen* 1999;**7**:442–52.
14. Armstrong DG, Jude EB. The role of matrix metalloproteinases in wound healing. *J Am Podiatr Med Assoc* 2002;**92**:12–8.
15. McCarty SM, Percival SL. Proteases and delayed wound healing. *Adv Wound Care (New Rochelle)* 2013;**2**:438–47.
16. Gottrup F. Oxygen in wound healing and infection. *World J Surg* 2004;**28**:312–5.
17. Demidova-Rice TN, Durham JT, Herman IM. Wound healing angiogenesis: innovations and challenges in acute and chronic wound healing. *Adv Wound Care (New Rochelle)* 2012;**1**:17–22.
18. Percival SL, McCarty S, Hunt JA, Woods EJ. The effects of pH on wound healing, biofilms, and antimicrobial efficacy. *Wound Repair Regen* 2014;**22**:174–86.
19. Panuncialman J, Falanga V. The science of wound bed preparation. *Surg Clin North Am* 2009;**89**(3):611–26.
20. Leaper DJ, Schultz G, Carville K, Fletcher J, Swanson T, Drake R. Extending the TIME concept: what have we learned in the past 10 years?(*). *Int Wound J* 2012;**9**(Suppl. 2):1–19.
21. Wagner FW. The dysvascular foot: a system for diagnosis and treatment. *Foot Ankle* 1981;**2**:64–122.
22. Lavery LA, Armstrong DG, Harkless LB. Classification of diabetic foot wounds. *J Foot Ankle Surg* 1996;**35**:528–31.

23. Falanga V, Saap LJ, Ozonoff A. Wound bed score and its correlation with healing of chronic wounds. *Dermatol Ther* 2006;**19**:383–90.

24. Moore K, Huddleston E, Stacey MC, Harding KG. Venous leg ulcers–the search for a prognostic indicator. *Int Wound J* 2007;**4**(2):163–72.

25. Flanagan M. Wound measurement: can it help us to monitor progression to healing? *J Wound Care* 2003;**12**:189–94.

26. Sheehan P, Jones P, Caselli A, Giurini JM, Veves A. Percent change in wound area of diabetic foot ulcers over a 4-week period is a robust predictor of complete healing in a 12-week prospective trial. *Diabetes Care* 2003;**26**(6):1879–82.

27. Lavery LA, Barnes SA, Keith MS, Seaman JW, Armstrong DG. Prediction of healing for postoperative diabetic foot wounds based on early wound area progression. *Diabetes Care* 2008;**31**(1):26–9.

28. Sarin S, Shami S, Shields DA, Scurr JH, Smith PD. Selection of amputation level: a review. *Eur J Vasc Surg* 1991;**5**(6):611–20.

29. Arsenault KA, Al-Otaibi A, Devereaux PJ, Thorlund K, Tittley JG, Whitlock RP. The use of transcutaneous oximetry to predict healing complications of lower limb amputations: a systematic review and meta-analysis. *Eur J Vasc Endovasc Surg* 2012;**43**(3):329–36.

30. Rhou Y, Henshaw FR, McGill MJ, Twigg SM. Congestive heart failure presence predicts delayed healing of foot ulcers in diabetes: an audit from a multidisciplinary high-risk foot clinic. *J Diabetes Complications* 2015;**29**(4):556–62.

31. Amir O, Liu A, Chang AS. Stratification of highest-risk patients with chronic skin ulcers in a Stanford retrospective cohort includes diabetes, need for systemic antibiotics, and albumin levels. *Ulcers* 2012;**2012**.

32. House SL. Psychological distress and its impact on wound healing: an integrative review. *J Wound Ostomy Continence Nurs* 2015;**42**(1):38–41.

33. Molnar JA, Underdown MJ, Clark WA. Nutrition and chronic wounds. *Adv Wound Care (New Rochelle)* 2014;**3**(11):663–81.

34. Glass GE, Murphy GF, Esmaeili A, Lai LM, Nanchahal J. Systematic review of molecular mechanism of action of negative-pressure wound therapy. *Br J Surg* 2014;**101**:1627–36.

35. Stoekenbroek RM, Santema TB, Legemate DA, Ubbink DT, van den Brink A, Koelemay MJ. Hyperbaric oxygen for the treatment of diabetic foot ulcers: a systematic review. *Eur J Vasc Endovasc Surg* 2014;**47**:647–55.

36. Vig S, Dowsett C, Berg L, Caravaggi C, Rome P, Birke-Sorensen H, et al. Evidence-based recommendations for the use of negative pressure wound therapy in chronic wounds: steps towards an international consensus. *J Tissue Viability* 2011;**20**(Suppl. 1):S1–18.

37. Peinemann F, McGauran N, Sauerland S, Lange S. Negative pressure wound therapy: potential publication bias caused by lack of access to unpublished study results data. *BMC Med Res Methodol* 2008;**8**:4.

38. Papanas N, Maltezos E. Becaplermin gel in the treatment of diabetic neuropathic foot ulcers. *Clin Interv Aging* 2008;**3**:233–40.

39. Barrientos S, Brem H, Stojadinovic O, Tomic-Canic M. Clinical application of growth factors and cytokines in wound healing. *Wound Repair Regen* 2014;**22**(5):569–78.

40. Blakytny R, Jude EB. Altered molecular mechanisms of diabetic foot ulcers. *Int J Low Extrem Wounds* 2009;**8**:95–104.

41. Dinh T, Tecilazich F, Kafanas A, Doupis J, Gnardellis C, Leal E, et al. Mechanisms involved in the development and healing of diabetic foot ulceration. *Diabetes* 2012;**61**:2937–47.

42. Zhang J, Li L, Li J, Liu Y, Zhang CY, Zhang Y, et al. Protein tyrosine phosphatase 1B impairs diabetic wound healing through vascular endothelial growth factor receptor 2 dephosphorylation. *Arterioscler Thromb Vasc Biol* 2015;**35**:163–74.

43. Eaglstein WH, Falanga V. Tissue engineering and the development of Apligraf, a human skin equivalent. *Adv Wound Care* 1998;**62**:1–8.

44. Hart CE, Loewen-Rodriguez A, Lessem J. Dermagraft: use in the treatment of chronic wounds. *Adv Wound Care (New Rochelle)* 2012;**1**:138–41.

45. Ehrenreich M, Ruszczak Z. Update on tissue-engineered biological dressings. *Tissue Eng* 2006;**12**:2407–24.

46. Greaves NS, Iqbal SA, Baguneid M, Bayat A. The role of skin substitutes in the management of chronic cutaneous wounds. *Wound Repair Regen* 2013;**21**:194–210.

47. Borenstein J, Megley K, Wall K, Pritchard E, Truong D, Kaplan D, et al. Tissue equivalents based on cell-seeded biodegradable microfluidic constructs. *Materials* 2010;**3**(3):1833–44.

48. Wu Y, Chen L, Scott PG, Tredget EE. Mesenchymal stem cells enhance wound healing through differentiation and angiogenesis. *Stem Cells* 2007;**25**:2648–59.

49. Falanga V, Iwamoto S, Chartier M, Yufit T, Butmarc J, Kouttab N, et al. Autologous bone marrow-derived cultured mesenchymal stem cells delivered in a fibrin spray accelerate healing in murine and human cutaneous wounds. *Tissue Eng* 2007;**13**:1299–312.

50. Wagner J, Kean T, Young R, Dennis JE, Caplan AI. Optimizing mesenchymal stem cell-based therapeutics. *Curr Opin Biotechnol* 2009;**20**:531–6.

51. Game FL, Hinchliffe RJ, Apelqvist J, Armstrong DG, Bakker K, Hartemann A, et al. A systematic review of interventions to enhance the healing of chronic ulcers of the foot in diabetes. *Diabetes Metab Res Rev* 2012;**28**(Suppl. 1):119–41.

52. Koenen P, Spanholtz TA, Maegele M, Stürmer E, Brockamp T, Neugebauer E, et al. Acute and chronic wound fluids inversely influence adipose-derived stem cell function: molecular insights into impaired wound healing. *Int Wound J* 2015;**12**:10–6.

53. Rennert RC, Sorkin M, Januszyk M, Duscher D, Kosaraju R, Chung MT, et al. Diabetes impairs the angiogenic potential of adipose derived stem cells by selectively depleting cellular subpopulations. *Stem Cell Res Ther* 2014;**5**:79.

54. Northcott PA, Jones DT, Kool M, Robinson GW, Gilbertson RJ, Cho YJ, et al. Medulloblastomics: the end of the beginning. *Nat Rev Cancer* 2012;**12**:818–34.

55. Emens LA. Trastuzumab: targeted therapy for the management of HER-2/neu-overexpressing metastatic breast cancer. *Am J Ther* 2005;**12**:243–53.

56. Lièvre A, Bachet JB, Le Corre D, Boige V, Landi B, Emile JF, et al. KRAS mutation status is predictive of response to cetuximab therapy in colorectal cancer. *Cancer Res* 2006;**66**:3992–5.

57. Fong PC, Boss DS, Yap TA, Tutt A, Wu P, Mergui-Roelvink M, et al. Inhibition of poly(ADP-ribose) polymerase in tumors from BRCA mutation carriers. *N Engl J Med* 2009;**361**:123–34.

58. Brem H, Stojadinovic O, Diegelmann RF, Entero H, Lee B, Pastar I, et al. Molecular markers in patients with chronic wounds to guide surgical debridement. *Mol Med* 2007;**13**:30–9.

59. Pastar I, Ramirez H, Stojadinovic O, Brem H, Kirsner RS, Tomic-Canic M. Micro-RNAs: new regulators of wound healing. *Surg Technol Int* 2011;**21**:51–60.

60. Zamboni P, Gemmati D. Clinical implications of gene polymorphisms in venous leg ulcer: a model in tissue injury and reparative process. *Thromb Haemost* 2007;**98**:131–7.

61. Park LK, Maione AG, Smith A, Gerami-Naini B, Iyer LK, Mooney DJ, et al. Genome-wide DNA methylation analysis identifies a metabolic memory profile in patient-derived diabetic foot ulcer fibroblasts. *Epigenetics* 2014;**9**:1339–49.

62. Rafehi H, El-Osta A, Karagiannis TC. Genetic and epigenetic events in diabetic wound healing. *Int Wound J* 2011;**8**:12–21.

63. Scales BS, Huffnagle GB. The microbiome in wound repair and tissue fibrosis. *J Pathol* 2013;**229**:323–31.

64. Bertesteanu S, Triaridis S, Stankovic M, Lazar V, Chifiriuc MC, Vlad M, et al. Polymicrobial wound infections: pathophysiology and current therapeutic approaches. *Int J Pharm* 2014;**463**:119–26.

65. Zhao G, Usui ML, Lippman SI, James GA, Stewart PS, Fleckman P, et al. Biofilms and inflammation in chronic wounds. *Adv Wound Care (New Rochelle)* 2013;**2**:389–99.

66. Percival S, Hill K, Williams D, Hooper S, Thomas D, Costerton J. A review of the scientific evidence for biofilms in wounds. *Wound Repair Regen* 2012;**20**(5):647–57.

67. Anderson GG, O'Toole GA. Innate and induced resistance mechanisms of bacterial biofilms. *Curr Top Microbiol Immunol* 2008;**322**:85–105.

68. Grant SS, Hung DT. Persistent bacterial infections, antibiotic tolerance, and the oxidative stress response. *Virulence* 2013;**4**:273–83.

69. Dhall S, Do D, Garcia M, Wijesinghe DS, Brandon A, Kim J, et al. A novel model of chronic wounds: importance of redox imbalance and biofilm-forming bacteria for establishment of chronicity. *PLoS One* 2014;**9**:e109848.

70. Jones CE, Kennedy JP. Treatment options to manage wound biofilm. *Adv Wound Care (New Rochelle)* 2012;**1**:120–6.

71. Bradley BH, Cunningham M. Biofilms in chronic wounds and the potential role of negative pressure wound therapy: an integrative review. *J Wound Ostomy Continence Nurs* 2013;**40**:143–9.

72. Kim PJ, Steinberg JS. Wound care: biofilm and its impact on the latest treatment modalities for ulcerations of the diabetic foot. *Semin Vasc Surg* 2012;**25**:70–4.

73. Velázquez-Velázquez JL, Santos-Flores A, Araujo-Meléndez J, Sánchez-Sánchez R, Velasquillo C, González C, et al. Anti-biofilm and cytotoxicity activity of impregnated dressings with silver nanoparticles. *Mater Sci Eng C Mater Biol Appl* 2015;**49**:604–11.

74. Price L, Liu C, Melendez J, Frankel Y, Engelthaler D, Aziz M, et al. Community analysis of chronic wound bacteria using 16S rRNA gene-based pyrosequencing: impact of diabetes and antibiotics on chronic wound microbiota. *PLoS One* 2009;**4**(7):e6462.

75. Ammons MC, Tripet BP, Carlson RP, Kirker KR, Gross MA, Stanisich JJ, et al. Quantitative NMR metabolite profiling of methicillin-resistant and methicillin-susceptible *Staphylococcus aureus* discriminates between biofilm and planktonic phenotypes. *J Proteome Res* 2014;**13**:2973–85.

76. Wiegand C, Schönfelder U, Abel M, Ruth P, Kaatz M, Hipler UC. Protease and pro-inflammatory cytokine concentrations are elevated in chronic compared to acute wounds and can be modulated by collagen type I in vitro. *Arch Dermatol Res* 2010;**302**:419–28.

77. Edsberg LE, Wyffels JT, Brogan MS, Fries KM. Analysis of the proteomic profile of chronic pressure ulcers. *Wound Repair Regen* 2012;**20**:378–401.

78. Ramsay S, Cowan L, Davidson J, Nanney L, Schultz G. Wound samples: moving towards a standardised method of collection and analysis. *Int Wound J* 2015 Jan 11. http://dx.doi.org/10.1111/iwj.12399 (epub ahead of print).

79. Broadbent J, Walsh T, Upton Z. Proteomics in chronic wound research: potentials in healing and health. *Proteomics Clin Appl* 2010;**4**(2):204–14.

80. Fernandez ML, Broadbent JA, Shooter GK, Malda J, Upton Z. Development of an enhanced proteomic method to detect prognostic and diagnostic markers of healing in chronic wound fluid. *Br J Dermatol* 2008;**158**(2):281–90.

81. Gibson DJ, Schultz GS. Molecular wound assessments: matrix metalloproteinases. *Adv Wound Care (New Rochelle)* 2013;**2**:18–23.

82. Gill SE, Parks WC. Metalloproteinases and their inhibitors: regulators of wound healing. *Int J Biochem Cell Biol* 2008;**40**:1334–47.

83. Chen P, Parks WC. Role of matrix metalloproteinases in epithelial migration. *J Cell Biochem* 2009;**108**:1233–43.

84. Veves A, Sheehan P, Pham HT. A randomized, controlled trial of Promogran (a collagen/oxidized regenerated cellulose dressing) vs standard treatment in the management of diabetic foot ulcers. *Arch Surg* 2002;**137**:822–7.

85. Serena TE. Development of a novel technique to collect proteases from chronic wounds. *Adv Wound Care (New Rochelle)* 2014;**3**:729–32.

86. Howangyin KY, Silvestre JS. Diabetes mellitus and ischemic diseases: molecular mechanisms of vascular repair dysfunction. *Arterioscler Thromb Vasc Biol* 2014;**34**:1126–35.

87. Arsenault KA, McDonald J, Devereaux PJ, Thorlund K, Tittley JG, Whitlock RP. The use of transcutaneous oximetry to predict complications of chronic wound healing: a systematic review and meta–analysis. *Wound Repair Regen* 2011;**19**(6):657–63.

88. McColl D, Cartlidge B, Connolly P. Real-time monitoring of moisture levels in wound dressings in vitro: an experimental study. *Int J Surg* 2007;**5**(5):316–22.

89. Ochoa M, Rahimi R, Ziaie B. Flexible sensors for chronic wound management. *IEEE Rev Biomed Eng* 2014;**7**:73–86.

90. Milne SD, Connolly P, Al Hamad H, Seoudi I. Development of wearable sensors for tailored patient wound care. *Conf Proc IEEE Eng Med Biol Soc* 2014;**2014**:618–21.

91. Mehmood N, Hariz A, Templeton S, Voelcker N. Calibration of sensors for reliable radio telemetry in a prototype flexible wound monitoring device. *Sens Bio-Sens Res* 2014;**2**:23–30.

92. Ausländer S, Fussenegger M. From gene switches to mammalian designer cells: present and future prospects. *Trends Biotechnol* 2013;**31**:155–68.

93. Karlsson M, Weber W. Therapeutic synthetic gene networks. *Curr Opin Biotechnol* 2012;**23**:703–11.

94. Pardee K, Green AA, Ferrante T, Cameron DE, DaleyKeyser A, Yin P, et al. Paper-based synthetic gene networks. *Cell* 2014;**159**:940–54.

95. Green AA, Silver PA, Collins JJ, Yin P. Toehold switches: de-novo-designed regulators of gene expression. *Cell* 2014;**159**:925–39.

96. Qi L, Lucks JB, Liu CC, Mutalik VK, Arkin AP. Engineering naturally occurring trans-acting non-coding RNAs to sense molecular signals. *Nucleic Acids Res* 2012;**40**:5775–86.

97. Lanahan AA, Lech D, Dubrac A, Zhang J, Zhuang ZW, Eichmann A, et al. PTP1b is a physiologic regulator of vascular endothelial growth factor signaling in endothelial cells. *Circulation* 2014;**130**:902–9.

98. Kutcher ME, Kolyada AY, Surks HK, Herman IM. Pericyte Rho GTPase mediates both pericyte contractile phenotype and capillary endothelial growth state. *Am J Pathol* 2007;**171**:693–701.

99. Lee S, Zeiger A, Maloney JM, Kotecki M, Van Vliet KJ, Herman IM. Pericyte actomyosin-mediated contraction at the cell-material interface can modulate the microvascular niche. *J Phys Condens Matter* 2010;**22**:194115.

100. Kotecki M, Zeiger AS, Van Vliet KJ, Herman IM. Calpain- and talin-dependent control of microvascular pericyte contractility and cellular stiffness. *Microvasc Res* 2010;**80**:339–48.

101. Durham JT, Surks HK, Dulmovits BM, Herman IM. Pericyte contractility controls endothelial cell cycle progression and sprouting: insights into angiogenic switch mechanics. *Am J Physiol Cell Physiol* 2014;**307**:C878–92.

102. Yang F, Tu J, Pan J-Q, Luo H-L, Liu Y-H, Wan J, et al. Light-controlled inhibition of malignant glioma by opsin gene transfer. *Cell Death Dis* 2013;**4**:e893.

Chapter 18

Cell Therapy for Cardiac Regeneration

Gabriela M. Kuster, Giacomo Della Verde, Ronglih Liao, Otmar Pfister

Key Concepts

1. The heart is capable of very limited cell replacement through cardiomyocyte division and differentiation of heart-resident progenitor cells, but these endogenous regenerative resources are insufficient to compensate for major cell loss, which is at the core of heart failure. Novel therapeutic approaches including cell transplantation and cellular reprogramming aiming at the regeneration of lost cardiac tissue are currently in preclinical and early clinical testing.
2. A variety of extracardiac and heart-resident stem and progenitor cells have been characterized, all of them having their inherent strengths and limitations. These cells are transplanted through either intramyocardial or intracoronary application.
3. Bone marrow-derived cells act mainly through paracrine mechanisms. Despite promising preclinical and early clinical results, more rigorously designed clinical trials employing surrogate endpoints failed to prove clear efficacy, and a larger phase 3 mortality trial is currently ongoing.
4. Heart-resident cardiac stem and progenitor cells have cardiomyogenic differentiation potential, and preclinical and first small clinical trials yielded encouraging results. However, their availability is limited and amplification time-consuming, preventing their use in acute cardiac injury (acute myocardial infarction). A first study testing allogeneic cardiosphere-derived cells in humans is currently ongoing.
5. Embryonic stem cells (ESCs) and inducible pluripotent stem cells (iPS) have high plasticity, however, the suitability of cardiogenic cells derived from ESCs and iPS may be limited by yet unsolved safety issues including immunogenicity, tumorigenicity, and genetic instability.

List of Abbreviations

BCRP Breast cancer resistance protein
BMC Bone marrow-derived cell
BMMNC Bone marrow mononuclear cell
BMSC Bone marrow stem cells
CCTRN Cardiovascular cell therapy research network
CD Cluster of differentiation
c-kit Receptor for SCF
CSC Cardiac stem cell
CXCR4 Chemokine receptor 4
DNA Deoxyribonucleic acid
EPC Endothelial progenitor cell
ESC Embryonic stem cell
G-CSF Granulocyte-colony stimulating factor
HSC Hematopoietic stem cell
iPS Induced pluripotent stem cell
Isl-1 Islet 1
KLF4 Kruppel like factor 4
MDR Multidrug resistance protein
MSC Mesenchymal stem cell
Oct-4 Octamer binding transcription factor 4
RNA Ribonucleic acid
Sca-1 Stem cell antigen 1
SCF Stem cell factor

Translating Regenerative Medicine to the Clinic. http://dx.doi.org/10.1016/B978-0-12-800548-4.00018-8

SDF1 Stromal cell-derived factor 1
SP Side population
Tbx5 T-box 5
VEGF Vascular endothelial growth factor

1. INTRODUCTION

Heart failure is a severe condition associated with high morbidity and mortality. As it can occur as the final common sequel of all cardiac diseases, it represents a leading cause of death and hospitalization.[1] In the Western world, an estimated 1–2% of the population suffers from heart failure and its prevalence exceeds 10% in people over 70 years of age.[2] According to the Heart Failure Association of the European Society of Cardiology in Europe alone, up to 14 million people are affected and the numbers keep rising.[3] Paradoxically, not only aging of the population, but also substantial advances in treatment of major cardiovascular diseases, such as coronary artery disease, contribute to the increase in the prevalence of heart failure. Especially, catheter-based techniques have markedly improved patient survival and tissue preservation after acute myocardial infarction, but as they still fail to completely restore the contractile competence of the severely injured heart, survivors may suffer from cardiac dysfunction and heart failure.

The magnitude of the contractile deficit of the injured heart directly correlates with the net loss of functional cardiac cells, in particular cardiomyocytes. In contrast to other muscle tissues, such as smooth muscle and skeletal muscle, the regenerative capacities of the adult myocardium are extremely limited and in itself insufficient to compensate for the cell loss arising from acute or chronic cardiac injury. Even state-of-the-art therapy of heart failure consisting of pharmacological neurohormonal blockade and devices, which aim at the stabilization of cardiovascular hemodynamics and the optimization of the myocardial remodeling process, fails to recover the loss of functional cardiomyocytes. Therefore, the quest for novel strategies to induce therapeutic regeneration of the injured heart is ongoing.

2. CONCEPTS AND STRATEGIES OF CARDIAC REGENERATION

Albeit limited, understanding of the cellular and molecular mechanisms responsible for intrinsic cardiac regeneration is seminal for the development of suitable strategies for therapeutic regeneration. The first compelling evidence of intrinsic cardiac cell renewal came from isotopic studies[4] and overturned a long-lasting paradigm of the heart as a strictly postmitotic organ. Bergmann and colleagues took advantage of the global exposure to ^{14}C released into the atmosphere during aboveground nuclear testing in the years of the Cold War, which was incorporated into the DNA and allowed determining the age of each cell in the body.[4] Although the overall amount of newly formed cardiomyocytes during adult life turned out to be limited and the precise origin of these cells remained unknown, their data confirmed cell turnover in the adult human heart as previously suggested by the discoveries of dividing cardiomyocytes after myocardial infarction and of cell-chimerism in transplanted hearts.[5,6] These earlier observations, together with subsequent animal studies, proposed two distinct cellular mechanisms of new cardiomyocyte formation: proliferation of preexisting cardiomyocytes[7–9] and myogenic differentiation of progenitor and immature precursor cells.[10,11]

The prospect of true myocardial regeneration has generated great enthusiasm in the cardiovascular field and ignited the search for strategies to therapeutically improve the heart's limited potential to replace lost cardiomyocytes. Various approaches including the stimulation of preexisting cardiomyocytes to reenter cell cycle and to proliferate, the activation or transplantation of endogenous or exogenous stem or progenitor cells with cardiomyogenic and angiogenic potential, and the in situ reprogramming of nonmyocytes, in particular fibroblasts, are currently under intense investigation. Whereas all these approaches have their own inherent limitations and intricacies, this chapter focuses on the use of stem and progenitor cells for cardiac regeneration discussing recent advances and backlashes as well as future challenges.

3. THE SEARCH FOR THE IDEAL CELL: EXTRACARDIAC SOURCES

During the past decade, multiple candidate cells, such as multipotent cells from extracardiac and—more recently—cardiac sources (see below) as well as pluripotent cells (embryonic stem cells (ESCs) and induced pluripotent stem cells (iPS)), have been proposed for cardiac regeneration. An overview of the various cell types is given in Figure 1. Cells from extracardiac sources mostly include endothelial progenitor cells (EPCs) and mesenchymal (MSCs) and hematopoietic stem cells (HSCs) originating from the peripheral blood, the adipose tissue, or the bone marrow, which, besides bone marrow mononuclear cells (BMMNCs) and skeletal myoblasts, all have been applied in clinical trials. More recent clinical trials using heart-resident progenitor cells such as c-kit$^+$ cells and cardiospheres (see below) are currently ongoing.[12,13]

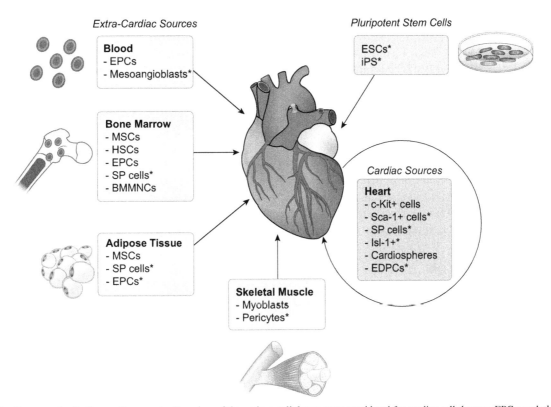

FIGURE 1 Sources of cells for cardiac repair. Overview of the various cellular sources considered for cardiac cell therapy. EPCs: endothelial progenitor cells; MSCs: mesenchymal stem cells; HSCs: hematopoietic stem cells; SP: side population; BMMNCs: bone marrow mononuclear cells; ESC: embryonic stem cells; iPS: induced pluripotent stem cells; c-kit: stem cell factor receptor; Sca-1: stem cell antigen 1; Isl-1: Islet-1; EDPCs: epicardial-derived progenitor cells. *not used in clinical trials.

3.1 The Bone Marrow as Reservoir of Somatic Stem Cells

One of the biggest and most extensively studied sources of extracardiac cells is the bone marrow. Together with the peripheral blood, the bone marrow represents an important reservoir of extracardiac cells that can be recruited and home to the injured myocardium. In particular, the ischemically injured heart produces a variety of cytokines and growth factors, which lead to the mobilization of endogenous bone marrow-derived cells (BMCs) and guide them to the site of injury.[14–18] Most of these factors are released in connection with the inflammatory response associated with myocardial ischemia and include granulocyte-colony stimulating factor, stem cell factor (SCF), vascular endothelial growth factor (VEGF), stromal cell-derived factor 1(SDF-1), and erythropoietin.[19] The inability to properly mobilize and home BMCs to the heart following myocardial infarction results in early cardiac failure and death, as shown in a proof-of-concept study using transgenic mice expressing a mutant of the SCF receptor c-kit, which interfered with cell mobilization.[16] Notably, restoration of the mobilization and homing capacities by transplantation of wild-type BMCs into these mice rescued the cardiomyopathy. Similarly, SDF-1 and its receptor CXCR4 are important for the mobilization of BMCs after myocardial infarction.[20] Together, these translational studies reveal an important functional interaction between the bone marrow and the heart in the setting of cardiac injury. Harvesting and transplanting BMCs to bypass the complicated processes of mobilization and homing and to increase the number of these apparently beneficial cells in the injured heart may therefore offer a therapeutic means to induce cardiac regeneration.

The possibility to functionally reconstitute a radio- or chemotherapy-ablated hematopoietic system through allogeneic or autologous BMCs has revolutionized the treatment of hemato-oncologic diseases and established bone marrow stem cells (BMSCs) as the epitome of adult stem cells. Early reports that these BMSCs may also transdifferentiate into nonhematologic cells such as skeletal muscle, hepatocytes, neurons, endothelial cells, and even cardiomyocytes further encouraged their use for cardiac regeneration.[21–23] In preclinical studies using myocardial infarction models of small and large animals, BMSCs expressing the stem cell receptor c-kit were isolated and intramyocardially injected into the infarct border zone.[24,25] This BMSC-based therapy resulted in impressive regeneration of lost myocardium, the formation of new capillaries, and improved cardiac function. Although later reports were unable to confirm true transdifferentiation of BMSCs into cardiomyocytes and suggested cellular fusion as underlying mechanism for the acquisition of a cardiomyogenic phenotype of BMSCs,[26] most animal studies consistently showed preservation of myocardial tissue and improvement of cardiac function after BMSC transplantation.[24,27] The realization that BMSCs may improve cardiac function after myocardial infarction in the absence of meaningful transdifferentiation into new cardiomyocytes gave rise to the *paracrine theory* of cardiac protection and regeneration. BMSCs and other adult stem cells possess profound paracrine activity, which includes the secretion of cardioprotective cytokines that inhibit apoptosis, of angiogenic factors (e.g., VEGF, basic fibroblast growth factor) that promote neovascularization, and of factors that activate resident cardiac stem cells (CSCs) and progenitor cells (e.g., insulin-like growth factor).[28] Although still poorly understood, the beneficial effects exerted by BMSCs may be explained by a combination of the following mechanisms of action: (1) preservation of viable cardiomyocytes in the infarct border zone due to antiapoptotic effects,[28] (2) induction of angiogenesis and vasculogenesis through the secretion of angiogenic factors and new capillary formation by EPCs,[29,30] (3) stimulation of resident CSCs and progenitor cells,[31] and (4) induction of cardiomyocyte mitosis.[32] The proposed mechanisms of BMSC-mediated cardioprotection and regeneration as they may also apply to other cell types are summarized in Figure 2.

3.2 BMCs: Clinical Trials

Inspired by the promising results from preclinical animal studies, the translation of BMSC-based therapy from the research laboratory to the clinical arena unfolded in a rapid pace. The implementation of BMSCs in clinical trials was facilitated by a certain convenience about their use, including the easy access to the bone marrow, which is renewable and contains a whole mixture of autologous cells with regenerative capacity. Most clinical studies utilized the mononuclear cell fraction (BMMNCs) for cardiac transplantation, because it contains the full array of BMSCs, progenitor, and precursor cells. Namely, these include HSCs, MSCs, EPCs, and side population (SP) cells, all of which were shown to improve cardiac function if transplanted into infarcted myocardium in various animal studies.[14,24,33–35] In humans, there are three major application routes for cell transplantation into the heart: (1) the intramyocardial injection trans-epicardially during open heart surgery (e.g., coronary artery bypass graft surgery)[36] or (2) trans-endocardially using catheter-based techniques[37] or (3) the direct intracoronary injection through an inflated over-the-wire

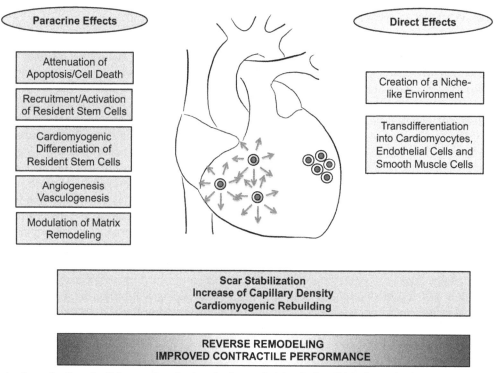

FIGURE 2 **Mechanisms of action of cell therapy.** Postulated mechanisms underlying the beneficial effects of cell therapy on myocardial remodeling and function include indirect, i.e., paracrine effects, as well as direct effects of supplied cells.

balloon catheter.[38] To date, most clinical BMSC-based trials were performed in acute myocardial infarction patients with direct intracoronary injection as the main application route.[39] After promising results from a variety of noncontrolled clinical trials,[38,40,41] larger randomized and controlled trials were initiated. However, although mostly well designed, in some cases even blinded, randomized clinical trials generated very mixed results with most of them yielding a neutral outcome.[42–47] These diverse findings were explained by the large variations in cell preparation,[48,49] the use of different cell types and cell numbers, and differences in patient selection and timing of cell administration.[50] In order to augment the mechanistic understanding of cell therapy and address major issues regarding clinical trial design, the *United States National Heart, Lung, and Blood Institute* established the *Cardiovascular Cell Therapy Research Network* (CCTRN).[51] Remarkably, none of the rigorously designed major clinical trials initiated by the CCTRN was able to confirm the beneficial effects of BMMNC transplantation reported in previous studies. The phase 2 randomized double-blind, placebo-controlled FOCUS-CCTRN trial using trans-endocardial injection of BMMNCs in chronic ischemic heart failure patients, for example, failed to show a significant benefit of such therapy on any of the prespecified endpoints.[52] Similarly, the TIME study program, which was launched to specifically clarify the open question of the ideal timing of cell therapy, produced sobering results. In the TIME study, patients were randomized to either receive intracoronary BMMNC-therapy or placebo at day 3 or day 7 after acute myocardial infarction. In the LateTIME study, patients were randomized to intracoronary BMMNC transplantation or placebo as late as 2–3 weeks postinfarct. Either trial failed to show any recovery of global or regional left ventricular function compared with placebo, regardless of the time point of cell therapy.[53,54] For an overview of the major preclinical and completed clinical trials using BMCs for myocardial regeneration, please see Tables 1 and 2.

Even today, after more than 10 years of clinical experience and more than 30 randomized controlled trials including over 2000 patients, the use of BMCs for cardiac regeneration remains highly controversial. Although this therapy has proven safe, its clinical efficacy appears rather limited as suggested by one of the latest meta-analyses.[39] There is no reduction of mortality, reinfarction, or rehospitalization, but it has to be noted that the overall event rate was low in these studies. At best, there may be a risk reduction for future revascularization and a modest improvement of left ventricular ejection fraction in the range of 3%.[39] In addition, a disconcerting more recent meta-analysis uncovered a high amount of factual discrepancies in reported BMC-based trials, with a high degree of discrepancy being associated with more favorable study outcome.[55] BMMNCs may indeed exhibit some inherent biological limitations that challenge their usefulness for successful regenerative therapy of the myocardium. Among these limitations are: (1) the poor retention and survival of transplanted cells in the injured myocardium, regardless of the route of delivery,[56,57] (2) the lack of meaningful cardiomyogenic differentiation,[26] and (3) the decline in paracrine function associated with age and cardiovascular risk factors.[58] As a consequence, new approaches to improve cell survival and preserve paracrine function are currently explored.[59,60] Still, trials using hard clinical endpoints will be needed to set the controversy at rest and provide a more definite answer as to the clinical effectiveness of BMCs. The BAMI trial, a phase 3 mortality trial investigating the effect of intracoronary injection of BMMNCs on all-cause mortality in acute myocardial infarction, is currently in its recruiting phase with the goal of enrolling 3000 patients and scheduled completion in 2018.[61] In view of the unresolved issues, results from this study will be awaited with great eagerness.

3.3 Bone Marrow-Derived MSCs

MSCs were originally identified by Friedenstein. Isolated from the bone marrow, these then so-called stromal cells adhered to plastic culture dishes, were clonogenic and capable to differentiate into mesoderm-derived tissues.[62] Due to their stem-like properties, they were later commonly referred to as MSCs. MSCs can be isolated from virtually all tissues[63] and are readily expandable in vitro. In the bone marrow, they are essential components of the microenvironment necessary for the proliferation and differentiation of HSCs. MSCs represent approximately 0.001–0.01% of bone marrow nucleated cells[64] and are a heterogeneous population of multipotent progenitor cells exhibiting various degrees of mesodermal lineage commitment.[65] Bone marrow-derived MSCs can differentiate into cardiomyogenic cells if properly stimulated in vitro[64,66,67] and cardiac engraftment and cardiomyogenic and angiogenic differentiation has been observed in large animal models of acute and chronic ischemic cardiomyopathy.[68–72] Preclinical studies using autologous or allogeneic MSCs to induce cardiac regeneration consistently reported improvement of left ventricular remodeling and ejection fraction, and reduction of scar size.[73–75] However, reminiscent of what has been observed with BMSCs, these beneficial effects occurred despite lack of long-term MSC engraftment and limited transdifferentiation of engrafted cells.[76,77] Therefore, similar as for BMSCs, the unique secretome of MSCs that includes cytokines and growth factors, which inhibit fibrosis and cell death and stimulate angiogenesis and differentiation of tissue-specific stem cells, is probably more important for the observed beneficial effects on cardiac remodeling and function than transdifferentiation of MSCs into functional cardiac cells.[78–80]

TABLE 1 Summary of Selected Proof of Concept Preclinical Studies

Cell Type	Animal	Infarct Model	Treated/Control, n	Delivery Route	Imaging Method	Increase in LVEF versus Control, Absolute %	Decrease in Scar Size versus Control, %	References
BMCs	Pig	Occlusion and reperfusion	20/10	Intracoronary	CMR	NS	3	149
BMCs	Pig	Occlusion and reperfusion	7/7	Intracoronary	CMR	NS	NS	150
MSCs	Pig	Occlusion and reperfusion	9/6	Intramyocardial	CMR	20	3.8	151
MSC/CSC mix	Pig	Occlusion and reperfusion	5/5	Intramyocardial	CMR	20	21	152
CSCs	Rat	Occlusion and reperfusion	24/17	Intracoronary	ECHO	10	29	153
CSCs	Pig	Occlusion and reperfusion	11/10	Intracoronary	ECHO	9	ND	154
CSCs (cardiospheres)	Pig	Occlusion and reperfusion	7/7	Intracoronary	CMR	NS	5	155
CSCs (cardiospheres)	Pig	Occlusion and reperfusion	18/10	Intramyocardial	ECHO	6	ND	156
iPS	Mice	Permanent occlusion	6/6	Intramyocardial	ECHO	13	ND	157
iPS	Mice	Permanent occlusion	25/13	Intramyocardial	CMR	5–10	25	158
iPS	Pig	Occlusion and reperfusion	9/9	Intramyocardial	SPECT	5.5	4	159
ESCs	Mice	Permanent occlusion	40/20	Intramyocardial	ECHO	10–15[a]	10–20	160

BMCs, bone marrow cells; MSCs, mesenchymal stem cells; CSCs, cardiac stem cells; iPS, induced pluripotent stem cells; ESCs, embryonic stem cells; CMR, cardiac magnetic resonance tomography; ECHO, echocardiography; SPECT, single-photon emission computed tomography; LVEF, left ventricular ejection fraction; ND, not determined; NS, nonsignificant.
[a]Percentage increase of fractional shortening.
Reprinted with minor adaptations from Pfister O, Della Verde G, Liao R, Kuster GM. Regenerative therapy for cardiovascular disease. *Transl Res* 2014;163(4), with permission from Elsevier.

TABLE 2 Summary of Selected Major Cell-Based Clinical Trials

Trial	Cell Type	Delivery Route	Treated/Control, n	Days after MI	Imaging Method	Change in LVEF versus Control, Absolute %	Decrease in Scar Size versus Control, %	References
Bone Marrow-Derived Cells								
BOOST	BMCs	Intracoronary	30/30	6	CMR	NS	ND	161
Janssens	BMCs	Intracoronary	33/34	1	CMR	NS	28	162
ASTAMI	BMCs	Intracoronary	47/50	6	SPECT CMR	NS	NS	163
REPAIR-AMI	BMCs	Intracoronary	101/103	3–7	LVA	2.5	ND	164
BONAMI	BMCs	Intracoronary	52/49	9	RNA CMR	NS	NS	165
HEBE	BMCs	Intracoronary	135/65	3–8	CMR	NS	NS	166
FOCUS-CCTRN	BMCs	Endomyocardial	61/31	ICMP	SPECT ECHO	NS	NS	167
TIME	BMCs	Intracoronary	78/41	3–7	CMR	NS	NS	168
LateTIME	BMCs	Intracoronary	58/29	14–21	CMR	NS	NS	169
Meta-analysis	BMCs	Intracoronary	1830			2.7	ND	170
Mesenchymal Stem Cells and Resident Cardiac Stem Cells								
C-CURE	Cardiopoietic MSCs	Endomyocardial	32/15	ICMP	ECHO	7	ND	171
POSEIDON	MSCs	Endomyocardial	31/no Control	ICMP	CCT	NS	33	172
SCIPIO	CSCs	Intracoronary	16/7	ICMP	ECHO CMR	8	24–30	173
CADUCEUS	CSCs (cardiospheres)	Intracoronary	17/8	45–90	CMR	NS	9	174

BMCs, bone marrow cells; MSCs, mesenchymal stem cells; CSCs, cardiac stem cells; CMR, cardiac magnetic resonance tomography; ICMF, ischemic cardiomyopathy; ECHO, echocardiography; SPECT, single-photon emission computed tomography; CCT, contrast-enhanced computer tomography; LVA, left ventricular angiogram; LVEF, left ventricular ejection fraction; RNA, radionuclide angiogram; ND, not determined; NS, nonsignificant.
Reprinted with minor adaptations from Pfister O, Della Verde G, Liao R, Kuster GM. Regenerative therapy for cardiovascular disease. *Transl. Res* 2014;163(4), with permission from Elsevier.

3.4 Clinical Trials Using MSCs

Given the heterogeneity of MSCs, differences in their isolation, expansion, and phenotypic characterization jeopardize the interpretability and comparability of study outcomes. The *Mesenchymal and Tissue Stem Cell Committee* of the *International Society for Cellular Therapy* addressed this problem by issuing the following criteria for a more standardized definition of human MSCs: (1) adherence to plastic culture dishes, (2) expression (≥95% positivity) of the surface markers CD73, CD90, and CD105 in the absence (≤2% positivity) of CD34, CD45, HLA-DR, CD14 or CD11b, CD79α or CD19 as assessed by fluorescence-activated cell sorter analysis, and (3) in vitro capacity to differentiate into mesoderm-derived tissues including chondroblasts, osteoblasts, and adipocytes.[81] Due to their unique immunophenotype with only moderate expression of major histocompatibility complex class I and lack of major histocompatibility complex class II as well as to their own immunomodulatory properties, human MSCs are immunoprivileged and therefore suited for allogeneic transplantation.[82] Clinical phase 1/2 trials have already been performed to investigate the safety and efficacy of intravenous, intracoronary, or endomyocardial injection of autologous or allogeneic MSCs in acute myocardial infarction and ischemic cardiomyopathy.[83–85] These early clinical trials documented the safety of MSC transplantation and reported reverse remodeling and improved regional contractility of the infarcted area.[85] Because of the limited cardiomyogenic differentiation potential of bone marrow-derived MSCs, recent approaches aimed at directing MSCs toward the cardiac lineage by exposing them to a cardiogenic cocktail to trigger expression and nuclear translocation of cardiac transcription factors.[86] In the multicenter randomized C-CURE (cardiopoietic stem cell therapy in heart failure) trial, this novel approach was applied to ischemic heart failure patients.[87] In the cell therapy arm, patients received primed autologous cardiopoietic MSCs by endomyocardial injection guided by NOGA® electromechanical mapping into areas of dysfunctional but viable myocardium, and outcome was compared to patients treated with standard therapy. After 6 months, patients randomized to cell therapy showed significant improvement of cardiac function and mitigated adverse ventricular remodeling compared to standard care.[87] Moreover, cardiopoietic MSC therapy was associated with better exercise tolerance and quality of life.[87] These promising results support the concept of lineage specification in stem cell therapy and provide the basis for larger, randomized controlled clinical trials to test this novel therapeutic approach.

Safety and feasibility of allogeneic MSC transplantation was recently demonstrated in the POSEIDON trial. In this small trial, 30 patients with ischemic cardiomyopathy were randomized to transendocardial injection of different amounts of either autologous or allogeneic MSCs.[88] Pooled analysis showed that MSC injection favorably affected patient functional capacity, quality of life, and ventricular remodeling. Importantly, allogeneic MSC transplantation was safe and not associated with adverse immune responses, hence opening the door for larger clinical trials using "off-the-shelf" allogeneic MSCs. Major preclinical and clinical trials using MSCs for myocardial regeneration are summarized in Tables 1 and 2, respectively.

4. HEART-RESIDENT STEM AND PROGENITOR CELLS

4.1 Endogenous CSCs and Progenitor Cells as Source of Cardiac Cells

In 2002, Hierlihy et al. were the first to describe a stem cell-like population in the postnatal heart.[89] Since then, various populations of CSCs and progenitor cells have been identified based on the expression pattern of specific surface markers or functional properties and some of them have been thoroughly characterized regarding their ability to differentiate into cardiac lineages. These populations include SP cells, cells expressing the SCF receptor c-kit, stem cell antigen-1 (Sca-1), or the transcription factor Islet-1 (Isl-1), as well as cardiospheres.[90–93] It is of note, however, that the true identity of Sca-1+ cells and the existence of a human Sca-1 homolog,[94] as well as the expression of Isl-1 in the adult working myocardium,[95] are still a matter of debate.

Recent work by Malliaras and colleagues showed that such endogenous CSCs and progenitor cells markedly contribute to new cardiomyocyte formation in response to injury.[96] However, although capable to differentiate into functional cardiac cells and to contribute to the myocardial cardiomyocyte mass, the limited amount of heart-resident stem and progenitor cells is overwhelmed in the setting of cardiac injury and their capacities are insufficient to meet major tissue loss. Therefore, harvesting and expansion of such cells represents an appealing alternative to the use of BMCs for cell therapy of the injured heart. Whereas identity and characteristics of some specific cell types are discussed in more detail below, an overview of reported CSCs and progenitor cells is given in Table 3.

4.2 Cardiac SP Cells

SP cells were first identified in 1996 in the bone marrow by Goodell and colleagues.[97] Isolated based on their functional ability to efflux the DNA binding dye Hoechst 33342, bone marrow SP cells were enriched in HSCs and contained the bulk of long-term repopulating cells. Since then, SP cells were isolated from a variety of solid organs including pancreas, pituitary, testis, mammary gland, lung, liver, skeletal muscle, and eventually the heart and shown to exhibit organ-specific progenitor cell properties.[98,99] They were also the first stem-like cell type described in the heart.[89] SP cells were named after their appearance in the fluorescence-activated cell sorting scatter plot, where they can be found in the Hoechst Blue-low

TABLE 3 Overview of Reported Cardiac Stem/Progenitor Cells

	Surface Markers	Isolation Method	In Vitro Differentiation	In Vivo Differentiation	Species	Developmental Stage	Impact on Infarction	References
SP+	Sca-1+; CD34+; CD31−a; CD45−;	FACS	CMCs, ECs, SMCs	CMCs, ECs, SMCs	Mice, pig, rat, human	Embryonic, neonatal, adult	Not determined	175–181
c-kit+	Sca-1+; Lin−; CD45−; CD31−; CD34−	FACS, MACS	CMCs, ECs, SMCs	CMCs, ECs, SMCs	Mice, pig, rat, dog, human	Embryonic, neonatal, adult	↑LVEF, ↑REF, ↓LVNM, ↑LVVM	173,182–185
Sca-1+	CD34−; Lin−; CD45−; c-kit−; SP	FACS, MACS	CMCs, ECs	CMCs, ECs	Mice, dog, human	Neonatal, adult	↑LVEF; ↓LVV	181,183, 186–189
Isl-1+	Lin−; CD45−; c-kit−; CD34−; CD31−; Sca-1−	FACS	CMCs, ECs, SMCs	Not determined	Mice, rat, human	Embryonic, neonatal, adult	Not determined	190,191
SSEA-1+	SSEA+; Oct3/4+; c-kit−; Sca-1−b	FACS	CMCs	CMCs, ECs	Rat, sheep	Embryonic, adult	↑LVEF	192,193
Cardiospheres	c-kit+; Sca-1+; CD34+; CD31+	Enzymatic digestion; culturing; handling	CMCs, ECs, SMCs	CMCs, ECs, SMCs	Mice, rat, pig, human	Neonatal, adult	↓Scar mass; ↑LVVM	155,194–196
EDPCs	c-kit+; CD34+; CD45−	Culturing; handling	CMCsc ECs, SMCs, FBs	CMCsc, ECs, SMCs, FBs	Mice, human	Fetal, adult	Not determined	197,198

SP, side population; Sca-1, stem cell antigen 1; c-kit, stem cell factor receptor; Isl-1, islet-1; SSEA-1, stage-specific embryonic antigen 1; EDPCs, epicardial-derived progenitor cells; FACS, fluorescence-activated cell sorting; MACS, magnetic-activated cell sorting; CMCs, cardiomyocytes; ECs, endothelial cells; SMCs, smooth vascular muscle cells; FBs, fibroblasts; LVEF, left ventricular ejection fraction; REF, regional ejection fraction; LVNM, left ventricular nonviable mass; LVVM, left ventricular viable mass; LVV, left ventricular volume.
aMartin et al. found no expression of CD31, whereas Pfister et al. reported two sub-populations of CD31+ and CD31− cells among cardiac SP. These differences may be due to differences in isolation procedures.
bInitially negative for c-kit and Sca-1.
cPossible differentiation into CMCs (expression of Nkx2.5, GATA4).
Reprinted with minor adaptations from Pfister O, Della Verde G, Liao R, Kuster GM. Regenerative therapy for cardiovascular disease. Transl Res 2014;163(4), with permission from Elsevier.

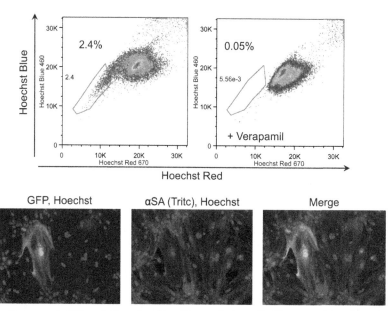

FIGURE 3 **Isolation and cardiomyogenic differentiation of cardiac side population cells.** *Upper panel, left:* Representative fluorescence-activated cell sorting plot of cardiac side population (SP) cells, which appear in the Hoechst Red-low and Hoechst Blue-low left corner, aside from the main population of cardiac mononucleated cells. *Right:* The Hoechst efflux is blocked by verapamil, prompting the SP to disappear. *Lower panel:* Cardiac SP cells were isolated from enhanced green fluorescence protein expressing mice and cocultured with rat cardiomyocytes for 3 weeks. The image series depicts a GFP+ former SP cell, which underwent cardiomyogenic differentiation and is now expressing α-sarcomeric actinin similar to the surrounding cardiomyocytes. GFP: green fluorescence protein; αSA: alpha-sarcomeric actinin; Tritc: tetramethylrhodamine.

and Hoechst Red-low corner, aside of the main population (Figure 3). Their ability to extrude Hoechst is based on enriched expression of ATP-binding cassette (ABC) transporters, such as ABCG2 (also known as breast cancer resistance protein 1, BCRP1) and ABCB1 (also known as multidrug resistance protein 1, MDR1), which allow for efficient efflux of cytotoxins.[100] In contrast to bone marrow SP cells, SP cells isolated from the myocardium do not express hematopoietic markers such as CD45 and CD34, but instead, expression of cardiac-specific transcription factors including Nkx2.5, Gata4, and Mef2 can be found.[101] Overall numbers of SP cells isolated from the heart are low and vary between 0.03 and 3.5% of the mononucleated cardiac cells.[102] Interestingly, a gradual decrease from neonatal (9.4 ± 1.2%) to adult age (0.8 ± 0.1%) can be observed in mice.[100] Murine SP cells can be propagated in vitro and stored frozen for further use.[100] Cardiac SP cells are capable of structural and functional cardiomyogenic differentiation including electromechanical coupling when cocultured with cardiomyocytes in vitro.[101] In vivo, SP cells are activated in response to myocardial infarction and proliferate within the infarct border zone.[15] In addition, upon intravenous injection, cardiac SP cells are home to the injured myocardium and differentiate into cardiomyocytes as demonstrated in a myocardial cryoinjury rat model.[103] Since the first description by Hierlihy in 2002, various independent groups consistently reported differentiation of cardiac SP cells into cardiomyocytes and endothelial cells in vitro and in vivo, hence establishing this unique cell population as a potential source of cardiac progenitor cells and suggesting that they may take part in the enigma of intrinsic cardiac regeneration.[101,103–106]

SP cells have also been isolated from the left atrium of human heart using similar techniques.[107] In a more indirect way, Emmert and colleagues recently identified cells positive for BCRP (ABCG2) but negative for the endothelial marker CD31 as potential cardiac progenitor cells in the human heart. They were mostly located in the atria and found enriched after myocardial ischemia.[108] Because SP cells from human myocardium still await more thorough characterization, they might not as yet be ready for clinical use.

4.3 c-Kit-Positive CSCs

In 2003, Beltrami and colleagues described a population of primitive c-kit+ cells in the heart that were clonogenic and capable of self-renewal and differentiation into all major cardiac lineages, including cardiomyocytes, endothelial cells, and vascular smooth muscle cells, hence fulfilling criteria for "stemness."[91] This seminal work proposed for the first time the existence of true CSCs in the adult myocardium. Subsequently, this concept was taken a step further by another group of investigators, who—based on different mouse and rat models of diffuse myocardial injury—suggested that these c-kit+ CSCs are both necessary and sufficient for cardiac regeneration based on the generation of new cardiomyocytes.[109] However, their study drew criticism, in particular as to the validity of the transgenic mouse model used for tracking of the c-kit$^+$ CSC fate in vivo,[110] and was challenged by recent work using two different genetic approaches to track c-kit$^+$ cells. These investigators found only minimal formation of new cardiomyocytes ($\leq$0.03%), but ample generation of new cardiac endothelial cells originating from c-kit$^+$ progenitor cells.[111] It is of note, however, that even these data, in principle, support the initial observation by Beltrami et al. regarding the general ability of c-kit$^+$ cells to generate cardiomyocytes in vivo. Regardless of the current controversy ensuing around the degree of the cardiomyogenic potential of c-kit$^+$ CSCs, several preclinical studies showed structural and functional improvement of the acutely or chronically injured myocardium after transplantation of rat or human c-kit$^+$ CSCs.[112–115]

4.4 Cardiospheres

First derived from mouse and human heart biopsies by Messina and colleagues in 2004,[92] cardiospheres represent microtissue rather than a single specific cell type. Harboring different cell types surrounded by extracellular matrix, they are grown from cultured tissue explants after mild enzymatic digestion. In culture, cardiospheres form multicellular clusters with more differentiated cells located in the outer layers and more primitive cells at their core. These more primitive cells express stem cell markers such as CD34, c-kit and Sca-1, and endothelial markers including CD31 and VEGF receptor 2 (Flk-1). Cardiospheres are clonogenic, able to self-renew, and their cells can differentiate into mature, beating cardiomyocytes and form microvascular structures. Similar as for c-kit$^+$ CSCs, several preclinical studies demonstrated their efficacy to improve myocardial remodeling and function in small and large animal models of myocardial infarction and ischemic cardiomyopathy.[116–119] As the technique for their isolation has gradually been refined, they are now ready and currently in use for clinical trials in humans.

4.5 Clinical Trials Using CSCs

Because of their cardiac origin and cardiomyogenic potential, in vitro expanded resident CSCs were soon considered a promising new cell type for cellular therapy in acute and chronic ischemic heart disease, and respective preclinical studies yielded promising results. An inherent disadvantage of autologous resident CSCs, however, is the time required for their ex vivo expansion, which can take several weeks until sufficient cell numbers are at hand for clinical application. Therefore, their use in acute myocardial infarction may be limited.

Recently, results from the first clinical phase 1 trials using autologous c-kit$^+$ CSCs (SCIPIO) or cardiosphere-derived cells (CADUCEUS) in patients with ischemic cardiomyopathy have been reported. The SCIPIO (Stem Cell Infusion in Patients with Ischemic cardiOmyopathy) trial enrolled patients with reduced left ventricular ejection fraction ($\leq$40%) due to ischemic cardiomyopathy undergoing aorto-coronary bypass surgery.[12] At the time of surgery, c-kit$^+$ CSCs were isolated from the right atrial appendage, which is routinely removed during bypass surgery, and expanded in vitro over several weeks up to a total cell number of approximately 2 million cells. In patients randomized to cell therapy, 0.5–1 million cells were intracoronary delivered 3–4 months after bypass surgery. An *interim* analysis of the first 23 randomized patients of this currently ongoing study showed significantly improved symptoms, quality of life, and left ventricular ejection fraction in CSC-treated patients at 4 months.[12] Interestingly, cardiac magnetic resonance tomography revealed a significant reduction of the myocardial scar size in these patients, which was even more pronounced after 12 months, suggesting possible myocardial regeneration within formerly infarcted areas.[120]

In the CADUCEUS (CArdiosphere-Derived aUtologous stem CElls to reverse ventricUlar dySfunction) trial, autologous cardiosphere-derived cells grown from endomyocardial biopsies were intracoronarily infused into the infarct-related artery from patients with postmyocardial infarction ventricular dysfunction (left ventricular ejection fraction $\leq$45%); and 6 months and 1 year data from 25 patients have been reported.[13,121] Also in this trial, cardiac magnetic resonance tomography showed significant reduction in scar size and improvement in regional cardiac function in cell-treated patients compared to control patients. However, contrasting to SCIPIO, cell therapy did not improve global cardiac function (left ventricular ejection fraction), symptoms, or quality of life.[13] In both trials—SCIPIO and CADUCEUS—intracoronary CSC transplantation was safe and did not lead to notable arrhythmias or tumor formation during follow-up. Despite impressive scar size reduction in both studies and improvement in cardiac function and symptoms in SCIPIO, these results should be interpreted with great caution given the low patient numbers and the lack of blinded study design. Larger randomized trials employing a placebo-controlled, double-blinded design are needed to confirm these results and to pave the way for CSC-based therapy into clinical practice. Major preclinical and clinical trials using CSCs for myocardial regeneration are summarized in Tables 1 and 2.

Given the time-consuming process of ex vivo expansion of autologous CSCs for therapeutic application, safety of allogeneic CSCs has recently been explored and confirmed in preclinical studies.[122,123] Based on these data, the phase 1/2 randomized double-blind and placebo-controlled ALLSTAR (Allogeneic heart stem cells to achieve myocardial regeneration) trial has been launched to address safety and efficacy of allogeneic cardiosphere-derived cell transplantation in patients with postinfarct cardiac dysfunction.[124] Particular attention will be paid to adverse immune responses to the donor cells with myocarditis associated with respective antibody formation as primary safety endpoint. The study aims at inclusion of 274 patients and initial data should be available toward the end of 2015.

5. PLURIPOTENT STEM CELLS

To enable easier access to potential donor cells, attempts to direct pluripotent stem cells into cardiomyogenic lineage prior to intracardiac transplantation are currently undertaken. ESCs and iPS are both cell types with high potential for cardiomyogenic differentiation upon proper in vitro manipulation.[125,126]

5.1 Human ESCs

Human ESCs can be differentiated into cardiomyocytes and several compounds supporting this process have been identified such as prostaglandin I_2, 5-azacytidine, and the small molecule p38 MAP kinase-inhibitor SB203580.[127] In a sophisticated protocol, Yang et al. applied a multistep procedure exposing human embryoid bodies to a combination of activin A, bone morphogenetic protein 4, and basic fibroblast growth factor, followed by VEGF and Dickkopf homolog 1 in the later stages of differentiation.[128] More recently, the induction of selected small noncoding RNAs (microRNAs) in ESCs has been proposed for cardiomyocyte production.[127] Although studies on ESC-derived cardiomyocytes confirmed their structural and functional differentiation including myofibrillar organization and gap junction formation,[125,129] compared to native cardiomyocytes they retain traits of a more immature phenotype,[130] which may allow for further proliferation of transplanted cells within the myocardium. In fact, as demonstrated by Laflamme and colleagues, human ESC-derived cardiomyocytes proliferate and structurally mature over time when transplanted into rat heart.[131] Overall, transplantation studies using murine or human ESC-derived cardiac-committed cells consistently showed a high engraftment rate and augmentation of cardiac function and capillary density.[132,133] Hence, human ESC-derived cardiomyocytes could be used as allogeneic cell source to be cryopreserved for easy availability when needed. However, their pluripotency and allogeneic origin may bear the danger of teratoma or tumor formation and immune rejection, making some form of immunosuppression mandatory.[134] Considering the enormous ethical hurdles for human ESC manipulation, it remains doubtful whether they will ever become an acceptable cell source for cardiac regeneration in the clinical setting.

5.2 Induced Pluripotent Stem Cells

Free from ethical concerns, iPS can be generated by reprogramming of autologous, terminally differentiated adult human fibroblasts through ectopic expression of two or more of the pluripotency transcription factors Oct4, Sox2 (both required), c-Myc, and KLF4.[135–137] Similar to ESCs, iPS can be differentiated into cardiomyocytes and feasibility and therapeutic efficacy of human iPS-derived cardiomyocyte sheets for regenerative therapy has recently been demonstrated in a porcine model of ischemic cardiomyopathy, in which cardiac remodeling and function could be improved.[138] Transplanted human iPS-derived cardiomyocytes were detectable as long as 8 weeks after transplantation, but long-term cell survival was poor. Despite these encouraging results, there are, however, major safety concerns associated with the potential clinical use of iPS-derived cardiomyocytes. Although autologous (syngeneic) iPS can be used, they may still provoke an immune response, which is in part caused by the expression of aberrant genes in reprogrammed cells and may require immunosuppression.[139] In particular, in the setting of autologous transplantation, also in immunosuppressed hosts, enhanced teratoma formation can occur.[140] Furthermore, genetic instability of reprogrammed cells is a major issue, especially when using viral vector-based techniques. These genetic alterations include chromosomal aberrations, single nucleotide mutations, and telomere instability.[141] Even with the use of nonintegrating technologies, an enhanced frequency of mutations can be observed, and incomplete or aberrant reprogramming may be associated with epigenetic alterations (e.g., DNA methylation).[141] Table 1 gives an overview of preclinical trials using ESCs and iPS for myocardial regeneration. Although reprogramming technologies open up new and promising avenues for cell-based therapies in cardiovascular medicine, further studies are needed and important safety issues have to be addressed before implementing this approach into clinics.

6. DIRECT REPROGRAMMING OF NONMYOCYTES

To avoid undesirable immune responses, in vivo reprogramming of cardiac fibroblasts may offer yet another promising strategy to induce cardiac regeneration. Using advanced cell technology, Ieda and colleagues were the first to reprogram cardiac fibroblasts into cardiomyocytes through transduction of three developmental transcription factors—Gata4, Mef2c, and Tbx5.[142] Two years later, Song and colleagues used a similar approach to directly reprogram endogenous fibroblasts into cardiomyocytes in vivo.[143] They showed that the forced expression of Gata4, Hand2, Mef2c, and Tbx5 by retroviral delivery reprogrammed dividing nonmyocytes into functional cardiomyocyte-like cells and improved cardiac remodeling and function after myocardial infarction in mice. Although with current technologies the efficiency of in vivo cardiomyogenic reprogramming is still low, this novel approach seems promising, because it would allow for endogenous cardiac regeneration without the necessity of cell transplantation.

7. UNRESOLVED ISSUES AND FUTURE PERSPECTIVES

From a clinical point of view, the ideal cell for cardiac regeneration should meet very specific criteria, including: (1) Easy accessibility and timely availability, (2) high cardioprotective, cardiomyogenic, and angiogenic potential, irrespective of patient age and cardiovascular risk factors, (3) persistence in hostile environment (ischemic myocardium), and (4) a high safety profile (low immunogenicity and tumorigenicity, high genetic stability). In light of these requirements, all of the above-mentioned cell types have their own individual advantages and shortcomings as summarized in Table 4. Despite enormous scientific achievements in the field during the last two decades, the need for improved understanding of the cellular and molecular mechanisms of cell-based cardiac regeneration is the longer the more urgent. Furthermore, the identification of the cell type and delivery route best suited for cellular therapy in the clinical setting is of highest priority. In a recent compilation of expert opinions on the topic,[144] Rosen and Myerburg put forward the idea of employing an adaptive design similar to the I-SPY2 trial in cancer research, which could be used as model trial for cardiac regeneration.[145] Such a setup would allow for testing of different strategies, in this case different cell types and/or application routes, simultaneously over different treatment arms, with appropriate interim analyses identifying the regimens holding most promise for continuation. In order to enhance cell-based therapeutic responses, combinations of two cell types with documented positive interactions might also be considered as a novel strategy. This has recently been applied in a porcine model of myocardial infarction, in which combination of human CSCs and MSCs was more effective in reducing infarct size and improving cardiac function than either cell type alone.[146]

Last but not least, because of limited proliferation, persistence, and cardiomyogenic differentiation of transplanted cells in the host myocardium, regardless of cell type and delivery route, new strategies to enhance cell engraftment and differentiation are needed. Cardiac-specific decellularized matrices and biopolymers that are currently in development may serve as attractive vehicles for cell delivery to the heart.[123] As an alternative cell-free strategy, transplantation of acellular matrices delivering cardiopoietic factors via sustained release may enhance endogenous paracrine effects and activate the intrinsic regenerative capacity of the heart.[147,148] However, such approach would also demand thorough characterization of the stem cell secretome and its changes during differentiation and maturation. Ultimately, from the clinician's point of view, cell-based therapies for cardiac regeneration will have to prove safe and effective in well-designed, randomized controlled, and double-blinded clinical trials before they are ready for clinical implementation.

Disclosures: There are no disclosures or conflict of interest to declare.

TABLE 4 Advantages and Disadvantages of Specific Cell Types for Cardiovascular Regeneration

	Availability in the Clinical Setting					Safety Concerns	
Cell Type	Availability	Autologous	Allogeneic	Off the Shelf	Efficacy	Need for Immunosuppression	Teratoma Formation
BMCs	+++	+	−	−	−/+	−	−
MSCs	+++	+	+	+++	+	−	−
CSCs	+	+	−	−	+(+)	−	−
iPS	+	+	+	+	++	−/+	++
ESCs	+	−	+	+	++	+	++

BMCs, bone marrow-derived cells; MSCs, mesenchymal stem cells; CSCs, cardiac stem cells; iPS, induced pluripotent stem cells; ESCs, embryonic stem cells.
Reprinted from Pfister O, Della Verde G, Liao R, Kuster GM. Regenerative therapy for cardiovascular disease. *Transl Res* 2014;163(4), with permission from Elsevier.

REFERENCES

1. Go AS, Mozaffarian D, Roger VL, Benjamin EJ, Berry JD, Blaha MJ, et al. Executive summary: heart disease and stroke statistics—2014 update: a report from the American Heart Association. *Circulation* 2014;**129**(3):399–410. Epub 2014/01/22.

2. McMurray JJ, Adamopoulos S, Anker SD, Auricchio A, Bohm M, Dickstein K, et al. Guidelines ESCCfP. ESC guidelines for the diagnosis and treatment of acute and chronic heart failure 2012: the task force for the diagnosis and treatment of acute and chronic heart failure 2012 of the European Society of Cardiology. Developed in collaboration with the Heart Failure Association (HFA) of the ESC. *Eur Heart J* 2012;**33**(14): 1787–847. Epub 2012/05/23.

3. Dickstein K. *Understanding heart failure: myths and facts about heart failure.* European Society of Cardiology, Heart Failure Association; 2014. http://www.heartfailurematters.org/en_GB/. [cited October 6, 2014].

4. Bergmann O, Bhardwaj RD, Bernard S, Zdunek S, Barnabe-Heider F, Walsh S, et al. Evidence for cardiomyocyte renewal in humans. *Science* 2009;**324**(5923):98–102. Epub 2009/04/04.

5. Beltrami AP, Urbanek K, Kajstura J, Yan SM, Finato N, Bussani R, et al. Evidence that human cardiac myocytes divide after myocardial infarction. *N Engl J Med* 2001;**344**(23):1750–7. Epub 2001/06/09.

6. Quaini F, Urbanek K, Beltrami AP, Finato N, Beltrami CA, Nadal-Ginard B, et al. Chimerism of the transplanted heart. *N Engl J Med* 2002;**346**(1):5–15.

7. Kikuchi K, Holdway JE, Werdich AA, Anderson RM, Fang Y, Egnaczyk GF, et al. Primary contribution to zebrafish heart regeneration by *gata4*(+) cardiomyocytes. *Nature* 2010;**464**(7288):601–5. Epub 2010/03/26.

8. Bostrom P, Mann N, Wu J, Quintero PA, Plovie ER, Panakova D, et al. C/EBPbeta controls exercise-induced cardiac growth and protects against pathological cardiac remodeling. *Cell* 2010;**143**(7):1072–83. Epub 2010/12/25.

9. Porrello ER, Mahmoud AI, Simpson E, Hill JA, Richardson JA, Olson EN, et al. Transient regenerative potential of the neonatal mouse heart. *Science* 2011;**331**(6020):1078–80. Epub 2011/02/26.

10. Hsieh PC, Segers VF, Davis ME, MacGillivray C, Gannon J, Molkentin JD, et al. Evidence from a genetic fate-mapping study that stem cells refresh adult mammalian cardiomyocytes after injury. *Nat Med* 2007;**13**(8):970–4. Epub 2007/07/31.

11. Hosoda T, D'Amario D, Cabral-Da-Silva MC, Zheng H, Padin-Iruegas ME, Ogorek B, et al. Clonality of mouse and human cardiomyogenesis in vivo. *Proc Natl Acad Sci USA* 2009;**106**(40):17169–74. Epub 2009/10/07.

12. Bolli R, Chugh AR, D'Amario D, Loughran JH, Stoddard MF, Ikram S, et al. Cardiac stem cells in patients with ischaemic cardiomyopathy (SCIPIO): initial results of a randomised phase 1 trial. *Lancet* 2011;**378**(9806):1847–57. Epub 2011/11/18.

13. Makkar RR, Smith RR, Cheng K, Malliaras K, Thomson LE, Berman D, et al. Intracoronary cardiosphere-derived cells for heart regeneration after myocardial infarction (CADUCEUS): a prospective, randomised phase 1 trial. *Lancet* 2012;**379**(9819):895–904. Epub 2012/02/18.

14. Jackson KA, Majka SM, Wang H, Pocius J, Hartley CJ, Majesky MW, et al. Regeneration of ischemic cardiac muscle and vascular endothelium by adult stem cells. *J Clin Invest* 2001;**107**(11):1395–402.

15. Mouquet F, Pfister O, Jain M, Oikonomopoulos A, Ngoy S, Summer R, et al. Restoration of cardiac progenitor cells after myocardial infarction by self-proliferation and selective homing of bone marrow-derived stem cells. *Circ Res* 2005;**97**(11):1090–2. Epub 2005/11/05.

16. Fazel S, Cimini M, Chen L, Li S, Angoulvant D, Fedak P, et al. Cardioprotective c-kit$^+$ cells are from the bone marrow and regulate the myocardial balance of angiogenic cytokines. *J Clin Invest* 2006;**116**(7):1865–77. Epub 2006/07/11.

17. Massa M, Rosti V, Ferrario M, Campanelli R, Ramajoli I, Rosso R, et al. Increased circulating hematopoietic and endothelial progenitor cells in the early phase of acute myocardial infarction. *Blood* 2005;**105**(1):199–206. Epub 2004/09/04.

18. Leone AM, Rutella S, Bonanno G, Contemi AM, de Ritis DG, Giannico MB, et al. Endogenous G-CSF and CD34(+) cell mobilization after acute myocardial infarction. *Int J Cardiol* 2006;**111**(2):202–8.

19. Vandervelde S, van Luyn MJ, Tio RA, Harmsen MC. Signaling factors in stem cell-mediated repair of infarcted myocardium. *J Mol Cell Cardiol* 2005;**39**(2):363–76. Epub 2005/07/05.

20. Abbott JD, Huang Y, Liu D, Hickey R, Krause DS, Giordano FJ. Stromal cell-derived factor-1alpha plays a critical role in stem cell recruitment to the heart after myocardial infarction but is not sufficient to induce homing in the absence of injury. *Circulation* 2004;**110**(21):3300–5.

21. Ferrari G, Cusella-De Angelis G, Coletta M, Paolucci E, Stornaiuolo A, Cossu G, et al. Muscle regeneration by bone marrow-derived myogenic progenitors. *Science* 1998;**279**(5356):1528–30.

22. Krause DS, Theise ND, Collector MI, Henegariu O, Hwang S, Gardner R, et al. Multi-organ, multi-lineage engraftment by a single bone marrow-derived stem cell. *Cell* 2001;**105**(3):369–77.

23. Mezey E, Chandross KJ, Harta G, Maki RA, McKercher SR. Turning blood into brain: cells bearing neuronal antigens generated in vivo from bone marrow. *Science* 2000;**290**(5497):1779–82.

24. Orlic D, Kajstura J, Chimenti S, Jakoniuk I, Anderson SM, Li B, et al. Bone marrow cells regenerate infarcted myocardium. *Nature* 2001; **410**(6829):701–5.

25. Fuchs S, Baffour R, Zhou YF, Shou M, Pierre A, Tio FO, et al. Transendocardial delivery of autologous bone marrow enhances collateral perfusion and regional function in pigs with chronic experimental myocardial ischemia. *J Am Coll Cardiol* 2001;**37**(6):1726–32.

26. Murry CE, Soonpaa MH, Reinecke H, Nakajima H, Nakajima HO, Rubart M, et al. Haematopoietic stem cells do not transdifferentiate into cardiac myocytes in myocardial infarcts. *Nature* 2004;**428**(6983):664–8.

27. Zeng L, Hu Q, Wang X, Mansoor A, Lee J, Feygin J, et al. Bioenergetic and functional consequences of bone marrow-derived multipotent progenitor cell transplantation in hearts with postinfarction left ventricular remodeling. *Circulation* 2007;**115**(14):1866–75. Epub 2007/03/29.

28. Uemura R, Xu M, Ahmad N, Ashraf M. Bone marrow stem cells prevent left ventricular remodeling of ischemic heart through paracrine signaling. *Circ Res* 2006;**98**(11):1414–21.

29. Kamihata H, Matsubara H, Nishiue T, Fujiyama S, Tsutsumi Y, Ozono R, et al. Implantation of bone marrow mononuclear cells into ischemic myocardium enhances collateral perfusion and regional function via side supply of angioblasts, angiogenic ligands, and cytokines. *Circulation* 2001;**104**(9):1046–52.

30. Walter DH, Haendeler J, Reinhold J, Rochwalsky U, Seeger F, Honold J, et al. Impaired CXCR4 signaling contributes to the reduced neovascularization capacity of endothelial progenitor cells from patients with coronary artery disease. *Circ Res* 2005;**97**(11):1142–51. Epub 2005/10/29.

31. Anversa P, Kajstura J, Rota M, Leri A. Regenerating new heart with stem cells. *J Clin Invest* 2013;**123**(1):62–70. Epub 2013/01/03.

32. Suzuki G, Iyer V, Cimato T, Canty Jr JM. Pravastatin improves function in hibernating myocardium by mobilizing CD133+ and cKit+ bone marrow progenitor cells and promoting myocytes to reenter the growth phase of the cardiac cell cycle. *Circ Res* 2009;**104**(2):255–64. 10 p. following 64. Epub 2008/12/20.

33. Asahara T, Masuda H, Takahashi T, Kalka C, Pastore C, Silver M, et al. Bone marrow origin of endothelial progenitor cells responsible for postnatal vasculogenesis in physiological and pathological neovascularization. *Circ Res* 1999;**85**(3):221–8.

34. Dai W, Hale SL, Martin BJ, Kuang JQ, Dow JS, Wold LE, et al. Allogeneic mesenchymal stem cell transplantation in postinfarcted rat myocardium: short- and long-term effects. *Circulation* 2005;**112**(2):214–23. Epub 2005/07/07.

35. Shake JG, Gruber PJ, Baumgartner WA, Senechal G, Meyers J, Redmond JM, et al. Mesenchymal stem cell implantation in a swine myocardial infarct model: engraftment and functional effects. *Ann Thorac Surg* 2002;**73**(6):1919–25; discussion 26.

36. Herreros J, Prosper F, Perez A, Gavira JJ, Garcia-Velloso MJ, Barba J, et al. Autologous intramyocardial injection of cultured skeletal muscle-derived stem cells in patients with non-acute myocardial infarction. *Eur Heart J* 2003;**24**(22):2012–20. Epub 2003/11/14.

37. Perin EC, Dohmann HF, Borojevic R, Silva SA, Sousa AL, Mesquita CT, et al. Transendocardial, autologous bone marrow cell transplantation for severe, chronic ischemic heart failure. *Circulation* 2003;**107**(18):2294–302. Epub 2003/04/23.

38. Strauer BE, Brehm M, Zeus T, Kostering M, Hernandez A, Sorg RV, et al. Repair of infarcted myocardium by autologous intracoronary mononuclear bone marrow cell transplantation in humans. *Circulation* 2002;**106**(15):1913–8. Epub 2002/10/09.

39. Zimmet H, Porapakkham P, Porapakkham P, Sata Y, Haas SJ, Itescu S, et al. Short- and long-term outcomes of intracoronary and endogenously mobilized bone marrow stem cells in the treatment of ST-segment elevation myocardial infarction: a meta-analysis of randomized control trials. *Eur J Heart Fail* 2012;**14**(1):91–105. Epub 2011/11/09.

40. Perin EC, Dohmann HF, Borojevic R, Silva SA, Sousa AL, Silva GV, et al. Improved exercise capacity and ischemia 6 and 12 months after transendocardial injection of autologous bone marrow mononuclear cells for ischemic cardiomyopathy. *Circulation* 2004;**110**(11 Suppl. 1):II213–8. Epub 2004/09/15.

41. Schachinger V, Assmus B, Britten MB, Honold J, Lehmann R, Teupe C, et al. Transplantation of progenitor cells and regeneration enhancement in acute myocardial infarction: final one-year results of the TOPCARE-AMI Trial. *J Am Coll Cardiol* 2004;**44**(8):1690–9. Epub 2004/10/19.

42. Wollert KC, Meyer GP, Lotz J, Ringes-Lichtenberg S, Lippolt P, Breidenbach C, et al. Intracoronary autologous bone-marrow cell transfer after myocardial infarction: the BOOST randomised controlled clinical trial. *Lancet* 2004;**364**(9429):141–8.

43. Schachinger V, Erbs S, Elsasser A, Haberbosch W, Hambrecht R, Holschermann H, et al. Intracoronary bone marrow-derived progenitor cells in acute myocardial infarction. *N Engl J Med* 2006;**355**(12):1210–21. Epub 2006/09/23.

44. Lunde K, Solheim S, Aakhus S, Arnesen H, Abdelnoor M, Egeland T, et al. Intracoronary injection of mononuclear bone marrow cells in acute myocardial infarction. *N Engl J Med* 2006;**355**(12):1199–209. Epub 2006/09/23.

45. Janssens S, Dubois C, Bogaert J, Theunissen K, Deroose C, Desmet W, et al. Autologous bone marrow-derived stem-cell transfer in patients with ST-segment elevation myocardial infarction: double-blind, randomised controlled trial. *Lancet* 2006;**367**(9505):113–21. Epub 2006/01/18.

46. Roncalli J, Mouquet F, Piot C, Trochu JN, Le Corvoisier P, Neuder Y, et al. Intracoronary autologous mononucleated bone marrow cell infusion for acute myocardial infarction: results of the randomized multicenter BONAMI trial. *Eur Heart J* 2011;**32**(14):1748–57. Epub 2010/12/04.

47. Hirsch A, Nijveldt R, van der Vleuten PA, Tijssen JG, van der Giessen WJ, Tio RA, et al. Intracoronary infusion of mononuclear cells from bone marrow or peripheral blood compared with standard therapy in patients after acute myocardial infarction treated by primary percutaneous coronary intervention: results of the randomized controlled HEBE trial. *Eur Heart J* 2011;**32**(14):1736–47. Epub 2010/12/15.

48. Seeger FH, Tonn T, Krzossok N, Zeiher AM, Dimmeler S. Cell isolation procedures matter: a comparison of different isolation protocols of bone marrow mononuclear cells used for cell therapy in patients with acute myocardial infarction. *Eur Heart J* 2007;**28**(6):766–72. Epub 2007/02/15.

49. Assmus B, Tonn T, Seeger FH, Yoon CH, Leistner D, Klotsche J, et al. Red blood cell contamination of the final cell product impairs the efficacy of autologous bone marrow mononuclear cell therapy. *J Am Coll Cardiol* 2010;**55**(13):1385–94. Epub 2010/03/27.

50. Bartunek J, Wijns W, Heyndrickx GR, Vanderheyden M. Timing of intracoronary bone-marrow-derived stem cell transplantation after ST-elevation myocardial infarction. *Nat Clin Pract Cardiovasc Med* 2006;**3**(Suppl. 1):S52–6. Epub 2006/02/28.

51. Hare JM, Bolli R, Cooke JP, Gordon DJ, Henry TD, Perin EC, et al. Phase II clinical research design in cardiology: learning the right lessons too well: observations and recommendations from the cardiovascular cell therapy research network (CCTRN). *Circulation* 2013;**127**(15):1630–5. Epub 2013/04/17.

52. Perin EC, Willerson JT, Pepine CJ, Henry TD, Ellis SG, Zhao DX, Cardiovascular Cell Therapy Research Network, et al. Effect of transendocardial delivery of autologous bone marrow mononuclear cells on functional capacity, left ventricular function, and perfusion in chronic heart failure: the FOCUS-CCTRN trial. *JAMA* 2012;**307**(16):1717–26. Epub 2012/03/27.

53. Traverse JH, Henry TD, Ellis SG, Pepine CJ, Willerson JT, Zhao DX, Cardiovascular Cell Therapy Research Network, et al. Effect of intracoronary delivery of autologous bone marrow mononuclear cells 2 to 3 weeks following acute myocardial infarction on left ventricular function: the LateTIME randomized trial. *JAMA* 2011;**306**(19):2110–9. Epub 2011/11/16.

54. Traverse JH, Henry TD, Pepine CJ, Willerson JT, Zhao DX, Ellis SG, Cardiovascular Cell Therapy Research Network, et al. Effect of the use and timing of bone marrow mononuclear cell delivery on left ventricular function after acute myocardial infarction: the TIME randomized trial. *JAMA* 2012;**308**(22):2380–9. Epub 2012/11/07.

55. Nowbar AN, Mielewczik M, Karavassilis M, Dehbi HM, Shun-Shin MJ, Jones S, et al. Discrepancies in autologous bone marrow stem cell trials and enhancement of ejection fraction (DAMASCENE): weighted regression and meta-analysis. *BMJ* 2014;**348**:g2688. Epub 2014/04/30.

56. Penicka M, Widimsky P, Kobylka P, Kozak T, Lang O. Images in cardiovascular medicine. Early tissue distribution of bone marrow mononuclear cells after transcoronary transplantation in a patient with acute myocardial infarction. *Circulation* 2005;**112**(4):e63–5. Epub 2005/07/27.

57. Muller-Ehmsen J, Krausgrill B, Burst V, Schenk K, Neisen UC, Fries JW, et al. Effective engraftment but poor mid-term persistence of mononuclear and mesenchymal bone marrow cells in acute and chronic rat myocardial infarction. *J Mol Cell Cardiol* 2006;**41**(5):876–84. Epub 2006/09/16.

58. Dimmeler S, Leri A. Aging and disease as modifiers of efficacy of cell therapy. *Circ Res* 2008;**102**(11):1319–30. Epub 2008/06/07.

59. Mohsin S, Siddiqi S, Collins B, Sussman MA. Empowering adult stem cells for myocardial regeneration. *Circ Res* 2011;**109**(12):1415–28. Epub 2011/12/14.

60. Jakob P, Landmesser U. Role of microRNAs in stem/progenitor cells and cardiovascular repair. *Cardiovasc Res* 2012;**93**(4):614–22. Epub 2011/12/03.

61. Mathur A. *BAMI. The effect of intracoronary reinfusion of bone marrow-derived mononuclear cells (BM-MNC) on all cause mortality in acute myocardial infarction.* U.S. National Institutes of Health; 2013. http://clinicaltrials.gov/show/NCT01569178. [cited October 9, 2014].

62. Friedenstein AJ, Chailakhjan RK, Lalykina KS. The development of fibroblast colonies in monolayer cultures of guinea-pig bone marrow and spleen cells. *Cell Tissue Kinet* 1970;**3**(4):393–403. Epub 1970/10/01.

63. da Silva Meirelles L, Chagastelles PC, Nardi NB. Mesenchymal stem cells reside in virtually all post-natal organs and tissues. *J Cell Sci* 2006; **119**(Pt 11):2204–13. Epub 2006/05/11.

64. Pittenger MF, Mackay AM, Beck SC, Jaiswal RK, Douglas R, Mosca JD, et al. Multilineage potential of adult human mesenchymal stem cells. *Science* 1999;**284**(5411):143–7.

65. Uccelli A, Moretta L, Pistoia V. Mesenchymal stem cells in health and disease. *Nat Rev Immunol* 2008;**8**(9):726–36. Epub 2009/01/28.

66. Makino S, Fukuda K, Miyoshi S, Konishi F, Kodama H, Pan J, et al. Cardiomyocytes can be generated from marrow stromal cells in vitro. *J Clin Invest* 1999;**103**(5):697–705.

67. Xu M, Wani M, Dai YS, Wang J, Yan M, Ayub A, et al. Differentiation of bone marrow stromal cells into the cardiac phenotype requires intercellular communication with myocytes. *Circulation* 2004;**110**(17):2658–65. Epub 2004/10/20.

68. Toma C, Pittenger MF, Cahill KS, Byrne BJ, Kessler PD. Human mesenchymal stem cells differentiate to a cardiomyocyte phenotype in the adult murine heart. *Circulation* 2002;**105**(1):93–8.

69. Quevedo HC, Hatzistergos KE, Oskouei BN, Feigenbaum GS, Rodriguez JE, Valdes D, et al. Allogeneic mesenchymal stem cells restore cardiac function in chronic ischemic cardiomyopathy via trilineage differentiating capacity. *Proc Natl Acad Sci USA* 2009;**106**(33):14022–7. Epub 2009/08/12.

70. Makkar RR, Price MJ, Lill M, Frantzen M, Takizawa K, Kleisli T, et al. Intramyocardial injection of allogenic bone marrow-derived mesenchymal stem cells without immunosuppression preserves cardiac function in a porcine model of myocardial infarction. *J Cardiovasc Pharmacol Ther* 2005;**10**(4):225–33. Epub 2005/12/31.

71. Yang YJ, Qian HY, Huang J, Li JJ, Gao RL, Dou KF, et al. Combined therapy with simvastatin and bone marrow-derived mesenchymal stem cells increases benefits in infarcted swine hearts. *Arterioscler Thromb Vasc Biol* 2009;**29**(12):2076–82. Epub 2009/09/19.

72. Dixon JA, Gorman RC, Stroud RE, Bouges S, Hirotsugu H, Gorman III JH, et al. Mesenchymal cell transplantation and myocardial remodeling after myocardial infarction. *Circulation* 2009;**120**(11 Suppl.):S220–9. Epub 2009/09/24.

73. Schuleri KH, Feigenbaum GS, Centola M, Weiss ES, Zimmet JM, Turney J, et al. Autologous mesenchymal stem cells produce reverse remodelling in chronic ischaemic cardiomyopathy. *Eur Heart J* 2009;**30**(22):2722–32. Epub 2009/07/10.

74. Perin EC, Silva GV, Assad JA, Vela D, Buja LM, Sousa AL, et al. Comparison of intracoronary and transendocardial delivery of allogeneic mesenchymal cells in a canine model of acute myocardial infarction. *J Mol Cell Cardiol* 2008;**44**(3):486–95. Epub 2007/12/07.

75. Halkos ME, Zhao ZQ, Kerendi F, Wang NP, Jiang R, Schmarkey LS, et al. Intravenous infusion of mesenchymal stem cells enhances regional perfusion and improves ventricular function in a porcine model of myocardial infarction. *Basic Res Cardiol* 2008;**103**(6):525–36. Epub 2008/08/16.

76. Iso Y, Spees JL, Serrano C, Bakondi B, Pochampally R, Song YH, et al. Multipotent human stromal cells improve cardiac function after myocardial infarction in mice without long-term engraftment. *Biochem Biophys Res Commun* 2007;**354**(3):700–6. Epub 2007/01/30.

77. Leiker M, Suzuki G, Iyer VS, Canty Jr JM, Lee T. Assessment of a nuclear affinity labeling method for tracking implanted mesenchymal stem cells. *Cell Transplant* 2008;**17**(8):911–22. Epub 2008/12/17.

78. Kinnaird T, Stabile E, Burnett MS, Lee CW, Barr S, Fuchs S, et al. Marrowderived stromal cells express genes encoding a broad spectrum of arteriogenic cytokines and promote in vitro and in vivo arteriogenesis through paracrine mechanisms. *Circ Res* 2004;**94**(5):678–85.

79. Gnecchi M, He H, Liang OD, Melo LG, Morello F, Mu H, et al. Paracrine action accounts for marked protection of ischemic heart by Akt-modified mesenchymal stem cells. *Nat Med* 2005;**11**(4):367–8. Epub 2005/04/07.

80. Gnecchi M, Zhang Z, Ni A, Dzau VJ. Paracrine mechanisms in adult stem cell signaling and therapy. *Circ Res* 2008;**103**(11):1204–19. Epub 2008/11/26.

81. Dominici M, Le Blanc K, Mueller I, Slaper-Cortenbach I, Marini F, Krause D, et al. Minimal criteria for defining multipotent mesenchymal stromal cells. The International Society for Cellular Therapy position statement. *Cytotherapy* 2006;**8**(4):315–7. Epub 2006/08/23.

82. Williams AR, Hare JM. Mesenchymal stem cells: biology, pathophysiology, translational findings, and therapeutic implications for cardiac disease. *Circ Res* 2011;**109**(8):923–40. Epub 2011/10/01.

83. Chen SL, Fang WW, Ye F, Liu YH, Qian J, Shan SJ, et al. Effect on left ventricular function of intracoronary transplantation of autologous bone marrow mesenchymal stem cell in patients with acute myocardial infarction. *Am J Cardiol* 2004;**94**(1):92–5. Epub 2004/06/29.

84. Hare JM, Traverse JH, Henry TD, Dib N, Strumpf RK, Schulman SP, et al. A randomized, double-blind, placebo-controlled, dose-escalation study of intravenous adult human mesenchymal stem cells (prochymal) after acute myocardial infarction. *J Am Coll Cardiol* 2009;**54**(24):2277–86. Epub 2009/12/05.

85. Williams AR, Trachtenberg B, Velazquez DL, McNiece I, Altman P, Rouy D, et al. Intramyocardial stem cell injection in patients with ischemic cardiomyopathy: functional recovery and reverse remodeling. *Circ Res* 2011;**108**(7):792–6. Epub 2011/03/19.

86. Behfar A, Yamada S, Crespo-Diaz R, Nesbitt JJ, Rowe LA, Perez-Terzic C, et al. Guided cardiopoiesis enhances therapeutic benefit of bone marrow human mesenchymal stem cells in chronic myocardial infarction. *J Am Coll Cardiol* 2010;**56**(9):721–34. Epub 2010/08/21.

87. Bartunek J, Behfar A, Dolatabadi D, Vanderheyden M, Ostojic M, Dens J, et al. Cardiopoietic stem cell therapy in heart failure: the C-CURE (Cardiopoietic stem Cell therapy in heart failURE) multicenter randomized trial with lineage-specified biologics. *J Am Coll Cardiol* 2013;**61**(23): 2329–38. Epub 2013/04/16.

88. Hare JM, Fishman JE, Gerstenblith G, DiFede Velazquez DL, Zambrano JP, Suncion VY, et al. Comparison of allogeneic vs autologous bone marrow-derived mesenchymal stem cells delivered by transendocardial injection in patients with ischemic cardiomyopathy: the POSEIDON randomized trial. *JAMA* 2012;**308**(22):2369–79. Epub 2012/11/03.

89. Hierlihy AM, Seale P, Lobe CG, Rudnicki MA, Megeney LA. The post-natal heart contains a myocardial stem cell population. *FEBS Lett* 2002;**530**(1–3):239–43.

90. Oh H, Bradfute SB, Gallardo TD, Nakamura T, Gaussin V, Mishina Y, et al. Cardiac progenitor cells from adult myocardium: homing, differentiation, and fusion after infarction. *Proc Natl Acad Sci USA* 2003;**100**(21):12313–8.

91. Beltrami AP, Barlucchi L, Torella D, Baker M, Limana F, Chimenti S, et al. Adult cardiac stem cells are multipotent and support myocardial regeneration. *Cell* 2003;**114**(6):763–76.

92. Messina E, De Angelis L, Frati G, Morrone S, Chimenti S, Fiordaliso F, et al. Isolation and expansion of adult cardiac stem cells from human and murine heart. *Circ Res* 2004;**95**(9):911–21.

93. van Vliet P, Roccio M, Smits AM, van Oorschot AA, Metz CH, van Veen TA, et al. Progenitor cells isolated from the human heart: a potential cell source for regenerative therapy. *Nether Heart J* 2008;**16**(5):163–9. Epub 2008/06/21.

94. Valente M, Nascimento DS, Cumano A, Pinto-do OP. Sca-1(+) cardiac progenitor cells and heart-making: a critical synopsis. *Stem Cells Dev* 2014;**23**(19):2263–73. Epub 2014/06/14.

95. Sussman MA. Myocardial Isl(+)land: a place with lots of rhythm, but no beat. *Circ Res* 2012;**110**(10):1267–9. Epub 2012/05/15.

96. Malliaras K, Zhang Y, Seinfeld J, Galang G, Tseliou E, Cheng K, et al. Cardiomyocyte proliferation and progenitor cell recruitment underlie therapeutic regeneration after myocardial infarction in the adult mouse heart. *EMBO Mol Med* 2013;**5**(2):191–209. Epub 2012/12/21.

97. Goodell MA, Brose K, Paradis G, Conner AS, Mulligan RC. Isolation and functional properties of murine hematopoietic stem cells that are replicating in vivo. *J Exp Med* 1996;**183**(4):1797–806.

98. Asakura A, Rudnicki MA. Side population cells from diverse adult tissues are capable of in vitro hematopoietic differentiation. *Exp Hematol* 2002;**30**(11):1339–45.

99. Challen GA, Little MH. A side order of stem cells: the SP phenotype. *Stem Cells* 2006;**24**(1):3–12.

100. Pfister O, Oikonomopoulos A, Sereti KI, Sohn RL, Cullen D, Fine GC, et al. Role of the ATP-binding cassette transporter *Abcg2* in the phenotype and function of cardiac side population cells. *Circ Res* 2008;**103**(8):825–35. Epub 2008/09/13.

101. Pfister O, Mouquet F, Jain M, Summer R, Helmes M, Fine A, et al. CD31⁻ but Not CD31⁺ cardiac side population cells exhibit functional cardiomyogenic differentiation. *Circ Res* 2005;**97**(1):52–61. Epub 2005/06/11.

102. Unno K, Jain M, Liao R. Cardiac side population cells: moving toward the center stage in cardiac regeneration. *Circ Res* 2012;**110**(10):1355–63. Epub 2012/05/15.

103. Oyama T, Nagai T, Wada H, Naito AT, Matsuura K, Iwanaga K, et al. Cardiac side population cells have a potential to migrate and differentiate into cardiomyocytes in vitro and in vivo. *J Cell Biol* 2007;**176**(3):329–41. Epub 2007/01/31.

104. Martin CM, Meeson AP, Robertson SM, Hawke TJ, Richardson JA, Bates S, et al. Persistent expression of the ATP-binding cassette transporter, *Abcg2*, identifies cardiac SP cells in the developing and adult heart. *Dev Biol* 2004;**265**(1):262–75. Epub 2003/12/31.

105. Yoon J, Choi SC, Park CY, Shim WJ, Lim DS. Cardiac side population cells exhibit endothelial differentiation potential. *Exp Mol Med* 2007;**39**(5):653–62. Epub 2007/12/07.

106. Liang SX, Tan TY, Gaudry L, Chong B. Differentiation and migration of Sca1⁺/CD31⁻ cardiac side population cells in a murine myocardial ischemic model. *Int J Cardiol* 2010;**138**(1):40–9. Epub 2009/03/04.

107. Sandstedt J, Jonsson M, Kajic K, Sandstedt M, Lindahl A, Dellgren G, et al. Left atrium of the human adult heart contains a population of side population cells. *Basic Res Cardiol* 2012;**107**(2):255. Epub 2012/03/01.

108. Emmert MY, Emmert LS, Martens A, Ismail I, Schmidt-Richter I, Gawol A, et al. Higher frequencies of BCRP⁺ cardiac resident cells in ischaemic human myocardium. *Eur Heart J* 2013;**34**(36):2830–8. Epub 2012/06/28.

109. Ellison GM, Vicinanza C, Smith AJ, Aquila I, Leone A, Waring CD, et al. Adult c-kit(pos) cardiac stem cells are necessary and sufficient for functional cardiac regeneration and repair. *Cell* 2013;**154**(4):827–42. Epub 2013/08/21.

110. Molkentin JD, Houser SR. Are resident c-Kit⁺ cardiac stem cells really all that are needed to mend a broken heart?. *Circ Res* 2013;**113**(9):1037–9. Epub 2013/10/12.

111. van Berlo JH, Kanisicak O, Maillet M, Vagnozzi RJ, Karch J, Lin SC, et al. c-kit⁺ cells minimally contribute cardiomyocytes to the heart. *Nature* 2014;**509**(7500):337–41. Epub 2014/05/09.

112. Dawn B, Stein AB, Urbanek K, Rota M, Whang B, Rastaldo R, et al. Cardiac stem cells delivered intravascularly traverse the vessel barrier, regenerate infarcted myocardium, and improve cardiac function. *Proc Natl Acad Sci USA* 2005;**102**(10):3766–71. Epub 2005/03/01.

113. Rota M, Padin-Iruegas ME, Misao Y, De Angelis A, Maestroni S, Ferreira-Martins J, et al. Local activation or implantation of cardiac progenitor cells rescues scarred infarcted myocardium improving cardiac function. *Circ Res* 2008;**103**(1):107–16. Epub 2008/06/17.

114. Bearzi C, Rota M, Hosoda T, Tillmanns J, Nascimbene A, De Angelis A, et al. Human cardiac stem cells. *Proc Natl Acad Sci USA* 2007;**104**(35): 14068–73. Epub 2007/08/22.

115. Tang XL, Rokosh G, Sanganalmath SK, Yuan F, Sato H, Mu J, et al. Intracoronary administration of cardiac progenitor cells alleviates left ventricular dysfunction in rats with a 30-day-old infarction. *Circulation* 2010;**121**(2):293–305. Epub 2010/01/06.

116. Smith RR, Barile L, Cho HC, Leppo MK, Hare JM, Messina E, et al. Regenerative potential of cardiosphere-derived cells expanded from percutaneous endomyocardial biopsy specimens. *Circulation* 2007;**115**(7):896–908. Epub 2007/02/07.

117. Takehara N, Tsutsumi Y, Tateishi K, Ogata T, Tanaka H, Ueyama T, et al. Controlled delivery of basic fibroblast growth factor promotes human cardiosphere-derived cell engraftment to enhance cardiac repair for chronic myocardial infarction. *J Am Coll Cardiol* 2008;**52**(23):1858–65. Epub 2008/11/29.

118. Johnston PV, Sasano T, Mills K, Evers R, Lee ST, Smith RR, et al. Engraftment, differentiation, and functional benefits of autologous cardiosphere-derived cells in porcine ischemic cardiomyopathy. *Circulation* 2009;**120**(12):1075–83. 7 p. following 83. Epub 2009/09/10.

119. Lee ST, White AJ, Matsushita S, Malliaras K, Steenbergen C, Zhang Y, et al. Intramyocardial injection of autologous cardiospheres or cardiosphere-derived cells preserves function and minimizes adverse ventricular remodeling in pigs with heart failure post-myocardial infarction. *J Am Coll Cardiol* 2011;**57**(4):455–65. Epub 2011/01/22.

120. Chugh AR, Beache GM, Loughran JH, Mewton N, Elmore JB, Kajstura J, et al. Administration of cardiac stem cells in patients with ischemic cardiomyopathy: the SCIPIO trial: surgical aspects and interim analysis of myocardial function and viability by magnetic resonance. *Circulation* 2012;**126**(11 Suppl. 1):S54–64. Epub 2012/09/22.

121. Malliaras K, Makkar RR, Smith RR, Cheng K, Wu E, Bonow RO, et al. Intracoronary cardiosphere-derived cells after myocardial infarction: evidence of therapeutic regeneration in the final 1-year results of the CADUCEUS trial (CArdiosphere-Derived aUtologous stem CElls to reverse ventricUlar dySfunction). *J Am Coll Cardiol* 2014;**63**(2):110–22. Epub 2013/09/17.

122. Malliaras K, Li TS, Luthringer D, Terrovitis J, Cheng K, Chakravarty T, et al. Safety and efficacy of allogeneic cell therapy in infarcted rats transplanted with mismatched cardiosphere-derived cells. *Circulation* 2012;**125**(1):100–12. Epub 2011/11/17.

123. Smith RR, Marban E, Marban L. Enhancing retention and efficacy of cardiosphere-derived cells administered after myocardial infarction using a hyaluronan-gelatin hydrogel. *Biomatter* 2013;**3**(1). Epub 2013/03/30.

124. Capricor Inc. *Allogeneic heart stem cells to achieve myocardial regeneration (ALLSTAR)*. U.S. National Institutes of Health; 2014. http://clinicaltrials.gov/ct2/show/NCT01458405. [cited October 9, 2014].

125. Kehat I, Kenyagin-Karsenti D, Snir M, Segev H, Amit M, Gepstein A, et al. Human embryonic stem cells can differentiate into myocytes with structural and functional properties of cardiomyocytes. *J Clin Invest* 2001;**108**(3):407–14. Epub 2001/08/08.

126. Mauritz C, Schwanke K, Reppel M, Neef S, Katsirntaki K, Maier LS, et al. Generation of functional murine cardiac myocytes from induced pluripotent stem cells. *Circulation* 2008;**118**(5):507–17. Epub 2008/07/16.

127. Bernstein HS, Srivastava D. Stem cell therapy for cardiac disease. *Pediatr Res* 2012;**71**(4 Pt 2):491–9. Epub 2012/03/21.

128. Yang L, Soonpaa MH, Adler ED, Roepke TK, Kattman SJ, Kennedy M, et al. Human cardiovascular progenitor cells develop from a KDR+ embryonic-stem-cell-derived population. *Nature* 2008;**453**(7194):524–8. Epub 2008/04/25.

129. Westfall MV, Pasyk KA, Yule DI, Samuelson LC, Metzger JM. Ultrastructure and cell-cell coupling of cardiac myocytes differentiating in embryonic stem cell cultures. *Cell Motil Cytoskelet* 1997;**36**(1):43–54. Epub 1997/01/01.

130. Jonsson MK, Vos MA, Mirams GR, Duker G, Sartipy P, de Boer TP, et al. Application of human stem cell-derived cardiomyocytes in safety pharmacology requires caution beyond hERG. *J Mol Cell Cardiol* 2012;**52**(5):998–1008. Epub 2012/02/23.

131. Laflamme MA, Gold J, Xu C, Hassanipour M, Rosler E, Police S, et al. Formation of human myocardium in the rat heart from human embryonic stem cells. *Am J Pathol* 2005;**167**(3):663–71. Epub 2005/08/30.

132. Tomescot A, Leschik J, Bellamy V, Dubois G, Messas E, Bruneval P, et al. Differentiation in vivo of cardiac committed human embryonic stem cells in postmyocardial infarcted rats. *Stem Cells* 2007;**25**(9):2200–5. Epub 2007/06/02.

133. Ardehali R, Ali SR, Inlay MA, Abilez OJ, Chen MQ, Blauwkamp TA, et al. Prospective isolation of human embryonic stem cell-derived cardiovascular progenitors that integrate into human fetal heart tissue. *Proc Natl Acad Sci USA* 2013;**110**(9):3405–10. Epub 2013/02/09.

134. Wakitani S, Takaoka K, Hattori T, Miyazawa N, Iwanaga T, Takeda S, et al. Embryonic stem cells injected into the mouse knee joint form teratomas and subsequently destroy the joint. *Rheumatology* 2003;**42**(1):162–5. Epub 2003/01/02.

135. Okita K, Ichisaka T, Yamanaka S. Generation of germline-competent induced pluripotent stem cells. *Nature* 2007;**448**(7151):313–7. Epub 2007/06/08.

136. Takahashi K, Tanabe K, Ohnuki M, Narita M, Ichisaka T, Tomoda K, et al. Induction of pluripotent stem cells from adult human fibroblasts by defined factors. *Cell* 2007;**131**(5):861–72. Epub 2007/11/24.

137. Lowry WE, Plath K. The many ways to make an iPS cell. *Nat Biotechnol* 2008;**26**(11):1246–8. Epub 2008/11/11.

138. Kawamura M, Miyagawa S, Miki K, Saito A, Fukushima S, Higuchi T, et al. Feasibility, safety, and therapeutic efficacy of human induced pluripotent stem cell-derived cardiomyocyte sheets in a porcine ischemic cardiomyopathy model. *Circulation* 2012;**126**(11 Suppl. 1):S29–37. Epub 2012/09/22.

139. Zhao T, Zhang ZN, Rong Z, Xu Y. Immunogenicity of induced pluripotent stem cells. *Nature* 2011;**474**(7350):212–5. Epub 2011/05/17.

140. Ahmed RP, Ashraf M, Buccini S, Shujia J, Haider H. Cardiac tumorigenic potential of induced pluripotent stem cells in an immunocompetent host with myocardial infarction. *Regen Med* 2011;**6**(2):171–8. Epub 2011/03/12.

141. Ronen D, Benvenisty N. Genomic stability in reprogramming. *Curr Opin Genet Dev* 2012;**22**(5):444–9. Epub 2012/10/09.

142. Ieda M, Fu JD, Delgado-Olguin P, Vedantham V, Hayashi Y, Bruneau BG, et al. Direct reprogramming of fibroblasts into functional cardiomyocytes by defined factors. *Cell* 2010;**142**(3):375–86. Epub 2010/08/10.

143. Song K, Nam YJ, Luo X, Qi X, Tan W, Huang GN, et al. Heart repair by reprogramming non-myocytes with cardiac transcription factors. *Nature* 2012;**485**(7400):599–604. Epub 2012/06/05.

144. Rosen MR, Myerburg RJ, Francis DP, Cole GD, Marban E. Translating stem cell research to cardiac disease therapies: pitfalls and prospects for improvement. *J Am Coll Cardiol* 2014;**64**(9):922–37. Epub 2014/08/30.

145. QuantumLeap. *I-SPY 2 TRIAL: neoadjuvant and personalized adaptive novel agents to treat breast cancer*. U.S. National Institutes of Health; 2009–2014. [cited October 9, 2014] http://clinicaltrials.gov/show/NCT01042379.

146. Williams AR, Hatzistergos KE, Addicott B, McCall F, Carvalho D, Suncion V, et al. Enhanced effect of combining human cardiac stem cells and bone marrow mesenchymal stem cells to reduce infarct size and to restore cardiac function after myocardial infarction. *Circulation* 2013;**127**(2):213–23. Epub 2012/12/12.

147. Suuronen EJ, Zhang P, Kuraitis D, Cao X, Melhuish A, McKee D, et al. An acellular matrix-bound ligand enhances the mobilization, recruitment and therapeutic effects of circulating progenitor cells in a hindlimb ischemia model. *FASEB J* 2009;**23**(5):1447–58. Epub 2009/01/13.

148. Eitan Y, Sarig U, Dahan N, Machluf M. Acellular cardiac extracellular matrix as a scaffold for tissue engineering: in vitro cell support, remodeling, and biocompatibility. *Tissue Eng Part C Methods* 2010;**16**(4):671–83. Epub 2009/09/29.

Abbreviated References from Tables 1-3

149. Moelker AD. *Eur Heart J* 2006.
150. De Silva. *Eur Heart J* 2008.
151. Schuleri KH. *Eur Heart J* 2009.
152. Williams AR. *Circulation* 2013.
153. Dawn B. *PNAS* 2005.
154. Bolli R. *Circulation* 2013.
155. Johnston PV. *Circulation* 2009.
156. Lee ST. *JACC* 2011.
157. Nelson TJ. *Circulation* 2009.
158. Mauritz C. *Eur Heart J* 2011.
159. Li X. *PLoS One* 2013.
160. Rajasingh J. *Circ Res* 2007.
161. Meyer GP. *Circulation* 2006.
162. Janssens S. *Lancet* 2006.
163. Lunde K. *N Engl J Med* 2006.
164. Schachinger V. *N Engl J Med* 2006.
165. Roncalli J. *Eur Heart J* 2011.
166. Hirsch A. *Eur Heart J* 2011.
167. Perin EC. *JAMA* 2012.
168. Traverse JH. *JAMA* 2011.
169. Traverse JH. *JAMA* 2012.
170. Zimmet H. *Eur J Heart Fail* 2012.
171. Bartunek J. *JACC* 2013.
172. Hare JM. *JACC* 2013.
173. Bolli R. *Lancet* 2011.
174. Malliaras K. *JACC* 2013.
175. Hierlihy AM. *FEBS Lett* 2002.
176. Martin CM. *Dev Biol* 2004.
177. Pfister O. *Circ Res* 2005.
178. Mouquet F. *Circ Res* 2005.
179. Meissner K. *J Histochem Cytochem* 2006.
180. Oyama T. *J Cell Biol* 2007.
181. Liang SX. *Int J Cardiol* 2010.
182. Beltrami AP. *Cell* 2003.
183. Linke A. *Proc Natl Acad Sci USA* 2005.
184. Chugh AR. *Circulation* 2012.
185. Ellison GM. *Cell* 2013.
186. Oh H. *Proc Natl Acad Sci USA* 2003.
187. Matsuura K. *J Biol Chem* 2004.
188. Wang X. *Stem Cells* 2006.
189. van Vliet P. *Nether Heart J* 2008.
190. Laugwitz KL. *Nature* 2005.
191. Moretti A. *Cell* 2006.
192. Ott HC. *Nat Clin Pract Cardiovasc Med* 2007.
193. Hou X. *J Clin Exp Cardiolog* 2012.
194. Messina E. *Circ Res* 2004.
195. Smith RR. *Circulation* 2007.
196. Makkar RR. *Lancet* 2012.
197. Limana F. *Circ Res* 2007.
198. Ruiz-Villalba A. *PLoS One* 2013.

Chapter 19

Cord Blood Transplantation in Hematological and Metabolic Diseases

Jessica M. Sun, Joanne Kurtzberg

Key Concepts
1. Umbilical cord blood is an established source of stem and progenitor cells for hematopoietic stem cell transplantation.
2. Compared to other stem cell sources, umbilical cord blood is associated with a longer time to engraftment, decreased incidence of GvHD, and less stringent HLA matching requirements.
3. Allogeneic umbilical cord blood transplantation can halt neurologic disease progression and extend life for decades in certain inherited metabolic diseases.
4. Umbilical cord blood is being investigated as a source of cellular therapies for acquired brain injuries.

Allogeneic hematopoietic stem cell transplantation (HSCT) has been established as an effective approach to curative therapy for both pediatric and adult patients with aggressive or recurrent malignancies, congenital immunodeficiency diseases, some genetic diseases, including inherited metabolic disorders (IMDs) and hemoglobinopathies, and congenital and acquired bone marrow failure syndromes. Traditionally, stem and progenitor cells have been obtained from bone marrow or mobilized peripheral blood. More recently, banked umbilical cord blood (CB) has emerged as an alternative source of stem and progenitor cells for transplantation. A major limitation for access to stem cell transplantation therapy is donor availability. Only 20–25% of patients in need of a transplant will have a human leukocyte antigen (HLA)-matched relative who can serve as their donor. Of those lacking a related donor, approximately 50% of Caucasian patients will identify an HLA-matched unrelated living bone marrow donor through the National Marrow Donor Program and other donor registries, but less than 10% of patients of ethnic minority backgrounds will find a suitably matched unrelated adult donor. For the remaining patients, a fully matched unrelated hematopoietic stem cell (HSC) donor cannot be identified. CB is readily available and can be transplanted across partially mismatched HLA barriers, increasing the availability of allogeneic stem cell donors for those patients lacking traditional HLA-matched related and unrelated donors.

Over the past two decades, CB has emerged as an alternative source of HSCs for use in HSCT. It is readily available, can be collected without risk to the mother or infant donor, and is significantly less likely to transmit infectious diseases transmissible through the blood. In addition, it is less likely to cause acute and chronic graft-versus-host disease (GvHD), a major obstacle to the success of allogeneic bone marrow transplantation. Recent data suggest that CB may confer protection against leukemic relapse in patients transplanted for recurrent leukemia. Current limitations of CB include a limited cell dose from a single donor unit, a smaller overall inventory of donors, decreased ability to utilize posttransplant donor-derived cellular therapy, and delayed immune reconstitution resulting in increased risk of posttransplant viral infections, particularly in the first year after transplant. New techniques, for example, ex vivo expansion, are being developed to overcome these obstacles, and CB is being increasingly utilized as a source of stem cells for HSCT.

1. UMBILICAL CB BANKING

In 1991, Dr Pablo Rubinstein established the first unrelated CB bank at the New York Blood Center supported by a pilot grant from the National Heart, Lung, and Blood Institute (NHLBI).[1] Since that time, more than 160 public CB banks have been established worldwide and there are approximately 720,000 CB units available for public use.[2] In general, public banks are nonprofit entities supported by third-party federal or private funding. Mothers may electively donate their infant's CB to one of these public banks, and the CB is banked and listed on a donor registry if the donor meets donor-screening criteria and the

Translating Regenerative Medicine to the Clinic. http://dx.doi.org/10.1016/B978-0-12-800548-4.00019-X

CB meets technical specifications. Not all donated CB units are used for transplantation but if selected for patient use, there is no contact between donor and recipient. Private CB banks, which store CB units for a particular family, usually for an up-front and yearly fee, are available worldwide and currently store an estimated 4 million CB units. Hybrid banks, banking for families and for the public, have also emerged. Pregnant women in some locations may have the option for either public or private CB banking or both. To aid in this decision, guidelines were established. In 2008, The American Society of Blood and Marrow Transplantation recommended donation to a public bank where possible, with the suggestion to review these recommendations in 5 years.[3] Both the American Academy of Pediatrics and the American Association of Obstetrics and Gynecology issued white papers recommending public donation unless there was a medical indication for autologous or related CB transplantation (CBT) in the donor's family.[4,5] Although private banks marketed promises of future uses of autologous CB in regenerative medicine, the evidence for these claims was felt to be insufficient to support endorsement of private banking at the time. Recently, there is renewed interest in private banking, with emerging data on use of autologous CB in neurologic diseases and regenerative medicine. In addition, CB licensure in the USA has increased the costs of public banking and unrelated donor CBT.

2. OVERVIEW OF BANKING TECHNOLOGY

The first CB collection for transplantation occurred in Salisbury, NC. Dr Gordon Douglas from NYU collected the CB dripping from the umbilical cord into a sterile plastic bottle containing preservative-free heparin. The CB was transported to Dr Hal Broxmeyer's laboratory where the unit was diluted with tissue culture media and DMSO, cryopreserved, and stored under liquid nitrogen. Dr Broxmeyer transported the CB in a dry shipper to Paris, France where Dr Elaine Gluckman performed the first CBT.

There have been enormous advances in the technical aspects of CB collection and banking since this first transplant. Initially manual techniques for CB collection and processing for public banking were developed.[1] Collection into the anticoagulant citrate-phosphate-dextrose quickly became standard practice in public banks and has been adopted by most private banks. While most banks depleted RBCs and plasma as a strategy for volume reduction during processing, some isolated purer populations of mononuclear cells or utilized plasma reduction alone.[6] Cryopreservation using 10% DMSO or 10% DMSO in 50% Dextran using controlled rate freezing was adopted. Methods for thawing and washing CB in Dextran 40 and 5% human serum albumin, and later, dilution without washing were developed and implemented by many transplant centers.[1] Unfortunately, controlled trials to determine the optimal anticoagulant for collection, cryoprotectant for long-term storage, or optimal thawing methods have never been conducted. CB units that are not RBC depleted should be washed to remove cellular debris and to prevent serious infusion reactions.[7]

The Cord Blood Transplantation (COBLT) study was the first prospective, open-label study of CB banking and transplantation in the world. In addition to establishing three additional public banks (at Duke, Children's Hospital of Orange County, and University of California at Los Angeles), it also created standard operating procedures for closed system CB collection, manual processing for RBC and plasma depletion and volume reduction, and tests for potency and viability.[8]

As banked unrelated CB was adopted as a source of cells for hematopoietic reconstitution, banking practices became more sophisticated. A series of devices for automated CB processing were manufactured, including robotic cryopreservation systems. Currently, any validated method of processing is accepted. A network of CB banks in Europe, Asia, Australia, and the USA, called Netcord, was established in 1997 and published the first standards for CB banking. These were adopted by the Foundation for Accreditation of Cellular Therapies and Joint Accreditation Committee of Europe and have been used for accreditation of CB banks for over 15 years. In 2004, the American Association of Blood Banks also published accreditation standards, and the FDA issued guidance for CB banking for unrelated transplantation in 2010. To date, five public banks in the USA have successfully completed the biologics licensure process with the FDA.

3. EARLY TRANSPLANT EXPERIENCE WITH UMBILICAL CBT

The first CB transplant was performed by Eliane Gluckman at the L'Hospital St. Louis in 1988 in a 6-year-old boy with bone marrow failure secondary to Fanconi anemia using CB from his HLA-matched sister. The patient, who was prepared for transplant with total lymphoid irradiation, cyclophosphamide and equine ATG, engrafted with donor cells on day +19, never developed any serious complications of transplantation, and did not experience acute or chronic GvHD.[9] More than 26 years later, he remains well and durably engrafted with donor cells. The success of this transplant paved the way for approximately 60 additional related CB transplants over the subsequent 5 years. Outcomes of these transplants demonstrated the feasibility of engraftment in children. Compared to results of allogeneic bone marrow transplantation from matched related donors, delayed times to engraftment of neutrophils and platelets and a reduced incidence of acute and chronic GvHD were observed after related CBT.[9–15] These encouraging results fueled the idea that CB could be used in the unrelated donor setting without full HLA matching. Accordingly, Dr Pablo Rubinstein, at the New York Blood Center, established the first unrelated donor CB bank in the USA with the support of the NHLBI at the NIH.

The first unrelated CBT was performed at Duke University in August 1993 in a child with refractory T-cell acute lymphoblastic leukemia utilizing an HLA-mismatched (4/6) unit from the New York Blood Center.[16] The Duke group subsequently published the first report of a series of 25 pediatric patients undergoing CBT.[16] A larger report from the New York Blood Center in 562 patients and all subsequent reports confirmed that despite partial HLA mismatching, CB could engraft in smaller (<40 kg) children, cell dose was critical, engraftment correlated with cell dose, and GvHD was reduced as compared to unrelated transplantation with adult donor cell sources.[17] Lower rates of engraftment were seen in diseases where resistance to engraftment was present such as acquired aplastic anemia, chronic myelogenous leukemia, and hemoglobinopathies. These early results were later confirmed by reports from Eurocord and the Center for International Blood and Marrow Transplant Research (CIBMTR).[18-20]

The COBLT study, conducted from 1997 to 2004, was the first prospective, multi-institutional study of unrelated CB banking and transplantation in the USA designed to examine the safety of unrelated CBT in infants, children, and adults with malignancies; children with congenital immunodeficiency disorders; and children with inborn errors of metabolism. Twenty-six transplant centers participated in the COBLT study, which also established three additional unrelated CB banks sponsored by the NHLBI. Common preparative regimens, prophylaxis against GvHD, and supportive care measures were used. Overall outcomes in children with malignant and nonmalignant conditions were favorable, with 2-year, event-free survivals of 55% in children with high-risk malignancies[21] and 78% survival in children with nonmalignant conditions.[22] Results in a very high-risk group of adults, however, were inferior to those seen in children and in individuals receiving unrelated bone marrow grafts. The cumulative incidence of engraftment by day 42 after transplantation was approximately 80% in all study strata including adults and children as well as children with malignant diseases, inborn errors of metabolism, and immunodeficiency syndromes. Factors adversely affecting engraftment or survival included lower cell doses, pretransplant cytomegalovirus seropositivity in the recipient, non-European ancestry, and greater HLA mismatching.

4. UMBILICAL CBT IN PEDIATRICS

Since the COBLT study, many studies have demonstrated benefit of CBT in children with hematological malignancies and further clarified the importance of cell dose and HLA matching. In a large registry study from the CIBMTR, outcomes of HSCT for children (<16 years) with acute leukemia were compared for patients who received single-unit HLA-matched (n=35), single-antigen mismatched (n=201) and two-antigen mismatched (n=267) CBT or matched (n=116) and mismatched (n=166) bone marrow. Compared to bone marrow, 5-year disease-free survival was slightly increased with matched CB and similar with 4/6 or 5/6 mismatched CB grafts. Relapse rates were lower after two-antigen HLA-mismatched unrelated CB transplants. Transplant-related mortality was higher after two-antigen mismatched CBT at any cell dose and one-antigen mismatched CBT at low cell doses (Total nucleated cell count (TNCC)<3×10^{-7}/kg), indicating that cell dose partially compensated for the degree of HLA mismatch.[20] The degree of HLA matching has also been further studied, with recent investigations suggesting that matching at HLA-C in addition to HLA-A,B, and DRB1 may reduce transplant-related mortality.[23] Recently, the Blood and Marrow Transplant Clinical Trials Network conducted a study to determine whether children with hematological malignancies would have improved survival after a double CBT, as compared to single CBT (BMT-CTN 0501). This study did not demonstrate an advantage for double CBT in children where a single CBT always provided an adequate cell dose.[24]

CBT has also been applied to transplantation of children with nonmalignant diseases including IMDs (discussed later in this chapter), primary immunodeficiencies, congenital marrow failure syndromes, and hemoglobinopathies. Unlike malignant conditions, in which complete donor engraftment is the goal for cure, a stable mixed chimerism state can be curative in many nonmalignant conditions. For that reason, reduced intensity conditioning (RIC), which carries lesser risks of transplant-related mortality but often leads to mixed chimerism, is the subject of investigation in many of these conditions. RIC in children with nonmalignant conditions is more frequently associated with graft rejection when CB is used as the source of donor cells, compared to adult cells which can be given in doses that are 10- to 100-fold higher than the dose contained in a single CB unit.

Allogeneic HSCT is curative for children with certain primary immunodeficiencies. The procedure has been being performed for this purpose since 1968, and successful outcomes using a matched related donor now exceed 90%.[25,26] In patients with certain types of severe combined immunodeficiency, matched related bone marrow may even be given without the need for preconditioning or GvHD prophylaxis. Haploidentical HSCT was introduced in the 1980s and is attractive since almost every child has a readily available haploidentical (parent) donor. However, haploidentical HSCT requires T-cell depletion of the graft, resulting in delayed immune reconstitution. In addition, only about 20% of recipients of haploidentical HSCT for primary immunodeficiency will have restoration of B-cell function.[26,27] CB grafts are readily available, do not require T-cell depletion, and have a decreased incidence of GvHD, making CB an attractive alternative donor source for patients with primary immunodeficiencies. In a recent study by Eruocord/EBMT comparing haploidentical HSCT versus CBT, survival rates were similar but the probability of discontinuing immunoglobulin supplementation was higher in patients receiving CB. Given the need to transplant patients with primary immunodeficiencies promptly, ideally before the onset of severe infection, CBT is a valuable alternative donor source if a matched related donor is not available.

As long as there are no pressing infectious issues, time to transplant is usually less hurried in the case of benign hematologic conditions such as inherited bone marrow failure syndromes, severe aplastic anemia, and hemoglobinopathies including thalassemia and sickle cell anemia. The majority of HSCTs for these diseases have utilized matched sibling donors, which continues to be the preferred donor source as it is associated with excellent outcomes, particularly in patients with beta thalassemia[28,29] and sickle cell disease.[30,31] Matched sibling CBT, after preparation with myeloablative chemotherapy, have also been successful.[32] However, most patients will not have a matched sibling bone marrow or CB donor available, requiring alternative donor sources. Initial experiences in unrelated donor CBT for sickle cell disease have been plagued by both transplant-related mortality and a high incidence of primary graft rejection.[33] While RIC regimens are attractive to minimize toxicity, they are also associated with increased risk of viral infections and may not be sufficient to overcome the immunologic barriers to engraftment in this population. In a recent retrospective study of unrelated donor CBT in patients with severe aplastic anemia, higher cell doses (TNCC > 3.9×10^7/kg) were associated with improved rates of engraftment, suggesting that higher cell doses may be necessary in patients with nonmalignant diseases.[34] Further studies are necessary to establish the most effective conditioning regimens, GvHD prophylaxis, graft criteria, and indications and timing of transplantation for these patients.

5. UMBILICAL CBT IN ADULTS

CBT was first extended to adult patients in the mid-1990s.[35] Initial studies of single CBT following myeloablative conditioning in high-risk, heavily pretreated adults were hampered by delayed engraftment and increased rates of primary graft failure, leading to high transplant-related mortality.[36] Similar to CBT in children, higher cell doses were associated with superior engraftment and event-free survival. As improvements have been made in patient selection, supportive care including use of growth factors, prophylactic and preemptive antiviral treatment, and selection of CBT units with higher nucleated cell doses/kg, outcome results have also improved.[37] More recent series in the USA, Europe, and Japan have indicated disease-free survivals of 40–70%, depending on patient age and disease status.[38–40]

Pretransplant comorbidities remain a significant source of transplant-related mortality in adults undergoing CBT. The use of RIC regimens has been employed to allow older patients and those with comorbid diseases to proceed safely to CBT. Disease-free survival at 1 year with the RIC approach has been reported at 40–60%.[41,42]

Cell dose has also been a limiting factor, as most single CB units do not contain enough cells to deliver the recommended minimum nucleated cell dose of > 2.5×10^7/kg for adult patients. Double CBT has been proposed as a method to overcome low cell doses for adults undergoing CBT. Compared to adult stem cell sources, double CBT has been associated with higher early transplant-related mortality, comparable overall survival, and significantly lower rates of relapse in adult patients with hematologic malignancies.[43–46] Despite multiple retrospective analyses, it is not clear whether single or double CBT is superior in adults.[47–49] Recent reviews show that outcomes of double and single CBT in adults are identical if the single unit provides an adequate cell dose (TNCC > 2.5×10^7/kg). The Spanish groups took a different approach to increase cell dose and pioneered the combination of a mismatched related or unrelated haploidentical peripheral blood CD34 cell donor with a single CBT. In this setting, initial neutrophil recovery is facilitated by the haploidentical donor while the CB donor provides durable engraftment.[50]

Novel strategies to improve engraftment and survival in adult CBT have included the use of ex vivo expansion, homing techniques, and infection prevention regimens. Elegant approaches to ex vivo expansion of CB cells are under development by several academic and biotech groups. These include expansion on Notch ligand, in the presence of cytokines plus nicotinamide, and on third-party mesenchymal stem cells.[51–53] Additional strategies to increase homing and migration of CB cells are also under development using prostaglandin E2, CD 26/dipeptidyl peptidase (DPP-IV), and fucosylation.[54–56] All of these approaches are in early clinical trials and showing promising results.[57] Strategies to support immune reconstitution are more challenging, but the emergence of new antiviral medications (Chimerix CMX001) and third-party cytotoxic T lymphocytes appear to have benefit in pilot clinical trials.[58]

6. HSCT AS A TREATMENT FOR IMDs

The IMDs are a heterogeneous group of genetic diseases, most of which involve a single gene mutation resulting in an enzyme defect. In the majority of cases, the enzyme defect leads to the accumulation of substrates that are toxic and/or interfere with normal cellular function. Oftentimes, patients may appear normal at birth but during infancy begin to exhibit disease manifestations, frequently including progressive neurological deterioration due to absent or abnormal brain myelination. The ultimate result is death in later infancy or childhood.

While enzyme replacement therapy (ERT) is available for selected IMDs and can be effective in ameliorating certain systemic disease manifestations, it does not cross the blood–brain barrier and therefore does not alter the progression of neurologic symptoms.[59,60] Currently, the only effective therapy to halt the neurologic progression of disease is allogeneic HSCT, which serves as a source of permanent cellular ERT.[61]

Following successful HSCT, the engrafted donor-derived HSCs provide a continuous endogenous source of the missing enzyme throughout the body, including the peripheral tissues as well as the central nervous system (CNS), for cross-correction of the defective metabolism.[62,63] Donor microglia cells of the brain, which are of myeloid origin, are thought to be the source of ERT after HSCT. These donor-derived cells not only act as normal scavengers in the CNS, but they can also secrete a portion of their lysosomal enzymes that can then be taken up by neighboring cells, thereby correcting the metabolic defect in affected host cells.[64–66] The timing of migration to, and engraftment of, donor-derived microglial cells after HSCT is not known but, based on clinical observations, is likely months after hematologic engraftment. It is also possible that donor cells exert anti-inflammatory and pro-neurogenic effects through paracrine signaling.

7. UMBILICAL CBT IN THE MUCOPOLYSACCHARIDOSES

The mucopolysaccharidoses (MPS), consisting of seven distinct clinical syndromes and numerous subtypes with a wide spectrum of clinical manifestations, represent about 35% of all lysosomal storage diseases. MPS are characterized by a progressive lysosomal accumulation of incompletely degraded glycosaminoglycans, previously termed mucopolysaccharides that typically cause disease manifestations including psychomotor retardation, musculoskeletal manifestations, vision and hearing impairment, and life-threatening cardiopulmonary failure.

Since 1980, when the first HSCT for a lysosomal storage disease was performed in a 1-year-old child with Hurler syndrome (MPS1) using bone marrow from his parents,[67] more than 500 HSCTs have been performed in Hurler syndrome worldwide, making it the most transplanted IMD. Numerous reports have demonstrated the effectiveness of HCST for Hurler syndrome including improvements in neurocognitive function, joint integrity, motor development, growth, hydrocephalus, corneal clouding, cardiac function, hepatosplenomegaly, hearing, visual and auditory processing, and overall survival following both bone marrow transplantation and CBT.[68–74] Factors associated with superior clinical outcomes include transplantation early in the course of the disease and achieving full-donor chimerism and normal enzyme levels posttransplant.[74–76]

In addition to CB's ready availability and more liberal HLA matching criteria, recent studies demonstrate that CB has additional advantages in the transplantation of MPS. Since most MPS are inherited in an autosomal recessive fashion, many related donors are heterozygous carriers and therefore have lower than normal levels of the affected enzyme. When CB is used as the donor source, potential units are screened and only used if they are noncarriers. A recent retrospective European Group for Blood and Marrow Transplantation (EBMT) study analyzed risk factors for graft failure in 146 with Hurler syndrome patients.[72] While this study showed no significant difference in survival and engraftment between the use of CB, bone marrow, and peripheral blood stem cells, significantly more patients receiving CB achieved full-donor chimerism (93% vs 67%) and normal enzyme levels (100% vs 72%) compared to patients receiving bone marrow or peripheral blood stem cells. Other studies using CB as an HSC source also showed high incidences of full-donor chimerism and normal enzyme levels in Hurler syndrome patients,[22,68,77,78] as well as in other lysosomal storage diseases.[22,77,79] Among patients receiving CBT for Hurler syndrome, a shorter interval between diagnosis and CBT (<4.6 months 82% vs >4.6 months 57%) and a conditioning regimen containing busulfan and cyclophosphamide (busulfan/cyclophosphamide 75% vs other 44%) are associated with a significantly higher event-free survival.[76]

Based on these observations, the EBMT developed transplantation guidelines for HSCT in MPS patients in 2005. In these guidelines, CB was prioritized as a donor source in the absence of a noncarrier matched sibling or fully matched unrelated donor, and myeloablative conditioning with busulfan/cyclophosphamide (later replaced with busulfan/fludarabine) with exposure-targeted intravenous busulfan was recommended. Since the introduction of these novel EBMT guidelines for HSCT in MPS patients, the survival and graft outcomes in patients with Hurler syndrome have been significantly improved with engrafted survival rates above 95% and low transplantation-related toxicity.[74] The use of CB in MPS patients has considerably increased, and fully matched CB grafts are now considered one of the most attractive sources, if not the most attractive source, for HSCT in MPS patients.

HSCT has been performed in MPS types other than Hurler syndrome, although often only as part of a larger heterogeneous cohort, or in small case reports.[22,77,80–87] Outcomes in these series have been variable, suggesting that some MPS diseases may be more sensitive to HSCT than others. Importantly, the observed high engrafted survival rates with low toxicity in Hurler patients may enable the extension of the HSCT indication to more "attenuated" but still severely affected MPS (sub)types.

8. UMBILICAL CBT IN THE LEUKODYSTROPHIES

The leukodystrophies are a group of disorders caused by genetic defects in the production or maintenance of myelin. Many affected children may appear normal at birth or early in infancy but then develop progressive deterioration in muscle tone, movements, gait, speech, ability to eat, vision, hearing, and behavior. HSCT has been shown to halt or slow disease progression in patients with Krabbe disease or globoid leukodystrophy, metachromatic leukodystrophy (MLD), adrenoleukodystrophy (ALD), and other leukodystrophies.[21–23,47–52] Stage of disease (e.g., presymptomatic, early symptomatic, or advanced), presentation (e.g., early infantile, late infantile, juvenile, or adult), and age of the patient all impact the prognosis and therefore the decision to proceed to transplantation.

More than 800 HSCTs have been performed for leukodystrophies worldwide since 1990, with the largest numbers of transplants in patients with ALD, followed by MLD, and lastly, Krabbe disease. A recent systematic review summarized the HSCT experience of 689 patients, including all donor sources.[88] Outcomes varied from poor to good, but follow-up in reports was short, demonstrating the need for long term, late outcome studies to be performed. In general, patients with presymptomatic or early disease had better outcomes with transplantation as compared to those with advanced and symptomatic disease at the time of transplantation, consistent with what has been described in disease-specific series.

The largest series of HSCT (including 53 CBT) for boys with ALD was reported from the University of Minnesota in 2011,[89] and Duke University described their experience in CBT for 12 boys with ALD, aged 2–11 years, in 2007.[90] In both series, boys with less disease manifestations (clinical and radiographic) at the time of transplantation had superior outcomes in terms of overall survival and neurologic outcomes, whereas symptomatic children demonstrated lower survival and rapid deterioration of neurologic function. Disease severity as demonstrated by Loes score on brain magnetic resonance imaging was also predictive of posttransplantation developmental outcome.

MLD is such a rapidly progressive disease that transplantation of babies with the early infantile form is only feasible if the diagnosis is known at the time of birth. The largest series of HSCT for children with MLD included 27 children, aged 4 months to 16 years old, with the late infantile (n=10) or juvenile (n=17) forms of the disease who underwent CBT at Duke University after myeloablative conditioning.[91] With a median follow-up of 5 years, 20/27 patients were survivors, for a long-term (5-year) survival probability of 74%. The late-infantile group had a 5-year survival probability of 60%, versus 82% in the juvenile group. Significant disease progression was noted in 10 patients (6 late-infantile onset, 4 juvenile onset). In the late-infantile onset group, only asymptomatic patients with minimal disease burden benefited from transplantation.

Krabbe disease, particularly the early infantile form, is a rapidly progressive and fatal condition. For that reason, most HSCTs for Krabbe disease have utilized CB as a graft source due to its ready availability. In 2005, the outcomes of 11 asymptomatic babies with Krabbe disease transplanted in the first month of life were reported along with 14 infants transplanted after the onset of symptoms.[79] With a median follow-up of 3 years, survival was dramatically increased in babies who underwent CBT prior to the development of symptoms (100% vs 42.8%). Symptomatic infants stabilized but did not demonstrate neurologic improvement. Newborns who underwent transplant with minimal-to-no symptoms of disease exhibited substantial neurodevelopmental gains in all areas of development compared to symptomatic infants and untreated patients. Nonetheless, some degree of gross motor function deficit became apparent in all the children.

In the USA, the success of CBT in infantile Krabbe disease and the need to intervene as early as possible led to the development of newborn screening programs for the disease, initiated in New York State in 2006. More than 2 million babies have been screened, with five cases (two siblings) of early infantile Krabbe disease detected. In addition, approximately 100 novel mutations in the GAL-C gene have been detected in babies that appear to be healthy at the present time. While newborn screening, leading to diagnosis in the presymptomatic state will enable more babies to have access to transplantation earlier in life, it has also highlighted the challenges of investigating false positive and indeterminate results.

As described above, results of CBT for IMDs suggest that greater benefit is likely when the transplant is performed early in the disease course prior to the development of clinical neurologic and other manifestations.[72,79,91] However, damage to the CNS occurs prenatally in some of these disorders. In addition, neurologic progression often occurs during and in the early months following HCST before sufficient numbers of donor cells engraft in the brain and produce adequate levels of the deficient enzyme. As a result, patients often experience a progressive loss of neurologic function for the first few months after transplantation before the disease stabilizes, and most patients are left with some residual and irreversible neurologic impairment. Additional approaches, such as augmented cellular therapies with CB- derived microglial-like cells (DUOC-01) and others,[92–94] gene therapies or supplemental enzyme therapy[95] or chaperone therapy will be needed alone or in combination with HSCT to normalize functional outcomes for these patients.

9. INVESTIGATIONS IN THE TREATMENT OF ACQUIRED BRAIN INJURIES WITH UMBILICAL CB

Observations using CB to treat children with genetic conditions led to the hypothesis that CB might also be beneficial in patients with acquired brain injuries. Over the past several years, much work has been done investigating CB cells in preclinical models of stroke, neonatal hypoxic-ischemic encephalopathy (HIE), traumatic brain injury, and spinal cord injury.

Numerous animal models have demonstrated neuroprotection,[96] neovascularization,[97] and neuronal regeneration[97] after CB administration leading to both neurological and survival benefits.[96,98–103] Based on these studies, early phase clinical trials of CB have begun in human patients with acquired brain injuries.

Clinical studies evaluating the use of CB in children with cerebral palsy are ongoing in both the USA and Korea. In a safety study of 184 infants and children with cerebral palsy (76%), congenital hydrocephalus (12%), and other brain injuries (12%), the only side effect associated with intravenous autologous CB infusions was a temporary hypersensitivity reaction (i.e., hives and/or wheezing) in approximately 1.5% of patients.[104] A randomized, double blind, placebo-controlled study is near completion to determine the efficacy of this approach. A similar study of allogeneic CB and erythropoietin was conducted in Korean children with cerebral palsy.[105] They reported greater improvements in cognitive and select motor functions in children who received CB and erythropoietin versus controls. There was no CB-only group for comparison.

In a phase I trial of newborns with HIE at birth conducted at Duke, fresh, noncryopreserved, volume- and RBC-reduced autologous CB was infused in 1, 2, or 4 doses of $1–5 \times 10^7$ nucleated cells/kg within the first 48–72 h of life in babies with moderate-to-severe encephalopathy qualifying for systemic hypothermia.[106] Babies who received CB infusions were compared to a concomitant group of babies treated at Duke who were cooled but did not receive CB cells. Infusions were found to be safe in these critically ill babies, and babies receiving cells had increased survival rates to discharge (100% vs 85%, p=0.20) and improved function at 1 year of age (74% vs 41% with development in the normal range, p=0.05). A phase II randomized trial is currently in development.

Intravenous infusion of autologous CB is also currently being investigated in young children with traumatic brain injury and autism. As most adults do not have an available autologous CB unit, studies of cell therapy for neurologic diseases in older patients have focused on bone marrow-derived cells. In an upcoming trial, allogeneic CB infusion will be investigated in adult patients who have suffered an acute stroke. Although the field of cell therapy as a possible treatment for acquired neurologic conditions is in its infancy, the relative availability, favorable safety profile, and pluripotential nature make CB a prime source of stem cells for such therapies.

10. SUMMARY

Over the past 25 years, CB has been established as a viable and, at times, preferred source of donor cells for hematopoietic reconstitution in allogeneic transplantation for all of the clinical indications treated with adult HSCs. CB expands access to transplantation for patients unable to find a fully matched donor and causes less acute and chronic GvHD whether or not full HLA matching is achieved. CBT is a unique option for treatment of children with IMDs affecting the brain and has potential for use in the treatment of acquired brain injuries. Looking ahead, CB also holds promise for use in the emerging fields of cellular therapies and regenerative medicine.

REFERENCES

1. Rubinstein P, Dobrila L, Rosenfield RE, et al. Processing and cryopreservation of placental/umbilical cord blood for unrelated bone marrow reconstitution. *Proc Natl Acad Sci USA* 1995;**92**(22):10119–22.
2. Bone Marrow Donors Worldwide. http://www.bmdw.org; 2014.
3. Ballen KK, Barker JN, Stewart SK, et al. Collection and preservation of cord blood for personal use. *Biol Blood Marrow Transpl* 2008;**14**(3):356–63.
4. Ecker JL, Greene MF. The case against private umbilical cord blood banking. *Obstetrics Gynecol* 2005;**105**(6):1282–4.
5. American Academy of Pediatrics Section on HO, American Academy of Pediatrics Section on AI, Lubin BH, Shearer WT. Cord blood banking for potential future transplantation. *Pediatrics* 2007;**119**(1):165–70.
6. Petz L, Jaing TH, Rosenthal J, et al. Analysis of 120 pediatric patients with nonmalignant disorders transplanted using unrelated plasma-depleted or -reduced cord blood. *Transfusion* 2012;**52**(6):1311–20.
7. Akel S, Regan D, Wall D, Petz L, McCullough J. Current thawing and infusion practice of cryopreserved cord blood: the impact on graft quality, recipient safety, and transplantation outcomes. *Transfusion* 2014;**54**(11):2997–3009.
8. Kurtzberg J, Cairo MS, Fraser JK, et al. Results of the cord blood transplantation (COBLT) study unrelated donor banking program. *Transfusion* 2005;**45**(6):842–55.
9. Gluckman E, Broxmeyer HA, Auerbach AD, et al. Hematopoietic reconstitution in a patient with Fanconi's anemia by means of umbilical-cord blood from an HLA-identical sibling. *N Engl J Med* 1989;**321**(17):1174–8.
10. Gluckman E. Hematopoietic stem-cell transplants using umbilical-cord blood. *N Engl J Med* 2001;**344**(24):1860–1.
11. Gluckman E, Rocha V, Boyer-Chammard A, et al. Outcome of cord-blood transplantation from related and unrelated donors. Eurocord Transplant Group and the European Blood and Marrow Transplantation Group. *N Engl J Med* 1997;**337**(6):373–81.
12. Kato S, Nishihira H, Sako M, et al. Cord blood transplantation from sibling donors in Japan. Report of the national survey. *Int J Hematol* 1998;**67**(4):389–96.
13. Wagner JE. Umbilical cord blood stem cell transplantation. *Am J Pediatr Hematol Oncol* 1993;**15**(2):169–74.

14. Wagner JE, Kernan NA, Steinbuch M, Broxmeyer HE, Gluckman E. Allogeneic sibling umbilical-cord-blood transplantation in children with malignant and non-malignant disease. *Lancet* 1995;**346**(8969):214–9.

15. Rocha V, Wagner Jr JE, Sobocinski KA, et al. Graft-versus-host disease in children who have received a cord-blood or bone marrow transplant from an HLA-identical sibling. Eurocord and International Bone Marrow Transplant Registry Working Committee on Alternative Donor and Stem Cell Sources. *N Engl J Med* 2000;**342**(25):1846–54.

16. Kurtzberg J, Laughlin M, Graham ML, et al. Placental blood as a source of hematopoietic stem cells for transplantation into unrelated recipients. *N Engl J Med* 1996;**335**(3):157–66.

17. Rubinstein P, Carrier C, Scaradavou A, et al. Outcomes among 562 recipients of placental-blood transplants from unrelated donors. *N Engl J Med* 1998;**339**(22):1565–77.

18. Gluckman E, Rocha V, Arcese W, et al. Factors associated with outcomes of unrelated cord blood transplant: guidelines for donor choice. *Exp Hematol* 2004;**32**(4):397–407.

19. Rocha V, Cornish J, Sievers EL, et al. Comparison of outcomes of unrelated bone marrow and umbilical cord blood transplants in children with acute leukemia. *Blood* 2001;**97**(10):2962–71.

20. Eapen M, Rubinstein P, Zhang MJ, et al. Outcomes of transplantation of unrelated donor umbilical cord blood and bone marrow in children with acute leukaemia: a comparison study. *Lancet* 2007;**369**(9577):1947–54.

21. Kurtzberg J, Prasad VK, Carter SL, et al. Results of the Cord Blood Transplantation Study (COBLT): clinical outcomes of unrelated donor umbilical cord blood transplantation in pediatric patients with hematologic malignancies. *Blood* 2008;**112**(10):4318–27.

22. Martin PL, Carter SL, Kernan NA, et al. Results of the cord blood transplantation study (COBLT): outcomes of unrelated donor umbilical cord blood transplantation in pediatric patients with lysosomal and peroxisomal storage diseases. *Biol Blood Marrow Transpl* 2006;**12**(2):184–94.

23. Eapen M, Klein JP, Sanz GF, et al. Effect of donor-recipient HLA matching at HLA A, B, C, and DRB1 on outcomes after umbilical-cord blood transplantation for leukaemia and myelodysplastic syndrome: a retrospective analysis. *Lancet Oncol* 2011;**12**(13):1214–21.

24. Wagner Jr JE, Eapen M, Carter S, et al. One-unit versus two-unit cord-blood transplantation for hematologic cancers. *N Engl J Med* 2014;**371**(18): 1685–94.

25. Gennery AR, Slatter MA, Grandin L, et al. Transplantation of hematopoietic stem cells and long-term survival for primary immunodeficiencies in Europe: entering a new century, do we do better?. *J Allergy Clin Immunol* 2010;**126**(3):602–10, e601–611.

26. Pai SY, Logan BR, Griffith LM, et al. Transplantation outcomes for severe combined immunodeficiency, 2000–2009. *N Engl J Med* 2014;**371**(5): 434–46.

27. Fernandes JF, Rocha V, Labopin M, et al. Transplantation in patients with SCID: mismatched related stem cells or unrelated cord blood? *Blood* 2012;**119**(12):2949–55.

28. Lawson SE, Roberts IA, Amrolia P, Dokal I, Szydlo R, Darbyshire PJ. Bone marrow transplantation for beta-thalassaemia major: the UK experience in two paediatric centres. *Br J Haematol* 2003;**120**(2):289–95.

29. Lucarelli G, Clift RA, Galimberti M, et al. Marrow transplantation for patients with thalassemia: results in class 3 patients. *Blood* 1996;**87**(5): 2082–8.

30. Bernaudin F, Socie G, Kuentz M, et al. Long-term results of related myeloablative stem-cell transplantation to cure sickle cell disease. *Blood* 2007;**110**(7):2749–56.

31. Walters MC, Storb R, Patience M, et al. Impact of bone marrow transplantation for symptomatic sickle cell disease: an interim report. Multicenter investigation of bone marrow transplantation for sickle cell disease. *Blood* 2000;**95**(6):1918–24.

32. Locatelli F, Rocha V, Reed W, et al. Related umbilical cord blood transplantation in patients with thalassemia and sickle cell disease. *Blood* 2003;**101**(6):2137–43.

33. Adamkiewicz TV, Szabolcs P, Haight A, et al. Unrelated cord blood transplantation in children with sickle cell disease: review of four-center experience. *Pediatr Transplant* 2007;**11**(6):641–4.

34. Peffault de Latour R, Purtill D, Ruggeri A, et al. Influence of nucleated cell dose on overall survival of unrelated cord blood transplantation for patients with severe acquired aplastic anemia: a study by eurocord and the aplastic anemia working party of the European group for blood and marrow transplantation. *Biol Blood Marrow Transpl* 2011;**17**(1):78–85.

35. Laporte JP, Gorin NC, Rubinstein P, et al. Cord-blood transplantation from an unrelated donor in an adult with chronic myelogenous leukemia. *N Engl J Med* 1996;**335**(3):167–70.

36. Laughlin MJ, Barker J, Bambach B, et al. Hematopoietic engraftment and survival in adult recipients of umbilical-cord blood from unrelated donors. *N Engl J Med* 2001;**344**(24):1815–22.

37. Ballen KK, Gluckman E, Broxmeyer HE. Umbilical cord blood transplantation: the first 25 years and beyond. *Blood* 2013;**122**(4):491–8.

38. Ruggeri A, Sanz G, Bittencourt H, et al. Comparison of outcomes after single or double cord blood transplantation in adults with acute leukemia using different types of myeloablative conditioning regimen, a retrospective study on behalf of Eurocord and the Acute Leukemia Working Party of EBMT. *Leukemia* 2014;**28**(4):779–86.

39. Sato A, Ooi J, Takahashi S, et al. Unrelated cord blood transplantation after myeloablative conditioning in adults with advanced myelodysplastic syndromes. *Bone Marrow Transplant* 2011;**46**(2):257–61.

40. Barker JN, Scaradavou A, Stevens CE. Combined effect of total nucleated cell dose and HLA match on transplantation outcome in 1061 cord blood recipients with hematologic malignancies. *Blood* 2010;**115**(9):1843–9.

41. Cutler C, Stevenson K, Kim HT, et al. Double umbilical cord blood transplantation with reduced intensity conditioning and sirolimus-based GVHD prophylaxis. *Bone Marrow Transplant* 2011;**46**(5):659–67.

42. Brunstein CG, Eapen M, Ahn KW, et al. Reduced-intensity conditioning transplantation in acute leukemia: the effect of source of unrelated donor stem cells on outcomes. *Blood* 2012;**119**(23):5591–8.

43. Barker JN, Fei M, Karanes C, et al. Results of a prospective multicentre myeloablative double-unit cord blood transplantation trial in adult patients with acute leukaemia and myelodysplasia. *Br J Haematol* 2015;**168**(3):405–12.

44. Brunstein CG, Gutman JA, Weisdorf DJ, et al. Allogeneic hematopoietic cell transplantation for hematologic malignancy: relative risks and benefits of double umbilical cord blood. *Blood* 2010;**116**(22):4693–9.

45. Gutman JA, Turtle CJ, Manley TJ, et al. Single-unit dominance after double-unit umbilical cord blood transplantation coincides with a specific CD8+ T-cell response against the nonengrafted unit. *Blood* 2010;**115**(4):757–65.

46. Ponce DM, Zheng J, Gonzales AM, et al. Reduced late mortality risk contributes to similar survival after double-unit cord blood transplantation compared with related and unrelated donor hematopoietic stem cell transplantation. *Biol Blood Marrow Transpl* 2011;**17**(9):1316–26.

47. Labopin M, Ruggeri A, Gorin NC, et al. Cost-effectiveness and clinical outcomes of double versus single cord blood transplantation in adults with acute leukemia in France. *Haematologica* 2014;**99**(3):535–40.

48. Ballen KK, Spitzer TR, Yeap BY, et al. Double unrelated reduced-intensity umbilical cord blood transplantation in adults. *Biol Blood Marrow Transpl* 2007;**13**(1):82–9.

49. Rocha V, Mohty M, Gluckman E, et al. Reduced-intensity conditioning regimens before unrelated cord blood transplantation in adults with acute leukaemia and other haematological malignancies. *Curr Opin Oncol* 2009;**21**(Suppl. 1):S31–4.

50. Bautista G, Cabrera JR, Regidor C, et al. Cord blood transplants supported by co-infusion of mobilized hematopoietic stem cells from a third-party donor. *Bone Marrow Transplant* 2009;**43**(5):365–73.

51. de Lima M, McNiece I, Robinson SN, et al. Cord-blood engraftment with ex vivo mesenchymal-cell coculture. *N Engl J Med* 2012;**367**(24):2305–15.

52. Delaney C, Heimfeld S, Brashem-Stein C, Voorhies H, Manger RL, Bernstein ID. Notch-mediated expansion of human cord blood progenitor cells capable of rapid myeloid reconstitution. *Nat Med* 2010;**16**(2):232–6.

53. Horwitz ME, Chao NJ, Rizzieri DA, et al. Umbilical cord blood expansion with nicotinamide provides long-term multilineage engraftment. *J Clin Invest* 2014;**124**(7):3121–8.

54. Cutler C, Multani P, Robbins D, et al. Prostaglandin-modulated umbilical cord blood hematopoietic stem cell transplantation. *Blood* 2013;**122**(17):3074–81.

55. Farag SS, Srivastava S, Messina-Graham S, et al. In vivo DPP-4 inhibition to enhance engraftment of single-unit cord blood transplants in adults with hematological malignancies. *Stem Cells Dev* 2013;**22**(7):1007–15.

56. Robinson SN, Thomas MW, Simmons PJ, et al. Fucosylation with fucosyltransferase VI or fucosyltransferase VII improves cord blood engraftment. *Cytotherapy* 2014;**16**(1):84–9.

57. Norkin M, Lazarus HM, Wingard JR. Umbilical cord blood graft enhancement strategies: has the time come to move these into the clinic? *Bone Marrow Transplant* 2013;**48**(7):884–9.

58. Hanley PJ, Bollard CM, Brunstein CG. Adoptive immunotherapy with the use of regulatory T cells and virus-specific T cells derived from cord blood. *Cytotherapy* 2015;**17**(6):749–55.

59. Shull RM, Kakkis ED, McEntee MF, Kania SA, Jonas AJ, Neufeld EF. Enzyme replacement in a canine model of Hurler syndrome. *Proc Natl Acad Sci USA* 1994;**91**(26):12937–41.

60. Tokic V, Barisic I, Huzjak N, Petkovic G, Fumic K, Paschke E. Enzyme replacement therapy in two patients with an advanced severe (Hurler) phenotype of mucopolysaccharidosis I. *Eur J Pediatr* 2007;**166**(7):727–32.

61. Krivit W, Peters C, Shapiro EG. Bone marrow transplantation as effective treatment of central nervous system disease in globoid cell leukodystrophy, metachromatic leukodystrophy, adrenoleukodystrophy, mannosidosis, fucosidosis, aspartylglucosaminuria, Hurler, Maroteaux-Lamy, and Sly syndromes, and Gaucher disease type III. *Curr Opin Neurol* 1999;**12**(2):167–76.

62. Di Ferrante N, Nichols BL, Donnelly PV, Neri G, Hrgovcic R, Berglund RK. Induced degradation of glycosaminoglycans in Hurler's and Hunter's syndromes by plasma infusion. *Proc Natl Acad Sci USA* 1971;**68**(2):303–7.

63. Knudson Jr AG, Di Ferrante N, Curtis JE. Effect of leukocyte transfusion in a child with type II mucopolysaccharidosis. *Proc Natl Acad Sci USA* 1971;**68**(8):1738–41.

64. Krivit W, Sung JH, Shapiro EG, Lockman LA. Microglia: the effector cell for reconstitution of the central nervous system following bone marrow transplantation for lysosomal and peroxisomal storage diseases. *Cell Transpl* 1995;**4**(4):385–92.

65. Unger ER, Sung JH, Manivel JC, Chenggis ML, Blazar BR, Krivit W. Male donor-derived cells in the brains of female sex-mismatched bone marrow transplant recipients: a Y-chromosome specific in situ hybridization study. *J Neuropathol Exp Neurol* 1993;**52**(5):460–70.

66. Neufeld EFMJ. The mucopolysaccharidoses. In: 8th ed. Scriver CR, Sly W, Valle D, editors. *The metabolic and molecular bases of inherited disease*, vol. III. McGraw-Hill; 2001. p. 3421–52.

67. Hobbs JR, Hugh-Jones K, Barrett AJ, et al. Reversal of clinical features of Hurler's disease and biochemical improvement after treatment by bone-marrow transplantation. *Lancet* 1981;**2**(8249):709–12.

68. Staba SL, Escolar ML, Poe M, et al. Cord-blood transplants from unrelated donors in patients with Hurler's syndrome. *N Engl J Med* 2004;**350**(19):1960–9.

69. Peters C, Shapiro EG, Anderson J, et al. Hurler syndrome: II. Outcome of HLA-genotypically identical sibling and HLA-haploidentical related donor bone marrow transplantation in fifty-four children. The Storage Disease Collaborative Study Group. *Blood* 1998;**91**(7):2601–8.

70. Souillet G, Guffon N, Maire I, et al. Outcome of 27 patients with Hurler's syndrome transplanted from either related or unrelated haematopoietic stem cell sources. *Bone Marrow Transpl* 2003;**31**(12):1105–17.

71. Bjoraker KJ, Delaney K, Peters C, Krivit W, Shapiro EG. Long-term outcomes of adaptive functions for children with mucopolysaccharidosis I (Hurler syndrome) treated with hematopoietic stem cell transplantation. *J Dev Behav Pediatr* 2006;**27**(4):290–6.

72. Boelens JJ, Wynn RF, O'Meara A, et al. Outcomes of hematopoietic stem cell transplantation for Hurler's syndrome in Europe: a risk factor analysis for graft failure. *Bone Marrow Transpl* 2007;**40**(3):225–33.

73. Aldenhoven M, Boelens JJ, de Koning TJ. The clinical outcome of Hurler syndrome after stem cell transplantation. *Biol Blood Marrow Transpl* 2008;**14**(5):485–98.

74. Aldenhoven M, Wynn RF, Orchard PJ, et al. Long-term outcome of Hurler syndrome patients after hematopoietic cell transplantation: an international multi-center study. *Blood* 2015;**125**(13):2164–72.

75. Boelens JJ, Aldenhoven M, Purtill D, et al. Outcomes of transplantation using various hematopoietic cell sources in children with Hurler syndrome after myeloablative conditioning. *Blood* 2013;**121**(19):3981–7.

76. Boelens JJ, Rocha V, Aldenhoven M, et al. Risk factor analysis of outcomes after unrelated cord blood transplantation in patients with hurler syndrome. *Biol Blood Marrow Transpl* 2009;**15**(5):618–25.

77. Prasad VK, Mendizabal A, Parikh SH, et al. Unrelated donor umbilical cord blood transplantation for inherited metabolic disorders in 159 pediatric patients from a single center: influence of cellular composition of the graft on transplantation outcomes. *Blood* 2008;**112**(7):2979–89.

78. Church H, Tylee K, Cooper A, et al. Biochemical monitoring after haemopoietic stem cell transplant for Hurler syndrome (MPSIH): implications for functional outcome after transplant in metabolic disease. *Bone Marrow Transplant* 2007;**39**(4):207–10.

79. Escolar ML, Poe MD, Provenzale JM, et al. Transplantation of umbilical-cord blood in babies with infantile Krabbe's disease. *N Engl J Med* 2005;**352**(20):2069–81.

80. Boelens JJ. Trends in haematopoietic cell transplantation for inborn errors of metabolism. *J Inherit Metab Dis* 2006;**29**(2–3):413–20.

81. Annibali R, Caponi L, Morganti A, Manna M, Gabrielli O, Ficcadenti A. Hunter syndrome (Mucopolysaccharidosis type II), severe phenotype: long term follow-up on patients undergone to hematopoietic stem cell transplantation. *Minerva Pediatr* 2013;**65**(5):487–96.

82. Guffon N, Bertrand Y, Forest I, Fouilhoux A, Froissart R. Bone marrow transplantation in children with Hunter syndrome: outcome after 7 to 17 years. *J Pediatr* 2009;**154**(5):733–7.

83. Jester S, Larsson J, Eklund EA, et al. Haploidentical stem cell transplantation in two children with mucopolysaccharidosis VI: clinical and biochemical outcome. *Orphanet J Rare Dis* 2013;**8**:134.

84. Peters C, Steward CG. National Marrow Donor P, International Bone Marrow Transplant R, Working Party on Inborn Errors EBMTG. Hematopoietic cell transplantation for inherited metabolic diseases: an overview of outcomes and practice guidelines. *Bone Marrow Transplant* 2003;**31**(4):229–39.

85. Tanaka A, Okuyama T, Suzuki Y, et al. Long-term efficacy of hematopoietic stem cell transplantation on brain involvement in patients with mucopolysaccharidosis type II: a nationwide survey in Japan. *Mol Genet Metabolism* 2012;**107**(3):513–20.

86. Vellodi A, Young E, Cooper A, Lidchi V, Winchester B, Wraith JE. Long-term follow-up following bone marrow transplantation for Hunter disease. *J Inherit Metabolic Dis* 1999;**22**(5):638–48.

87. Vellodi A, Young E, New M, Pot-Mees C, Hugh-Jones K. Bone marrow transplantation for Sanfilippo disease type B. *J Inherit Metabolic Dis* 1992;**15**(6):911–8.

88. Musolino PL, Lund TC, Pan J, et al. Hematopoietic stem cell transplantation in the leukodystrophies: a systematic review of the literature. *Neuropediatrics* 2014;**45**(3):169–74.

89. Miller WP, Rothman SM, Nascene D, et al. Outcomes after allogeneic hematopoietic cell transplantation for childhood cerebral adrenoleukodystrophy: the largest single-institution cohort report. *Blood* 2011;**118**(7):1971–8.

90. Beam D, Poe MD, Provenzale JM, et al. Outcomes of unrelated umbilical cord blood transplantation for X-linked adrenoleukodystrophy. *Biol Blood Marrow Transpl* 2007;**13**(6):665–74.

91. Martin HR, Poe MD, Provenzale JM, Kurtzberg J, Mendizabal A, Escolar ML. Neurodevelopmental outcomes of umbilical cord blood transplantation in metachromatic leukodystrophy. *Biol Blood Marrow Transpl* 2013;**19**(4):616–24.

92. Tracy E, Aldrink J, Panosian J, et al. Isolation of oligodendrocyte-like cells from human umbilical cord blood. *Cytotherapy* 2008;**10**(5):518–25.

93. Tracy ET, Zhang CY, Gentry T, Shoulars KW, Kurtzberg J. Isolation and expansion of oligodendrocyte progenitor cells from cryopreserved human umbilical cord blood. *Cytotherapy* 2011;**13**(6):722–9.

94. Kurtzberg J, Buntz S, Gentry T, et al. Preclinical characterization of DUOC-01, a cell therapy product derived from banked umbilical cord blood for use as an adjuvant to umbilical cord blood transplantation for treatment of inherited metabolic diseases. *Cytotherapy* 2015;**17**(6):803–15.

95. Li Y, Sands MS. Experimental therapies in the murine model of globoid cell leukodystrophy. *Pediatr Neurol* 2014;**51**(5):600–6.

96. Vendrame M, Cassady J, Newcomb J, et al. Infusion of human umbilical cord blood cells in a rat model of stroke dose-dependently rescues behavioral deficits and reduces infarct volume. *Stroke* 2004;**35**(10):2390–5.

97. Taguchi A, Soma T, Tanaka H, et al. Administration of CD34+ cells after stroke enhances neurogenesis via angiogenesis in a mouse model. *J Clin Invest* 2004;**114**(3):330–8.

98. Chen J, Sanberg PR, Li Y, et al. Intravenous administration of human umbilical cord blood reduces behavioral deficits after stroke in rats. *Stroke* 2001;**32**(11):2682–8.

99. Meier C, Middelanis J, Wasielewski B, et al. Spastic paresis after perinatal brain damage in rats is reduced by human cord blood mononuclear cells. *Pediatr Res* 2006;**59**(2):244–9.

100. Nan Z, Grande A, Sanberg CD, Sanberg PR, Low WC. Infusion of human umbilical cord blood ameliorates neurologic deficits in rats with hemorrhagic brain injury. *Ann N. Y. Acad Sci* 2005;**1049**:84–96.

101. Lu D, Sanberg PR, Mahmood A, et al. Intravenous administration of human umbilical cord blood reduces neurological deficit in the rat after traumatic brain injury. *Cell Transpl* 2002;**11**(3):275–81.

102. Zhao ZM, Li HJ, Liu HY, et al. Intraspinal transplantation of CD34+ human umbilical cord blood cells after spinal cord hemisection injury improves functional recovery in adult rats. *Cell Transpl* 2004;**13**(2):113–22.

103. Nishio Y, Koda M, Kamada T, et al. The use of hemopoietic stem cells derived from human umbilical cord blood to promote restoration of spinal cord tissue and recovery of hindlimb function in adult rats. *J Neurosurg Spine* 2006;**5**(5):424–33.

104. Sun J, Allison J, McLaughlin C, et al. Differences in quality between privately and publicly banked umbilical cord blood units: a pilot study of autologous cord blood infusion in children with acquired neurologic disorders. *Transfusion* 2010;**50**(9):1980–7.

105. Min K, Song J, Kang JY, et al. Umbilical cord blood therapy potentiated with erythropoietin for children with cerebral palsy: a double-blind, randomized, placebo-controlled trial. *Stem Cells* 2013;**31**(3):581–91.

106. Cotten CM, Murtha AP, Goldberg RN, et al. Feasibility of autologous cord blood cells for infants with hypoxic-ischemic encephalopathy. *J Pediatr* 2014;**164**(5):973–9.

Chapter 20

Mobilizing Endogenous Stem Cells for Retinal Repair

Honghua Yu, Mays Talib, Thi H. Khanh Vu, Kin-Sang Cho, Chenying Guo, Dong F. Chen

Key Concepts

1. Retinal degenerative diseases cause retinal neuron death that leads to permanent loss of vision.
2. Stem cell-based therapy offers a promise for restoring vision in retinal degenerative diseases through a cell replacement or neuron regenerative approach.
3. As retinal stem-like cells exist in the adult mammalian retina, mobilizing endogenous stem cells may provide a less invasive method than the stem cell transplantation strategy.
4. Directed differentiation of stem-like cells into desired retinal neurons and establishment of functional connectivity remain as major challenges for stem cell therapy to regain vision.

List of Abbreviations

α-AA α-Aminoadipic acid
AC Amacrine cell
AMD Age-related macular degeneration
BMCs Bone marrow-derived cells
BP Bipolar cell
CE Ciliary epithelium
CMZ Ciliary marginal zone
CNS Central nervous system
Dnmts DNA methyltransferases
EGF Epidermal growth factor
ESCs Embryonic stem cells
FGF Fibroblast growth factor
H3K27me3 Histone H3 lysine 27 trimethylation
HC Horizontal cell
HDAC1 Histone deacetylatase 1
hESCs Human embryonic stem cells
IGF Insulin-like growth factor
iPSCs Induced pluripotent stem cells
MC Müller cell
MSCs Mesenchymal stem cells
NFL Nerve fiber layer
PCs Progenitor cells
PR Photoreceptor
PRCs Polycomb repressive complexes
RGCs Retinal ganglion cells
RPCs Retinal progenitor cells
RPE Retinal pigment epithelium

Translating Regenerative Medicine to the Clinic. http://dx.doi.org/10.1016/B978-0-12-800548-4.00020-6

1. INTRODUCTION

The retina, as the most accessible part of the central nervous system (CNS), is susceptible to degeneration as a result of genetic mutation or acquired conditions. A variety of diseases can cause retinal neurodegeneration, leading to irreversible blindness. These include conditions that cause photoreceptor death, such as age-related macular degeneration (AMD), retinitis pigmentosa, and cone or rod dystrophy, or damage to the optic nerve and retinal ganglion cells (RGCs), such as glaucoma and optic neuritis. These diseases share common pathophysiological features like permanent loss of retinal neurons.

Recent advancements in pharmacological therapies, for example, the antiangiogenic treatment for patients with neovascular AMD,[1,2] have been successful in slowing down the progression of certain retinal diseases or prevent further deterioration of function. However, no treatments are available to completely halt neurodegeneration or enable regeneration and reestablishment of retinal functions in patients once the neurons are lost.

With recent progress, stem cell therapy either by transplanting stem cells or by recruiting endogenous stem cell populations is emerging as a new approach that has the potential to reverse vision loss after retinal degeneration or damage. Attempts have been made in human trials to replace those lost through harvesting and transplanting donor stem cells into the eyes of patients with retinal degenerative diseases, and several clinical trials are in progress.[3,4] The exciting findings in successful restoration of sight in both human and animal models suggest the feasibility of reversing vision loss through a regenerative approach. To this end, new neurons may originate either from an engrafted or from an endogenous source of stem/progenitor cells. Cell transplantation is still a complex multistep process, even though transplanted stem cells have the capacity to proliferate, differentiate into various cell lineages, and repopulate the host retina. Drug-based regenerative therapy that aims at mobilizing the endogenous progenitor cell population to repair the retina may offer many advantages over the transplantation approach. These include less concerns about immune rejection, neuron integration, tumor formation, and disease transmission by implanted cells. The idea of retinal repair through mobilizing endogenous stem cells presents an attractive approach that intends to relieve vision loss in patients by generating and preserving the disease-afflicted cells with their own cells. The eye being a relatively small organ presents a special advantage in this approach as it reduces the number of cells required for regenerative therapies—a critical barrier to cell-based approach. To date, the field is rapidly advancing with encouraging results.

2. SOURCES OF ENDOGENOUS STEM/PROGENITOR CELLS

The concept that the adult mammalian CNS contains populations of resident neural stem/progenitor cells was accepted two decades ago.[5,6] Emerging evidence suggests that Müller cells (MCs) are dormant stem-like cells found throughout the retina and serve as a source of progenitor cells to regenerate retinal neurons after injury.[7,8] In addition, ciliary epithelia-derived cells, retinal pigment epithelium (RPE), and bone marrow-derived cells (BMCs) have also been reported as potential sources of progenitor cells that can be mobilized to the injured retina (Figures 1 and 2).

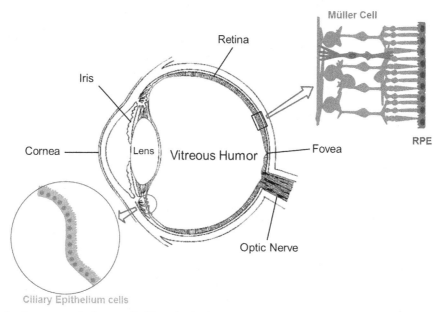

FIGURE 1 **Sources of endogenous retinal stem cells.** The retina has been shown to contain a few populations of endogenous stem cells, including Müller cells in the neural retina, retinal pigment epithelium cells, and ciliary epithelium cells in the ciliary margin zone. RPE, retinal pigment epithelium.

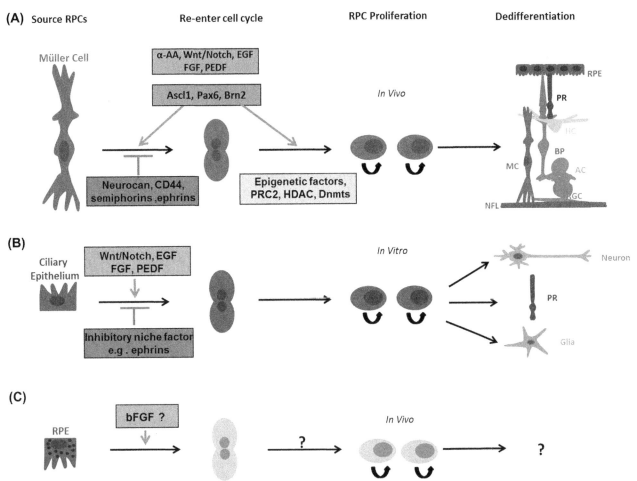

FIGURE 2   Müller cells (MCs), retinal pigment epithelium, and ciliary epithelium cells are sources of retinal stem cells with the potential for retinal self-repair in mammals. (A) Latent MCs are stimulated or suppressed to reenter the cell cycle by different factors (green (light dark gray in print version) or red (dark gray in print version) boxes). MCs can be induced to proliferate by growth factors and transcription factors listed in the green boxes and differentiate into various retinal cell types. Modification of epigenetic factors may also contribute to the regulation of the stem cell potency of MCs. (B) It has been shown that in vitro, ciliary epithelium cells isolated from mature retina can be stimulated (green (light dark gray in print version) box) or suppressed (red (dark gray in print version) box) to reenter the cell cycle, differentiating into new neurons, including photoreceptors (PR), and glia cells (GC). (C) Retinal pigment epithelium has also been shown to possess the ability to proliferate and transdifferentiate into retinal neurons, but the neurogenic potential of these cells in mammals are fairly restricted and appear to be deficient in the regulatory elements that are required for induction of transdifferentiation. EGF: epidermal growth factor; FGF: fibroblast growth factor; PEDF: pigment epithelium-derived factor; α-AA: α-aminoadipic acid; Ascl1: achaete–scute complex-like 1; Pax6: paired box gene 6; Brn2: brain-2; PRC2: polycomb repressive complex 2; HDAC: histone deacetylatase 1; Dnmts: DNA methyltransferases; RPC: retinal progenitor cell; PR: photoreceptor; RPE: retinal pigment epithelium; HC: horizontal cell; BP: bipolar cell; MC: Müller cell; AC: amacrine cell; RGC: retinal ganglion cell; NFL: nerve fiber layer.

Lower vertebrates, such as fish and amphibian, are capable of regenerating the retina, and MCs are thought to serve as the primary source of retinal progenitors.[9] After injury, quiescent MCs reenter the cell cycle and dedifferentiate to form multipotent progenitors that subsequently generate all retinal neuron types that repair the retina and restore visual function.[10–16] Over the past decade, efforts have been placed to investigate whether retinal neuroregeneration can be induced from MCs of adult mammals, such as mouse and rat.

MCs of adult mammals share many properties of retinal progenitor cells (RPCs). They express the same neurogenic genes, such as Notch and Wnt, as those found in the fish,[17,18] and can be reprogrammed in a dish to become retinal neural or photoreceptor progenitors.[19,20] In vivo, it has been shown that, by targeting specific signaling pathways through administering fibroblast growth factor (FGF),[21] Notch,[22,23] Wnt,[24–26] or sonic hedgehog,[27] a significant number of MCs can be induced to reenter the cell cycle and display properties of retinal progenitors. While transcription factor Ascl1a was shown to be required for retinal regeneration in the fish,[12,14,28] recent report indicates that overexpressing a single transcription factor, Ascl1, is also sufficient to induce a neurogenic state of mature MCs of mice.[29] These results suggest that some part of the regenerative program occurring in nonmammalian vertebrates remains in the MCs of mammalian retina, which may be induced for retinal repair in patients with retinal degeneration.

The ciliary marginal zone of lower vertebrates, such as teleost and amphibians, is also known to harbor a pool of RPCs capable of producing new retinal neurons throughout life.[30] A population of multipotent RPCs has been isolated from the ciliary epithelium (CE) of adult rodents and humans that shows the capacity to generate various retinal cell types in vitro.[31,32] However, their ability to proliferate and generate new retinal neurons, such as photoreceptors, appears to be limited in vivo.[33,34] Mitogens, including basic FGF, insulin, Wnt3a, and pigment epithelium-derived factor, are found to promote the proliferative potential of CE-derived RPCs.[35–39] Transcription factors, such as OTX2, Crx, and Chx10, increase the photoreceptor progeny of CE-derived RPCs.[40] Nevertheless, the neurogenic potential of these cells in birds and mammals is fairly restricted. There has been scarce evidence suggesting that these cells contribute to retinal regeneration after injury in adult mammals or birds.

In urodela of amphibians, such as salamanders, RPE cells located between the retina and choroid are capable of transdifferentiating into neurons and regenerating the entire retina.[41,42] The regenerative process usually starts with the dedifferentiation of pigmented cells, which then proceed to depigmentation, reentrance of the cell cycle, and expression of progenitor cell genes.[43] In mammals, RPE of embryonic rats has also been shown to possess the ability to transdifferentiate into retinal neurons and develop into neural retina, but only during the earliest developmental stage.[44] In addition, peripheral RPE cells of adult rats retain the capacity to enter the cell cycle and complete cellular division in vivo, although they divide at a low cycling rate.[45] Interestingly, RPE cells from adult humans are reported as being capable of generating stable RPE and differentiating into mesenchymal lineages in vitro.[46] Besides, it has been shown that cultured human RPE cells can differentiate into neurons that are positive for beta-III tubulin, MAP2, and neurofilament proteins; whereas, no photoreceptor or glial marker positive cells were observed in these cultures.[47] Mammalian RPEs appear to be deficient in the regulatory elements that are required for induction of transdifferentiation.[45] The neurogenic potential of these cells in birds and mammals is fairly restricted. Evidence suggesting that these cells contribute to retinal regeneration after injury in adult mammals or birds is limited.[48]

Some studies have described the ability of BMCs to cross lineage boundaries and express tissue-specific proteins in different organs.[49–51] In mice, it has been shown that endogenous BMCs can migrate to the subretinal space in the damaged retina, presumably to initiate or participate in neural repair.[52–54] However, no evidence has been suggested that they can transdifferentiate into cells with anatomical or functional characteristics of retinal neurons.[53,54] To date, MCs are the best characterized mammalian cell type that shows RPC properties and generates new retinal neurons after injury in adult mammals.[55–57]

3. NICHE SIGNALS AND STEM CELL POTENTIAL

Neuroregenerative potential of RPCs depends on the intrinsic properties of both neural stem cells and the environment, or "niche," in which stem cells reside. The regenerative properties of MCs are evolutionarily conserved. In contrast to lower vertebrates, mammals have lost the ability to regenerate retinal neurons, likely due to the constraints of the nonneurogenic environment of the adult retina.[58] In rodent retinas, for example, the MCs become reactive and hypertrophic in response to injury, but few reenter the cell cycle—a first step toward MC transdifferentiation into RPCs.[58] Treating the retina with exogenous activating factors after damage has been shown to induce proliferation of endogenous RPCs. For example, Wnt3a,[24] epidermal growth factor (EGF),[59] FGF,[60] insulin-like growth factor (IGF),[61] retinoic acid,[24] Notch,[62] N-methyl-N-nitrosourea,[63] and α-aminoadipic acid (α-AA)[55] have all been reported to stimulate the proliferation from at least a subpopulation of MCs.

MC proliferation has been studied in a variety of species, both in vitro and in vivo, and some of these mitogenic factors are better characterized. In the posthatch chick, a combination of insulin and FGF causes a large percentage of the MCs to proliferate. P2Y-receptor activation stimulates MCs proliferation in guinea pig, as does the activation of the platelet-derived growth factor receptor.[64] Perhaps the best-studied mitogenic factor for MCs is EGF, which stimulates MCs proliferation of mice,[59] rats,[65] rabbits,[66] guinea pigs,[64] and humans.[67] Intraocular injection of EGF significantly increases the number of BrdU-positive MCs in adult rats after light damage.[65] Wnt3a also induces BrdU-positive MCs in retinal explant cultures, where retinal damage inevitably occurs during culture preparation.[24]

Several groups have studied retinal damage-induced MC proliferation in mice[25,55,56] and reported that mouse MCs could be induced to proliferate when neurotoxic damage was coupled with growth factor stimulation. Interestingly, mouse MCs can be stimulated to proliferate in the absence of neural death by a subtoxic dose of glutamate or alpha-aminoadipate (α-AA).[55] Together, these observations implicate that the nonneurogenic environment of adult mammals may present an inhibitory niche that suppresses the regenerative potential of MCs. Similar to MCs, CE-derived RSCs have also been found to be capable of reentering the cell cycle in the presence of certain mitogens, including basic fibroblast growth factor, insulin, Wnt3a, and pigment epithelium-derived factor.[35–39] Likely, there is a large overlap in the molecular pathways that regulate the proliferative and regenerative potentials of RPCs of different sources.

Recent studies have also begun to unveil the components of the negative niche factors that inhibit the regenerative potential of the adult retina and brain. Among them, ephrin-A and EphA receptors are thought to act as important players.[68–70] For example, adult neurogenesis is detected in two restricted areas, the subgranular zone (SGZ) of the hippocampus and the subventricular zone (SVZ) of the CNS; however, neural stem cells are found widely distributed throughout the adult CNS, including those that are considered the nonneurogenetic areas.[71] The astrocytes in areas outside of the neurogenic SGZ and SVZ of adult mice express high levels of ephrin-A2 and -A3, which present an inhibitory niche negatively regulating neural stem cell growth.[72] Adult mice lacking ephrin-A2 and -A3 display active ongoing neurogenesis throughout the CNS.[72] Interactions of ephrin-A/EphA family with neurogenic signals, such as Wnt and FGF, have been documented,[73–75] further supporting a role of ephrin-As in the regulation of neural stem cell behavior. Recent reports indicate that ephrin-As also play a key inhibitory role in the developing retina and adult CE to suppress stem cell proliferation and retinogenesis via suppressing Wnt signaling.[76] Together, these data indicate a novel mechanism associated with ephrin-A/EphAs as endogenous modulators in the control of neurogenesis and regeneration in the adult CNS.

Studies in the brain and retina have suggested that the limitation of retinal neuroregeneration can be attributed to an assortment of factors, including the presence of inhibitory extracellular matrix or cell adhesion molecules. Among these factors, injury-induced inhibitory molecules, such as chondroitin sulfate proteoglycans, neurocan, hyaluronan-binding glycoprotein CD44, semaphorins, and ephrins, are particularly abundant in the retina.[76–79] These molecules have previously been shown to function as inhibitory cues for neurite and axonal extension and block the regeneration of nerve fibers as well as neural stem cell growth.[77,80] The retina, as part of the CNS, upregulates these growth inhibitory molecules after injury. Both neurocan and CD44 are expressed in the normal retina, and injury stimulates the increased production of these proteins in reactive retinal astrocytes and MCs.[81] Upregulation of these inhibitory molecules leads to an inhospitable environment for the proliferation of RPCs and neuron regeneration,[76] and hinders successful RPC transplantation by blocking donor–host integration.[82] In addition, the most recent study showed that BMP2, BMP4, and sFRP2 secreted by the adult lens and cornea inhibit CE-derived RPCs and contribute to adult RPCs quiescence in the adult mammalian eye.[83] Thus, to induce regeneration and restore visual function through transplantation, neutralization of these inhibitory proteins may be necessary.

Together, both stimulating the growth pathways and blocking the effects of extracellular inhibitory molecules of neurons have been proved to be beneficial for promoting retinal regeneration after diseases or injury. Although the detailed mechanisms involved remain to be determined, a two-pronged approach might be the most efficient way of promoting retinal regeneration and repair after injury.

4. INTRACELLULAR SIGNALS AND TRANSCRIPTIONAL REGULATION

Niche factors must signal through intracellular pathways to regulate stem cell behavior. The self-renewal of stem cells is tightly controlled through concerted actions of intrinsic transcription factors and networks. Moreover, it has been shown that aging generally has a negative impact on the proliferation and regenerative potential of neural stem cells. Adult neurogenesis in the brain decreases with aging, as a result of both reduced proliferation and differentiation of newborn cells; in parallel with it, an age-associated decline in cognitive performance is often observed. The age of the animal at the time of testing also seems to be a crucial determinant of the ability of MCs to reenter the cell cycle. There is a large decline in the response of MCs to mitogens during the second postnatal week in mice.[56] This loss in responsiveness to EGF correlates with the downregulation of EGF receptor that occurs over the same period. These studies support the notion that the failure of retinal neurons to regenerate in the adult is not a mere outcome of changes in the extraneuronal milieus; rather, the intrinsic nature of RPCs also contributes critically to their regenerative capacity.

In recent years, some key intracellular signals and transcription factors that control the intrinsic growth programs of retinal progenitors have been identified. Ascl1 is emerging as a key regulator determining the neuronal fate of glial lineage neural stem cells. Forced expression of Ascl1 induced a neurogenic state of mature MCs of mice,[29] suggesting it as a potential target for stimulating retinal neuroregenerative therapy after disease or injury. Moreover, it is evident that expression of Ascl1 along with two other transcription factors, Brn2 and Myt1l, using viral gene delivery directly converted fibroblasts into neurons,[84] suggesting a crucial involvement of Ascl1 in neuronal fate determination.

Pax6 gene encodes a transcription factor controlling retinal neurogenesis and regeneration.[85–87] In the vertebrate retina, Pax6 is highly expressed during retinal development to maintain the multipotency and proliferation capacity of retinal progenitors.[88] Pax6 is also detected in a subpopulation of MCs of adult mice where it may be involved in the molecular response to retinal injury.[11,18] It has been showed that photoreceptor injury induces migration of Pax6-positive MC nuclei toward the outer retinal layers. These cells express markers of cell cycle, implicating their potential to reenter the cell cycle similarly to that seen in lower vertebrates.[89] Moreover, Pax6 is upregulated in the MCs of mice after *N*-methyl-D-aspartate glutamate induced neural damage or Wnt3a treatment, as are components of the Notch pathway, Dll1, Notch1, and Nestin.[17,24,56] Studies with the human retina have indicated that many progenitor cell genes (e.g., Pax6 and Sox2) are reactivated in the mammalian MCs after damage and mitotic stimulation.[67]

Crx, NeuroD, Nrl, and Nr2e3 are the major transcription factors known to be involved in photoreceptor genesis during development thus far. Crx and NeuroD are expressed in photoreceptors of the developing and mature retina, and are essential for precise differentiation and maturation.[90] Nrl is exclusively expressed in rod photoreceptors and is essential for their development and maintenance.[91] Ex vivo studies show that Nr2e3 acts synergistically with Nrl and inhibits the activation of cone genes by Crx.[92,93]

Despite the increased understanding of the roles of the molecular signals in the regulation of retinal regeneration, to date, successful repair of the damaged or diseased retina remains a challenge. The critical issue hampering our understanding of the mechanisms controlling retinal regeneration lies in the complexity of the problem and its potential involvement of multiple factors. In order to develop clinically feasible and applicable therapies, studies are needed to further elucidate the interactive effects of these factors as well as the mechanisms underlying the regulation of the proliferation and regenerative behavior of RPCs.

5. EPIGENETIC REGULATION OF STEM CELL POTENTIAL

Epigenetics is one of the most promising and expanding fields in the current biomedical research landscape. The term generally refers to chromatin modifications that persist from one stage of cell division to the next stage. It involves heritable alterations of gene expression without changes in DNA sequence, and contributes to the diversity of gene expression and memory of cell lineage. Epigenetics is believed to play a major role in retinal development and cell specification, partly through stabilizing transcriptional programs in embryonic progenitors and differentiated descendants, and establishing and maintaining gene expression in RPCs in the postnatal life. Thus, epigenetic mechanism is a likely avenue which should be explored to change the plasticity of RPCs and enhance the endogenous regenerative potential of the retina.

Epigenetic regulation includes histone modifications, DNA methylation, and other mechanisms, which work together to establish and maintain the global and local condensed or decondensed chromatin states to determine gene expression.[94–96] Disruption of epigenetic machineries is known to provoke aberrant gene expression patterns that give rise to developmental defect. Histone modifications, including histone methylation and acetylation, are areas of intensive interest. In part, this is because chemical compounds that manipulate these processes have been recently identified and some have been shown to affect retinal neurons survival.[97,98]

The histone methyltransferase complex, termed polycomb repressive complexes (PRCs), controls key steps in developmental transitions and cell fate choices.[97,99] PRC2 methyltransferase activity, for example, catalyzes the addition of histone H3 lysine 27 trimethylation (H3K27me3) to specific genomic loci, which act as docking sites for recruiting additional repressive complexes. PRC2 regulates the progression of retinal progenitors from proliferation to differentiation. In *Xenopus*, the PRC2 core components are enriched in retinal progenitors and downregulated in differentiated cells. Knockdown of the PRC2 core component Ezh2 leads to reduced retinal progenitor proliferation, in part due to upregulation of the Cdk inhibitor p15Ink4b. In addition, although PRC2 knockdown does not alter eye patterning or the expression of retinal progenitor genes, such as Sox2, it does cause suppression of proneural basic helix-loop-helix gene expression. These studies indicate that PRC2 is crucial for the initiation of neural differentiation in the retina. Consistent with these observations, knocking down or blocking PRC2 function constrains the generation of most retinal neural cell types and promotes a Müller glial cell fate decision.[97]

It is thought that histone acetylation promotes a more open chromatin structure that allows for gene transcription, while histone methylation in general stabilizes transcriptional programs in progenitor cells and their differentiated descendants. Thus, histone modification is crucial for establishing and maintaining gene expression in cell's postnatal life.[100,101] Accordingly, alterations in histone acetylation may cause behavioral changes of RPCs, such as death and aberrant differentiation. In contrast, mutations associated with histone methylation are likely to result in long-term consequences on cell survival and function. Histone deacetylation generally represses gene promoters and is used to silence genes during differentiation. Loss of histone deacetylase 1 (HDAC1) in zebrafish leads to retinal overproliferation and inhibition of differentiation through activation of the Wnt and Notch pathways.[102] In mouse retinal explants, HDAC inhibition resulted in defects in rod differentiation but also, unlike in zebrafish, in a reduction in proliferation.[103]

Another major form of epigenetic regulation is DNA methylation. Although the role of cell-specific DNA methylation in the retina is still unclear, one potential mechanism may be that it helps direct proper lineage decisions and differentiation of retinal precursor cells. Recent study demonstrates that DNA methyltransferases (Dnmts) are involved in development of the vertebrate eyes. High levels of mouse Dnmts expression are observed during early stages of retinal differentiation.[104] In the zebrafish embryo, knockdown of the maintenance Dnmts by a translation-blocking antisense morpholino results in a profound disorganization of all retinal layers.[105]

Epimutation, defined as abnormal transcriptional repression or activation of genes caused by mutations in epigenetic modulators,[106] is generally considered reversible in comparison to genetic mutations. The potential to pharmacologically modify gene expression through the manipulation of epigenetic regulation is currently an area of intense interest. The research in this field may unveil novel pathways underlying stem cell regulation and lead to new epigenetic drug targets for boosting the regenerative potential of endogenous stem cells for treating and reversing vision loss.

6. FUNCTIONAL RESTORATION OF RETINAL NEURONS

Translation of stem cell biology into clinical application for retinal degenerative disorders via endogenous stem cells must overcome three major obstacles: The first obstacle concerns developing methods to mobilize endogenous stem cells to sufficiently proliferate and restore the lost cell numbers; the second obstacle regards directing the targeted differentiation of endogenous stem cells into retinal progenitors capable of regenerating desired retinal cell types in vivo; finally, enabling restoration of sight newly generated neurons must integrate into the neural circuitry, form synaptic contacts with the existing neurons, and establish functional connectivity. By repopulating an injured retina with newly generated neurons in fish, these cells have been shown to develop functional connections with the existing circuitry and restored sight.[107] Emerging studies on stem cell-based therapy for targeted neuron replacements, such as photoreceptors and RGCs, indicate that directed neurodifferentiation by endogenous stem cells could also be achieved in adult mammals.

Photoreceptors are photosensitive cells, and their degeneration is a major cause of blindness worldwide, partly due to their incapability of regeneration or self-repair. Stem cell therapy for photoreceptor replacement provides an exciting prospect for restoring sight in those whose vision is significantly impaired by retinal disease affecting primarily the photoreceptors. Currently, no treatments are available that can effectively reverse vision loss due to photoreceptor degeneration. Various sources of cells, including neural precursors,[108] embryonic and postnatal RPCs,[109] neural stem cell lines,[110] and bone marrow stem cells,[111] have been tested for their ability to differentiate and replace photoreceptors. An appropriate source of precursor cells is a key for photoreceptor cell replacement therapy. Postmitotic photoreceptor precursor cells could easily be derived from RPCs isolated from neonatal mouse retinas (P1–P5).[112] However, equivalent human RPCs would have to be derived from second-trimester fetus.[112] Aside from ethical considerations, such tissues are in limited provision and might not provide a consistent source of cells for retinal cell transplantation.[112]

Considerable progress has been made in differentiating embryonic stem cells (ESCs) in vitro toward a neural retinal precursor phenotype that is competent to generate photoreceptor-like cells.[109,113] Opsin- and rhodopsin-positive cells are obtained after subretinal grafting of human ESCs, indicating the potential of human ESCs to differentiate into retinal cells, while the subretinal microenvironment supports their differentiation toward a photoreceptor cell fate.[114] New rod and cone photoreceptors have also been successfully generated from ESCs from mouse, monkey, and human.[115–120] Most recent study has demonstrated that retinal stem cells isolated from the adult retina have the potential of producing functional photoreceptor cells that can integrate into the retina, morphologically resembling endogenous photoreceptors, and forming synapses with resident retinal neurons.[121] Both structural integration of grafted cells and improvement of pupillary reflex have been reported after transplantation of photoreceptor precursors into a mouse model of retinal degeneration.[122]

Currently, many laboratories have reported an increase in proliferation of mammalian MCs—an endogenous source of RPCs—and their migration into the injured areas of the retina.[17,25,55,123] However, it remains unclear if the newly developed neurons can integrate and allow restoration or improvement of visual function. A number of studies using a transplantation approach further support the extraordinary potential of cell replacement therapy in functionally refurbishing damaged retinal tissues. These include the studies directed toward the creation of new photoreceptors[116,124] or RPE by grafted stem cells.[125–127] A variety of different cell types have been tested in their ability to restore retinal function. ESCs, RPCs, and photoreceptor precursor cells have all been shown to form new functioning photoreceptors and improve retinal function following transplantation into the degenerative retinal hosts.[116,122] Moreover, there has been evidence that transplanted cells are capable of forming synaptic contacts with local retinal neurons, suggesting that functional communication between newly generated cells with native retinal neurons can be developed for improvement of visual function.

Compared with photoreceptors, replacing lost RGCs is a more challenging task. This is because successful replacement of RGCs requires not only the survival, migration, and integration of donor cells into the ganglion cell layer and differentiation into RGC-like cells. These cells must also extend long axons which navigate through the optic disc, entering the optic nerve through the lamina cribrosa. Newly generated RGC axons must be properly myelinated in the optic nerve and continuously extend and make the right way (cross or not cross) into the chiasm, and finally establish functional and topographical connections to the central visual targets. Due to these challenges, efforts at transplantation-based replacement of RGCs are still lagging behind.

ESCs,[128,129] RPCs,[130] and MC-derived stem cells[131] have been investigated for their potential of replacing RGCs in treating retinal degenerative diseases caused by RGC dysfunction. Interestingly, these results suggest that the stem cells are capable of migrating and integrating into the retina depleted of RGCs or populated by apoptotic RGCs, expressing RGC markers and extending neurites.[132] iPSCs-derived RGC-like cells with electrophysiological properties similar to RGCs have also been generated from mouse fibroblasts through adenoviral gene delivery.[133] However, these cells showed limited ability to integrate into the normal retina after transplantation. Some evidence suggests that a glaucomatous retina presents even a less permissive environment to the integration of transplanted cells.[134] Mesenchymal stem cells,[135] on the other hand, failed completely to migrate into the injured eye after intravenous engraftment; nevertheless, some neuroprotective effect was observed following transplantation into a rat model of glaucoma.[136,137] To date, evidence for synaptic integration and functional improvement by stem cell-derived RGCs remains elusive.

7. CONCLUSIONS AND FUTURE DIRECTIONS

Intense efforts and substantial progress have been made in the field of retinal neuroregeneration in the last decade. Emerging evidence suggests that the mammalian retina contains subpopulations of stem-like cells, including primarily MCs and CE-derived RPCs, in addition to RPE and BMC populations, which may retain the capacity for neuronal regeneration under certain conditions. Limitation of retinal regeneration in adult mammals reflects both intrinsic inability of retinal neurons to reinitiate robust regeneration and lack of a permissive environment for such growth.

The regenerative strategy by mobilizing endogenous stem cells to participate in retinal repair has several advantages over cell transplantation therapy as it does not need to face the shortage of donor cells nor diseases or disorders that may be transmitted via implant; it avoids potential immune rejections. Moreover, endogenously derived RPCs are generally thought to be better programmed to differentiate into retinal neurons and integrate into the existing neural circuitry than exogenously transplanted stem cells. Recent research has shown that adult human and mouse MCs can be induced to reenter the cell cycle and regenerate new neurons in vitro and in vivo following stimulation by a single compound, although the number and diversity of regenerative neurons is still limited. Practically, a drug therapy for stimulating residential RPCs derived from the patient's own retina would be clinically more viable than transplanting exogenous cells. This is because the (drug therapy) method reduces the concerns over the ethical issues associated with the use of ESCs, while injection of a drug solution into the eye is a clinically established procedure and considered less invasive than cell transplantation. Patient's native MCs (stem-like) are likely to be competent to generate retinal-specific cells. Currently, the primary challenge of inducing retinal regeneration from Müller progenitors falls onto the limited number of MCs that can be activated to reenter the cell cycle and participate in regeneration and repairing process. The tumorogenic potential of Müller progenitors is much less of an issue as compared to transplanted hESCs and iPSCs. To our knowledge, there has not been any report of tumor formation by Müller progenitors in vitro or in vivo upon mitogen stimulation. Although barriers to regenerative cell survival, migration, and integration as well as long-term efficacy and safety concerns remain to be overcome, endogenous retinal repair is progressing at a rapid pace and may soon turn the endogenous stem cell approach into a viable therapy.

ACKNOWLEDGMENT

All authors have read the journal's policy on conflicts of interest and have none to declare.

REFERENCES

1. Rosenfeld PJ. Bevacizumab versus ranibizumab for AMD. *N Engl J Med* May 19, 2011;**364**(20):1966–7.
2. Martin DF, Maguire MG, Ying GS, Grunwald JE, Fine SL, Jaffe GJ. Ranibizumab and bevacizumab for neovascular age-related macular degeneration. *N Engl J Med* May 19, 2011;**364**(20):1897–908.
3. Cramer AO, MacLaren RE. Translating induced pluripotent stem cells from bench to bedside: application to retinal diseases. *Curr Gene Ther* April 2013;**13**(2):139–51.
4. Medina RJ, Archer DB, Stitt AW. Eyes open to stem cells: safety trial may pave the way for cell therapy to treat retinal disease in patients. *Stem Cell Res Ther* 2011;**2**(6):47.
5. Reynolds BA, Weiss S. Generation of neurons and astrocytes from isolated cells of the adult mammalian central nervous system. *Science* March 27, 1992;**255**(5052):1707–10.
6. Richards LJ, Kilpatrick TJ, Bartlett PF. De novo generation of neuronal cells from the adult mouse brain. *Proc Natl Acad Sci USA* September 15, 1992;**89**(18):8591–5.
7. Karl MO, Reh TA. Regenerative medicine for retinal diseases: activating endogenous repair mechanisms. *Trends Mol Med* April 2010;**16**(4): 193–202.
8. Fischer AJ, Reh TA. Potential of Muller glia to become neurogenic retinal progenitor cells. *Glia* July 2003;**43**(1):70–6.
9. Jones BW, Watt CB, Frederick JM, et al. Retinal remodeling triggered by photoreceptor degenerations. *J Comp Neurol* September 8, 2003;**464**(1): 1–16.
10. Raymond PA, Barthel LK, Bernardos RL, Perkowski JJ. Molecular characterization of retinal stem cells and their niches in adult zebrafish. *BMC Dev Biol* 2006;**6**:36.
11. Bernardos RL, Barthel LK, Meyers JR, Raymond PA. Late-stage neuronal progenitors in the retina are radial Muller glia that function as retinal stem cells. *J Neurosci* June 27, 2007;**27**(26):7028–40.

12. Fausett BV, Gumerson JD, Goldman D. The proneural basic helix-loop-helix gene ascl1a is required for retina regeneration. *J Neurosci* January 30, 2008;**28**(5):1109–17.

13. Thummel R, Kassen SC, Enright JM, Nelson CM, Montgomery JE, Hyde DR. Characterization of Muller glia and neuronal progenitors during adult zebrafish retinal regeneration. *Exp Eye Res* November 2008;**87**(5):433–44.

14. Ramachandran R, Fausett BV, Goldman D. Ascl1a regulates Muller glia dedifferentiation and retinal regeneration through a Lin-28-dependent, *let-7* microRNA signalling pathway. *Nat Cell Biol* November 2010;**12**(11):1101–7.

15. Fausett BV, Goldman D. A role for alpha1 tubulin-expressing Muller glia in regeneration of the injured zebrafish retina. *J Neurosci* June 7, 2006;**26**(23):6303–13.

16. Ramachandran R, Reifler A, Parent JM, Goldman D. Conditional gene expression and lineage tracing of tuba1a expressing cells during zebrafish development and retina regeneration. *J Comp Neurol* October 15, 2010;**518**(20):4196–212.

17. Das AV, Mallya KB, Zhao X, et al. Neural stem cell properties of Muller glia in the mammalian retina: regulation by Notch and Wnt signaling. *Dev Biol* November 1, 2006;**299**(1):283–302. Epub 2006 July 29.

18. Roesch K, Jadhav AP, Trimarchi JM, et al. The transcriptome of retinal Muller glial cells. *J Comp Neurol* July 10, 2008;**509**(2):225–38.

19. Garcia M, Forster V, Hicks D, Vecino E. Effects of muller glia on cell survival and neuritogenesis in adult porcine retina in vitro. *Invest Ophthalmol Vis Sci* December 2002;**43**(12):3735–43.

20. Giannelli SG, Demontis GC, Pertile G, Rama P, Broccoli V. Adult human Muller glia cells are a highly efficient source of rod photoreceptors. *Stem Cells* February 2011;**29**(2):344–56.

21. Fischer AJ, McGuire CR, Dierks BD, Reh TA. Insulin and fibroblast growth factor 2 activate a neurogenic program in Muller glia of the chicken retina. *J Neurosci* November 1, 2002;**22**(21):9387–98.

22. Del Debbio CB, Balasubramanian S, Parameswaran S, Chaudhuri A, Qiu F, Ahmad I. Notch and Wnt signaling mediated rod photoreceptor regeneration by Muller cells in adult mammalian retina. *PLoS One* 2010;**5**(8):e12425.

23. Hayes S, Nelson BR, Buckingham B, Reh TA. Notch signaling regulates regeneration in the avian retina. *Dev Biol* December 1, 2007;**312**(1):300–11.

24. Osakada F, Ooto S, Akagi T, Mandai M, Akaike A, Takahashi M. Wnt signaling promotes regeneration in the retina of adult mammals. *J Neurosci* April 11, 2007;**27**(15):4210–9.

25. Liu B, Hunter DJ, Rooker S, et al. Wnt signaling promotes Muller cell proliferation and survival after injury. *Invest Ophthalmol Vis Sci* January 2013;**54**(1):444–53.

26. Sanges D, Romo N, Simonte G, et al. Wnt/beta-Catenin signaling triggers neuron reprogramming and regeneration in the mouse retina. *Cell Rep* July 25, 2013;**4**(2):271–86.

27. Wan J, Zheng H, Xiao HL, She ZJ, Zhou GM. Sonic hedgehog promotes stem-cell potential of Muller glia in the mammalian retina. *Biochem Biophys Res Commun* November 16, 2007;**363**(2):347–54.

28. Ramachandran R, Zhao XF, Goldman D. Ascl1a/Dkk/beta-catenin signaling pathway is necessary and glycogen synthase kinase-3beta inhibition is sufficient for zebrafish retina regeneration. *Proc Natl Acad Sci USA* September 20, 2011;**108**(38):15858–63.

29. Pollak J, Wilken MS, Ueki Y, et al. ASCL1 reprograms mouse Muller glia into neurogenic retinal progenitors. *Development* June 2013;**140**(12):2619–31.

30. Kubota R, Hokoc JN, Moshiri A, McGuire C, Reh TA. A comparative study of neurogenesis in the retinal ciliary marginal zone of homeothermic vertebrates. *Brain Res Dev Brain Res* March 31, 2002;**134**(1–2):31–41.

31. MacNeil A, Pearson RA, MacLaren RE, Smith AJ, Sowden JC, Ali RR. Comparative analysis of progenitor cells isolated from the iris, pars plana, and ciliary body of the adult porcine eye. *Stem Cells* October 2007;**25**(10):2430–8.

32. Xu H, Sta Iglesia DD, Kielczewski JL, et al. Characteristics of progenitor cells derived from adult ciliary body in mouse, rat, and human eyes. *Invest Ophthalmol Vis Sci* April 2007;**48**(4):1674–82.

33. Gualdoni S, Baron M, Lakowski J, et al. Adult ciliary epithelial cells, previously identified as retinal stem cells with potential for retinal repair, fail to differentiate into new rod photoreceptors. *Stem Cells* June 2010;**28**(6):1048–59.

34. Cicero SA, Johnson D, Reyntjens S, et al. Cells previously identified as retinal stem cells are pigmented ciliary epithelial cells. *Proc Natl Acad Sci USA* April 21, 2009;**106**(16):6685–90.

35. Zhao X, Das AV, Soto-Leon F, Ahmad I. Growth factor-responsive progenitors in the postnatal mammalian retina. *Dev Dyn* February 2005;**232**(2):349–58.

36. Abdouh M, Bernier G. In vivo reactivation of a quiescent cell population located in the ocular ciliary body of adult mammals. *Exp Eye Res* July 2006;**83**(1):153–64.

37. Kubo F, Nakagawa S. Hairy1 acts as a node downstream of Wnt signaling to maintain retinal stem cell-like progenitor cells in the chick ciliary marginal zone. *Development* June 2009;**136**(11):1823–33.

38. Inoue T, Kagawa T, Fukushima M, et al. Activation of canonical Wnt pathway promotes proliferation of retinal stem cells derived from adult mouse ciliary margin. *Stem Cells* January 2006;**24**(1):95–104.

39. De Marzo A, Aruta C, Marigo V. PEDF promotes retinal neurosphere formation and expansion in vitro. *Adv Exp Med Biol* 2010;**664**:621–30.

40. Inoue T, Coles BL, Dorval K, et al. Maximizing functional photoreceptor differentiation from adult human retinal stem cells. *Stem Cells* March 31, 2010;**28**(3):489–500.

41. Ikegami Y, Mitsuda S, Araki M. Neural cell differentiation from retinal pigment epithelial cells of the newt: an organ culture model for the urodele retinal regeneration. *J Neurobiol* February 15, 2002;**50**(3):209–20.

42. Susaki K, Chiba C. MEK mediates in vitro neural transdifferentiation of the adult newt retinal pigment epithelium cells: is FGF2 an induction factor? *Pigment Cell Res* October 2007;**20**(5):364–79.

43. Araki M. Regeneration of the amphibian retina: role of tissue interaction and related signaling molecules on RPE transdifferentiation. *Dev Growth Differ* February 2007;**49**(2):109–20.

44. Zhao S, Thornquist SC, Barnstable CJ. In vitro transdifferentiation of embryonic rat retinal pigment epithelium to neural retina. *Brain Res* April 24, 1995;**677**(2):300–10.

45. Al-Hussaini H, Kam JH, Vugler A, Semo M, Jeffery G. Mature retinal pigment epithelium cells are retained in the cell cycle and proliferate in vivo. *Mol Vis* 2008;**14**:1784–91.

46. Salero E, Blenkinsop TA, Corneo B, et al. Adult human RPE can be activated into a multipotent stem cell that produces mesenchymal derivatives. *Cell Stem Cell* January 6, 2012;**10**(1):88–95.

47. Amemiya K, Haruta M, Takahashi M, Kosaka M, Eguchi G. Adult human retinal pigment epithelial cells capable of differentiating into neurons. *Biochem Biophys Res Commun* March 26, 2004;**316**(1):1–5.

48. Wohl SG, Schmeer CW, Isenmann S. Neurogenic potential of stem/progenitor-like cells in the adult mammalian eye. *Prog Retin Eye Res* May 2012;**31**(3):213–42.

49. Orlic D, Kajstura J, Chimenti S, et al. Bone marrow cells regenerate infarcted myocardium. *Nature* April 5, 2001;**410**(6829):701–5.

50. Krause DS, Theise ND, Collector MI, et al. Multi-organ, multi-lineage engraftment by a single bone marrow-derived stem cell. *Cell* May 4, 2001;**105**(3):369–77.

51. LaBarge MA, Blau HM. Biological progression from adult bone marrow to mononucleate muscle stem cell to multinucleate muscle fiber in response to injury. *Cell*. November 15, 2002;**111**(4):589–601.

52. Li Y, Reca RG, Atmaca-Sonmez P, et al. Retinal pigment epithelium damage enhances expression of chemoattractants and migration of bone marrow-derived stem cells. *Invest Ophthalmol Vis Sci* April 2006;**47**(4):1646–52.

53. Li Y, Atmaca-Sonmez P, Schanie CL, Ildstad ST, Kaplan HJ, Enzmann V. Endogenous bone marrow derived cells express retinal pigment epithelium cell markers and migrate to focal areas of RPE damage. *Invest Ophthalmol Vis Sci* September 2007;**48**(9):4321–7.

54. Machalinska A, Klos P, Baumert B, et al. Stem Cells are mobilized from the bone marrow into the peripheral circulation in response to retinal pigment epithelium damage–a pathophysiological attempt to induce endogenous regeneration. *Curr Eye Res* July 2011;**36**(7):663–72.

55. Takeda M, Takamiya A, Jiao JW, et al. alpha-Aminoadipate induces progenitor cell properties of Muller glia in adult mice. *Invest Ophthalmol Vis Sci* March 2008;**49**(3):1142–50.

56. Karl MO, Hayes S, Nelson BR, Tan K, Buckingham B, Reh TA. Stimulation of neural regeneration in the mouse retina. *Proc Natl Acad Sci USA* December 9, 2008;**105**(49):19508–13.

57. Ooto S, Akagi T, Kageyama R, et al. Potential for neural regeneration after neurotoxic injury in the adult mammalian retina. *Proc Natl Acad Sci USA* September 14, 2004;**101**(37):13654–9.

58. Bringmann A, Iandiev I, Pannicke T, et al. Cellular signaling and factors involved in Muller cell gliosis: neuroprotective and detrimental effects. *Prog Retin Eye Res* November 2009;**28**(6):423–51.

59. Ueki Y, Reh TA. EGF stimulates Muller glial proliferation via a BMP-dependent mechanism. *Glia* May 2013;**61**(5):778–89.

60. Romo P, Madigan MC, Provis JM, Cullen KM. Differential effects of TGF-beta and FGF-2 on in vitro proliferation and migration of primate retinal endothelial and Muller cells. *Acta Ophthalmol* May 2011;**89**(3):e263–8.

61. Ikeda T, Waldbillig RJ, Puro DG. Truncation of IGF-I yields two mitogens for retinal Muller glial cells. *Brain Res* July 17, 1995;**686**(1):87–92.

62. Wang Z, Sugano E, Isago H, Murayama N, Tamai M, Tomita H. Notch signaling pathway regulates proliferation and differentiation of immortalized Muller cells under hypoxic conditions in vitro. *Neuroscience* July 12, 2012;**214**:171–80.

63. Wan J, Zheng H, Chen ZL, Xiao HL, Shen ZJ, Zhou GM. Preferential regeneration of photoreceptor from Muller glia after retinal degeneration in adult rat. *Vis Res* January 2008;**48**(2):223–34.

64. Milenkovic I, Weick M, Wiedemann P, Reichenbach A, Bringmann A. P2Y receptor-mediated stimulation of Muller glial cell DNA synthesis: dependence on EGF and PDGF receptor transactivation. *Invest Ophthalmol Vis Sci* March 2003;**44**(3):1211–20.

65. Close JL, Liu J, Gumuscu B, Reh TA. Epidermal growth factor receptor expression regulates proliferation in the postnatal rat retina. *Glia* August 1, 2006;**54**(2):94–104.

66. Sagar SM, Edwards RH, Sharp FR. Epidermal growth factor and transforming growth factor alpha induce c-fos gene expression in retinal Muller cells in vivo. *J Neurosci Res* August 1991;**29**(4):549–59.

67. Bhatia B, Jayaram H, Singhal S, Jones MF, Limb GA. Differences between the neurogenic and proliferative abilities of Muller glia with stem cell characteristics and the ciliary epithelium from the adult human eye. *Exp Eye Res* December 2011;**93**(6):852–61.

68. Theus MH, Ricard J, Bethea JR, Liebl DJ. EphB3 limits the expansion of neural progenitor cells in the subventricular zone by regulating p53 during homeostasis and following traumatic brain injury. *Stem Cells* July 2010;**28**(7):1231–42.

69. Depaepe V, Suarez-Gonzalez N, Dufour A, et al. Ephrin signalling controls brain size by regulating apoptosis of neural progenitors. *Nature* June 30, 2005;**435**(7046):1244–50. Epub 2005 May 15.

70. Holmberg J, Armulik A, Senti KA, et al. Ephrin-A2 reverse signaling negatively regulates neural progenitor proliferation and neurogenesis. *Genes Dev* February 15, 2005;**19**(4):462–71.

71. Jiao J, Chen DF. Induction of neurogenesis in nonconventional neurogenic regions of the adult central nervous system by niche astrocyte-produced signals. *Stem Cells* May 2008;**26**(5):1221–30.

72. Jiao JW, Feldheim DA, Chen DF. Ephrins as negative regulators of adult neurogenesis in diverse regions of the central nervous system. *Proc Natl Acad Sci USA* June 24, 2008;**105**(25):8778–83.

73. Stolfi A, Wagner E, Taliaferro JM, Chou S, Levine M. Neural tube patterning by Ephrin, FGF and Notch signaling relays. *Development* December 2011;**138**(24):5429–39.

74. Yokote H, Fujita K, Jing X, et al. Trans-activation of EphA4 and FGF receptors mediated by direct interactions between their cytoplasmic domains. *Proc Natl Acad Sci USA* December 27, 2005;**102**(52):18866–71.

75. Lim BK, Cho SJ, Sumbre G, Poo MM. Region-specific contribution of ephrin-B and Wnt signaling to receptive field plasticity in developing optic tectum. *Neuron* March 25, 2010;**65**(6):899–911.

76. Fang Y, Cho KS, Tchedre K, et al. Ephrin-A3 suppresses Wnt signaling to control retinal stem cell potency. *Stem Cells* February 2013;**31**(2):349–59.

77. Busch SA, Silver J. The role of extracellular matrix in CNS regeneration. *Curr Opin Neurobiol* February 2007;**17**(1):120–7.

78. Silver J, Miller JH. Regeneration beyond the glial scar. *Nat Rev Neurosci* February 2004;**5**(2):146–56.

79. Kita EM, Bertolesi GE, Hehr CL, Johnston J, McFarlane S. Neuropilin-1 biases dendrite polarization in the retina. *Development* July 2013;**140**(14):2933–41.

80. Ponta H, Sherman L, Herrlich PA. CD44: from adhesion molecules to signalling regulators. *Nat Rev Mol Cell Biol* January 2003;**4**(1):33–45.

81. Krishnamoorthy R, Agarwal N, Chaitin MH. Upregulation of CD44 expression in the retina during the rds degeneration. *Brain Res Mol Brain Res* April 14, 2000;**77**(1):125–30.

82. Zhang Y, Klassen HJ, Tucker BA, Perez MT, Young MJ. CNS progenitor cells promote a permissive environment for neurite outgrowth via a matrix metalloproteinase-2-dependent mechanism. *J Neurosci* April 25, 2007;**27**(17):4499–506.

83. Balenci L, Wonders C, L.K-T B, Clarke L, van der Kooy D. Bmps and Sfrp2 maintain the quiescence of adult mammalian retinal stem cells. *Stem Cells* July 10, 2013;**31**:2218–30.

84. Vierbuchen T, Ostermeier A, Pang ZP, Kokubu Y, Sudhof TC, Wernig M. Direct conversion of fibroblasts to functional neurons by defined factors. *Nature* February 25, 2010;**463**(7284):1035–41.

85. Nabeshima A, Nishibayashi C, Ueda Y, Ogino H, Araki M. Loss of cell-extracellular matrix interaction triggers retinal regeneration accompanied by Rax and Pax6 activation. *Genesis* June 2013;**51**(6):410–9.

86. Gehring WJ, Ikeo K. Pax 6: mastering eye morphogenesis and eye evolution. *Trends Genet* September 1999;**15**(9):371–7.

87. Insua MF, Simon MV, Garelli A, de Los Santos B, Rotstein NP, Politi LE. Trophic factors and neuronal interactions regulate the cell cycle and Pax6 expression in Muller stem cells. *J Neurosci Res* May 15, 2008;**86**(7):1459–71.

88. Ashery-Padan R, Gruss P. Pax6 lights-up the way for eye development. *Curr Opin Cell Biol* December 2001;**13**(6):706–14.

89. Joly S, Pernet V, Samardzija M, Grimm C. Pax6-positive Muller glia cells express cell cycle markers but do not proliferate after photoreceptor injury in the mouse retina. *Glia* July 2011;**59**(7):1033–46.

90. Furukawa T, Morrow EM, Cepko CL. Crx, a novel otx-like homeobox gene, shows photoreceptor-specific expression and regulates photoreceptor differentiation. *Cell.* November 14, 1997;**91**(4):531–41.

91. Mears AJ, Kondo M, Swain PK, et al. Nrl is required for rod photoreceptor development. *Nat Genet* December 2001;**29**(4):447–52.

92. Chen J, Rattner A, Nathans J. The rod photoreceptor-specific nuclear receptor Nr2e3 represses transcription of multiple cone-specific genes. *J Neurosci* January 5, 2005;**25**(1):118–29.

93. Haider NB, Jacobson SG, Cideciyan AV, et al. Mutation of a nuclear receptor gene, NR2E3, causes enhanced S cone syndrome, a disorder of retinal cell fate. *Nat Genet* February 2000;**24**(2):127–31.

94. Muhchyi C, Juliandi B, Matsuda T, Nakashima K. Epigenetic regulation of neural stem cell fate during corticogenesis. *Int J Dev Neurosci* October 2013;**31**(6):424–33.

95. Hu XL, Wang Y, Shen Q. Epigenetic control on cell fate choice in neural stem cells. *Protein Cell* April 2012;**3**(4):278–90.

96. Gao Y, Jammes H, Rasmussen MA, et al. Epigenetic regulation of gene expression in porcine epiblast, hypoblast, trophectoderm and epiblast-derived neural progenitor cells. *Epigenetics* September 1, 2011;**6**(9):1149–61.

97. Aldiri I, Moore KB, Hutcheson DA, Zhang J, Vetter ML. Polycomb repressive complex PRC2 regulates *Xenopus* retina development downstream of Wnt/beta-catenin signaling. *Development* July 2013;**140**(14):2867–78.

98. He S, Li X, Chan N, Hinton DR. Review: epigenetic mechanisms in ocular disease. *Mol Vis* 2013;**19**:665–74.

99. Stojic L, Jasencakova Z, Prezioso C, et al. Chromatin regulated interchange between polycomb repressive complex 2 (PRC2)-Ezh2 and PRC2-Ezh1 complexes controls myogenin activation in skeletal muscle cells. *Epigenetics Chromatin* 2011;**4**:16.

100. Barth TK, Imhof A. Fast signals and slow marks: the dynamics of histone modifications. *Trends Biochem Sci* November 2010;**35**(11):618–26.

101. Lee S, Lee SK. Crucial roles of histone-modifying enzymes in mediating neural cell-type specification. *Curr Opin Neurobiol* February 2010;**20**(1):29–36.

102. Yamaguchi M, Tonou-Fujimori N, Komori A, et al. Histone deacetylase 1 regulates retinal neurogenesis in zebrafish by suppressing Wnt and Notch signaling pathways. *Development* July 2005;**132**(13):3027–43.

103. Chen B, Cepko CL. Requirement of histone deacetylase activity for the expression of critical photoreceptor genes. *BMC Dev Biol* 2007;**7**:78.

104. Nasonkin IO, Lazo K, Hambright D, Brooks M, Fariss R, Swaroop A. Distinct nuclear localization patterns of DNA methyltransferases in developing and mature mammalian retina. *J Comp Neurol* July 1, 2011;**519**(10):1914–30.

105. Rai K, Nadauld LD, Chidester S, et al. Zebra fish Dnmt1 and Suv39h1 regulate organ-specific terminal differentiation during development. *Mol Cell Biol* October 2006;**26**(19):7077–85.

106. Banno K, Kisu I, Yanokura M, et al. Epimutation and cancer: a new carcinogenic mechanism of Lynch syndrome (review). *Int J Oncol* September 2012;**41**(3):793–7.

107. Fleisch VC, Fraser B, Allison WT. Investigating regeneration and functional integration of CNS neurons: lessons from zebrafish genetics and other fish species. *Biochim Biophys Acta* March 2011;**1812**(3):364–80.

108. Ng L, Lu A, Swaroop A, Sharlin DS, Forrest D. Two transcription factors can direct three photoreceptor outcomes from rod precursor cells in mouse retinal development. *J Neurosci* August 3, 2011;**31**(31):11118–25.

109. Gonzalez-Cordero A, West EL, Pearson RA, et al. Photoreceptor precursors derived from three-dimensional embryonic stem cell cultures integrate and mature within adult degenerate retina. *Nat Biotechnol* August 2013;**31**(8):741–7.

110. Hambright D, Park KY, Brooks M, McKay R, Swaroop A, Nasonkin IO. Long-term survival and differentiation of retinal neurons derived from human embryonic stem cell lines in un-immunosuppressed mouse retina. *Mol Vis* 2012;**18**:920–36.

111. Zhang Y, Wang W. Effects of bone marrow mesenchymal stem cell transplantation on light-damaged retina. *Invest Ophthalmol Vis Sci* July 2010;**51**(7):3742–8.

112. West EL, Pearson RA, MacLaren RE, Sowden JC, Ali RR. Cell transplantation strategies for retinal repair. *Prog Brain Res* 2009;**175**:3–21.

113. Ikeda H, Osakada F, Watanabe K, et al. Generation of Rx+/Pax6+ neural retinal precursors from embryonic stem cells. *Proc Natl Acad Sci USA* August 9, 2005;**102**(32):11331–6.

114. Banin E, Obolensky A, Idelson M, et al. Retinal incorporation and differentiation of neural precursors derived from human embryonic stem cells. *Stem Cells* February 2006;**24**(2):246–57.

115. Lamba DA, Karl MO, Ware CB, Reh TA. Efficient generation of retinal progenitor cells from human embryonic stem cells. *Proc Natl Acad Sci USA* August 22, 2006;**103**(34):12769–74.

116. Lamba DA, Gust J, Reh TA. Transplantation of human embryonic stem cell-derived photoreceptors restores some visual function in Crx-deficient mice. *Cell Stem Cell* January 9, 2009;**4**(1):73–9.

117. Lamba DA, McUsic A, Hirata RK, Wang PR, Russell D, Reh TA. Generation, purification and transplantation of photoreceptors derived from human induced pluripotent stem cells. *PLoS One* 2010;**5**(1):e8763.

118. Osakada F, Ikeda H, Mandai M, et al. Toward the generation of rod and cone photoreceptors from mouse, monkey and human embryonic stem cells. *Nat Biotechnol* February 2008;**26**(2):215–24.

119. Ramsden CM, Powner MB, Carr AJ, Smart MJ, da Cruz L, Coffey PJ. Stem cells in retinal regeneration: past, present and future. *Development* June 2013;**140**(12):2576–85.

120. Homma K, Okamoto S, Mandai M, et al. Developing rods transplanted into the degenerating retina of Crx-knockout mice exhibit neural activity similar to native photoreceptors. *Stem Cells* June 2013;**31**(6):1149–59.

121. Li T, Lewallen M, Chen S, Yu W, Zhang N, Xie T. Multipotent stem cells isolated from the adult mouse retina are capable of producing functional photoreceptor cells. *Cell Res* June 2013;**23**(6):788–802.

122. MacLaren RE, Pearson RA, MacNeil A, et al. Retinal repair by transplantation of photoreceptor precursors. *Nature* November 9, 2006;**444**(7116):203–7.

123. Nickerson PE, McLeod MC, Myers T, Clarke DB. Effects of epidermal growth factor and erythropoietin on Muller glial activation and phenotypic plasticity in the adult mammalian retina. *J Neurosci Res* July 2011;**89**(7):1018–30.

124. Klassen H, Sakaguchi DS, Young MJ. Stem cells and retinal repair. *Prog Retin Eye Res* March 2004;**23**(2):149–81.

125. Okamoto S, Takahashi M. Induction of retinal pigment epithelial cells from monkey iPS cells. *Invest Ophthalmol Vis Sci* 2011;**52**(12):8785–90.

126. Park UC, Cho MS, Park JH, et al. Subretinal transplantation of putative retinal pigment epithelial cells derived from human embryonic stem cells in rat retinal degeneration model. *Clin Exp Reprod Med* December 2011;**38**(4):216–21.

127. Juuti-Uusitalo K, Vaajasaari H, Ryhanen T, et al. Efflux protein expression in human stem cell-derived retinal pigment epithelial cells. *PLoS One* 2012;**7**(1):e30089.

128. Aoki H, Hara A, Niwa M, Motohashi T, Suzuki T, Kunisada T. Transplantation of cells from eye-like structures differentiated from embryonic stem cells in vitro and in vivo regeneration of retinal ganglion-like cells. *Graefes Arch Clin Exp Ophthalmol* February 2008;**246**(2):255–65.

129. Jagatha B, Divya MS, Sanalkumar R, et al. In vitro differentiation of retinal ganglion-like cells from embryonic stem cell derived neural progenitors. *Biochem Biophys Res Commun* March 6, 2009;**380**(2):230–5.

130. Cho JH, Mao CA, Klein WH. Adult mice transplanted with embryonic retinal progenitor cells: new approach for repairing damaged optic nerves. *Mol Vis* 2012;**18**:2658–72.

131. Singhal S, Bhatia B, Jayaram H, et al. Human Muller glia with stem cell characteristics differentiate into retinal ganglion cell (RGC) precursors in vitro and partially restore RGC function in vivo following transplantation. *Stem Cells Transl Med* March 2012;**1**(3):188–99.

132. Mellough CB, Cui Q, Harvey AR. Treatment of adult neural progenitor cells prior to transplantation affects graft survival and integration in a neonatal and adult rat model of selective retinal ganglion cell depletion. *Restor Neurol Neurosci* 2007;**25**(2):177–90.

133. Meng F, Wang X, Gu P, Wang Z, Guo W. Induction of retinal ganglion-like cells from fibroblasts by adenoviral gene delivery. *Neuroscience* July 13, 2013.

134. Bull ND, Limb GA, Martin KR. Human Muller stem cell (MIO-M1) transplantation in a rat model of glaucoma: survival, differentiation, and integration. *Invest Ophthalmol Vis Sci* August 2008;**49**(8):3449–56.

135. Zwart I, Hill AJ, Al-Allaf F, et al. Umbilical cord blood mesenchymal stromal cells are neuroprotective and promote regeneration in a rat optic tract model. *Exp Neurol* April 2009;**216**(2):439–48.

136. Johnson TV, Bull ND, Martin KR. Identification of barriers to retinal engraftment of transplanted stem cells. *Invest Ophthalmol Vis Sci* February 2010;**51**(2):960–70.

137. Johnson TV, Bull ND, Hunt DP, Marina N, Tomarev SI, Martin KR. Neuroprotective effects of intravitreal mesenchymal stem cell transplantation in experimental glaucoma. *Invest Ophthalmol Vis Sci* April 2010;**51**(4):2051–9.

Chapter 21

Experimental Cell Therapy for Liver Dysfunction

Bo Wang, Jason A. Wertheim

Key Concepts

In this chapter, we summarized the current development and challenges of hepatocytes transplantation in the following aspects: cell source, donor liver preservation, and cell engraftment real-time monitoring.

1. INTRODUCTION

The liver performs several critical functions necessary for survival including protein synthesis and metabolism, detoxification, and nutrient storage within the body. Liver transplantation is the established and widely applicable treatment for end-stage inherited hepatic disorders and liver failure, both acute and chronic. However, there is a shortage of donor organs for transplantation, and nearly 16,000 patients are currently waiting for a liver transplant in the United States. In addition to the high risks associated with this complex surgical procedure, associated financial costs, and need for lifelong immunosuppressive therapy to reduce the risk of rejection, the shortage of donor organs further limits the availability of liver transplantation to only a subset of patients who require a liver for transplantation.[1]

Human hepatocyte transplantation has been proposed as an experimental alternative therapy for liver transplantation and also considered as a "bridge" therapy for patients waiting for an orthotopic liver transplant. The first animal experiment using isolating hepatocytes from the liver was introduced by Berry and Friend in 1969[2] and the first clinical trial was conducted by Mito et al. in 1992.[3]

Theoretically, there are several advantages to hepatocyte transplantation compared to whole liver transplantation, which include (1) a relatively less invasive procedure whereby cells are injected into the portal vasculature, (2) lower cost compared to liver transplantation, (3) ability to recover hepatocytes from marginal or extended criteria donor livers, (4) potential to cryopreserve cells until clinically needed, (5) retention of the native liver, which may provide additional time for hepatic recovery in acute cases, and (6) experimental ability to correct genetic defects through gene editing to improve specific hepatocyte functionality prior to transplantation.[4–6] Hepatocyte transplantation has been proposed as an option for patients who are waiting for liver transplantation as a temporary metabolic support prior to definitive organ transplantation or for patients with fulminant hepatic failure to facilitate liver regeneration.[1] However, in many cases of clinical hepatocyte transplantation, only partial liver correction was observed and function of the transplanted cells declined at ~9 months after implantation leading to graft loss and need for definitive liver transplantation. Several hurdles still need to be overcome to improve the efficacy of hepatocyte transplantation before the procedure becomes routine clinical practice.[6] In this chapter, we describe the current state of hepatocyte transplantation, liver preservation, and cell fate after hepatocyte transplantation.

2. HUMAN HEPATOCYTES

Human hepatocytes are typically recovered from donated liver tissues that are not used or are unsuitable for liver transplantation, including livers from brain dead and non-heart-beating donors.[7] Cells are isolated using a two-step perfusion method with a sequence of buffer and collagenase.[8,9] The overall quality of many liver tissues used for hepatocyte isolation is lower than donor livers used for whole organ transplantation, and may be slightly damaged or have a degree of underlying liver disease.[10] The major limitation to isolating hepatocytes from discarded livers is the time-dependent viability, limited availability, and decreased engraftment potential after cell transplantation.[7,11]

Translating Regenerative Medicine to the Clinic. http://dx.doi.org/10.1016/B978-0-12-800548-4.00021-8

After isolation, hepatocytes may be either transplanted directly into the recipient patient or cryopreserved for emergency application or planned treatment for liver-based metabolic disorders.[6] The potential advantage of cryopreservation of hepatocytes is the ability to bank cells for long-term preservation to generate a sufficient supply and to perform quality control analysis to predict suitability for transplantation.[12] However, viability, protein synthesis capacity, and P450 activity currently limit optimal clinical outcomes using cryopreserved cells, which are much lower compared to the fresh hepatocytes due to formation of intracellular ice crystals during the cryopreservation process causing cell dehydration, rupture, apoptosis, low cell attachment, mitochondrial damage, and decreased adenosine triphosphate (ATP, the primary energy-storing molecule that transfers energy within cells) production during the thawing process.[7,13]

To improve the viability and function of cryopreserved hepatocytes, recent studies have focused on optimizing the cryopreservation process by modifying the composition of freezing medium, the rate and duration of freeze/thaw cycles,[14–16] modifying cell density, and type of cytoprotectant used.[17–20]

3. ALTERNATIVE CELL SOURCES

The shortage of donor organs is the main hurdle for both strategies of whole liver transplantation and hepatocyte transplantation. To increase the ability to perform hepatocyte transplantation and other cell-based therapies for liver diseases, it is essential to develop a sustainable and readily available source of hepatocytes or hepatocyte-like cells.

To date, multiple cell types have been studied as candidates to replace human hepatocytes, including human hepatocyte cell lines,[21] fetal liver cells,[22] and stem cell-derived hepatocytes.[23] Both tumor-derived hepatocyte cell lines and immortalized cells have shown their advantages in easy proliferation, storage, and maintenance, yet these cells have reduced metabolic functionality and have the potential to develop tumors in vivo, which is unsuitable for clinical applications.[24,25] Human fetal hepatocytes are isolated from fetal liver tissues and can undergo multiple cell divisions and differentiate into more mature hepatic cells in vivo.[26] Interestingly, a clinical case report described hepatic recovery in a patient with acute liver decompensation and underlying fatty liver disease after delivery of fetal hepatocytes.[27] However, most studies indicate that the fetal hepatocytes have lower levels of maturity and hepatic functionality compared to primary hepatocytes and may also display tumorigenic phenotypes.[28,29]

Stem cells, including adult, embryonic, and induced pluripotent stem (iPS) cells, can divide for unlimited passages in culture and have high potential to differentiate into multiple cell types, including hepatocytes.[30–33] Adult stem cells, such as bone marrow mesenchymal (MSCs) and hematopoietic stem cells, may transdifferentiate into hepatocyte-like cells and have lower risk of teratoma formation and ethical issues compared with fetal hepatocytes and embryonic stem cells (ESCs) though further experimental study is needed in this area.[34,35] Animal studies showed that MSCs transplanted into the damaged liver could generate hepatocyte-like cells[36,37] and MSCs.[38] The major limitations of adult stem cells are their low efficiency of hepatic differentiation and potential to differentiate into myofibroblasts at the site of liver injury.[39,40] ESCs and iPS cells are capable of engrafting in the recipient liver and express hepatocyte and bile duct cell-specific markers.[41–45] Another recent published cell source is the biliary tree stem/progenitor cell (BTSC) that is isolated from the biliary epithelium of fetal or adult large intrahepatic and extrahepatic bile ducts.[46] The BTSCs have shown self-replicate ability in culture and typical markers expressing as the primitive endoderm and biliopancreatic progenitors, and are able to differentiate into several cell types such as hepatocytes, cholangiocyte, or pancreatic β-cells both in vitro and in vivo.[47,48] A recent clinical study transplanted human BTSCs through hepatic artery in two patients with liver cirrhosis. Both patients demonstrated a constant improvement of liver function during the first 6 months postsurgery and the second patient maintained steady liver function up to 12 months.[49]

However, the cell sources introduced above for hepatocyte transplantation therapy still retain several technical obstacles including their immature or fetal-stage status and low repopulation efficiency.[50,51] Due to the limitations of these cell types, caution and additional experimentation are needed to enhance functionality of these cells and evaluate the potential of teratoma formation before these cells can be translated into clinical use.

4. MACHINE PERFUSION FOR LIVER PRESERVATION

Simple cold storage is the universal method for liver preservation before transplantation and the low temperature (0–4 °C) can decrease liver metabolism and lengthen the time for liver preservation. The disadvantages of this method includes ATP depletion, cellular swelling, mitochondrial calcium overload, damage to sinusoidal endothelial and Kupffer cells, and electrolyte disturbances that can lead to further graft injuries after liver reperfusion.[52–54] To better preserve donor livers for transplantation or hepatocyte isolation, machine perfusion (as an alternative to static cold storage) has been developed to provide near-physiological conditions for liver preservation.[55,56] Various techniques of extracorporeal machine perfusion for liver preservation have been investigated during the past two decades, including pulsatile or dual perfusion, retrograde or antegrade, double or single vessel perfusion, and various protocols in delivering physiologic pressures, flows, temperature, or oxygen during perfusion.[55,57–59]

Hypothermic (HMP) and normothermic (NMP) machine perfusion are the two primary protocols that are commonly studied. HMP has been shown to improve the process of cold storage by adding a continuous oxygen supply and metabolic substrates into the organ perfusate, while reducing liver metabolic demands and increasing energy stores compared to the static cold storage.[60] Several animal and early clinical experiments show a certain level of reduction in biliary complications and improvement in early allograft function.[61–64] The approach of NMP minimizes preservation injury by mimicking near-physiological conditions for liver perfusion, and preservation and viability of the donor livers during storage can be easily assessed.[65,66]

Machine perfusion is considered a valid method for liver preservation and similar strategies have been applied in other organs used for transplantation, such as kidneys.[67,68] Before wide clinical use, machine perfusion was criticized for its limitations in logistics, complex operation, poor clinical practicability, and possible damage to vital structures during perfusion.[54] The development of liver preservation protocols may improve the outcome for patients waiting for a liver transplantation, allow for rehabilitation of suboptimal allografts that were previously considered unsuitable for clinical transplantation, and may result in decompression of transplant waiting lists.

5. MONITORING CELL ENGRAFTMENT

Due to the low viability and integration of transplanted hepatocytes into diseased livers, noninvasive imaging techniques are needed to track transplanted cells in the recipient livers, determine cell fate, migration, and contribution to overall liver metabolic function.

Early imaging methods used to label cells with radioactive indium for nuclear medicine-based imaging in the acute phase after transplantation to demonstrate hepatocyte distribution.[69] However, indium has a short half-life and lack of defined anatomical resolution that limit the application of this imaging technique.

Recently, magnetic resonance imaging (MRI) has been widely used as a real-time imaging tool to track cell delivery and assess initial engraftment patterns via labeling the cells with iron oxide prior to transplantation.[70–73] MRI offers excellent anatomical detail of the body especially for soft tissue detection by using powerful magnetic fields. However, normal MRI protocols cannot distinguish transplanted cells from host liver tissues. Therefore, recent studies are focused on investigating new contrast agents to label cells that result in safe and fully functional cells for clinical application.

Labeling of cells with supraparamagnetic iron oxide particles is the most common method used for clinical application. These iron oxide particles increase signal intensity of labeled cells that allow them to be easily distinguished from surrounding tissues. Iron is biodegradable and is compartmentalized in lysosomes, metabolized, and recycled into the normal iron pool.[69,70,74] However, there remains some debate about the biological effects of this strategy and some studies show that iron oxide particles inhibit cell proliferation and morphology of certain cell types.[75–77]

Another strategy is to use a fluorine-based cellular tracer agent to detect transplanted cells with fluorine-19 (19F) MRI. The advantage of the 19F MRI approach in hepatocyte transplantation is its ability to selectively and quantitatively detect transplanted cells since the human body lacks a 19F background.[78] Recent experimental studies have focused on adding multiple 19F signal channels to visualize and distinguish different cell types.

6. CONCLUSION

Hepatocyte transplantation remains a developing and experimental therapy as a clinical procedure for the treatment of liver disease. Despite the limited availability, adult hepatocytes are still the primary cell source. With continued experimental studies in liver storage, cell isolation, stem cell biology, and techniques to improve cell engraftment, this form of cell therapy may have a wider application in the treatment of liver disease.

REFERENCES

1. Ito M, et al. Review of hepatocyte transplantation. *J Hepatobiliary Pancreat Surg* 2009;**16**(2):97–100.
2. Berry MN, Friend DS. High-yield preparation of isolated rat liver parenchymal cells: a biochemical and fine structural study. *J Cell Biol* 1969;**43**(3):506–20.
3. Mito M, Kusano M, Kawaura Y. Hepatocyte transplantation in man. *Transplant Proc* 1992;**24**(6):3052–3.
4. Ohashi K, Park F, Kay MA. Hepatocyte transplantation: clinical and experimental application. *J Mol Med* 2001;**79**(11):617–30.
5. Hughes RD, Mitry RR, Dhawan A. Current status of hepatocyte transplantation. *Transplantation* 2012;**93**(4):342–7.
6. Jorns C, et al. Hepatocyte transplantation for inherited metabolic diseases of the liver. *J Intern Med* 2012;**272**(3):201–23.

7. Hughes RD, et al. Isolation of hepatocytes from livers from non-heart-beating donors for cell transplantation. *Liver Transpl* 2006;**12**(5):713–7.

8. Seglen PO. Preparation of isolated rat liver cells. *Methods Cell Biol* 1976;**13**:29–83.

9. Strom SC, et al. Isolation, culture, and transplantation of human hepatocytes. *J Natl Cancer Inst* 1982;**68**(5):771–8.

10. Sundback CA, Vacanti JP. Alternatives to liver transplantation: from hepatocyte transplantation to tissue-engineered organs. *Gastroenterology* 2000;**118**(2):438–42.

11. Kawahara T, et al. Factors affecting hepatocyte isolation, engraftment, and replication in an in vivo model. *Liver Transpl* 2010;**16**(8):974–82.

12. Donato MT, et al. Functional assessment of the quality of human hepatocyte preparations for cell transplantation. *Cell Transplant* 2008;**17**(10–11):1211–9.

13. Stephenne X, et al. Cryopreservation of human hepatocytes alters the mitochondrial respiratory chain complex 1. *Cell Transplant* 2007;**16**(4):409–19.

14. Loretz LJ, et al. Optimization of cryopreservation procedures for rat and human hepatocytes. *Xenobiotica* 1989;**19**(5):489–98.

15. Chesne C, et al. Viability and function in primary culture of adult hepatocytes from various animal species and human beings after cryopreservation. *Hepatology* 1993;**18**(2):406–14.

16. Jamal HZ, Weglarz TC, Sandgren EP. Cryopreserved mouse hepatocytes retain regenerative capacity in vivo. *Gastroenterology* 2000;**118**(2):390–4.

17. Dhawan A, et al. Human hepatocyte transplantation: current experience and future challenges. *Nat Rev Gastroenterol Hepatol* 2010;**7**(5):288–98.

18. Miyamoto Y, et al. Improvement of hepatocyte viability after cryopreservation by supplementation of long-chain oligosaccharide in the freezing medium in rats and humans. *Cell Transplant* 2006;**15**(10):911–9.

19. Katenz E, et al. Cryopreservation of primary human hepatocytes: the benefit of trehalose as an additional cryoprotective agent. *Liver Transpl* 2007;**13**(1):38–45.

20. Luebke-Wheeler JL, et al. E-cadherin protects primary hepatocyte spheroids from cell death by a caspase-independent mechanism. *Cell Transplant* 2009;**18**(12):1281–7.

21. Roberts EA, et al. Characterization of human hepatocyte lines derived from normal liver tissue. *Hepatology* 1994;**19**(6):1390–9.

22. Monga SP, et al. Mouse fetal liver cells in artificial capillary beds in three-dimensional four-compartment bioreactors. *Am J Pathol* 2005;**167**(5):1279–92.

23. Soto-Gutierrez A, et al. Reversal of mouse hepatic failure using an implanted liver-assist device containing ES cell-derived hepatocytes. *Nat Biotechnol* 2006;**24**(11):1412–9.

24. Nyberg SL, et al. Primary hepatocytes outperform Hep G2 cells as the source of biotransformation functions in a bioartificial liver. *Ann Surg* 1994;**220**(1):59–67.

25. Palakkan AA, et al. Liver tissue engineering and cell sources: issues and challenges. *Liver Int* 2013;**33**(5):666–76.

26. Wege H, et al. Telomerase reconstitution immortalizes human fetal hepatocytes without disrupting their differentiation potential. *Gastroenterology* 2003;**124**(2):432–44.

27. Khan AA, et al. Peritoneal transplantation of human fetal hepatocytes for the treatment of acute fatty liver of pregnancy: a case report. *Trop Gastroenterol* 2004;**25**(3):141–3.

28. Poyck PP, et al. Functional and morphological comparison of three primary liver cell types cultured in the AMC bioartificial liver. *Liver Transpl* 2007;**13**(4):589–98.

29. Diekmann S, Bader A, Schmitmeier S. Present and future developments in hepatic tissue engineering for liver support systems: state of the art and future developments of hepatic cell culture techniques for the use in liver support systems. *Cytotechnology* 2006;**50**(1–3):163–79.

30. Takayama K, et al. Efficient generation of functional hepatocytes from human embryonic stem cells and induced pluripotent stem cells by HNF4alpha transduction. *Mol Ther* 2012;**20**(1):127–37.

31. Cho CH, et al. Homogeneous differentiation of hepatocyte-like cells from embryonic stem cells: applications for the treatment of liver failure. *FASEB J* 2008;**22**(3):898–909.

32. Stutchfield BM, Forbes SJ, Wigmore SJ. Prospects for stem cell transplantation in the treatment of hepatic disease. *Liver Transpl* 2010;**16**(7):827–36.

33. Szkolnicka D, et al. Pluripotent stem cell-derived hepatocytes: potential and challenges in pharmacology. *Annu Rev Pharmacol Toxicol* 2013;**53**:147–59.

34. Stock P, et al. Hepatocytes derived from adult stem cells. *Transplant Proc* 2008;**40**(2):620–3.

35. Saulnier N, et al. Mesenchymal stromal cells multipotency and plasticity: induction toward the hepatic lineage. *Eur Rev Med Pharmacol Sci* 2009;**13**(Suppl. 1):71–8.

36. Sgodda M, et al. Hepatocyte differentiation of mesenchymal stem cells from rat peritoneal adipose tissue in vitro and in vivo. *Exp Cell Res* 2007;**313**(13):2875–86.

37. Sato Y, et al. Human mesenchymal stem cells xenografted directly to rat liver are differentiated into human hepatocytes without fusion. *Blood* 2005;**106**(2):756–63.

38. Parekkadan B, Milwid JM. Mesenchymal stem cells as therapeutics. *Annu Rev Biomed Eng* 2010;**12**:87–117.

39. Baertschiger RM, et al. Fibrogenic potential of human multipotent mesenchymal stromal cells in injured liver. *PloS One* 2009;**4**(8):e6657.

40. Piryaei A, et al. Differentiation of bone marrow-derived mesenchymal stem cells into hepatocyte-like cells on nanofibers and their transplantation into a carbon tetrachloride-induced liver fibrosis model. *Stem Cell Rev* 2011;**7**(1):103–18.

41. Kumashiro Y, et al. Enrichment of hepatocytes differentiated from mouse embryonic stem cells as a transplantable source. *Transplantation* 2005;**79**(5):550–7.

42. Gouon-Evans V, et al. BMP-4 is required for hepatic specification of mouse embryonic stem cell-derived definitive endoderm. *Nat Biotechnol* 2006;**24**(11):1402–11.

43. Heo J, et al. Hepatic precursors derived from murine embryonic stem cells contribute to regeneration of injured liver. *Hepatology* 2006;**44**(6):1478–86.

44. Si-Tayeb K, et al. Highly efficient generation of human hepatocyte-like cells from induced pluripotent stem cells. *Hepatology* 2010;**51**(1):297–305.
45. Yi F, Liu GH, Izpisua Belmonte JC. Human induced pluripotent stem cells derived hepatocytes: rising promise for disease modeling, drug development and cell therapy. *Protein Cell* 2012;**3**(4):246–50.
46. Carpino G, et al. Biliary tree stem/progenitor cells in glands of extrahepatic and intraheptic bile ducts: an anatomical in situ study yielding evidence of maturational lineages. *J Anat* 2012;**220**(2):186–99.
47. Cardinale V, et al. Multipotent stem/progenitor cells in human biliary tree give rise to hepatocytes, cholangiocytes, and pancreatic islets. *Hepatology* 2011;**54**(6):2159–72.
48. Carpino G, et al. Evidence for multipotent endodermal stem/progenitor cell populations in human gallbladder. *J Hepatol* 2014;**60**(6):1194–202.
49. Cardinale V, et al. Transplantation of human fetal biliary tree stem/progenitor cells into two patients with advanced liver cirrhosis. *BMC Gastroenterol* 2014;**14**(1):204.
50. Russo FP, Parola M. Stem cells in liver failure. *Best Pract Res Clin Gastroenterol* 2012;**26**(1):35–45.
51. Vitale AM, Wolvetang E, Mackay-Sim A. Induced pluripotent stem cells: a new technology to study human diseases. *Int J Biochem Cell Biol* 2011;**43**(6):843–6.
52. Mendes-Braz M, et al. The current state of knowledge of hepatic ischemia-reperfusion injury based on its study in experimental models. *J Biomed Biotechnol* 2012;**2012**:298657.
53. Guibert EE, et al. Organ preservation: current concepts and new strategies for the next decade. *Transfus Med Hemother* 2011;**38**(2):125–42.
54. Monbaliu D, Brassil J. Machine perfusion of the liver: past, present and future. *Curr Opin Organ Transplant* 2010;**15**(2):160–6.
55. Pienaar BH, et al. Seventy-two-hour preservation of the canine liver by machine perfusion. *Transplantation* 1990;**49**(2):258–60.
56. Yamamoto N, et al. Seventy-two-hour preservation of porcine liver by continuous hypothermic perfusion with UW solution in comparison with simple cold storage. *J Surg Res* 1991;**51**(4):288–92.
57. Yanaga K, et al. A new liver perfusion and preservation system for transplantation research in large animals. *J Invest Surg* 1990;**3**(1):65–75.
58. Xu H, et al. Prolonged hypothermic machine perfusion preserves hepatocellular function but potentiates endothelial cell dysfunction in rat livers. *Transplantation* 2004;**77**(11):1676–82.
59. Vairetti M, et al. Correlation between the liver temperature employed during machine perfusion and reperfusion damage: role of Ca^{2+}. *Liver Transpl* 2008;**14**(4):494–503.
60. Belzer FO, Southard JH. Principles of solid-organ preservation by cold storage. *Transplantation* 1988;**45**(4):673–6.
61. Guarrera JV, et al. Hypothermic machine preservation in human liver transplantation: the first clinical series. *Am J Transplant* 2010;**10**(2):372–81.
62. Jamieson NV, et al. Preservation of the canine liver for 24–48 hours using simple cold storage with UW solution. *Transplantation* 1988;**46**(4):517–22.
63. de Rougemont O, et al. One hour hypothermic oxygenated perfusion (HOPE) protects nonviable liver allografts donated after cardiac death. *Ann Surg* 2009;**250**(5):674–83.
64. Guarrera JV, et al. Hypothermic machine perfusion of liver grafts for transplantation: technical development in human discard and miniature swine models. *Transplant Proc* 2005;**37**(1):323–5.
65. Schon MR, et al. Liver transplantation after organ preservation with normothermic extracorporeal perfusion. *Ann Surg* 2001;**233**(1):114–23.
66. Brockmann J, et al. Normothermic perfusion: a new paradigm for organ preservation. *Ann Surg* 2009;**250**(1):1–6.
67. Kievit JK, et al. Outcome of machine-perfused non-heart-beating donor kidneys, not allocated within the Eurotransplant area. *Transpl Int* 1998;**11**(Suppl. 1):S421–3.
68. Reznik ON, et al. Machine perfusion as a tool to select kidneys recovered from uncontrolled donors after cardiac death. *Transplant Proc* 2008;**40**(4):1023–6.
69. Kudo M. Diagnostic imaging of hepatocellular carcinoma: recent progress. *Oncology* 2011;**81**(Suppl. 1):73–85.
70. Bulte JW, Kraitchman DL. Iron oxide MR contrast agents for molecular and cellular imaging. *NMR Biomed* 2004;**17**(7):484–99.
71. Heyn C, et al. In vivo MRI of cancer cell fate at the single-cell level in a mouse model of breast cancer metastasis to the brain. *Magn Reson Med* 2006;**56**(5):1001–10.
72. Shapiro EM, et al. MRI detection of single particles for cellular imaging. *Proc Natl Acad Sci USA* 2004;**101**(30):10901–6.
73. Medarova Z, Moore A. MRI as a tool to monitor islet transplantation. *Nat Rev Endocrinol* 2009;**5**(8):444–52.
74. Briley-Saebo K, et al. Hepatic cellular distribution and degradation of iron oxide nanoparticles following single intravenous injection in rats: implications for magnetic resonance imaging. *Cell Tissue Res* 2004;**316**(3):315–23.
75. Kostura L, et al. Feridex labeling of mesenchymal stem cells inhibits chondrogenesis but not adipogenesis or osteogenesis. *NMR Biomed* 2004;**17**(7):513–7.
76. Nohroudi K, et al. In vivo MRI stem cell tracking requires balancing of detection limit and cell viability. *Cell Transplant* 2010;**19**(4):431–41.
77. Pisanic TR, 2nd, et al. Nanotoxicity of iron oxide nanoparticle internalization in growing neurons. *Biomaterials* 2007;**28**(16):2572–81.
78. Morawski AM, et al. Quantitative "magnetic resonance immunohistochemistry" with ligand-targeted (19)F nanoparticles. *Magn Reson Med* 2004;**52**(6):1255–62.

Chapter 22

Microfluidic-Based 3D Models of Renal Function for Clinically Oriented Research

Natalia Sánchez-Romero, Patricia Meade, Ignacio Giménez

Key Concepts

1. A large fraction of our knowledge on renal epithelial cells biology and disease is based on far from physiological models, which seriously affects our ability to understand or predict the in vivo responses. There is an increasing demand for 3D culture renal epithelial models that incorporate all physical and chemical microenvironmental cues, including luminal flow, cell–extracellular matrix interactions, and cell–cell interactions.
2. Different strategies are taken to ensure availability of high numbers of highly differentiated cells required for regenerative medicine application, like immortalization of isolated primary cells or differentiation of stem cells toward the renal phenotype. Maintenance of their phenotype will rely on appropriate culture in serum-free hormonally defined media and mimicking the in vivo physical and chemical 3D microenvironment.
3. Synthetic biopolymers with significant bioactive substances functionalization are the future direction to simulate the cell–extracellular matrix interaction and allow for cell–cell interactions through coculture with representative cells. It will also solve the dependence on animal products.
4. Microfabrication techniques like microfluidics or microelectronic sensors are being applied to the fabrication of complex biomimicking environments and long-term maintenance and monitoring of renal cell activity.
5. A better understanding of renal epithelial cell physiology will have a tremendous positive impact on our ability to reproduce renal function for replacement therapies and to better exploit the intrinsic regenerative potential of renal cells.

List of Abbreviations

AQP-1 Aquaporin-1
BM Basal Membrane
BAK Bioartificial Kidney
CD Collecting Duct
COC Cyclic Olefin Copolymer
DCT Distal Convoluted Tubule
DN Distal Nephron
EMT Epithelial-to-Mesenchymal Transition
ECM Extracellular Matrix
GGT-1 Gamma Glutamyl Transferase
hPTC human Proximal Tubule Cells
hTERT human Telomerase
HA Hyaluronic Acid
MM Metanephric Mesenchyme
mIMCD Mouse Inner Medullary Collecting Duct cells
PDMS Polydimethylsiloxane
PMMA Poly(methylmethacrylate)
PC Polycarbonate
PKD Polycystic Kidney Disease
PES Polyethersulfone
PS Polystyrene
PT Proximal Tubule

Translating Regenerative Medicine to the Clinic. http://dx.doi.org/10.1016/B978-0-12-800548-4.00022-X

PTC Proximal Tubule Cells
RPTEC Renal Proximal Tubule Epithelial Cells
TAL Thick Ascending Limb
TEER Transepithelial Electrical Resistance
UB Ureteric Bud
ZO-1 Zonula Occludens type 1

1. INTRODUCTION

Human kidneys are extremely powerful and efficient organs able to handle 180 L of plasma filtrate every day, to finally excrete 1/100 of that volume in the form of urine containing all the waste products required to keep our blood and extracellular fluid clean of excess ions and toxic organic substances and drugs. Unfortunately, it is precisely their high metabolic rate and high exposure to cytotoxic agents that result in a very high incidence of acute and chronic kidney diseases. Renal replacement therapies (RRT) (dialysis, kidney transplant) are associated with undesired adverse effects and have a marked impact on social and labor terms. Besides, renal transplant is only available for a limited number of patients due to organ donor shortage.

Strategies to regenerate renal function using administration of cell-based therapies or de novo organ generation will have to overcome significant hurdles derived from the intricate structural and functional complexity of the kidney, making it possibly the hardest organ to regenerate. Interesting approaches to take advantage of renal cell functions in order to improve or complement dialysis/hemofiltration-based RRT (bioartificial kidney (BAK), see Box 1) have been challenged by the apparent lack of knowledge on renal cell physiology. The same can be said for current nephrotoxicity screenings, since there is a weak correlation between the results obtained in in vitro models with those obtained in in vivo testing, thus delaying the validation process and increasing the need for research animals.

This chapter reviews current efforts aimed to fill the gap existing in in vitro modeling of renal epithelial function. Generation of 3D culture devices based on microfabricated techniques, incorporating advanced biocompatible, functionalized biopolymer hydrogels and immortalized, highly differentiated renal epithelial cells is expected to enable renal organotypic culture and to revolutionize our ability to understand and predict clinically relevant renal epithelial functions and responses for their application in kidney disease.

2. CELL SOURCES FOR IN VITRO KIDNEY MODELS

Perhaps the most critical challenge in developing any in vitro or artificial model of renal function is to be able to cultivate large numbers of cells with the specific phenotype. This challenge is complicated because of several reasons. First, the kidney is one of the most complex organs in the body: aside from cell types common to every organ, its functional unit, the nephron, is made up of most than a dozen distinct epithelial cell types. Along the nephron cells exhibit very large differences in morphology, protein expression patterns, and very specific activities (Figure 1). Such heterogeneity enables the kidney to independently maintain several equilibriums in the body. There is a clear work division along the nephron, starting with plasma filtration at the renal corpuscle. In the proximal tubule (PT), the bulk of ion and water reabsorption takes place, important substrates like glucose and protein (amino acids) are reabsorbed, and organic anions and cations—including drugs and toxicants, are secreted into the lumen (see[1] for a recent comprehensive review). PT is also the place

---

Box 1 Clinical Applications of In Vitro Kidney Function Models.

The increasing incidence of end-stage renal disease, a shortage of kidney organ donors, and the significant impact on patient's life of current dialysis and hemofiltration techniques, generates an urgent need for alternative renal replacement therapies. The concept of bioartificial kidney (BAK) was developed in the 1980s with the goal to complement dialysis treatment (reviewed in Refs 30,143). Dialysis and hemofiltration are efficient in clearing toxins from blood but other important substances are lost and must be replenished. They also fail to provide a kidney's metabolic, endocrine, and immunomodulatory functions, resulting in significant long-term complications. A BAK is a cell therapy based on in vitro culture of renal cells.

Initially, renal cells were grown inside hollow fiber cartridges connected to hemofiltration devices to reproduce their reabsorptive function. Success in preclinical studies led to clinical trials in the 2000s. However, phase II trials had to be interrupted due to undesired adverse effects and technical issues, despite some clinical improvement had been observed through phase I trials. Recent approaches are focusing in replacing the metabolic, endocrine, and immunomodulatory functions. Protocols to freeze cells inside the device solve the issue of transport and long-term storage of bioartificial devices. Another critical aspect is to reduce the device size as to make it wearable or even implantable.

for important metabolic and endocrine activities. High metabolic rates and exposure to toxins make the PT more exposed to hypoxia and chemical insult than other nephron segments. Accordingly, most in vitro models of renal function have focused on reproducing PT function. Starting at the thick ascending limb (TAL) of the loop of Henle and extending through the distal convoluted tubule (DCT), connecting tubule, and collecting ducts (CDs), the distal nephron (DN) is where the fine-tuning of ion and water handling by the kidney takes place. Several layers of regulation ensure proper maintenance of whole body parameters like blood pressure, plasma osmolarity, and pH. The widely used classes of loop diuretic and thiazide drugs have their targets in the DN.

A second challenge faced by those trying to reproduce renal function in vitro is differentiated renal cells quickly lose their very specific activities upon being cultured in vitro. To solve this issue, one would need to ask first about the factors determining renal cell differentiation, causing such abrupt changes in cell morphology and function as those seen in the nephron. Figure 1 helps to understand the four interactions that a given renal tubule cell could use to maintain its full phenotype: (1) mechanical stimulus by luminal flow, (2) mechanical and chemical interactions with the extracellular protein matrix, (3) interactions with neighboring cells, and (4) delivery of blood-borne oxygen, nutrients, and hormones. This chapter reviews current efforts being made in an attempt to reproduce the physical and chemical environment of a renal tubule cell. It starts by discussing the available sources of cells with defined renal phenotypes, with special focus on cells reproducing PT phenotype.

2.1 Primary Culture of Isolated Nephron Segments

Primary cell cultures are defined as those growing from tissue explants or from individualized cells obtained from the original tissue. Their phenotype is closely related to that of the cells making the tissue. As such, primary culture is the gold standard for renal function studies. Successful in vitro culture has been accomplished for most nephron segments, and for a variety of mammal species, including humans.[2–4] We already mentioned the PT is the best studied segment from a clinical perspective. Thus, periodically there are reports describing new or improved methods to isolate and grow proximal tubule cells (PTC), with recent focus on generating cultures of human origin because of their need for cell therapy applications.[5–11]

While renal epithelial cells can be grown out of renal slices or small explants,[12] the use of an enzymatic digestion step is a more common approach.[13,14] Mild treatment with collagenase or related enzymes allows for single tubule individualization. Tubules can then be microdissected with the help of a stereoscope, and manually selected using morphological criteria.[15] High purity of the starting material compensates for the painstaking procedure. Yield is low, though, implying it will take more than 2 weeks for the cells to start growing exponentially. Most often, enzymatic digestion is extended until the tubules or cells have been completely liberated, followed by application of different purification techniques. Density gradient[13,14] or size-based sieving has been shown to be efficient to isolate TAL or PT tubules.[5–8] More sophisticated purification procedures involve the use of immunoseparation. Successful use of immunomagnetic separation has been reported for PT,[16–18] TAL–DCT,[4,19] and CDs.[16,18] Immunoseparation combined with Fluorescence-Activated Cell Sorting has also

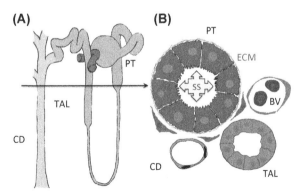

FIGURE 1 Functional 3D interactions for a renal tubule. (A) Nephrons are the kidney functional units. They consist of longitudinal tubule structures made of epithelial cells. Every nephron starts at the renal corpuscle, where it meets the vasculature and plasma is filtrated through the glomerular capillaries. Then, ultrafiltrate flows along the proximal tubule (PT), the loop of Henle (TAL), and the distal nephron, eventually draining into the CD. Abrupt morphological changes occur at every segment transition, reflecting very different functions. (B) Cells within each tubule are exposed to four independent interactions. Luminal flow creates a shear stress (SS) over the apical surface. Binding to surrounding protein matrix (ECM, fibers) serves to sense mechanical and biochemical cues important for tubule differentiation and function regulation. Neighboring interstitial and tubular cells send paracrine signals to modulate function and repair process (TAL, CD). Finally, oxygen, nutrients, and hormones supply is ensured by blood perfusion through surrounding capillaries (BV).

been employed in purifying different nephron segments.[10,20] More recently, the use of a complex object parametric analyzer and sorter has allowed for the isolation of very pure population of isolated DCT tubules.[21]

The possibility of isolating renal epithelial cells from human urine has recently received significant attention. Growing PTC, or any other nephron segment, from an individual's urine would enable personalized clinical studies. A few groups have reported successful protocols to isolate and grow PT-like cells from human urine.[11,22,23] A new cell line retaining numerous phenotypic characteristics of PT, ciPTEC, was recently generated by immortalization of PTC isolated from the urine of a healthy volunteer.[24]

Once isolated, tubules or cells require an attachment substrate and nutrient and oxygen support. Conventional culture techniques have been proven to suffice for the maintenance of epithelial renal cells, whereas the use of defined extracellular matrix (ECM)-like substrates (discussed in the next section) has been shown to improve their differentiated state. A serum-free hormonally defined culture medium suited for growing renal epithelial cells was defined in the 1980s by Mary Taub[25] and is based on Dulbecco's Modified Eagle Medium: Nutrient Mixture F-12 1:1 mix supplemented with insulin, selenium, transferrin, prostaglandin E1, triiodothyronine, epidermal growth factor, and hydrocortisone. Renal epithelial growth medium (REGM) is a commercial solution with similar components, which is usually added 0.5% serum. Although these media were formulated to preserve PT phenotype, they also serve to culture other nephron segments in low or serum-free conditions. Defining serum-free culture conditions is important, if clinical application is the final goal, since most regulatory agencies will oppose to the use of animal-derived serum in products intended for human applications.

Epithelial cells in culture exhibit a characteristic morphology generally described as a cobblestone pavement appearance. However, careful phenotyping is mandatory to establish both a cultured cell real identity and the level of cell heterogeneity. Transcript and protein expression levels of epithelial and specific nephron segment markers, together with functional assays to show specific cell activities characteristic of the nephron segment under study, are the common approaches to evidence renal cell phenotypes[5–11] (Tables 1 and 2). However, studies looking into a specific nephron segment phenotype, usually contain a scant analysis about the absence of other nephron segments markers or the presence of nonepithelial, nonrenal markers, which makes difficult to evaluate the real degree of purity and differentiation of the cells under study.

Dedifferentiation and senescence are the main limitations for the use of primary cultures. For instance, commercial human PTC are characterized by the expression of 1 or 2 markers, and guaranteed only for 10–15 doublings. Considering a conservative 48 h doubling time it results in a usable period of only a few weeks. Most authors report that their primary cultures are able to maintain their phenotype for less than 10 passages,[10,26] before they start dedifferentiating, remain quiescent, or die. Primary cells activate gene expression programs required to respond to the isolation procedure and to survive under the new physical and chemical environment imposed by conventional culture techniques. The process of growing as primary culture has indeed been described as an epithelial-to-mesenchymal transition (EMT).[12,27] Also, primary cells are nontransformed adult cells with a biological age, so they enter into senescence and stop proliferating after a few passages. These limitations mean that a continuous source of tissue material is required. In the case of human tissue, besides ethical constraints, it implies a large degree of interindividual genetic and epigenetic variability will be introduced in the study.

2.2 Immortalized Cell Lines of Renal Origin

In order to overcome the limitations of primary cultures, a number of continuous cell lines of renal origin have been generated and characterized. These cell lines, originated as spontaneous or induced transformation of renal cell primary cultures, have been extensively used in studies related to epithelial and nephron function. A common immortalization method consists in transform cells with the SV40(T) large antigen (reviewed in Refs 28,29). Transformed cells acquire the ability to proliferate indefinitely; however, in most cases these cells had already suffered some dedifferentiation allowing them to grow under artificial conditions. More importantly, immortalization does not guarantee cells will retain their phenotype indefinitely, so proper care must be taken to often check their phenotype. The gold standard continues to be primary cultures of human PT, because immortalization and long-term passaging have been shown to cause genetic changes.[28]

For more than 50 years, renal continuous cell lines have been obtained through spontaneous or induced immortalization of renal cells isolated from mammalian kidney. Cell lines like Madin-Darby Canine Kidney Epithelial Cells (canine), LLC-PK1 (porcine), normal rat kidney NRK-52 (rat), or OK (opossum) have been extremely useful for in vitro research of normal and altered renal epithelial function. These cells retain enough phenotypic parameters to allow for studying specific characteristics or activities, and have lesser requirements and proliferate indefinitely, unlike primary culture. Careful isolation, purification, and characterization have allowed for the generation of specific cell lines with adequate preservation of characteristic functional markers of defined nephron segments.[14] However, in general they exhibit some degree of dedifferentiation and transdifferentiation, which hinders their usefulness in understanding integrated in vivo responses of specific nephron segments. Moreover, the use of renal cells in artificial kidney devices or other cell-based therapies requires cells of human origin.[30]

TABLE 1 Expression Markers for Phenotypic Characterization of Renal Epithelial Cells

Phenotype	Marker	Technique of Choice
Epithelial	E-cadherin	IF, WB
	N-cadherin	IF
	Ksp-cadherin	IF
	Cytokeratin 8	IF, WB
	Cytokeratin 18	IF, WB
	Zonula occludens-1 (ZO-1)	IF, WB
Podocyte	Synaptopodin	IF, WB
	Podocalyxin	IF, WB
Proximal tubule	Megalin	ICC, IF
	Gamma glutamyl transferase-1 (GGT-1)	ICC, WB
	Alkaline phosphatase	ICC, WB
	N-acetyl-beta-D-glucosaminidase (NAG)	ICC, WB
	Glutamine synthetase	ICC, WB
	Sodium–phosphate transporter (NaPi)	IF, WB
	Aquaporin-1 (AQP1)	IF, WB
	Alanine aminopeptidase M (CD13)	IF, FC
	Dipeptidyl peptidase IV, DPPIV (CD26)	IF, FC
	Tetragonolobus agglutinin	IHC, IF
	PNA lectin	IHC, IF
	Sodium glucose cotransporter (SGLT2)	IF, WB
	Aminopeptidase A (APA)	IHC, IF
Thick ascending limb of Henle	NKCC2	IF, WB
	Tamm–Horsfall (THP)	ICC, IF
Distal convoluted tubule/connecting tubule	Calbindin 28 (CB28)	ICC, IF
	NCC	IF, WB
Cortical collecting duct	Aquaporin 2 (AQP2)	IF, WB
Metanephric mesenchyme	Six2	ICC, WB
(Nephrogenesis markers)	Wilms Tumor gene 1(Wt1)	ICC, WB
	Cadherin-6	ICC, WB
	Glial-derived neurotrophic factor (Gdnf)	ICC, WB
	c-kit	ICC, WB
	Sall1	ICC, WB
	CD24	IF, FC
	CD133 (prominin-1)	IF, FC
	CD29 (beta1-integrin)	IF, FC
	CD44	IF, FC

IF, immunofluorescence; WB, Western blotting; ICC, immunocytochemistry; IHC, immunohistochemistry; FC, flow cytometry.

TABLE 2 Functional Markers for Phenotypic Characterization of Renal Epithelial Cells

Phenotype	Marker	Function
Epithelial	TEER (transepithelial electric resistance)	Monitoring epithelial tight junction permeability, surrogate for epithelial health
Proximal tubule	Albumin uptake	Endocytosis through the apical membrane
	Gamma-glutamyl transferase-1 (GGT)	Brush border enzyme
	Alkaline phosphatase	Brush border enzyme
	N-acetyl-beta-D-glucosaminidase	Lysosomal enzyme
	Glutamine synthetase	Cytosolic enzyme
	Hormone-induced cAMP production	cAMP production dependent on expression of hormone receptor
	Glucose uptake	Sodium-dependent glucose
	Sodium uptake	Cotransported with glucose, amino acids, bicarbonate,...
Thick ascending limb of Henle	Na–K–Cl cotransporter	Loop diuretic sensitive cell uptake of Na$^+$, K$^+$, or Cl$^-$
	Vasopressin stimulation	cAMP production dependent on expression of hormone receptor
Distal convoluted tubule/connecting tubule	Na–Cl cotransporter	Thiazide-sensitive cell uptake of Na$^+$, Cl$^-$
	Aldosterone stimulation	
Cortical collecting duct	Vasopressin stimulation	cAMP production dependent on expression of hormone receptor

HK-2, a PT continuous cell line obtained by expression of genes E6/E7 from papilloma virus in human kidney cells,[31] has been extensively used in renal physiology and pharmacology studies for the last two decades.[32,33] However, among other defects, HK-2 cells do not express most of the molecular machinery required for xenobiotic transport and metabolism.[34] It is also not very useful in long-term studies, since it apparently does not tolerate well confluent conditions.[35] Cell-based therapies like the BAK require large numbers of well-differentiated, proliferative PTC of human origin that will stand long-term culture as confluent monolayers. This need has prompted the search for new strategies to generate immortalized human PTC. An alternative method to viral transformation is inducing expression of human telomerase (hTERT). Telomerase activity blocks the senescence process and appears to be more amenable and less-intrusive procedure for renal epithelial cells than viral transformation. TERT-expressing cell lines have been established for PT[24,36,37] and principal cells of the CD.[38] TERT-immortalized PTC exhibited PT functions such as hormone-mediated cAMP production, pH-dependent ammonia production, sodium-dependent phosphate uptake, and protein endocytosis.[36] Another human PT cell line (ciPTEC) has been generated that exhibits a PT phenotype after more than 45 passages, including expression of drug transport proteins involved in drug metabolism, what makes this cell line a promising tool for nephrotoxicity studies.[11,24] Immortalization by hTERT in NKi-2 cells was shown to not interfere with renal cells ability to form tubules in 3D hydrogels.[39]

Despite numerous efforts to isolate and purify the starting material, cultured cell lines generated from renal epithelial tissue lose their characteristics and trans- or dedifferentiate to some extent. This has to be understood as a normal consequence of the culture process. When taken out of their environment, cells change their expression patterns in order to respond to the aggression and to adapt to the new conditions, otherwise they would die. PTC are known to possess the ability to regenerate the damaged tubule through a process involving dedifferentiation, proliferation, and redifferentiation.[40] In vitro culture might in fact represent a model for the first two steps of the process.[12,27] It remains to be proven that cultured cells, once the right conditions are provided, are able to redifferentiate and acquire the full phenotype characteristic of in vivo cells. Thus, focus has switched over efforts to improve in vitro culture technologies with the goal to provide the right conditions, the subject we will discuss throughout the rest of this chapter.

2.3 Differentiation of Progenitor Cells toward a Renal Phenotype

Before we move on to discuss the current strategies aimed to avoid cell dedifferentiation by providing better biomimicking environments, we will briefly comment on the opposite approach: to take advantage of regenerative or developmental programs to generate specific renal cells by inducing differentiation of pluripotent proliferative cells. As in other organs, several niches containing cells with regenerating ability have been identified in different regions of the kidney: PT interstitium,[41–44] tubular epithelial cells,[40,45] human Bowman's capsule,[46] the corticomedullary region,[47] and the renal papilla.[48,49]

Upon tubular damage, kidney-regenerating cells are able to proliferate and differentiate into different cell types.[40] Cells within the PT wall exhibit markers of progenitor cells and have been shown to proliferate and regenerate the PT and the parietal and visceral walls of the Bowman's capsule after moderate insults.[45] Renal progenitor cells have been isolated and grown in culture for subsequent in vivo implantation, resulting in tubule integration[41] and renal function improvement.[46] However, these approaches do not appear to be the more straightforward for the generation of large numbers of differentiated human renal epithelial cells.

The ability to direct the differentiation of pluripotent stem cells toward renal lineage cells opens the possibility to generate renal cells on demand, allowing for individualized in vitro studies of kidney function and drug susceptibility, and paving the way to personalized therapies. All types of extrarenal stem cells (ESC, BMDSC, AFDSC) can be derived into partially or totally committed renal epithelial cells.[30,50,51] Cells with PT phenotypic characteristics have been derived from human embryonic stem cells (hESCs).[52,53] Kidney organoids and tubules have been generated from human pluripotent stem cells.[54,55] Organ development was recapitulated from human ESC.[56]

To overcome the limitations of ESCs (ethical issues and teratogenesis) the future is in reprogramming adult cells. iPSCs (induced pluripotent stem cells) have been generated from human PTC,[57] human kidney mesangial cells,[58] and urine-derived renal cells.[59] iPSCs may potentially provide an inexhaustible source of patient- and tissue-specific stem cells. Xia and colleagues[60] were able to differentiate iPSCs into ureteric bud (UB) progenitor-like cells that were lately used to generate kidney organoids.[54] iPSCs retain the epigenetic signature of the parent cells, facilitating targeted, organ-specific differentiation while exhibiting less potential for abnormal tissue formation than embryonic or bone marrow-derived stem cells (BMDSC). On the other hand, iPSCs have been shown to induce immune reaction in syngenic mice.

One of the main advantages of deriving renal cells from pluripotent stem cells is that it would virtually yield unlimited number of cells. It also raises the possibility of generating different nephron segments from a single originating cell type, allowing for re-creation of complex renal tissue. Indeed, the use of appropriately reprogrammed cells could be directed to the emulation of the developmental program and the generation of full renal structures.[54,56] Such structures may be used to investigate complex integrated functions, or can be further exploited toward regenerative strategies.

3. MODELING RENAL TUBULES COMPLEX 3D INTERACTIONS

Renal tubules are formed by intercellular unions through specialized proteins like those found in tight junctions. Tubular structure is encased in the BM, a thin layer made of nonfibrillar collagen IV, laminin, and fibronectin, which is largely secreted by the tubule itself. Nephrons, blood vessels, and other elements found in kidney parenchyma are held together by a stromal matrix, constituted primarily by fibrous collagen I, proteoglycans, and their constituent glycosaminoglycans.[61] The stromal matrix forms the tissue interstitial compartment that includes sparse fibroblast numbers (Figure 1). This ECM is not a fixed scaffold but it must be ready for remodeling to allow and even guide, tubule morphogenesis during development or repair.[62] Through the BM, tubules are in contact with the interstitium, other renal tubules, and the vascular network comprised of peritubular capillaries and vasa recta (Figure 1). To fully understand, and be able to reproduce, normal or altered renal function, one has to take into account all these relationships.

3.1 Role of ECM in Renal Cell Differentiation and Function

HK-2 cells cultured on micro-scaffolds obtained by decellularizing 300-μm fragments of renal stroma have been shown to improve their PT phenotype.[63] This finding illustrates the importance of cell–ECM interactions in driving differentiation toward a particular phenotype.[64] The BM and the interstitium provide a number of physical and chemical cues that cells need for proper self-recognition and differentiation. As in other tissues, renal cells recognize the roughness and hardness of the BM.[65] This trait is being exploited in attempts to recreate artificial ECM substrates,[66,67] where it is being appreciated that the topology offered by the polymeric structures can actually be more important than the bioactive signals they provide.[68–70] The protein matrix surrounding the tubule also acts as a reservoir for biologically active substances like growth factors and cytokines.[30,71]

Nevertheless, epithelial cells' most important influence from the ECM arises from biochemical interactions with the BM.[62] Binding of integrins and other transmembrane receptors to BM components such as collagen IV, laminin, and fibronectin triggers intracellular signaling pathways important for differentiation. Integrins not only contribute to development and tissue morphogenesis but also may play a role in epithelial function and regulation. For instance, deletion of integrin beta-1 changes the paracellular permeability of PTC toward a tighter epithelium.[72] Importance of cell–ECM interactions is further illustrated by embryonic or perinatal lethality resulting from integrin gene disruption. In addition to integrins, polycystin-1, another transmembrane protein localized to focal adhesions and the gene product that is mutated in autosomal-dominant polycystic kidney disease (PKD), has been involved in cell–matrix interactions. Expression of polycystin-1 C-terminal fragment was shown to be sufficient to induce tubulogenesis in kidney epithelial cells.[73] Diabetic nephropathy is a paradigmatic example of how disruption of ECM homeostasis can explain an altered renal function.[74]

This section discusses the use of ECM substitutes to promote attachment and differentiation of renal cell cultures in vitro. Strategies aimed to reconstruct kidney structures using decellularized organs have been reviewed elsewhere in this book.

3.2 Two-Dimensional Renal Cell Culture on ECM-Coated Surfaces

The simplest approach to mimicking the cell–ECM interactions consists in coating the culture container surface with a thin layer of ECM-like material. Some cells, like human bronchial epithelial cells, require such treatment to ensure proper attachment to the culture vessel. Although this is not the case with renal epithelial cells, the chemical nature and physical properties of the coating protein matrix have been shown in many cases to be critical for an adequate differentiation state.[63,75]

Type I collagen is the ECM-like substrate most commonly used in cell culture procedures, possibly because it is more accessible and affordable than other alternatives. Collagen I is the main interstitial ECM component but is not present in the composition of the BM directly in contact with the tubule cells. Substrates like collagen IV, laminin, or fibronectin present in BM have been more sparsely used.[76,77] An alternative is the use of complex matrices extracted from stromal tumors, such as Matrigel.[78] Matrigel has the advantage of providing a mixture of BM components, including collagen IV and laminin, and also contains growth factors. A common approach in 3D culture models consists in mixing collagen I 1:1 with Matrigel, as discussed below.[39,79,80]

Daniel Zink's group has compared the differentiated state of primary PTC cultured onto several ECM substrates. PTC were found to exhibit higher levels of ZO-1 (zonula occludens), a marker of polarized epithelial cells, when cultured in the presence of collagen type IV and laminin.[77,81,82] In the same study, common attachment substrates like poly-lysine or pronectin failed to promote ZO-1 expression. On the other hand, HK-2 cells appeared to perform better on fibronectin-coated polycarbonate (PC) membranes compared to laminin or Matrigel.[70] Matrigel offered the best performance in a comparison of PT primary cell proliferation, attachment ratio, and monolayer formation on PC membranes coated with one of five different ECM substrates (Matrigel, gelatin, collagen I, poly-L-lysine, laminin).[76] Narayanan et al. have recently analyzed in depth the conditions for PTC differentiation from hESCs. Differentiation was induced by incubation in REGM. Under these conditions, Matrigel outperformed fibronectin, laminin, and collagen IV on cell differentiation tests based on expression of aquaporin-1 (AQP-1), gamma-glutamyl transferase, and cytokeratin 18. Differentiated hESCs formed confluent epithelia showing the typical ZO-1 chicken-wire staining pattern. In addition, cells formed networks of cords, which consisted of rows of single cells, and multicellular tubule-like structures when cultivated on Matrigel.[52]

Complex ECM extracts like Matrigel usually provide the best results in terms of preserving or inducing the differentiated state. However, their cost and batch-to-batch composition variability intrinsic to tissue extracts may limit their use. There have been significant efforts, especially in the BAK field, into designing and fabricating artificial ECMs with optimal physical properties that could substitute for natural ECM. A good synthetic biomaterial for renal epithelium scaffolding will be one whose composition can be tailored to provide significant physical and biochemical signals and that exhibits low degradation under the conditions expected in a BAK.[83] Supramolecular micro- or nanostructured polymers can be generated by electrospinning, incorporating bioactive molecules that are able to support growth and differentiation of PTC.[84] In this study, human PTC cultured for 19 days on a supramolecular polymer functionalized with bioactive ECM peptides showed tight monolayer formation and expression of functional markers indicative of a PT phenotype. A synthetic polyamide matrix has been successfully employed to support normal growth of NRK cells in the presence of serum.[67] Since the main advantage in the use of synthetic substrates is the ability to tailor the hydrogel polymer physical properties and to functionalize them with appropriate chemical signals, it looks the future for in vitro models of renal epithelia resides in artificial substrates.

3.3 Three-Dimensional Renal Cell Culture in Hydrogel and Transwell Devices

Interestingly, cells of renal origin, either primary cultures or continuous cell lines, retain their ability to form tubular structures when given the appropriate conditions. There are numerous reports on MDCK or primary PTC forming

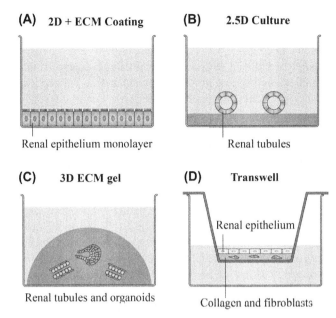

(A) 2D + ECM Coating

Renal epithelium monolayer

(B) 2.5D Culture

Renal tubules

(C) 3D ECM gel

Renal tubules and organoids

(D) Transwell

Renal epithelium

Collagen and fibroblasts

FIGURE 2 In vitro renal cell culture models. (A) Two-dimensional (2D) culture. Cells are typically cultured onto a highly rigid substrate coated with an ECM substitute. Renal epithelial cells form polarized monolayers but do not stand long-term confluent culture (B) Two and one-half-dimensional (2.5D) culture. Cells are cultured on top of a thin, organized layer of ECM, and medium is supplemented with growth factors to induce tubulogenesis. (C) Three-dimensional (3D) cell culture embedded within an ECM gel. Cells are uniformly exposed to an organized ECM and can further remodel and restructure the ECM over time. This format is not well suited for imaging techniques. A variation of these formats involves the coculturing of an epithelial monolayer on top of a gel with embedded interstitial cells. (D) Culture onto permeable supports—the Transwell system. The three types of in vitro renal cell culture models described above can be performed on a tissue culture insert which bottom is a permeable support. Immersion of the insert into a well creates two different fluid compartments. Renal epithelium growing as a monolayer will create a functional barrier between an apical, "luminal" compartment and a basal, "blood" compartment, allowing renal epithelial cells to recreate in vivo functions. Endothelial or stromal cells can be cocultured in the dish compartment to study paracrine effects without direct physical contact between cell types. *Adapted by permission from Macmillan Publishers Ltd: Figure 2 in Nat Rev Mol Cell Biol 15:10, 647–64, copyright 2014.*

tubular structures by simply seeding them on top of a thin ECM gel. This culture system has been termed 2.5D culture[85] (Figure 2), and it has been very useful in studying basic principles of tubule morphogenesis. There have also been reports of nephrotoxicity models based on this approach. A fully 3D culture system, on the other hand, consists in embedding the cells in an ECM gel, which is achieved by mixing the isolated renal cells with a liquid ECM matrix at the time of seeding[85,86] (Figure 2). Renal cells have been shown to spontaneously form tubular structures when cultured inside collagen or Matrigel gels.[87,88] In this model, nutrient delivery and functional assays depend on solute diffusion through the gel from a static fluid compartment placed above the gel. Clinical significance of 3D tubulogenesis models is based on the fact that the repair process observed after acute tubular damage shares similar cell and molecular mechanisms. Surviving tubular cells undergo partial EMT, proliferate, and then redifferentiate. These models can also contribute to successful implementation of BAKs by helping to understand the minimum requirements for PTC health.

Studying tubular morphogenesis has been useful in establishing the role of secreted growth factors, cell–cell interactions, and cell–matrix interactions in renal cell differentiation.[89] One of the best studied models is the tubule formation of MDCK cells in collagen I gels.[90] Since the original observation by Montesano et al. that MDCK cultured inside collagen I gels would form tubules when exposed to fibroblasts or hepatocyte growth factor (HGF),[91,92] tubulogenesis has also been shown to be stimulated by EGF[93] or TGFa, and downregulated by TGFb, and BMPs 2 and 7.[89] HK-2 cells can also form tubules inside an ECM matrix, improving their expression of N-cadherin, which in humans is distinctly expressed in PTC.[94] Tubules generated by inclusion of PTC (NKi-2 immortalized cell line) in collagen I–Matrigel gels have been compared to 2D cultures as cytotoxicity models. Cells from 3D cultures exhibited better expression of functional markers like hormone-mediated cAMP production, and higher sensitivity to nephrotoxins that could be explained by differential expression of proteins involved in drug transport in PT in vivo.[39] A fibrosis model consisting of a coculture of human PTC (HKC-8) on top of a collagen I gel with embedded fibroblasts has been used to evaluate fibrosis marker expression after cisplatin exposure.[95] The tubulogenesis 3D model has also been successfully achieved with primary renal cells. Rat primary kidney cells seeded on a collagen–Matrigel (10%) gel led to the formation of glomerular and tubular-like structures, which stained positive for cytokeratin 18.[96] Inclusion in collagen I gels of human renal cells obtained from discarded kidney tissue led to tubule formation after 10 days. Importantly, the phenotype of different nephron segments was present in the system, since tubules were positively stained for markers of proximal tubule (NEP), distal tubule (EMA), or thick ascending limb (Tamm–Horsfall) in an exclusive fashion.[9,97] Thus, the ability of renal cells to form tubules appears to be independent of species, or nephron segment, and is conserved in immortalized cell lines. Moreover, tubulogenesis consistently improves the differentiation state and allows for more representative studies of normal and altered function. It has also been shown to be instrumental in inducing differentiation of embryonic cells toward renal lineages.[39,98,99]

An important limitation of collagen hydrogels is fast degradation due to cellular remodeling and relatively low mechanical integrity, reducing its resistance to shear forces and limiting their utility for long-term culture. To overcome this limitation, Subramanian et al. have combined collagen gels with 3D porous synthetic silk polymer scaffold to increase their resistance. Cells embedded in a collagen–Matrigel gel were added to the scaffold. Tubule- and cyst-like structures lasted longer than 2 weeks, exhibiting hormone-dependent cAMP production, drug transport activity, and positive staining for tubular and polarization markers.[80] This model has been successfully applied to study the effects of polycystin-1 silencing on cell–ECM interactions in mouse inner medullary collecting duct cells.[100]

Materials other than collagen and Matrigel have been successfully employed to grow renal cells in 3D scaffolds and stimulate tubulogenesis. Astashkina and colleagues have developed a 96-well format 3D culture model of PTC for nephrotoxicity studies based on culturing mouse primary PTC inside hyaluronic acid (HA) matrix. Under these conditions, cells form tubular structures that remain viable and express PT functional markers for at least 2 weeks. As could be expected, when compared to immortalized cells LLC-PK1 and HEK-1 growing as monolayers directly on plastic, the PT 3D system exhibited a more clinically relevant response to nephrotoxicants.[101] This model has been used recently to evaluate nanoparticle nephrotoxicity.[102] HA has also been shown to promote differentiation of both metanephric mesenchyme and the UB.[99] Custom functionalization of artificial hydrogels is a powerful approach to reveal cell–ECM interactions necessary for normal and altered states of tubule morphogenesis.[103]

Another approach employed in renal 3D cell culture is the use of microcarriers, microstructured particles made from biocompatible materials. Human primary PTC were cultured on alginate microcarriers coated with gelatin (75–150 μm, Global Eukaryotic Microcarriers, GEMS™). When compared to 2D culture on collagen-treated dishes or transwells, cells exhibited an improved morphology, a large increase in microvilli and the associated protein villin and an enhanced sodium uptake. PTC can be kept confluent for several months since they remain quiescent.[23] A variety of renal cells (HK-2, LLC-PK1, primary cultures) were efficiently grown on fibrin microcarriers prepared from human fibrinogen and thrombin (diameter range of 100–300 μm). These particles maintain cell growth for several weeks before they degrade.[104] A main advantage in using microcarriers is that they are pipettable and allow for production of high yields of differentiated cells. Microcarriers and artificial scaffolds like those described above[80] also facilitate perfusion of medium and coculture with additional cell types.

Induction of tubule formation in 3D hydrogels has helped understanding tubule morphogenesis and branching, and it has provided some examples of utility in studies of renal function and disease. However, their main limitation is the lack of access to the tubule lumen. In fact, the luminal compartment is filled with the same solution filling the ECM matrix or, if tubules are closed, their composition is determined by tubule cell secretion, unknown and not experimentally controlled. Additional limitations are the difficulty to obtain high-resolution imaging, sampling for biochemical studies, and medium manipulation for functional and drug application. Another approach to recreating the 3D structure of a renal tubule consists in growing the renal epithelium onto a permeable support placed between two different liquid compartments. Up to now, this 3D culture model has been realized by growing cells on the so-called Transwell devices (Figure 2).[85] Transwell platform has been traditionally used in 6- or 24-well plate format, although it is already available in 96-well plate format. Working with transwells is costly and laborious, what may explain why it has not been more widely adopted. Notwithstanding, the Transwell system ensures proper polarization of epithelial cells, simultaneous experimental access to both compartments, and facilitates coculturing of additional cell types in independent compartments.[105,106] Additionally, the benefits of growing the cells onto or inside ECM gels are perfectly attainable in the Transwell format. For example, Miya et al. developed a model of PTC and endothelial cell coculture to study the regulation of tubulogenesis by surrounding capillaries. Human primary PTC were embedded in a 1:1 mix of Matrigel and atelocollagen I (a protease-derived soluble collagen) and seeded into 0.4-μm pore Transwells. Tubulogenesis was stimulated by HGF in a dose-dependent manner, but not by other growth factors (EGF, BMP-7, IGF-I, or follistatin). Coculture with HUVEC or HUVEC-conditioned medium resulted in increased tubulogenesis, gel invasion, and migration, although the paracrine factors involved could not be identified.[79]

Growing renal cells in 3D culture systems, either as tubules embedded in a collagen–Matrigel gel or as monolayers growing onto an ECM-coated permeable membrane in a Transwell, appears to be a good solution to mimic the cell–ECM interactions that are critical for complete differentiation and proper function of renal tubules. It also allows for coculturing with relevant cells like fibroblasts or endothelial cells. However, an important component of in vivo microenvironment is missing. While the Transwell approach allows for clear distinction of apical versus basolateral compartments, it is generally not amenable to individualized media perfusion (although Minucell is a commercial solution to this problem). In the next section, we review recent strategies to incorporate microfluidic technology into renal cell culture and how it could finally evolve into the definitive biomimicking system.

4. RENAL ORGANOTYPIC CULTURE IN MICROFLUIDIC DEVICES

4.1 Justification for the Use of Microfluidic Devices in Renal Epithelial Cell Culture

Renal tubular cells are continuously exposed to a plasma ultrafiltrate flowing along the luminal compartment (Figure 1(B)). Luminal flow generates a shear stress (SS) force over the apical surface of the cells. Cells can sense SS magnitudes through mechanical bending of primary cilium -or of brush-border microvilli in PTC. Intracellular signaling triggered by luminal SS is a key physiological stimulus for renal tubular cells. For example, luminal flow dynamics regulate ion transport in the TAL and the CD.[107] Loss of this sensing mechanism leads to severe pathological states (e.g., PKD).[108] Despite luminal flow relevance in nephron function and disease, there are relatively few in vitro studies of cultured renal cells under conditions of luminal flow. Actually, such studies are designed to specifically investigate the effect of flow/SS. In other words, luminal flow is currently not considered a requisite for an in vitro model of renal tubular function, quite surprisingly.

Equally surprising is the fact that a highly polarized cell that, in vivo, is exposed to two compartments with very different dynamic characteristics (slow flow, small distribution volume in the luminal compartment vs a much wider basolateral compartment) is commonly modeled under basically opposite conditions (very small fixed basolateral compartment and a huge apical compartment which is entirely renewed every 3–5 days). 2D culture conditions (Figure 2(B)) eliminate the possibility to reproduce real tubular function, which consists in concentrating or diluting solutes in the luminal fluid. Keeping adequate transepithelial gradients (i.e., blood to urine) is critical for maintaining chemical equilibriums in the extracellular fluid compartment to which the basolateral side of the tubule is connected. Ironically, the formation of domes is many times referred to as an indication of a renal cell monolayer healthy status, acknowledging vectorial transepithelial transport is a key function of renal cells.

It is thus evident that conventional 2D culture techniques fail to reproduce the actual environment of renal tubule cells, which may explain the difficulty in translating in vitro results to in vivo applications. But, why have these two important in vivo features of renal tubular function been obviated in the majority of studies employing in vitro renal culture? Growing cells on permeable supports (Transwells) has been available for more than 25 years. However, their low throughput probably does not compensate for their cost and more difficult handling. On the other hand, the introduction of flow as an experimental condition has been limited to rather cumbersome, single sample devices to perform experiments under a microscope (parallel plates flow chambers) and, in their almost entire majority, over the course of only a few hours.[109] Recognizing this issue, there is a current surge of intense cooperation between technical engineering and biological laboratories interested in developing microfluidic devices for their use in studies of renal epithelium. This section reviews such efforts aimed to produce a reliable, user-friendly dispositive for the in vitro culture of polarized renal cells under continuous flow conditions.

4.2 Design and Fabrication of Microfluidic Cell Culture Devices

Microfluidics studies the properties and applications of small fluid volumes (10^{-9}–10^{-18} L), usually flowing through rectangular channels within the range of 100 nm–100 μm transverse section.[110] It started to develop in the mid-1970s and is being intensively applied in biomedical sciences through the concept of Lab-on-a-chip, where biochemical reactions are carried out inside a microfluidic channel in order to reduce sample and reagent volumes, among other possible advantages. Microfluidics has also caught the attention of the cell culture field because it allows for introducing flow in the culture system, with two large impacts: (1) enabling prolonged culture of cells without the need for subpassaging or repeated media changes and (2) incorporating mechanical stimulation that can be interesting for those cells or tissues that endure such stimulation in vivo and respond to it by acquiring specific phenotypes. Tissue engineering rapid adoption of microfluidics is based on these two important advantages. In the particular case of the nephron, the size of a microfluidic channel is very well suited to recreate the flow inside a renal tubule lumen, which usually ranges in the 20–60 μm diameter. SS forces depend on the dimensions of the channel but also on flow magnitudes. Wider channels would need higher flows to recreate physiological SS. This has a downside, though, since higher flows will significantly reduce changes in solute concentration brought about by epithelium transport activity (it would also dilute the concentration of paracrine substances). Most solute reabsorption is dependent on ion gradients built up through the activity of membrane ion transport. On the other hand, lowering flow would not provide enough SS. The ideal solution is thus to try to reproduce the geometry of a renal tubule as much as possible, something within the capabilities of microfluidics.

The most employed technique to generate a microchannel pattern for tissue culture is polydimethylsiloxane (PDMS) molding using soft photolithography.[105,106,111–127] PDMS is the material of choice for microfluidic bioengineers. Besides its properties (high compliance, optical clarity, low cost, reproducibility), it is clearly more advantageous over other materials for rapid device prototyping and fabrication.[110,128,129] This is particularly important in biological applications that usually require extensive optimization of the prototype design through sequential iterations. On the other hand, PDMS presents significant differences with regard to polystyrene (PS), the material most cell culture disposables are fabricated with, but a material not well suited for microfabrication techniques. After providing justifications for the respective use of PDMS and PS in the engineering and biology fields, Berthier and Beebe[129] hinted the engineering field should move toward materials biologists are most used to. According to them, five limitations of PDMS for microfluidic cell culture systems are (1) deformation: causes variable responses to SS forces, (2) evaporation: modifies composition of media (osmolarity),

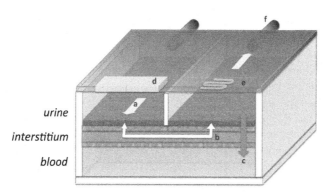

urine

interstitium

blood

FIGURE 3 Conceptual design of a microfluidic-based organotypic culture of renal tubule cells. Three different compartments are fabricated, separated by two permeable surfaces. Renal cells are grown at the bottom of the "urine" compartment. They form a polarized confluent monolayer exposed to shear stress (a) resulting from fluid flow through the "urine" compartment. The epithelium establishes a functional barrier between the "urine" compartment and the compartments situated below. Immediately under the cells, there is an interstitial compartment filled with ECM. Optimal design would allow for epithelium to grow directly on top of the ECM gel. This interstitium can harbor mesenchymal or stem cells to study their paracrine effects on epithelial cells (b). Different renal cell types seeded on independent fluidic channels could also interact through this pathway, or by luminal paracrine signals if growing in connected channels. Finally, a "blood" compartment consisting in a fluidic channel is used to maintain culture homeostasis and could also accommodate an endothelial monolayer to complete a 3D functional environment of a renal tubule (c). Elements for optical (d) or microelectrode-based monitoring (e), and for biochemical sampling (f) are shown in the device.

introduces bubbles, (3) absorption of less hydrophilic small soluble factors reduces their effects, (4) leaching of uncross-linked oligomers may interfere with cell-membrane signaling and trafficking, and (5) hydrophobic recovery hampers cell attachment. Also, the fabrication, shipping, and storage of large numbers of PDMS devices required to make them available to the general laboratory is not as easy as with plastic materials like PS. Alternatives to PDMS are thermoplastics such as poly(methylmethacrylate) (PMMA or Lucite),[130] PS, PC, cyclic olefin copolymer,[131] and teflon.

The simplest microfluidic pattern design includes a single cell culture chamber with inlet and outlet fluidic channels connected for culture medium delivery and removal. A proper renal epithelium model would demand a second fluidic compartment, separated by a thin layer of porous material, usually a porous membrane, able to support cell attachment and proliferation (Figure 3). The same biocompatible materials used previously in the Transwell open chamber system can be bonded to PDMS. Renal cells have been successfully cultured inside microfluidic devices onto translucent track-etched PC,[114,123,132] optically clear polyester (polyethylene terephthalate),[106,122,124] polyethersulfone[121] membranes, and directly on PDMS porous membranes.[111]

Microfluidic flexible patterning allows for a myriad of complex designs aimed to provide useful characteristics such as exposure to solute gradients, migration tests,[133] hydrogel confinement, and even MEMS-driven pumping,[134] and to include actuators imitating specialized organ structure and function.[110,135] Eric Leclerc's group incorporated a microstructure array into the single culture chamber to promote cell adhesion.[117,118,127] In a following design, a second chamber was added for liver cell coculture. Serial connection of liver and kidney chambers allowed studying the cytotoxic effect on MDCK renal cells of isofosfamide metabolites generated by the liver cells.[112] Wang and colleagues placed a microsized barrier inside the culture chamber in the path of the fluid flow to generate within it a wide range and modes of SS.[125] After modeling the shear forces, they showed the effects of 24 h constant flow on MDCK morphology and actin cytoskeleton rearrangements. In an approach to study the interactions of renal epithelial cells with surrounding interstitial cells, Huang and colleagues devised a cup-like coculture system consisting of two independent but connected compartments.[105] Adipose stem cells (ASCs) were embedded in a collagen gel and injected into the central channel. After polymerization of the first gel, a suspension of MDCK cells was introduced in the peripheral channel and formed a monolayer. Exposure to ASCs markedly promoted MDCK cells epithelial differentiation. A similar design was employed to study PTC invasion into the ECM substrate after induced EMT.[136]

Probably the most ambitious goal for microfluidic-based cell culture devices is to achieve long-term culture under automatized maintenance and experimentation interventions. Current culture procedures are far from simulating physiological homeostasis and introduce significant variability (nutrient and waste concentrations change in opposite directions over the course of 3–7 days between culture medium renewals); they also increase the risk for contamination. Culture medium perfusion inside microfluidic chambers must ensure adequate oxygen and nutrients supply to maintain culture growth and survival. The simplest approach relies on passive, gravity-driven movement of fluid between two reservoirs.[113,122,126] While being simple and affordable, this method provides a variable flow and requires continued media replacement for long-term cultures. Syringe pumps are another common solution.[105,106,114,115,119,120,124,130,136] Continuous, higher flow rates over longer period of times can be achieved by peristaltic pumps,[112,117,118,121,123,125,127] although they create pulsatile flow. Levels of flow-mediated SS achieved by using these impulsion methods cover a wide range (0.025–10 dyne cm^{-2}). SS forces in renal tubules are one order of magnitude lower than those found in blood vessels. For PT, SS can be estimated through calculations from glomerular filtration rates to be around 1 dyne cm^{-2}.[109] However, a wide range of SS has been employed in isolated tubule microperfusion experiments (e.g., as low as 0.2 dyne cm^{-2}).[115] Syringe pumps or microinjectors will have less trouble working under such low flow range requirements. Flow exposure in the reviewed works ranged between 1 and 72 h. Here the advantage is for peristaltic pumps because they allow for medium recirculation and thus could easily enable long-term culture conditions. Pneumatic pressure-driven pumps are also employed in microfluidic devices (Table 3). Independently of the perfusion system employed, flow is usually initiated after cells have reached confluence under static conditions.

4.3 Experimentation with Renal Cells Cultured in Microfluidic Devices

A variety of cells of renal origin have been grown in microfluidic devices: primary human PT,[106,114,123] primary inner medullary CD,[124] HK-2,[114,115,119,136] MDCK,[105,112,116–118,120–123,125,127,130] OK,[132] COS7,[130] and HEK (although not an epithelial cell).[113] Cell culture surfaces (PDMS or porous membranes) were coated with ECM-like substrates: collagen I,[105] Matrigel,[136] collagen IV,[114,115,123] fibronectin,[105,106,112,117,118,121,122,127] or just primed with the culture medium.[123,125]

Optical properties of PDMS or glass employed in most microfluidic devices allow for real-time monitoring of cell morphology, viability, and activity. Optical probes can be used to assess specific cell activities. For example, fluorescent microscopy monitoring of calcein uptake can be used as a cell viability test[121] or to demonstrate functional expression of the Pgp transporter.[106] Fluorescence microscopy was also employed, using Fluo-4AM indicator, to measure calcium release in MDCK cells following exposure to ifosfamide and cloroacetaldehyde.[112] Oxygen consumption is a critical parameter indicative of a healthy cell metabolism that can be determined optically in microfluidic devices.[137,138] Living cells can also be assayed in situ by means of integrated microelectronic sensors. Transepithelial electrical resistance (TEER) is an estimate of epithelial monolayer tight junctions integrity calculated from impendanciometry measurements that has already been successfully implemented in microfluidic devices.[122,123]

Cells growing inside microfluidic devices can be fixed and subjected to immunofluorescence and immunocytochemistry staining. Limitations arising from the optical properties of materials used for fabrication (autofluorescence) and from the particular available microscope setup need to be considered during the designing phase. Immunostaining is a common method to determine the differentiation state of renal cells cultured inside microfluidic devices (Table 1),[105,106,114,115,119,123–125,130,136] and to evaluate viability in toxicity assays.[106]

A potential advantage in using microfluidic devices consists in saving expensive reagents because of the low volumes running through the microfluidic chambers. On the other hand, the risk exists that this small sampling volume or the small number of cells growing in the chambers, will not suffice to feed current assay instrumentation. However, the number of biochemical assays already performed on samples from cells growing in microfluidic devices does not support this concern—on the contrary, they suggest microfluidic devices can easily replace conventional culture platforms. Metabolite chromatographic analysis and full metabolomic studies using samples from cells growing in small area microfluidic chambers have been performed.[112,116] A high-throughput small-molecule screening metabolomic study was carried out by ^{1}H-NMR spectroscopy on samples from renal cells exposed to toxicants like ammonia.[116] More conventional biochemical assays have allowed measuring metabolic activities like glucose[113,127] or glutamine consumption,[117,118] lactate production,[113] ammonia production,[117,118,127] alkaline phosphatase activity,[106] lactate dehydrogenase activity release,[106,127] albumin uptake,[106,109,132] fluorescein isothiocyanate-inulin transport,[123] sodium transport,[124] changes in medium osmolarity,[124] or reactive oxygen species production.[126] Cell numbers do not appear to be a problem either for conventional techniques used in nucleic acid and protein analysis, such as quantitative PCR,[113] western blotting,[106,113,132] or even analysis by microarray technologies.[117,118] Cells growing on microfluidic devices have also been subjected to molecular biology procedures such as siRNA or plasmid transfections.[105,119,120,124]

Thus, although most of the work done so far with microfluidic devices has consisted in proof of concept studies to demonstrate the approach feasibility, available evidence supports the idea that this technology will definitely provide a stronger model for the study of renal function and disease in a short future. Fundamental questions applied to significant clinical problems like renal toxicology,[106,112,117,118] EMT and proper cell differentiation,[136] albumin handling,[132] stone formation,[119] and metabolomics[116] have already been addressed by using such devices. Indeed, several commercial cell culture devices incorporating microfluidics have been introduced in recent years (Table 3). Some of them address the need for high-throughput 3D culture experimentation or include the possibility of exposing cells to two fluidic compartments. Their use in renal research has been very limited yet, likely because it implies a large investment in a still not very mature technology. It is foreseen that as they demonstrate their usefulness and more laboratories adopt them, this technology will become more affordable.

TABLE 3 Commercial Devices Providing Microfluidic Solutions for Cell Culture

Company	Applications	[a]	Permeable	Impulsion Solutions	Geometries	Publications
Cellix Ltd http://www.cellixltd.com/	Cell rolling, adhesion, migration, chemotaxis, shear stress	8	No	Syringe pump peristaltic	Parallel longitudinal channels	Endothelium Blood cells Cancer cells
Cytoo www.cytoo.com	Renal proximal tubule model (closed lumen)	100s	No	N/A	Micropatterned chips and well plates	Cell biology Polarization Mechanotransduction
Ebers www.ebersmedical.com	Chemical gradients Shear stress Cell polarization	3	Yes	Incubators integrate peristaltic pumps	Parallel channels	Cell biology Epithelial biology Cancer research
Fluxion Biosciences http://www.fluxionbio.com/	Aggregation, adhesion, cell rolling, shear stress	24	No	Electropneumatic pump	Well plates	Microbiology Cell biology
	Patch clamp	100s	No	Passive (cell trapping)	Well plates	Electrophysiology
Gradientech www.gradientech.se	Migration, chemotaxis, morphogenesis	1	No	Syringe pump	2D and 3D chemical gradients	Cancer biology Immunology
IBIDI www.ibidi.com	Cell migration, chemotaxis, angiogenesis, shear stress	6	No	Air pressure pump Syringe pump	Parallel channels	(85, 110, 140)
Kirkstall http://kirkstall.org/	Organ models (skin, cornea, respiratory epithelium)	1	Yes	Peristaltic pump	Individual chambers	Cancer Stem cell Drug discovery
Merck Millipore www.emdmillipore.com	Chemotaxis/migration, drug screening, hypoxia, shear stress	4	No[b]	Pneumatic pump	Parallel chambers	Cell Biology, Microbiology Cancer
microLIQUID www.microliquid.com	Shear stress, chemical gradients	1	No	N/A	Customized	Neurosciences
Micronit Microfluidics http://www.micronit.com/	Organ on a chip	1	Yes	Pneumatic (on chip)	Customized	Cell Biology
Mimetas http://www.mimetas.com	Organ on a chip	100	No		Multichamber	
Minucells http://www.minucells.de/	Gradient culture container for a single tissue carrier	1–6	Yes	Peristaltic pump	Single channel	(141, 142)
Nortis www.nortisbio.com	Organ on a chip/vascular biology	12	Yes	Pneumatic pump	3D hydrogels	(143)

[a]*High-throughput screening, number of samples per chip/device.*
[b]*In the CellASIC platform (Millipore) cells are separated from perfusion channel by a perfusion barrier, but do not grow onto it.*

5. CURRENT LIMITATIONS AND FUTURE DIRECTIONS IN IN VITRO KIDNEY RESEARCH

Application of microfabrication techniques to the in vitro study of renal epithelial cells is just a drop in the ocean of microfluidics and MEMS applications in cell biology. However, the technology has not reached enough maturity to be readily accessible to the general laboratory. Commercially available solutions are very young in the market and have not been widely adopted yet. One important limitation is the need for specific equipment like incubators with integrated perfusion to ensure adequate long-term perfusion that simply has not been developed yet or is offered by very limited suppliers. There are already solutions aimed to provide high-throughput assays based on microfluidic plates (Table 3). However, for the case of complex tissue architecture like that of the renal tubule, perfusion of a high number of independent compartments might represent a significant challenge.

As technical solutions become available, extensive assessment of previous knowledge based on conventional techniques will become a prerequisite for general adoption of these models. There are already reports that the materials employed for rapid prototyping in microfluidic devices for cell culture can have a negative impact on cells, suggesting that a transition toward materials more common in cell biology applications would be advisable. A first report on the effects of long-term culture under conditions of microfluidic channels stresses the notion that it is a completely different environment for cells.[113] It remains to be established whether it really helps to close the gap with the in vivo reality.

Of course, the expectations are high that a better differentiation and enriched cell–cell interactions will result in more representative experimentation. Physiological knowledge with applications in BAK development will be generated. Drug screening and preclinical studies of cell-based therapies will as a result be simplified and the use of animals greatly reduced.

GLOSSARY

3D culture Artificial environment containing more than one cell type, which are growing with its surrounding and it generates formation of a complex ECM. A 3D culture represents a more physiologically relevant environment.

Bioartificial kidney A BAK consists in the combination of a hemofilter in series with a bioreactor unit containing renal epithelial cells, termed a renal assist device. It is used like RRT.

Biomimicking Reproduce the conditions of the cells in its natural environment.

Coculture The combination of growing two or more cell types in a culture.

Dedifferentiation Cellular process by which a specialized cell or tissue lose its specific function to an unspecialized form.

Explant culture Technique used for the isolation and culture of cells starting from a piece of tissue.

Extracellular matrix The extracellular matrix is a complex network of material secreted locally. ECM provides structural, adhesive, and biochemical signaling support.

Extrarenal stem cells Cells can go through a transdifferentiation process into specialized kidney cells supporting the inherent renal regenerative capacity.

Hydrogel polymer A 3D network constituted by polymeric materials capable of holding large amounts of water.

iPSCs Technological process, where adult cells are genetically reprogrammed to generate pluripotent cells.

Matrigel Trade name of a basement membrane matrix constituted by structural proteins. The real composition of matrigel is not well defined.

Microfluidics Science that deals with the flow of liquid through channels of micrometer size. The volume of the flow is in the order of nanoliters or picoliters.

Organoids A piece of tissue that works like an organ in vitro.

Organotypic culture Cell culture model that mimics in vivo tissue architecture through interactions of heterotypic cell types and extracellular matrices.

PDMS A silicon-based organic polymer with rheological properties.

Pluripotent stem cell Descendants of totipotent cells, can differentiate into nearly all cells in an organism, but not a whole organism.

Primary cell culture Cell culture, where cells are isolated from the parental tissue and proliferate under the appropriate conditions.

Redifferentiation Process by which a group of differentiated cells acquire a specialized form.

Senescence Process of aging, where the cells lose the capacity of division, as a result of DNA damage or a shortening of telomeres.

Transdifferentiation Conversion of one differentiated cell type into another cell type.

Transwell Inserts with permeable support for the study of both anchorage-dependent and anchorage-independent cell lines.

Tubulogenesis Process by which cells form tubules.

ACKNOWLEDGMENT AND DISCLAIMER

This work was supported by Spain's Ministerio de Economía y Competitividad through a grant DPI-2011-28262-C04-02 (IG) and a predoctoral fellowship BES-2012-059562 (to NS). One of the companies listed in Table 3 (EBERS) has been an observer to our project, although no direct funding or support has been received, nor are we stakeholders in the company.

REFERENCES

1. Masereeuw R, Mutsaers HA, Toyohara T, Abe T, Jhawar S, Sweet DH, et al. The kidney and uremic toxin removal: glomerulus or tubule? *Semin Nephrol* March 2014;**34**(2):191–208.
2. Trifillis AL. Isolation, culture and characterization of human renal proximal tubule and collecting duct cells. *Exp Nephrol* 1999;**7**(5–6):353–9.
3. Glynne PA. Primary culture of human proximal renal tubular epithelial cells. *Methods Mol Med* 2000;**36**:197–205.
4. Baer PC, Geiger H. Human renal cells from the thick ascending limb and early distal tubule: characterization of primary isolated and cultured cells by reverse transcription polymerase chain reaction. *Nephrol Carlt* June 2008;**13**(4):316–21.
5. Vesey DA, Qi W, Chen X, Pollock CA, Johnson DW. Isolation and primary culture of human proximal tubule cells. *Methods Mol Biol* 2009;**466**:19–24.
6. Valente MJ, Henrique R, Costa VL, Jerónimo C, Carvalho F, Bastos ML, et al. A rapid and simple procedure for the establishment of human normal and cancer renal primary cell cultures from surgical specimens. *PLoS One* 2011;**6**(5):e19337.
7. Presnell SC, Bruce AT, Wallace SM, Choudhury S, Genheimer CW, Cox B, et al. Isolation, characterization, and expansion methods for defined primary renal cell populations from rodent, canine, and human normal and diseased kidneys. *Tissue Eng Part C Methods* March 2011;**17**(3):261–73.
8. Sharpe CC, Dockrell ME. Primary culture of human renal proximal tubule epithelial cells and interstitial fibroblasts. *Methods Mol Biol* 2012;**806**:175–85.
9. Guimaraes-Souza NK, Yamaleyeva LM, AbouShwareb T, Atala A, Yoo JJ. In vitro reconstitution of human kidney structures for renal cell therapy. *Nephrol Dial Transpl* August 2012;**27**(8):3082–90.
10. Van der Hauwaert C, Savary G, Gnemmi V, Glowacki F, Pottier N, Bouillez A, et al. Isolation and characterization of a primary proximal tubular epithelial cell model from human kidney by CD10/CD13 double labeling. *PLoS One* 2013;**8**(6):e66750.
11. Jansen J, Schophuizen CM, Wilmer MJ, Lahham SH, Mutsaers HA, Wetzels JF, et al. A morphological and functional comparison of proximal tubule cell lines established from human urine and kidney tissue. *Exp Cell Res* April 15, 2014;**323**(1):87–99.
12. Winbanks CE, Darby IA, Kelynack KJ, Pouniotis D, Becker GJ, Hewitson TD. Explanting is an ex vivo model of renal epithelial-mesenchymal transition. *J Biomed Biotechnol* 2011;**2011**:212819.
13. Boogaard PJ, Nagelkerke JF, Mulder GJ. Renal proximal tubular cells in suspension or in primary culture as in vitro models to study nephrotoxicity. *Chem Biol Interact* 1990;**76**(3):251–91.
14. Bens M, Vandewalle A. Cell models for studying renal physiology. *Pflugers Arch* October 2008;**457**(1):1–15.
15. Glaudemans B, Terryn S, Gölz N, Brunati M, Cattaneo A, Bachi A, et al. A primary culture system of mouse thick ascending limb cells with preserved function and uromodulin processing. *Pflugers Arch* July 26, 2013;**466**(2):343–56.
16. Baer PC, Nockher WA, Haase W, Scherberich JE. Isolation of proximal and distal tubule cells from human kidney by immunomagnetic separation. Tech note. *Kidney Int* November 1997;**52**(5):1321–31.
17. Sanechika N, Sawada K, Usui Y, Hanai K, Kakuta T, Suzuki H, et al. Development of bioartificial renal tubule devices with lifespan-extended human renal proximal tubular epithelial cells. *Nephrol Dial Transpl* September 2011;**26**(9):2761–9.
18. Luttropp D, Schade M, Baer PC, Bereiter-Hahn J. Respiration rate in human primary renal proximal and early distal tubular cells in vitro: considerations for biohybrid renal devices. *Biotechnol Prog* 2011;**27**(1):262–8.
19. Pizzonia JH, Gesek FA, Kennedy SM, Coutermarsh BA, Bacskai BJ, Friedman PA. Immunomagnetic separation, primary culture, and characterization of cortical thick ascending limb plus distal convoluted tubule cells from mouse kidney. *Vitro Cell Dev Biol* May 1991;**27A**(5):409–16.
20. Helbert MJ, Dauwe S, De Broe ME. Flow cytometric immunodissection of the human nephron in vivo and in vitro. *Exp Nephrol* 1999;**7**(5–6):360–76.
21. Markadieu N, San-Cristobal P, Nair AV, Verkaart S, Lenssen E, Tudpor K, et al. A primary culture of distal convoluted tubules expressing functional thiazide-sensitive NaCl transport. *Am J Physiol Ren Physiol* September 15, 2012;**303**(6):F886–92.
22. Price KL, Hulton SA, van't Hoff WG, Masters JR, Rumsby G. Primary cultures of renal proximal tubule cells derived from individuals with primary hyperoxaluria. *Urol Res* June 2009;**37**(3):127–32.
23. Gildea JJ, McGrath HE, Van Sciver RE, Wang DB, Felder RA. Isolation, growth, and characterization of human renal epithelial cells using traditional and 3D methods. *Methods Mol Biol* 2013;**945**:329–45.
24. Wilmer MJ, Saleem MA, Masereeuw R, Ni L, van der Velden TJ, Russel FG, et al. Novel conditionally immortalized human proximal tubule cell line expressing functional influx and efflux transporters. *Cell Tissue Res* February 2010;**339**(2):449–57.
25. Taub N, Livingston D. The development of serum-free hormone-supplemented media for primary kidney cultures and their use in examining renal functions. *Ann N Y Acad Sci* 1981;**372**:406–21.
26. Cummings BS, Lasker JM, Lash LH. Expression of glutathione-dependent enzymes and cytochrome p450s in freshly isolated and primary cultures of proximal tubular cells from human kidney. *J Pharmacol Exp Ther* May 2000;**293**(2):677–85.
27. Baer PC, Bereiter-Hahn J. Epithelial cells in culture: injured or differentiated cells? *Cell Biol Int* May 14, 2012;**36**(9):771–77.
28. Lipps C, May T, Hauser H, Wirth D. Eternity and functionality - rational access to physiologically relevant cell lines. *Biol Chem* December 2013;**394**(12):1637–48.
29. Salmon P. Generation of human cell lines using lentiviral-mediated genetic engineering. *Methods Mol Biol* 2013;**945**:417–48.
30. Jansen J, Fedecostante M, Wilmer MJ, van den Heuvel LP, Hoenderop JG, Masereeuw R. Biotechnological challenges of bioartificial kidney engineering. *Biotechnol Adv* November 15, 2014;**32**(7):1317–27.
31. Ryan MJ, Johnson G, Kirk J, Fuerstenberg SM, Zager RA, Torok-Storb B. HK-2: an immortalized proximal tubule epithelial cell line from normal adult human kidney. *Kidney Int* January 1994;**45**(1):48–57.
32. Wu Y, Connors D, Barber L, Jayachandra S, Hanumegowda UM, Adams SP. Multiplexed assay panel of cytotoxicity in HK-2 cells for detection of renal proximal tubule injury potential of compounds. *Toxicol Vitro* September 2009;**23**(6):1170–8.
33. Poveda J, Sanchez-Niño MD, Glorieux G, Sanz AB, Egido J, Vanholder R, et al. P-Cresyl sulphate has pro-inflammatory and cytotoxic actions on human proximal tubular epithelial cells. *Nephrol Dial Transpl* January 2014;**29**(1):56–64.

34. Jenkinson SE, Chung GW, van Loon E, Bakar NS, Dalzell AM, Brown CD. The limitations of renal epithelial cell line HK-2 as a model of drug transporter expression and function in the proximal tubule. *Pflugers Arch* December 2012;**464**(6):601–11.

35. Hoppensack A, Kazanecki CC, Colter D, Gosiewska A, Schanz J, Walles H, et al. A human in vitro model that mimics the renal proximal tubule. *Tissue Eng Part C Methods* July 2014;**20**(7):599–609.

36. Wieser M, Stadler G, Jennings P, Streubel B, Pfaller W, Ambros P, et al. HTERT alone immortalizes epithelial cells of renal proximal tubules without changing their functional characteristics. *Am J Physiol Ren Physiol* November 2008;**295**(5):F1365–75.

37. Ellis JK, Athersuch TJ, Cavill R, Radford R, Slattery C, Jennings P, et al. Metabolic response to low-level toxicant exposure in a novel renal tubule epithelial cell system. *Mol Biosyst* January 2011;**7**(1):247–57.

38. Steele SL, Wu Y, Kolb RJ, Gooz M, Haycraft CJ, Keyser KT, et al. Telomerase immortalization of principal cells from mouse collecting duct. *Am J Physiol Ren Physiol* December 2010;**299**(6):F1507–14.

39. DesRochers TM, Suter L, Roth A, Kaplan DL. Bioengineered 3D human kidney tissue, a platform for the determination of nephrotoxicity. *PLoS One* 2013;**8**(3):e59219.

40. Kusaba T, Lalli M, Kramann R, Kobayashi A, Humphreys BD. Differentiated kidney epithelial cells repair injured proximal tubule. *Proc Natl Acad Sci USA* January 28, 2014;**111**(4):1527–32.

41. Bussolati B, Bruno S, Grange C, Buttiglieri S, Deregibus MC, Cantino D, et al. Isolation of renal progenitor cells from adult human kidney. *Am J Pathol* February 2005;**166**(2):545–55.

42. Kitamura S, Yamasaki Y, Kinomura M, Sugaya T, Sugiyama H, Maeshima Y, et al. Establishment and characterization of renal progenitor like cells from S3 segment of nephron in rat adult kidney. *FASEB J* November 2005;**19**(13):1789–97.

43. Lindgren D, Boström AK, Nilsson K, Hansson J, Sjölund J, Möller C, et al. Isolation and characterization of progenitor-like cells from human renal proximal tubules. *Am J Pathol* February 2011;**178**(2):828–37.

44. Hansson J, Hultenby K, Cramnert C, Pontén F, Jansson H, Lindgren D, et al. Evidence for a morphologically distinct and functionally robust cell type in the proximal tubules of human kidney. *Hum Pathol* February 2014;**45**(2):382–93.

45. Maeshima A, Sakurai H, Nigam SK. Adult kidney tubular cell population showing phenotypic plasticity, tubulogenic capacity, and integration capability into developing kidney. *J Am Soc Nephrol* January 2006;**17**(1):188–98.

46. Angelotti ML, Ronconi E, Ballerini L, Peired A, Mazzinghi B, Sagrinati C, et al. Characterization of renal progenitors committed toward tubular lineage and their regenerative potential in renal tubular injury. *Stem Cells* August 2012;**30**(8):1714–25.

47. Kim K, Lee KM, Han DJ, Yu E, Cho YM. Adult stem cell-like tubular cells reside in the corticomedullary junction of the kidney. *Int J Clin Exp Pathol* 2008;**1**(3):232–41.

48. Oliver JA, Maarouf O, Cheema FH, Martens TP, Al-Awqati Q. The renal papilla is a niche for adult kidney stem cells. *J Clin Invest* September 2004;**114**(6):795–804.

49. Ward HH, Romero E, Welford A, Pickett G, Bacallao R, Gattone VH, et al. Adult human CD133/1(+) kidney cells isolated from papilla integrate into developing kidney tubules. *Biochim Biophys Acta* October 2011;**1812**(10):1344–57.

50. Salvatori M, Peloso A, Katari R, Orlando G. Regeneration and bioengineering of the kidney: current status and future challenges. *Curr Urol Rep* January 2014;**15**(1):379.

51. Nowacki M, Kloskowski T, Pokrywczyńska M, Nazarewski Ł, Jundziłł A, Pietkun K, et al. Is regenerative medicine a new hope for kidney replacement? *J Artif Organs* June 2014;**17**(2):123–34.

52. Narayanan K, Schumacher KM, Tasnim F, Kandasamy K, Schumacher A, Ni M, et al. Human embryonic stem cells differentiate into functional renal proximal tubular-like cells. *Kidney Int* April 2013;**83**(4):593–603.

53. Li Y, Kandasamy K, Chuah JK, Lam YN, Toh WS, Oo ZY, et al. Identification of nephrotoxic compounds with embryonic stem-cell-derived human renal proximal tubular-like cells. *Mol Pharm* July 7, 2014;**11**(7):1982–90.

54. Xia Y, Sancho-Martinez I, Nivet E, Rodriguez Esteban C, Campistol JM, Izpisua Belmonte JC. The generation of kidney organoids by differentiation of human pluripotent cells to ureteric bud progenitor-like cells. *Nat Protoc* November 2014;**9**(11):2693–704.

55. Lam AQ, Freedman BS, Morizane R, Lerou PH, Valerius MT, Bonventre JV. Rapid and efficient differentiation of human pluripotent stem cells into intermediate mesoderm that forms tubules expressing kidney proximal tubular markers. *J Am Soc Nephrol* June 2014;**25**(6):1211–25.

56. Takasato M, Er PX, Becroft M, Vanslambrouck JM, Stanley EG, Elefanty AG, et al. Directing human embryonic stem cell differentiation towards a renal lineage generates a self-organizing kidney. *Nat Cell Biol* January 2014;**16**(1):118–26.

57. Montserrat N, Ramírez-Bajo MJ, Xia Y, Sancho-Martinez I, Moya-Rull D, Miquel-Serra L, et al. Generation of induced pluripotent stem cells from human renal proximal tubular cells with only two transcription factors, OCT4 SOX2. *J Biol Chem* July 13, 2012;**287**(29):24131–8.

58. Song B, Smink AM, Jones CV, Callaghan JM, Firth SD, Bernard CA, et al. The directed differentiation of human iPS cells into kidney podocytes. *PLoS One* 2012;**7**(9):e46453.

59. Zhou T, Benda C, Dunzinger S, Huang Y, Ho JC, Yang J, et al. Generation of human induced pluripotent stem cells from urine samples. *Nat Protoc* December 2012;**7**(12):2080–9.

60. Xia Y, Nivet E, Sancho-Martinez I, Gallegos T, Suzuki K, Okamura D, et al. Directed differentiation of human pluripotent cells to ureteric bud kidney progenitor-like cells. *Nat Cell Biol* December 2013;**15**(12):1507–15.

61. Kuraitis D, Giordano C, Ruel M, Musarò A, Suuronen EJ. Exploiting extracellular matrix-stem cell interactions: a review of natural materials for therapeutic muscle regeneration. *Biomaterials* January 2012;**33**(2):428–43.

62. Clause KC, Barker TH. Extracellular matrix signaling in morphogenesis and repair. *Curr Opin Biotechnol* October 2013;**24**(5):830–3.

63. Finesilver G, Bailly J, Kahana M, Mitrani E. Kidney derived micro-scaffolds enable HK-2 cells to develop more in-vivo like properties. *Exp Cell Res* March 10, 2014;**322**(1):71–80.

64. Lelongt B, Ronco P. Role of extracellular matrix in kidney development and repair. *Pediatr Nephrol* August 2003;**18**(8):731–42.

65. Kim DH, Provenzano PP, Smith CL, Levchenko A. Matrix nanotopography as a regulator of cell function. *J Cell Biol* April 30, 2012;**197**(3):351–60.

66. Nur-E-Kamal A, Ahmed I, Kamal J, Schindler M, Meiners S. Three-dimensional nanofibrillar surfaces promote self-renewal in mouse embryonic stem cells. *Stem Cells* February 2006;**24**(2):426–33.

67. Schindler M, Ahmed I, Kamal J, Nur-E-Kamal A, Grafe TH, Young Chung H, et al. A synthetic nanofibrillar matrix promotes in vivo-like organization and morphogenesis for cells in culture. *Biomaterials* October 2005;**26**(28):5624–31.

68. Kim MH, Sawada Y, Taya M, Kino-Oka M. Influence of surface topography on the human epithelial cell response to micropatterned substrates with convex and concave architectures. *J Biol Eng* 2014;**8**:13.

69. le Digabel J, Ghibaudo M, Trichet L, Richert A, Ladoux B. Microfabricated substrates as a tool to study cell mechanotransduction. *Med Biol Eng Comput* October 2010;**48**(10):965–76.

70. Sciancalepore AG, Sallustio F, Girardo S, Gioia Passione L, Camposeo A, Mele E, et al. A bioartificial renal tubule device embedding human renal stem/progenitor cells. *PLoS One* 2014;**9**(1):e87496.

71. Yue B. Biology of the extracellular matrix: an overview. *J Glaucoma* 2014;**23**(8 Suppl 1):S20–3.

72. Elias BC, Mathew S, Srichai MB, Palamuttam R, Bulus N, Mernaugh G, et al. The integrin β1 subunit regulates paracellular permeability of kidney proximal tubule cells. *J Biol Chem* March 21, 2014;**289**(12):8532–44.

73. Nickel C, Benzing T, Sellin L, Gerke P, Karihaloo A, Liu ZX, et al. The polycystin-1 c-terminal fragment triggers branching morphogenesis and migration of tubular kidney epithelial cells. *J Clin Invest* February 2002;**109**(4):481–9.

74. Kolset SO, Reinholt FP, Jenssen T. Diabetic nephropathy and extracellular matrix. *J Histochem Cytochem* December 2012;**60**(12):976–86.

75. Forino M, Torregrossa R, Ceol M, Murer L, Della Vella M, Del Prete D, et al. TGFbeta1 induces epithelial-mesenchymal transition, but not myofibroblast transdifferentiation of human kidney tubular epithelial cells in primary culture. *Int J Exp Pathol* June 2006;**87**(3):197–208.

76. Gao X, Tanaka Y, Sugii Y, Mawatari K, Kitamori T. Basic structure and cell culture condition of a bioartificial renal tubule on chip towards a cell-based separation microdevice. *Anal Sci* 2011;**27**(9):907–12.

77. Zhang H, Lau SF, Heng BF, Teo PY, Alahakoon PK, Ni M, et al. Generation of easily accessible human kidney tubules on two-dimensional surfaces in vitro. *J Cell Mol Med* June 2011;**15**(6):1287–98.

78. Kleinman HK, Martin GR. Matrigel: basement membrane matrix with biological activity. *Semin Cancer Biol* October 2005;**15**(5):378–86.

79. Miya M, Maeshima A, Mishima K, Sakurai N, Ikeuchi H, Kuroiwa T, et al. Enhancement of in vitro human tubulogenesis by endothelial cell-derived factors: implications for in vivo tubular regeneration after injury. *Am J Physiol Ren Physiol* August 2011;**301**(2):F387–95.

80. Subramanian B, Rudym D, Cannizzaro C, Perrone R, Zhou J, Kaplan DL. Tissue-engineered three-dimensional in vitro models for normal and diseased kidney. *Tissue Eng Part A* September 2010;**16**(9):2821–31.

81. Zhang H, Tasnim F, Ying JY, Zink D. The impact of extracellular matrix coatings on the performance of human renal cells applied in bioartificial kidneys. *Biomaterials* May 2009;**30**(15):2899–911.

82. Ni M, Teo JC, Ibrahim MS, Zhang K, Tasnim F, Chow PY, et al. Characterization of membrane materials and membrane coatings for bioreactor units of bioartificial kidneys. *Biomaterials* February 2011;**32**(6):1465–76.

83. Petkau-Milroy K, Brunsveld L. Supramolecular chemical biology; bioactive synthetic self-assemblies. *Org Biomol Chem* January 14, 2013;**11**(2):219–32.

84. Dankers PY, Boomker JM, Huizinga-van der Vlag A, Wisse E, Appel WP, Smedts FM, et al. Bioengineering of living renal membranes consisting of hierarchical, bioactive supramolecular meshes and human tubular cells. *Biomaterials* January 2011;**32**(3):723–33.

85. Shamir ER, Ewald AJ. Three-dimensional organotypic culture: experimental models of mammalian biology and disease. *Nat Rev Mol Cell Biol* October 2014;**15**(10):647–64.

86. Desrochers TM, Palma E, Kaplan DL. Tissue-engineered kidney disease models. *Adv Drug Deliv Rev* April 2014;**69-70**:67–80.

87. Zegers MM, O'Brien LE, Yu W, Datta A, Mostov KE. Epithelial polarity and tubulogenesis in vitro. *Trends Cell Biol* April 2003;**13**(4):169–76.

88. Schlüter MA, Margolis B. Apical lumen formation in renal epithelia. *J Am Soc Nephrol* July 2009;**20**(7):1444–52.

89. Karihaloo A, Nickel C, Cantley LG. Signals which build a tubule. *Nephron Exp Nephrol* 2005;**100**(1):e40–5.

90. Kim M, O'Brien LE, Kwon SH, Mostov KE. STAT1 is required for redifferentiation during madin-darby canine kidney tubulogenesis. *Mol Biol Cell* November 15, 2010;**21**(22):3926–33.

91. Montesano R, Schaller G, Orci L. Induction of epithelial tubular morphogenesis in vitro by fibroblast-derived soluble factors. *Cell* August 23, 1991;**66**(4):697–711.

92. Montesano R, Matsumoto K, Nakamura T, Orci L. Identification of a fibroblast-derived epithelial morphogen as hepatocyte growth factor. *Cell* November 29, 1991;**67**(5):901–8.

93. Taub M, Wang Y, Szczesny TM, Kleinman HK. Epidermal growth factor or transforming growth factor alpha is required for kidney tubulogenesis in matrigel cultures in serum-free medium. *Proc Natl Acad Sci USA* May 1990;**87**(10):4002–6.

94. Kher R, Sha EC, Escobar MR, Andreoli EM, Wang P, Xu WM, et al. Ectopic expression of cadherin 8 is sufficient to cause cyst formation in a novel 3D collagen matrix renal tubule culture. *Am J Physiol Cell Physiol* July 2011;**301**(1):C99–105.

95. Moll S, Ebeling M, Weibel F, Farina A, Araujo Del Rosario A, Hoflack JC, et al. Epithelial cells as active player in fibrosis: findings from an in vitro model. *PLoS One* 2013;**8**(2):e56575.

96. Lü SH, Lin Q, Liu YN, Gao Q, Hao T, Wang Y, et al. Self-assembly of renal cells into engineered renal tissues in collagen/matrigel scaffold in vitro. *J Tissue Eng Regen Med* November 2012;**6**(10):786–92.

97. Joraku A, Stern KA, Atala A, Yoo JJ. In vitro generation of three-dimensional renal structures. *Methods* February 2009;**47**(2):129–33.

98. Sakurai H, Barros EJ, Tsukamoto T, Barasch J, Nigam SK. An in vitro tubulogenesis system using cell lines derived from the embryonic kidney shows dependence on multiple soluble growth factors. *Proc Natl Acad Sci USA* June 10, 1997;**94**(12):6279–84.

99. Rosines E, Schmidt HJ, Nigam SK. The effect of hyaluronic acid size and concentration on branching morphogenesis and tubule differentiation in developing kidney culture systems: potential applications to engineering of renal tissues. *Biomaterials* November 2007;**28**(32):4806–17.

100. Subramanian B, Ko WC, Yadav V, DesRochers TM, Perrone RD, Zhou J, et al. The regulation of cystogenesis in a tissue engineered kidney disease system by abnormal matrix interactions. *Biomaterials* November 2012;**33**(33):8383–94.

101. Astashkina AI, Mann BK, Prestwich GD, Grainger DW. A 3-D organoid kidney culture model engineered for high-throughput nephrotoxicity assays. *Biomaterials* June 2012;**33**(18):4700–11.

102. Astashkina AI, Jones CF, Thiagarajan G, Kurtzeborn K, Ghandehari H, Brooks BD, et al. Nanoparticle toxicity assessment using an in vitro 3-D kidney organoid culture model. *Biomaterials* August 2014;**35**(24):6323–31.

103. Chung IM, Enemchukwu NO, Khaja SD, Murthy N, Mantalaris A, García AJ. Bioadhesive hydrogel microenvironments to modulate epithelial morphogenesis. *Biomaterials* June 2008;**29**(17):2637–45.

104. Shimony N, Gorodetsky R, Marx G, Gal D, Rivkin R, Ben-Ari A, et al. Fibrin microbeads (FMB) as a 3D platform for kidney gene and cell therapy. *Kidney Int* February 2006;**69**(3):625–33.

105. Huang HC, Chang YJ, Chen WC, Harn HI, Tang MJ, Wu CC. Enhancement of renal epithelial cell functions through microfluidic-based coculture with adipose-derived stem cells. *Tissue Eng Part A* September 2013;**19**(17–18):2024–34.

106. Jang KJ, Mehr AP, Hamilton GA, McPartlin LA, Chung S, Suh KY, et al. Human kidney proximal tubule-on-a-chip for drug transport and nephrotoxicity assessment. *Integr Biol (Camb)* May 3, 2013.

107. Weinbaum S, Duan Y, Satlin LM, Wang T, Weinstein AM. Mechanotransduction in the renal tubule. *Am J Physiol Ren Physiol* December 2010;**299**(6):1220–36.

108. Ferkol TW, Leigh MW. Ciliopathies: the central role of cilia in a spectrum of pediatric disorders. *J Pediatr* March 2012;**160**(3):366–71.

109. Raghavan V, Rbaibi Y, Pastor-Soler NM, Carattino MD, Weisz OA. Shear stress-dependent regulation of apical endocytosis in renal proximal tubule cells mediated by primary cilia. *Proc Natl Acad Sci USA* June 10, 2014;**111**(23):8506–11.

110. Huh D, Hamilton GA, Ingber DE. From 3D cell culture to organs-on-chips. *Trends Cell Biol* December 2011;**21**(12):745–54.

111. Huh D, Kim HJ, Fraser JP, Shea DE, Khan M, Bahinski A, et al. Microfabrication of human organs-on-chips. *Nat Protoc* November 2013;**8**(11):2135–57.

112. Choucha-Snouber L, Aninat C, Grsicom L, Madalinski G, Brochot C, Poleni PE, et al. Investigation of ifosfamide nephrotoxicity induced in a liver-kidney co-culture biochip. *Biotechnol Bioeng* February 2013;**110**(2):597–608.

113. Su X, Theberge AB, January CT, Beebe DJ. Effect of microculture on cell metabolism and biochemistry: do cells get stressed in microchannels? *Anal Chem* February 5, 2013;**85**(3):1562–70.

114. Frohlich EM, Alonso JL, Borenstein JT, Zhang X, Arnaout MA, Charest JL. Topographically-patterned porous membranes in a microfluidic device as an in vitro model of renal reabsorptive barriers. *Lab Chip* June 21, 2013;**13**(12):2311–9.

115. Frohlich EM, Zhang X, Charest JL. The use of controlled surface topography and flow-induced shear stress to influence renal epithelial cell function. *Integr Biol (Camb)* January 2012;**4**(1):75–83.

116. Shintu L, Baudoin R, Navratil V, Prot J-M, Pontoizeau C, Defernez M, et al. Metabolomics-on-a-Chip and predictive systems toxicology in microfluidic bioartificial organs. *Anal Chem* January 13, 2012;**84**(4):1840–8.

117. Choucha Snouber L, Jacques S, Monge M, Legallais C, Leclerc E. Transcriptomic analysis of the effect of ifosfamide on MDCK cells cultivated in microfluidic biochips. *Genomics* July 2012;**100**(1):27–34.

118. Snouber LC, Letourneur F, Chafey P, Broussard C, Monge M, Legallais C, et al. Analysis of transcriptomic and proteomic profiles demonstrates improved Madin-Darby canine kidney cell function in a renal microfluidic biochip. *Biotechnol Prog* 2012;**28**(2):474–84.

119. Wei Z, Amponsah PK, Al-Shatti M, Nie Z, Bandyopadhyay BC. Engineering of polarized tubular structures in a microfluidic device to study calcium phosphate stone formation. *Lab Chip* October 21, 2012;**12**(20):4037–40.

120. Rahimzadeh J, Meng F, Sachs F, Wang J, Verma D, Hua SZ. Real-time observation of flow-induced cytoskeletal stress in living cells. *Am J Physiol Cell Physiol* September 2011;**301**(3):C646–52.

121. Ramello C, Paullier P, Ould-Dris A, Monge M, Legallais C, Leclerc E. Investigation into modification of mass transfer kinetics by acrolein in a renal biochip. *Toxicol Vitro* August 2011;**25**(5):1123–31.

122. Douville NJ, Tung YC, Li R, Wang JD, El-Sayed ME, Takayama S. Fabrication of two-layered channel system with embedded electrodes to measure resistance across epithelial and endothelial barriers. *Anal Chem* March 15, 2010;**82**(6):2505–11.

123. Ferrell N, Desai RR, Fleischman AJ, Roy S, Humes HD, Fissell WH. A microfluidic bioreactor with integrated transepithelial electrical resistance (TEER) measurement electrodes for evaluation of renal epithelial cells. *Biotechnol Bioeng* November 1, 2010;**107**(4):707–16.

124. Jang KJ, Suh KY. A multi-layer microfluidic device for efficient culture and analysis of renal tubular cells. *Lab Chip* January 7, 2010;**10**(1):36–42.

125. Wang J, Heo J, Hua SZ. Spatially resolved shear distribution in microfluidic chip for studying force transduction mechanisms in cells. *Lab Chip* January 21, 2010;**10**(2):235–9.

126. Lo JF, Sinkala E, Eddington DT. Oxygen gradients for open well cellular cultures via microfluidic substrates. *Lab Chip* September 21, 2010;**10**(18):2394–401.

127. Baudoin R, Griscom L, Monge M, Legallais C, Leclerc E. Development of a renal microchip for in vitro distal tubule models. *Biotechnol Prog* 2007;**23**(5):1245–53.

128. Sia SK, Whitesides GM. Microfluidic devices fabricated in poly(dimethylsiloxane) for biological studies. *Electrophoresis* November 2003;**24**(21):3563–76.

129. Berthier E, Young EW, Beebe D. Engineers are from PDMS-land, biologists are from polystyrenia. *Lab Chip* April 7, 2012;**12**(7):1224–37.

130. Rydholm S, Frisk T, Kowalewski JM, Andersson Svahn H, Stemme G, Brismar H. Microfluidic devices for studies of primary cilium mediated cellular response to dynamic flow conditions. *Biomed Microdevices* August 2008;**10**(4):555–60.

131. van Midwoud PM, Janse A, Merema MT, Groothuis GM, Verpoorte E. Comparison of biocompatibility and adsorption properties of different plastics for advanced microfluidic cell and tissue culture models. *Anal Chem* May 1, 2012;**84**(9):3938–44.

132. Ferrell N, Ricci KB, Groszek J, Marmerstein JT, Fissell WH. Albumin handling by renal tubular epithelial cells in a microfluidic bioreactor. *Biotechnol Bioeng* March 2012;**109**(3):797–803.

133. Wu MH, Huang SB, Lee GB. Microfluidic cell culture systems for drug research. *Lab Chip* April 21, 2010;**10**(8):939–56.

134. Huang SB, Wu MH, Wang SS, Lee GB. Microfluidic cell culture chip with multiplexed medium delivery and efficient cell/scaffold loading mechanisms for high-throughput perfusion 3-dimensional cell culture-based assays. *Biomed Microdevices* June 2011;**13**(3):415–30.

135. Huh D, Torisawa YS, Hamilton GA, Kim HJ, Ingber DE. Microengineered physiological biomimicry: organs-on-chips. *Lab Chip* June 21, 2012;**12**(12):2156–64.

136. Zhou M, Ma H, Lin H, Qin J. Induction of epithelial-to-mesenchymal transition in proximal tubular epithelial cells on microfluidic devices. *Biomaterials* February 2014;**35**(5):1390–401.

137. Mehta G, Mehta K, Sud D, Song JW, Bersano-Begey T, Futai N, et al. Quantitative measurement and control of oxygen levels in microfluidic poly(dimethylsiloxane) bioreactors during cell culture. *Biomed Microdevices* April 2007;**9**(2):123–34.

138. Grist SM, Chrostowski L, Cheung KC. Optical oxygen sensors for applications in microfluidic cell culture. *Sensors (Basel)* 2010;**10**(10):9286–316.

139. Wang WC, Liu SF, Chang WT, Shiue YL, Hsieh PF, Hung TJ, et al. The effects of diosgenin in the regulation of renal proximal tubular fibrosis. *Exp Cell Res* May 1, 2014;**323**(2):255–62.

140. Fliedl L, Manhart G, Kast F, Katinger H, Kunert R, Grillari J, et al. Novel human renal proximal tubular cell line for the production of complex proteins. *J Biotechnol* April 2014;**20**(176):29–39.

141. Minuth WW, Denk L, Glashauser A. Cell and drug delivery therapeutics for controlled renal parenchyma regeneration. *Adv Drug Deliv Rev* June 15, 2010;**62**(7–8):841–54.

142. Kelly EJ, Wang Z, Voellinger JL, Yeung CK, Shen DD, Thummel KE, et al. Innovations in preclinical biology: ex vivo engineering of a human kidney tissue microperfusion system. *Stem Cell Res Ther* 2013;**4**(Suppl 1):S17.

143. Humes HD, Buffington D, Westover AJ, Roy S, Fissell WH. The bioartificial kidney: current status and future promise. *Pediatr Nephrol* March 2014;**29**(3):343–51.

Index

Note: Page numbers followed by "f" indicate figures, "t" indicate tables, and "b" indicate boxes.

Printed in the United States
By Bookmasters